12th International Conference on Compressors and their Systems (Compressors 2021)

IOP Conference Series: Materials Science and Engineering
Volume 1180

London, United Kingdom
6 - 8 September 2021

ISBN: 978-1-7138-4640-6
ISSN: 1757-8981

Printed from e-media with permission by:

Curran Associates, Inc.
57 Morehouse Lane
Red Hook, NY 12571

CURRAN ASSOCIATES INC. proceedings.com

Some format issues inherent in the e-media version may also appear in this print version.

Institute of Physics
Dirac House, Temple Back
Bristol BS1 6BE UK

Phone: 44 1 17 929 7481
Fax: 44 1 17 920 0979

techtracking@iop.org

Additional copies of this publication are available from:

Curran Associates, Inc.
57 Morehouse Lane
Red Hook, NY 12571 USA
Phone: 845-758-0400
Fax: 845-758-2633
Email: curran@proceedings.com
Web: www.proceedings.com

TABLE OF CONTENTS

IOP Conf. Series: Materials Science and Engineering 1180 (2021) 011001 doi:10.1088/1757-899X/1180/1/011001

Foreword

It is my great pleasure to welcome you to our 12th International Conference on Compressors and their Systems, held at City, University of London.

This series of biennial conferences started 22 years ago, in 1999 as the result of a collaborative effort by the Fluid Machinery Group of the Institution of Mechanical Engineers (IMechE) and City, with the aim of bringing together researchers, academics and professionals in the field of compressors and their systems from across the globe. Despite the present need to hold it online, as a result of the difficulties and restrictions created by the CORONA Virus, its success continues.

This year, it is being held from 6th to 8th September 2021 to be followed, immediately after, by the 5th Short course/Forum on CFD in positive displacement rotating machines.

It has been organised in collaboration with our esteemed partners; the IMechE, the Institute of Refrigeration (IOR), and the International Institute of Refrigeration (IIR), who are joined, this year, by two new partners, namely; the Manufacturing Technologies Association (MTA) and Machines, an open access journal by MDPI. On behalf of the Organising Committee, I would like to thank them for their support in preparing and promoting this event.

This year, the theme of the conference is sustainability and in the first keynote address, Professor Ian Arbon will draw on his decades of international experience in the compressor industry and in sustainable development, renewable energy and climate change mitigation/adaptation, to define what this term really means.

On the second day Dr Andy Pearson will present a keynote address on challenges facing compressors, in heat pump applications, while Professor Ruzhu Wang will talk about the role of heat pumps in meeting the Carbon Neutral Target.

The third day keynote address will be on compressors and industrial digitisation and will be delivered by Dr Billy Milligan of Howden.

Since the conference is being held online, we have decided to deliver these keynotes twice each day, once in the mornings and the second time in the afternoon to allow participants from all over the world to listen to and interact with the speakers at convenient times.

Social responsibility is an extremely important issue for a sustainable future. In his talk titled "The Engineer in Society", Dr Collin Brown, who served as Chief Executive Officer of IMechE, will reflect on the multiple perceptions that engineering can generate, including its ability to do harm as well as good.

I would like to thank all of them for sharing their enormous knowledge and experience with us.

The two industry sessions and panel discussions are an important part of the programme and will include 9 presentations in which you can hear from world leading experts on topics related to the sustainability of compressors and their systems, and novelties in materials and lubrication.

The review committee consisted of 75 world experts in this field, led by Dr Matthew Read. They accepted 69 papers. All of these will be presented virtually in the technical sessions. The attendees of the conference will have full access to them through Conftool and the mobile app Conference4Me, while the entire proceedings will be published, by the IOP (Institute of Physics). I would like to take this opportunity to express my deepest gratitude to all the reviewers and, especially, to Dr Read.

Let me also draw your attention to the 5th Short course/forum on Computational Fluid Dynamics in Rotating Positive Displacement Machines which is held on 9th and 10th September 2021. Our focus this year is on Multiphysics simulations including, ever more important, conjugate heat transfer studies in rotating positive displacement machines. Thirteen talks will be delivered by leading experts including Professor Hrvoje Jasak, of the university of Cambridge, who is one of the founders of OpenFOAM.

I would like to extend my special thanks to our sponsors: Holroyd PTG, Howden, Fetu, Gamma Technologies, PDM Analysis and Megger as well as the exhibitor Victrex for their generous support.

On this occasion, I must also thank all those who have helped to make this occasion memorable. Special thanks go to Lucy Bennett of the City Events team, Dr Matthew Read - Chair of the Reviewing and programme Committee, Dr Sham Rane for organising the Short Course and Dominique Spence-Polin, the conference secretary, for their great efforts in arranging and promoting the conference and making us all feel at home at City, despite only virtual attendance. Not to be forgotten are the session Chairs and Co-chairs, namely; Dr Thibaud Plantagenet, Thibault Tam, Aleksander Krupa and Brijesh Patel. In addition, I wish to extend my special thanks to the members of the Organising Committee and the International Liaison Committee for their constant support and guidance.

Last but not least, I would like to thank all the authors and participants for their contribution.

I hope you will all thoroughly enjoy this occasion and that you find it to be truly worthwhile attending!

Professor Ahmed Kovacevic

Conference Chairman

Centre for Compressor Technology
School of Mathematics, Computer Science and Engineering
City, University of London, United Kingdom

Sponsors

City, University of London would like to thank the following sponsors and exhibitors

Platinum Sponsors

PTG HOLROYD MACHINE TOOLS & PRECISION COMPONENTS

Howden

Gold Sponsor

fetu ROTICULATE

Silver Sponsors

GT Gamma Technologies

pdm analysis

Social Event and Prize Award Sponsor

Megger.

Exhibitors

victrex

IOP Conf. Series: Materials Science and Engineering 1180 (2021) 011001 doi:10.1088/1757-899X/1180/1/011001

Conference Committees

Organising Committee 2021

Professor Ahmed Kovacevic, Conference Chair, City, University of London
Dr Matthew Read, City, University of London
Dr Sham Rane, University of Oxford
Louise Gordon, City, University of London
Lucy Bennett, City, University of London
Dominique Spence-Polin, City, University of London
Professor Ian Smith, City, University of London
Miriam Rodway, Institute of Refrigeration
Didier Coulomb, International Institute of Refrigeration
Deonie Lambert, International Institute of Refrigeration
Robert Porter, Institution of Mechanical Engineers
Ivor Rhodes, ACE Cranfield
Professor Christohper S. Holmes, Holroyd
Dr Hamid R Hazby, PCA Engineers Limited/IMechE Fluid Machinery Group
Donald lawson, Howden Compressors, United Kingdom
Joshua Dugdale, The Manufacturing Technologies Association

International Liaison Committee 2021

Professor Andreas Bruemmer, TU Dortmund University, Germany
Professor Annamalai Mani, Indian Institute of Technology, India
Dr Ashvin Dhunput, ADC
Professor Cesar J. Deschamps , University of Santa Catarina, Brazil
Professor Eckhard Groll, Purdue University, Indiana, USA
Dr Guy Hundy, Aleph Zero, United Kingdom
Professor Ian Arbon, Engineered Solutions, United Kingdom
Professor Joaquim Rigola, Polytechnic University of Catalonia, Spain
Professor Kazuyoshi Miyagawa, Waseda University, Japan
Kenji Toyo, Waseda University, Japan
Dr Kristina V. Soldatova, St Petersburg University, Russia
Dr Kim Tiow Ooi, Nanyang Technology University, Singapore
Professor Liansheng Li, Qingdao University of Science and Technology, China
Dr. Markus Digruber HOERBIGER Austria
Matthew Cambio, Ingersoll Rand/Trane, USA
Matthew Irons, Emerson Climate Technologies GmBH, Belgium
Dr Mitsushiro Fukuta, Shizuoka University, Japan
Professor Nikola Stosic, City, University of London, United Kingdom
Professor Pega Hrnjak, University of Illinois, US
Professor Vincent Lemort, University of Liège, Belgium
Dr Vishnu Sishtla, UTC, USA
Professor Vladimir Yusha, Omsk State Technical University, Russia
Dr Xiaolin Wang, University of Tasmania, Australia
Ziwen Xing, Xi'an Jaotong University, China

IOP Conf. Series: Materials Science and Engineering 1180 (2021) 011001 doi:10.1088/1757-899X/1180/1/011001

Scientific Review Committee 2021

Dr. Stefano Murgia Ing. Enea Mattei SpA
Dr. Timur Mustafin Kazan National Research Technological University
Kai Nadler Leybold GmbH
Alexander Nikolov TU Dortmund University
Dr. Tsutomu Nozaki Hitachi
Dr. Ali Bahadir Olcay Yeditepe University
Prof. Kim Tiow Ooi Nanyang Technological University
Dr. Michael Perevozchikov Emerson Climate Technologies
Dr. Suxin Qian Xi'an Jiaotong University
Dr. Sham Rane City, University of London
Dr. Matthew Read City, University of London
Ivor Rhodes Engineering Consultancy
Dr. Melanie Rigby SMT
Prof. Joaquim Rigola CTTC – UPC
Dr. Christopher Seeton Shrieve Chemical
Dr. Tadayoshi Shoyama Panasonic Corporation
Ernane Silva UFSC
Prof. Ian K Smith City, University of London
Dr. Kristina Soldatova Peter the Great St.Petersburg Polytechnic University
Prof. Nikola Stosic City, University of London
Dr. Shuaihui Sun Xi'an University of Technology
Dr. Kok Ming Tan Sanden International Singapore Pte Ltd
Dr. Chengzhi Tang Helmer Scientific
Dr. Hong Tao shanghai Hitachi Electrical Appliances Co. Ltd
Dr. Christiane Thomas TU Dresden
Dr. Muhammad Waqas Tofique voestalpine Precision Strip AB, Sweden
Kenji Tojo TOJO R&D Design Office / Waseda University
Prof. Gianluca Valenti Politecnico di Milano
Dr. Kacani Vasillaq Leobersdorfer Maschinenfabrik GmbH & Co.KG
Dr. Junfeng Wang Simerics Inc.
Dr. Xianfa Wang PTG Holroyd Precision Ltd.
Prof. Xiaolin Wang University of Tasmania
Dr. Martin White City, University of London
Prof. Weifeng Wu Xian Jiaotong University
Prof. Ziwen Xing Xi'an Jiaotong University
Prof. Vladimir Leonidovich Yusha Omsk State Technical University
Prof. yuanyang zhao Qingdao University of Science and Technology
Dr. Davide Ziviani Purdue University
Dr. Yang Zou Lubrizol

IOP Conf. Series: Materials Science and Engineering 1180 (2021) 011002 doi:10.1088/1757-899X/1180/1/011002

Peer review declaration

All papers published in this volume of *IOP Conference Series: Materials Science and Engineering* have been peer reviewed through processes administered by the Editors. Reviews were conducted by expert referees to the professional and scientific standards expected of a proceedings journal published by IOP Publishing.

- **Type of peer review: Single-blind**
- **Conference submission management system: ConfTool**
- **Number of submissions received: 72**
- **Number of submissions sent for review: 71**
- **Number of submissions accepted:**
  69
- **Acceptance Rate (Number of Submissions Accepted / Number of Submissions Received X 100): 96%**
- **Average number of reviews per paper: 2-3 (minimum 2)**
- **Total number of reviewers involved: 71**
- **Any additional info on review process:**

  Reviewers were asked to consider the following key points in red, related to scientific content, quality and presentation:
- General criteria
  •Comply with the template for an IOP Publishing publication. (please refer to page 2)
  •Length of the papers to be less or equal to 10 pages.
  •Written in good style and with appropriate use of English language.
- Technical criteria
  •Scientific merit: notably scientific rigour, accuracy and correctness.
  •Clarity of expression; communication of ideas; readability and discussion of concepts.
  •Sufficient discussion of the context of the work, and suitable referencing.
- Quality criteria
  •Originality: is the work relevant and novel?
  •Motivation: does the problem considered have a sound motivation? All papers should clearly demonstrate the scientific interest of the results.
  •Repetition: have significant parts of the manuscript already been published?
  •Length: is the content of the work of sufficient scientific interest to justify its length?
- Presentation criteria

IOP Conf. Series: Materials Science and Engineering 1180 (2021) 011002 doi:10.1088/1757-899X/1180/1/011002

•Title: is it adequate and appropriate for the content of the article?
•Abstract: does it contain the essential information of the article? Is it complete? Is it suitable for inclusion by itself in an abstracting service?
•Diagrams, figures, tables and captions: Are they essential and clear?
•Text and mathematics: are they brief but still clear? If you recommend shortening, please suggest what should be omitted.
•Conclusion: does the paper contain a carefully written conclusion, summarizing what has been learned and why it is interesting and useful? A 3rd review was conducted where there was a significant difference between first 2 reviewer scores.

- **Contact person for queries: Matthew Read (matthew.read.3@city.ac.uk)**

IOP Conf. Series: Materials Science and Engineering 1180 (2021) 012001 doi:10.1088/1757-899X/1180/1/012001

Bi-Directional System Coupling for Conjugate Heat Transfer and Variable Leakage Gap CFD Analysis of Twin-Screw Compressors

S Rane, A Kovačević, N Stošić and I K Smith

City, University of London, Centre for Compressor Technology, United Kingdom.
Email: sham.rane@city.ac.uk

Abstract. Oil-free twin-screw compressors are essential in various industrial applications where clean compressed gas is required. Due to the absence of the cooling oil in these machines, thermal deformations are large. Hence, design clearances are generally set at a relatively large value of more than 150 μm. Leakage through these clearances are the primary source of flow loss. It is essential to predict the change of the gap size in operation accurately so that the design clearances can be minimised, allowing reliable operation and maximising compressor efficiency. To achieve this, CFD and Structural solvers were combined. The CFD model uses a single domain deforming grid of the twin-screw rotors generated in SCORG grid generator, together with ANSYS CFX flow solver. The thermal model of the rotors and housing uses ANSYS Structural solver. Two modelling systems were coupled bi-directionally to obtain variation in the radial leakage gap size for calculation of performance in the CFD model. The predicted compressor performance thus obtained was compared with measurements of flow, internal pressure-rise, power, specific power, volumetric and adiabatic efficiency. For the test case, three variations of radial gap size were evaluated, two of them with the uniform gap size of 10 μm and 160 μm and the third one with a variable gap size as predicted by the bi-directionally coupled model. The coupled model predicted this gap size to vary from 24 to 117 μm, thus predicting an improved flow and volumetric efficiency by 8.2%, lower indicated power by 2.5% and a higher adiabatic efficiency by 5.5%, in comparison to the design gap size of 160 μm. These predicted gap sizes could be used to improve the design clearances of the compressor by reducing them from 160 to 120 μm which would result in a better performance during operation.

1. Introduction

Oil-free twin-screw air compressors are essential in various industrial applications where clean compressed gas is required. As a result of the absence of oil within the working chamber, during compression, the internal temperature of the gas may rise up to 130°C. This causes large thermal deformation of the rotors and casing. Accordingly, the design clearances are generally set at a value of about 150 μm, which is much larger than in oil flooded machines. Leakage through clearances being the primary source of loss, it is essential to predict the resulting gap size variation accurately to enable safe design clearances to be set at a minimum reliable value and thereby improve the compressor efficiency. *Kovačević et al.* [1, 2] presented various techniques for CFD modelling of twin-screw machines and in the chapter on fluid-solid interaction, the analysis of rotor deformation is discussed for several cases such as a 127 mm oil injected air compressor operating from 1.0 to 7.0 bar, dry air operation from 1.0 to 3.0 bar, and a high-pressure CO_2 refrigeration compressor with suction conditions of 30.0 bar at 0°C and a discharge pressure of 90.0 bar at 40 °C. Deformation levels in the range of 70 to 100 μm were reported in these studies. The Comet solver for computational continuum mechanics was employed for both fluid and structural calculations. In their three-part literature from 2006, *Sauls et al.* [3, 4 5] proposed a procedure whereby an initial thermodynamic chamber model computational result is used to define boundary conditions for intermediate thermal analysis of refrigeration compressor components, followed by calculation of deformation and clearance variation along the rotor-to-rotor mesh line and in the rotor-to-housing gaps. A similar approach that couples the chamber model KaSim to the ANSYS solver was described by *Nikolov et al.* [6] to estimate the influence of rotor and casing thermal deformation on the performance of a screw expander in an ORC system, based on the iterative coupling of a thermodynamic chamber model and 3D FEM thermal analysis. More recently, *Ding et al.* [7] used an approach similar to [1, 2] to solve Conjugate Heat Transfer (CHT) and deformation of solid parts of a dry air compressor. The Simerics-MP+ solver was used in conjugation with the SCORG grid generator. *Buckney* [8] applied a procedure to calculate CHT and clearance variation in SCORG chamber models which was extended to 3D FEM analysis by *Husak et al.* [9].

Table 1. Specification of the dry air twin-screw compressor

Rotor profile	N Type		Rotor Data	
Axis distance	93.00 mm		Main Rotor	Gate Rotor
Rotor length	203.913 mm	**Lobes**	3	5
Built-in volume index	1.8	**Outer diameter**	127.324 mm	120.262 mm
Interlobe gap	160 microns	**Wrap angle**	285.00°	171.00°
Axial gap	160 microns	**Radial gap**	160 microns	160 microns

Figure 1. The N-profile rotor pair.

A bi-directional system coupling approach that utilizes CFD and Structural solvers in co-simulation is presented here. The CFD model uses a single domain deforming grid of the twin-screw rotors generated in SCORG and ANSYS CFX flow solver. The thermal model of the rotors and housing uses the ANSYS Structural solver. System coupling by means of CHT analysis is used to map the fluid-solid boundaries where the local heat transfer coefficient, wall adjacent fluid temperature and heat flux are exchanged in a co-simulation loop. Additionally, a considerable saving in computational time has been achieved by attributing a large time-scale factor to solid heat conduction. The compressor specification is listed in Table 1 and a photograph of the N-profile [10] rotor pair is shown in Figure 1, taken from the suction side axial end. Estimates of the housing temperature distribution, exit air temperature, compressor performance and internal pressure distribution were compared with measured values for validation of the model. The bi-directional system coupling approach used, improved the reliability of the performance results obtained from the computational models.

2. Methodology of the bi-directional system coupling

As shown in Figure 2, the bi-directional system is comprised of five analytical models, which are coupled and managed via the ANSYS Workbench platform. Each model is identified by an alphabetical letter and its elements are numbered as cells. For example, C5 describes the setup cell of the structural analysis model. The compressor design data, the rotor CAD model and the grids required for the CFD model are generated externally by SCORG. The system coupling procedure is an iterative process that starts with an initial CFD solution to get an estimate of the temperature field. In the current work, the initial CFD solution was obtained with an assumed radial leakage gap of 10 μm. This is derived from a combined flow and CHT analysis with uniform leakage gap sizes. However, the component E estimates the leakage gap size, based on the deformation data received from component C during the simulation process. ANSYS CFX solver with a Finite Volume Method (FVM) is used for both models. Once the initial flow field and temperature distribution are known, the housing and rotor surface temperature data are transferred as boundary conditions to the component B which defines the thermal analysis model. The thermal analysis model uses a Finite Element Method (FEM) to solve temperature distribution in the metal components and the results are transferred as body loads to the transient structural analysis model in the component C. Components C and E are connected via the system coupling component D. In the current work, the housing deformation has been accounted for only in components C and E. For simplification, the rotor deformation is not considered. Accordingly, in order to account for rotor deformation, the leakage gap size is set at a lower value of 30 μm in the model and is updated by the housing deformation data as the simulation progresses.

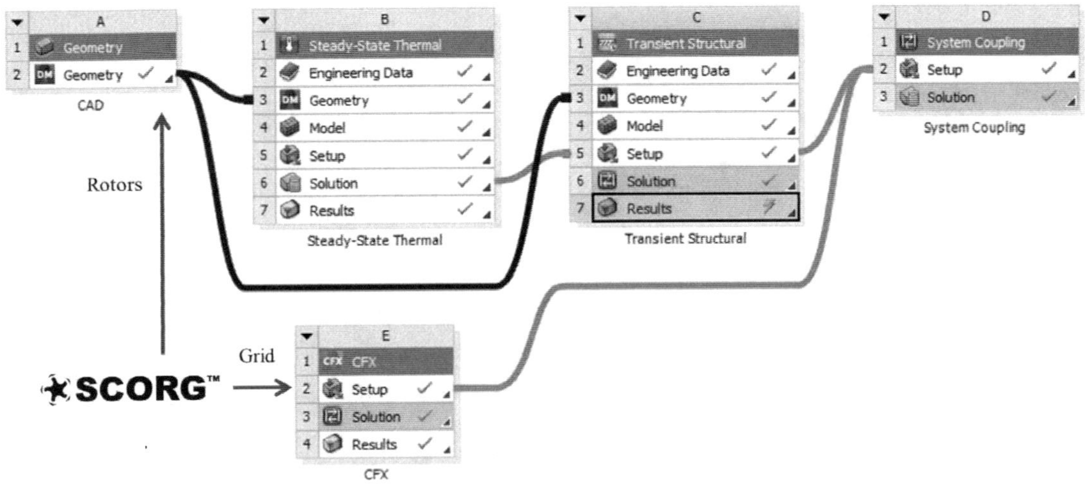

Figure 2. Components of the bi-directional system coupling analysis.

3. Numerical modelling of the system components

CFD modelling of a twin-screw compressor is challenging due to the complex deformation of the working chamber. To achieve a reliable numerical model, the in-house grid generator SCORG is used as shown in Figure 2. Details of the methodology used for CFD analysis in this paper can be found in examples of the use of the SCORG grid generator in references [11 - 18].

3.1. Screw compressor CFD model with CHT

The CFD model of the twin-screw compressor is detailed in Figure 3. The whole compressor block is split into several domains as shown in Figure 3a.

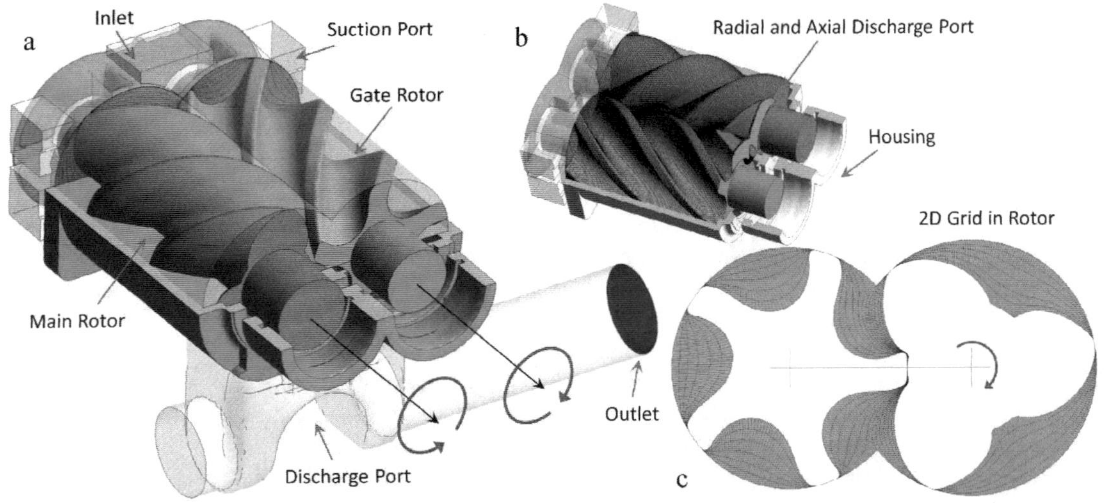

Figure 3. CFD model of the twin-screw compressor with CHT, a) domains and boundaries, b) 3D rotor grid and discharge port, c) 2D rotor grid at one position.

Namely, these are the suction port, the rotor fluid domain, the discharge port, the main rotor solid domain, the gate rotor solid domain and the housing solid domain. High-pressure axial end leakage has not been included in the present study. In Figure 3b, the housing and the radial and axial discharge port areas are highlighted. in Figure 3c a cross-section of the rotor grid generated by SCORG is shown at one rotor position. The rotor domain is a fully hexahedral cell structure with both rotors contained in a single grid [12, 13]. All other domains are generated using a tetrahedral cell structure. The domains are connected in the CFD solver through non-conformal interfaces. The compressed gas is air as an ideal gas, with a molar mass of 28.96 kg/kmol, a specific heat capacity at constant pressure of $1.0044e^{03}$ J/kg K, a dynamic viscosity of $1.831e^{-05}$ kg/m s and a thermal conductivity of $2.61e^{-02}$ W/m K. Both the rotor and housing components, are assumed to be made of structural steel with a density of 7850 kg/m^3, a coefficient of thermal expansion of $1.2e^{-05}$ units/°C, a tensile yield strength 250 MPa, Young's modulus 200 GPa, Poisson's ratio 0.3 and thermal conductivity 60.5 W/m K.

Table 2. Specification of the numerical setup in the ANSYS CFX solver.

Mesh deformation	User defined nodal displacement	**Advection scheme**	High Resolution
Mesh in ports	Tetrahedral (ANSYS Mesh)	**Inner loop coefficients**	Up to 10 iterations per time step
Outlet boundary condition	Opening (Static pressure 2.0 bar, backflow acts as total pressure and temperature)	**Transient scheme**	Second order Backward Euler $\Delta t = 5.55556e^{-05}$ sec CHT solid time scale $\tau = 10^6$
Inlet boundary condition	Opening (Total pressure 1.0 bar and temperature 25°C)	**Convergence criteria**	r.m.s residual level $1e^{-03}$
Turbulence model	SST – k Omega (Standard Wall Functions)	**Relaxation parameters**	Solver relaxation fluids 0.1 Solver relaxation scalar 0.2

In Table 2, the main numerical setup parameters of the CFD solver have been specified. It is noted that in the transient scheme and solver time-step definition, a large scaling factor of 10^6 has been applied to accelerate the conjugate heat transfer analysis. By this setting, the flow solver progresses in time with a time-step size of $5.5556e^{-05}$ sec while the heat transfer in the solid parts of the model progresses with an average diffusion timescale close to 600 sec.

Table 3. CFD model grid size and quality metrices.

	Nodes	Cells	Orthogonal Angle		Expansion Factor		Aspect Ratio	
			Average	Minimum	Average	Maximum	Average	Maximum
Rotor	684398	594696	67.23	10.8	1.88	4406	78.88	1182
Suction Port	34001	169807	65.73	26.2	3.26	1056	2.46	98
Discharge Port	58584	303416	66.11	27.0	3.06	405	2.20	59
Main Rotor	426170	2260951	66.58	43.7	2.95	16	2.16	7
Gate Rotor	198584	1000546	66.44	45.6	2.82	16	2.14	6
Housing	227788	1179811	66.25	26.9	2.93	157	2.10	51

Grid size and quality factors in various domains have been reported in Table 3. The average quality of cells in the rotor domain is critical for the stability of the solver. The rotor domain grid with a very high expansion factor and high aspect ratio of cells could be processed using a double precision calculation.

3.2. Housing thermal model

From the CHT calculations of the initial CFD model, with a radial gap size of 10 μm, the data of the time averaged housing bore surface temperature is transferred to system B to serve as a thermal load for the deformation calculation. The housing thermal model calculates heat conduction in the metal and provides volumetric temperature data for the housing deformation model of system C. The bore temperatures imported from the CFD solution and the housing thermal model setup are shown in Figure 4a, while the calculated housing temperature and transferred to system C is illustrated in Figure 4b.

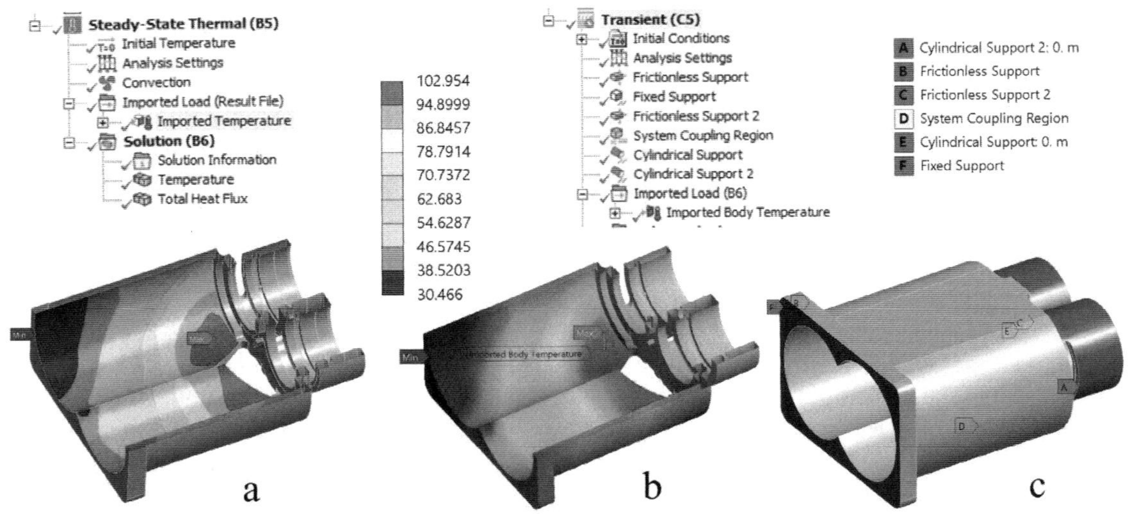

Figure 4. Thermal and structural model of the compressor housing, a) Bore temperature from CFD, b) housing temperature, c) supports and system coupling boundaries.

3.3. Coupled CFD and housing deformation model

System C presented in Figure 4c is a transient structural model of the housing which is coupled with the CFD system E through the system coupling D. In the structural model both local surface forces acting on the bore are received from the CFD system and the volumetric temperature is received from system B. Thus, the deformation of the housing is calculated under both the net force and the thermal load. The force load is dependent on the position of the rotors and gets updated every coupling iteration. The instantaneous deformation of the housing bore is transferred to the CFD system E as an incremental boundary displacement, and this changes the radial leakage gap size with time and space. Other supports defined in the structural model are also represented in Figure 4c.

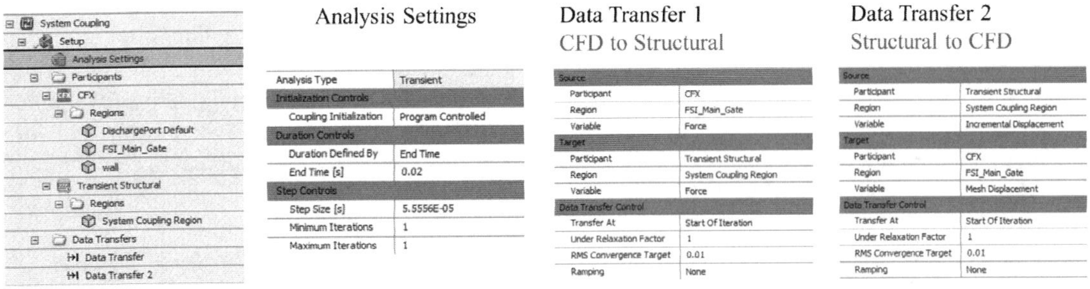

Figure 5. Data exchange routine of the bi-directional system coupling.

3.4. Routine for Data exchange

Figure 5 illustrates the setup used for bi-directional data transfer during calculations. Two data transfer units are defined, one for the force due to compression pressure rise from CFD model to the structural model and another for deformation due to thermal and force load from the structural to the CFD model. The flow solver leads the loop, and the time step size is defined in order to set the main rotor speed at 6000 rpm for the set rotor grid advancement per step. In order to reduce computational cost only one coupling iteration is defined per time step with r.m.s convergence level of 0.01.

4. Results and discussion

The above methodology has been applied to solve flow, CHT and housing deformation in a dry air compressor at 6000 rpm speed operating at 1.0 bar suction pressure, 24°C suction temperature and 2.0

bar discharge pressure. Measurement results at this operating condition were recorded at the City, University of London test facility and are available as compressor performance data and compression chamber pressure cycle data [12, 13]. These data have been used for comparison with three numerical cases; Case 1 is with the uniform radial leakage gap size of 10 μm, used as an initial CHT solution, Case 2 is with the design specified uniform radial leakage gap size of 160 μm and Case 3 is with the variable radial leakage gap obtained using the bi-directional system coupling with the housing deformation and an initial gap size of 30μm.

4.1. Conjugate heat transfer analysis

When the solid components of the screw compressor unit are included in the CFD model, heat flux conservation is achieved between the fluid and solid parts and the assumption of adiabatic walls at the housing and the rotor surfaces could be eliminated. This helps to achieve more accurate heat transfer results that consequently impact the compression process. The temperature distribution on rotors, in the compression chamber and in the discharge port is presented in Figure 6. The results are compared between the three cases. With a 10 μm radial leakage gap, the peak temperature reaches to 133.7°C. With a 160 μm radial leakage gap, the leakage is substantially increased and hence the peak air temperature reaches 156.9°C. While, with the variable radial leakage gap the peak air temperature is around 151.4°C.

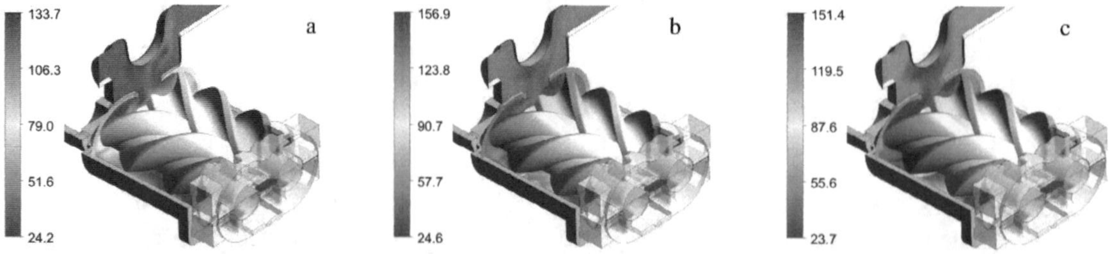

Figure 6. Instantaneous air temperature distribution, a) radial gap 10 μm, b) radial gap 160 μm, c) variable radial gap.

On the rotor and housing components, a larger time scale factor was applied to achieve close to steady operating temperature results. A time averaged temperature distribution is presented in the rotor and housing section in Figure 7 after 20 compression cycles. The differences in air temperature observed in Figure 6 influence the CHT results of the metal temperature here. With a 10 μm radial leakage gap, peak time averaged temperature in the solid reaches 98°C. With a 160 μm radial leakage gap, the time averaged temperature in the solid increase to 129°C. While, with the variable radial leakage gap the time averaged temperature in the solid is around 120°C.

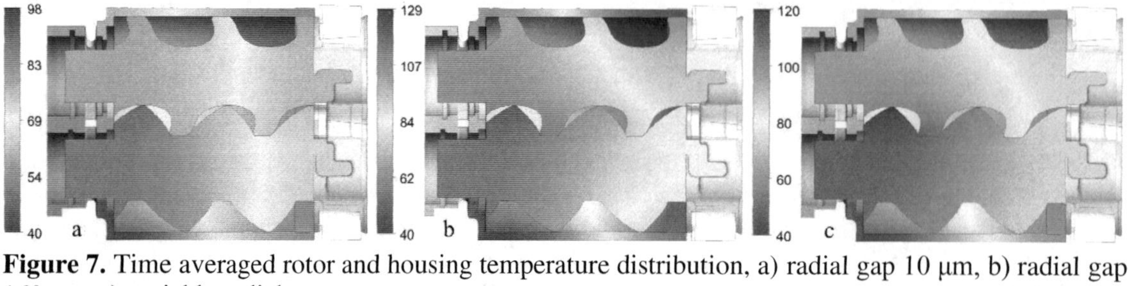

Figure 7. Time averaged rotor and housing temperature distribution, a) radial gap 10 μm, b) radial gap 160 μm, c) variable radial gap.

4.2. Variation of radial leakage gaps

Due to the thermal non-uniformity on the housing and the rotors, and high metal temperature, the housing deforms. This deformation causes the clearance between the rotor outer diameters and the housing bore diameter to vary significantly from their design values. Results of the calculated variation in the radial leakage gap size after 20 compression cycles are presented in Figure 8. It can be observed that the gap size distribution is highly non-uniform, and ranges from 24 to 117 μm over both the main and gate rotor bores. It should be noted that the rotor deformation has not been accounted in the current study and hence an initial gap size of 30 μm was specified in the radial leakage clearance. The obtained non-uniform gap size distribution is an impression of the housing deformation on this initial uniform 30 μm clearance.

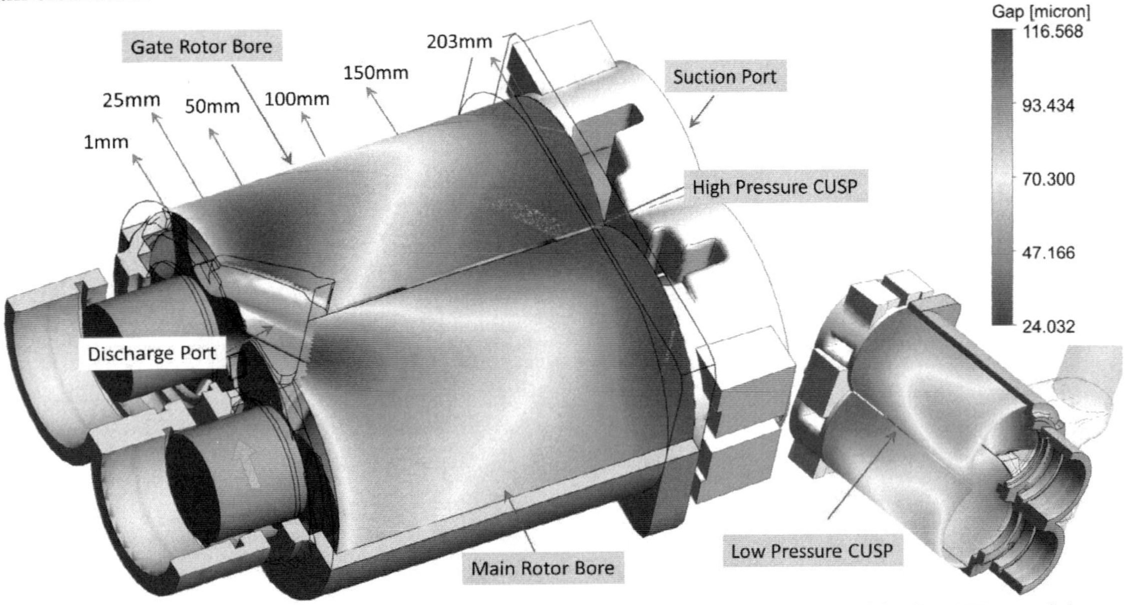

Figure 8. Non-uniform variation of the radial leakage gap size, as captured in the CFD model.

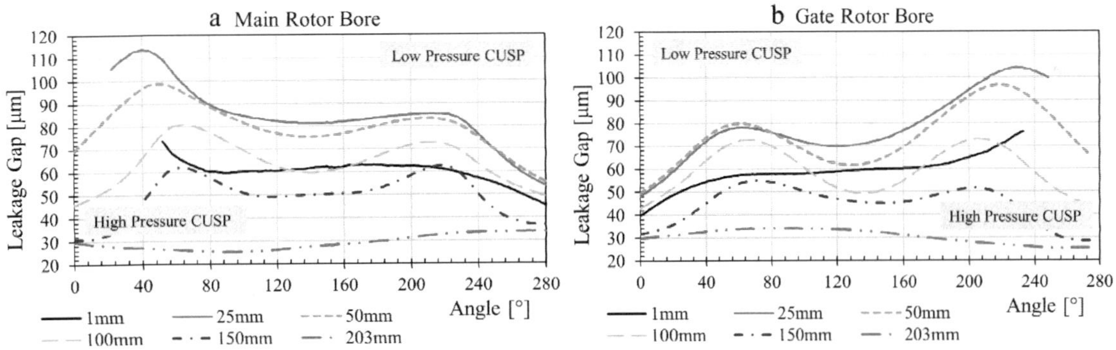

Figure 9. Variation of the radial leakage gap size along the housing bore axis, a) main rotor bore, b) gate rotor bore.

A detailed evaluation of the radial gap size variation in presented in Figure 9. Figure 9a is for a set of axial locations on the main rotor bore, while Figure 9b is for the same axial locations on the gate rotor bore. These axial locations have been defined in Figure 8. It can be observed that at 25 mm from the discharge end, the housing deflection is the highest and here the leakage gap varies from 55 to 115 μm, from the low-pressure CUSP (intersection of the bore diameter) to the high-pressure CUSP, respectively on both housing bores. CUSP locations have been indicated in Figure 8. The gap variation at the suction end is small due to the relatively low temperature and a fixed support boundary applied on the intake flange. At 1 mm from the discharge end, the gap variation is lower than at 25 mm due to higher housing stiffness provided by the end wall. At axial locations, between 25 and 150 mm, the gap variation shows a gradual change with lower gaps at the low-pressure CUSP as compared to the high-pressure CUSP for both housing bores. This nature of leakage gap size distribution results in lower volumetric efficiency and higher specific power of the compressor. The data could be used to redesign the housing with cooling jackets and stiffness elements.

4.3. Pressure distribution

The pressure-angle diagram of the compression cycle is compared between the three cases in Figure 10a, and the instantaneous distribution of pressure on the rotors is presented in Figure 10b. The pressure-angle diagram is compared with measurements at the same operating condition.

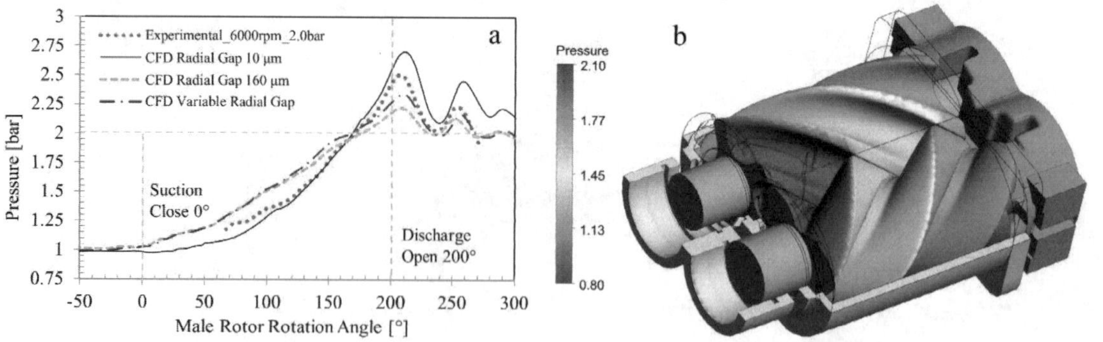

Figure 10. Compression cycle pressure distribution, a) pressure-angle diagram, b) instantaneous pressure in the rotors and suction port.

With a radial gap size of 10 μm, the initial pressure rise is more gradual and lower than the measured value. After about 150° main rotor angle, the pressure rises more steeply, and the peak pressure exceeds that of the measured value, to reach about 2.71 bar. This is followed by pressure pulsations of a higher magnitude in the remaining cycle. In the case of a radial gap size of 160 μm, the initial pressure rise exceeds that of the measured value, up to a rotor angle of 175°, but the peak pressure, at about 2.2 bar, is lower than the measured value. The variable leakage gap size also estimates a similar initial pressure rise, higher than the measured value and the peak pressure is about 2.3 bar.

4.4. Housing deformation and stress distribution

The distribution of deformation magnitude on the housing is presented in Figure 11a. It can be observed that the highest deflection is about 115 μm which correlates in magnitude as well as location to the radial leakage gap variation presented in Figure 8 and Figure 9. These results confirm validity of the bi-directional coupling setup and accuracy of the coupling solver.

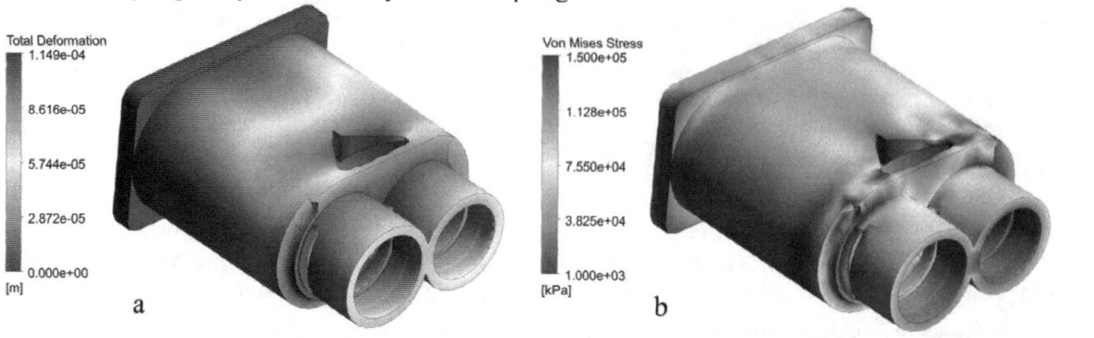

Figure 11. Housing structural analysis results, a) deformation, b) Von Mises equivalent stress.

The distribution of the Von Mises equivalent stress in the housing material is presented in Figure 11b. For the specified thermal and force load and the locations of the supports such as fixed axis distance, axial end face planar displacement and fixed mounting on the suction flange, produce locally high stress in the discharge end of the housing. The maximum stress magnitude is about 150 MPa which is within the material tensile strength of 350 MPa.

4.5. Compressor performance

A comparison of the compressor performance parameters is presented in Table 4. The measured performance data at 6000 rpm was available at 2.0 bar discharge pressure [12, 13]. With a fixed radial gap clearance of 10 μm the flow was 17% higher than that measured. With a larger radial gap of 160 μm, the flow was reduced considerably and was 18% lower than the measured value. With a variable radial gap that ranged from 30 to 120 μm, the flow was reduced and about 11% lower than that measured.

Table 4. Comparison of compressor performance at 6000 rpm.

No	Case	Flow	Torque	Indicated Power	Specific Power	Discharge Temperature	Volumetric Efficiency	Adiabatic Efficiency
		m3/min	*Nm*	*kW*	*kW/m3/min*	°C	%	%
1	Measured	7.036	25.397	15.15	2.153	129.4	68.98	67.59
2	Radial Gap 10 μm	8.244	24.355	15.30	1.856	114.1	80.83	78.39
3	Radial Gap 160 μm	5.751	23.849	14.98	2.606	135.1	56.38	55.84
4	Variable Radial Gap	6.223	24.454	15.36	2.469	132.5	61.01	58.94

The torque obtained from the CFD model relates to the indicated power only. However, the measured torque includes the mechanical losses. Hence a mechanical efficiency of 95% has been assumed here in order to compare the calculated results with the measured values. At 10 μm gap, (Figure 10a), the internal pressure rise is higher, hence the torque and indicated power is 1% higher than that measured. With 160 μm clearance, the indicated power is 1% lower than the measured and with a variable gap, the indicated power is 1.4% lower than that measured. These results can be correlated with the nature of the pressure distribution represented in Figure 10a. The discharge air temperature is 12% lower with 10 μm gap, 4.4% higher with 160 μm gap and 2.4% higher with the variable gap model. Based on the flow results, the volumetric efficiency is 11.85 units higher with 10 μm, 12.6 units lower with 160 μm and 8 units lower with the variable gap model in comparison to the measured values. Based on the indicated power, the adiabatic efficiency is 10.8 units higher with a 10 μm gap, 12 units lower with a 160 μm gap and 8.6 units lower with the variable gap model. Between the three models, the difference in specific power is of the same order as that of the flow, as the indicated power is very close in all three of them.

5. Conclusions

High temperature of the gas during compression, and the thermal deformation resulting from it, limits the maximum pressure that can be achieved in a dry air twin-screw compressor. Due to these factors, the leakage gap sizes tend to be over specified in the design, leading to lower volumetric efficiency and higher specific power. In order to improve the accuracy of computational models used, a bi-directional system coupling approach that solves CFD and structural models in co-simulation has been presented here. The following conclusions can be drawn from the study:

- Adding compressor metal components in the CFD model eliminates the requirement to specify a boundary condition at the fluid-solid interfaces for the conjugate heat transfer analysis.
- Bi-directional system coupling can be used to predict a variable radial leakage gap size, thereby improving the accuracy of performance prediction. For the test case the gap size was found to vary from 24 to 117 μm.
- A comparison between 10 μm, 160 μm and a variable radial leakage gap calculation indicated that time averaged peak temperature in the solid reaches 98°C, 129°C and 120°C respectively.
- The coupled model predicted an improved flow and volumetric efficiency by 8.2%, lower indicated power by 2.5% and a higher adiabatic efficiency by 5.5%, in comparison to the design gap size of 160 μm.
- The presented solution was a conceptual test, as rotor deformation was not included, and housing deformation was used to vary an initial 30 μm radial leakage gap. The system needs to be developed further to account for rotor deformation.
- Interlobe and high-pressure axial end leakage gaps are also required to be incorporated, to further improve the numerical model.

Such an improved CHT analysis with bi-directional system coupling can be used to accurately estimate the thermal deformation and leakage gap changes in the compressor during operation and improve the reliability of the performance results obtained from the computational models.

Acknowledgments

Funding for this research was received from Royal Academy of Engineering, UK, towards the project Smart Efficient Compression: Reliability & Energy Targets (SECRET).

References

[1] Kovačević A, Stošić N and Smith I K 2007 *Screw compressors - Three dimensional computational fluid dynamics and solid fluid interaction*, Springer-Verlag Berlin Heidelberg New York, ISBN 3-540-36302-5.

[2] Kovačević A, Stošić N and Smith I K 2002 Numerical Simulation of Fluid Flow and Solid Structure in Screw Compressors, Proceedings of ASME Congress, Symposium on the Analysis and Applications of Heat Pump and Refrigeration Systems, IMECE2002-33367, pp. 409-416, doi: 10.1115/IMECE2002-33367

[3] Sauls J, Powel G and Weathers B, 2006 Transient Thermal Analysis of Screw Compressors, Part I- Use of Thermodynamic Simulation to determine Boundary Conditions for Finite Element Analysis. Proc. Int. Compressor Conf. at Purdue, Paper 1811.

[4] Weathers B, Sauls J and Powel G 2006 Transient Thermal Analysis of Screw Compressors, Part II - Transient Thermal Analysis of a screw Compressor to determine Rotor-to-housing Clearances. Proc. Int. Compressor Conf. at Purdue, Paper 1812.

[5] Powel G, Weathers B and Sauls J 2006 Transient Thermal Analysis of Screw Compressors, Part III-Transient Thermal Analysis of a screw Compressor to determine Rotor-to-Rotor Clearances. Proc. Int. Compressor Conf. at Purdue, Paper 1813.

[6] Nikolov A, Huck C and Brummer A 2012 Influence of Thermal Deformation on the Characteristics Diagram of a Screw Expander in Automotive Application of Exhaust Heat Recovery. Proc. Int. Compressor Conf. at Purdue, pp. 1447.

[7] Ding H, Jiang Y and Dhar S 2019 CFD Modelling of Coupled Heat Transfer between Solid and Fluid in a Twin Screw Compressor, IOP Conf. Ser.: Mater. Sci. Eng. **604**, 012005, doi: 10.1088/1757-899X/604/1/012005

[8] Buckney D 2017 Clearance management in twin screw compressors. Doctoral thesis, City, University of London. https://openaccess.city.ac.uk/id/eprint/18235/

[9] Husak E, Kovačević A and Karabegovic I 2019 Calculation of clearances in twin screw compressors. IOP Conf. Ser.: Mater. Sci. Eng. 604 012001 https://doi.org/10.1088/1757-899X/604/1/012001

[10] Stošić N, Smith I K and Kovačević A 2005 *Screw Compressors: Mathematical Modeling and Performance Calculation*, Springer Verlag, Berlin, ISBN: 3-540-24275-9.

[11] Rane S, Kovačević A, Stošić N and Kethidi M 2013 Grid Deformation Strategies for CFD Analysis of Screw Compressors, *Int Journal of Refrigeration*, **36**, 7, 1883-1893. Doi: 10.1016/j.ijrefrig.2013.04.008

[12] Rane S, Kovačević A, Stošić N and Kethidi M 2014 Deforming grid generation and CFD analysis of variable geometry screw compressors. *Comput. Fluids*, **99**, 124–141. doi: 10.1016/j.compfluid.2014.04.024

[13] Rane S and Kovačević A 2017 Algebraic generation of single domain computational grid for twin screw machines. Part I – Implementation, *Adv. Eng. Softw.*, vol **107**, pp. 38–50, doi: 10.1016/j.advengsoft.2017.02.003

[14] Kovačević A and Rane S 2017 Algebraic generation of single domain computational grid for twin screw machines Part II – Validation, *Adv. Eng. Softw.*, vol **109**, pp. 31–43, doi: 10.1016/j.advengsoft.2017.03.001

[15] Kovačević A and Rane S 2013 3D CFD analysis of a twin screw expander, 8th International conference on compressors and their systems, London, p. 417-429. Cambridge: Woodhead Publishing. ISBN 9781782421696.

[16] Rane S and Kovačević A 2017 Application of numerical grid generation for improved CFD analysis of multiphase screw machines. IOP Conf. Ser.: Mater. Sci. Eng. **232**, 012017. Doi: 10.1088/1757-899x/232/1/012017

[17] Basha N, Kovačević A and Rane S 2019 Analysis of Oil-Injected Twin-Screw Compressor with Multiphase Flow Models, *Designs*, vol **3**, 4, p. 54, doi: 10.3390/designs3040054

[18] Rane S, Kovačević A, Stošić N and Smith I 2021 Analysis of real gas equation of state for CFD modelling of twin screw expanders with R245fa, R290, R1336mzz(Z) and R1233zd(E). *Int Journal of Refrigeration*. **121**, 313-326. Doi: 10.1016/j.ijrefrig.2020.10.022

IOP Conf. Series: Materials Science and Engineering 1180 (2021) 012002 doi:10.1088/1757-899X/1180/1/012002

A Thermodynamic Chamber Modelling Approach for Oil Free and Oil Injected Twin Screw Compressors

G Ramchandran[1] and J Harrison[1]

[1] Gamma Technologies, LLC., Westmont IL 60559, USA

E-mail: g.ramchandran@gtisoft.com

Abstract. As computational modelling becomes an increasingly reliable and key component in accelerating the design process for twin screw machines, the goals for engineers now include developing faster running and physically accurate component models to optimize machine performance and efficiency, minimize internal leakage, reduce unwanted noise and pulsations, and properly size bearing supports in the machine. Accurately capturing these aspects via physical models helps in analyzing operating points that were not tested as well as in understanding how the machine will perform in a surrounding system. Thereafter, engineers can find an optimal design in a timely manner for the fastest speed to market as well as reduce physical testing to keep development costs low. This paper presents the use of a multi-physics modelling platform - GT-SUITE - in conjunction with SCORG – a well-established tool for the design and analysis of twin screw machines – to explore meeting the aforementioned goals. Two case studies are presented for a 3/5 oil free air compressor and a 4/5 oil injected air compressor. Comparisons to the mass flow rates of the gas and oil, temperatures, indicated power and the instantaneous chamber pressure vs rotation angle were made against test data available from the Centre for Compressor Technology at City University. The sensitivity of oil injection timing on the discharge temperature and power is shown and an optimum timing was found. The validated chamber models may be integrated into a system as well as used for further optimization to improve the original compressor performance.

1. Introduction

Rotary twin screw compressors are positive displacement machines that are widely used in light to heavy duty industrial applications to supply compressed air at a desired, continuous rate. Both oil free and oil injected variants are predominant in the market. Modern design of screw compressors involves the use of computational tools to analyze and optimize the performance of existing designs. Many researchers have investigated lumped parameter modelling approaches for screw compressors in the years past due to the advantages they offer in terms of computational speed and accuracy compared to more detailed 3D CFD based approaches. While 3D CFD provides more detailed insights into the operation of screw machines, it can be time consuming and not particularly convenient for geometrical optimization. Some of the original numerical models for oil free and oil injected screw compressors [1-10] were based on the approach of solving the equations of conservation of mass and energy applied to a control volume. Geometrical considerations such as the change in the chamber volume with rotation angle and the estimation of various leakage gaps are important aspects of these models to predict the performance of these machines with an acceptable degree of accuracy. Recent advances in computational capabilities and the demand for quicker designs at low cost has spawned more advanced modelling efforts for screw compressors. Some of these include the works in [11-15].

In this paper, a commercial system modelling tool, GT-SUITE, is used in conjunction with an industry established screw compressor geometry modelling tool, SCORG, to formulate an effective alternative to existing thermodynamic chamber models for screw compressors. GT-SUITE has been used to model swashplate type refrigerant-based compressors [16], scroll compressors [17], and screw expanders [18-22]. This is the first study showcasing the ability to model twin screw air compressors using GT-SUITE. Two case studies are presented comparing GT-SUITE's model results to SCORG's thermodynamics solver and test data for oil free and oil injected screw compressors. Also shown is a use case for optimizing screw compressor designs using GT-SUITE.

2. Modelling workflow

A complete model considers all flow paths present within the system. GEM3D (a GT-SUITE pre-processor) was used to import the CAD of the flow path, discretize this into multiple pipes and flowsplits, and finally export the flow system into the GT-ISE modelling environment with the screw machine template. SCORG was then used to generate the chamber volume and port/leakage area profiles. Seamless information transfer from SCORG to GT-SUITE greatly reduces model setup time. Figure 1 illustrates the modelling workflow followed. Details are provided in the subsequent sections.

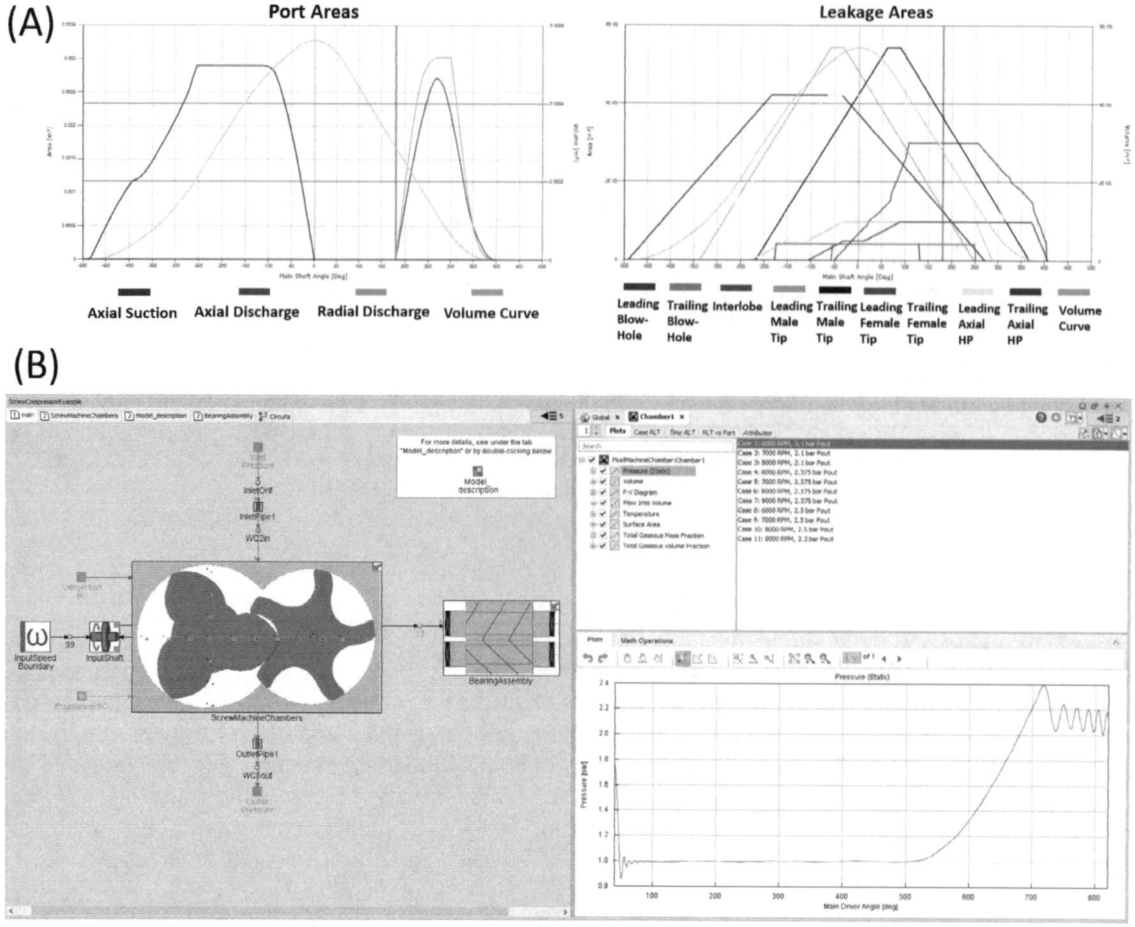

Figure 1. (a) Geometrical parameters and profiles generated by SCORG and written into text files, (b) GT-SUITE model run with pre-processed inputs.

2.1. *Model setup and pre-processing in SCORG*

SCORG V5.8.3 was utilized for the first step in the modelling process which involves setting up the rotor profile geometry, defining ports, thermodynamic operating conditions, working fluid, and oil injection (if present). Once geometry calculations are run, a text file containing the chamber volume, inlet and outlet port areas (radial/axial/both) and leakage areas for each rotation angle of the main rotor is generated for use by the GT-SUITE chamber model. An additional text file containing parameters such as boundary pressures, temperatures, compressor speed, geometrical features for the inlet/outlet pipes, port volumes, etc. is generated for the GT model. Optionally, force calculations may be run in SCORG to generate another text file with the axial and radial bearing loads vs angle in order to take advantage of the detailed journal and roller bearing models available in GT-SUITE to evaluate compressor rotordynamics, modal analysis, friction losses and bearing performance. However, these topics are out of the scope of the current study which is focused on the thermodynamic performance.

2.2. *Thermodynamic chamber modelling in GT-SUITE*

Once the geometry calculations are complete, the GT-SUITE solver may be accessed either from within SCORG itself or via the GT-SUITE user interface. Templates with various combinations of the number of working chambers are available in GT-SUITE. Running the GT-SUITE solver from within the SCORG interface automatically picks the GT template with the expected number of chambers.

The GT templates contain chamber models built from flow components and connections as shown in Figure 2. Pipe templates are used for the inlet and outlet piping. Every working chamber is linked to the inlet and outlet port volumes that are modeled as flowsplits. Individual links are present for the axial and radial ports, as well as for the various leakage paths between chambers. The compressor speed is imposed on the shaft. The number of chambers in the model is evaluated based on the product of the number of main rotor lobes and the integral number of shaft rotations in one working cycle.

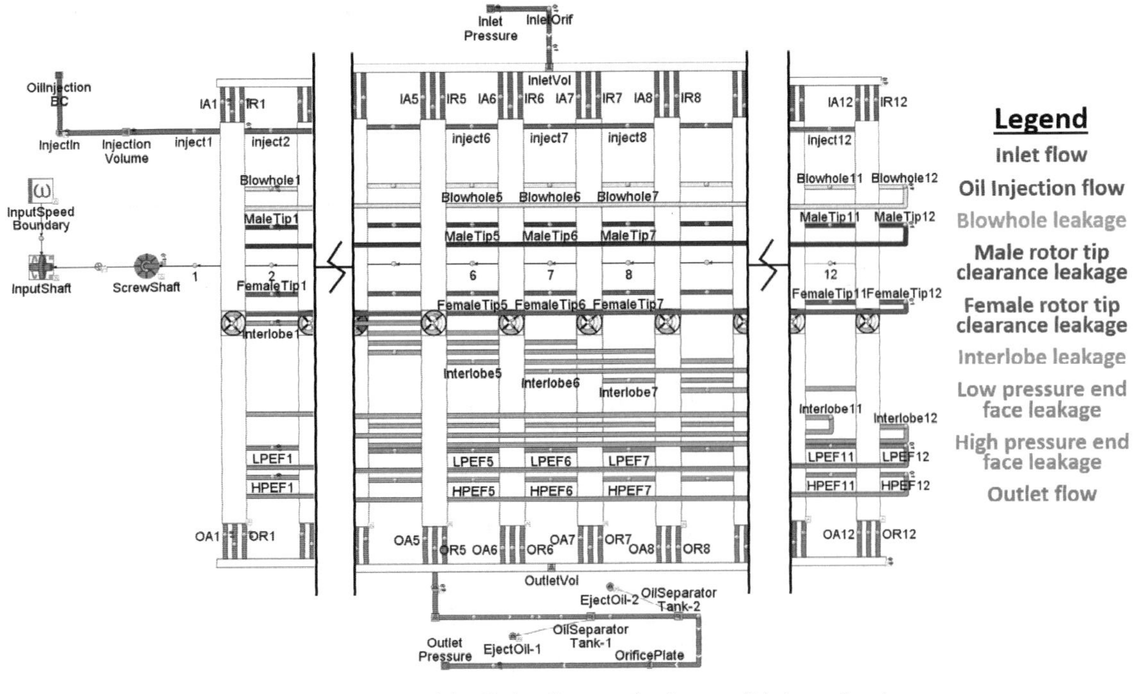

Figure 2. GT-SUITE model map with all the flow paths in an oil injected twin screw compressor

The pre-processor GEM3D is a useful tool where the CAD of the flowsystem (pipe volumes) is imported, cutting planes are used to separate the flowsystem into different flow components, and a simple conversion process is used to convert each shape into a flow component. The 1D flow model consists of the compressor flow system which is discretized into many volumes comprised of pipes (which are further discretized into subvolumes) and flowsplits (single volumes). A staggered 1D grid approach is adopted (as indicated in Figure 3) where scalar quantities like pressure, internal energy, temperature, density, etc. are solved at the center of each subvolume, and vector quantities like mass flow rate, velocity, species flow rate, etc. are solved at the boundary between subvolumes.

GT-SUITE solves equations for the conservation of continuity, momentum, energy and species in 1D, which means all quantities are uniform within each pipe subvolume. The timestep size used by the explicit flow solver is adaptive and always small enough to satisfy the Courant condition. This allows for the ability to capture phenomena such as unsteady flow, pressure pulsation and high frequency wave dynamics. Although the code is nominally 1D in pipes, flowsplits are specifically designed to account for conservation of momentum in 3D. Scalar terms are solved at the volume's center while the solution of momentum equation (vector quantities) is carried out separately at each volume opening (boundary). For the momentum equation, the flowsplit geometry is characterized for each boundary by its expansion diameter (diameter into which flow expands after entering flowsplit), characteristic length (distance from boundary plane to opposite side of flowsplit) and boundary orientation (relative angle of each boundary). A characteristic velocity vector is calculated for the flowsplit based on the contributions of momentum flux from all of its boundaries. The momentum flux out of the flowsplit is calculated by using the component of the characteristic velocity in the direction of the boundary. This adequately captures flow through boundaries with arbitrary orientations which is not possible with conventional 1D treatments or typical lumped parameter modelling approaches.

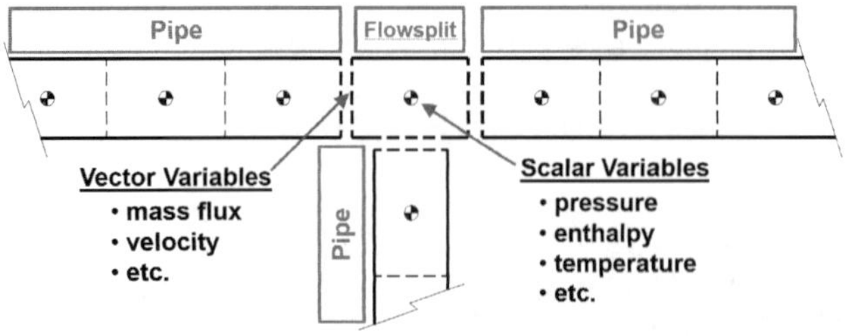

Figure 3. Staggered grid approach in GT-SUITE to solve conservation equations and species transport

In the current setup, the walls of the flow components are assumed to be adiabatic, so no heat transfer is considered between the fluid and the wall. However, the option to consider heat transfer is available and may be investigated in future studies. Flow through leakage paths and between adjacent subvolumes are modelled through orifices. These are defined via diameters or areas and discharge coefficients as inputs. The momentum equation is solved to calculate mass flow rate through the orifices. For gases, the mass flow rate is calculated based on the isentropic nozzle relationships for subsonic and choked flow regimes (Equations 1 and 2 respectively).

$$\dot{m} = C_D A_R \rho_o \left(\frac{P_{2,Static}}{P_{1,Total}}\right)^{1/\gamma} \sqrt{RT_o} \left\{\frac{2\gamma}{\gamma-1}\left[1-\left(\frac{P_{2,Static}}{P_{1,Total}}\right)^{\frac{\gamma-1}{\gamma}}\right]\right\}^{1/2} \tag{1}$$

$$\dot{m} = C_D A_R \rho_o \left(\frac{2}{\gamma+1}\right)^{\frac{1}{\gamma-1}} \sqrt{\gamma RT_o} \left\{\frac{2}{\gamma+1}\right\}^{1/2} \tag{2}$$

Here, $\dot{m}$ is the mass flow rate, ρ_o is the upstream stagnation density, C_D is the discharge coefficient, A_R is the reference flow area, $P_{1,Total}$ is the total pressure upstream of restriction, $P_{2,Static}$ is the static pressure downstream of restriction, R is the gas constant, T_o is the upstream stagnation temperature, and γ is the specific heat ratio.

In the case of liquids, the flow rate is based on the incompressible Bernoulli flow equation through an orifice plate (Equation 3). Here, D is the orifice diameter and ρ is the liquid's density.

$$\dot{m} = C_D \rho \left(\frac{\pi D^2}{4}\right) \sqrt{\frac{2(P_{1,Total} - P_{2,Static})}{\rho}} \tag{3}$$

The transient species mass fraction is solved for within each subvolume and this is important in the case of oil injected compressors where the oil and air are modeled as a homogeneous mixture within each subvolume. Flow is based on the species mass fraction in the upstream subvolume.

In the case studies discussed in the following section, the models built have between 41 to 65 flow volumes depending on the total number of chambers. The average timestep sizes are 1.5e-5 seconds and 6.2e-6 seconds for the oil free and oil injected compressor models, respectively. Most operating conditions converge in 6 to 7 cycles, requiring less than a minute of run time on a modern PC.

3. Case studies

Two screw compressors for which test measurements are available in the public forum were chosen to validate the GT-SUITE model predictions. Measurements at a few operating points were carried out on the air compressor test rig at the Centre for Compressor Technology in City University of London. Details regarding the test setup such as the various sensors (orifice plates, pressure transducers, thermocouples, torque meters) for measuring flow, pressures, the instantaneous chamber pressure in the interlobes, temperatures, and the compressor torque are available for the dry [23] and oil injected air compressors [24]. Figure 4 shows the test schematics for both compressors. Additional steady state measurements for the dry compressor are available in [13]. To highlight GT-SUITE as an effective alternative to one of the predominantly used chamber models in the industry, comparisons were also made to SCORG's thermodynamic solver in addition to test data.

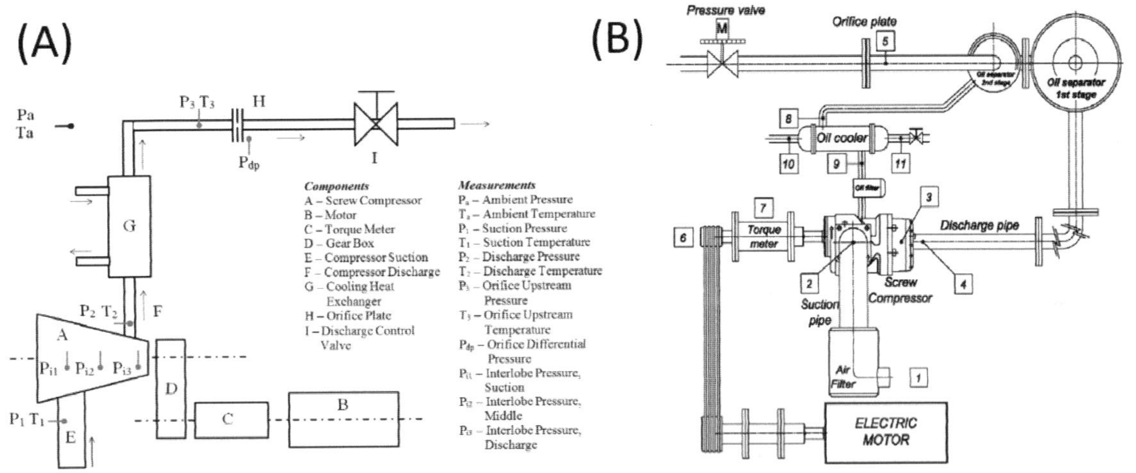

Figure 4. Test schematic for (a) oil free compressor from [23], (b) oil injected compressor from [24].

Table 1 contains the range of operating conditions presented in these comparisons. Table 2 provides information regarding the geometry of the two compressors being modeled. The interlobe, radial, and end axial clearances were reduced by 5 to 20% from the nominal values in the models in order to better correlate to the test measurements. This was deemed acceptable since the operational clearances vary non-uniformly from nominal clearances based on the operating conditions.

Table 1. Operating Conditions Modelled for Comparison to Measurement

	Units	Oil Free Machine	Oil Injected Machine
Compressor Speed	RPM	6000 to 9000	3000 to 6000
Suction Pressure	bar	1.0	1.0
Suction Temperature	K	298.15	298.15
Discharge Pressure (Pd)	bar	2.0 to 2.5	6.0 to 8.0
Oil Injection Pressure	bar	-	Pd - 0.5

IOP Conf. Series: Materials Science and Engineering 1180 (2021) 012002 doi:10.1088/1757-899X/1180/1/012002

Table 2. Screw Compressor Geometry (with 'N' rotor profiles)

	Units	Oil Free Type	Oil Injected Type
Main Rotor Lobes	-	3	4
Gate Rotor Lobes	-	5	5
Axis Center Distance	mm	93	93
L/D Ratio	-	1.6	1.55
Main Rotor Diameter	mm	127	127
Wrap Angle	deg	306	306
Built-in Volume Index	-	1.9	3.6
Displacement	cc/rev	1787	1512

3.1. Case study 1 – oil free screw compressor

In this section, some of the GT-SUITE model results and comparisons to test data and the SCORG thermodynamics solver are presented for the oil free twin screw compressor.

3.1.1. Pressure-angle diagram. Figure 5 shows the variation of the chamber pressure with the angle of rotation of the main rotor at two different speeds and for different pressure ratios. While the general trend and GT-SUITE comparisons to test measurement are good, there are some differences noticed close to the opening of the discharge port near the pressure peak. Additionally, the pressure wave dynamics occurring during the compression of the working chamber is captured reasonably well compared to test and is an improvement to SCORG's prediction where the oscillations appear to be over-damped. The peak pulsation is also captured more accurately as compared to SCORG.

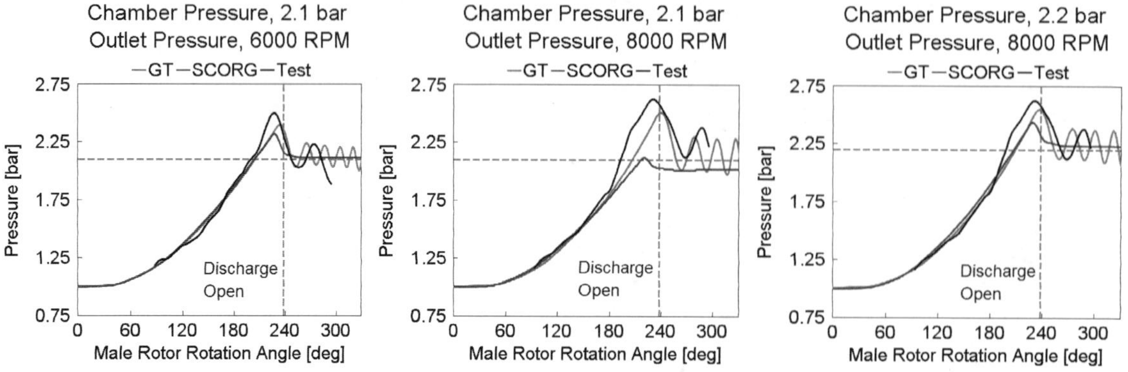

Figure 5. Comparisons of chamber pressure variation for the oil free screw compressor.

3.1.2. Mass flow rates. Figure 6 shows the comparison of the mass flow rate predictions against test data at different compressor speeds and pressure ratios. By adjusting clearances, a good match was found between test and GT-SUITE. This tracks well with what is also predicted by the SCORG model.

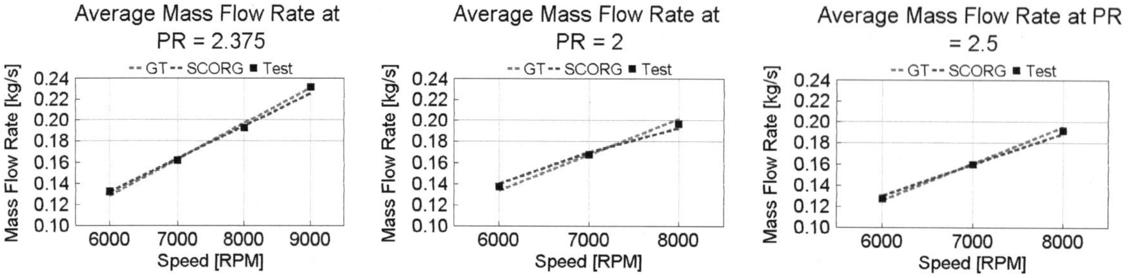

Figure 6. Comparisons of average mass flow rates at various speeds for the oil free screw compressor.

3.1.3. Indicated power. Figure 7 shows the comparison of the GT-SUITE indicated power predictions against test data and SCORG at different compressor speeds for a pressure ratio of 2.0. Since the power was measured on the shaft, a constant mechanical efficiency of 95% was assumed for the gearbox driving the main rotor in the test data across all speeds. While probably not the best estimate since the mechanical efficiency of the gearbox typically varies with compressor speed, there is a good agreement in the results with the model predictions within 8% of the test measurements. The GT-SUITE results show a closer prediction to test data compared to SCORG.

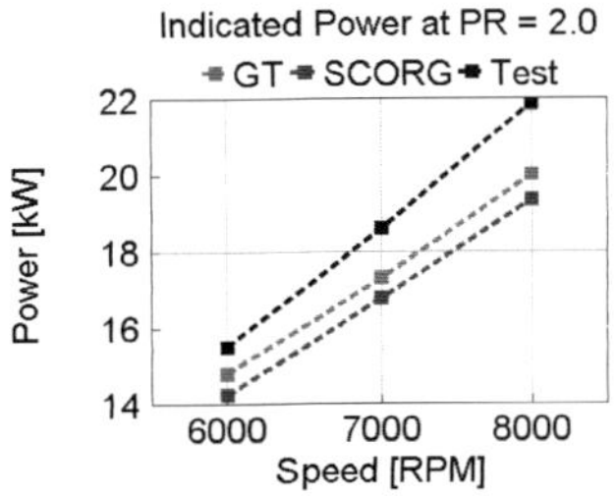

Figure 7. Comparisons of the indicated power at various speeds for the oil free screw compressor.

3.2. Case study 2 – oil injected screw compressor
In the next study, model comparisons to available test data for the case of the oil injected twin screw compressor are discussed.

3.2.1. Mass flow rates. Figure 8 shows the comparison of the normalized mass flow rates predicted by the model compared to available test data. The model trend compares well to the test data with the predicted values being within 7% of the test results.

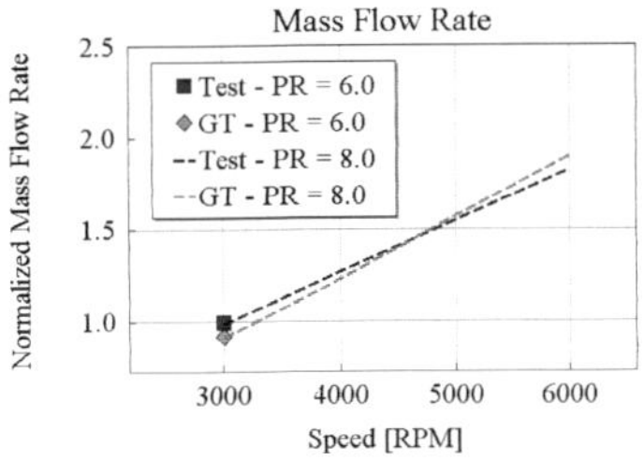

Figure 8. Comparison of the normalized average mass flow rates at different steady state speeds to test data for the oil injected screw compressor.

3.2.2. Indicated power. Figure 9 shows the comparison of the normalized indicated power predicted by the model compared to available test data at various compressor speeds for the same operating conditions. A constant mechanical efficiency of 70% was applied to the measured power to estimate the experimental indicated power. For a better comparison against the actual measured power, the built-in GT-SUITE friction model for seals, bearings and rotor drag may be used. Nevertheless, with the current assumption, the model predictions are within 9% of the test data.

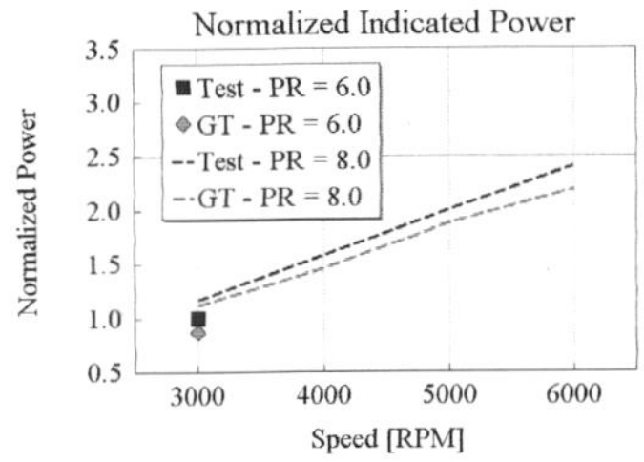

Figure 9. Comparison of the normalized indicated power at different steady state speeds to test data for the oil injected screw compressor.

3.2.3. Additional observations. Figure 10 shows a few other model predictions. It can be seen that the temperature never exceeds 335 K even at 6000 RPM and a pressure ratio of 8.0. From the P-V diagram, it is clear that the amplitude of the pressure pulsations in the chamber increase with both speed and discharge pressure. The compressor appears to be under-compressed at the operating conditions analysed, i.e., the pressures in the chambers are less than the targeted discharge pressures at the time of the discharge port opening. The peak over-compression can be attributed to throttling and dynamic losses occurring near the outlet port. This restriction is captured by the model with an orifice downstream of the discharge port. A 3D CFD analysis of the same compressor [25] confirms a similar trend. The model also predicts the mass fraction of oil being injected into the working chamber.

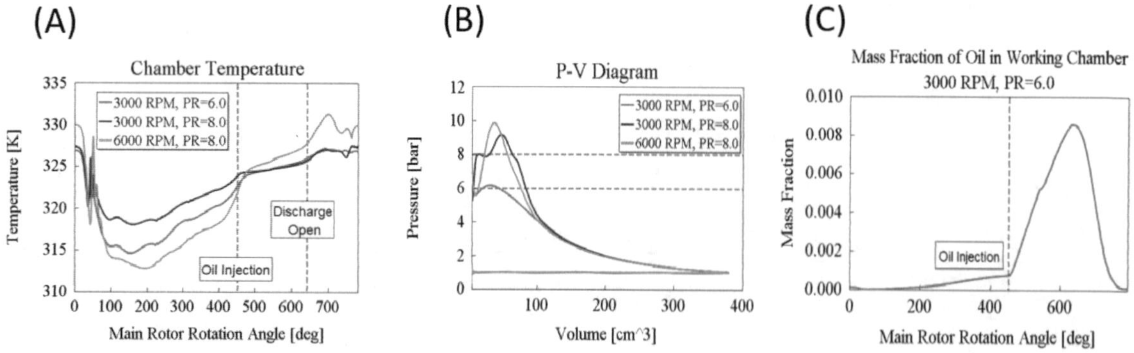

Figure 10. (a) Comparison of chamber temperatures, (b) Comparison of the P-V diagrams, (c) Model predicted instantaneous mass fraction of oil in the working chamber

3.2.4. Optimizing oil injection. An exercise was carried out to study the sensitivity of varying oil injection timing and injection port areas on the discharge temperature and indicated power of the compressor. The built-in design optimizer in GT-SUITE was used and a genetic algorithm was chosen to conduct this study. The objective of this design optimization was to find pareto-optimal designs that minimized the discharge temperature of the compressor while maximizing its indicated power. The objective functions for the average discharge temperature and the indicated power were created by equally weighting the two operating conditions: (3000 RPM, PR = 6) and (6000 RPM, PR = 8.0). 150 designs were evaluated by the genetic algorithm. Figure 11 shows the contour plots of the discharge temperature and indicated power plotted for different combinations of timing shift (base design angle shift = 0) and injection port areas (base design injection area multiplier = 1). It is evident that as more oil is injected into the compressor, temperature in the working chamber and subsequently discharge temperature is closer to the oil injection temperature. With more injected oil combined with earlier oil injection (negative angle shift) in the working cycle, a better design is achievable. The pareto plot in Figure 12 indicates all the combinations of optimized designs available to choose from.

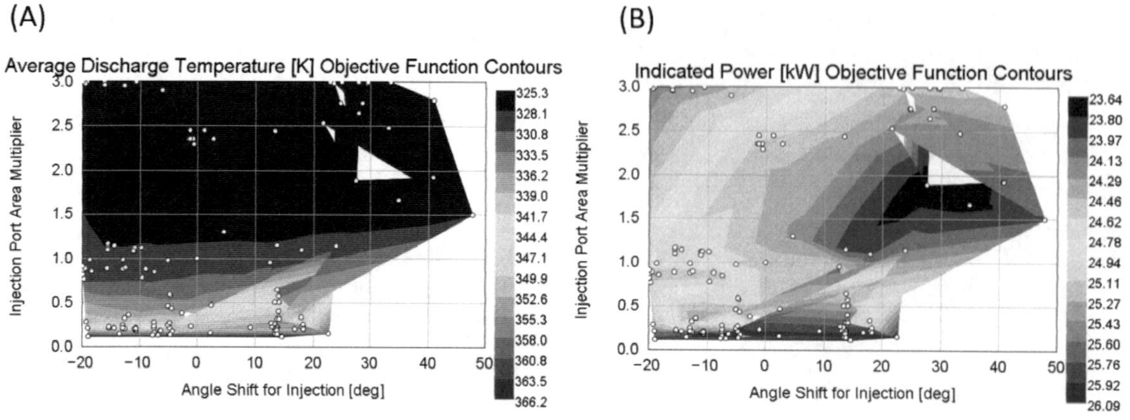

Figure 11. Contour plots of (a) average discharge temperature, and (b) average indicated power, (both equally weighted for the two operating conditions considered) for designs varying the injection timing and injection port area magnitude.

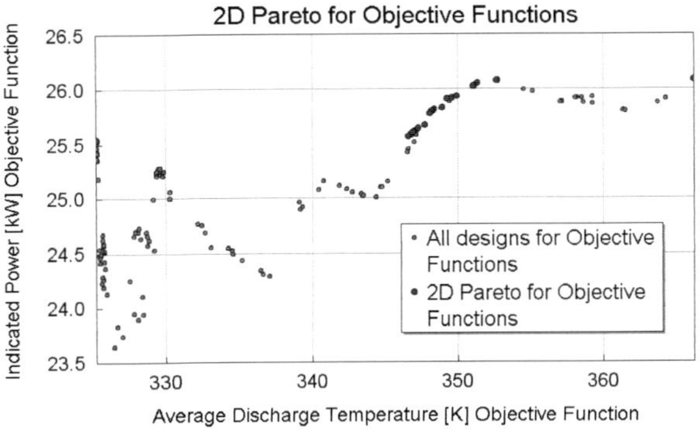

Figure 12. Pareto optimal designs obtained by maximizing indicated power and minimizing average discharge temperature.

4. Conclusions

Thermodynamic chamber models were formulated for oil free and oil injected twin screw air compressors using GT-SUITE. The overall objective of this work has been to provide screw compressor designers with a toolset to find optimal designs in a timely manner for the fastest speed to market while reducing physical testing and keeping development costs low.

- Model comparisons to test measurements of mass flow rates, instantaneous chamber pressures, and indicated powers of both test compressors from available literature show strong agreement.
- The validated model may be used to predict performance at operating conditions difficult to test.
- A design analysis was conducted to evaluate the optimal injection port area as well as the injection timing. A set of pareto-optimal designs were identified, and it was observed that by increasing the oil injection port area and advancing the injection timing, designs offering a lower discharge temperature along with higher indicated powers may be obtained.
- Future work may involve adding a friction model to evaluate operational power accurately and break down the different contributors to friction power loss (bearings, seals, rotor drag, etc.).
- The model may be extended to study the operation of these compressors when present in a system.
- Further studies may also be conducted to minimize leakages, noise and pulsation (muffler design), size bearing supports for sustainable operation, and model shaft rotordynamics in GT-SUITE.

Acknowledgments

The authors would like to thank Dr. Ahmed Kovacevic and Dr. Sham Rane from City University of London for their support during model development by providing crucial inputs necessary to build the compressor models and experimental data for validation.

References

[1] Fujiwara M, Mori H, and Suwama T 1974 Prediction of Oil Free Screw Compressor Performance Using Digital Computer *International Compressor Engineering Conference at Purdue* 119

[2] Fujiwara M, Kasuya K, Matsunaga T, and Watanabe M 1984 Computer Modeling for Performance Analysis of Rotary Screw Compressor *International Compressor Engineering Conference at Purdue* 503

[3] Fukazawa Y and Ozawa U 1980 Small Screw Compressors for Automobile Air Conditioning Systems *International Compressor Engineering Conference at Purdue* 351

[4] Sangfors B 1982 Analytical Modeling of Helical Screw Machine for Analysis and Performance Prediction *International Compressor Engineering Conference at Purdue* 144

[5] Sangfors B 1984 Computer Simulation of the Oil Injected Twin Screw Compressor *International Compressor Engineering Conference at Purdue* 528

[6] Singh P J and Onuschak A D 1984 A Comprehensive Computerized Method For Twin Screw Rotor Profile Generation and Analysis *International Compressor Engineering Conference at*

Purdue 501
[7] Dagang X, Xiong Z, and Yu Y 1986 The Computer Simulation of Oil-Flooded Refrigeration Twin-Screw Compressor *International Compressor Engineering Conference at Purdue* 537
[8] Kauder K, and Harling H B 1994 Visualisierung der Olverteilung in Schraubenkompressoren *Proc. VDI Tagung "Schraubenmaschinen 94"* Dortmund VDI Berichte 1135
[9] Fujiwara M and Osada Y 1995 Performance Analysis of Oil Injected Screw Compressors and their Application *Int. J. Refrig* **18**, 4
[10] Janicki M 2007 Modellierung und Simulation von Rotationsverdrängermaschinen PhD Disseration TU Dortmund
[11] Hanjalic K and Stosic N 1997 Development and Optimization of Screw Machines with a Simulation Model - Part II, Thermodynamic Performance Simulation and Design *ASME Transactions Journal of Fluids Engineering* **119** 664
[12] Venumadhav K, Stosic N, Kovacevic A, and Smith I K 2001 The Design of a Family of Screw Compressors for Oil–Flooded Operation *International Conference Compressors and Their Systems at London*
[13] McCreath P, Stosic N, Kovacevic A, and Smith I. K2001 The Design of Efficient Screw Compressors for Delivery of Dry Air *International Conference Compressors and Their Systems at London*
[14] Stosic N, Smith I K and Kovacevic A 2003 Optimisation of Screw Compressors *Applied Thermal Engineering* **23** pp. 1177-1195
[15] Stosic N, Smith I. K. and Kovacevic A 2005 *Screw Compressors, Mathematical Modeling and Performance Calculation, Monograph* (Berlin: Springer Verlag)
[16] Ramchandran G, Bhatia K, and Aihara R 2018 Simulation of 1D Flow Coupled with 3D Multi-Body Dynamics Model of a Double-Acting Swashplate Compressor *International Compressor Engineering Conference at Purdue* 2623
[17] Harrison J N, Koester S, Aihara R, and Ratner D 2018 From CAD to 1D: A Direct Approach to Modeling Scroll Compressors with Multi-Physics Simulation *International Compressor Engineering Conference at Purdue* 2539
[18] Bianchi G, Kennedy S, Zaher O, Tassou S. A, Miller J, and Jouhara H 2018 Numerical modeling of a two-phase twin-screw expander for Trilateral Flash Cycle applications *Int. J. Refrig* **88** 248-259
[19] Bianchi G, Marchionni M, Kennedy S, Miller J, and Tassou S 2019 One-Dimensional Modelling of a Trilateral Flash Cycle System with Two-Phase Twin-Screw Expanders for Industrial Low-Grade Heat to Power Conversion *Designs* **3** 41
[20] Bianchi G, Marchionni M, Miller J and Tassou S 2020 Modelling and off-design performance optimisation of a trilateral flash cycle system using two-phase twin-screw expanders with variable built-in volume ratio *Applied Thermal Engineering* **179** 115671
[21] Vimalakanthan K, Read M, and Kovacevic A 2020a Numerical Modelling and Experimental Validation of Twin-Screw Expanders *Energies* **13** 4700
[22] Vimalakanthan K, Read M, and Kovacevic A, 2020b Performance Prediction and Optimisation of Twin-Screw Expander *Proceedings of the IIR International Rankine 2020 Conference-Heating, Cooling and Power Generation*
[23] Rane S, 2015 Grid Generation and CFD Analysis of Variable Geometry Screw Machines Thesis City University of London
[24] Mujic E 2009 A Numerical and Experimental Investigation of Pulsation Induced Noise in Screw Compressors PhD Thesis City University London
[25] Rane S R, Kovacevic A and Stosic N 2016 CFD Analysis of Oil Flooded Twin Screw Compressors *International Compressor Engineering Conference at Purdue* 2392

Special feature of a screw vacuum-compressor 'chamber' model development

T N Mustafin, R R Yakupov, M S Khamidullin, I G Khisameev, V A Alyaev

Kazan National Research Tehnological University, 68 Karl Marx str., Kazan, 420015, Russian Federation

Email: mustimur@rambler.ru

Abstract. Wildly using of screw vacuum pumps and vacuum-compressors (which discharge pressure is higher than atmosphere pressure) can be explained by their easy service, reliability, excellent energetic and capacity characteristic and excellent vacuum, they produce. Their future optimization can be reached by using of the working process mathematic modelling, which can takes the place of expensive experimental research (can reduce the count of it). In an engineer point of view, chamber models of the positive displacement machine are still popular and wildly used due to simple of the initial data entering, small calculation time and simple of the results' explanation. Therefore, their development and updating are still actual goal. The chamber model accuracy are determined generally by the accuracy of leakage flow calculation through the gaps and ports and by the accuracy of heat exchange calculation inside the working chamber. This is especially important for the vacuum machines, which minimal vacuum pressure are determined by the internal leakages. The vacuum inside the working chamber can lead to not only viscosity flow, but also to molecular and mixed flows. Therefore, using only viscosity flow calculation methodises leads to the significant errors. The vacuum also leads to change the heat exchange conditions inside the working chamber. Therefore, those facts should be taken into account. This paper is dedicated to the presenting the way of the solving the problems described above.

Keywords: vacuum compressor, 'chamber' model, compressor characteristics

1. Introduction

The object of the study is a screw vacuum compressor, the structural diagram of which is given in figure 1.

The screw vacuum compressor [1] falls within the class of machines of the positive displacement principle of operation – that is, machines in which the processes of suction, transfer, compression and displacement of the aspirated medium take place in chambers with a time-variable volume. The main working bodies are rotors 1 and 2, which are coarse pitch screws with the pitch variable along the length. Working cavities are formed in their depressions and are limited by their surfaces as well as by the surface of the pump housing. Suction is performed through nozzle 3 and the suction port located on the inner part of the pump housing. The gas is released through the discharge port, also located on the inner part of the housing, through check valve 4. This valve prevents backflow of the medium into the pumped out volume. The pump rotors are in constant engagement with each other, while the torque transfer from drive rotor 1 to driven rotor 2 is carried out by means of synchronising gears 6, ensuring the absence of direct contact between the rotors during the pump operation. The pump is driven by electric motor 5, the internal windings of which are directly mounted at the end of drive rotor 1.

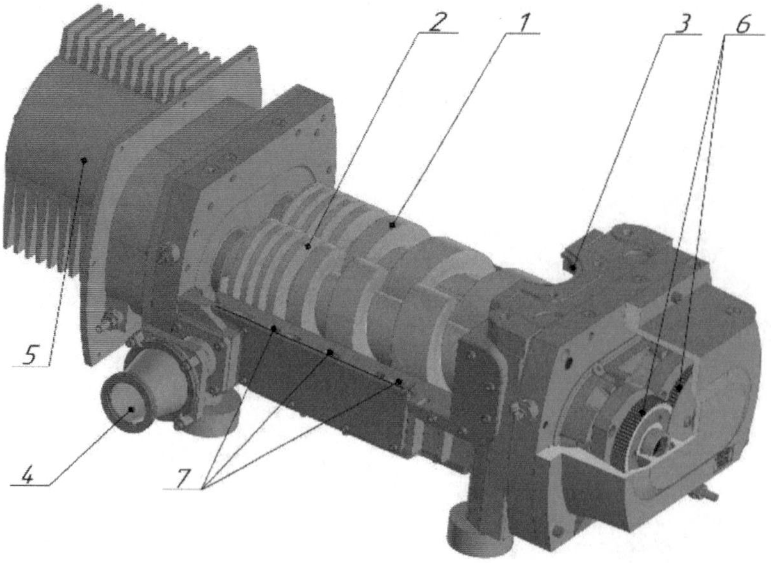

Figure 1. Structural diagram of the screw vacuum pump:

1 – main rotor, 2 – driven rotor, 3 – suction nozzle, 4 – discharge nozzle with check valve, 5 – electric motor, 6 – synchronising gears, 7 – water channels

One of the features of vacuum pumps and positive displacement vacuum compressors in comparison with similar compressors is the extremely large values of the compression ratios at low pressure drops in the cavities. This leads to significant heating of the pumped-out medium, the heat from which is partially removed with the cooling water flowing in channels 7 of the compressor housing, which washes over it from all four sides.

2. Mathematical model of the working process

For the theoretical study of the vacuum compressor under consideration, a 'chamber' model has been selected based on the law of conservation of energy and the equation of state and presented in the form of two differential equations of the first order [2, 3]:

$$\begin{cases} \dfrac{dP}{d\varphi} = \dfrac{(k-1)}{V\cdot\omega}\cdot\left(\dfrac{dQ}{d\tau} + \sum_{i=1}^{n}\left(h_{in.i}\cdot\dot{m}_{in.i}\right) - \sum_{j=1}^{m}\left(h_{ex.i}\cdot\dot{m}_{ex.i}\right) - \dfrac{\omega\cdot k\cdot P}{(k-1)}\cdot\dfrac{dV}{d\varphi}\right) \\ \dfrac{dT}{d\varphi} = \dfrac{(k-1)\cdot T}{P\cdot V\cdot\omega}\cdot\left(\dfrac{dQ}{d\tau} + \sum_{i=1}^{n}\left(\left(h_{in.i} - h\right)\cdot\dot{m}_{in.i}\right) + h\cdot\left(\dfrac{k-1}{k}\right)\cdot\left(\sum_{i=1}^{n}\left(\dot{m}_{in.i}\right) - \sum_{j=1}^{m}\left(\dot{m}_{ex.i}\right)\right) - P\cdot\omega\cdot\dfrac{dV}{d\varphi}\right), \end{cases} \tag{1}$$

where φ is the angle of rotation of the drive element (drive rotor); P and T are the pressure and the temperature in the working chamber, respectively; V is the current volume of the working chamber; ω is the angular speed of the drive element (drive rotor); $h = c_P\cdot T$ is the enthalpy of the compressed medium; c_P is the specific heat capacity at constant pressure of the compressed medium; c_V is the specific heat capacity at constant volume of the compressed medium; $k = \dfrac{c_P}{c_V}$ is the adiabatic index of the compressed medium; $\frac{dQ}{d\tau}$ is the heat gain supplied or removed from the compressed medium due to heat exchange with the walls of the housing and the surface of the rotors; $\dot{m}$ is the mass flow rate of inflow or outflow gas from the working cavity; without the index, the parameters refer to parameters in the working cavity; n and m are the numbers of slots and cavities through which inflow and outflow occur; the index '$_{in.}$' relates to the parameters of the inflow medium; '$_{ex.}$' relates to the parameters of the outflow medium.

The system (1) is a mathematical model of the process in the working cavity of 'dry' machines with the positive displacement principle of operation, based on the following assumptions:

- The medium in the volume under consideration is homogeneous and continuous.
- The parameters of the medium vary throughout the volume simultaneously and in the same way.
- The change in the kinetic energy and energy of the medium associated with its position in the volume under consideration is disregarded.
- The medium in the reference volume is considered an ideal gas obeying the Mendeleev–Clapeyron equation.

The dependences of the volume of the working chamber, the geometric parameters of the ports of the vacuum compressor as well as the dimensions of the slots formed in the working chamber as a function of the angles of rotation of the drive element are determined based on the analysis of the design, and the dimensions, configuration and profile of the rotor screws. In fact, these dependencies are determined initially and are the initial data along with the parameters of the investigated operating mode for the calculation of the operating mode. They can only be refined iteratively, taking into account the thermal strains of the main parts of the vacuum compressor and calculated based on the parameters of the working process in a given operating mode.

The mathematical model describes the phenomena and processes of a real object approximately, which is mainly governed by the assumptions made during the derivation of each of the equations used, the assumptions of the mathematical model development and the errors of the numerical values used.

The development of the mathematical model of the screw vacuum compressor at this stage has been performed subject to the following assumptions:

1. The pressure in the suction and discharge nozzles is assumed to be constant – that is, on suction and discharge, the pump is connected to the cavities of infinitely large volume, and the discrete gas supply does not affect the pressure change in them.

2. The working fluid within the cavity under consideration is assumed to be homogeneous – that is, its parameters are the same for any point, and mixing of the incoming gas and the gas in the working cavity occurs instantly.

3. Since the processes in the working cavities occur in the same way and are displaced along the angle of rotation of the rotor, only one working cavity is under consideration. This creates the need for an iterative calculation using the indicator and temperature diagram of the previous approximation. The parameters in adjacent cavities in the calculations of each approximation are taken to be the same as in the given cavity at the respective angle of rotation in the previous approximation.

The following are taken as initial conditions:

1. The pressure and the temperature at the end of the suction process are taken as the initial pressure and temperature of the working process.

2. The temperature of the working fluid in the discharge nozzle is initially set in sufficient degree arbitrarily; therefore, the discharge temperature in the first approximation is selected based on the adiabatic compression condition. At the moment of equalisation of pressures in the working cavity and the discharge nozzle, the discharge temperature is redistributed and taken as equal to that in the working cavity. This predetermines the need to calculate the parameters of the gas upon discharge in an iterative way.

3. Indicator and temperature diagrams of the first approximation are set from the condition of constant pressure and temperature in the suction and discharge processes and their instant change when the working cavity is connected to the discharge port, the compression and expansion processes are taken as polytropic.

We shall highlight the following stages of the working process, which differ both in terms of the nature of the phenomena occurring and in terms of the pattern of leaks: suction; gas compression; discharge.

The processes of suction, compression and discharge are themselves subdivided into processes that differ in the pattern of leakage through the slot. There are several cavities communicating with each other through the slotted gaps in the machine simultaneously, in which various stages of the working process occur. They are displaced relative to each other along the angle of rotation of the rotor by the angles that are multiples of periodisation. It is enough to conduct the simulation for one working cavity since the rest represent the same cavity at other angles of rotation of the drive rotor. Therefore, the parameters in adjacent cavities are determined from the previous iteration. The calculation is carried out until the convergence of the iterative values of temperature and pressure is less than 2%.

The volume delivery of the compressor is determined by integrating the indicator and temperature diagrams:

$$S = \frac{1}{\omega_1} \cdot \frac{\int_{\varphi 1}^{\varphi 2} m_{\mathrm{W}} \cdot \mathrm{d}\varphi}{\rho_{\mathrm{S}}} \cdot n \cdot z, \tag{2}$$

where ρ_{S} is the density of the medium in the suction nozzle; m_{W} is the instantaneous mass flow through the port; $\varphi 1$ is the discharge start angle; $\varphi 2$ is the discharge start angle; z is the number of cavities involved in the discharge process per one rotor revolution; n is the rotation frequency.

The work done on the gas within one cycle is determined on the basis of the indicator and temperature diagrams of the working cavity, obtained by integrating the differential equations of the working process of the vacuum pump, by integrating the differential equation:

$$\frac{\mathrm{d}L}{\mathrm{d}\varphi} = \frac{(k-1)}{\omega} \cdot \left(\frac{\mathrm{d}Q}{\mathrm{d}\tau} + \sum_{i=1}^{n} \left(h_{\mathrm{in.i}} \cdot \dot{m}_{\mathrm{in.i}} \right) - \sum_{j=1}^{m} \left(h_{\mathrm{ex.i}} \cdot \dot{m}_{\mathrm{ex.i}} \right) - \frac{\omega \cdot k \cdot P}{(k-1)} \cdot \frac{\mathrm{d}V}{\mathrm{d}\varphi} \right), \tag{3}$$

The indicator power is determined as:

$$N_{\mathrm{ind}} = Z_1 \cdot L \cdot \frac{\omega}{2 \cdot \pi}, \tag{4}$$

where Z_1 is the number of cavities (teeth) of the drive rotor.

3. Calculation of gas leaks in the working chamber of the vacuum compressor

To solve equation (1), it is required to determine the gas flow rate through the slots formed in the working chamber. The calculation method from [4, 5, 6] was taken as a basis. The mass flow is calculated using the following formula:

$$\dot{m} = \frac{U(P_1 - P_2)}{RT}, \tag{5}$$

where P_1 and P_2 are the pressure values upstream and downstream of the channel, T is the temperature upstream of the channel, R is the universal gas constant, and U is the channel conductivity. The operating pressure range of non-contact pumps can cover more than eight orders of magnitude (from 10 to 10^5 Pa). Accordingly, the flow regime in the channels can be either molecular, transient or viscous. The geometry of the slotted channels of the rotor mechanism of non-contact vacuum pumps and vacuum compressors is complicated and depends on the type of profile, the size of the gaps and the relative position of the rotors. Of greatest interest are the four types of slots shown in figure 2 since almost all channels can be reduced to one of these types. The following formula was used to calculate the conductivity:

$$U = \begin{cases} U_V, \, if \; Kn \le 0.01 \\ U_V + zU_M, \, if \; 0.1 > Kn > 0.01, \\ U_M, \, if \; Kn \ge 0.1 \end{cases} \tag{6}$$

where Kn is the Knudsen number, z is the empirical coefficient, U_M is the conductivity of the element in the molecular flow regime; U_V is the conductivity of the element in a viscous mode with a flow without sliding on the walls.

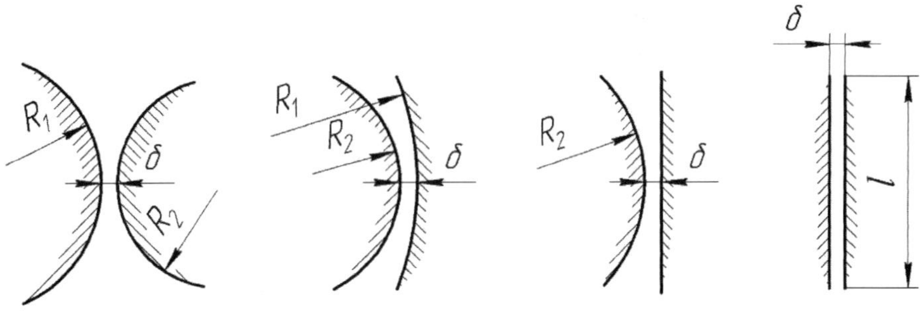

Figure 2. Geometry of channels

For channels 1–3, the following equation has been used

$$U_V = \frac{\sqrt{2RT}\xi(1+\tau)\delta L}{9\pi\left(1+\sqrt{1+\alpha\left(1-\tau^2\right)\xi^2}\right)}, \tag{7}$$

where $\alpha = c_0 + c_1 x_k + c_2 x_k^2 + c_3 x_k^3 - 0{,}00129\tau^{0{,}397\xi}$, $x_k = \frac{\ln(\tau)}{\ln(\tau)-1}$, $\xi = \frac{\delta^2 P_1}{\eta\sqrt{RT} l_{EQ}}$, $\tau = \frac{P_1}{P_2}$, η is the coefficient of dynamic viscosity, $l_{EQ} = \sqrt{\frac{\delta R_1 R_2}{R_1 \pm R_2}}$ is the effective channel length, $c_0 = \frac{0{,}0008}{1+0{,}00013\xi^3} + \frac{775}{10^6+0{,}202\xi^3} + \frac{0{,}00088\xi^3}{10^6+0{,}362\xi^3}$, $c_1 = -0{,}00153 + \frac{0{,}0103\xi^2}{2544+\xi^2+0{,}0000275\xi^4}$, $c_2 = \frac{284{,}6}{9086+\xi^2} + \frac{0{,}0105\xi^2}{58153+\xi^2} + \frac{0{,}0137\xi^3}{1{,}11\cdot 10^7+\xi^3}$, $c_3 = -\left(\frac{0{,}00805\xi^2}{36345+\xi^2} + \frac{39823}{1{,}646\cdot 10^6+\xi^3} + \frac{0{,}0129\xi^3}{6{,}93\cdot 10^6+\xi^3}\right)$.

When determining the effective channel length, the plus sign is selected for channel 1; the minus sign, for channel 2; and $l_{EQ} = \sqrt{\delta R_2}$ is adopted for channel 3. For slot 4:

$$U_V = \frac{\delta^3 L P_1(1+\tau)}{12 l \eta\left(1+\sqrt{1+\frac{\alpha\left(1-\tau^2\right)P_1^2\delta^4}{6 l^2 R T \eta^2}}\right)}, \tag{8}$$

where $\alpha = c_0 + c_1 ln(\tau + 0{,}03) + c_2 \ln^2(\tau + 0{,}03)$, $c_0 = 0{,}0687 - 0{,}00581\ \ln(1/\xi) - 0{,}001\ \ln^2(1/\xi)$, $c_1 = -0{,}264 - 0{,}0677\ \ln(1/\xi) - 0{,}00452\ \ln^2(1/\xi)$,

$c_2 = -0{,}0711 - 0{,}0163\ \ln(1/\xi) - 0{,}000142\ \ln^2(1/\xi)$, $\xi = \dfrac{\delta^2 P_1}{l\eta\sqrt{RT}}$.

Analysis of the results of numerical calculation of the mass flow rate of gas has made it possible to obtain an expression for determining the critical pressure ratio, which for a flat slot is written in the following form

$$\tau_{cr} = 0{,}528\left(\frac{1}{2} + \frac{1}{\pi}\text{arctg}\left[0{,}65(\ln\xi - 1{,}94) + 0{,}012(\ln\xi - 1{,}94)^3\right]\right)^2. \quad (9)$$

For calculation in the molecular mode U_M, it is proposed to use the well-known Clausing expression, presented in the following form:

$$U_M = \frac{C}{4} L\delta K_3, \quad (10)$$

where C is the arithmetic mean velocity of gas molecules; K_3 is the coefficient of conductivity reduced to the minimum gap; δ is the size of the minimum gap; and L is the slot height. For channels 1–3, it is recommended to use the following expressions:

$$K_3^{I} = \exp\left[0{,}7884 + 0{,}4443x + \frac{3{,}27}{x} + \frac{4{,}807}{x^2} + \frac{2{,}883}{x^3} + 0{,}0238 arctg\left(8{,}95\frac{R_2}{R_1}\right)\right],$$

$$K_3^{II} = K_3^{I}\exp\left[-\exp[2y]\left(0{,}057y + 0{,}0304\sqrt{y} + 0{,}0067y^{0{,}16}x\right)\right],$$

where $x = \ln\left[\delta\left(\dfrac{1}{R_2} \pm \dfrac{1}{R_1}\right)\right]$, "+" sign for channel 1, "–" sign for channel 2, $y = R_2 / |R_1|$.

Finally, we have:

$$K_3 = \begin{cases} K_3^{I},\ for\ channel\ 1{,}3 \\ K_3^{II},\ for\ channel\ 2 \\ \dfrac{\delta}{L}\ln\left(\dfrac{L}{\delta}\right),\ for\ channel\ 4. \end{cases} \quad (11)$$

Coefficient z for the variable cross-section channels 1–3 can be calculated as:

$$z = \frac{1 + \delta P_{CP}\sqrt{M(RT)^{-1}}\eta^{-1}}{1 + (1{,}24 + 0{,}5(1-\tau)^2)\delta P_M\sqrt{M(RT)^{-1}}\eta^{-1}}, \quad (12)$$

where M is the molar mass of the compressible medium, and $P_M = \dfrac{P_1 + P_2}{2}$ is the average pressure. Coefficient z for the variable cross-section channel 4 can be calculated as:

$$z = \frac{1 + dP_M\sqrt{M(RT)^{-1}}\eta^{-1}}{1 + 1{,}24dP_M\sqrt{M(RT)^{-1}}\eta^{-1}}, \quad (13)$$

where $d=\sqrt{\frac{4F}{\pi}}$ is the hydraulic diameter of the channel, and F is the cross-sectional area of the channel.

The mass exchange of the working cavity through the ports with the cavities of the suction and discharge nozzles as well as through the 'triangular' slots is determined by the formula

$$m_{\mathrm{W}}=\mu F_{\mathrm{W}} W_{\mathrm{S}} \rho_2, \tag{14}$$

where the port discharge coefficient μ, which takes into account the difference between real and theoretical flow conditions, is determined in the first approximation as for an orifice flowmeter; F_{W} is the passage area. The rate of adiabatic outflow of gas W_{S} is determined by the Saint-Venant–Wantzel formula:

$$W_{\mathrm{S}}=\begin{cases}\sqrt{2RT_2\dfrac{k}{k-1}\left[1-\left(\dfrac{p_1}{p_2}\right)^{\frac{k-1}{k}}\right]}, if\ \dfrac{p_1}{p_2}\geq\beta=\left(\dfrac{2}{k+1}\right)^{\frac{k}{k-1}} \\ \sqrt{2RT_2\dfrac{k}{k+1}}, if\ \dfrac{p_1}{p_2}<\beta=\left(\dfrac{2}{k+1}\right)^{\frac{k}{k-1}}.\end{cases} \tag{15}$$

The equation (14) has been used for the viscous mode of the flow. For the molecular and mixed flow regime, the equations (5), (6), (10) and (13) have been used. It should be noted, that those kinds of the vacuum compressor gaps have short length in the flow direction, therefore their coefficient of conductivity can be taken as $K_3=1$.

4. Heat exchange calculation

The heat gain supplied or removed from the compressed medium due to heat exchange with the walls of the housing and the surface of the rotors is calculated using the Newton–Richmann formula:

$$\frac{\mathrm{d}Q}{\mathrm{d}\tau}=\alpha F_{\mathrm{W}}\left(T_{\mathrm{W}}-T\right), \tag{16}$$

where $\alpha=\frac{Nu\cdot\lambda}{d_{\mathrm{eq}}}$ is the heat transfer coefficient; F_{W} is the area of the surface that limits the working chamber; T_{W} is the average temperature of the surface that limits the working chamber; Nu is the Nusselt number; λ is the coefficient of thermal conductivity of the compressible medium; $d_{\mathrm{eq}}=\frac{4F}{P}$ is the equivalent diameter; F is the cross-sectional area of the channel; P is the channel perimeter.

When considering the heat transfer processes in compressors, it is assumed that the gas is a continuum. However, at low absolute pressures, the phenomenon of heat transfer can be explained only if the molecular structure of the substance is taken into account. The flow of rarefied gases is characterised by the speed and temperature jumps on the surface. Quantitatively, these phenomena are characterised using slip and accommodation coefficients. To calculate the heat transfer of a rarefied gas, the following formula was used:

$$Nu=\frac{Nu_0}{1+Nu_0\cdot Kn\cdot c}, \tag{17}$$

where Nu_0 is the Nusselt number for the heat transfer of non-rarefied gas under the same heat transfer conditions; c is the auxiliary coefficient:

$$c = \frac{2-a}{a} \cdot \frac{2 \cdot k}{k+1} \cdot \frac{1}{\Pr}, \tag{18}$$

where Pr is the Prandtl number; a is the accommodation coefficient introduced by Knudsen, which characterises the proportion of true energy exchange in comparison with the maximum possible one. For air and processed metals, the range of the accommodation coefficient is 0.95 to 0.97. However, in calculations, the accommodation coefficient is most commonly taken as equal to 1.

To calculate the heat transfer in channels in a turbulent mode, the following formula is proposed in the paper [2]:

$$Nu_0 = 0.023 \cdot \mathrm{Re}^{0.8} \cdot \mathrm{Pr}^{0.4} \left(1 + 1.77 \frac{d_{\mathrm{eq}}}{R_{\mathrm{R}}}\right), \tag{19}$$

where Re is the Reynolds number, R_{R} is the rotor cavity radius. The average speed of the contact points of the rotors was chosen as the determining speed. To take into account the influence of rarefaction, a correction to the value of the dynamic viscosity of the gas was also introduced:

$$\eta = \frac{\eta_0}{1 + \beta \cdot Kn}, \tag{20}$$

where η_0 is the dynamic viscosity of the gas at atmospheric pressure, and the value β is determined from the solution of the system of equations:

$$\begin{cases} Kn = z - \dfrac{A}{(2-c_2)(3-c_2)z^{2-c_2}} \\ \beta = c_1 z^{c_2} + \dfrac{A}{c_1 c_2 (2-c_2)(3-c_2) z}, \end{cases} \tag{21}$$

where z is the parametric variable; $A = 0.15$; $c_1 = 1.479952$; $c_2 = 0.1551753$.

5. Simulation results

As a result of mathematical simulation, the calculated pump characteristics of the vacuum compressor under study were obtained. They are given below (figure 3) for the operating mode in the vacuum pump mode since the preliminary test data are available for this mode.

The qualitative analysis of the obtained characteristics evidences that the design mode corresponding to the internal compression ratio in the screw vacuum compressor occurs at a residual pressure of slightly less than 10 kPa (upon discharge into the atmosphere). In modes with a residual pressure below 10 kPa, a characteristic pattern of 'external compression' of the evacuated air is observed; and in modes with a residual pressure of 10 kPa and above, a pattern of 'internal additional compression' of the evacuated air (at 10 kPa, the pattern of 'internal additional compression' is already present, but not notably).

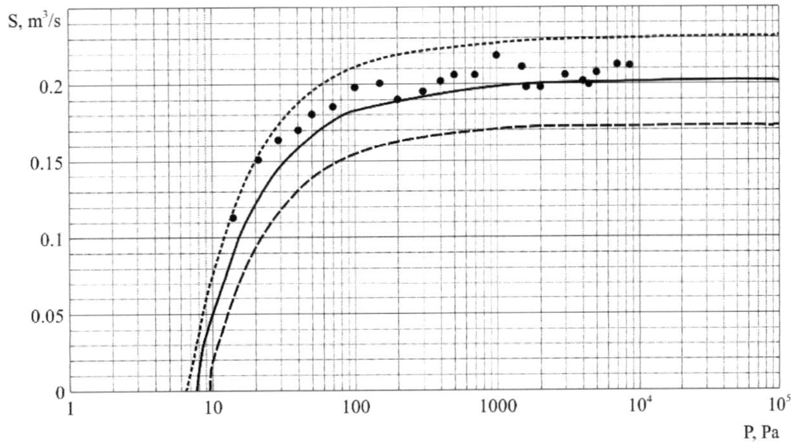

Figure 3. Pumping characteristics

6. Conclusion

The obtained characteristics have the greatest discrepancy with the experimental data in the area of low pressures and minimum productivity (no more than 15%) – that is, in those areas where internal leaks have a dominant influence, while a molecular flow regime is observed. At the same time, it should be noted that the mobility of the walls has a significant effect on leaks in it and requires further correction of the above methodology. However, it can already be recommended for preliminary calculations in the presented form, and the method for calculating leaks in the viscous mode can be extended to a wide range of positive displacement compressor machines without coolant injection.

References

[1] Moesch T W, Janitschke T, Thomas C and Hesse U 2019 Analysis of cycloid type vacuum compressors for water vapor compression systems at sub-atmospheric pressures *IOP Conf. Ser.: Mater. Sci. Eng.* **604** 012008

[2] Burmistrov A, Salikeev S, Raykov A and Fomina M 2017 Mathematical model of working process of oil free scroll vacuum pump *Vakuum in Forschung und Praxis* **29 (6)** 28-31 10.1002/vipr.201700663

[3] Vizgalov S V, Ibraev A M, Volkov M V, Idiyatullin B F and Khisameev I G 2020 Modeling the working process of the roots blower with an ejector attachment and gas bypass through profiled windows on the housing *AIP Conference Proceedings* **2285** 030029

[4] Salikeev S, Burmistrov A, Bronshtein M, Fomina M and Raykov A 2014 Conductance calculation of channels in laminar gas flow regime at an arbitrary pressure difference *Vacuum* **107** 178-183

[5] Burmistrov A, Salikeev S, Bronshtein M, Fomina M and Raykov A 2015 Conductance calculation of slot channels with variable cross section *Vakuum in Forschung und Praxis* **27 (1)** 36-40.

[6] Burmistrov A 2006 *Development and Investigation of Contact-free Vacuum Pumps. Thesis* (Kazan) 451 p (in Russian)

Application of Path Homotopy in Twin Screw Compressor Rotor Profile Design

Sumit Patil[1,2], Nikola Stosic[1], Ahmed Kovacevic[1], Ian Smith[1], Neeraj Asati[2]

[1]Centre for Compressor Technology, City, University of London, Northampton Square, Clerkenwell, London EC1V 0HB, UK

[2]Kirloskar Pneumatic Co. Ltd., Plot No.1, Hadapsar Industrial Estate, Hadapsar, Pune, Maharashtra 411013, India

E-mail: Sumit.Patil@city.ac.uk

May 2021

Abstract. This work explores application of an idea called 'Path Homotopy' from Topology in twin screw compressor rotor profile design. It is well known that twin screw rotor profiles impact the overall performance of screw compressors to a great extent. Almost all the well known profiles that are commercially used in twin screw compressors are generated out of analytical curves such as circles, lines and conic sections. Analytical representation has its own advantages in profile design but it limits the search space of all possible shapes that can be incorporated in profiles. This work aims at proposing an approach for rotor profiling that sticks to the analytical nature of curve representation in profiles while simultaneously extending the search space of curves that can be deployed in the design. Path homotopy has been identified as one of such approaches. Profile elements can be defined as homotopies between well known analytical curves (circles, lines and conic sections) and the homotopy parameter can directly influence these curves morphologically in their analytical form itself. In this way, shapes other than those used traditionally in profiling can be explored in search of the better rotor profiles.

Keywords: Homotopy, Rotor Profile, Screw Compressor

1. Introduction

Screw rotor is a heart of screw compressor. Any improvement in it directly impacts the screw compressor efficiency and reduces the power consumption for an unit flow of compressed gas under given conditions. Their shapes conventionally defined as rotor profiles on cross section perpendicular to their length, have largest and direct impact on thermodynamic efficiency of compression, leakage areas and hence volumetric efficiency of the compressor. The art of choosing the right curves for a rotor profile in order to achieve the maximum possible performance is termed as rotor profiling. It has a history of close to 90 years since the invention of first screw compressor by Lysholm and William (1938).

The earliest rotor profiles were symmetric wherein main and gate rotor lobes had a single circular arc featured on them (Figure1). Symmetric profiles had large blow hole areas (leakage area at the cusp formed between rotor tips and casing). Rotor profiling was revolutionized by the invention of SRM-A profile by Schibbye (1970) which was asymmetric in nature and reduced the blow hole area by 90%. It was a very practical profile with a good manufacturability. Stosic et al. (2011) lauds SRM-A for making screw compressors commercially viable for the first time.

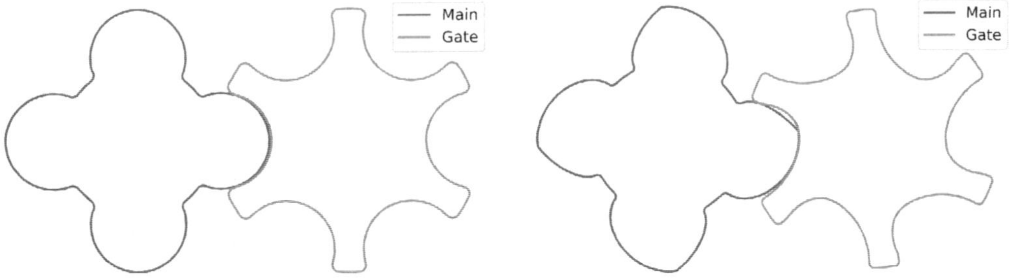

(a) (b)

Figure 1: (a) Circular Symmetric Profile (b) SRM-A Profile

These profiles were designed and generated by precise mathematical rules and procedures but no literature on theory of profile generation was published until Sakun (1960) who introduced the envelope method from differential geometry to screw rotor profile generation. Following this, Andreev (1961) published his work that elaborated on tooling and manufacturing aspects of screw rotor profiles. Post this period, multiple successful rotor profiles were invented, such as SIGMA profile (Bammert, 1982), SRM-D (Astberg, 1984), CYCLONE profile (Hough et al., 1987) and HYPER profile (Chen, 1995). A particular development in the method of generation of rotor profiles- rack generation- first proposed by Menssen (1977) led to many successful rotor profiles. These include Rinder (1987) profile, N-Profile by Stosic (1997), Cavatorta and Tomei (2014) profile and N-Silent profile by Stosic (2017). This family of profiles is rack generated while those mentioned before them are rotor generated. Rack generation has definitive advantage over rotor generation in terms of ease and error-less generation of profiles. Most of the modern successful rotor profiles are based on rack generation and calculation procedures developed in Stosic and Hanjalic (1997).

This work particularly aims is aimed at identifying the prime challenge in rotor profiling and proposing a promising method priori unused in profiling. The immediate next section lays out the details of this prime challenge in profiling as well as evaluates different attempts in the literature to tackle it. Based on the evaluation of the problem, in the following section, 'path homotopy' (Armstrong, 2013) is proposed as an approach to design better rotor profiles. Conclusions following this section are based on the literature review and the proposal to use this novel method in profiling.

2. The problem of choosing the right curves

The prime challenge in profiling is the perfect choice of curves to make-up the shape of rotors best suited for a desired application. In the most early days of rotor profiling, this issue of right choice of curves was sidelined simply because, even if there was any wish to test new types of curves in profiles, the formalism to calculate conjugate shapes of complex curves was not in place which made entire process a trial and error way to incremental improvement. For this reason, only the most simple curves such as circles, lines and single points are used in most of the earliest profiles in literature. Only after a fairly decent formalism at place established by Sakun (1960), Litvin (1989), Stosic and Hanjalic (1997) and others, one can see the use of curves like ellipses, parabolas, hyperbolas and cycloids in rotor profiles used either for their own benefits or simply to obtain a patent. Still the range of analytical curves used in rotor profiling are limited to conic sections, involutes and trochoids.

The reason for this limited use might be either that the profiles generated with these curves alone are sufficiently good; or that the inherent benefits despite of the challenges with incorporating analytical curves into design procedure that searches the best suitable shape of profile outweigh the ease at doing the same thing with numerical shape functions such as splines or Bezier curves. Kauder et al. (2002) and Helpertz (2003) explored the efficacy of later approach wherein Bezier curves whose shapes can be manipulated by tweaking control points were used to design a profile. Looking at this problem of finding the best shape of profile as an optimization problem, they proposed genetic algorithm with selection criteria on simple geometric features such as blow hole area, sealing line length and rotor inertia to quicken the calculations and convergence. Though this approach looks promising there are few issues here (as reported by authors themselves) such as -

- In optimizations problems like such which involve a large number of technical criteria, authors state that even minor deviations in true/normal forms of Bezier curve lead to major faults in meshing conditions.

- Even if the profiles that don't confer to theoretical meshing conditions are excluded from further generations of optimization process, profiles represented by simple splines seriously deviate from optimized solution.
- A complex consistency check must be performed after each iteration to ensure that all segments (discrete shape functions) are tangent after minor variations. This impedes the computation time and convergence of solution.

Hauser et al. (2008) addressed some of the challenges in previous attempts of using this approach by suggesting NURBS (Non-Rational Uniform Rational B-Spline) for representing the curves. Authors claimed advantages such as high quality of shape representation with relatively less information content and complex consistency checks after each iteration can be skipped. Still, the issue of being an indirect approach and the feasibility in real applications is not fully addressed. Švígler (2010) has commented in his book on use of NURBS in screw rotor profiling - he says, "though spline curves are advantageous for curve formation or representation, they are very inconvenient for calculating contact points of conjugate surfaces".

A more practical approach to optimizing the shape of profiles was demonstrated by Stosic et al. (2003) who stuck to analytical representation of profile curves and optimized within the already known and tested N-profile to take advantage of all the features inherently built in its design through years of experience. The algorithm used for this optimization "Box-Complex method" searches for the optimized solution in the vicinity of the base profile based on thermodynamic performance of different variants. The representation of curves in N-profile offers a distinct advantage too. The general circle arc equation-

$$ax^p + by^q = 1 \tag{1}$$

enables to represent wide range of curves (all conic sections and lines) along with ability to parametrically tweaking them using a, b, p and q. Proof of this approach being practical is found in the same article Stosic et al. (2003) where it has been reported that screw compressors for different operating conditions and applications were designed using this approach and tested too.

Coming to the more basic point of the challenge of choice of right curves for profile, almost all of the successful and patented profiles till date follow the tested path of analytically represented curves. In order to improve or incorporate other desirable features such as varying radius of curvature along the curve (which is not possible with circles as radius of curvature is fixed with its radius) for better distribution of contact stress or reducing blow hole area by sharpening tips at the end, inventions like Kim and Lee (2004) have introduced better representations of the common analytical curves such as hyperbolae in a parametric way so as to control flatness or roundness of the hyperbola. This provides advantage of already stated analytical representations as well as control over curve features.

Based on all the considerations and indications from literature review of different methods, one thing is clear- there is a need for method that can handle analytical curves like those in conventional profiles while providing better control over curve features like that of recent approaches based on numerical shape functions. This work proposes 'path homotopy' as such a method lying on the middle-ground. It is found in the abstract science of topology and is not yet used in mechanical engineering applications. Only one known application of this idea in engineering is found in computer graphics by Dym et al. (2015).

3. Path Homotopy

As evident from the literature review, the strategy to stick to analytical representation of profile curves have definitive advantages. However, it is also pointed out that this condition of analyticity must be met along with considerable control over curve features in order to explore wide range of shapes for a better profile. In order to achieve this goal, the current methods in literature such as that of using general circle arc equation in N-Profile or ingeniously setting up analytical curves in parametric way are pretty effective. Though, in the process of choosing right curves for a profile, one desires greater local control over curve features and parametric tweaking is expected to be less cumbersome (less number of control parameters are desirable).

In order to do so, an alternate strategy from Topology named 'Homotopy Deformations' (Armstrong, 2013) priori unused in rotor profiling or other mechanical shape optimization problems turns out to be more suitable than the existing methods. This is a concept that allows for the continual deformations of an analytical curve from one curve (initial) to the other (final). It can reach all the intermediate curves with single parameter variation. What is required is just a definition of initial and final state of curve for the deformations to take place in between. Mathematically, homotopy between two curves $Curve_1$ and $Curve_2$ with respect to homotopy parameter or deformation parameter q is written as-

$$\mathscr{H} = (q)(Curve_2) + (1-q)(Curve_1) \tag{2}$$

The continuous deformation is graphically demonstrated in Figure2. Here, the $Curve_1$ is set as the initial state of homotopy and $Curve_2$ is set as the final state of the deforming homotopy. When $q = 0$, $\mathscr{H} = Curve_1$; and when $q = 1$, $\mathscr{H} = Curve_2$. The homotopy parameter 'q' continually deforms the $Curve_1$ into the $Curve_2$ while

IOP Conf. Series: Materials Science and Engineering 1180 (2021) 012004 doi:10.1088/1757-899X/1180/1/012004

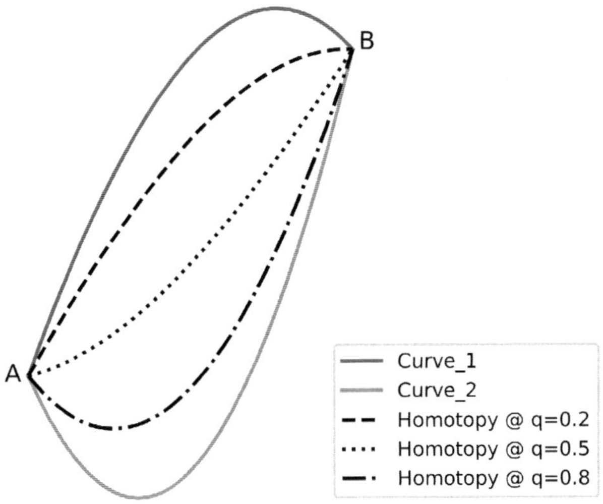

Figure 2: Homotopy demonstration between any two curves defined in between points 'A' and 'B'

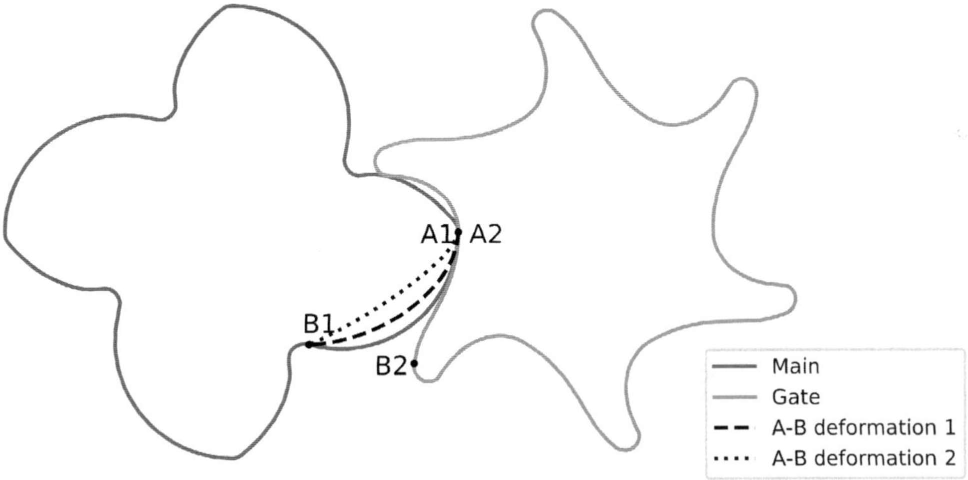

Figure 3: Example of the application of homotopy in Demonstrator profile (Stosic et al., 2005)

taking intermediate shapes as q goes smoothly from 0 to 1 (three such instances depicted as $q = 0.2$, $q = 0.5$ and $q = 0.8$).

In rotor profiling, such homotopic constructions could be used in place of certain profile sections. One such example is demonstrated using the 'Demonstrator profile' (Stosic et al., 2005) in Figure3. It is a rotor generated profile entirely made of circles defined primarily on main rotor. The main rotor lobe section marked as '*A1-B1*' in Figure3 is also a circular arc with a radius r and a center at (x_0, y_0) which can be defined as-

$$y = \sqrt{r^2 - (x - x_0)^2} + y_0 \tag{3}$$

This curve '*A1-B1*' can be replaced by a homotopy that deforms according to a paired curve in homotopy construction. If the other curve is chosen to be a straight line between points $A1 = (x_1, y_1)$ and $B1 = (x_2, y_2)$-

$$y = \left(\frac{y_2 - y_1}{x_2 - x_1}\right)(x - x_1) + y_1 \tag{4}$$

The homotopy hinged on points $A1$ and $B1$ is constructed using (3) and (4) as -

$$\mathscr{H} = (q)\left(\left(\frac{y_2 - y_1}{x_2 - x_1}\right)(x - x_1) + y_1\right) + (1 - q)\left(\sqrt{r^2 - (x - x_0)^2} + y_0\right) \tag{5}$$

The homotopy parameter q can be tweaked to deform the circular arc into a straight line. Two such instances of deformations are depicted in Figure3. These deformed states of the profile section *A1-B1* still have an analytical representation. The generation of the corresponding curve *A2-B2* on gate rotor can be followed from solving the conjugacy condition derived using Euler's envelope theory as presented in Stosic and Hanjalic (1997). Alternatively, the homotopy on main rotor can be transformed onto a rack and the respective rack profile can then be transformed to the corresponding section on gate rotor.

Following are the key points why this novel strategy could prove useful in rotor profiling-

- It brings definitiveness to the search space of probable curves with determination of initial and final curves beforehand. These could be chosen based on already known best shapes in certain portions of the rack and/or rotor profiles. Then the designer can optimize around those shapes using the homotopic deformations.
- This method is so general that one can write homotopy between any two arbitrary analytical curves without any restriction to certain form or family of curves.
- The method also provides provision to choose our initial and final states of the curves in such a way that the curve features such as radii of curvature at certain points could be controlled more efficiently.
- Designer gets the advantage of analytical representation (avoiding intrinsic problems in numerical shape function approaches) while gaining a wider search space (avoiding the constraint of strictly analytical approach).

- The method is general enough to be seamlessly used in any type of profile design be it rotor generated, rack generated or a rotor-rack generated. Homotopies can be defined either on rotors or rack and transformed suitably using appropriate meshing conditions.

4. Conclusions

Literature review of the most successful screw rotor profiles and the methods of rotor profiling indicate that the foremost challenge for a designer is the choice of most appropriate curves on a profile. The choice of curves should create the shape having maximum displacement and minimum leakages for a typical screw compressor. There are two widely known approaches of designing rotor profiles. First of these is a traditional approach where strictly analytically represented curves such as conic sections and involutes are chosen as the rotor profile elements. In second approach, profile shapes are generated out of numerical shape functions such as Bezier curves or B-splines and tweaked iteratively in the direction of better profile characteristics. First approach is robust in terms of profile generation and avoiding errors but lacks the relative control over curve features and achievable shapes through them. On the other hand, second approach provides tremendous control over curve features and the range of shapes that can be generated out of it. But it lacks the robustness of first approach. Analytical representation has its definitive practical advantage in rotor profiling.

A compromised approach as a middle-way that can pick the pros of both these approaches and avoid the cons of both may lead to better rotor profiles. 'Path Homotopy' is found to be one of such approaches. Its application in rotor profiling is advocated in this work by demonstrating it through an example.

Acknowledgments

Authors would like to thank Kirloskar Pneumatic Company Limited, India for financially supporting this research.

Bibliography

Andreev, P. A.
1961. *Vintovie kompressornie mashinii [Screw Compression Machines].* SUDPROM Leninngrad.

Armstrong, M. A.
2013. *Basic topology.* Springer Science & Business Media.

Astberg, A.
1984. Screw rotor machine and rotor profile therefor. US Patent 4,435,139.

Bammert, K.
1982. Intermeshing screw rotor machine with specific thread profile. US Patent 4,350,480.

Cavatorta, P. and U. Tomei
2014. Screw compressor having male and female rotors with profiles generated by enveloping a rack profile. US Patent 8,702,409.

Chen, C.-H.
1995. Screw compressor with rotors having hyper profile. US Patent 5,454,701.

Dym, N., A. Shtengel, and Y. Lipman
2015. Homotopic morphing of planar curves. In *Computer Graphics Forum*, volume 34, Pp. 239–251. Wiley Online Library.

Hauser, J., A. Bruemmer, and K. Kauder
2008. Rotor profile generation and optimization of screw machines using nurbs.

Helpertz, M.
2003. *Methode zur stochastischen Optimierung von Schraubenrotorprofilen.* PhD thesis, Universität Dortmund.

Hough, D., S. J. Morris, and A. D. Barber
1987. Screw rotor machines with specific tooth profiles. US Patent 4,636,156.

Kauder, K., M. Helpertz, B. Reusch, and S. Berlik
2002. Optimisation methods for rotors of twin-screw compressors. *VDI BERICHTE*, 1715:29–50.

Kim, J. S. and J. H. Lee
2004. Rotor profile for screw compressors. US Patent 6,779,993.

Litvin, F. L.
1989. *Theory of gearing.*

Lysholm, A. and B. G. K. William
1938. Rotary compressor. US Patent 2,111,568.

IOP Conf. Series: Materials Science and Engineering 1180 (2021) 012004 doi:10.1088/1757-899X/1180/1/012004

BIBLIOGRAPHY

Menssen, E.
1977. Screw compressor with involute profiled teeth. US Patent 4,028,026.

Rinder, L.
1987. Screw rotor profile and method for generating. US Patent 4,643,654.

Sakun, I. A.
1960. *Screw Compressors [in Russian]*. Mashgiz Moscow.

Schibbye, L.
1970. Improvements in and relating to rotary positive displacement machines of the intermeshing screw type and rotors therefor. GB Patent App. 3,421,766 A.

Stosic, N. and K. Hanjalic
1997. Development and optimization of screw machines with a simulation model—part i: Profile generation.

Stosic, N., I. Smith, and A. Kovacevic
2005. *Screw compressors: mathematical modelling and performance calculation*. Springer Science & Business Media.

Stosic, N., I. K. Smith, and A. Kovacevic
2003. Optimisation of screw compressors. *Applied Thermal Engineering*, 23(10):1177–1195.

Stosic, N., I. K. Smith, A. Kovacevic, and E. Mujic
2011. Review of mathematical models in performance calculation of screw compressors. *International Journal of Fluid Machinery and Systems*, 4(2):271–288.

Stosic, N. R.
1997. Plural screw positive displacement machines. International Patent WO1997043550A1.

Stosic, N. R.
2017. Reduced noise screw machines. US Patent 9,714,572.

Švígler, J.
2010. *A treatise on the theory of screw machines*. Západočeská univerzita.

Numerical evaluation of an Oil free Screw Compressor's Suction Port design using Ansys CFX and SCORG

S Tankhiwale[1*], A Dagwar[1], S Rane[2], A Kovacevic[2], A Birari[1], S Abdan[1,2] and N Asati[1]

[1] Kirloskar Pneumatic Company Limited, Pune, India
[2]Centre for Compressor Technology, City, University of London, London, U.K.

E-mail: shubhankar.tankhiwale@kirloskar.com

*Corresponding author

Abstract. Quasi one-dimensional chamber models and experimental methods are commonly adopted by researchers to predict and validate the performance of the oil free screw compressors. However, to understand the scope for improvement in performance as impacted by compressor ports, one needs to know the details of physical processes happening in compression chamber. This can be done by understanding flow field such as internal temperature and pressure with the help of CFD model.
With this CFD setup, comparison of specific power consumption of oil free screw compressor between the axial and radial suction flange is presented in this paper. A hexahedral numerical mesh is used to represent flow field which is generated by SCORG™. Flow solver ANSYS CFX has been used for study with these single domain deforming rotor grids.
This paper presents the methods used for setting up the CFD model and analysis of specific power consumption for two different suction port geometries, an axial and the radial suction flange. At the design operating condition, although axial suction flange seems to be more favourable in terms of lower suction pressure drops at high operating speed, the radial suction flange when simulated using CFD model predicts almost the similar performance. However, radial suction flange gives an added advantage for ready adaptability with the suction elements like the suction filter. The results of the simulation show that the radial suction flange can be adopted for better compatibility with the oil free package and the CFD model used for the analysis can be used for optimization of the air-end at various operating conditions.

Keywords: CFD, Suction port, CFX, SCORG, Oil free screw compressor

1. Introduction.

The flow analysis of Positive displacement (PD) machines is highly complex mainly due to 3D flow structure involving turbulence, complexity in mesh due to micron size gap and unsteadiness. In recent years, a growing availability of computational resources and progress in the accuracy of numerical methods brought PD machines Computational Fluid Dynamics (CFD) methods from pure research work into the competitive industrial markets. Computational Fluid Dynamics (CFD) can provide a useful means of modeling the oil free machines and study the physical phenomenon in detail in order to understand the complete behavior of oil free twin screw compressors.

The numerical computation of screw machines requires accurate grid generation because the fluid flow is transient and depends on the rotor position. Kovačević [4] [5] made a breakthrough to successfully use an algebraic grid generation method with boundary adaptation and transfinite interpolation which has been implemented in the program called SCORG. This aims to generate high quality grids in a short time as a result of which CFD can now be used very frequently in industrial applications of PD machines.

An oil-free twin screw compressor the working chamber is free from oil. This is opposite to the oil injected machines, where the oil is used for lubrication, temperature control and reduced internal leakages in the working chamber. Oil-free machines are typically used in industries where the working fluid must be free of any containment. Eliminating oil from the rotor chamber adds in own complexities and challenges, such as sealing arrangements, increased operating temperatures etc. Additionally, as oil is absent from the working chamber, the induced efficiency losses that occur due to oil shear are avoided, meaning the oil-free screw compressor can operate at significantly higher tip speeds.

The suction port is a part of the suction element which influences the compressor's performance. The use of CFD allows the analysis of the flow inside the suction port which is impossible to investigate experimentally but it helps influencing the performance of a screw compressor. The Suction port's design optimization is quite difficult to take place until and unless the exact boundary conditions are known. Hence incorporating the gap between rotors and its casing in the simulation is must to improve the suction port design. VandeVoorde et al. (2005) [2] used an algorithm for generating block structured mesh from the solution of the Laplace equation for twin screw compressors and pumps using differential methods. Reports on analysis of oil free air screw compressors and twin screw expanders with real gas models are available in literature using these techniques (Paper et al., 2013) [1]. Recently, Arjeneh et al. (2014) [3] have presented the analysis of flow through the suction port of a screw compressor with water injection. With both deforming rotor domains and multiphase models it was reported to be difficult to stabilize the solver in a full 3D analysis. Although several attempts have been made in the recent past to extend the CFD technology to oil injected compressors, it has proven to be difficult to achieve the desired grid structure and the modelling conditions that can provide stability to the numerical solvers. An extensive study undertaken by Rane [6] [8] presents a high level of accuracy between CFD modeling and test data is achievable when using a single domain rotor grid. An attempt has been made in the presented work to utilize the latest developments in CFD technology for design optimization of suction port of an oil free twin screw compressor.

2. Approach for modelling suction ports of the oil free screw compressors

2.1 Modelling Methodologies

The performance attributes of the machine can be determined numerically, through CFD modelling or the uniquely designed thermodynamic chamber model tools - such as that used in SCORGTM [1].There are extreme differences in modelling the 1D and 3D cases. The greatest is the processing time and power required to obtain the results using the specified boundary conditions. Where the chamber model can output predictions in seconds, the CFD model would take days, ultimately due to complex and quantity of equations required to be calculated. Additionally, to perform these calculations, a large amount of high specification processing and enough number of cores is required to minimize the computational time, whereas the chamber model can operate on a single core. Further to this, the CFD model will provide an extensive range of results, including flow phenomena that can be numerically and visually investigated, as opposite to the chamber model which outputs single calculated values relating to performance. This is not to say a particular model is superior, simply their use should be considered on a case to case basis.

2.2 Thermodynamic Chamber Model (SCORG^{TM})

One-dimensional mathematical modelling of screw machines has been used to great effect for many years. An accurate prediction of many aspects of machine performance can be obtained from the models which simultaneously solve thermodynamic processes, based on ideal or real gas laws, with the geometric parameters of the machine, including chamber volume, port areas and sealing line lengths.

From the many modelling methodologies, that presented by Stosic et al [7] forms the foundation of the model used in this study. Described as multi-chamber thermodynamic model, the screw machine is divided into sub-domains representing each stage of the compression process. The challenge of computing time dependent flow variables within the control volumes is overcome by using a combination of differential equations to solve the rate of change of mass and internal energy, and algebraic equation to solve other internal flow behaviour such as velocity and leakages. From a user perspective, the complexities involved in the computation of the abovementioned flow variables are hidden. As much as it is recommended to have a theoretical understanding of the model, developers have excelled in creating a tool that is as much user friendly as it is reliable. This study will utilize the thermodynamic model inherent to SCORG^{TM}.

2.3 Rotor grid generation

SCORG^{TM} is used to generate the 3D rotor fluid domain.

The tool allows the user to define certain mesh parameters to calculate a suitable mesh. For that applying the principles of analytical grid generation through transfinite interpolation with adaptive meshing, the authors have derived a general, fast and reliable algorithm for automatic numerical mapping of arbitrary twin screw machine geometries. The procedure of analytical grid generation of screw machine working domain is explained in Kovačević (2002) [7]. In order to achieve a conformal single domain mesh, a new approach of background blocking has been presented in Rane (2015) [6]. In this procedure, outer boundary in each background block is defined as a combination of the rack segment and the casing circle segment. The rack segment stretches between the lower and higher cusp points and is closed by the casing as shown in Figure 1. The distribution obtained on the outer boundaries of the two blocks is used as the reference for the rotor profile distribution. Since the blocks of the main and gate rotors are obtained separately, the intersection points obtained on the common rack curve from the two blocks can be different. If different the resulting mesh has a non-conformal map between the two rotor blocks. Using the blocking approach it is possible to transform this boundary map into a conformal boundary map. The surface mesh on the interlobe surface mostly follows the axial grid lines with small transverse movements in the vicinity of the cusp points. The movements of points are along the surface of the interface and do not cause any irregularity of cells. This implementation allows for a fully conformal interface with the equal index around cusp points to ensure accurate capturing is the blow hole leakage area. The blocking approach is in more detail explained in Rane (2015) [6]. It allows both conformal and non-conformal boundary map to produce fully hexahedral 3D grid with both, the main and the gate rotor surfaces smoothly captured. The surface mesh on the casing is of the highest quality with the regular quadrilateral cells.

Table 1 Rotor grids parameters

Parameter	Description	Input
A	Radial Divisions	7
B	Angular Divisions	40
C	Interlobe Divisions	50
D	Circumferential Divisions	40
E	Distribution type	Casing to rotor conformal

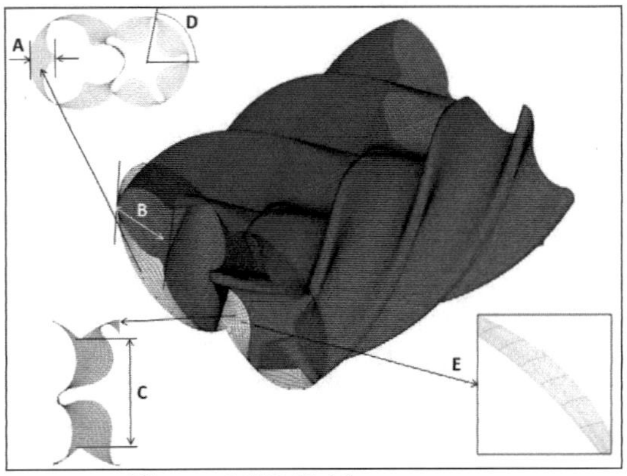

Figure 1 3D Grid parameters

With above setting shown in Table 1, the produced mesh indicates a good overall quality rotor mesh, with <1% of the cells in the 3D mesh not meeting the recommended Orthogonality and expansion factor quality criteria as shown in Table 2. Minimum acceptable orthogonality angle is 20°. Whereas an aspect ratio for 3D mesh is 100% in good category of the quality criteria. CFX uses a Fortran interface called 'junction box routine' to exchange external meshes with the solver.

2.4 Stationary domain mesh
Figure 2 and Figure 3 shows the CFD model flow domain in which the mesh lines are highlighted with rendering. Table 2 summarizes the statistics of the 3D domain. All stationary domains have a combination of tetrahedral and hexahedral cell structure. Figure 3 shows an extended geometry called adapter which is connected with the axial suction flange by fluid-fluid interfaces. The rotor domain itself is a deforming grid, generated using the procedure described in Section 2.3

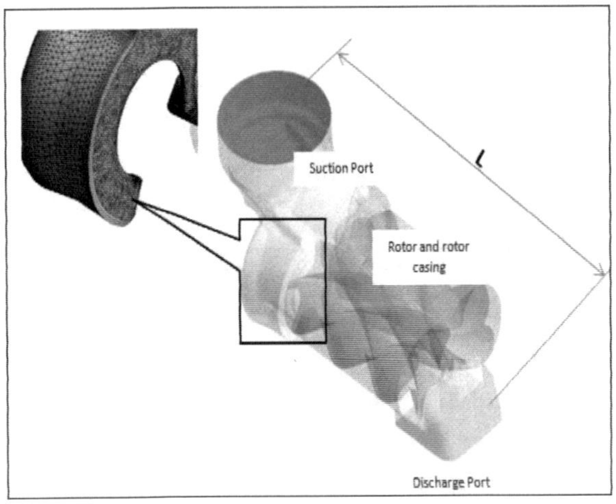

Figure 2 Flow domain of screw compressor with Radial suction

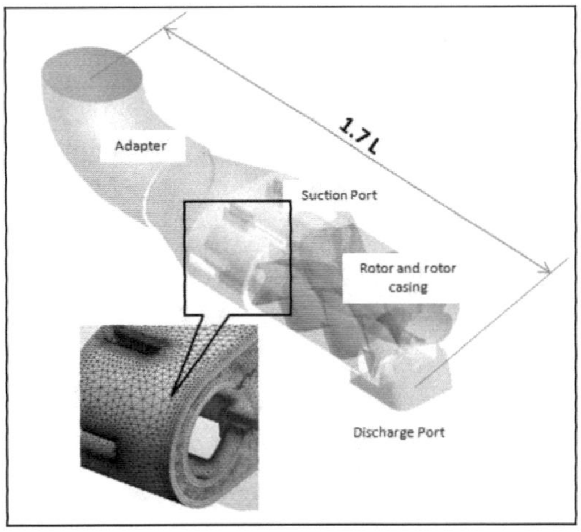

Figure 3 Flow domain of screw compressor with Axial suction and adapter

Table 2 Mesh statistics

Domain	Cell structure	Node count	Cell count	Orthogonality angle (Min)	Expansion factor	Aspect ratio
Rotor	Hex	482636	426258	8.2	586	389
Suction Port Radial	Tetra+Hex	277464	931693	28	388	7.17
Suction Port Axial	Tetra+Hex	165342	504516	34.2	365	10.55
Adapter	Tetra+Hex	43742	190007	52.4	8	5.48
Discharge Port	Tetra+Hex	53337	156042	20.3	54	11.54

2.5 CFD Model

CFD modeling is rapidly being adopted in all areas of modern engineering practice to solve complex flow and heat transfer problems. However, the challenges involved in modeling and computing the complex physics and transient nature inherent to the machine have proven difficult to overcome. Nonetheless, through extensive and committed research, the development of design tools and users obtaining a greater understanding of the problems, it is now possible to undertake these simulations with a certain degree of confidence that a valid solution can be obtained. To obtain this, CFX as a solver is used for this study. CFX is vertex-centred solver for which the spatial domain is firstly discretised into a mesh by using an external grid generator like SCORG™. This mesh is then used to construct virtual control volumes within the solver.

There are two CFX case setups, one is for axial suction port and other one is radial suction port. In Axial suction flange the adapter is an additional component required to connect vertically with suction filter. Whereas in radial type suction port the suction filter will be connected directly on suction flange. The male rotor has three lobes and the female rotor has five lobes with 'N' profiles. The built-in volume index V_i of the compressor is 2.5. The wrap angle is 306°. Main and gate rotors has 3 and 5 lobes respectively. Operating speed of the machine is 18918.4 rpm.

2.5.1 CFD boundary conditions

In both the cases, air as a single phase is defined as an ideal gas state with molar mass of 28.96 kg/kmol, specific heat capacity 1.0044e03 J/kg K, dynamic viscosity 1.831e-05 kg/m s and thermal conductivity 2.61e-02 W/m K. The boundary condition is summarized in Table 3.

Table 3 Operating boundary condition for both CFD models.

Speed (rpm)	Working Fluid	Suction Pressure (bar A)	Suction Temperature (K)	Discharge Pressure (bar A)	Discharge Temperature (K)
18918.4	Air	1	298.15	3	393

2.5.2 Solver parameters

Numerical setup for CFX is shown in Table 4. The Shear Stress Transport k-omega model is used in both the cases. k-epsilon model is used for away from wall treatment or free stream treatment whereas k-omega is used for near wall treatment of the case. The SST model compiles the best elements from both k-epsilon and k-omega models for predicting the best flow behaviour. CFX requires less number of iterations to approach the converged solution.

Table 4 Numerical setup for CFX solver.

Criteria	CFX-controls
Turbulence Model	SST k-omega
Inlet Boundary Condition	Opening (specified pressure and temperature)

Outlet Boundary Condition	Opening (specified pressure and temperature)
Pressure Velocity Coupling	Co-located layout
Turbulence Scheme	First order upwind
Transient Scheme	Second order backward Euler
Transient Inner loop coefficient	10 iterations per time steps
Convergence Criteria	1e-4
Relaxation Parameters	0.1

3. Results
The results present the performance calculations, applying the boundary conditions listed above, from both the models. From here the variance are calculated and a clear comparison can be drawn.

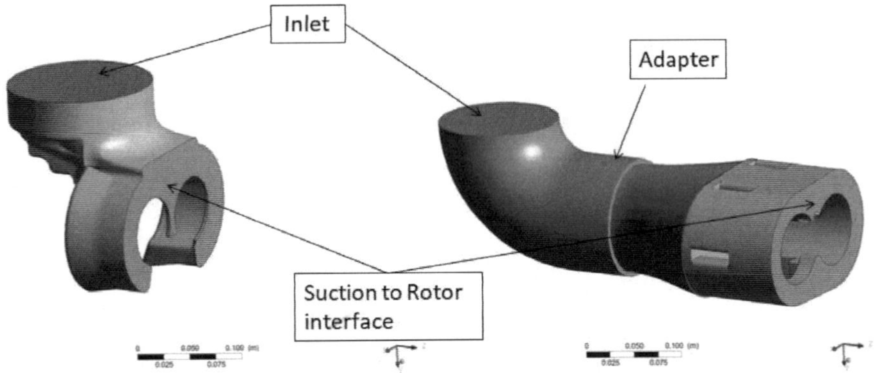

Figure 4 Fluid volume of a) Radial suction port and b) Axial suction port with adapter.

Results are mainly focused on two designs of the suction port shown in Figure 4. In case of radial suction port, air will flow in curvilinear passage from radial inlet to suction to rotor interface. Whereas in axial suction port design, air will flow in the same manner as radial design but here it has more area to travel between the same two surfaces.

3.1 Pressure distribution.
A pressure contour is plotted at X = 0.04 m in YZ plane for male rotor speed at 18918.4 rpm on the flow domain shown in Figure 4 to observe the pressure distribution in both the cases. Figure 5 shows the similar pressure characteristics at the cross section. Comparatively lower suction pressure drop is observed on axial type suction port.

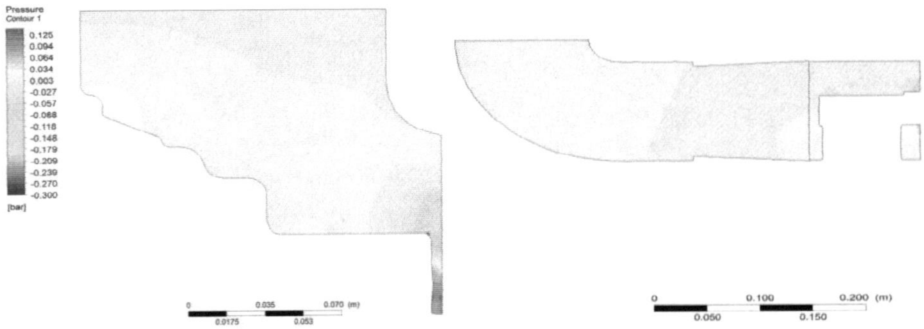

Figure 5 Pressure contour plot **a)** Axial Suction port with adapter and **b)** Radial Suction Port at time step = 820 for the operating speed of 18918.4 rpm

Figure 6 shows the suction port pressure drop against male rotor rotation angle for axial and radial type suction ports. It is clear from the graph that the pressure drop for axial suction port is lower than radial type suction port design. This is due to an additional equipment called adapter is attached to the axial suction port. Adapter reduces the pressure drop due to its larger area. Whereas in case of radial suction port the pressure drop is more due to smaller area is available for air to pass from suction flange to the rotor domain as shown in Figure 4. As per standard formula of minor loss [9], the estimated loss coefficient for radial port is 28% higher than that of axial port.

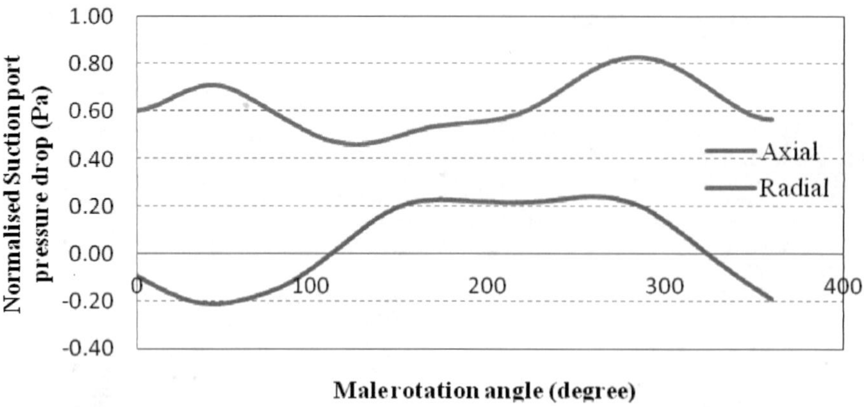

Figure 6 Graph of Suction port pressure drop against male rotor rotation angle for axial and radial suction port design

3.2 Specific Power

Specific power is nothing but the Free air (m3/min) delivered per kW of power. Overall performance results from both the simulations are averaged for each cycle to achieve performance values of specific power. Percentage difference in the specific power for both the designs is as low as below 1 %.

3.3 Velocity distribution

Velocity vectors are plotted on the YZ plane of the radial and axial suction ports as shown in Figure 7. Higher velocities are observed at the end of both the suction ports. More recirculation can be observed towards suction to rotor interface in case of axial port which leads to the loss of free air delivery.

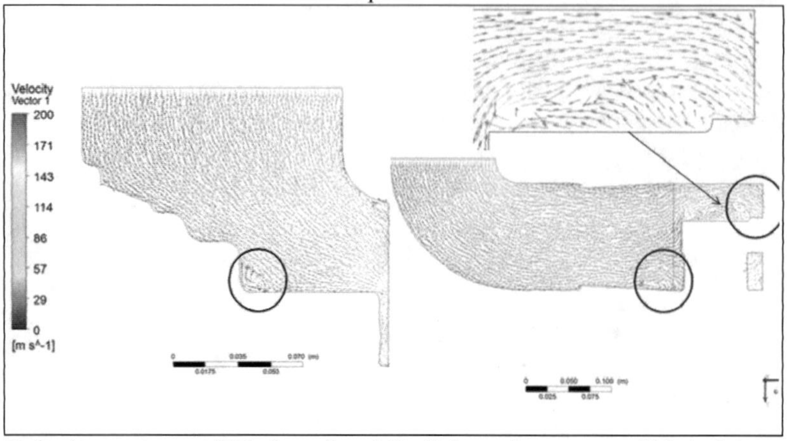

Figure 7 Velocity vectors at YZ plane for a) Radial suction port and b) Axial suction port

Figure 8 shows velocity contours on the surface of the suction port. Higher velocities are observed on the suction to rotor interface.

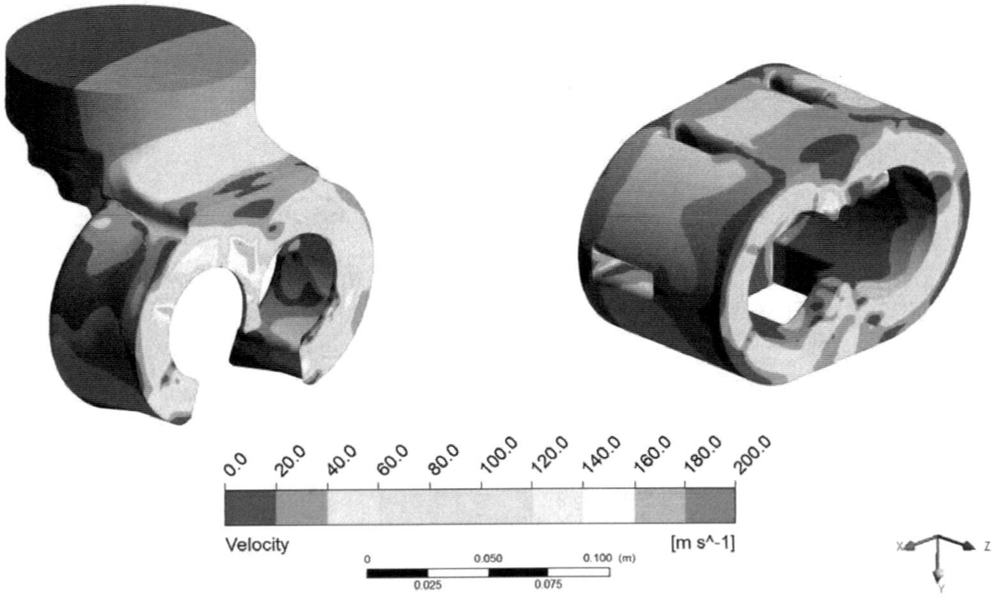

Figure 8 Velocity contour plot **a)** Radial Suction port and **b)** Axial Suction Port at time step = 820

4. Conclusion

This paper presents the numerical evaluation of the suction port of an oil free twin screw compressors at design operating condition. A CFD modelling approach has been used for the calculation of performance and pressure drop across the suction port. A single domain structured numerical mesh of the flow domain was generated by SCORG™ using recently developed boundary blocking and differential grid generation procedure. CFX case with Axial as well as Radial suction port design is modelled, solved and studied in this paper and comparison is made between both the cases to conclude the results.

- Cycle averaged specific power is very small between the two designs
- Despite of having a little high pressure drop at radial suction port, it can be used for its better compatibility with oil free package due to the size difference observed in Figure 2 and Figure 3.

The detailed performance characteristics have been studied with varying operating speeds of the screw compressor for further optimization of the radial suction port.

5. References

[1] Papes, I., Degroote, J. and Vierendeels, J. 2013 3D CFD analysis of an oil injected twin screw expander. Int. Mechanical Engineering Congress and Exposition,ASME IMECE at San Diego, USA.

[2] Vierendeels, Voorde Vande J. 2005 A grid manipulation algorithm for ALE calculations in screw compressors. Canada : AIAA 2005-4701, 2005. 17th AIAA Computational Fluid Dynamics Conf.

[3] Arjeneh M., Kovačević A., Gavaises M. and Rane S. 2014. Study of Multiphase Flow at the Suction of Screw Compressor at Purdue paper 1353.

[4] Kovačević, A. 2005, Boundary Adaptation in Grid Generation for CFD Analysis of Screw Compressors. Int. journal for numerical methods in engineering.
[5] Kovačević, A. 2002. Three-Dimensional Numerical Analysis for Flow Prediction in Positive Displacement Screw Machines, Thesis. City University London.
[6] Rane S. 2015 Grid Generation and CFD analysis of Variable Geometry Screw Machines. City university London
[7] Stošic N., Smith I.K. and Kovacevic A 2005 Screw Compressors: Mathematical Modelling and Performance Calculation Berlin ISBN: 3-540-24275-9..
[8] N Basha, A Kovacevic and S Rane. 2019 User defined nodal displacement of numerical mesh for analysis of screw machines in FLUENT IOP Publishing Ltd Int. Conf. on Compressors and their Systems.
[9] Rajput, R.K. (2013). *A Text book of fluid mechanics and hydraulic machine in SI Unit*. New Delhi, S. Chand and company limited.

Investigation of Screw Compressors for Low Pressure Ratio Applications

Y Lu[1], A Kovacevic[1], M Read[1]*, D Buckney[2], Y Endo[3]

[1] City, University of London, United Kingdom
[2] Glasgow Technical Centre, Mayekawa UK Ltd.
[3] Mayekawa MFG. Co., Ltd.

*Corresponding author: m.read@city.ac.uk

Abstract. Screw compressors have been widely investigated for many applications including air and process gas compression, and refrigeration systems. There is however a surprising lack of literature for low pressure ratio application of these machines, defined here as application requiring volume ratios in the range of 1-1.5. The aim of this paper is to characterise the loss mechanisms for oil injected screw compressors with low volume index (defined as the ratio of maximum to minimum volumes during the internal compression process), V_i. This knowledge will be applied in the future to identify appropriate optimisation of rotor profiles and porting geometry for a range of low-pressure ratio applications. The current study involves the use of chamber models to investigate the influence of wrap angles and porting. This initial analysis will be developed in future work to allow a detailed parametric study of the factors that limit the performance of low V_i oil-injected screw compressors.

1. Background

Positive displacement machines are commonly used to achieve compression of gases. There are many applications where such compressors are required to operate with a low-pressure ratio. This can either be due to the nature of the process, such as in an engine supercharger, or due to variable operating conditions, such is in gas booster systems and refrigeration systems, where low pressure ratio may be required during some operating regimes.

Roots blowers are an example of positive displacement device commonly used for low pressure ratio applications in industrial systems and engine supercharging. Significant work has been done in the past on characterising the rotor geometry and leakage characteristics of these machines [1, 2]. More recent work has focused on improving aspects such as flow pulsations [3] and using experimental and CFD methods to investigate the flow field [4]. These machines have no internal compression, and can be thought of as 'gas pumps'; pressurisation occurs due to working chambers containing low pressure fluid being exposed to the high-pressure fluid, resulting in some degree of backflow. As such, the efficiency of these machines is maximum close to a pressure ratio of 1, and drops rapidly as pressure ratio increases due to leakage at the rotor-to-rotor and rotor-to-casing clearance gaps and dynamic pressure losses during filling and discharge. Helical profiles are sometimes applied in roots blowers in order to achieve progressive engagement between the rotors and reduce vibrations and noise [5]. This leads to an increase in leakage path lengths, but the wrap angles involved are small and this does not significantly affect the operation of the machine.

Twin screw compressors use helical rotors with higher wrap angles, allowing internal compression to occur within the machine. The ability to match the working chamber and downstream pressures before discharge leads to higher efficiency at higher pressure ratio than can be achieved with a roots blower. These machines have been widely investigated for many applications including air and process gas compression, and refrigeration systems [6-8]. There is however a surprising lack of literature for low pressure ratio application of these machines, defined here as requiring a volume ratio of less than 1.5. When compared to simple positive displacement machines such as roots blowers, screw compressors have improved performance, but the increased manufacturing complexity is a drawback. Dynamic compressors are also an option which have some advantages in performance over screw compressors at design point, but have different challenges in terms of machine design, operation, and off-design performance.

This study is motivated by the identification of new industrial refrigeration applications requiring high efficiency compressor operation in this low-pressure ratio region, and hence research is necessary to understand both the fundamental loss mechanisms for screw machines in this operating regime, and the challenges of cost-effective design and optimisation.

2. Performance of a conventional screw compressor with low V_i

A conventional screw compressor design has been analysed using the compressor geometry generation and thermodynamic chamber modelling features of the software package SCORG [9]. This screw compressor is designed to operate with V_i values in the range of 1.8-5, with a relatively high-pressure difference between suction and discharge conditions. This configuration does however provide a useful base case configuration for investigating the influence of various parameters on the performance of screw machines when operating with a low-pressure ratio. The modelled geometry is based on a Mayekawa 400L class UD series refrigeration compressor [10], with outlet ports adapted for low V_i operation. The compressor configuration is 4/6 lobes on main and gate rotors respectively, with a maximum operating speed of 3600rpm and a corresponding theoretical displacement of 11700m^3/h.

2.1. *Low pressure ratio operating conditions*

The chamber model calculations using the SCORG software package have been performed for a range of V_i values, which are achieved by varying the geometry of the high-pressure discharge port of the machine. The operating conditions considered in this analysis are for an NH_3 gas compressor with oil injection. An ideal gas model is used with the following parameters:

- P_{suc} = 3.55 bar, T_{suc} = -5°C
- P_{dis} = 4.62 bar
- Pressure ratio = 1.3
- Oil injection angle = 0°
- Oil supply temperature = 290K
- Oil density at supply temperature = 1.60 mPa.s
- Main rotor speed = 3600 rpm

The chamber model allows calculation of the working chamber pressure profile and the resulting indicated power requirement. The additional mechanical losses relating to the bearings, shaft seal and oil drag losses are then estimated based on empirical relationships between the operating speed, oil viscosity and machine geometry. The resulting net power requirement can then be calculated as shown in Figure 1, with the corresponding results for efficiency and oil-to-gas mass ratio shown in Figure 2.

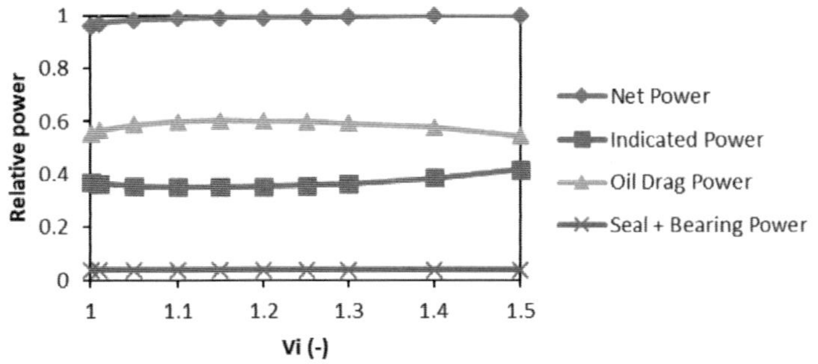

Figure 1 – Breakdown of power requirement as a function of the volume index for the base case compressor

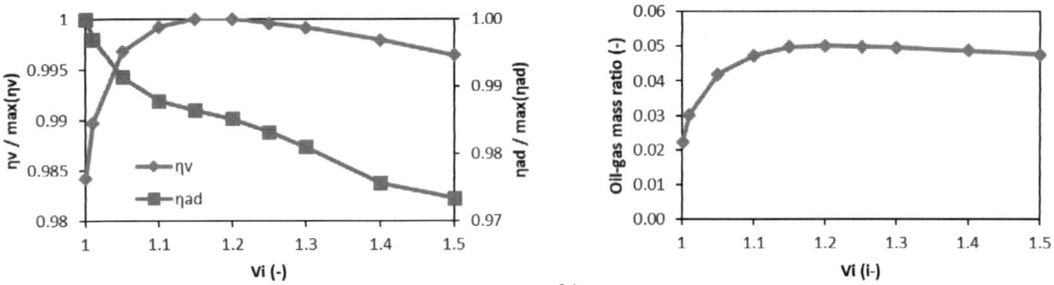

a) b)

Figure 2 – a) Normalised volumetric and adiabatic efficiency, and b) oil-to-gas mass ratio as function of the volume index for the base-case compressor

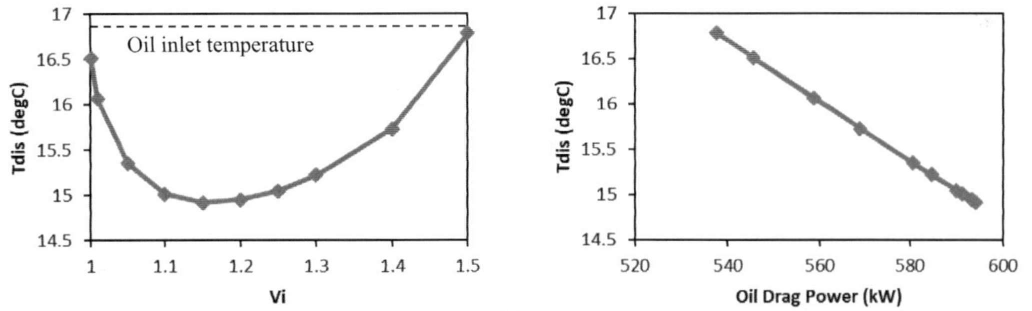

a) b)

Figure 3 – a) Discharge temperature as function of volume index for the base-case compressor, and b) illustration of oil drag temperature dependence

This initial study of screw compressor operation and performance has indicated that oil drag is likely to be a key factor in overall power consumption for low pressure ratio applications. This is due to the low indicated power requirement relative to the machine size and operating speed. Drag losses at rotor end faces appear to be the dominant oil drag loss [11] and more detailed characterization of this mechanism is necessary to better understand the influence on specific power and adiabatic efficiency. The selection of oil is also an important factor, due to the influence of viscosity on these losses. Due to the low pressure ratio, there is very little temperature rise during the compression process, and the oil temperature at inlet and discharge are similar (Figure 3a). The effect of viscosity is shown in the relationship between discharge temperature and oil drag losses in Figure 3b. A further important factor to consider is the mass flow rate of oil. The results in Figure 2b shown that the oil-to-gas mass ratio varies between 2-5% when operating with such low discharge pressure with the oil injection supply pressurized to the same level. This is lower than would be expected for operation at higher pressure ratios. However, due to the low pressure-ratio, the gas temperature rise is relatively small and the oil is not required to achieve cooling of the gas. The small pressure difference between suction and discharge conditions also leads to relatively low leakage flow rate, reducing the need for oil to be present in clearance gaps to achieve sealing. The main requirement for oil is therefore to lubricate the rotors, and the oil-to-gas ratio can be substantially lower than in screw compressors operating at more conventional conditions with higher pressure ratios. The effects of this reduced oil mass flow rate on oil drag losses should also be considered, along with the potential for oil-free compressor operation, although this has the drawback of requiring more complex bearing and seals design.

3. Influence of geometric parameters on performance of machine with low V_i

The wrap angle of the rotors is a geometric parameter that has an influence on both the swept volume of the machine and the size of the leakage areas, particularly that of the rotor tip-to-casing clearances, during the compression cycle. The influence of this geometrical parameter on the performance of the machine has been investigated by keeping all other machine geometry fixed and varying the wrap angle from 150-300 degrees. The influence of this on three important leakage paths, the rotor tip-to-casing radial leakage areas (A_{rad}) of the main and gate rotors and the blow hole (A_{bh}) are shown in Figure 4. The general trend is that decreasing the wrap angle leads to an increased blow hole area, but decreased max radial leakage areas for both rotors. The change in the total of these three leakage areas is seen to decrease with decreasing wrap angle, suggesting leakage flow rates between working chamber may be reduced. Decreasing the wrap angle also reduces the duration of the internal compression phase so there is less time for leakage to occur. This is supported by the results of the chamber modelling, which show that the volumetric efficiency is increased by decreasing the wrap angle (Figure 5a).

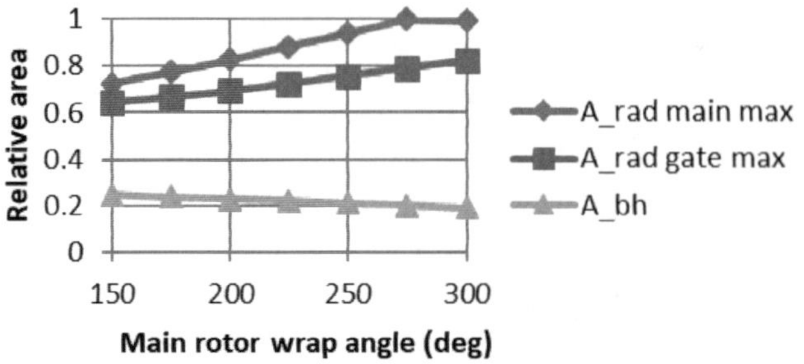

Figure 4 – Influence of rotor wrap angle on maximum radial and blow-hole leakage areas ($A_{rad,max}$ and A_{bh} respectively). Notes areas are normalised by the largest value of $A_{rad,max}$

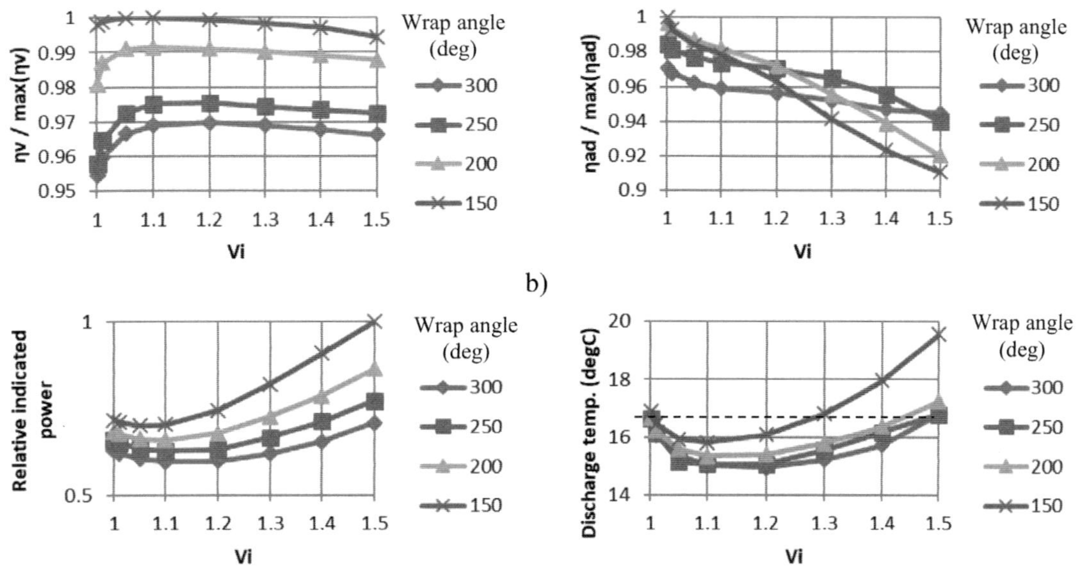

a) b) c) d)

Figure 5 – Results showing normalised a) volumetric and b) adiabatic efficiency, c) normalised indicated power and d) discharge temperature for fixed screw machine geometry with a range of wrap angle values

The results for volumetric efficiency suggest that leakage is minimized with a V_i value of between 1.1-1.2 in all cases (Figure 5a). The results for adiabatic efficiency (Figure 5b) suggest that where under-compression occurs (corresponding to $V_i < 1.15$), this is also improved by reducing the wrap angle, while when over-compression occurs ($V_i > 1.15$) this is no longer the case. For the significantly over-compressed case when $V_i = 1.5$ the maximum adiabatic efficiency is seen to occur with the highest value of wrap angle (300deg). The cause of this reduction in adiabatic efficiency with increasing wrap angle can be explained by considering the indicated power, as shown in Figure 5c. Reducing the wrap angle leads to a higher maximum rate of change of volume in the working chamber, while the maximum inlet and discharge port areas remain similar; the fluid must pass through the ports over a shorter time period, resulting in higher velocities and increased pressure losses during the filling and discharge phases, increasing the indicated power required for the compression process. Figure 6 shows a comparison of the working chamber volume and port areas for cases with $V_i = 1.1$ and wrap angles of 150 and 300 degrees, while Figure 7 shows a comparison of the resulting $p - V$ diagrams for the same cases, clearly showing the significantly higher indicated power for the lower wrap angle. The higher discharge temperature achieved with lower wrap angle does tend to reduce the oil drag power, and the slightly higher maximum working chamber volume combined with the higher volumetric efficiency results in 6% higher mass flow rate for this case. The net effect is a small increase in the adiabatic efficiency when the wrap angle is decreased from 300° to 150° for machines with $V_i = 1.1$ (as seen in Figure 5b).

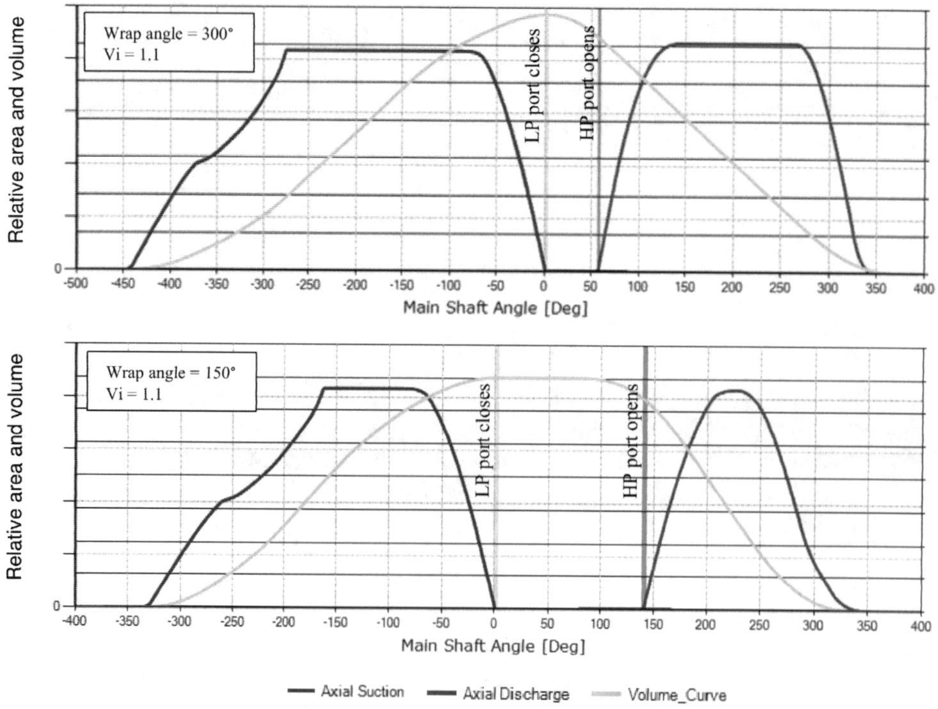

Figure 6 – Working chamber volume and port areas for cases with $V_i = 1.1$ and wrap angles of 300° and 150° respectively

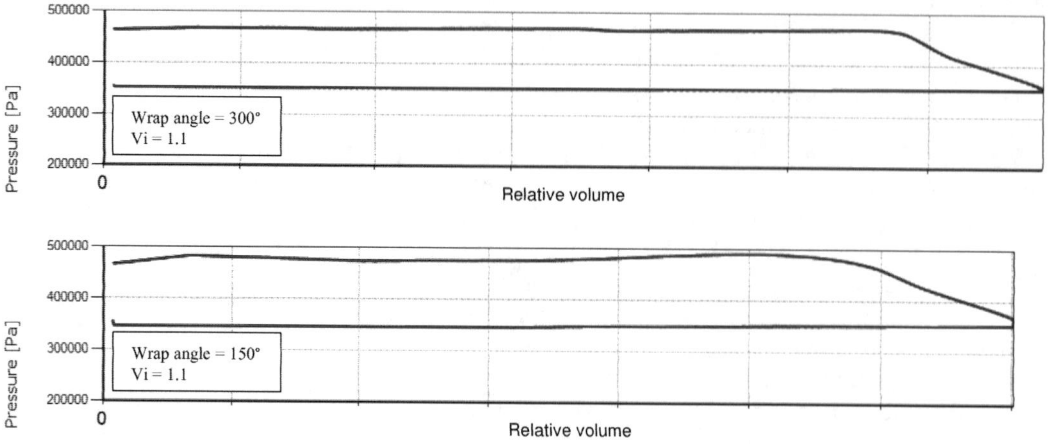

Figure 7 – Working chamber pressure vs. relative volume for cases shown in Figure 6

4. Conclusions

The main conclusions from the current study are as follows:

- Mechanical losses are an important consideration for screw machines operating at low pressure ratio due to the relatively low indicated power requirement.
- Oil drag losses appear to be a major factor in determining the adiabatic efficiency of the machine, and further consideration of the oil properties and the oil injection system is necessary.
- Oil mass flow rate should be provided in order ensure sufficient lubrication of rotors, but the sealing and cooling effects are less important for low pressure ratio operation. The optimum oil-to-gas mass ratio is therefore expected to be significantly lower than in normal screw compressor operation.
- The wrap angle of the rotors is shown to affect the volumetric efficiency and indicated power. For the case considered here, decreasing wrap angle from 300deg to 150deg increases adiabatic efficiency as long as the fluid is not over-compressed ($V_i < 1.15$)

Future work will focus on more detailed characterization of the loss mechanisms, and investigate the influence of a wider range of geometrical parameters such as rotor profile, lobe number and rotor length. It is also planned to use experimental test data to improve and validate these empirical models for the loss mechanisms at low V_i operation. The aim is then to carry out a detailed parametric study of the factors that limit the performance of low V_i oil-injected screw compressors, and identify cost-effective design solutions to improve efficiency of these machines.

References

[1] Ritchie, J.B. and Patterson, J., 1968. Geometry and leakage aspects of involute rotors for the Roots blower. Proceedings of the Institution of Mechanical Engineers, 183(1), pp.707-724.

[2] Mcdougald, S., Imrie, B.W. and Cole, B.N., 1974. An Investigation of the Volumetric. Efficiency of a Roots Blower. International Compressor Engineering Conference. Paper 127.

[3] Sun, S.K., Jia, X.H., Xing, L.F. and Peng, X.Y., 2018. Numerical study and experimental validation of a Roots blower with backflow design. Engineering Applications of Computational Fluid Mechanics, 12(1), pp.282-292.

[4] Singh, G., Sun, S., Kovacevic, A., Li, Q. and Bruecker, C., 2019. Transient flow analysis in a Roots blower: Experimental and numerical investigations. Mechanical Systems and Signal Processing, 134, p.106305.

[5] Mimmi, G. C. and Pennacchi, P. E., 1999. Dynamic loads in the three-lobe supercharger. J. Mech. Des., 121(4), 602–605.

[6] Stosic N., Smith I.K., Kovacevic A. ,2005. Screw Compressors: Mathematical Modeling and Performance Calculation, Springer Verlag, Berlin.

[7] Stošić, N., 2004. Screw compressors in refrigeration and air conditioning. *HVAC&R Research*, *10*(3), pp.233-263.

[8] Kovačević, A., Stošić, N. and Smith, I.K., 2006. Numerical simulation of combined screw compressor–expander machines for use in high pressure refrigeration systems. Simulation Modelling Practice and Theory, 14(8), pp.1143-1154.

[9] https://www.scorg.info/, Accessed 15/02/2021

[10] https://mayekawa.com/mycom/screw_compressors/, Accessed 15/02/2021

[11] Gräßer, M. and Brümmer, A., 2015, July. Influence of liquid in clearances on the operational behaviour of twin screw expanders. In IOP Conference Series: Materials Science and Engineering (Vol. 90, No. 1, p. 012060). IOP Publishing.

Reducing rotors clearance - a way to increase the performance of a screw compressor

M Nitulescu[1], C Slujitoru[1], V Petrescu[1], V Silivestru[2], G Fetea [3], S Tomescu[1]
[1]Dipl. Engineer COMOTI, *220 D Iuliu Maniu Bd., sector 6, cod 061126, OP 76, CP174 Bucharest, Romania*
[2]Scientific Researcher 1, Dr. Engineer President General Director, COMOTI, *220 D Iuliu Maniu Bd., sector 6, cod 061126, OP 76, CP174 Bucharest, Romania*
[3]Dr. Dipl. Engineer COMOTI, Technical Director, *220 D Iuliu Maniu Bd., sector 6, cod 061126, OP 76, CP174 Bucharest, Romania*

Abstract. The screw compressors due to their constructive simplicity and the known advantages in operation is increasingly demanded on the market. Many companies are producing this type of compressor. The influence of leakages – conditioned by rotors clearances – on compressor performance is well known. The literature addresses the problem of coatings, but for dry compressors. Our intention is to apply this technology in the case of oil injected screw compressors (when the profile deviation is large or in the case of rotors with slight wear). Tests were performed to evaluate the performance of a screw compressor with uncoated rotors and performance of compressor assembled with coated rotors (both pairs of rotors have the same profile deviations – the profile accuracy measured is -8 ÷ -10 μm). The test aim to highlight only the influence of the coating layer – by reducing clearances – on the performance of the compressor, without performing further quality test of the coated layer. The quality of the coating process is the subject of additional research. We intend to use this technology for oil injected screw compressors both for new manufactured rotors (with large profile tolerances) and rotors with slight.

1. Introduction

Within this article, NRDI COMOTI aims to present an innovative method – at least in Romania – to reduce the high costs related to the manufacture of component parts of high-pressure discharge compressors with very tight tolerances, using a modern coating technology.

It's significant to emphasize that in the open literature, the rotor's coating process is presented as being applicable **only** for the dry (oil free) screw compressors types. The arguments specified in open literature and the advantages of applying this type of coating process have led us to consider that this coating process can also be applied to rotors used in oil injected screw compressors in the two aforementioned circumstances. Thus, the possibility to practically verify the stated considerations was offered by a concrete situation: COMOTI ordered the manufacture of rotor pairs to a partner company. From the profile diagrams it resulted that the recorded profile's deviations fit into the range -8 ÷ -11μm. The pairs of rotors that will be coated were chosen accordingly to figures 1 and 2, respectively the pairs of rotors without coating accordingly to figures 3 and 4.

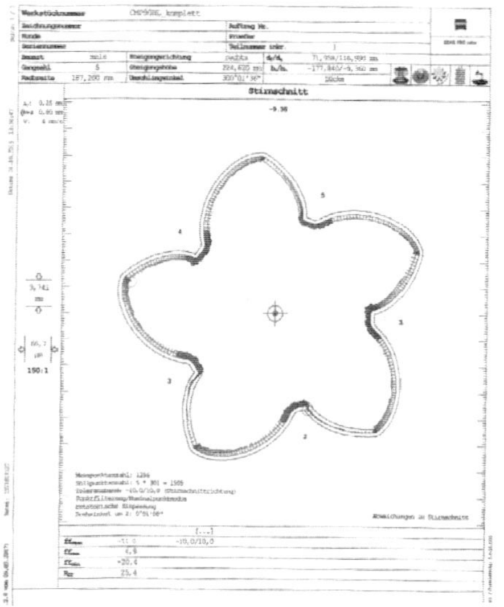

Figure 1. Profile diagram of male rotor chosen to be coated

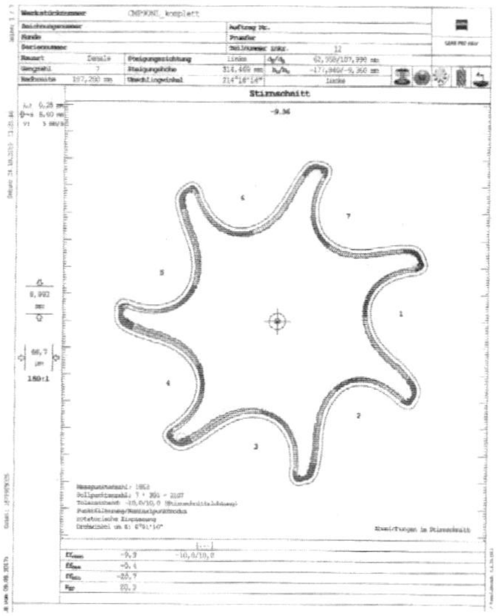

Figure 2. Profile diagram of female rotor chosen to be coated

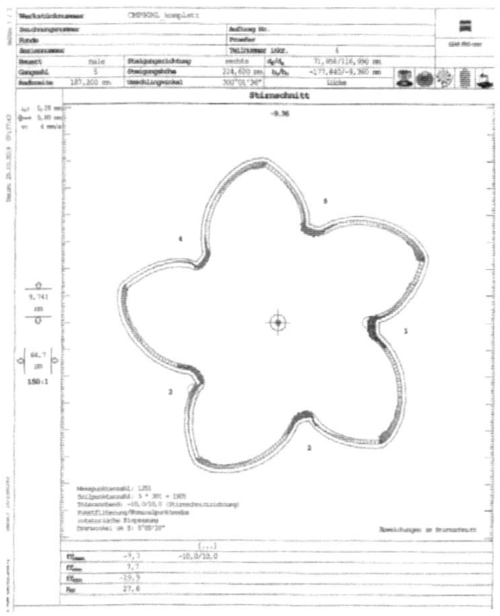

Figure 3. Profile diagram of male rotor chosen to be uncoated

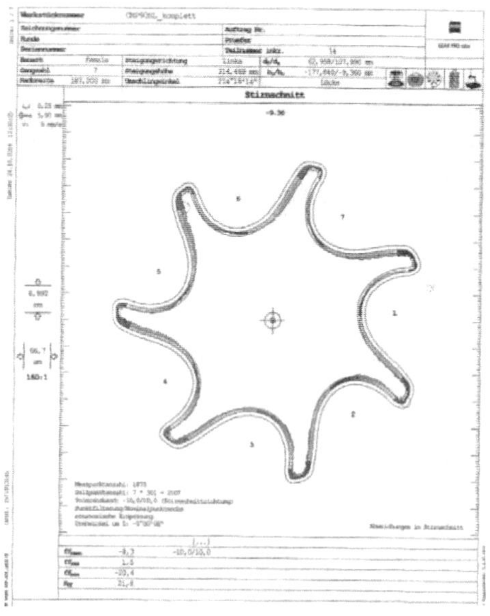

Figure 4. Profile diagram of male rotor chosen to be uncoated

By coating, the clearances between male and female rotors were reduced with 20 μm and clearances between rotors and housing were reduced with 10 μm.

Thus, figure 5 shows a sample rotor's pair of a screw compressor developed by COMOTI, for each rotor that is coated with XILAN 1052 CTG ON LAS, the thickness of the deposited layer fits into the range 0.009 ÷ 0.012 mm. This was a first attempt to test the possibility of coating rotors, checking only the uniformity of the thickness of the deposited layer (see figure 6). Subsequent research will determine the quality of the layer during operation period (checking adhesion, elasticity, oil resistance, resistance to operational stresses, aso).

Figure 5. Rotors coated with XILAN 1052 CTG ON LAS

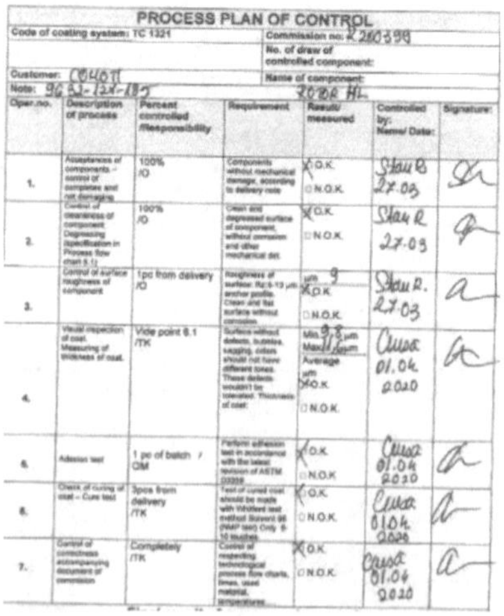

PROCESS PLAN OF CONTROL

Code of coating system: TC 1321
Commission no: R 280398
No. of draw of controlled component:
Customer: COMOTI
Name of component:
Note: 9651-124-185 ROTOR HL

Oper.no.	Description of process	Percent controlled /Responsibility	Requirement	Result/ measured	Controlled by: Name/ Date:	Signature:
1.	Acceptances of components – control of completes and not damaging	100% /O	Components without mechanical damage, according to delivery note	X O.K. ☐ N.O.K.	27.03	
2.	Control of cleanliness of component Degreasing (specification in Process flow chart 5.1)	100% /O	Clean and degreased surface of component, without corrosion and other mechanical det.	X O.K. ☐ N.O.K.	27.03	
3.	Control of surface roughness of component	1pc from delivery /O	Roughness of surface: Rz 5-13 µm anchor profile. Clean and flat surface without corrosion	µm 9 X O.K. ☐ N.O.K.	27.03	
4.	Visual inspection of coat. Measuring of thickness of coat.	Vide point 6.1 /TK	Surface without defects, bubbles, sagging, colors should not have different tones. These defects wouldn't be tolerated. Thickness of coat:	Min. 9.8 µm Max. 11.6 µm Average µm X O.K. ☐ N.O.K.	01.04 2020	
5.	Adesion test	1 pc of batch / QM	Perform adhesion test in accordance with the latest revision of ASTM D3359	X O.K. ☐ N.O.K.	01.04 2020	
6.	Check of curing of coat – Cure test	3pcs from delivery /TK	Test of cured coat should be made with Whitford test method Solvent 96 (NMP test) Only 5-10 touches	X O.K. ☐ N.O.K.	01.04 2020	
7.	Control of correctness accompanying document of commission	Completely /TK	Control of respecting technological process flow charts, times, used material, temperatures	X O.K. ☐ N.O.K.	01.04 2020	

Figure 6. Measurement sheet for deposited layer

Figures 7 and 8 suggestively presents the pair of rotors both the one with coated rotors and the one with uncoated rotors mounted in the compressor assembly before test and in figure 9 the coated rotors pair after nearly 30 testing hours.

Figure 7. Compressor with coated rotors

Figure 8. Compressor with uncoated rotors

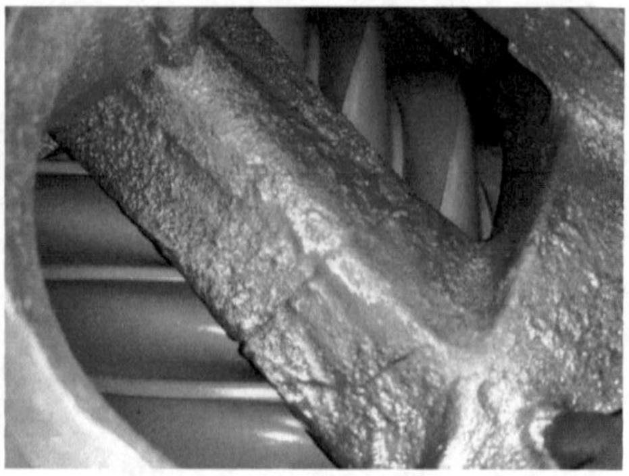

Figure 9. Compressor with coated rotors after nearly 30 testing hours

So, it seems that any reduction of the clearance leads to the improvement of the performances. In recent time, a modern machine tool can manufacture helical rotors obtaining a tolerance to the rotor profile around +/- 5 μm, which ultimately leads to a value of less than 15 μm of the clearance between the rotors. With such small gap, collision between the rotors is certitude and therefore the profile and distribution of the clearance along the profile must be generated so as to avoid damage if this collision occurs. It should be mentioned that the manufacture of rotors with tolerances so tight is expensive. A solution would be the acceptance of wider tolerances, which will be adjusted by coating procedures, to comply with the initially prescribed value of clearance between the moving parts.

This article does not intend to analyze the principles and methods of designing the rotor profile / profile tolerances / clearance distribution also, but to analyze solutions to improve the technological process of manufacturing / reconditioning the rotors regarding the following aspects:

- the possibility of reconditioning rotors that have worked in heavy duty and shows wear of profile that leads to the increase of the clearance and the alteration of the performances;

- the reduction of the costs related to the manufacture of rotors with tight tolerances, by using the coating procedures.

In order to limit clearances, precise manufacturing of rotors and housings is essential, but even in the case of using precise CNC machines, the ideal case – zero tolerance – is practically unattainable. For example, for small compressors, the recommended clearances are up to 50 μm, values that are difficult to obtain, a solution being the rotor pairing to obtain the optimal clearances. But these operations raise the cost of manufacturing. One solution would be treating of the surface, to reduce clearance and minimize friction. Examples of such solutions would be: hard anodizing, chrome plating, plating or PTFE coating – see figure 10.

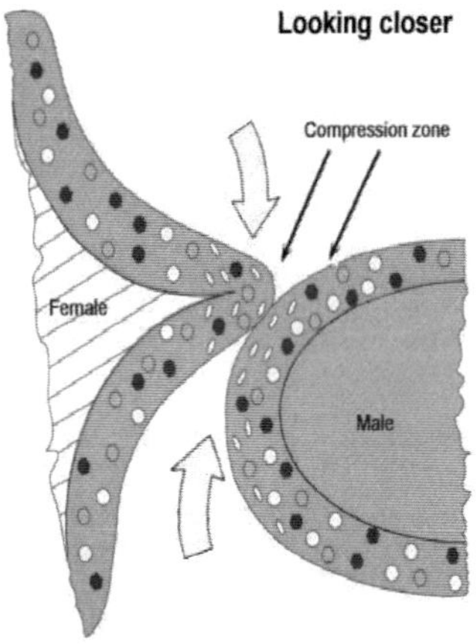

Figure 10. Coating of the male / female rotors' surfaces

In the article [1] it is stated that compressor efficiency is limited by geometry. Air leaks through the gaps between the rotors, between the housing and rotor tips, and at the output ends of the rotors. And the higher is the pressure, the more likely air is to escape back to the low-pressure side of the compressor. All that leakage is lost efficiency and lost power. In fact, pumping efficiency is related to the clearances between surfaces that push air. The answer seems easy: the tighter the clearance the better. But that isn't easily attained. Because metal surfaces in these screw-type compressors are a three-dimensional spiral slope, machining them to precise dimensions is difficult. Manufacturers usually match each male-female set manually to get the best fit.

In the article [2] it is stated that screw compressor efficiency is improved by reducing internal leakage and this can be affected by minimizing the clearance between the rotors and the casing. The effect on performance of three factors which influence the working clearance was analyzed. These are: the interlobe clearance, adjusted to the rotor temperature change, rotor contact on the lobe flat side and displacement of the discharge bearing centers.

In the article [3] it is stated that designing twin-screw compressors to safely operate at higher-than-normal temperatures poses a challenge as the compressor must accommodate larger peak thermal distortions while maintaining efficiency at nominal operating conditions. This paper presents a case study of an oil-injected compressor tested at elevated discharge temperatures with original and revised clearances. A procedure is presented to use boundary conditions derived from a chamber model to approximate component temperature distributions that are then used to predict possible thermal distortions and the resulting effect on clearance gaps. The original and revised clearance designs are evaluated and performance penalties incurred due to the modifications are discussed.

2. Tests

2.1. General conditions

The configuration/instrumentation layout of the test bench is shown in figure 11. The parameters to be measured for the estimated performance of the compressor will be: compressor suction/discharge temperature and pressure; speed of the male rotor; power of the electric motor driver; pressure/temperature of oil after the oil cooler; differential pressure, pressure and temperature across the differential devices (designed according to ISO 5167-2); in/out temperature for water, in the oil cooler; oil flow rates, for each oil point injection (seal, suction bearings, injection, discharge bearings); pressure of oil at each oil injection point; atmospheric pressure, atmospheric temperature and the relative humidity.

In figure 12 there is a photo of the test bench, with the compressor installed on it.

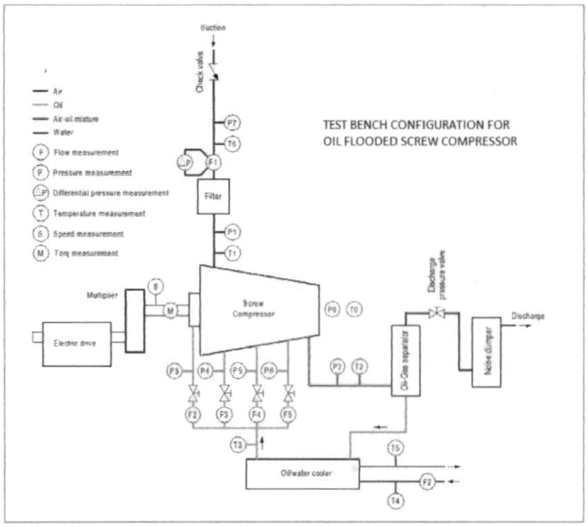

Figure 11. Test bench configuration

Figure 12. Compressor on test bench

2.2. Measured parameters. Measuring equipment. Precisions

The equipment, precision and measurement methods are in accordance with the recommendations of ISO 1217, ASME PTC9, relating to the determination of the performance of a compressor during the test, delivered flow or required power.

We emphasize once again that for licensed compressors, the performance test - with air - is done for a point of operation indicated by the licensing company. Depending on the type of compressor, respectively volumetric ratio Vi, the licensor indicates the specific test parameters:

- suction pressure / suction temperature;
- discharge pressure;
- flow;
- speed.
- During the test, you measure:
- the gas flow rate;
- the power;
- discharge temperature (controlled by oil flow).

For compressors developed by COMOTI, the test conditions follow the recommendations in Annex E-ISO 1217- concerning the acceptance test for volumetric compressors driven by variable speed electric motors.

At the same time, ISO 1217 recommends that the measuring equipment that may affect test results be calibrated at the specified range to ensure the accuracy/reliability of the measured information. Measurements of parameters used in performance calculations are made with instruments / devices that meet the following conditions (for the COMOTI stand):

- for pressure – pressure transducer, signal 4...20mA, line precision ±0,2%;
- for temperature – thermocouple, signal 4...20mA, line precision ±0,5%;
- for oil flow - flow meter, signal 4...20mA, line precision ±0,2%
- for gas flow - differential pressure transducer, signal 4...20mA, instrument precision ±0,1%, line precision ±1%.

The flow rate is measured at the suction of the compressor accordingly to ISO 5167-2, the flow meter being mounted on the pipe according to the installation conditions. For calculating the corrections and displaying the flow in Nm3, the acquisition system picks up atmospheric pressure and ambient temperature signals (with transducers mounted in the vicinity of the suction port).

- for power – torque converter mounted on the drive shaft of the compressor, type HBM-T40 precision class 0,05%;
- for speed - speed transducer, signal 4...20mA, line precision ±0,2%c

During the tests, the vibrations are measured with VIBER X5-see figure13 and thermal imprint with Fluke –see figure14.

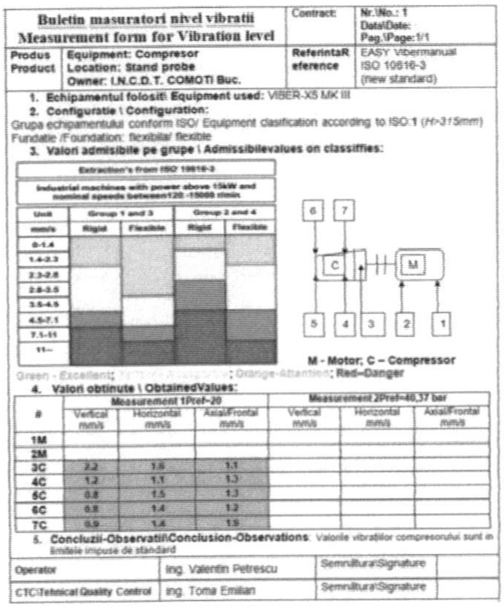

Buletin masuratori nivel vibratii Measurement form for Vibration level		Contract:	Nr.\No.: 1 Data\Date: Pag.\Page: 1/1
Produs Product	Equipment: Compresor Location: Stand probe Owner: I.N.C.D.T. COMOTI Buc.	ReferintaR eference	EASY Vibermanual ISO 10816-3 (new standard)

1. Echipamentul folosit\ Equipment used: VIBER-X5 MK III
2. Configuratie \ Configuration:

Grupa echipamentului conform ISO/ Equipment clasification according to ISO:1 (H>315mm)
Fundatie /Foundation: flexibila/ flexible

3. Valori admisibile pe grupe \ Admissibilevalues on classifies:

Extraction's from ISO 10816-3

Industrial machines with power above 15kW and nominal speeds between120 -15000 r/min

Unit	Group 1 and 3		Group 2 and 4	
mm/s	Rigid	Flexible	Rigid	Flexible
0-1.4				
1.4-2.3				
2.3-2.8				
2.8-3.5				
3.5-4.5				
4.5-7.1				
7.1-11				
11--				

M - Motor; C – Compressor

Green - Excellent; … ; Orange-Attention; Red–Danger

4. Valori obtinute \ ObtainedValues:

#	Measurement 1Pref-20			Measurement 2Pref=40,37 bar		
	Vertical mm/s	Horizontal mm/s	Axial/Frontal mm/s	Vertical mm/s	Horizontal mm/s	Axial/Frontal mm/s
1M						
2M						
3C	2.2	1.6	1.1			
4C	1.2	1.1	1.3			
5C	0.8	1.5	1.3			
6C	0.8	1.4	1.2			
7C	0.9	1.4	1.9			

5. Concluzii-Observatii\Conclusion-Observations: Valorile vibratiilor compresorului sunt in limitele impuse de standard

Operator	Ing. Valentin Petrescu	Semnatura\Signature	
CTC\Tehnical Quality Control	Ing. Toma Emilian	Semnatura\Signature	

Figure 13. Vibration analysis

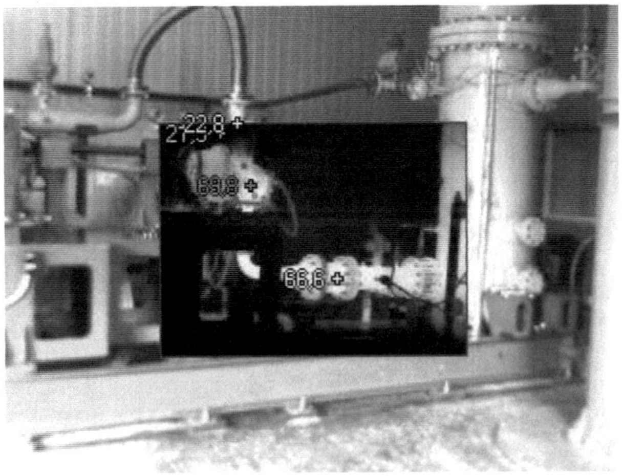

Figure 14. Thermal image

When testing a new compressor developed by the COMOTI, the tests in the stand are detailed, aiming at the acquisition of data allowing the development of the operating diagrams (previously mentioned: flow/speed and power/speed chart for a given suction pressure, a given volume ratio Vi and different discharge pressures, in operating range).

The scope of our tests is to compare the performance of the two compressors:

- with coated rotors;
- with uncoated rotors.

We keep the other parts – housings – unchanged to determine only the influence of the coating on the performance (keep constant axial clearances), and made tests in the same conditions (similar suction pressure, discharge pressure, discharge temperature, male rotor speed).

2.3. Recorded parameters. Diagrams

We specify that the equipment, precision and measurement methods are in accordance with the recommendations of ISO 1217, ASME PTC9, relating to the determination of the performance of a compressor during the test, delivered flow or required power. Tests were performed to evaluate the performance of a screw compressor with uncoated rotors and performance of compressor assembled with coated rotors (both pairs of rotors have the same profile deviations – the profile accuracy measured is -8 ... -10 μm). The test aim to highlight only the influence of the coating layer – by reducing clearances – on the performance of the compressor. As a procedure, we kept the discharge pressure constant and gradually decreased the speed (keeping it constant until the discharge temperature stabilized) – see figures 15, 16.

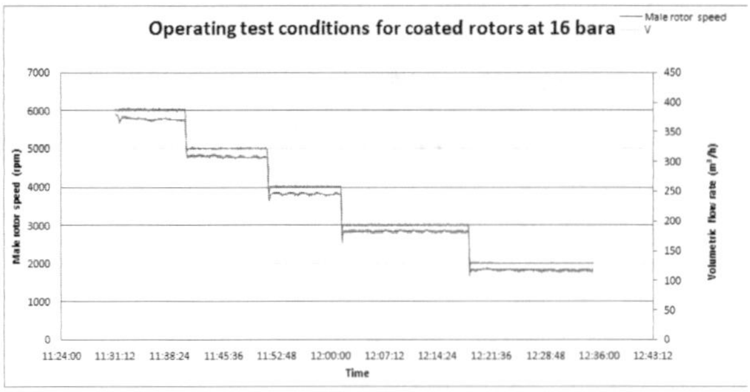

Figure 15. Operating test conditions for coated rotors at 16 bar abs.

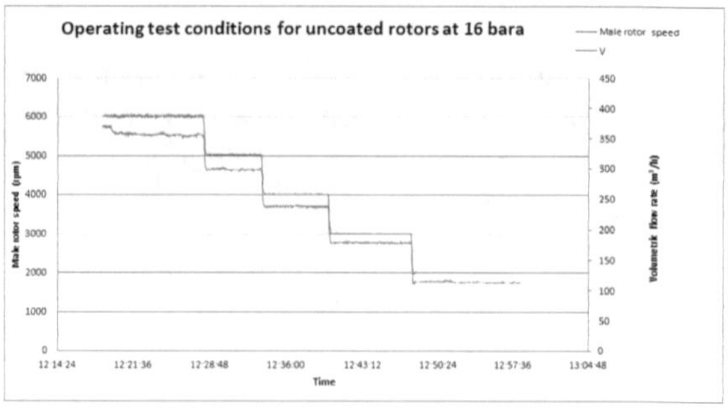

Figure 16. Operating test conditions for uncoated rotors at 16 bar abs

We attach print-screen with a sample of recorded parameters for male rotor speed at 6000 rpm– see figure 17 and table 1 (the notations of the measured parameters are the same for the print-screen and for the table 1).

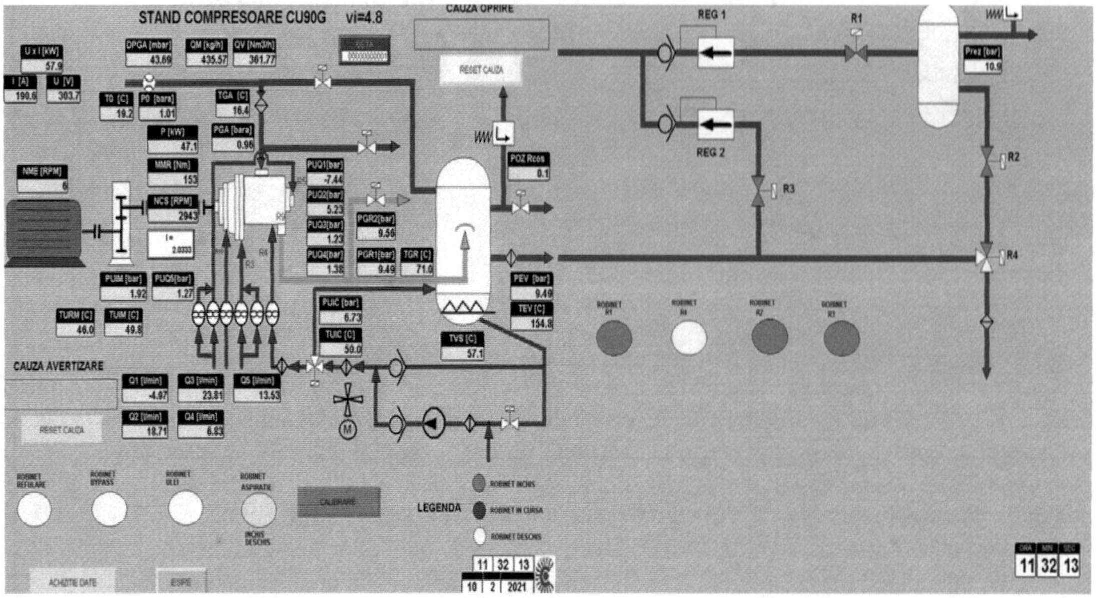

Figure 17. Sample of print-screen for recorded data at 11 bar abs. – uncoated rotors

Table 1. Sample of recorded parameters for uncoated (UC) and coated (C) rotor's pair at 11 and 16 bar abs.

Rotor's pair UC or C-p (bar abs)	Differential pressure suction diaphragm DPGA (mbar)	Momentum MMR (Nm)	Driver speed NCS (RPM)	Ambient pressure P0 (bar abs)	Suction pressure PGA (bar abs)	Discharge pressure PGR1 (bar)	Lube oil pressure PUIC (bar)	Suction temperature TGA (°C)	Discharge temperature TGR (°C)	Ambient temperature T0 (°C)	Oil injection temperature TUIC (°C)	Mass flow rate QM (kg/h)	Volumetric flow rate QV (Nm^3/h)	Power consumption P_MMR (kW)
UC-11	43,69	152,88	2943,28	1,01	0,96	9,49	6,73	16,4	70,95	19,2	49,95	435,57	361,77	47,12
UC-16	41,53	194,13	2944,69	1,01	0,96	14,21	12,21	16,7	71,6	18,3	49,85	426,5	354,24	59,86
C-11	45,47	158,5	2954,53	1,01	0,96	9,49	7,48	13,75	68,4	16,5	50,25	448,64	372,62	49,04
C-16	44,34	192,25	2981,25	1,01	0,95	14,08	11,01	9,5	70,85	11,05	49,95	446,12	370,53	60,02

So, applying the recommendations about computation of tests result specific diagrams were plotted. For an easy understanding, we synthesized the test results, presenting, in a diagram, the measured and corrected values at a certain speed/discharge pressure (Pref) – for coated rotor/uncoated rotor – see figures 18÷21.

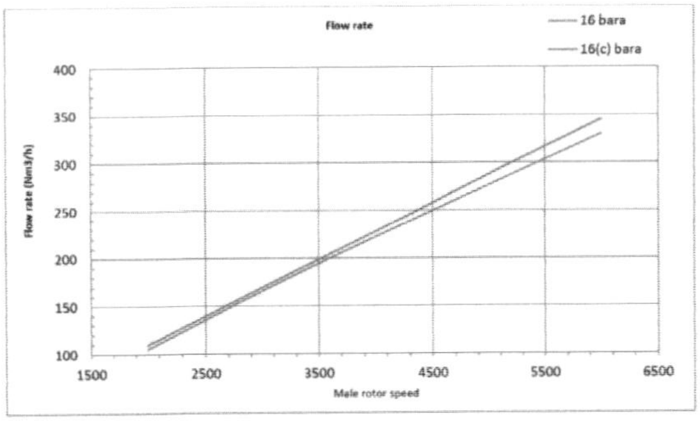

Figure 18. Flow rate for Pref =16 bar abs.

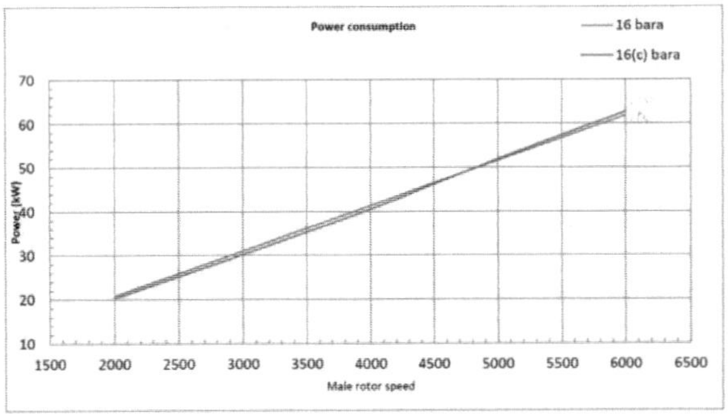

Figure 19. Power for Pref =16 bar abs.

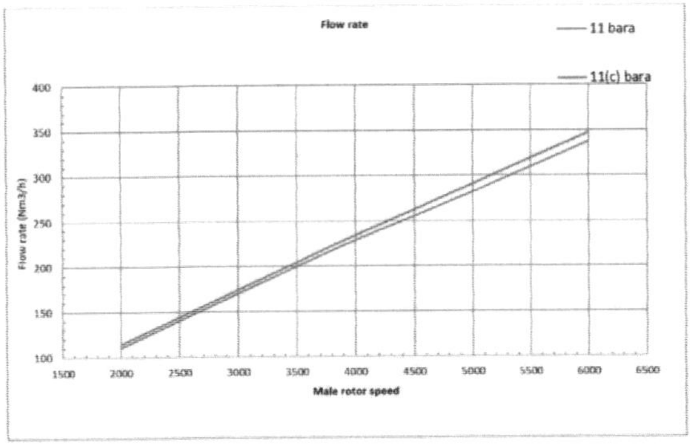

Figure 20. Flow rate for Pref =11 bar abs.

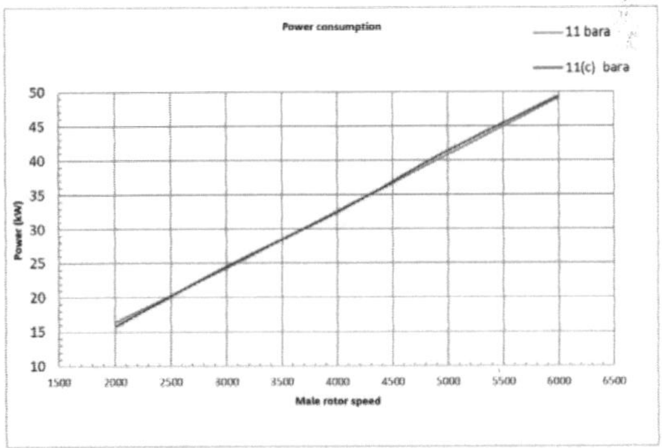

Figure 21. Power for Pref =11 bar abs.

Starting from recorded parameters also we achieve following diagrams – figures 22 and 23 – that show – in percentages – the rapport between flow rate (Vcr - volumetric flow rate coated rotors, Vur - volumetric flow rate uncoated rotors) and power (Pcr - power consumption for coated rotors, Pur - power consumption for uncoated rotors) at different speed, for coated and uncoated rotors.

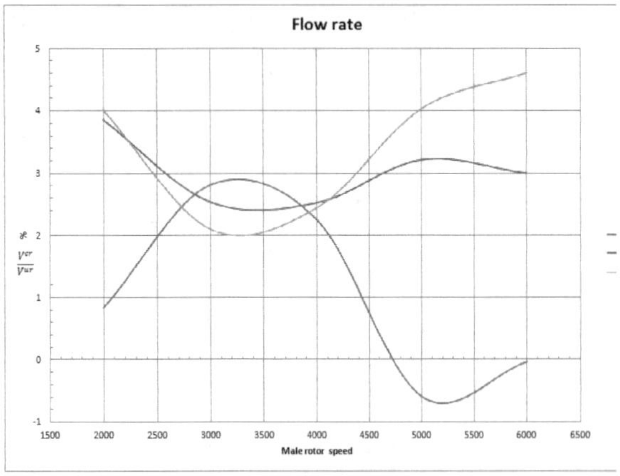

Figure 22. V^{cr}/V^{ur} diagram

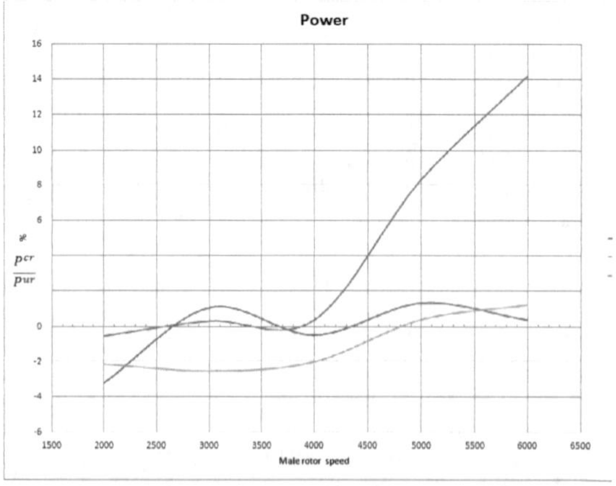

Figure 23. P^{cr}/P^{ur} diagram

It is also suggestive to show in the tables 2, 3 and 4 the values of the volumetric efficiency, for compressor with coated and uncoated rotor, and the increase of volumetric efficiency -%-. Practically, concluding, analyzing the power and flow diagrams it can be stated that, by coating the rotors, an increase of the flow is obtained, in the conditions of a practically constant power. So, the specific consumption kWh / 1000Nm3 decreases (with approx.3%).

Table 2. Volumetric efficiency for uncoated rotors

Speed (RPM)	Pressure at 6 bar abs	Pressure at 11 bar abs	Pressure at 16 bar abs
2000	0.806314	0.776078	0.740404
3000	0.80887	0.796326	0.776858
4000	0.826	0.802018	0.781326
5000	0.830231	0.792316	0.777829
6000	0.818061	0.789565	0.773123

Table 3. Volumetric efficiency for coated rotors

Speed (RPM)	Pressure at 6 bar abs	Pressure at 11 bar abs	Pressure at 16 bar abs
2000	0.713112	0.806042	0.770096
3000	0.831638	0.81652	0.793172
4000	0.844572	0.822248	0.800414
5000	0.825293	0.817781	0.809175
6000	0.817807	0.813257	0.808689

Table 4. Volumetric efficiency increase from uncoated rotors to coated rotors

Speed (RPM)	Pressure at 6 bar abs	Pressure at 11 bar abs	Pressure at 16 bar abs
2000	0.843057 %	3.860977 %	4.010283 %
3000	2.814755 %	2.535852 %	2.100056 %
4000	2.248338 %	2.522376 %	2.443014 %
5000	-0.59475 %	3.217659 %	4.029867 %
6000	-0.031102 %	3.000666 %	4.600234 %

IOP Conf. Series: Materials Science and Engineering 1180 (2021) 012007 doi:10.1088/1757-899X/1180/1/012007

3. Conclusions

Following the tests results we can draw the following conclusions:

- tests were suited to evaluate the performance of a screw compressor with uncoated rotors and performance of compressor assembled with coated rotors (both pairs of rotors have the same profile deviations – the profile accuracy measured is -8 ÷ -10 μm);
- applying this technology are reduced the costs related to the manufacture of rotors with tight tolerances, (by using the coating procedures); also, by coating, the price of the compressor unit rises with nearly 2% (for tested type compressor)
- the test show that, for coated rotors the compressor performance has improved;
- for discharge pressure 11 bara and 16 bara, the average flow increase is approx. 3% (the power remaining practically constant);
- applying this technology, the specific consumption in kWh / 1000Nm3 decreases (if we compare the parameters of two compressors – with/without coating layer), with nearly 3%;
- we specify that quality of the coating process is the subject of additional research. But after approximately 30 hours of tests, a visual analysis of coating looks well, without detachments or other traces (see figure 9).

4. References

[1] Bruce Nesbitt: *Conformable coating narrows compressor air gaps.* https://www.machinedesign.com/archive/conformable-coating-narrows-compressor-air-gaps-boosts-efficiency

[2] Nikola Stosic, Ian K. Smith, Ahmed Kovacevic: *Improving screw compressor performance* https://www.researchgate.net/publication/289206577_Improving_screw_compressor_performance

[3] Buckney D, Kovacevic A, Stosic N 2016 *Design and evaluation of rotor clearances for oil* injected screw compressors, proceedings of the institution of Mechanical Engineers, Part E: Journal of Process Mechanical Engineering 231 26-37

[4] ISO10440 Rotary-type Positive Displacement Compressors for Petroleum, Petrochemical and Natural Gas Industries (API 619)

[5] ISO 1217 – Displacement compressors – Acceptance tests (PTC 9)

[6] ISO 5167-2 – Measurement of fluid flow by means of pressure differential devices in circular cross-section conduits running full – Part 2 Orifice plates

Design, Testing and Feasibility Analysis of an Oil-Free Twin Screw Compressor with In-Chamber Flash Cooling

G Stupple
Jäcklin GmbH, Unterer Talweg 50, 86179 Augsburg, Germany

stupple@jaecklin.de

Abstract. With the current aim of reducing CO2 emissions worldwide compressor efficiency is more in focus than ever before. An estimated 20% of all electricity generated worldwide is used for compressing air. Oil injected twin screw compressors form a large percentage of the total. The injected oil is used for cooling, as well as for sealing and lubrication of the rotors. Typical pressure ratios lie between 8:1 and 16:1. There are many investigations being conducted into oil-free screw technology for specific applications requiring no oil carry over. In this presentation we will take a close look into the possibility of producing a single stage oil-free twin-screw compressor operating at a pressure ratio of 8:1 and using evaporated water for cooling. The main objective in this case is to improve compressor efficiency. Additionally, we will compare efficiency results with a state-of-the-art single stage oil injected twin screw compressor. Previous results from an earlier paper by the author and team from City University of London on CFD analysis of a similar system are taken as the basis. Further simulated calculations using SCORG for the actual case in question allow preliminary performance analysis using the multi-domain chamber model. From simulation we can see that the oil drag force acting on the rotors forms a significant part of the total power requirement. It is therefore suggested that an oil-free solution will achieve improved specific power. To confirm this concept, verify the simulated calculations and understand the heat transfer process in the compression chamber more fully, experimental analysis was undertaken.

1. Introduction

The object of this work is to design a compressor which will produce better efficiency over the specified operating range than is currently achievable from a typical oil injected compressor.
This is a market driven requirement and is considered to be an important development step for the screw compressor by Jaecklin GmbH. During this stage of the project we would like to produce a stable running prototype providing data to be used for further development. This is a big challenge and the design will be a compromise between high efficiency, low manufacturing cost and high reliability. Additional targets are low noise, low rpm, high start/stop capability, low operational temperature and good air quality.

Certain design decisions were made in advance taken from the above list of requirements. A single stage solution is the only commercially acceptable possibility. The size of the compressor will be in the range of 22 kW up to 45 kW. Direct drive is required with variable speed up to 6000 rpm. The designed pressure ratio will initially be 8:1 which may be extended up to 11:1 during future development.

Since cooling during compression will be a major challenge in the oil free screw compressor various alternatives were considered. The possibility of using impinging jet rotor cooling commonly referred to as flash cooling has an advantage of reducing compressor temperature at the same time reducing fluid injection quantity. It was suggested that a reduction in injected fluid induced rotor drag might lead to reduced power and possibly to higher compressor efficiency.

The challenge of removing oil from an oil-injected compressor is expected to provide the following effects. Possible increase in the leakage between rotors and housing may result. Reduced lubrication properties between the rotors may cause reliability issues. The suggested advantage of reduced drag generated between the rotors and housing was numerically simulated using SCORG multi-domain chamber model and the results indicate a power saving of 17% at the target operating condition. The increase in leakage will be measured by comparing the volumetric efficiency of both the oil injected and oil free versions and the results used for designing optimised rotors for future evaluation. For the comparative tests both oil injected and oil free versions use exactly the same rotors and the clearances in both compressors are identical. At this stage it was not clear how effective the flash cooling would be. The SCORG simulation model predicted a gas temperature at the outlet of 156°C.

A two stage dry running solution was also considered and rejected on the basis of increased complexity and manufacturing cost. A dry running single stage twin screw compressor is normally limited to a pressure ratio of 3:1 and for this reason was not considered to be a suitable solution for the specified 8:1 pressure ratio. The high gas temperature produced during the compression process deforming the rotors and housing, and at a pressure ratio of 8:1 leading to damage or eventual seizure. This was confirmed during a CFD analysis commissioned by Jaecklin GmbH at an earlier stage of the project. Nevertheless, an interesting development taken from dry running compressors was considered as a possible solution for cooling the oil free rotors in our case. Referring to a previous paper, *Stosic et al 2011, Numerical and experimental research in heat transfer to screw compressor rotors,* we can see that the pressure ratio of a dry running single stage compressor could be extended considerably by injecting minute quantities of a volatile liquid such as water into the housing at the high-pressure port end in any circumferential position. Experimental results from the paper confirm that the theory of flash cooling works effectively up to a pressure ratio of nearly 6:1. The conclusion from this previous work tells us that the liquid would then impinge on the rotors at a rate which would be instantly evaporated on contact with them. This concept was taken over and analysed more fully as a suggested solution for a single stage oil-free compressor with an operating pressure ratio of 8:1.

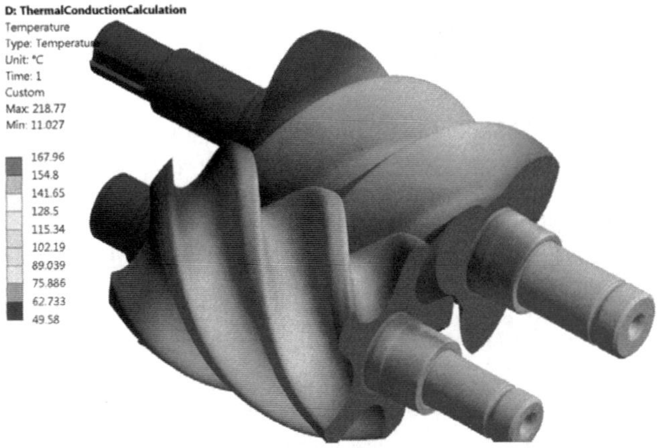

Figure 1: Shows the CFD thermal conduction calculation at a pressure ratio of 12:1. The impinging jet cooling is directed at both rotors as close to the discharge end as possible.

The results of a more recent paper, *Stosic et al 2018 numerical investigation of water evaporation in twin screw compressors* provide a numerical study into the feasibility of using flash cooling to operate a single stage compressor up to a pressure ratio of 12:1 by using a very small amount of water injection, just enough to provide cooling by evaporation. A numerical procedure was used to estimate the mass of water required to produce saturated air at delivery pressure. A multiphase CFD model was set up to solve air and water-liquid flow along with a simplified evaporation model to account for the latent heat of evaporation. The main physical effect is rotor cooling from heat exchange produced by the impinging water jet. The conclusion tells us that flash cooling operates more efficiently at 4500 rpm than at 6000 rpm due to the longer time that the injected water mass has to provide heat transfer. Nevertheless, the numerical results from this paper show that when the water mass is selected to provide full evaporation at a pressure ratio of 12:1 the exit temperature exceeds 300°C therefore thermal expansion in the machine would lead to rotor damage unless the injected water amount was higher than that required for evaporation only. It should be emphasised that the main difference between impinging jet rotor cooling and fluid flooding is the quantity of fluid used and the exact injection position. The minute quantity of water used evaporates on contact with the rotors. The rotors are therefore not subjected to a fluid induced drag force during operation. This effect was confirmed in CFD calculations and during comparison of numerical models in SCORG.

The compressor chosen for the comparative tests was an oil injected machine designed by Jäcklin GmbH. This particular machine had been previously measured in the CUL test laboratory in London.

Table 1: The compressor unit used for this experiment was 4/5 male driven configuration with the following dimensions:

Male rotor diameter	130 mm
Length/diameter ratio	1.36
Female rotor diameter	102 mm
Centre distance	90 mm

By using known test results as input data for SCORG an accurate case for the oil injected compressor was modelled. After completing and verifying the oil injected SCORG model, the oil injection was removed and replaced with flash water cooling in the multi chamber model. No other changes were made in the numerical model at this time. In particular the simulated influence of water drag on the rotors compared to oil drag, the calculated leakage rates and the predicted volumetric efficiency without oil would be investigated. The measured results from the oil free compressor would be compared to the results taken from the oil injected compressor of identical design. To eliminate any other design influences from the comparison both reference machines use the same rotors, the same clearances, and the same basic housing design. Modifications were made to the oil free compressor to provide flash cooling. Sealing of the compression chamber from the bearing chamber at the high-pressure end using proprietary shaft seals was considered necessary to prevent the ingress of water vapour into the bearings.

2. Oil injected compressor validation of numerical simulation model.

Table 2: Shows the measured results taken over a speed range from 1500 rpm up to 6000 rpm. The measurements were made in accordance with ISO 1217. Inlet conditions were 1000 mbar and 20°C. The pressure ratio was 8:1.

Shaft Speed (RPM)	Air Flow (m3/min)	Power (kW)	Specific Power (kW/m3/min)
1500	1.72	10.18	5.90
2000	2.31	13.59	5.89
2500	2.95	17.42	5.91
3000	3.60	20.93	5.81
3500	4.19	24.53	5.86
4000	4.80	28.40	5.91
4500	5.39	32.34	6.00
5000	6.05	37.04	6.12
5500	6.63	41.51	6.26
6000	7.29	46.60	6.39

The measurement was made in the CUL laboratory in London. The results were used for verification of the SCORG multi-chamber model during the modelling and simulation process. All input data for the simulation model was taken from the actual compressor and measured data. The numerical results from the SCORG simulation case show an extremely good correlation with the measured results taken from the oil injected compressor over the entire performance range of this machine (Figures 2&3).

Table 3: Simulated performance calculation for the identical operating conditions used during testing.

Speed rpm	Vol m3/m	M kg/min	Vol Eff.	Power kW	Pspec	Ad Eff.	Psuc bar	Tsuc degC	Pdis bar	Tdis degC	Moil kg/sec
1500	1.66	1.98	0.81	10.10	6.05	0.78	1	20	8	63.41	1.12
2000	2.28	2.72	0.83	13.60	5.94	0.79	1	20	8	64.26	1.14
2500	2.91	3.45	0.85	17.23	5.92	0.79	1	20	8	65.12	1.15
3000	3.53	4.19	0.86	20.99	5.94	0.79	1	20	8	65.98	1.15
3500	4.15	4.94	0.86	24.86	5.98	0.79	1	20	8	66.83	1.16
4000	4.78	5.68	0.87	28.83	6.02	0.78	1	20	8	67.68	1.16
4500	5.40	6.24	0.87	32.89	6.08	0.77	1	20	8	68.53	1.16
5000	6.03	7.17	0.88	37.04	6.13	0.77	1	20	8	69.37	1.16
5500	6.66	7.92	0.88	41.26	6.19	0.76	1	20	8	70.20	1.16
6000	7.29	8.67	0.88	45.54	6.24	0.75	1	20	8	71.03	1.16

In the right-hand column, the quantity of injected oil in kg/sec can be seen. This corresponds to 66 l/min at 4500 rpm. The measured amount was 69.8 l/min at 4500 rpm. This is a reasonable comparison.

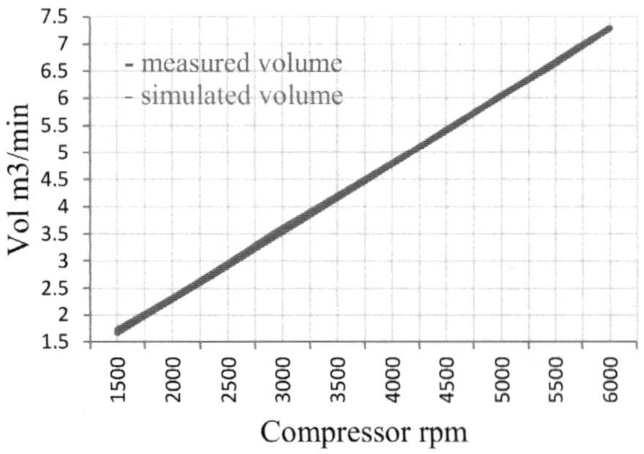

Figure 2: The air flow delivery in m³/min.

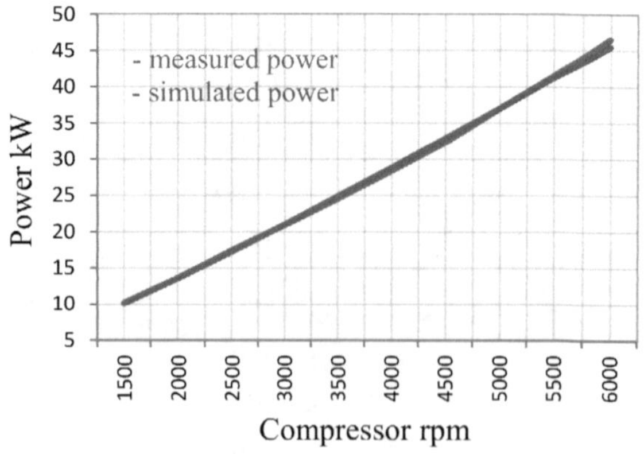

Figure 3: The power comparison in kW.

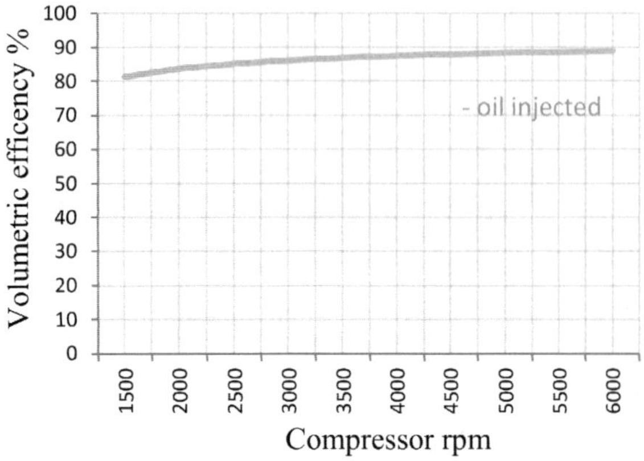

Figure 4: The simulated volumetric efficiency.

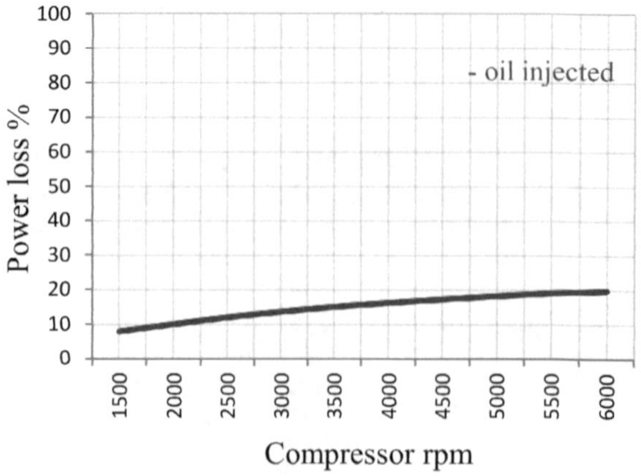

Figure 5: The simulated power loss (%) due to oil drag.

In figure 4 the volumetric efficiency taken from the simulated model of the oil injected compressor is illustrated. At 4500 rpm the measured value was 83.6 % the simulated model shows 87.9 % at the same point. The difference will be investigated.

The total shaft power in kW calculated in SCORG is the sum of the indicated power, shaft seal power, bearing power and oil drag power. In the oil injected compressor, the oil drag power (kW) forms a considerable part of the total sum. By calculating the oil drag value in kW for each rpm step from 1500 rpm up to 6000 rpm and comparing it to total shaft power (kW) the percentage difference can be determined and plotted in figure 5.

The result illustrates that the oil drag expressed as a percentage of the total shaft power increases from 10% at 2000 rpm up to 20% at 6000 rpm. It can be seen that the drag losses increase with speed so the power saving between 4500 rpm and 6000 rpm is particularly interesting. From the previous CFD numerical study *Stosic et al 2018 numerical investigation of water evaporation in twin screw compressors* we know that flash cooling is likely to be more effective at 4500 rpm than at 6000 rpm due to the longer time that the injected water mass has to provide heat transfer. In consideration of these two factors it was decided to concentrate on 4500 rpm for the initial comparison of prototypes on test. This decision was also influenced by a restriction on test resources. The oil free test rig has a maximum motor rpm limit of just over 4500.

3. The oil free compressor numerical simulation case.

The procedure followed for simulation of the oil free compressor was to copy the oil injected model and remove the oil injection completely replacing it with flash cooling. All other input parameters remain unchanged and a reasonable result was achieved. Due to uncertainty of actual clearances required it was decided to leave interlobe clearance, radial clearance and axial clearance at 0.05 mm, the same values as in the oil injected machine. At this time no form of optimising of the model was undertaken. The comparative results after testing would be used for gathering data to be included in the final design. A fully optimised oil free version will be designed at the next stage of this project.

It was anticipated that oil free cooling of the rotors would be a challenge and it would be required to keep the compression temperature low to avoid thermal deformation and contact between rotors and housing. The flash water injection in the model uses 0.340 l/min at the test conditions. The modelled result produces 160°C outlet port temperature and a volumetric efficiency of 91% which compares favourably with the 88% modelled and verified in the oil injected version. Why the oil free model shows less leakage is unclear, but this will be examined during later verification.

Table 4: Simulated performance calculation for the oil free version.

Speed rpm	Vol m3/m	M kg/min	Vol Eff.	Power kW	Pspec	Add Eff.	Psuc bar	Tsuc degC	Pdis bar	Tdis degC	Mwat kg/sec
1500	1.59	1.89	0.77	10.60	6.64	0.71	1	20	8	167.07	0.0054
2000	2.27	2.69	0.82	14.04	6.19	0.76	1	20	8	159.59	0.0054
2500	2.94	3.49	0.86	17.45	5.93	0.79	1	20	8	157.63	0.0055
3000	3.60	4.27	0.87	20.81	5.78	0.81	1	20	8	161.38	0.0056
3500	4.27	5.07	0.89	24.22	5.67	0.83	1	20	8	157.69	0.0057
4000	4.93	5.86	0.90	27.64	5.59	0.84	1	20	8	155.79	0.0057
4500	5.59	6.65	0.90	31.08	5.55	0.85	1	20	8	156.46	0.0057
5000	6.25	7.43	0.91	34.54	5.52	0.85	1	20	8	158.91	0.0057
5500	6.91	8.22	0.91	38.02	5.49	0.86	1	20	8	162.45	0.0058
6000	7.56	8.99	0.92	41.50	5.48	0.86	1	20	8	168.86	0.0058

In the right-hand column, the quantity of injected water in kg/sec can be seen. At 4500 rpm this corresponds to 0.34 l/min and the specific power (Pspec) is 5.5 kW/m^3/min. By comparison the validated oil injected model indicates a specific power of 6.1 kW/m^3/min under the same conditions.

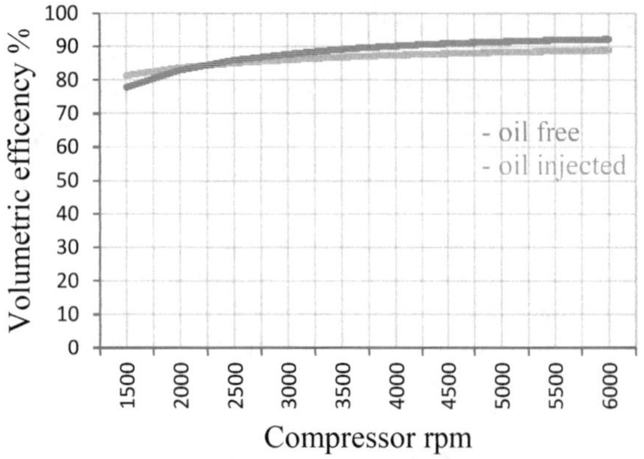

Figure 6: The simulated volumetric efficiencies for both cases.

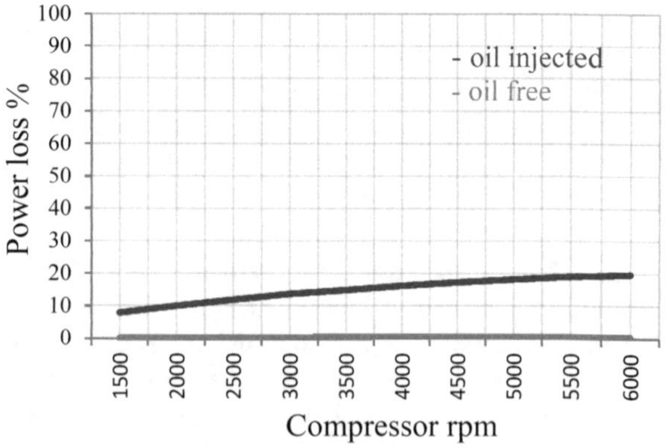

Figure 7: Simulated fluid drag for both cases.

4. Experimental investigation.

Table 5: Measured results comparison, the difference between the specific power (kW/m³/min) measured at the block outlet in comparison to the compressor outlet is due to the oil separation pressure loss.

	Oil injected version	Oil free version
Input speed	4500 rpm	4500 rpm
Volume delivery	5.39 m³/min. @20°C /1000 mbar	5.0 m³/min. @20°C /1000 mbar
Shaft power	32.34 kW	29.2 kW
Working pressure	8 bar(a)	8 bar(a)
Suction pressure	1 bar(a)	1 bar(a)
Pressure ratio	8:1	8:1
Outlet temperature	76°C	100°C
Specific power block only	6.0 kW/m³/min.	5.8 kW/m³/min.
Specific power at compressor outlet	6.75 kW/m³/min.	6.0 kW/m³/min.

Comparison of results and discussion.

Volumetric efficiency: Comparing the results at 4500 rpm which is 30 m/sec tip speed the air delivery of the oil free version is 7% less than the oil injected version. The reason for this could be increased leakage between rotors and housing due to removing the oil. The simulation results show a different situation so this point will need to be examined further. In the oil free version, the clearances are more critical in the sense that dimensional accuracy influences leakage fluctuation. Reducing clearances without compromising reliability will be investigated in the next prototype.

Cooling: In this experiment 0.0057 kg/sec water was injected into the compression chamber. In comparison the oil injected version uses 1.1645 kg/sec oil. The flash cooling effect was stronger than expected. The exit gas measured 100°C whereas the simulation model predicted 165°C. The difference between simulated and test results will be investigated and used to verify SCORG for these conditions. A possible explanation is that in the simulation not all of the injected water actually evaporates. At 7 bar evaporation occurs at 170°C and the test compressor is running cooler than this. During operation all compressor parts record a temperature of less than 100°C so there is no risk of damage due to inadequate clearance. The maximum temperature occurs at the exit port close to the temperature sensor. This is as close as possible to the point of maximum compression. It is assumed that the rotor temperature is less than the gas temperature due to the impinging jet water cooling.

The hot gas/water mixture exiting through the compressor outlet pipe goes straight into an aftercooler and the water content condenses. Because the injected water mass is extremely low compared to a fluid flooded compressor the separation process is much easier and requires no special separation equipment. The external housing temperature was between 45°C and 65°C and the bearing cover reaches a maximum of 85°C in the area directly adjacent to the exit gas. Further away from the exit port the temperature reduces to 65°C.

Power saving: A reduction of 10% at 30 m/s tip speed was measured. The simulation shows the power saving is greater as the tip speed increases. It was not possible to test this effect at higher rpm due to the test rig limitation. The numerical simulation indicates that power loss due to water drag in the flash cooled compressor is close to zero over the entire speed range. This has not been confirmed in the provisional test results to date.

Specific power: The specific power improvement predicted from simulation was - 6.08 reducing to 5.55 kW/m³/min. an improvement of 8% for the bare block only. In fact, the measured value was 6.0 reducing to 5.8 kW/m³/min an improvement of 3%. Comparison of results measured at the compressor outlet show 6.75 reducing to 6.0 kW/m³/min an improvement of 11%. Despite the measured reduction in power the volumetric efficiency also reduced in the experiment so the predicted advantage was not fully achieved. The net result is a significant gain in overall efficiency. More testing time for further optimisation is required. Considerable potential remains to reduce the clearances and increase the volumetric efficiency which in turn will improve the adiabatic efficiency. These experiments will continue during 2021. A new prototype is being produced with the intention of correcting these issues. Comparison over the entire rpm range and verification of results will then be possible.

Uncertainty of measurement: The oil injected compressor was measured in the City University Laboratory and the results reported in accordance with ISO 1217 for bare screw compressors. The simulation case was built around these results. The correlation between measured results and numerical simulation was good and provides an accurate basis for this work.

Figure 8: The oil free prototype on test.

5. Conclusions

The result confirms that flash cooling can be used most efficiently in a screw compressor up to a pressure ratio of 8:1. A reduction in power has also been confirmed. A measured efficiency improvement of 3% has been achieved. The numerically predicted 8.7% specific performance improvement has not been achieved at this time. The likely reason for this is higher leakage rate in the experimental case than numerically predicted. The numerical study indicates increased oil drag with increasing speed. The leakage reduces with increasing speed so higher rpm may improve efficiency. Testing up to 6000 rpm is planned for confirmation.

There are two advantages which have been confirmed by testing. An additional 10% improvement in efficiency for the whole compressor unit is the result of having no oil separation equipment. The second advantage is that any condensation produced after cooling is reused by the flash cooling system. There is no need to separate out and collect the condensation. The separation system in an oil injected compressor is normally required to prevent the oil from becoming contaminated by condensation produced during the compression and cooling process. The condensation produced in an oil injected compressor normally contains a small amount of oil carry over which makes disposal in line with environmental restrictions costly. In the flash cooled compressor these issues do not exist and there is no negative environmental impact. Certain applications for Jäcklin GmbH customers may be interested in this advantage.

6. Acknowledgements

Jaecklin GmbH

References

[1] Stosic N, Kovacevic A, Rane S and Stupple G. 2018: Analysis of water evaporation in twin screw compressors using Eulerian multiphase approach in CFD, ID:1358 Purdue Conferences USA.This reference has two entries but the second one is not numbered (it uses the 'Reference (no number)' style.
[2] Stosic N, Smith I. K. and Kovacevic A. 2011: Numerical and experimental research in heat transfer to screw compressor rotors ASME/JSME AJTEC2011-44255 Hawaii USA.
[3] Stosic N, Smith I. K. and Kovacevic A. 2004: Estimation and control of heat transfer in screw compressor rotors ASME IMECE2004-60561
[4] Kovacevic A, Rane S, 2020: SCORGTM Help Manual
[5] Rinder L, 1979: Schraubenverdichter (Screw Compressors), Springer Verlag, New York
[6] Stosic N, Smith I. K. and Kovacevic A, 2005: Screw Compressors: Mathematical Models in Computer Aided Design, Springer Verlag
[7] Stosic N, Smith I K, Kovacevic A and Mujic E, 2011: Geometry of screw compressor rotors and their tools, Journal of Zhejiang University-SCIENCE A (Applied Physics & Engineering) ISSN 1673-565X (Print); ISSN 1862-1775 (Online), DOI:10.1631/jzus.A1000393
[8] Stosic N, Smith I K, Kovacevic A and Mujic E, 2011a: Review of Mathematical Models in Performance Calculation of Screw Compressors, International Journal of Fluid Machinery and Systems, Vol.4, No.2, April-June, 2011, ISSN (Online): 1882-9554, DOI: 10.5293/IJFMS.2011.4.2.200
[9] Stosic N, Smith I K and Kovacevic A, 2019: Improving Screw Compressor Displacement and Efficiency by Increasing the Rotor Profile Depth, 11th International Conference on Compressors and Their Systems, London, September 9-11th, 2019 (Online), DOI: 10.1088/1757-899x/604/1/011012

IOP Conf. Series: Materials Science and Engineering 1180 (2021) 012009 doi:10.1088/1757-899X/1180/1/012009

Modelling of inhomogeneous chamber states in rotary positive displacement vacuum pumps

Heiko Pleskun[1], Timo Jünemann[2], Tobias Bode[3] and A. Brümmer[4]

TU Dortmund University, Chair of fluidics, Germany

E-mail: [1]heiko.pleskun@tu-dortmund.de, [2]timo.juenemann@tu-dortmund.de, [3]tobias.bode@tu-dortmund.de, [4]andreas.bruemmer@tu-dortmund.de

September 2021

Abstract. Chamber model simulation is a common approach to simulate rotary positive displacement vacuum pumps. Therefore the pump is abstracted into working chambers and connecting clearances, whereby the clearance leakages can be identified as the major loss mechanism in such machines. The clearance mass flow rates are calculated with respect to the thermodynamic states in the adjacent chambers, which are normally considered to be homogeneous. In this work it is shown, that the chamber state is inhomogeneous for rarefied gases due to the movement of the pistons which causes a pressure gradient within the chamber. This effect increases with higher Knudsen numbers, because of the increasingly dominant friction. An efficient way to model these inhomogeneous states with a one-dimensional approach for geometrically abstracted chambers is shown. The approach is applied to chamber model simulations of a screw spindle vacuum pump (SSVP) and the results are compared to simulations with homogeneous chamber states and to measurements.

1. Introduction

Dry running vacuum pumps have become an important part of vacuum systems due to their oil-free working principle. The need for a clean vacuum, for instance in the semiconductor industry, has pushed the development of dry running positive displacement vacuum pumps, such as screw and roots pumps (see Ref. [1]). Due to contactless operation of the pistons the sealing between the working chambers is realized by small clearances. Regarding figure 1 the effective mass flow rate of a screw machine can be calculated by the delivered mass flow rate $\dot{m}_d$ of a working chamber driven by the rotation of the pistons and reduced by the clearance mass flows $\dot{m}_{clearance}$. Clearances are located between both rotary pistons like the radial clearance, the inter-lobe clearance and the blow hole and between the rotary pistons and the housing. In order to calculate the thermodynamic operation of these pumps, the knowledge of these clearance mass flow rates is essential, which represent the main loss mechanism. A common approach

Table 1. List of most important symbols

symbol	explanation	unit	symbol	explanation	unit
A	cross section area	m^2	c_m	most probably speed	m/s
D	diameter	m	f	rotational speed	Hz
G	reduced flow rate	-	h	chamber/clearance height	m
L	chamber Length	m	m	mass	kg
$\dot{m}$	mass flow rate	kg/s	p	pressure	Pa
R	specific gas constant	J/(kg K)	s	rotor lead	m
T	temperature	K	U	velocity	m/s
V	volume	m^3	w	chamber width	m
z	number of lobes	-	δ	gas rarefaction parameter	-
η	cross section ratio	-	μ	dynamic viscosity	Pa s
ρ	density	kg/m^3	ψ	wrap angle	-

to calculate the thermodynamic operation of vacuum pumps is chamber model simulation where the pump is divided in a network of chambers that transport gas from low pressure port to high pressure port and connecting clearances which cause a mass flow rate in opposite direction [2–4]. The thermodynamic state of the gas within the chambers is normally considered to be homogeneous and the clearance boundary conditions are taken from mean values within the adjacent chambers.

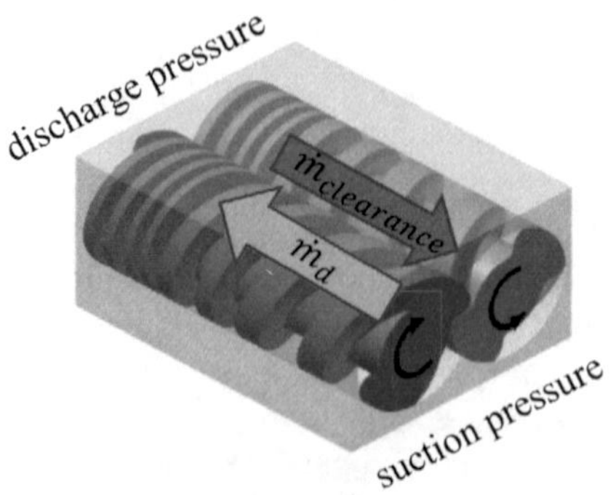

Figure 1. Sketch of an SSVP with 2-lobed cycloid profile.

The calculation of clearance flows for rarefied gases including moving boundaries has previously been investigated in references [5–7]. In this paper it is shown, that these chamber states are not homogeneous for rarefied gases but a pressure gradient occurs in tangential direction of the rotor lobe.

Figure 2 shows the chamber model abstraction for a screw spindle vacuum pump (SSVP) with the sketched network of chambers and connecting clearances. In this representation pressure p related to the chamber's mean pressure p_{ch} is sketched. The inhomogeneous pressure distribution causes the clearances to not only connect two adjacent chambers, but the high pressure side and a low pressure side of adjacent chambers. Thus the boundary conditions of the clearances have to be corrected for better agreement of simulation results to the pumps' operation performance.

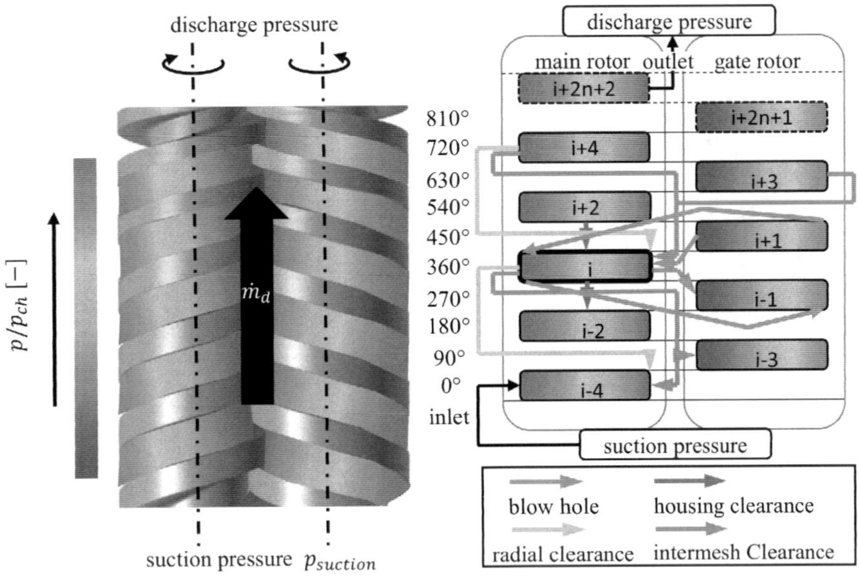

Figure 2. Sketch of an SSVP and the investigated impact of inhomogeneous chamber states on the boundaries of adjacent clearances similar to [3, 8].

2. Theoretical approach

2.1. Kinematics

In order to show why the pressure gradient occurs within the chambers it is necessary to take the kinematics into account. Figure 3 shows the sketch of an unwind cylinder cut of one rotor of an SSVP, where the chamber and the profile are alternately placed along the rotor axis. The rotor lead and the piston's rotation causes a chamber to move from low pressure port to high pressure port along the rotor axis with a velocity

$$U_{ax} = s \cdot f \tag{1}$$

where f is the rotational speed and s is the rotor lead. This is the velocity of the relative system. Furthermore each point on the rotor has a movement in circumferential direction

$$U_{circ} = \pi \cdot D \cdot f \tag{2}$$

which is the absolute velocity, where D is twice the distance from the rotor axis to the considered rotor point. The relation between the absolute velocities of the rotor points due to the rotation and the movement of the relative system can be obtained by the angle φ.

Therefore the magnitude of the relative movement U_r of rotor points within the chamber's fixed coordinate system can be obtained by

$$U_r = \frac{\pi D \cdot f}{\cos(\varphi)} \qquad \varphi(D) = \tan^{-1}\left(\frac{s}{\pi D}\right) \qquad D_i \leq D \leq D_o \tag{3}$$

Inhomogeneous chamber states in vacuum pumps

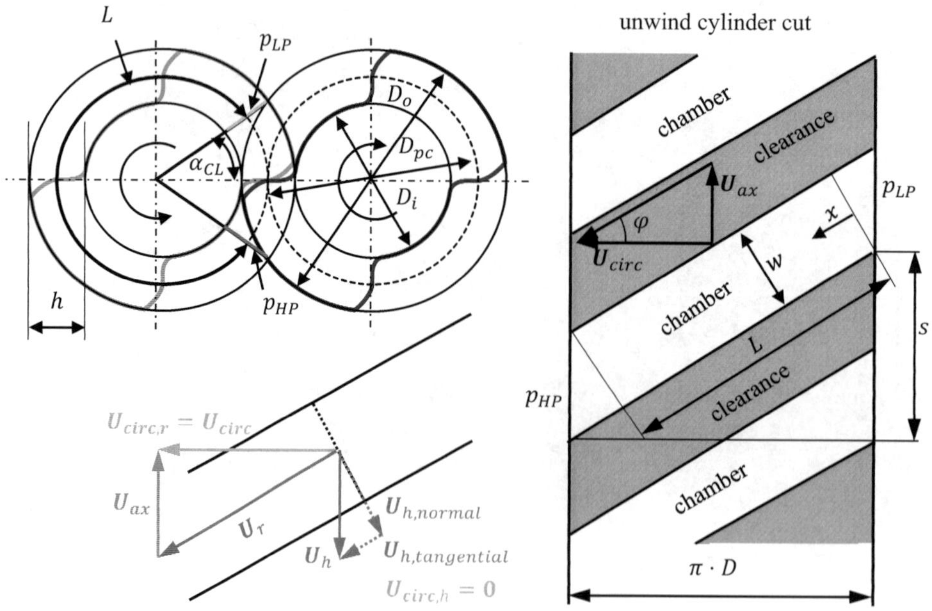

Figure 3. Geometry of a 2-lobed cycloid profile and boundary velocities in the chamber fixed relative system of screw vacuum pumps.

where D_o is the vacuum pump's outer rotor diameter and D_i the inner rotor diameter respectively.

The housing (h) does not move, so the housing's relative velocity $U_h = -U_{ax}$ is negative to the movement of the relative system. This can be divided into a part tangential and normal to the rotor lead:

$$U_{h,tangential} = s \cdot f \cdot \sin(\varphi(D_o)) \qquad U_{h,normal} = s \cdot f \cdot \cos(\varphi(D_o)) \tag{4}$$

So in every cross section of a chamber there are four bounding surfaces causing a Couette flow in the same direction tangential to the rotor lobe and the housing which also causes a flow in direction normal to the rotor lobe. Because of the chamber's encapsulation due to the other rotor and the surrounding housing this flow is dammed up. This leads to a pressure gradient which causes a Poiseuille flow in opposite direction of the Couette flow caused by the moving boundaries.

Geometrically the chamber can be abstracted as a channel with rectangular cross section with

$$h = \frac{D_o - D_i}{2} \qquad w = \frac{s}{2z} \cos\left(\varphi(D_{pc})\right) \tag{5}$$

the channel height and width is the linearized mean distance between two lobes. D_{pc} is the pitch circle diameter and mean value of inner and outer diameter. The chamber length is defined as

$$L = (\pi - \alpha_{CL}) \cdot D_{pc} \cdot \sqrt{1 + \left(\frac{s}{\pi D_{pc}}\right)^2} \qquad \alpha_{CL} = \tan^{-1}\left(\frac{\sqrt{D_o^2 - D_{pc}^2}}{D_{pc}}\right) \tag{6}$$

Inhomogeneous chamber states in vacuum pumps

with the angle of the cut lense defined in figure 3. This is the length of the helix, at the pitch circle. The end chamber geometry is defined between the intersection lines of rotor axis and cusps of the cut lense which is the intersection area of the outer diameters of both rotors (see the yellow and red lines in figure 3). The mean pressure in these intersection lines are defined as the chambers high pressure p_{HP} and low pressure p_{LP} respectively, which are used as inlet and outlet boundary values of adjacent clearances. Thus the chamber is defined with constant cross section area for constant rotor lead s. The impact of a non-constant rotor lead on the pressure distribution within the chamber is neglected in this work.

2.2. Governing equations

In order to calculate the pressure distribution within the chamber the approach of an isothermal channel flow model is used similar to the approach in [5, 9–11]. Thus in steady state the clearance mass flow rate $\dot{m}_{cl,t}$ tangential to the lobe can be expressed as the sum of the mass flow rate $\dot{m}_C$ caused by a Couette flow and the mass flow rate $\dot{m}_P$ caused by a Poiseuille flow:

$$\dot{m}_{cl,t} = \dot{m}_C + \dot{m}_P \tag{7}$$

The tangential clearance mass flow rate is the sum of the mass flow rates of the blow hole, the radial clearance and the intermesh clearance connected to the chamber. Furthermore when there is a difference between inflowing and outflowing housing clearance mass flow rate, this could be modelled as mass sources along the chamber length and thus $\dot{m}_{cl,t}$ is not constant. The mass flow rate caused by a Couette flow in a rectangular channel is the sum of the mass flow rate caused by each moving wall:

$$\dot{m}_C = \rho A \cdot \sum_{j=1}^{4} G_{C,j}(\delta, \eta) \cdot U_{w,j} \tag{8}$$

ρ is the density, $A = hw$ the cross section area, $U_{w,i}$ is the mean velocity of the respective wall and $G_{C,i}$ is the shape factor depending on the geometric ratio $\eta = \tilde{w}^2/A$, with $\tilde{w}$ the width of the respective moving wall within the cross section and

$$\delta = \frac{\min(h, w) \cdot p_{ch}}{\mu \cdot c_m} \tag{9}$$

is the gas rarefaction parameter with the dynamic viscosity μ and

$$c_m = \sqrt{2RT} \tag{10}$$

the molecules' most probable speed depending on the specific gas constant R and temperature T. For $\eta \geq 1$ the following formula for the shape factor is used:

$$G_C(\delta, \eta) = G_{C,mol}(\eta) + k_1 \ln(\eta) \tan^{-1}(k_2 \delta \eta + k_3 \delta) \exp(k_4 \ln^2(k_5 \eta)) \tag{11}$$

$$k_1 = 86.74 \; k_2 = 2.66 \; k_3 = -1.07 \; k_4 = -23.08 \cdot 10^{-3} \; k_5 = 46.08 \cdot 10^6 \tag{12}$$

with the analytical molecular solution with diffuse wall scattering

$$G_{C,mol}(\eta) = \frac{1}{4\pi} \left[\left(\eta - \eta^{-1}\right) \ln\left(\eta^2 + 1\right) - 2\eta \ln(\eta) + 4 \tan^{-1}(\eta) \right] \tag{13}$$

Inhomogeneous chamber states in vacuum pumps

The factors in equation 12 are results of a non linear regression with the Levenberg Marquardt algorithm which hold a maximum error smaller than 1% and an average error of 0.27% to analytical solutions of the hydrodynamic regime ($\delta \to \infty$) and the free molecular regime ($\delta = 0$) and solutions of the DSMC method in a large variety of $0.01 \le \delta \le 100$ and $1 \le \eta \le 100$ in between. [12]

For $\eta < 1$ the relation

$$G_C(\delta, \eta) = \frac{1}{2} - G_C(\delta, \eta^{-1}) \quad \forall \quad \eta < 1 \tag{14}$$

is used. The mass flow rate of the Poiseuille flow reads as follows

$$\dot{m}_P = -\frac{\min(h, w) \cdot A}{c_m} \cdot G_P(\delta, \eta) \cdot \frac{\partial p}{\partial x} \tag{15}$$

and is related to the pressure gradient in flow direction (in this case the tangential lobe direction, see the local coordinate system in figure 3). The dimensionless flow rates G_P used in this model are described in a previous work [5] and the values are related from references [11, 13, 14]. Note, that due to symmetry $G_P(\delta, \eta) = G_P(\delta, \eta^{-1})$ holds.

Inserting equations 8 and 15 in equation 7 leads to a differential equation in tangential lobe direction which is solved by an explicit Euler method

$$p_{i+1} = p_i + \frac{2\Delta x}{G_P(\delta_i, \eta) \min(h, w) c_m} \left(p_i \sum_{i=j}^{4} G_{C,j}(\delta_i, \eta) \cdot U_{w,j} - \frac{\dot{m}_{cl,t} RT}{hw} \right) \tag{16}$$

where the local rarefaction parameter $\delta_i(p_i)$ is used. To solve this equation the tangential clearance mass flow rate $\dot{m}_{cl,t}$ needs to be known, which can principally be used from the last time step within chamber model simulation. In this work it is assumed to be small compared to the couette flow within the chamber and is therefore neglected.

3. Chamber model simulation

3.1. Geometric and thermodynamic model

As sketched in figure 2 for the chamber model the vacuum pump is divided into chambers that transport mass from low pressure port to high pressure port and connecting clearances. Therefore the geometric properties of the chambers and clearances as a function of the rotation angle are needed. The geometric model used in this paper is adapted from references [15] and [16]. As a new feature the clearances are not simply connected with two chambers, but with a certain chamber position. Therefore the blow hole connects the high pressure side of two chambers, while the radial and inter-lobe clearance connect the high pressure side of the chamber at higher rotation angle with the low pressure side of the chamber with lower rotation angle considering the rotation angle to start at the low pressure port. The boundary conditions of the housing clearance are taken from mass averaged mean values of the certain chamber. The mass flow rates for given geometric and thermodynamic boundary conditions are obtained by tabulated data of measurements and simulations of the specific clearance shapes investigated and described in references [3] and [17].

Inhomogeneous chamber states in vacuum pumps

Due to the chamber's geometric abstraction for the 1D model, there is a discrepancy between the 1D model's chamber volume V_{1D} and the total chamber volume V_t, as the volume within the cut lense is neglected. Thus for chambers, that are not connected to the high or low pressure port, the total chamber volume is divided as follows

$$V_t = V_s + V_{1D} + V_e \qquad V_{1D} = L \cdot A \tag{17}$$

where $V_s = V_e$ are the chamber's start and end volume respectively. This classification is performed as the total mass m_t within the chamber is divided on these three volumes with the linearization, that the pressure within the end volumes is constant with

$$p(V_s) = p_{LP} \quad \text{and} \quad p(V_e) = p_{HP} \tag{18}$$

being the high and low pressure boundaries calculated by the 1D model. Thus the pressure distribution calculated with equation 16 has to match the constraint

$$m_t = \frac{1}{RT}\left(V_s p_{LP} + A\int_0^L p\,\mathrm{d}x + V_e p_{HP}\right) \tag{19}$$

where p_{LP} is found by a shooting method according to reference [18]. The integral can be solved numerically

$$m_{1D} = \frac{A}{RT}\int_0^L p\,\mathrm{dx} \approx \frac{A}{RT}\sum_{i=1}^{N} p_i \Delta x_i \tag{20}$$

where N is the number of increments used to solve the differential equation 16. For a chamber which is open to low pressure or high pressure port the effective chamber length is reduced, because the pressure at inlet or outlet area is fixed. The new model is implemented in the chamber simulation software *KaSim* developed at the Chair of Fluidics of TU Dortmund University described in reference [19].

3.2. Comparison to measurements

In order to work out the impact of the new investigated model for inhomogeneous chamber states, simulations for a test machine are performed and compared to measurements. This machine is an isochoric two-lobed cycloid profiled SSVP. The dimensionless data is given in table 2, where L_R is the rotor length, h_h, h_{ilc} and h_{rc} are clearance heights of the housing clearance, the inter-lobe clearance and the radial clearance respectively and ψ_t is the machines total wrap angle. As the machine is cooled it is assumed to be fully isothermal at room temperature $T = 293K$. The machine neither has a radial inlet area nor a pressure relief valve.

Table 2. Data of the test machine measured in reference [3]

D_o/D_i	D_o/s	D_o/L_R	D_o/h_h	D_o/h_{ilc}	D_o/h_{rc}	ψ_t	z
1.571	5.641	0.667	880	733.33	733.33	3474°	2

Inhomogeneous chamber states in vacuum pumps

Figure 4 shows the ratio of the suction speed $\dot{V}$ to the theoretical suction speed $\dot{V}_{th}$

$$\dot{V} = \frac{\dot{m}_{effective} \cdot R \cdot T}{p_{suction}} \qquad \dot{V}_{th} = 2 \cdot V_{ch,max} \cdot f \cdot z \tag{21}$$

as a function of the suction pressure $p_{suction}$. $\dot{m}_{effective}$ is the effective mass flow rate of the machine. The discharge pressure is always 10^5 Pa. Comparing the results of *KaSim* without the new model (red line) to the measurements, a good agreement is obtained for large values of the suction pressure. But with the decrease of the suction pressure those two curves diverge. The measurements show that the machine cannot evacuate the low pressure volume lower than about 6 Pa at the given rotational speed, while the simulated suction speed is still at a high value. Regarding the new model (blue line) the curve is not affected in the high pressure regime, but collapses at low suction pressures in qualitatively good agreement with the measurements. But the new model in the current implementation slightly seems to overrate the effect, as the results indicate a worse suction speed than the measurements.

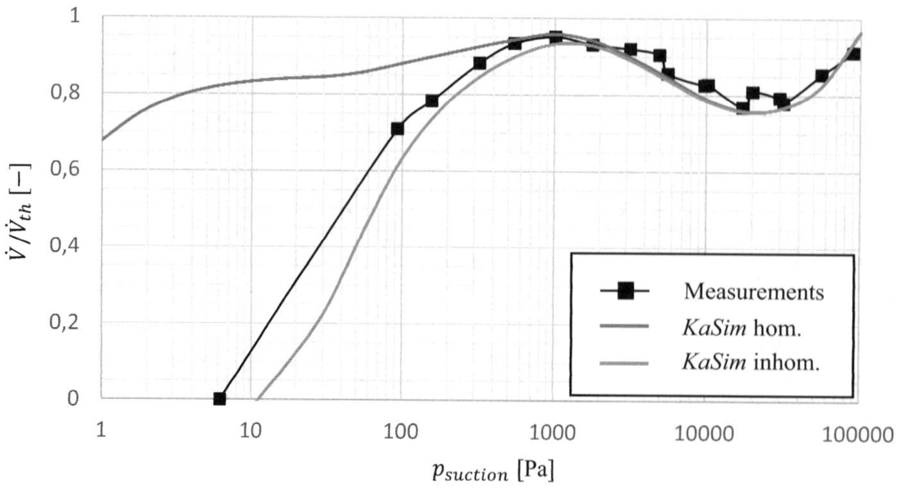

Figure 4. Comparison of the suction speed curves of measurements and chamber model simulation with homogeneous and inhomogeneous pressure distribution within the chambers for a rotational speed $f = 83.33$ Hz at constant temperature $T = 293$ K.

Figure 5 shows the pressure distribution within the machine as a function of the wrap angle ψ for a suction pressure $p_{suction} = 10$ Pa and $p_{suction} = 100$ Pa. The straight lines show the pressure distribution of the old model with homogeneous chamber states while the dashed lines show the mass averaged mean pressure distribution of the new model. The dotted lines show the chamber's low and high pressure distribution respectively. As expected all lines converge to the same line during the transport phase when the pressure is high enough. Big differences are obtained at the low pressure regime especially during the suction phase. As the model with homogeneous chamber states delivers a mean chamber pressure which is not significantly lower than the

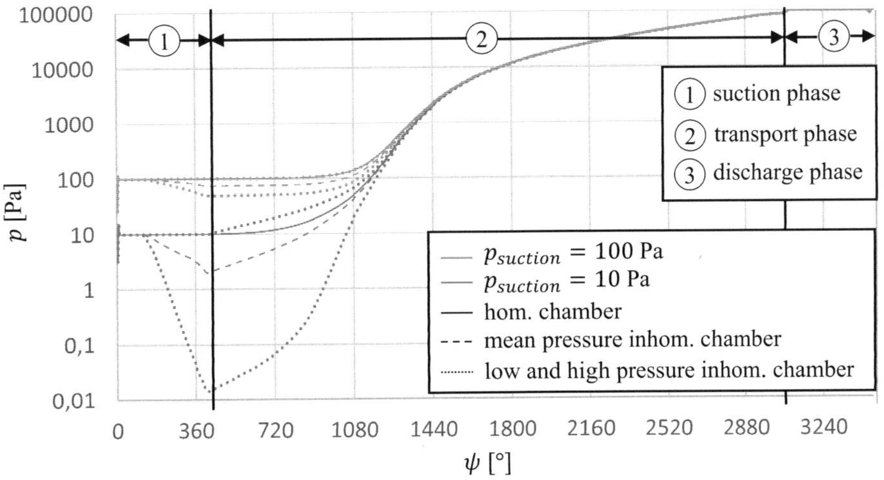

Figure 5. Comparison of the pressure distribution within the vacuum pump for homogeneous and inhomogeneous pressure distribution within the chambers.

suction pressure, the new model simulates a pressure drop until the chamber is fully encapsulated. Thus in this regime an immense suction throttling is depicted with this approach. Nevertheless the maximum pressure ratios within the chamber seem very high. Especially in the suction phase an improvement is expected when the impact of the clearance mass flows on the pressure distribution is considered. In the current implementation the clearances just affect the chamber's mass and therefore the mean pressure but the last term in equation 16 is neglected.

4. Conclusion

In this work a one-dimensional model in order to calculate the pressure distribution within the chambers of an SSVP has been investigated. This model has been implemented into the chamber model simulation software *KaSim*. A comparison with measurements shows a great improvement of the simulation results due to the new model as especially the suction phase at low suction pressure is highly affected by inhomogeneous chamber states. In order to have a further improvement of the chamber model simulation, the impact of clearance mass flow rates on the chamber's pressure distribution should be included. Furthermore a better accuracy is expected if the mass flow rates of the clearances are calculated by the model presented in reference [6]. In order to validate the one-dimensional approach of the new model in detail three-dimensional CFD simulations including Maxwell velocity slip boundary conditions as described in reference [20] should be performed.

Acknowledgments

Supported by:

Federal Ministry for Economic Affairs and Energy

on the basis of a decision by the German Bundestag

Supported by the Federal Ministry for Economic Affairs and Energy (BMWi) on the basis of a decision by the German Bundestag, AiF, IGF No.: 19859 N/1.

References

[1] Jousten K 2018 *Wutz - Handbuch der Vakuumtechnik* vol 12 (Springer Vieweg)
[2] Huck C, Brümmer A and Sachs R 2014 Thermodynamical operation behaviour of screw vacuum pumps with cycloid- and quimby-toothed rotors *9th International Conference on Screw Machines: VDI Bericht 2228* pp 331–346
[3] Stratmann D 2010 *Kennfeldverhalten zweiwelliger Rotationsverdrängervakuumpumpen* Dissertation Technische Universität Dortmund URL `http://d-nb.info/1007720948`
[4] Rohe A 2005 *Wärmehaushalt von Schraubenspindel-Vakuumpumpen* Dissertation Technische Universität Dortmund URL `http://d-nb.info/1011533766`
[5] Huck C, Pleskun H and Brümmer A 2018 *J. Vac. Sci. Technol. A* **36** 031606
[6] Jünemann T and Brümmer A 2020 *J. Vac. Sci. Technol. B* **38** 044201
[7] Müller R 2013 *Spaltströmung mit Wärmeübertragung in Vakuumpumpen* Dissertation Universität Kaiserslautern URL `http://d-nb.info/1044891858`
[8] Pfaller D 2012 *Einfluss der Verdrängergeometrie auf die Auslegung von Schraubenspindel-Vakuumpumpen* Dissertation Technische Universität Dortmund
[9] Fukui S and Kaneko R 1988 *J. Tribol.* **110** 253
[10] Sharipov F, Fahrenbach P and Zipp A 2005 *J. Vac. Sci. Technol., A* **23** 1331
[11] Graur I and Ho M T 2014 *Vacuum* **101** 328
[12] H Pleskun T Bode A B Calculation of the couette flow in a long rectangular channel in the whole range of the gas rarefaction under isothermal conditions unpublished
[13] Sharipov F 1999 *J. Vac. Sci. Technol., A* **17** 3062
[14] Titarev V A and Shakhov E M 2010 *Comput. Math. Math. Phys.* **50** 1221
[15] Jünemann T and Brümmer A 2018 *IOP Conf. Ser.: Mater. Sci. Eng* **425** 012028
[16] Nadler K, Brümmer A and Sachs R 2010 A method for the geometrical analysis of twin-shaft rotary displacement machines *8th International Conference on Screw Machines: VDI Bericht 2101* pp 333–344
[17] Wenderott D 2001 *Spaltströmungen im Vakuum* Dissertation Technische Universität Dortmund URL `http://d-nb.info/963650505`
[18] Press W H 2007 *Numerical recipes: The art of scientific computing* vol 3 (Cambridge university press)
[19] Janicki M 2007 *Modellierung und Simulation von Rotationsverdrängermaschinen* Dissertation Technische Universität Dortmund URL `http://d-nb.info/997683996`
[20] Jünemann T, Pleskun H and Brümmer A 2021 Maxwell velocity slip and smoluchowski temperature jump boundary condition for ansys cfx *In: 12th International Conference on Compressors and their Systems*

Estimation of radial shaft seal, oil drag and windage loss in twin screw oil injected compressor

S Abdan[1,2*], N Stosic[1], A Kovacevic[1], I Smith[1] and N Asati[2]

[1] Centre for Compressor Technology, City, University of London, London, U.K.
[2] Kirloskar Pneumatic Company Limited, Pune, India.

E-mail: suraj.abdan@city.ac.uk, n.stosic@city.ac.uk, a.kovacevic@city.ac.uk, i.k.smith@city.ac.uk, neeraj.asati@kirloskar.com

*Corresponding author

Abstract. The objective of this paper is to study the mechanical power loss arising from radial shaft seal, oil drag, and windage within the twin-screw oil-injected screw compressor. The previous papers in the series presented the evaluation of power loss due to the gears and bearings where it was found that the calculated total power loss was underestimated when compared to the experimental results for different sizes of compressors and different operating conditions. This indicates that few additional contributing elements were missing and needed due consideration to enhance the prediction. Literature review of the basic fluid dynamics and transmission gears confirms that oil drag loss and windage loss can also contribute to the power loss considerably.
This paper presents a review of a few methods for the prediction of the radial shaft seal, oil drag, and windage loss. Predictions with the use of these methods and their comparison with the experimental results are analysed in this paper. Experimental results of three different sizes of the compressors tested in laboratory conditions were taken as reference for the comparison at different operating speeds and pressure ratios. The analysis helps to understand the contribution of each element of the power loss which can help a designer to optimally design a screw compressor.

Keywords: Oil drag, Radial shaft seal, Windage loss, Screw Compressor

Nomenclature

B	width of rotor tip	m
B_θ	body forces	N
d	diameter	m
F	gear width/rotor length	m
M	gear module/ratio of pitch to number of lobes	m
n	rotational speed	rpm
P_{drag}	power loss in drag	W

$P_{windage}$	power loss in windage	W
p	pressure	Pa
r	radius	m
r_i	inner radius/rotor tip radius	m
r_o	outer radius/housing inner radius	m
t	time	s
u_r	velocity in radial direction	m/s
u_θ	velocity in angular direction	m/s
V_t	tip velocity	m/s

Greek letters

θ	angle	degrees
λ	gearbox space function	
μ	dynamic viscosity	Pa.s
ρ	density	kg/m^3
τ_i	shear stress on rotor tip	N/m^2
ϕ	oil mixture function	
ω	angular velocity	rad/s

1. Introduction

With the increasing demand for more energy-efficient machines, one of the ideas in the screw compressor to focus on is to reduce the power loss. To do this, one first needs to understand the elements of the mechanical power loss in the screw compressor and their contribution to different sizes of the compressors at different operating conditions. Although a few methods are available in the literature to independently predict the individual element's power loss, a comprehensive model suitable for the screw compressor application is unavailable. Realisation of major contributors of power loss elements in the screw compressor can help a designer to optimally design the screw compressor as per the need.

A Sankey diagram representing elements of power loss in a screw compressor package is shown in Figure 1. Only a part of the total input power to the screw compressor package is available for the compression where the remaining part is lost in overcoming prime mover loss, transmission loss, and bearing, seal, and oil drag loss within the bare compressor.

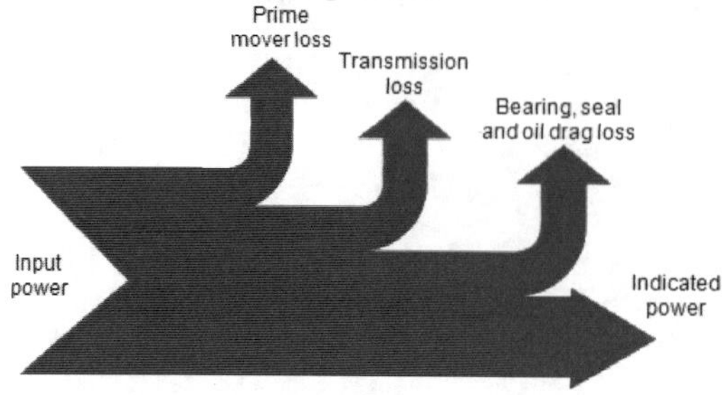

Figure 1 Sankey diagram of power loss in screw compressor package

Referring to previous studies in this series on bearing power loss estimation, the study presented here includes power loss estimation of radial shaft seals, oil drag, and windage loss. A comparison of the SKF and Harris model for the estimation of bearing power loss has been presented in Abdan et al. (2019) [1]. Frölich et al. (2104) [2] and Engelke et al. (2011) [3] have presented semi-analytical and

IOP Conf. Series: Materials Science and Engineering 1180 (2021) 012010 doi:10.1088/1757-899X/1180/1/012010

experimental studies on the radial shaft seal power loss. A rotating Couette flow model along with a correlation proposed by Cicchitti et al. (1959) [4], for calculation of the two-phase flow properties is used for the prediction of the oil drag loss. Figure 2 shows a general arrangement of the shaft seal position and oil injection for an oil-flooded, twin-screw compressor block (Image: Kirloskar Pneumatic Company Limited).

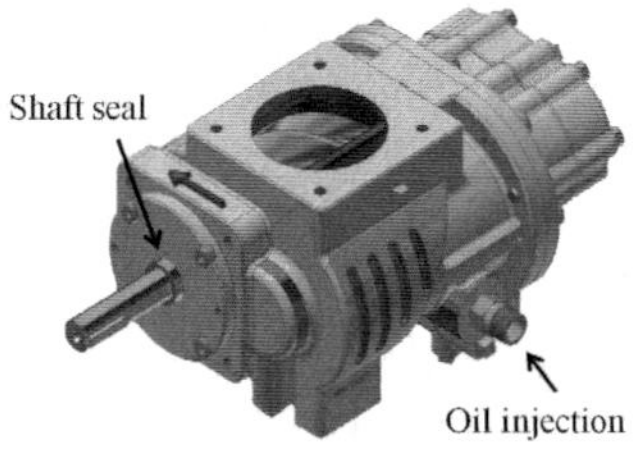

Figure 2 General arrangement of the shaft seal position and oil injection

Dawson (1984) [5] presented an approximation for the calculation of windage loss in power transmission gears which is referred to and adopted with a modification for prediction of windage loss in screw compressors. The experimental results of three different sizes were taken as a reference for comparison and results of the proposed method were compared with the experimental results.

2. Elements of mechanical loss

2.1 Bearing

Rolling element bearings are preferred over journal bearings because of their low starting friction, intermittent operation and are more adaptable for handling misalignment between shafts.

The Harris model for bearing power loss estimation shows better agreement with the experimental observations as presented by Tu (2016) [6]. According to the Harris model, the frictional loss of the ball bearings is classified into two categories, one due to load and another due to lubricant. For roller bearings, an additional loss because of sliding between roller ends and ring flange is taken into account as given in Haris et al. (2006) [7]. For ball bearings, total frictional torque is divided into two categories, load-dependent, and load-independent. The details of the model and approximations are presented in Abdan et al. (2019) [1] with verification.

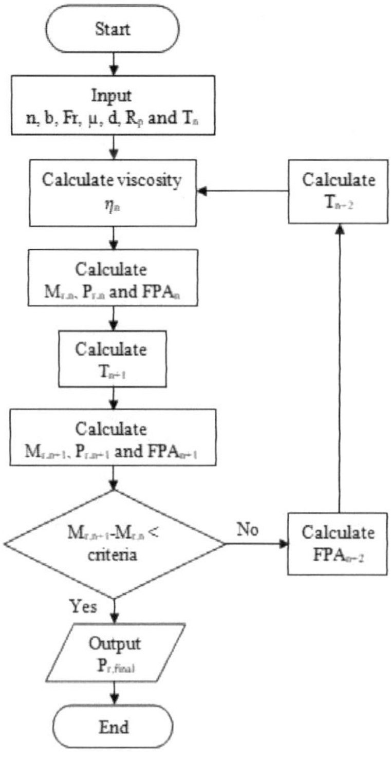

Figure 3 Flow chart

2.2 Shaft seal loss

The radial shaft seal is an integral part of the screw compressors whose function is to prevent oil leakage out of the compressor. Since, the lip of the stationary seal tightly sits on the rotating shaft, thus creating friction and ultimately loss of power. Frölich et al. (2104) [2] presents a semi-analytical approach for the calculation of contact temperature and an empirical approach for the calculation of friction. The experimental measurements of the influence of elastomeric lubricant combination on the operating performance of radial shaft seals are presented by Engelke et al. (2011) [3] with an algorithm to estimate the seal frictional loss.

The method to calculate the radial shaft seal power loss is the combination of iterative and semi-analytical approaches. The value of friction coefficient, radial force on the seal and seal-shaft contact surface width is assumed and this assumption is validated by Engelke et al. (2011) [3]. The shaft diameter, shaft surface roughness, rotating speed and oil temperature are used as input parameters. In the first iteration, the friction torque value is calculated for the assumed temperature while for the next iteration, the seal surface temperature is raised by 16.5 K for 1 W/mm^2 sealing contact area as presented in Engelke et al. (2011) [3]. Once the friction torque for raised surface temperature is calculated, the difference in the last two time steps for friction torque is calculated and compared with the convergence criteria. A flow chart for the calculation of shaft seal power loss is given in Figure 3.

There are two assumptions in the model, one is the radial force per unit circumferential length and the second one is the coefficient of friction. The experimental measurements have indicated the values of the radial force per unit circumferential length and the coefficient of friction for combinations of the seal material and the seal-shaft material respectively by Engelke et al. (2011) [3].

2.3 Oil drag loss

The oil flow in rotor tip-housing clearance can be equated to rotating Couette flow where flow is dominated by viscous effects and inertial effects are negligible. In a Cylindrical coordinate system, this is compared to the flow in annular space. A two-dimensional Navier-Stokes equation that can be solved exactly by analytical techniques with a number of significant assumptions is as follows.

$$\rho\frac{\partial u_\theta}{\partial t}+u_r\frac{\partial u_\theta}{\partial r}+\frac{u_\theta}{r}\frac{\partial u_\theta}{\partial r}+\frac{u_r u_\theta}{r}=-\frac{1}{r}\frac{\partial p}{\partial \theta}+B_\theta+\mu\left(\frac{1}{r}\frac{\partial\left(r\frac{\partial u_\theta}{\partial r}\right)}{\partial r}-\frac{u_\theta}{r^2}+\frac{1}{r^2}\frac{\partial^2 u_\theta}{\partial \theta^2}+\frac{2}{r^2}\frac{\partial u_r}{\partial \theta}\right) \tag{1}$$

While assuming the flow is steady, body forces are negligible, flow is incompressible and pressure gradient is negligible, using continuity equation, the above equation reduces to the following form

$$r^2\frac{\partial^2 u_\theta}{\partial r^2}+r\frac{\partial u_\theta}{\partial r}-u_\theta=0 \tag{2}$$

This is a second-order, ordinary differential equation. After applying the following boundary conditions, at $r=r_i$ (on rotor tip), $u_\theta=V_t$ and at $r=r_o$ (on housing inner surface) and $u_\theta=0$, the equation reduces to

$$u_\theta=\frac{V_t r_i}{r_o^2-r_i^2}\left(\frac{r_o^2}{r}-r\right) \tag{3}$$

This is the velocity profile in the clearances. The shear stress can be calculated by differentiating the velocity.

$$\tau_i=-\frac{2\mu V_t r_o^2}{\left(r_o^2-r_i^2\right)r_i} \tag{4}$$

The product of shear stress, area, and speed gives the power loss through drag.

$$P_{drag}=\tau B r_i \omega \tag{5}$$

2.4 Windage loss

The force created by the friction between the air and the object is referred to as windage. The higher the relative velocity between the rotors and the air, the higher is the frictional force and ultimately the loss. An equation for approximating the windage loss in gears is given by Dawson (1984) [5]. The windage power loss is a function of the rotational speed of the gear (n), gear width (F), tooth module (M), the amount of oil mist present inside the casing, and the diameter of the gears.

$$P_{windage} = n^{2.9}\left(0.16d^{3.9} + d^{2.9}F^{0.75}M^{1.15}\right)\times10^{-20}\phi\lambda \tag{6}$$

The same equation has been used to predict the windage loss through the compression chamber of the screw compressor. The factor ϕ represents the oil mixture function while the factor λ represents a gearbox space function. However, the predictions of the model did not match in good agreement with the experimental results. Hence, the equation proposed by Dawson (1984) [5] is modified to make it suitable for the screw compressor application. The modified form of the equation is as below

$$P_{windage,m} = n^{2}\left(0.16d^{5.5} + d^{2.9}F^{2.3}M^{1.16}\right)\times10^{-20}\phi\lambda \tag{7}$$

The results of the equation for different sizes of the compressor are presented in further sections.

3. Results and Discussion

Three sizes of the twin-screw, oil-injected compressors used for 15-30 kW (size 1), 37-55 kW (size 2), and 75-160 kW (size 3) with "N" rotor profile, 4/5 lobe combination, and L/D=1.55 were considered for the calculation. These compressors were tested in laboratory conditions at different pressure ratios for the measurement of shaft power and airflow.

Harris model is used for the calculation of power loss through the anti-friction bearings. The results and verification of the calculation for a different set of bearings were presented by Abdan et al. (2019) [1]. With the use of methods mentioned in the previous section for calculation of power loss through shaft seal, oil drag, and windage loss, the total power loss through the above three sizes of screw compressors was estimated.

Table 1 lists the anti-friction bearings used in the compressors with their location. The radial and axial loads generated during compression were taken from DISCO (Design Integration for Screw COmpressors), a software package developed through continuous research and development at City, University of London, UK.

Table 1 Bearings used in screw compressors

Location	Size 1	Size 2	Size 3
Male-Radial-Suction	NU 205 ECP	NU 307 ECP	NU 213 ECP
Female-Radial-Suction	NU 203 ECP	NU 2304 ECP	NU 1009 ECP
Male-Radial-Discharge	NU 205 ECP	NU 2207 ECP	NU 211 ECP
Female-Radial-Discharge	NU 205 ECP	NU 2207 ECP	NU 211 ECP
Male-Axial-Discharge	7305 BEP	7407 BEP	7311 BECBP
Female-Axial-Discharge	7205 BEP	7207 BEP	7211 BEP

The radial shaft seals of sizes 25x32x4 mm, 32x42x4 mm, and 60x80x8 mm were used for sizes 1, 2, and 3 respectively. It is assumed that the friction coefficient between seal and shaft is 0.35, the average radial force on seal lip is 145.9 N/m based on experimental measurements presented by Engelke et al. (2011) [3] while the shaft surface roughness is 0.8 micrometers. Vogel model is used for the calculation of temperature-dependent viscosity as presented by Knežević et al. (2006) [8].

A correlation proposed by Cicchitti et al. (1959) [4] was used for the evaluation of the two-phase thermo-physical properties of the air-oil mixture. A constant oil-air mass ratio of 4 was assumed with an oil density of 950 kg/m^3 and viscosity of 68 cSt at 40°C. The air mass flow rate varies with respect to the rotor shaft speed and operating pressure ratio which is obtained from DISCO. From the total peripheral area of the rotor housing, the cusp area was deducted using the Cosine law of trigonometry. Similar to the Couette flow model in a Cylindrical coordinate system used for calculation of drag loss in radial clearances between the rotor and the housing, a Couette flow model in the Cartesian coordinate system is used for the calculation of drag loss in the axial clearances.

The experimental results were obtained for three sizes, sizes 1, 2, and 3, at different pressure ratios (6.5-12.5) and rotor tip speeds (10-40 m/s) at laboratory conditions. The compressor inlet and outlet temperatures were measured by Platinum Resistance Thermometers with errors within ± 0.50^{o}C. Oil temperatures were measured by K-type thermocouples with errors within ± 1.0°C. All pressures were measured with transducers with errors within ± 0.6%. The compressor speed was measured by a shaft encoder with ± 2.7% error. The compressor torque was measured by the torque meter with a strain gauge transducer with an accuracy of ± 0.25%. A comparison between the experimental results, DISCO estimation, and calculation from the above-mentioned methods is plotted in Figure 4.

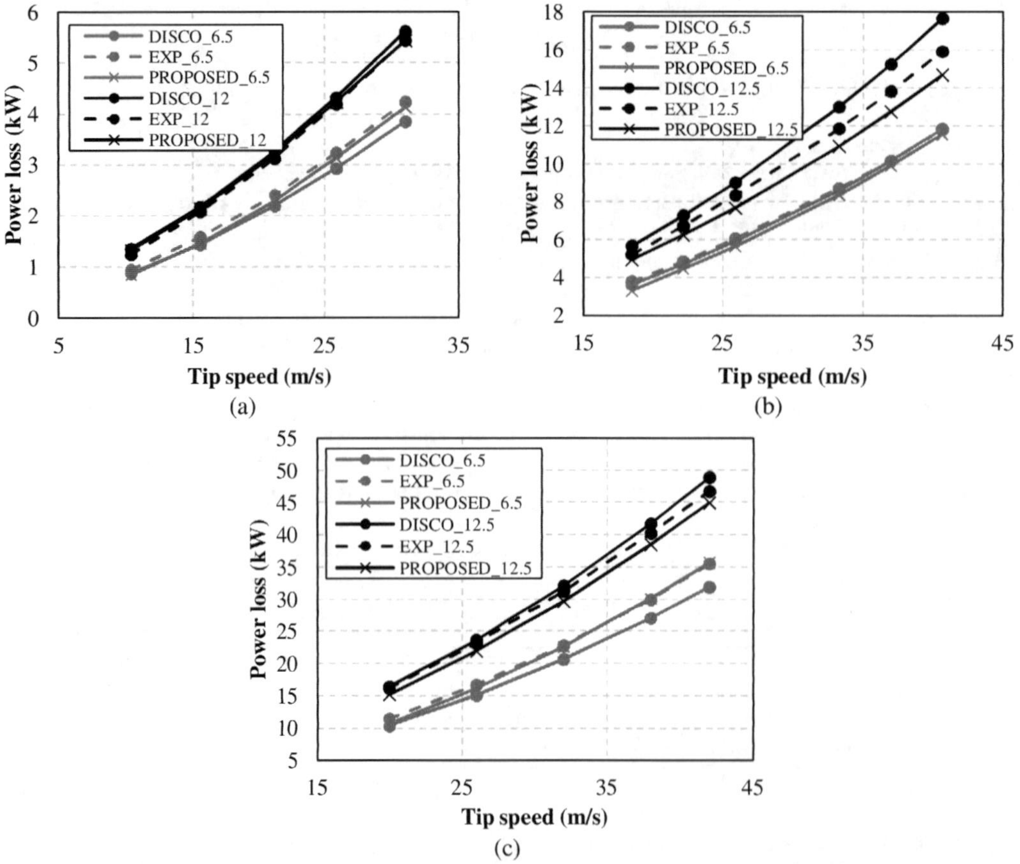

Figure 4 Comparison of total power loss in compressor size 1 (a), 2 (b) and 3 (c)

Although total shaft power was measured using the torque meter at laboratory conditions, the experimental total mechanical power loss is calculated from the measured shaft power by deducting the indicated power as predicted by the DISCO programme. Figure 4 clearly shows that the prediction for total power loss with the use of the proposed method matches closely with the experimental results for different sizes of screw compressors at a pressure ratio of 6.5. However, for a pressure ratio of 12.5, the proposed model slightly under-predicts the mechanical loss as compared to the experimental measurements as can be seen in Figure 4 (b) and (c).

Upon quantification of the results, the effect of shaft speeds and pressure ratios on the individual elements of the power loss is presented in Figure 5 and Figure 6.

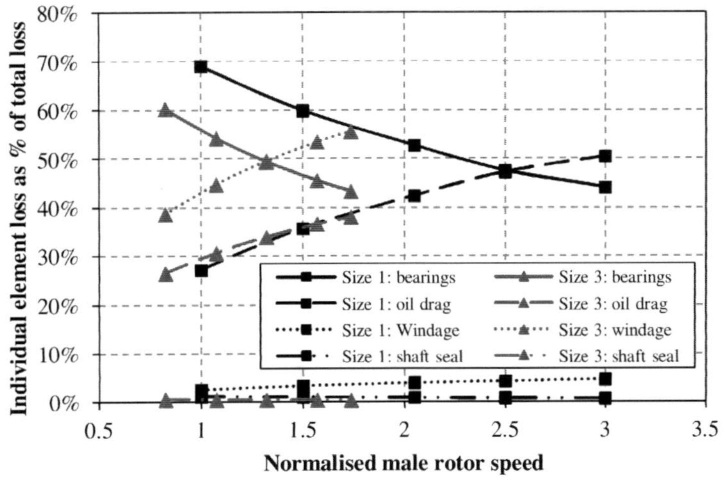

Figure 5 Effect of tip speed on the elements of power loss

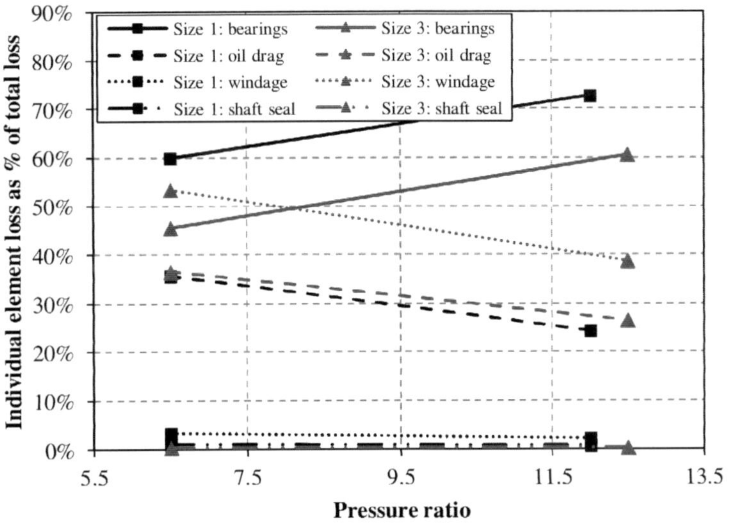

Figure 6 Effect of pressure ratio on the elements of power loss

Effect clearly shows that the anti-friction bearings are the main contributors to the power loss followed by oil drag loss, windage, and radial shaft seals in decreasing order of their contribution. However, with the increase in the size of the compressor, windage loss becomes more dominant and can contribute substantially to the power loss.

4. Conclusions

Different elements contributing to the power loss within the air screw compressor with oil injection and twin screw rotor configuration have been studied. The elements are anti-friction bearings, shaft seal, oil drag, and windage. The details of different power loss prediction models for anti-friction bearings and their comparison with available experimental results have been presented in the previous work along with a recommendation to use the Harris model.

A combination of semi-analytical approach and iterative procedure proposed by Frölich et al. (2104) [2] and Engelke et al. (2011) [3] is used for the prediction of power loss through radial shaft seal. For the prediction of power loss through oil drag in radial and axial clearance gaps, a rotating Couette flow model is used with suitable boundary conditions. A modified Dawson's correlation is used for the prediction of windage loss. With the use of the above-mentioned models, total power loss prediction for three different sizes of the compressor, 15-30 kW, 37-55 kW, and 75-160 kW at different speeds and operating pressures found to be in good agreement with the experimental data as indicated in Figure 4.

A parametric analysis shows that anti-friction bearings are the biggest contributors to the power loss followed by oil drag loss, windage loss, and the radial shaft seal. The percentage contribution of anti-friction bearing reduces while that of oil drag loss, windage, and radial shaft seal loss increase with respect to the tip speed as shown in Figure 5. The trend reverses when the pressure ratio is increased for the same tip speed as seen in Figure 6.

Acknowledgement
We gratefully thank Kirloskar Pneumatic Company Limited, Pune, India for sponsorship and Prof. Stosic and Prof. Kovacevic of City, University of London for their continued support and guidance.

References

[1] S. Abdan, N. Stosic, A. Kovacevic, I. Smith and N. Asati, "Analysis of rolling bearing power loss models for twin screw oil injected compressor," in *International Conference on Compressors and their Systems*, London, 2019.

[2] D. Frölich, B. Magyar and B. Sauer, "A comprehensive model of wear, friction and contact temperature in radial shaft seals," Wear, 2014.

[3] T. Engelke, "Einfluss der Elastomer-Schmierstoff-Kombination auf das Betriebsverhalten von Radialwellendichtringen," Gottfried Wilhelm Leibniz Universität, Hannover, 2011.

[4] A. Cicchitti, C. Lombardi, M. Silvestri, G. Soldaini and R. Zavattarelli, "Two-phase cooling experiments: pressure drop, heat transfer and burnout measurements," in *Centro Informazioni Studi Esperienze*, Milan, 1959.

[5] P. H. Dawson, "Windage loss in larger high-speed gears," in *Proceedings of the Institution of Mechanical Engineers*, 1984.

[6] M. Tu, "Validation and modeling of power losses of NJ 406 cylindrical roller bearings," KTH Industrial Engineering and Management, Machine Design, Stockholm, 2016.

[7] M. Harris, N. Kotzalas and A. Tedric, Essential Concepts of Bearing Technology, New York: Taylor & Francis, 2006.

[8] D. Knežević and V. Savić, "Mathematical modeling of changing of dynamic viscosity, as a function of temperature and pressure, of mineral oils for hydraulic systems," in *Facta Universitatis. Series. Mechanical Engineering*, Serbia, 2006.

Study on screw rotor thermal machining method of single screw compressor

S Y Wang, Z L Wang*, H W Shi, Z M Wang, M M Hao, J Wang

College of Chemical Engineering, China University of Petroleum (East China), Qingdao 266580, China
Email address: wangzengli25@ upc.edu.cn
* Corresponding Author

Abstract. Single screw compressor (SSC) is a type of rotary compressor with excellent performance. However, under high temperature conditions, the hot deformation of the screw rotor and the star wheel cause the change of the gap of the meshing pairs, which seriously degrades the compressor performance. In order to solve the problem of the gap variation of the meshing pairs, a new hot-state machining method is proposed to study the deformation characteristics of the screw rotor in the hot-state machining process, which provides some reference data for the pre-deformation of the rotor in the machining process. Numerical calculation model of thermal machining process for screw rotor was established by using finite element method. With the validated model, the deformation characteristics in the process of hot working were studied by changing the number of heating holes, diameter of heating holes and temperature of heating holes. Results show that screw rotor deformation will increase with the increase of heat source temperature and diameter, and the deformation distribution of the screw rotor is not significantly affected by the heat source temperature and diameter, but will be more uniform with the increase of heat source quantity. The research work in this paper can provide theoretical basis for hot machining of screw rotor for SSC.

1. Introduction

The single-screw compressor (SSC) was invented by the French engineer Zimmern in the 1950s [1], which is composed of a screw and two symmetrically arranged star wheels as shown in the figure1. SSC has the advantages of small volume, good force balance performance, high volume utilization rate, has been widely used in refrigeration and air conditioning, aerodynamics, process gas compression and other fields. Nowadays, with the rapid development of contemporary society, the carbon emissions are constantly rising, resulting in increasingly serious environmental problems [2-4], improving energy efficiency is of great significance to alleviate the environmental problem. In recent years, in order to improve energy efficiency, low temperature heat-pump, mechanical vapor recompression (MVR) system have been rapidly developed [5-8]. Considering the great mechanical property and wide operation range, single screw compressor (SSC) has received increasing attention in these fields. As shown in figure 1, SSC relies on the meshing movement between the star-wheel and the screw rotor to form a periodically changed working chamber to achieve gas suction, compression and discharge. So the meshing pair is the core working component of the SSC. However, the axis of the star-wheel and the screw rotor is spatially perpendicular, which makes the meshing profile characteristics become very complex. The thermal deformation of SSC during operation causes the deviation of the center position of the star wheel and screw, which has a complex influence on the mating position of the meshing pair, directly affects the meshing relationship of the meshing pair, and reduces the reliability of the meshing pair. Therefore, it is of great significance to study the thermal deformation mechanism of screw rotor under hot machining condition.

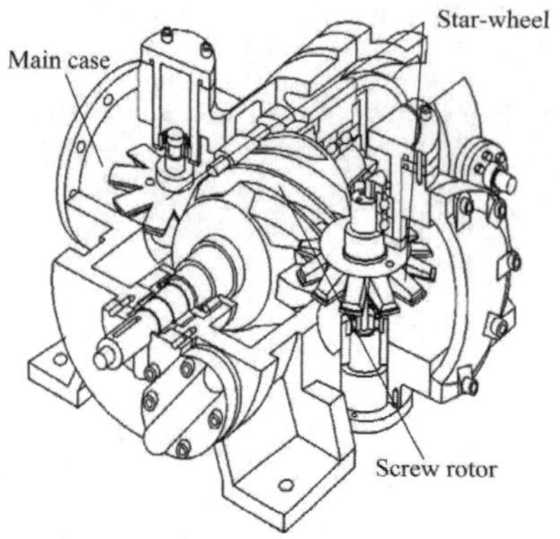

Figure 1. The structure of a single screw compressor

In order to research the thermodynamic deformation mechanism of meshing pair tooth profile under the action of multiple coupling fields,a series of studies have been carried out for the deformation characteristics theoretical analysis, deformation predict and the deformation compensation design on the compressor by the scholars at home and abroad. Diao and Yang et al. calculated the thermal and force deformations of oil-free twin screw compressors, the research results show that the force deformations were much smaller than the thermal deformations [9,10]. Hsieh et al. proposed a mathematical model and a calculation procedure to efficiently calculate the temperature distributions in the male and female rotors of the oil-injected screw compressors. The temperature distributions of the screw rotor is calculated in their study [11]. Mustafin et al. analyzed the formation and factors influencing the profile clearances of screw compressors [12-15]. As for the single screw compressor, Zhou comprehensively analyzed the working process of the single screw compressor, and calculated the temperature field and thermal deformation of the main components of single screw compressor by using the boundary element numerical technology. Based on research results, the influences of thermal deformation on the leakage and on the fit clearance of the compressor under the working state were simulated [16,17].

According to the above literatures, scholars at home and abroad have analyzed the factors causing the thermal deformation of the compressor, but have not compensated for the change of the clearance caused by the thermal deformation, resulting in the thermal deformation of the compressor has not been well controlled. To this end, this paper present a thermal machining method of screw rotor for single screw compressor. Which use the simulation of the screw rotor under the hot processing condition, the deformation of the screw rotor was obtained to compensate for the gap change of the mesh pair caused by the hot deformation under the load condition of the single screw compressor, so as to improve the hot meshing characteristics of the single screw compressor.

2. Theoretical basis of thermal expansion

In this paper, the single screw steam compressor used in high temperature heat pump has been selected as the physical model to investigate the thermal machining process of the screw rotor. The material property of the screw rotor are shown in table 1.

Table 1. Material properties of brass

Parameter	Brass (Analytical model)	QT600-3 (Validation model)
Coefficient of expansion /K^{-1}	1.85×10^{-5}	1.2×10^{-5}
Modulus of elasticity /GPa	90	170
Density /$kg\cdot m^{-3}$	8600	7300
Poisson ratio	0.36	0.27
Thermal conductivity / $W\cdot m^{-1}\cdot K^{-1}$	120	35.5

2.1 The geometry model

The geometry model of the screw rotor hot machining process used in finite element method (FEM) is show in figure 2 (take the hot machining model with 3 heat sources for example). In this paper, the geometry model of the screw rotor hot machining process was modified with different thermal machining parameters as follows:

1) Adding 1, 2, 3 and 4 heat sources to the screw rotor model separately.
2) Changing the diameter of heat sources to 10mm, 12mm, 14mm, 16mm,18mm and 20mm.
3) The heat sources established in (1) was loaded with thermal loads of 160°C, 180°C and 200°C respectively.
4) Adding a PATH channel.

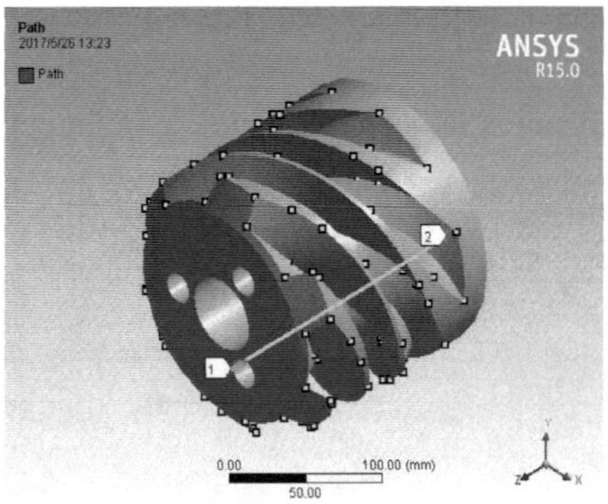

Figure 2. The position of three heating holes in the screw rotor

Figure 3 show the meshing diagram of the screw rotor with three heating holes of 18mm diameter. Because of the complex structure of these components, the tetrahedral mesh has been chosen. Different mesh sizes are numerically checked for the gird dependency of temperature of lines on the screw rotor. Considering the calculation data, cost of simulation time and the grid independence test, the final selected element numbers of screw rotor are 259846, the minimum size is 2.5mm. The average grid cell mass of screw rotor is 0.805. And the mesh pass the grid independence test. So the element quality meets the calculation requirements.

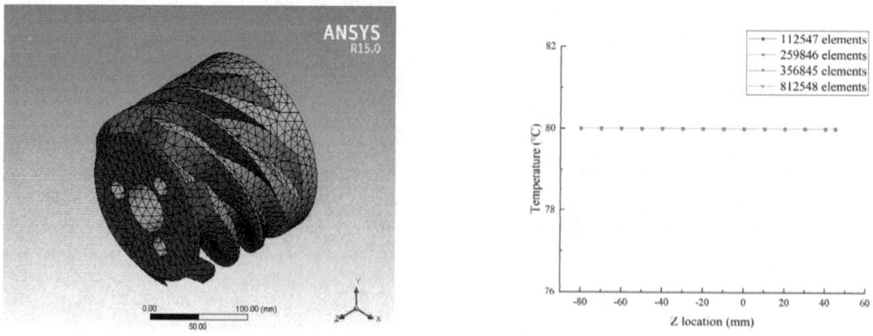

(a) Meshing diagram (b) Grid independence test

Figure 3. Mesh details

2.2 Boundary conditions

When the temperature load is applied to the screw rotor, the thermal deformation of the screw rotor will be induced. In the prototype, the screw rotor and the screw shaft were fitted through interference mode. The screw rotor can be fixed and supported axially by two angular contact ball bearings installed on the shaft in exhaust side. But in the suction side, the screw rotor can be fixed and supported by the cylindrical roller bearing, the screw shaft can freely adjustable in the axial. Therefore, the screw rotor should be constrained on the exhaust end with fixed constraint, and be constrained on the section end with the axial displacement of the constraints, keep the axial degree of freedom. In this paper, two different thermostatic heat sources are set on the surface of the screw rotor heating hole and the whole screw rotor, so that a constant temperature is obtained on the surface of the heating hole, and the screw rotor is thus heated. The screw rotor with three heat sources was taken as an example. The surface temperature of the heating hole was set at 180°C and the ambient temperature was set at 22°C, as shown in the figure 4.

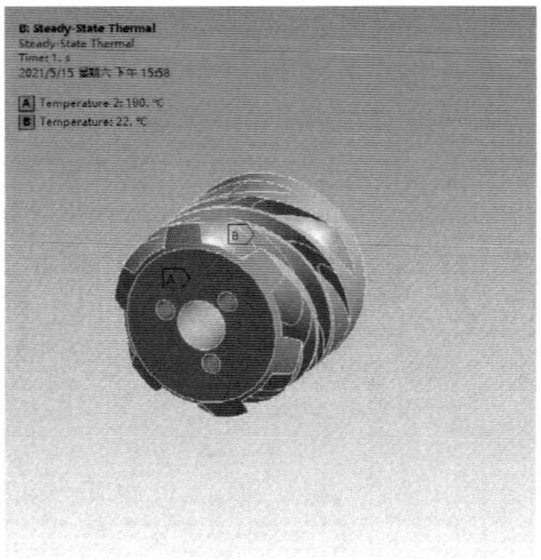

Figure 4. The setting of temperature load

3. Result and discussions

3.1 The parameters of the SSRC

Combining with the geometric model of the screw rotor thermal machining process and the thermal deformation analysis model of screw rotor describing the thermal processing characteristics of the SSC, the Numerical calculation model used to analysis the thermal machining process of the SSC can be set up. In order to solve and verify this model, a typical single screw compressor with the structure parameters as shown in table 2 was developed in this paper as an example for the theoretical and experimental study. By using this FEM, the thermal processing characteristics of the SSC are investigated and the influence of the thermal machining parameters on the preset thermal deformation characteristics are also discussed in this section.

Table 2 Structure parameters of the SSC with MEMP

Structure parameters	Value	Structure parameters	Value
Screw rotor diameter /m	0.181	Suction seal angle α_{in} /rad	-0.681
Screw rotor length /m	0.1557	Central distance /m	0.1448

3.2 The Path deformation analysis

By using the verified model, the thermal machining characteristics of the SSC with the structure parameters shown in table 2 were studied. For investigating the influences of the thermal machining on the thermal machining characteristics, further researches are presented to evaluate the effects on the thermal deformation of the screw groove profile. The concentrated heat load, quantity loaded and the diameter of the heat source were selected as the main operating parameters to be examined.

Figure 5 shows the thermal deformation of different heat sources under the condition that the heat source temperature is equal to 180°C and the heat source diameter is equal to 20mm. As shown in the figure 5, the thermal deformation near the inlet and outlet ends is large, and the minimum deformation is in the middle. It can be seen from the variation tendency of the thermal deformation along the path from the inlet end to the outlet end that a deformation peak is reached quickly, and then the deformation begins to decrease until the intermediate position reaches the minimum value, and then begins to rise to reach the second peak and also the maximum deformation near the outlet end, and then there is a downward trend before reaching the outlet end. By changing the number of heat sources, the deformation distribution of the screw rotor did not change much, but with the increase of heat source, the deformation increased, but the thermal deformation is more uneven.

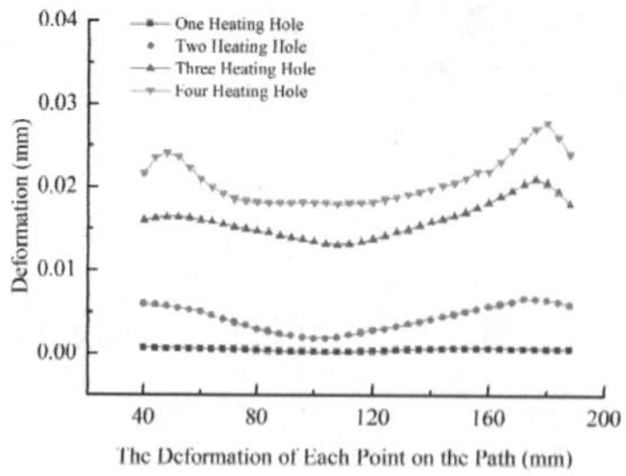

Figure 5. The deformation of the path of the screw rotor with different heat sources

Figure 6 shows the deformation of different heat source diameters when there are three heat sources and the heating temperature is equal to 180°C. As shown in this figure, the deformation near the inlet and outlet ends is large, and the minimum deformation is in the middle. It can be seen from the variation tendency of the thermal deformation along the path from the intake end to the outlet end that, the thermal deformation variation tendency with different heat source diameters is the same with each other. By changing the heat source diameters of the screw rotor thermal machining process, the Uniformity of the thermal deformation of screw rotor did not change greatly. But the thermal deformation value of the screw rotor increased with the increase of the heat source diameters. When the heat source is doubled in diameter, the deformation increases by 43%.

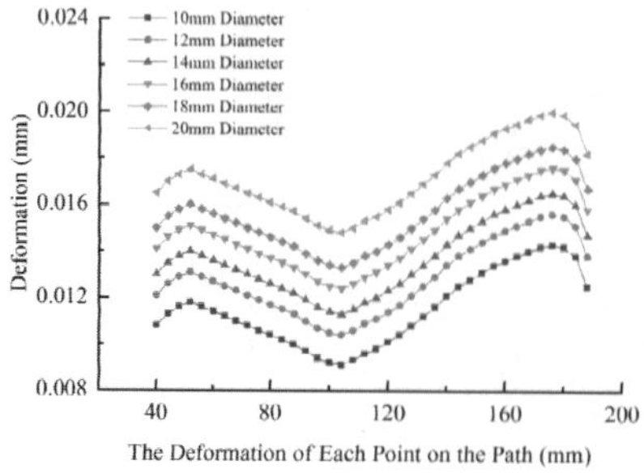

Figure 6. Deformation of the path of screw rotor with different heating hole diameters

Figure 7 shows the deformation at different heat source temperature when there are three heat sources and the heating hole diameter is equal to 20mm. As shown in the figure 7, the variation tendency of the thermal deformation along the path from the intake end to the outlet end under different heat source temperature is the same with each other.. By changing the heat source temperature of the screw rotor, the deformation distribution of the screw rotor changed little, but with the increase of heat source temperature, the deformation increased, When the heat source temperature be increased by 25%, the deformation will be increased by 30.5%. By comparing the influence of heat source diameter and temperature on thermal deformation, the influence of heat source diameter on thermal deformation is greater than that of heat source temperature.

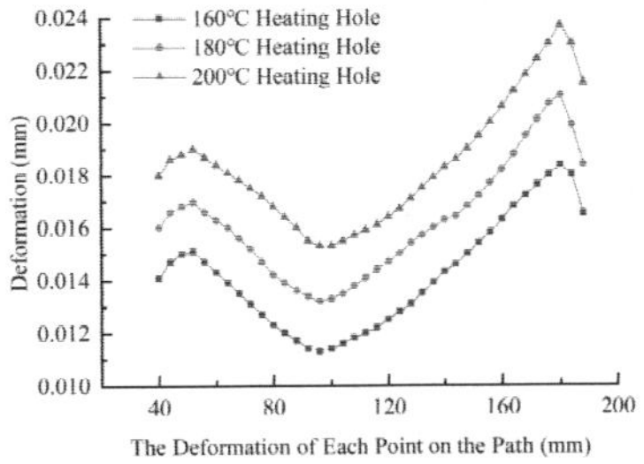

Figure 7. The deformation of the path of the screw rotor with different heat source temperature

The uniformity of the thermal deformation of the screw groove at the same star wheel rotation angle is an important standard to measure the thermal machining characteristics. So the thermal deformations of the screw rotor groove profile along the circumference were analyzed. The results are as follows:

Figure 8 and figure 9 shows the thermal deformation under different heat sources when the heat source temperature is equal to 180°C and the heat source diameter is equal to 20mm. As shown in the figure 8 and figure 9, when there is one heat source, the side of the screw groove with the heat source has the maximum deformation, but the other part of the screw rotor without heat source basically does not have the deformation. When there are two heat sources, the maximum deformation is located near the plane composed of the central line of the screw rotor axial and the central line of the heat source. The deformation near the two heat sources is roughly the same. The deformation is distributed symmetrically along the axis with the central line of the heat source. When there are three and four heat sources, the maximum deformation also occurs near the installation location of heating source, and the deformation is roughly rotational symmetry. When there are multiple heat sources, the deformation decreases rapidly from the heat source installation location until the farthest distance from the heat source reaches the minimum, and then begins to increase to the position of the next heat source. The deformation is distributed symmetrically on the axis of the heat source. At the same time, the heat source is different, the place with the greatest deformation is all outside the heat source. By changing the heat source number acted on the screw rotor, the deformation distribution of the screw rotor did not change much, but with the increase of heat source, the deformation increased.

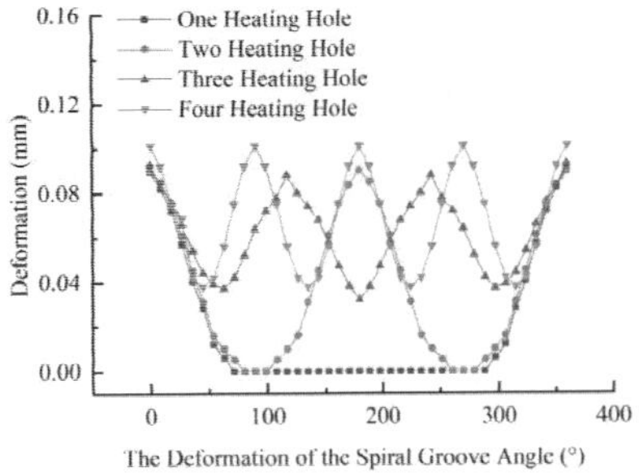

Figure 8. The deformation of the spiral groove of the screw rotor with different heat sources

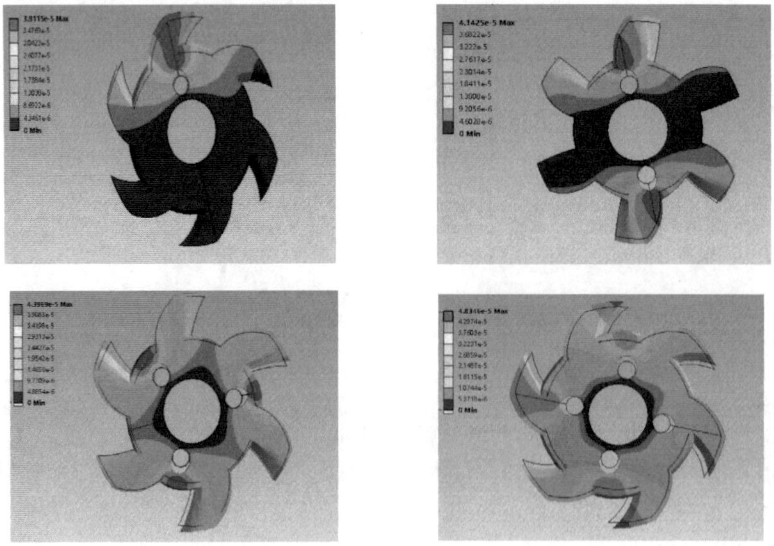

Figure 9. The contour of the spiral groove of the screw rotor with different heat sources

Figure 10 shows the thermal deformation of the screw rotor with different heat source diameters when there are three heat sources and the heating temperature is equal to 180°C. As shown in the figure 10, the deformation was mostly located near plane composed of the central line of the screw rotor axial and the central line of the heat source, and outside the heat source. The deformation is distributed symmetrically on the axis of the heat source. By changing the heat source diameter acted on the screw rotor, the deformation distribution of the screw rotor did not change greatly, but the deformation increased with the increase of the heating hole diameters. The cures in figure 10 were also compared and analyzed to research the effect of the heat source diameter on the thermal deformation uniformity of the screw rotor during the thermal machining process. By contrast of the curves in figure 9, the amplitude variation range of thermal deformation increases gradually with the increase of the of heat source diameter. So the thermal deformation uniformity of screw rotor during the thermal machining process will be increased with the increase of the heat source diameter.

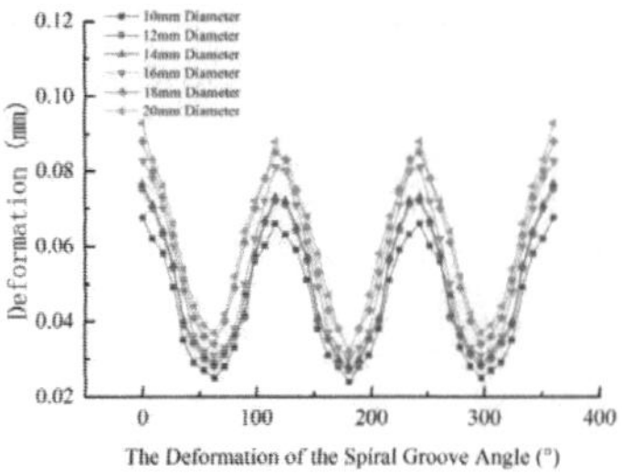

Figure 10. Deformation of the spiral groove of screw rotor with different heating hole diameters

Figure 11 shows the thermal deformation of the screw rotor at different heat source temperature when there are three heat sources and the heating hole diameter is equal to 20mm. As shown in the figure 11, the thermal deformation distribution of the screw rotor changed little when the heat source temperature of the screw rotor increased, but with the increase of heat source temperature, the thermal deformation of the screw rotor increased. The contrast of the curves in figure 11 shows that the amplitude variation range of thermal deformation remain nearly constant with the increase of the of heat source temperature. So the thermal deformation uniformity of screw rotor during the thermal machining process will be remain nearly constant when the heat source temperature increases.

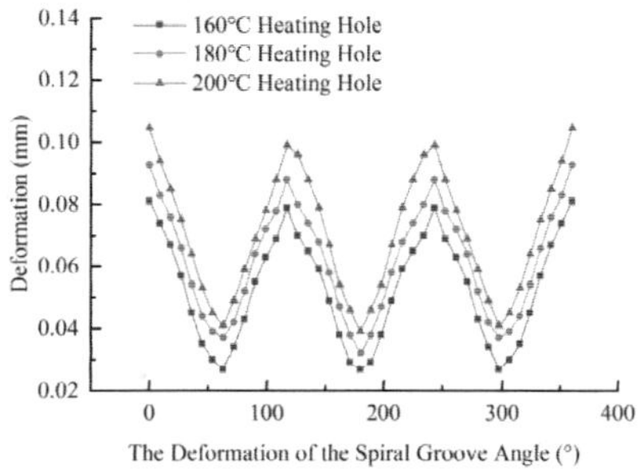

Figure 11. The deformation of the spiral groove of the screw rotor with different heat source temperature

4. Conclusion

In this paper, Numerical calculation model of thermal machining process for the screw rotor in SSC was established, and the thermal deformation law of screw rotor and the effect of thermal machining parameters on the deformation characteristics of the screw groove profile were investigated for SSC. Through analysis, the following conclusions can be drawn:

1) The heat source number has influences on the thermal machining characteristics of screw rotor. the thermal deformation and its uniformity along the peripheral direction increases gradually with the increase of the of heat source number. But the thermal deformation is more uneven along the screw rotor axis when the heat source number increases.
2) The heat source diameter is a factor that affects the machining characteristics of screw rotor. The thermal`deformation value of the screw rotor increases with the increase of the heat source diameters. When the heat source is doubled in diameter, the deformation increases by 43%. The amplitude variation range of thermal deformation increases gradually with the increase of the of heat source diameter. So the thermal deformation uniformity of screw rotor during the thermal machining process will be increased with the increase of the heat source diameter.
3) The heat source temperature is also an important factor affecting the machining characteristics of screw rotor, the deformation increased with the increase of heat source temperature, When the heat source temperature be increased by 25%, the deformation will be increased by 30.5%. The amplitude variation range of thermal deformation remain nearly constant with the increase of the of heat source temperature. By comparing the influence of heat source diameter and temperature on thermal deformation, the influence of heat source diameter on thermal deformation is greater than that of heat source temperature.

Acknowledgements

This work is supported by National Key R&D Program of China [NO. 2020YFB2010002], National Natural Science Foundation of China [NO. 51706247, NO. 51975585], the Fundamental Research Funds for the Central Universities and the Opening Fund of National Engineering Laboratory of Offshore Geophysical and Exploration Equipment [NO. 20CX02312A].

References

[1] Zimmern B. Worm rotary compressors with liquid joints. USA, 3180565[P]. 1965.
[2] N. America. BP energy outlook 2035. UK: British Petroleum, 2012.
[3] Dincer I. Renewable energy and sustainable development: a crucial review[J]. Renew Sustain Energy Rev. 2000, 4(2): 157–75.
[4] U. Zafar, T.U. Rashid, A.A. Khosa, et al. An overview of implemented renewable energy policy of Pakistan[J]. Renew Sustain Energy Rev. 2018, (82): 654-665.
[5] Iodice P, Langella G, Amoresano A. Modeling and energetic-exergetic evaluation of a novel screw expander based direct steam generation solar system[J]. Appl Therm Eng. 2019, 155: 82-95.
[6] Iodice P, Langella G, Amoresano A. Energy performance and numerical optimization of a screw expander-based solar thermal electricity system in a wide range of fluctuating operating conditions[J]. International Journal of Energy Research. 2020, 44(3): 1858-74.
[7] Tian Y, Shen J, Wang C, Xing Z, Wang X. Modeling and performance study of a water-injected twin-screw water vapor compressor[J]. Int J Refrig. 2017, 83: 75-87.
[8] Jamil MA, Zubair SM. Effect of feed flow arrangement and number of evaporators on the performance of multi effect mechanical vapor compression desalination systems[J]. Desalination. 2018, 429: 76-87.
[9] B.C. Yang, J.F. Li, Z.W. Xing. Analysis of rotor deformation of oil-free screw air compressor[J]. Compress. Technolo. 2005, (01): 8-11.
[10] A.N. Diao, B. Zhou, B.S. Lu. Optimal design of assembly clearance of oil-free screw compressor[J]. Compress. Technolo. 2013, (06): 29-32.
[11] S.H. Hsieh, Y.C. Shih, W.H. Hsieh, F.Y. Lin, M.J. Tsai. Calculation of temperature distributions in the rotors of oil-injected screw compressors[J]. Int. J. Therm. Sci. 2011, (50):1271-1284.
[12] T.N. Mustafin, R.R. Yakupov, et al. The analysis of typical profile clearances formation in meshing rotors of the screw compressor,AIP Conference Proceedings. 2017, (1876):20-24.
[13] T.N. Mustafin, R.R. Yakupov, et al. Analysis of the screw compressor rotors' non-uniform thermal field effect on transmission error[J]. IOP Conference Series: Mater. Sci. and Eng. 2015, (90): 4-12.
[14] T.N. Mustafin, R.R. Yakupova, et al. Analysis of influence of screw compressor construction parameters and working condition on rotor temperature fields[J]. Procedia Engineering. 2016, (152): 423-433.
[15] T.N. Mustafin, R.R. Yakupov, et al. Determining of actual profile clearances and screw compressor rotor positions depending on working conditions[J]. IOP Conference Series: Mater. Sci. and Eng. 2017, (232):12-23.
[16] L. Zhou. Study on meshing pair thermoelastic deformation and dynamic meshing characteristics of single screw compressor[D]. J. of Xi'an Jiaotong Uni. 1999.
[17] L. Zhou, G.X. Jin. Research on meshing accuracy of single screw compressor[J]. Fluid Machinery. 1998, (7): 37-39.

IOP Conf. Series: Materials Science and Engineering 1180 (2021) 012012 doi:10.1088/1757-899X/1180/1/012012

Numerical study of intake flow field inside the high-speed refrigeration scroll compressor with spiral suction channels in the motor rotor

G X Bu[1], X R Li[1], X W Li[1], Y Zhang[1], W F Wu [1*], X L Wang[2] and X P Zhang[3]

1 School of Energy and Power Engineering, Xi'an Jiaotong University, Xian, China
2 School of Engineering, University of Tasmania, Hobart, Australia
3 Xinxiang Aviation Industry (Group) Co., Ltd, Xinxiang, China

*Correspondence: weifengwu@mail.xjtu.edu.cn

Abstract: To control the temperature of the electric motor, the suction gas is introduced flowing through the motor in hermetic scroll compressors. For high-speed compressors, this flow resistance is high, which results in decreasing of the efficiency. In this paper, spiral flow channels were designed inside the motor rotor to decrease the suction flow resistance and increase the refrigerant flow rate inside the scroll compressor. Simulation of the suction flow through the motor was carried out. Two structure of straight and spiral intake channels were brought into the simulation model, under 3000 rpm, 5000 rpm, 7000 rpm and 9000 rpm. The results showed that the spiral channels could increase the refrigerant flow rate inside the motor rotor and the suction pressure of the compressor, and with the increase of the motor speed, the effect of the intake supercharging improved. At 9000 rpm, the suction pressure at the suction port of the compressor increased by about 4% by using spiral channels, while the mass flow rate increased by about 6%. The increases of total refrigerant flow rate in spiral channels are mainly due to the increase in unblocked channels. The size and the layout of the balance weight have a great effect on intake flow.

1. Introduction

Recently, the development of high-speed compressor has gained increased attention because of smaller size and lighter weight. The high-speed scroll compressors have been widely used in mobile platforms, such as electric vehicles [1]. High-speed scroll compressors are usually hermetic. In the hermetic compressors, it is impossible to observe the internal temperature, pressure and flow directly. Therefore, the cooling of motor is always a problem. Many scholars have reported the heat dissipation of the motor.

For compressor motor, suction cooling, a kind of air cooling, is the most widely used cooling method. The refrigerant enters the compressor, cools the motor first, and then flows into the compressor to operate. The inlet temperature of the refrigerant is lower than the motor temperature, and the refrigerant flows through the internal flow channels of the motor to reduce the motor temperature [2].

At the same time, in order to increase the heat exchange of the motor, the channels are usually set inside the motor. Huang et al. studied the axial cooling channels at the edges of housing or/and stator back, and carried out thermal analysis with rectangular, oval and elliptical flow channels [3]. He et al. developed a comprehensive model of the flow and thermal characteristics of a semi hermetic twin-screw refrigeration compressor. Straight cooling channels were set inside the stator and the rotor [4]. Tang et al. used CFD to analyze the influence of increasing rotor ventilation channel and optimizing stator cutting on motor cooling performance of a scroll compressor [5]. Yang et al. introduced a structure design of an external rotor permanent magnet synchronous torque motor, which adopted spiral magnetic pole and set cooling channels on the surface of the outer rotor and inside the stator core [6]. K. Rönnberg et al. compared different modelling approaches, and mainly focused on the stator side, where the straight cooling channels are arranged [7].

So far, the research about motor cooling only focuses on the heat dissipation effect of motor, and the research object is only the motor. However, for the motor used in the compressor, there is no report concern about the cooling structure and its impact on compressor performance. With the increase of heat transfer, the flow channels in the motor will bring resistance loss. Thanks to the high rotation speed, structures similar to turbine could be efficiently introduced to enhance the suction flow of the high-speed scroll compressor. At present, most cooling channels in the motor rotor or stator are straight. Therefore, in this paper, for high-speed scroll compressor, spiral inlet channels set in the motor rotor is proposed. In order to discuss the influence of refrigerant flow in the spiral channels on the performance of the compressor, the spiral channels with 240mm pitch and the straight channels are compared using CFD software. The simulation is carried out at 3000 rpm, 5000 rpm, 7000 rpm and 9000 rpm.

2.Motor structure and computational model

2.1. Geometric model and spiral channels of the motor

In order to increase the suction pressure of the compressor and the refrigerant flow rate of the refrigeration system, spiral rotor channels are arranged inside the motor rotor. Figure 1 shows the cross section of the internal structure of the scroll compressor. The hermetic scroll compressor mainly divided into two parts: the motor and the compressor driven by the motor. As shown in Figure 1, all the refrigerant flow through the motor by the channels in the rotor and the gap between the rotor and the stator. After cooling the motor, the refrigerant then enters the suction port of the scroll compressor. Hence, we choose the flow passage of the motor part as the calculation domain, as shown in the Figure 1. The outlet pressure of the computational domain is the suction pressure of the compressor. The performance of the compressor will be improved with the increase of the outlet pressure and mass flow rate of the computational domain.

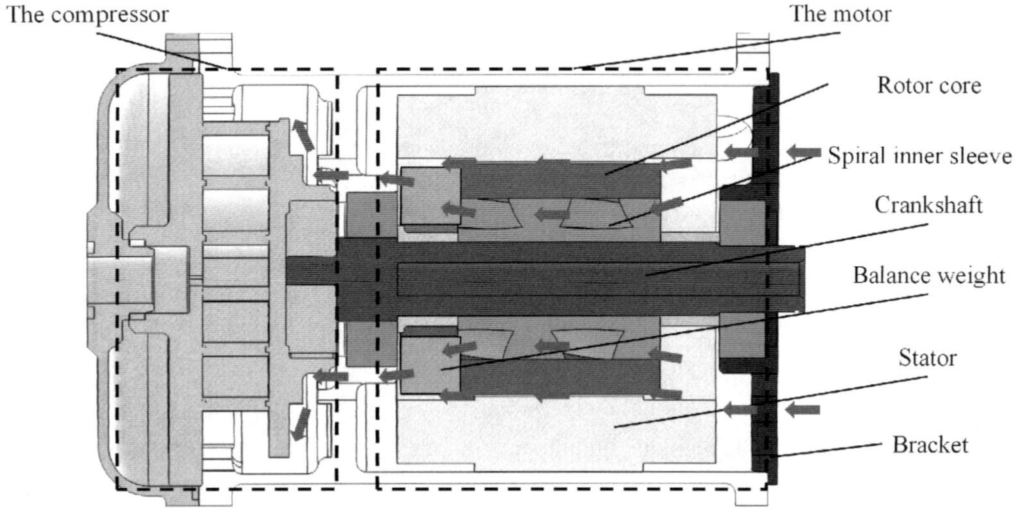

Figure 1. Cross section of internal structure of the scroll compressor.

The main structure of the motor shows in Figure 2. The motor is permanent magnet synchronous motor. The motor is composed of a crankshaft, a spiral inner sleeve, a rotor core, a balance weight and a stator. The inner rotor consists of rotor core and magnets. The balance weight is located on the rotor core. It is used to balance the rotating inertia force of the orbiting scroll and have an influence on the flow rate of the refrigerant. Accordingly, it's very important to design the size and shape of the balance weight. Axle plays an important part in the motor and supports the whole motor body, so its mechanical strength and rigidity must meet a high standard.

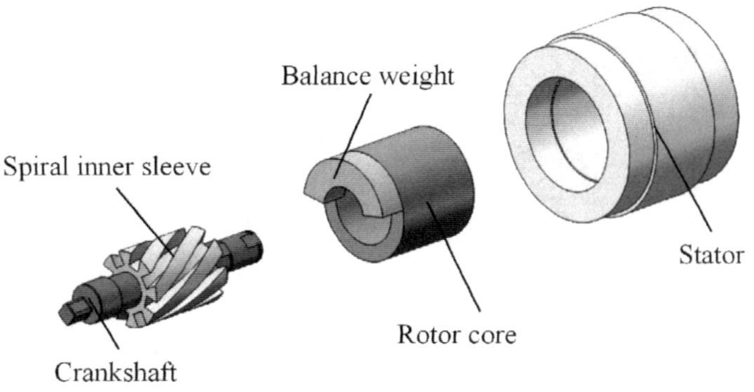

Figure 2. Main components of the motor.

As is shown in Figure 3, after the spiral inner sleeve added inside the rotor core, the refrigerant flow channels (F1-F8) are formed between the spiral inner sleeve and the rotor core. The spiral inner sleeve and the rotor core adopt interference fit to make rotation movement under the action of the crankshaft. The rotation direction of the spiral inner sleeve is set according to the rotation direction of the crankshaft. In this study the spiral inner sleeve rotates counterclockwise when viewed from the balance weight direction, as showed in Figure 3. However, due to the existence of the balance weight, some channels (F2-F4) are still blocked, so the design principle of the inner sleeve is to minimize the flow loss and ensure the maximum outlet area of channels.

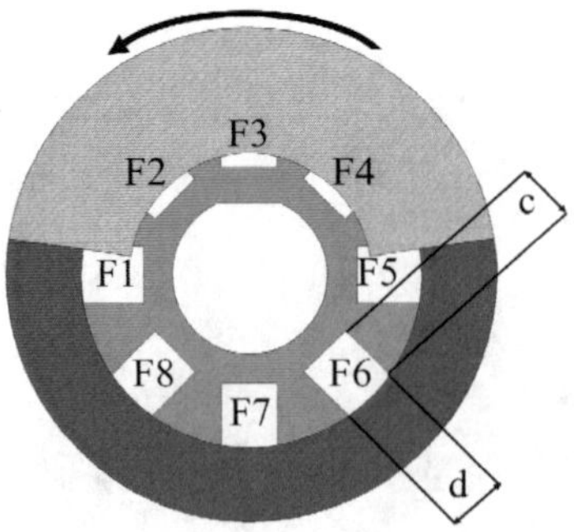

Figure 3. Rotor cooling channel dimensions.

The radial dimension, the outside diameter and the inside diameter, and the axial length of the rotor core are decided by the structure of the motor. In this paper, the geometrical dimensions of the spiral inner sleeve we designed are shown in the Table 1 below. The outside diameter is 48 mm, the inside diameter is 22 mm, the axial length is 60 mm. Table 1 also show the d, c, the pitch and the total ventilation area, which are formed by the sleeve and the rotor core.

Table 1. The geometric dimensions of the spiral inner sleeve.

Geometric dimensions	Value
outside diameter	48 mm
inside diameter	22 mm
axial length	60 mm
d	8 mm
c	8.66 mm
pitch	240
total ventilation area	568.88 mm^2

2.2. Internal flow field model of the motor

In order to study the influence of the motor rotor channels on compressor performance, the internal runner of the motor was selected as the computational domain. The motor dashed box in Figure 1 shows the computational fluid domain. So, the calculation domain starts from the refrigerant entering the motor and ends at leaving the motor. It mainly includes the channels in the rotor core, the air gap and the gap between balance weight and crankshaft.

Figure 4 shows the computational model with straight channels, where (a) is the geometric model and (b) is the mesh of the geometric model. The computational model with spiral channels is show in Figure 5, where (a) shows the geometric model and (b) is the mesh of the geometric model. The motor rotor, the inner sleeve and the balance weight is rotating when the compressor is operating. Hence, the static domain is adopted for the gap between the rotor and the stator and the refrigerant inlet and outlet area, while the rotating domain is adopted for the internal flow channels of rotor and the gap between balance weight and crankshaft. In order to discuss the influence of refrigerant flow in the spiral channels on the performance of the compressor, the spiral channels with a pitch of 240mm and the straight channels are compared and analyzed. The number of the two types of channels are both 8. Besides, the cross-sectional area, shape and outlet location of the channels are all same.

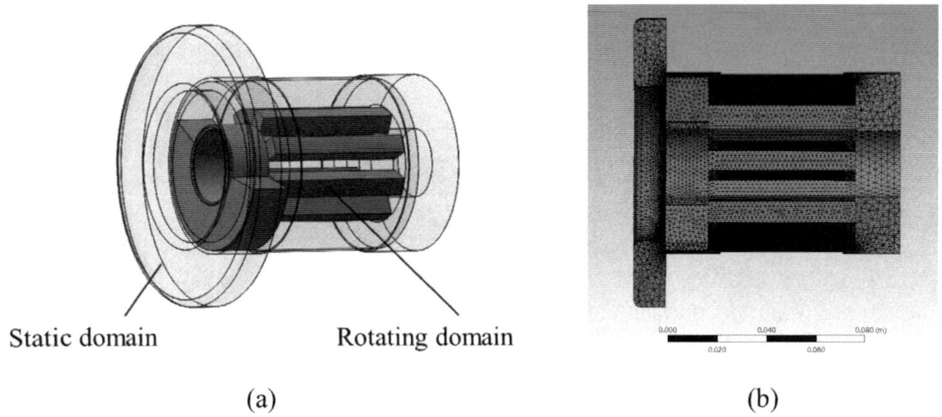

Figure 4. The geometric mode (a) and the section view of mesh (b) of the computational model with straight channels.

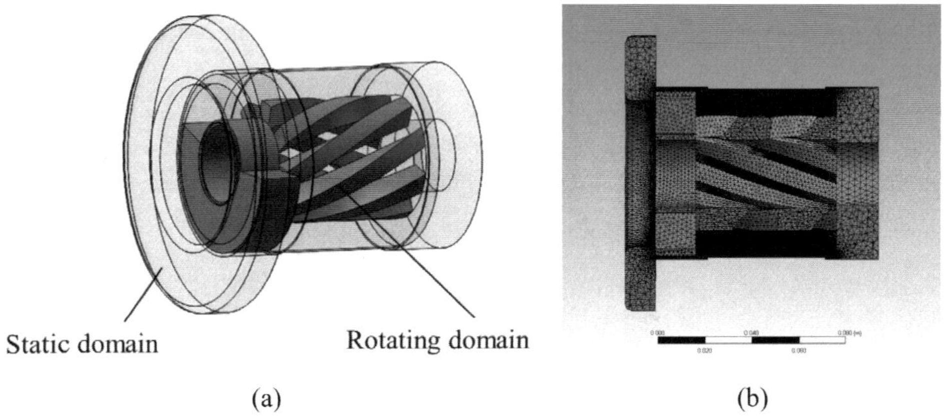

Figure 5. The geometric mode (a) and the section view of mesh (b) of the computational model with spiral channels.

2.3. Computational Model

The simulation method and calculation conditions are presented in Table 2. The actual refrigerant R134a gas model is used for calculation. And the first-order upwind scheme is used to solve the momentum and energy equations. The convergence results are obtained by monitoring the residual of mass, energy and heat transfer equation. When the residual converges to 10^{-4}, the calculation stops.

Table 2. Simulation method and calculation condition.

Computation method		Operating conditions			
software	turbulent model	inlet temperature	inlet pressure	rotational speed (rpm)	outlet flow velocity (m/s)
CFX17.2	k-Epsilon	288.15K	3.4966bar	3000	4.3067
				5000	7.4789
				7000	10.1106
				9000	12.4621

In order to discuss the influence of refrigerant flow in the spiral channels on the performance of the compressor, the spiral channels and the straight channels are compared and analyzed. The simulation is carried out at 3000 rpm, 5000 rpm, 7000 rpm and 9000 rpm. In addition, the grid independence verification is show in Table 3. After grid independence verification, the grid quantity of case 3 meets the quality requirement.

Table 3. Grid independence verification.

Type	Case	Mesh number	Outlet pressure (bar)
the computational domain with straight channels	1	3801056	336.085
	2	5391963	338.917
	3	6806930	338.638
	4	8189592	338.857
the computational domain with spiral channels	1	3384108	321.054
	2	4834654	324.767
	3	6297658	324.053
	4	7836120	324.671

The heat transfer between the surface of the motor and the refrigerant gas is convection heat transfer. In the setting of the wall boundary conditions, the outside temperature is assumed at 40°C, and the calculation of the heat transfer coefficient is relatively complex, and empirical formula is usually used to calculate it.

(1) heat transfer coefficient of inner gap between stator and rotor

There is an annular gap between the rotating rotor, and the stationary stator. Nerg et al. [8] in their study pointed out heat transfer characteristics in such inner gap can be derived as:

$$\begin{aligned} &Nu_g = 2, T_a^* < 1700 \\ &Nu_g = 0.128(Ta^*)^{0.367}, 1700 < Ta^* < 10^4 \\ &Nu_g = 0.409(Ta^*)^{0.241}, 10^4 < Ta < 10^7 \end{aligned} \tag{1}$$

$$Ta^* = \frac{16997(0.0056 + 0.0571\zeta^2)(1 - \dfrac{\delta}{2r_o})Ta}{\pi^4 \zeta} \tag{2}$$

$$Ta = \frac{\rho^2 \omega^2 r_o \delta^3}{\mu^2} \tag{3}$$

$$\zeta = \frac{2r_o - 2.304\delta}{2r_o - \delta} \tag{4}$$

Where: Ta^* is the modified Taylor number, δ is the inner gap length (m), r_o is the radius of the rotor outer surface, ω is the rotation angular velocity (rad/s), μ is the dynamic viscosity (Pa·s) and ρ is the density(kg/m^3).

(2) heat transfer coefficient of the rotor inner channels

The rotor inner channels circumferential distribution around the rotor axis, rotating along with the rotor and spiral sleeve. The heat transfer characteristics can be calculated as follow [4]:

$$N_{uc} = \frac{0.043P_r}{(P_r)^{0.2} - 0.050} \frac{R_e}{\Gamma^{0.1}} \left[1 + \frac{0.061}{(\frac{R_e}{\Gamma^2})^{0.2}} \right], \tag{5}$$

$$\Gamma = \frac{R_e^{\frac{18}{11}}}{(G_r P_r^{\frac{2}{3}})^{\frac{10}{11}}}, \tag{6}$$

$$G_r = \frac{H\omega^2 \beta(\tau \cdot \frac{d_c}{2})\rho^2 d_c^3}{\mu^2}, \tag{7}$$

Where: H is the distance from cooling hole axis to rotor axis (m), β is the thermal expansion coefficient and τ is the temperature gradient along the cooling hole axis, d_c is the equivalent diameter (m).

(3) heat transfer coefficient of end-winding space [9]:

$$\alpha_w = \frac{(1 + 0.04v)}{0.045}, \tag{8}$$

Where: v is the outer diameter straight velocity of the motor rotor(m/s).

(4) heat transfer coefficient of the inner and outer surface of the stator domain contacting with frame

The inner and outer of the stator domain contacted with the frame and the bearing at the area close to the outlet. Therefore, the refrigerant is heated by the compressor components. The heat transfer on the inner and outer surface can be treated as natural convection around a horizontal cylinder.

3. Results and discussion

3.1. Discussion on the compressor inlet mass flow rate and pressure

According to Figure 6, the outlet pressures of computational domain, the suction pressures of the compressor, with two type rotor channels are calculated. It is found that with the increase of rotation speed, the pressure loss increases due to the increase of refrigerant flow rate. However, under the same compressor operating condition, the outlet pressure of the spiral channel is higher than that of the straight channel, and the pressure loss is smaller. The pressure increases by 0.63% at 3000 rpm, 1.79% at 5000 rpm, 2.26% at 7000 rpm, and 4.35% at 9000 rpm. Therefore, the spiral channel can reduce the intake resistance of the compressor and the suction pressure is increased. And with the increase of rotation speed, the difference of outlet pressure is larger. Accordingly, in the high-speed compressor the spiral inner sleeve of the motor rotor is effective.

Under different rotational speeds of the compressor, the simulation results of the total refrigerant mass flow rate, the compressor suction mass flow rate, are shown in Figure 7. It can be seen from the figure that the flow rate of spiral channels is greater than that of straight channels at the same rotation speed. With the increase of the rotation speed, the difference of mass flow rate between the straight channels and spiral channels increases. The outlet mass flow rate increases by 3.94% at 3000 rpm, 4.19%

at 5000 rpm, 6.86% at 7000 rpm, and 6.87% at 9000 rpm. It is obvious that the spiral channels help to increase the refrigerant flow and the refrigeration capacity of the compressor system.

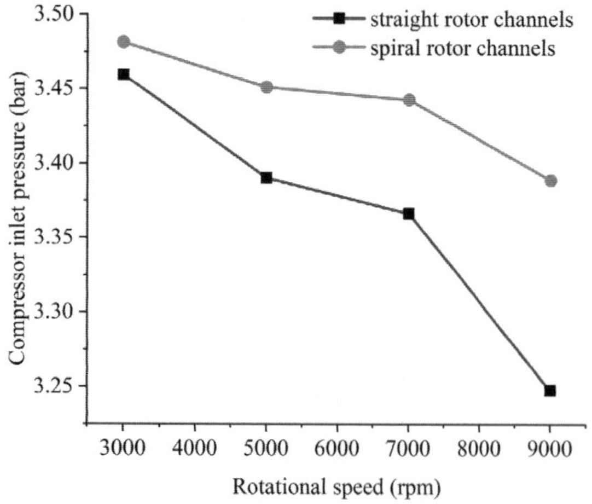

Figure 6. Compressor inlet pressure of rotor channels at different rotation speeds.

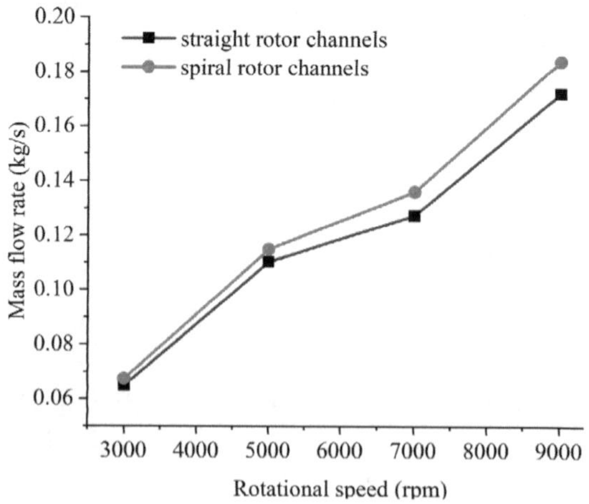

Figure 7. Mass flow rate of rotor channels at different rotation speeds.

3.2. Discussion on refrigerant distribution

The outlet position of the rotor channels is shown in Figure 3. At 3000 rpm, 5000 rpm, 7000 rpm and 9000 rpm, the refrigerant mass flow distribution shows the same law. Take 9000 rpm as an example. Figure 8 (a) (b) shows the sectional drawing of velocity vector of the computational domain with straight and spiral channels. The refrigerant flows into the gap between balance weight and crankshaft through the air gap and the channels in the rotor, and finally reaches the outlet. It can be seen that the velocity of refrigerant in unblocked channels is larger. By comparing the straight channel with the spiral channel, it can be found that in the unblocked channel, the refrigerant flow rate is higher, which results in larger mass flow rate of the spiral channels.

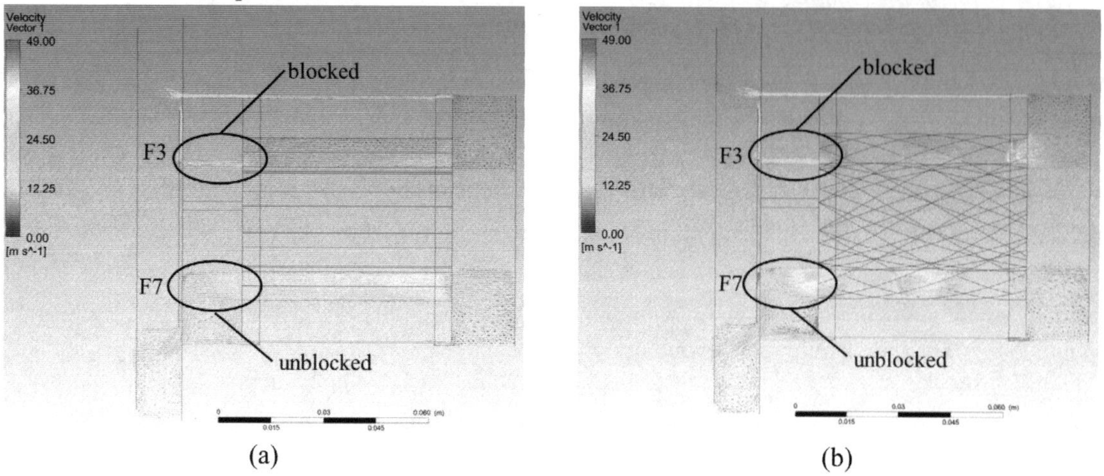

Figure 8. Cross section of velocity vector (a) with straight channels (b) with spiral channels at 9000 rpm.

Table 4 shows the percentage of mass flow in the rotor channel F1-F8 in the total flow at 9000 rpm. In different types of channels, the mass flow rate of refrigerant increases with the increase of the rotation speed. And the refrigerant distribution in each channel is uneven, among which, F1-F5 channels are blocked to varying degrees at the outlet, and F6-F8 are directly connected with the rotating domain of the gap between balance weight and crankshaft. Among them, the straight channels have higher refrigerant flow rate than the spiral at F1-F5 channel, while the spiral channels have higher refrigerant flow rate than the straight at F6-F8 channel. The increases of total refrigerant flow rate in spiral channels are mainly due to the increase of F6-F8 flow rate.

Table 4. The percentage of mass flow in the rotor channels (F1-F8) in the total flow at 9000rpm.

Rotor channels	Percentage of mass flow rate in rotor channels		Deviation
	Straight channels	Spiral channels	
F1	15.58%	14.54%	-1.04%
F2	4.27%	2.96%	-1.31%
F3	4.21%	3.31%	-0.90%
F4	4.71%	3.97%	-0.74%
F5	18.68%	18.65%	-0.03%
F6	18.52%	19.96%	1.44%
F7	17.28%	18.72%	1.44%
F8	16.74%	17.88%	1.14%

The main reason for the great difference of flow rate is the existence of balance weight of scroll compressor. Due to the existence of the balance weight, the refrigerant flow and flow rate in the F1-F5 channels are reduced. Therefore, in order to further increase the refrigerant flow in the motor of the scroll compressor, the location of the balance weight of the scroll compressor can be optimized. On the premise of ensuring that the balance weight meets the requirements of the compressor, the volume is reduced and the layout should be reasonable, so as to increase the refrigerant flow and reduce the pressure loss of the channels of the motor rotor.

4. Conclusion

In this paper, a kind of spiral channel in the motor rotor of the high-speed scroll compressor is proposed. The spiral inner sleeve is added inside the rotor core, the spiral refrigerant flow channels are formed between the spiral inner sleeve and the rotor core. In order to discuss the influence of the rotor channels structure on the performance of the compressor, the spiral channels with a pitch of 240 mm and the straight channels are compared and analyzed. The simulation is carried out at 3000 rpm, 5000 rpm, 7000 rpm and 9000 rpm.

It is found that under the same compressor operating condition, the outlet pressure of the spiral channels is higher than that of the straight channels and the refrigerant mass flow rate of spiral channels also is greater. The pressure increases by 0.63% at 3000 rpm, 1.79% at 5000 rpm, 2.26% at 7000 rpm, and 4.35% at 9000 rpm. And, at different rotational speeds, the outlet mass flow increases by 3.94% at 3000 rpm, 4.19% at 5000 rpm, 6.86% at 7000 rpm, and 6.87% at 9000 rpm. With the increase of the motor speed, the supercharging effect of spiral channels in rotor is more obvious.

Due to the existence of balance weight of scroll compressor, the refrigerant distribution in each channel is very uneven. The refrigerant flow velocity and mass flow rate are larger in the unblocked channels than blocked channels. And in the unblocked channels, the spiral channels show better flow characters. The increases of total refrigerant flow rate in spiral channels are mainly due to the increase in unblocked channels. Therefore, in order to further increase the refrigerant mass flow in the motor of the scroll compressor, the size of balance weight is reduced and the layout should be reasonable.

Acknowledgements: This paper is supported by the National Natural Science Foundation of China (item number: 52076162).

References

[1] Tian Z and Gu B 2019 Analyses of an integrated thermal management system for electric vehicles *Int. J. Energy Res.* **43** 5788–802
[2] Chen W, Xing Z, Shen J and He Z 2013 Thermal modeling for the motor in semi-hermetic screw refrigeration compressors under part-load conditions *Int. J. Refrig.* **36** 1874–82
[3] Huang Z, Marquez F, Alakula M and Yuan J 2012 Characterization and application of forced cooling channels for traction motors in HEVs *Proceedings - 2012 20th International Conference on Electrical Machines, ICEM 2012* pp 1212–8
[4] He Z, Xing Z, Chen W and Wang X 2013 Thermal and hydraulic analysis on the flow around the motor in semi-hermetic twin screw refrigeration compressors *Appl. Therm. Eng.* **58** 114–24
[5] Tang W, Wang L, Yao L and Lin J 2016 The Improvement of Motor Cooling Through Stator Profile Optimization Using CFD Analysis in Hermetic Scroll Compressors Danfoss (Tianjin) Ltd ., Danfoss Commercial Compressors 2 . ELECTRIC MOTOR LOSSES AND COOLING ANALYSIS *Int. Compress. Eng. Refrig. Air Cond. High Perform. Build. Conf.* 1–10
[6] Kai Y and Yaojing F 2010 Design of novel spiral magnetic poles and axial-cooling structure of outer-rotor PM torque motor *IEEE Transactions on Applied Superconductivity* vol 20 pp 838–41
[7] Ronnberg K and Beniakar M E 2018 Thermal Modelling of Totally Enclosed Fan Cooled motors *Proceedings - 2018 23rd International Conference on Electrical Machines, ICEM 2018* (Institute of Electrical and Electronics Engineers Inc.) pp 2619–25
[8] Nerg J, Rilla M and Pyrhönen J 2008 Thermal analysis of radial-flux electrical machines with a high power density *IEEE Trans. Ind. Electron.* **55** 3543–54
[9] Chen Q, Shao H, Huang J, Sun H and Xie J 2019 Analysis of Temperature Field and Water Cooling of Outer Rotor In-Wheel Motor for Electric Vehicle *IEEE Access* **7** 140142–51

Simulation of the Deformation and Contact of Scrolls in the Scroll Compressor

C Wang[1], S Zhang[1], J M Cheng[2], B W Lei[1], Y J Du[1], J H Wu[1*]

1 School of Energy and Power Engineering, Xi'an Jiaotong University, Xi'an, China

2 Guangdong Meizhi Compressor Co. Ltd., Shunde, China

*Corresponding author, E-mail: jhwu@xjtu.edu.cn

Abstract. The control of the clearances in a scroll compressor is crucial to ensure the reliability and efficiency of compressor, especially for the compressors without radial compliance mechanism. In this study, the fluid-solid-thermal simulation is conducted to get the temperature field, the pressure load distribution on scroll wraps. Then, the influences of the temperature, pressure and the inertial force on the deformation of scroll wraps are investigated. On these bases, the variation of gap size and potential contact between scroll wraps is analysed under various orbiting radii. It is shown that the deformation caused by temperature is more significant than the pressure and inertial force. The contact occurs even though orbiting follows the theoretical orbiting radius due to the positive radial deformation of scroll wrap and negative radial deformation of inner wall of fixed scroll.

1. Introduction

The scroll compressor is widely used in the refrigeration system for its high energy efficiency, quite operation and low vibration. The control of the clearances in a scroll compressor is crucial to ensure the reliability and efficiency of compressor. The axial and radial clearances between orbiting scroll and fixed scroll greatly affect the internal leakage between working chambers [1]. Generally, intermediate chamber for orbiting scroll is used to balance the axial gas force and to control the axial clearance. For the radial gaps, some scroll compressors use radial compliance mechanism to ensure the scroll wraps radially engaged while some scroll compressors use journal bearings without radial compliance mechanism [2], which possess higher request on the control on manufacturing tolerances and on scroll wraps deformation to avoid radial contact or oversized gaps between scroll wraps [3]. Additionally, in this case, the real orbiting radius of orbiting scroll is not only determined by the geometry but also influenced by the operating conditions because the radial clearance of journal bearings allows the orbiting scroll slightly deviate from theoretical orbiting radius.

The radial gaps are simultaneously influenced by the thermal deformation, pressure deformation, inertial force deformation as well as the motion of orbiting scroll (OS). Computational Fluid Dynamics (CFD) simulation is generally used to obtain the temperature and pressure distribution [4]. Recently, many useful technologies such as structural dynamic mesh and fluid-structural-interaction (FSI) are introduced into the CFD simulation of scroll compressor to get a faster and more accurate result. Wang et al. [5] developed a novel structured dynamic mesh generation method for the scroll compressors and indicated that the structured mesh is much more accurate than the unstructured mesh. Hesse et al. [6] developed a software to generate dynamic mesh for scroll machines. Their simulation results show good agreements with experiments. Sun et al. [7] presented the sharp decrease of pressure and temperature when leaked fluid flowed through the radial gaps with dynamic grids. Hsieh et al. [8] simulated the internal flow fields, thermal deformation of an oil-free scroll vacuum pump and compressor with fluid-solid-thermal coupling analysis. In conclusion, plenty of research has been done on the temperature-distribution on scroll wraps and its influence on the axial thermal deformation. Furthermore, few devotions have been made in the radial gaps and the possible contact between OS and fixed scroll (FS) under multi-field load.

Based on previous studies, this paper mainly focuses on the transient radial deformation of scroll wraps and possible contact between scroll wraps. The temperature and pressure fields are firstly obtained by the fluid-solid-thermal coupling simulation of compression chamber and scrolls. Then, temperature, pressure and inertial force loads are applied to scroll walls with one-way coupling method, where both the axial and radial deformation distribution of scroll wraps are determined. Moreover, the variation of orbiting radius of OS is taken into consideration to investigate the variation of radial gap and possible contact behavior between OS and FS wraps with finite element analysis (FEA).

2. Model and methodology

2.1. Geometry

The schematic of the scroll compressor is presented in Figure 1. The schematic of the scroll compressor is show in figure 1. In particular, models for fluid and solid domains are simplified in this paper. For the fluid domain: the working chambers, inlet, outlets (in order to avoid over-compression, eight relief valves are installed near the main discharge valve), upper-axial gap and lower-axial gap are considered while the intermediate chamber and high-pressure discharge chamber are ignored in the CFD simulation. For the solid domain, only the models of FS and OS is established in the fluid-solid-thermal coupling simulation and in the FEA simulation, while the shell, crank shaft and the main frame are not taken into consideration. The scroll wraps of OS and FS enclosed several cylindrical working chambers.Specific parameters of the scroll compressor are listed in Table 1. The materials properties are the same as those of the real scroll parts, both the OS or FS is made by HT300.

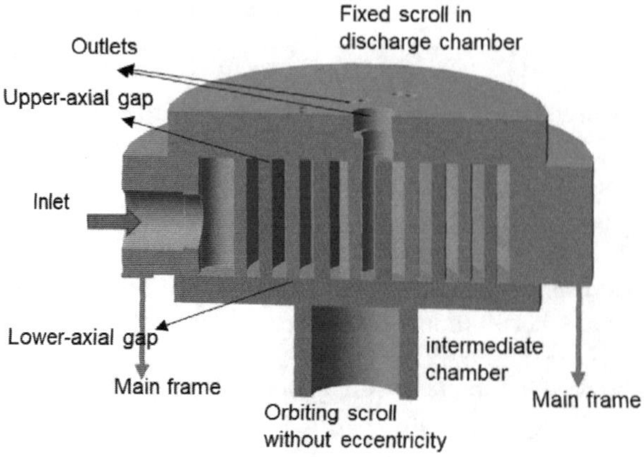

Figure 1 Schematic of the scroll compressor

Table 1 the parameters of the scroll compressor

Parameter	Value
Type of scroll profile	algebraic spiral
Height of scroll wrap [mm]	41.4
Stroke volume [mL/rev]	96
Theoretical orbiting radius [mm]	4.52
Numbers of relief valves	8

Refrigerant	R410a
Suction pressure – Absolute [MPa]	0.998
Discharge pressure –Absolute [MPa]	3.35
Designed intermediate pressure – Absolute [MPa]	1.4
Suction temperature [°C]	18
Rotational speed [rpm]	1800 to 7200
Mass of orbiting scroll [kg]	1.50

2.2. *CFD model*

The commercial software ANSYS is used to solve the complex fluid-solid-thermal coupling simulation. As Figure 2 shows, the three-dimensional grids for inlet, outlets, upper axial gaps are stationary region, while the dynamic mesh strategy is used for working chambers and lower axial gap that fixed with OS. The radial clearance of scroll wraps is set as low as 20 μm to avoid negative volume and to reduce the leakage between chambers in present work. Additionally, axial gaps are also included in the fluid domain and its height is 10 μm, where the refrigerant could radially leak to low-pressure chamber from high-pressure working chamber. Moreover, a pressure inlet boundary at is applied to the outside of lower-axial gap to realize the effect of the intermediate pressure chamber.

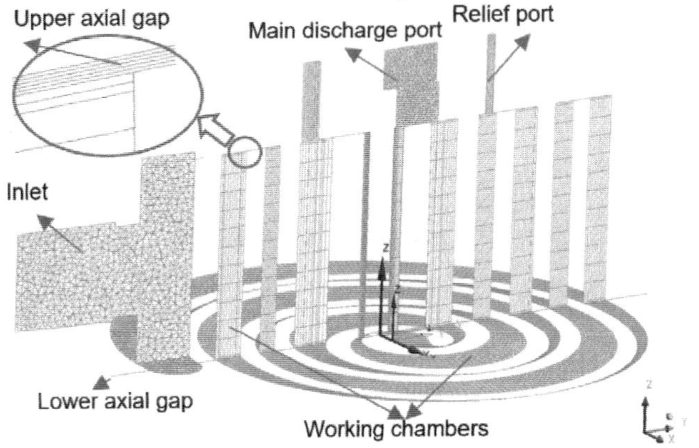

Figure 2. Illustration of the grids for fluid domain

Groups of grids are tested as a grid-independence check, where the total number of fluid domains varies from 300,000 to 1,000,000 with various layers in radial and axial directions for working chambers. The results of volume efficiency of the scroll compressor stand at about 84.8% under given operating condition, showing no significant changes with different grid densities. Thus, about 500,000 of total number of fluid domain elements is used in the fluid-solid-thermal simulation, taking the computational cost and performance into consideration.

The solid domain is composed of OS and FS with their discs respectively. The motion of OS is constrained properly as that in the reality. Fluid-structure interfaces (FSI) are established to handle the heat transfer between fluid and solid. Thus, the FSI boundary is employed for scroll wraps and discs of FS and OS as well as the inner wall of FS. An assumption is made that the temperature is fixed at 88 °C for outer side of the FS because the average discharge temperature is 88 °C. The adiabatic boundary is used for outer wall of OS. The grid-independence check for solid domain is not done and the total number elements is 100,000.

2.3. *FEA model*

Since the temperature and pressure distribution of FS and OS are obtained, FEA simulation is firstly done to investigate the temperature, pressure and inertial force deformation respectively. The temperature and pressure distributions are exported as external date at each timestep and then import into structural analysis. The results of CFD simulation are imported into FEA module respectively. Additional discharge pressure boundaries are manually applied at the outer walls of FS and OS, because they are not given in the results of CFD simulation. The inertial force of scroll wrap of OS is calculated by the rotational speed and it is manually added too.

As the main frame, the crankshaft and the shell are ignored to reduce grid size, additional boundary conditions are needed for the OS and FS. In the deformation analyses, all the degree-of-freedom of FS is constrained at the red solid lines in Figure 3, while the axial and radial displacement of OS is fixed at zero at the red solid and dot lines. The influences of temperature, pressure and inertial force are calculated respectively. Then, in contact analysis, the multi-field results are imported simultaneously. The radial displacement at dot lines of OS is set according to the eccentricity, which is decided by the shaft angle (θs) and the orbiting radius of OS (Ror), at the red dot line to simulate the drive of crankshaft. Specifically, only the contact between side walls of scroll wraps is considered in the contact analysis.

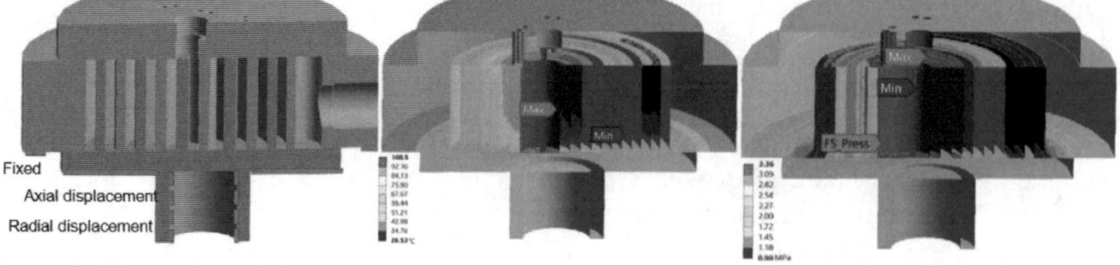

Figure 3. Illustration of displacement boundary condition and imported temperature and pressure boundary conditions from CFD results for the FEA simulation

3. Results and discussions

3.1. CFD results

The gas force on OS could be obtained by integrating the pressure on the scroll wrap walls and the discs. The forces with θs are presented in Figure 4. Lateral forces consist of radial force (Fr) and tangential force (Ft) in cylindrical coordinate (and Fx and Fy in cartesian coordinate). The value of Fr nears constant, standing at about -400 N, and the Ft fluctuates from 4650 N to 5200 N in a cycle. However, the Fz is much large than lateral force, almost reaching -19,500 N, which needs to be balanced by the force of intermediate pressure chamber. It should also be noticed that the Fr is larger than the eccentric force (240 N) of OS when rotational speed is 1800 rpm, which could lead to the separating tendency between OS and FS. Fortunately, the limited clearances of journal bearings could provide extra constrain for OS to avoid radial clearances out of control.

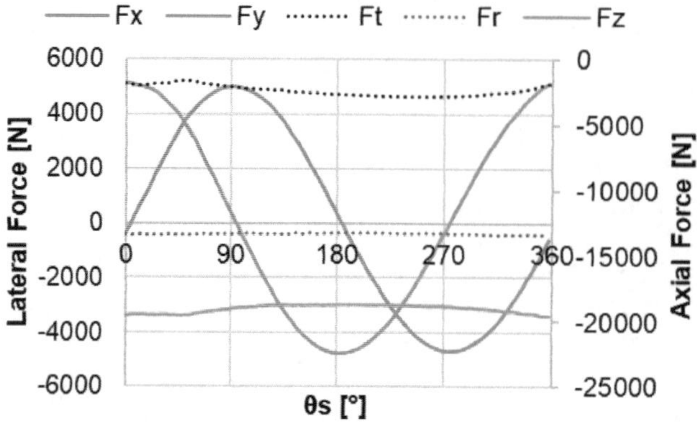

Figure 4: Gas forces on orbiting scroll

Since asymmetrical chamber design is used, the pressure variations are also asymmetrical for different working chambers as Figure 5 shows. As the compression angle (θc, θc = θs-137°) for outer of OS with inner of FS is 180° earlier than that of inner of OS with outer of FS, the pressure of working chamber formed by them is higher at the same θs. It should be also pointed out that the pressure distribution on the lower axial gap is affected by the working chamber and the intermediate chamber. Similarly, the upper axial gap is also affected by the discharge ports, as the pressure remains unchanged when the scroll wrap hides the relief ports.

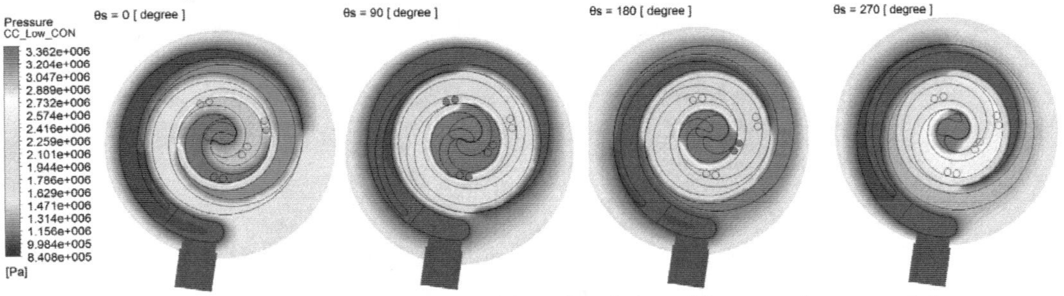

Figure 5. Pressure distribution on working chamber and lower and upper axial gaps

The temperature distribution on the middle cross-section of working chamber, OS and FS is shown in Figure 6. In a single chamber, the uneven temperature distribution is caused by the leakage high-temperature fluid and heat transfer of solid. The temperature at center of scroll wraps is the highest, reaching 110 °C, and it gradually decreases along the wrap. The temperature distribution of FS is greatly affected by the environmental temperature and the minimum temperature of FS is about 45 °C, while the minimum temperature of OS is 18 °C. Overall, the variation of temperature distribution on solid is insignificant in a cycle.

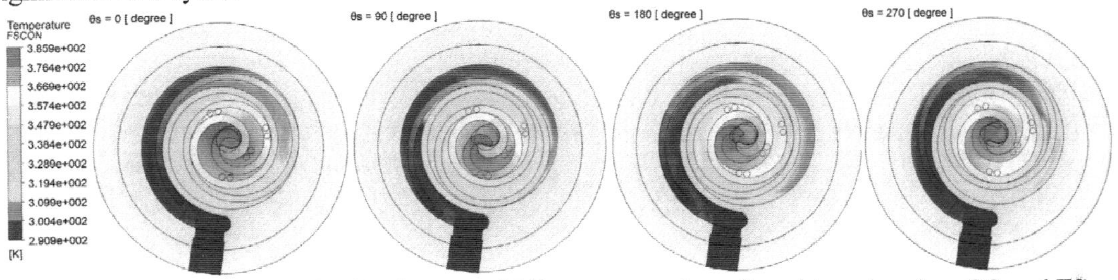

Figure 6. Temperature distribution on middle cross-section of working chamber, OS and FS

3.2. *FEA results*

Since the temperature and pressure fields are obtained by CFD simulation, the transient results are exported and imported into structural module as external loads. Other boundary conditions are also applied as the section 2.3 described. In this section, a cylindrical coordination origin in the center of FS and OS is established to illustrate the radial and the axial deformation of solid domain.

3.2.1. Temperature deformation. Figure 7 presents the radial (dR) and axial (dZ) deformation of FS and OS respectively. Generally, the scroll wraps show a tendency to expand in radial direction and positive deformation tendency in axial direction. The maximum radial deformation of FS wrap occurs near the inlet, reaching about 30 μm, which is as about twice larger as that of OS wrap. However, the negative dR of FS inner wall at bottom reaches -22 μm, which is different from the wraps. The maximum axial deformation of scroll wrap occurs at the center of OS and it gradually decreases at the tail of scroll wrap, while the dZ of FS is quite uniform along its wrap. The dZ of discs shows opposite trend. The dZ of OS disc remains near zero because the constrain is close. The maximum dZ of FS disc, being 85.8 μm, occurs at the center of FS and it decreases radially as the constrain is applied on the bottom surface in the FS. The transient deformation is not given as the variation of temperature loads on FS and OS nearly remains unchanged in a cycle as described in section 3.1.

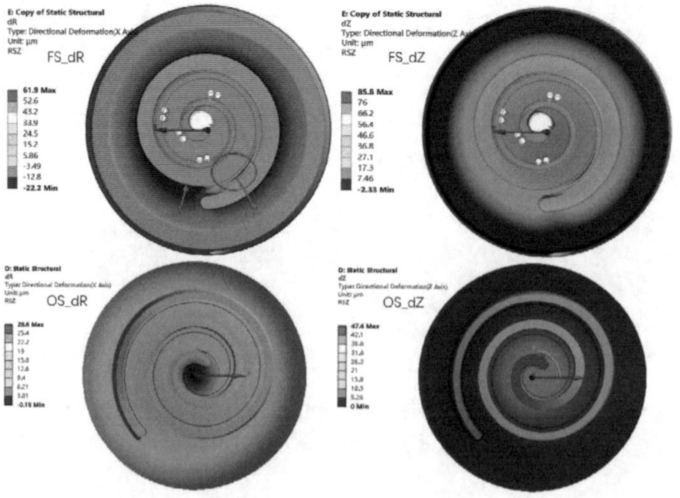

Figure 7. Temperature deformation of FS and OS in radial and axial directions

3.2.2. Pressure deformation. Figure 8 shows the deformations in radial and axial directions of FS and OS over the time respectively. The pressure deformation is slightly small than that caused by temperature load, especially for the dZ. The largest dR of scroll wrap is about 10 μm at the top due to the cantilever effect of wrap. Generally, positive and negative dR occur around the engage line of OS and FS. Taking the pressure distribution into consideration, positive dR is caused by the pressure difference between inside and outside of the wrap chambers. The negative dR could not only be influenced by the nearby deformed scroll wrap region, but also the pressure difference on symmetric chambers of inner and outer OS, which is caused by the phase difference of θc for theses chambers. Thus, both positive and negative dR occur in scroll wrap under pressure load. As the pressure distribution changes with the orbiting of OS, mirror distribution dR is obtained when θs is 360°. However, similar deformation tendency of both FS and OS could be concluded as those of 180°.

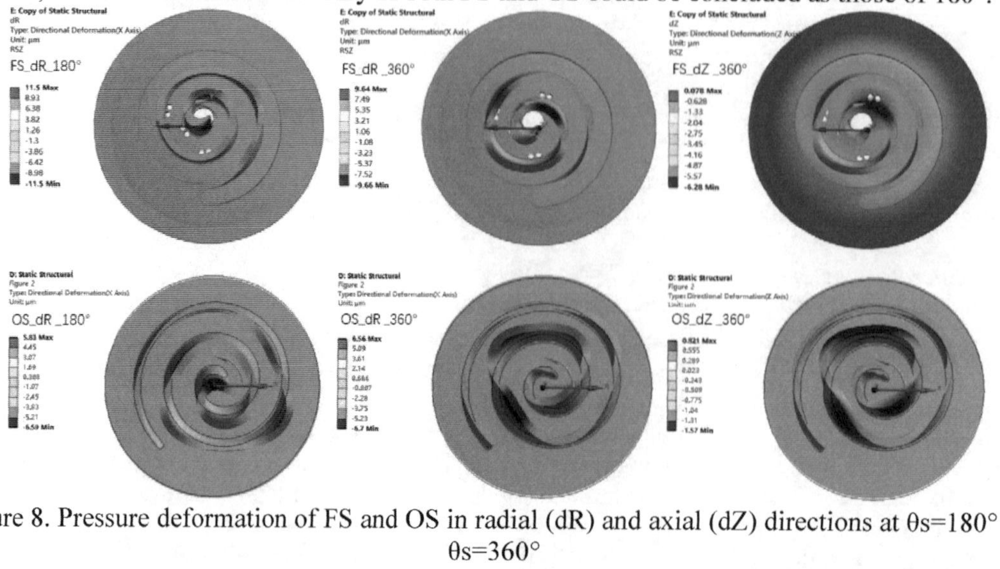

Figure 8. Pressure deformation of FS and OS in radial (dR) and axial (dZ) directions at θs=180° and θs=360°

3.2.3. Inertial force deformation. Figure 9 illustrates the total deformation of OS driven by inertial force at various θs when rotational speed is 1800 rpm. It seems that the deformation mainly occurs at the tail region of scroll wrap. The value of deformation caused by inertial force is fairly smaller than those caused by temperature and pressure load as the rotational speed is small.

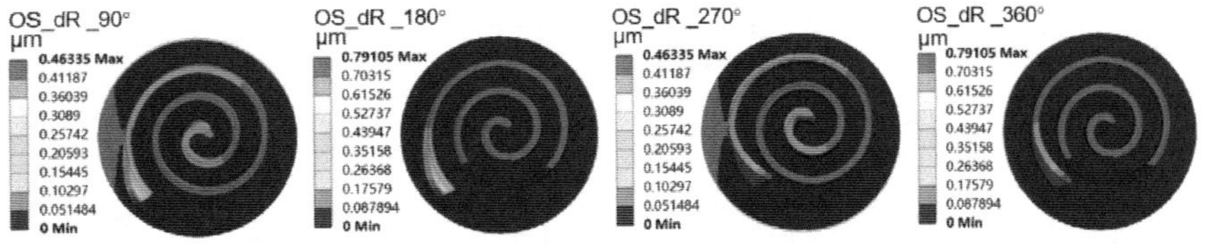

Figure 9. Inertial force deformation of OS when rotational speed is 1800 rpm

3.2.4. Coupling deformation. The coupling deformation under temperature, pressure and inertial loads when θs=360° is shown in Figure 10. It should be noticed that only the loads are applied while the radial displacement boundary of OS is set to zero for a clear view of the deformation distribution. As have been noticed in 3.2.1, the negative dR of inner wall of FS attracts our attention again, while the positive dR of scroll wrap pairs are relatively uniform in the cross-section. The negative dR of inner wall of FS as well as the positive dR of scroll wrap could provide extra constrain to the motion of OS, even though the OS have not reached theoretical orbiting radius. Though the deformation of OS wrap tends to reduce the axial gap, the maximum deformation of FS disc reaches 80 µm that is much larger than that of OS. Thus, the axial gap is increased under multi-field loads. It is also shown the uneven axial gap distributes along the wraps. The axial gap decreases from the center region to the outer region of scroll wraps.

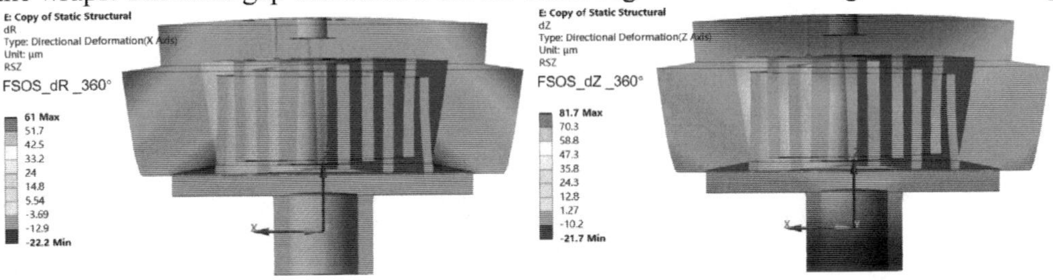

Figure 10. Coupling deformation of FS and OS

3.2.5. Contact analysis. The results of contact analysis without or with multi-field loads on solid are shown in Figure 11. When loads and theoretical Ror are applied, the contact only occurs at the bottom outer wall of OS and inner wall of FS on the suction side of scroll wraps, which agrees well with the deformation results discussed in section 3.2.4. The maximum contact pressure is 485 MPa. Moreover, since the contact occurs on limited region of engaging lines, the sealing of chambers is not geometrically granted now. However, it is noted that, in reality, the lubricating oil in the working chamber could improve the sealing between chambers by filling in the gaps and reduce the contact pressure by hydrodynamic effect.

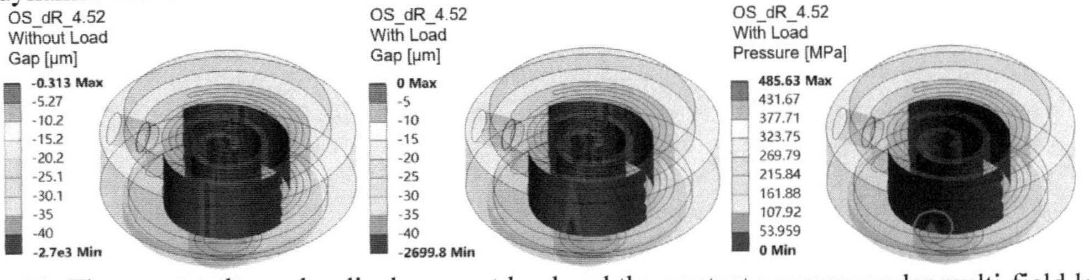

Figure 11. The gap results under displacement load and the contact pressure under multi-field loads at Ror =4.52 mm and θs =360°

The variation of contact force with the change of Ror and θs is shown in Figure 12. The orientation of contact forces is normal to the rotation direction. The average contact force is 9100 N in a cycle when Ror keeps the theoretical value, being 4.52 mm. Since the eccentric force of OS is much smaller than 9100 N, the deformation and contact of wraps could also cause the deviated motion of OS as well as the tilting crankshaft. As the OS and FS have separating tendency by the contact force, the full sealing of the engage line could not be secured. Additionally, potential reliability risks could be raised due to the misalignment of shaft in journal bearings on this condition. There is a drastic decrease of contact force to 400 N when θs increase to 120°, where then the ending profiles for outer of OS and inner of FS start to engage. The drastic decrease might be caused by the positive dR of FS wrap near suction port as Figure 7 shows. Thus, a shock load could occur on the tail of OS wrap during its engagement according to the simulation results, especially at a higher rotational speed.

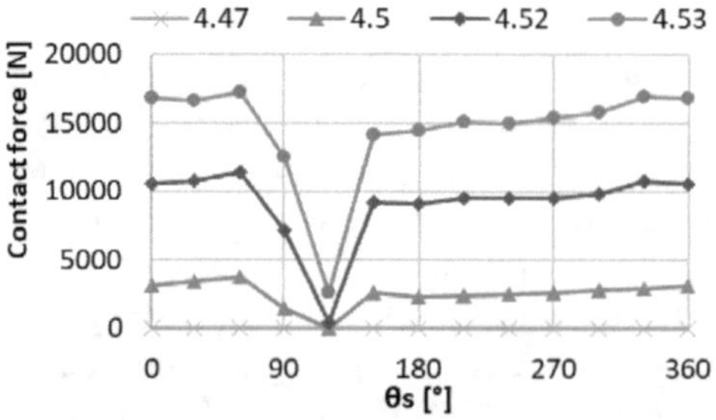

Figure 12. The variation of contact force with different Ror in a cycle

4. Conclusion

In this paper, the fluid-solid-thermal CFD simulation is conducted to obtain the pressure and temperature distributions of an asymmetric algebraic scroll compressor. Then the multi-field loads are imported into fixed scroll and orbiting scroll in the FEA analysis to get the deformation results. At last, the contact behavior under various orbiting radius of OS is analyzed. Following conclusion can be made:

- The average radial gas force exerted on OS is -400 N while the eccentric force of OS at 1800 rpm is about 240 N. The larger radial gas force could lead to the separating tendency of OS to FS at lower rotational speed.
- Additional axial gap is caused by the axial deformation of scroll wraps and discs, and its height decreases from the center region to the outer region of scroll wraps.
- In radial direction, the deformation of solid causes the 'expanded scroll wraps' and 'shrinked FS inner wall', which lead to the contact behavior even though OS follows the theoretical orbiting radius. In this case, the radial gap could be increased and there is a drastic decrease of contact force when θs increase to 120°.

Moreover, several boundary conditions could be improved in current work to obtain more precise results, including the temperature distribution of outer wall of FS, the axial constrain of OS and radial motion of OS driven by crankshaft. Further research would be done in these aspects in the future.Firstly, the temperature distribution of outer wall of FS should be also obtained with fluid-solid-thermal CFD analysis or experiment. Besides, the OS is not constrained in Z-direction in reality but pushed by the back pressures of OS. The motion of OS in R-direction is coupled with the motion of crankshaft.

Acknowledgement

This work was financially supported by the Guangdong Meizhi Compressor Co. Ltd. in China, for whom the authors would like to express their sincere gratitude.

References

[1] J. W. Bush and J. P. Elson, "Scroll compressor design criteria for residential air conditioning and heat pump applications. Part I: Mechanics.," in International Compressor Engineering Conference School, 1988, p. 605.
[2] S. Ahn et al., "Development of Compact and High-efficient Scroll Compressor with Novel Bearing Structure," in International Compressor Engineering Conference, 2016, p. 2483.
[3] D. Gross and D. Genevois, "Scroll Wrap Additional Stress in Variable Speed Application," in International Compressor Engineering Conference, 2018, p. 2584.

[4] M. Morini, C. Pavan, M. Pinelli, E. Romito, and A. Suman, "Analysis of a scroll machine for micro ORC applications by means of a RE / CFD methodology," Appl. Therm. Eng., vol. 80, pp. 132–140, 2015, doi: 10.1016/j.applthermaleng.2015.01.039.
[5] J. Wang, Y. Song, Q. Li, and D. Zhang, "Novel structured dynamic mesh generation for CFD analysis of scroll compressors," Proc. Inst. Mech. Eng. Part A J. Power Energy, vol. 229, no. 8, pp. 1007–1018, Dec. 2015, doi: 10.1177/0957650915601927.
[6] J. Hesse, R. Andres, C. F. X. Berlin, and S. Gmbh, "Cfd Simulation of a Dry Scroll Vacuum Pump Including Leakage Flows," Int. Compress. Eng. Refrig. Air Cond. High Perform. Build. Conf., pp. 1–10, 2016.
[7] S. Sun, K. Wu, P. Guo, and J. Yan, "Analysis of the three-dimensional transient flow in a scroll refrigeration compressor," Appl. Therm. Eng., vol. 127, no. 1, pp. 1086–1094, Dec. 2017, doi: 10.1016/j.applthermaleng.2017.08.115.
[8] S. Hsieh, J. Zhang, S. Chou, and C. Huang, "A Comparison of Thermal Deformation of Scroll Profiles inside Oil-free Scroll Vacuum Pump and Compressor via CAE / CFD Analysis," Int. Compress. Eng. Conf. Purdue, pp. 1–8, 2018.

Optimization of the tip profile of orbiting scroll in an asymmetry suction chamber scroll compressor

SUN Shuaihui[1,2], Wang Zhe[1,2], GUO Pengcheng[1,2], ZHANG bo[3], MAO Zhenkai[3]

1. Xi'an University of Technology, Xi'an, Shaanxi, 710048, China
2. State Key Laboratory of Eco-hydraulics in Northwest Arid Region of China, Xi'an, Shaanxi, 710048, China
3. Power China Northwest Engineering Corporation Limited, Xi'an, 710065, Shaanxi, PR China.

E-mail address: shs@xaut.edu.cn

Abstract: The scroll compressors with asymmetrical suction chamber have large discharge loss, as the pressure in outside chamber is far larger than the discharge pressure when the discharge process begins. It is necessary to optimize the scroll profile around the discharge port to reduce the discharge loss. The theoretical method is to redesign the tip profile to obtain the same volume ratio of outside working chamber with the inner working chamber. However, as it is difficult to predict the discharge loss through the flank gap in discharge chamber, the optimum design cannot be obtained. In this paper, we establish an optimization platform based on the CFD method. The profile of orbiting scroll tip is parametrized with CAD software, then the CFD calculation will be done and the performance parameters will be extracted. The MOST optimization algorithm is used to search the best performance parameters and control the parameterized value of scroll profile. After iterations, the best profile with the modification angle of 66.3° was obtained. Compared to the original model, the volumetric efficiency of the compressor increased by 4.11%; and the isentropic efficiency increased by 4.91%. Afterwards, the flow field are analysed to study the characteristics of the optimized scroll compressor. The results show that the backflow, the secondary vortex in discharge chamber and the discharge mass flow pulsation were reduced, which led to the decrease of the discharge loss and the improvement of the isentropic efficiency.

1. Introduction

Scroll compressors usually have several pairs of crescent-shaped working chambers which are formed by the engagement between the orbiting scroll and the fixed scroll. As the orbiting scroll orbits, the volume of each pair of working chamber changes, and a series of continuous processes of suction, compression and discharge happen for pressurizing the gas. In a hermetical scroll refrigeration compressor with a high-pressure shell, only one inlet port is designed to reduce the costs. As shown in Figure 1, during the suction process of symmetric suction chamber compressor (SSC), the fluid which flows into the suction chamber Suc_B has to bypass the suction flow passage. If the flow passage is not designed well, the flow loss in the passage is large, and even the fluid is blocked by the scroll vane of orbiting scroll[1], which results in the high pulsation of mass flow rate and the low volumetric efficiency.

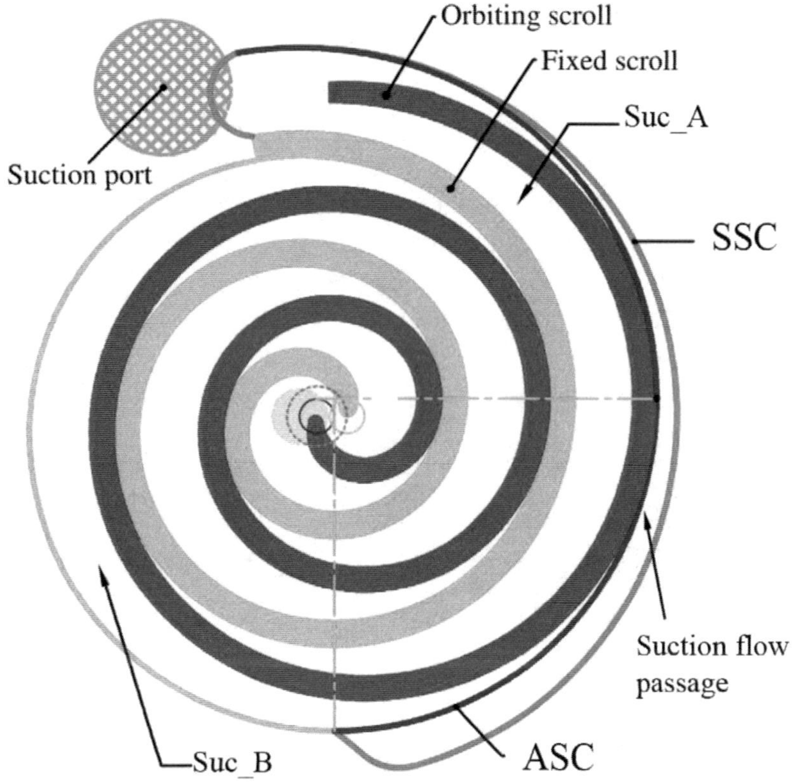

Figure 1. Schematic diagram of SSC and ASC

In order to solve the problem, the asymmetric suction chamber compressors(ASC) were designed by extending the end angle of the fixed scroll profile by 180°. In this case, the original suction flow passage was changed into the suction chamber Suc_B, which was different from the suction chamber Suc_A in Figure 1[2]. The termination angle of fixed scroll profile in asymmetric suction chamber compressor(ASC) is larger than that of orbiting scroll, thereby the suction chamber Suc_A is asymmetrical with the Suc_B, as shown in Figure1. Suefuji et al.[3] proposed the patent of ASC firstly, where the terminal angle of fixed scroll was about 60° to 120° larger than that of orbiting scroll. Qiao et al.[4] calculated the force and moment on the orbiting scroll in an asymmetric scroll compressor and pointed out the force and moment increased while their pulsation amplitudes decreased after the asymmetric structure was used. Shaffer et al.[5] developed a geometric model for an ASC using a control volume approach and proposed the calculation equations for the area of working chambers under different crank angles. The authors[6]established a transient numerical simulation model of scroll compressors with different flow passage and analyze the suction process of these compressors.

The results indicated that the suction mass flow rate increased and the suction pulsation was reduced significantly as the asymmetric suction chamber was used; but the discharge loss increased due to the asymmetric pressure distribution between working chambers. Cao et al.[7] modified the scroll profile around the discharge port of an ASC and calculated its performance; the results indicated that the pressure difference between two working chambers and the discharge losses were reduced after the modification. Wang[8] modified the orbiting scroll profile to ensure the volume ratio of the outside working chamber is equal to that of the inner working chamber. The isentropic efficiency increased but the under-compression and backflow happened. Therefore, the modification angle was not the optimum one. Recently, the optimization technology based on CFD method had been applied to modify the shape of blade of centrifugal pump and wind turbine[8-9], and obtained good modification schemes to improve the efficiency or reliability of these machines. Cavazzini combined the scroll compressor parameterized model, CFD method and optimization algorithm in the MATLAB to study the impact of different parameters on compressor efficiency [10].

Therefore, this paper established the optimization model based on the CFD method and ISIGHT software. By parameterizing the orbiting scroll tooth profile, the fluid domain of the discharge chamber was modified after the performance data was obtained by the CFD simulation. The MOST algorithm was used for searching the best modification angle of the profile of orbiting scroll. Finally, the flow field of the scroll compressor with the best modification angle was analyzed.

2. Mathematical Model

2.1. Physical Model

A scroll refrigeration compressor was taken as the research object in this paper. The main parameters of the scroll compressor are shown in Table 1. Figure 2. shows the modification angle of the orbiting scroll according to the equal built-in volume ratio[11]. In order to eliminate the working fluid backflow and improve the efficiency, the modification angle would be optimized based on CFD and optimization model.

Table 1. The main parameters of the original scroll refrigeration compressor

Parameters	Value
Thickness of scroll vane /mm	3.6
Profile height /mm	4
Orbiting radius ρ /mm	4.5
Scroll warp height H/mm	36
Base circle radius a /mm	2.578
Initial angle of involute α /rad	0.698
Determination angle of fixed scroll /rad	21.99
Determination angle of orbiting scroll /rad	18.85
flank clearance /mm	0.030
Axial clearance /mm	0.015

2.2. Numerical simulation model

The numerical simulation model was established using commercial software PumpLinx[6]. The hexahedral grids were generated in working chambers. Ten layers of grids were generated in the flank clearance and five layers were generated in the axial clearance. As the orbiting scroll rotated, all grids were regenerated at every timestep during the numerical simulation. The turbulence model was adopted RNG k-e model. The adiabatic wall was used for wall boundary conditions. Refrigerant R22 was used as the working fluid and its real properties were obtained from NIST Chemistry WebBook[12]. The R22 property tables including density, enthalpy, thermal conductivity and viscosity were added into PumpLinx.

The pressures of inlet and outlet were set to be 0.627MPa and 2.146MPa, respectively. The inlet temperature was 307.7K. The rotational speed was set to be 2880 $r \cdot min^{-1}$. When the averaged mass flow rate difference between the inlet and outlet less than 1% was considered convergence. The accuracy of the numerical model was verified by the experimental results. The comparisons results can be found in Ref. [6]

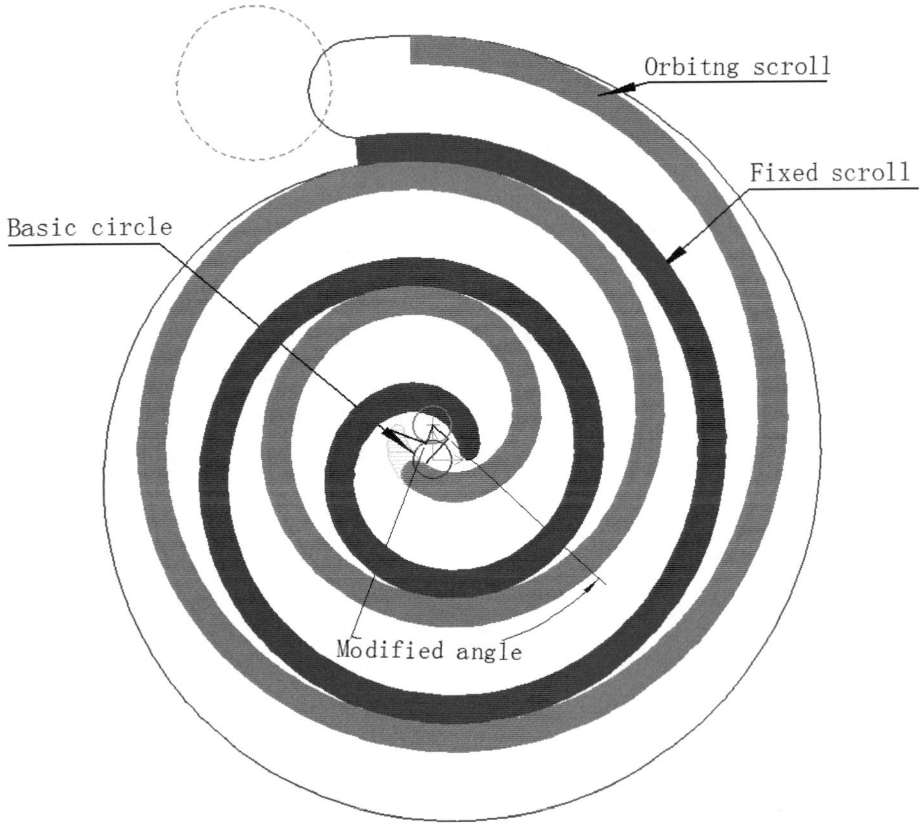

Figure 2. Schematic diagram of modified profile of orbiting scroll tip

2.3. Optimization procedure

The optimization program was built through the isight optimization platform which contained the following parts[13], as shown in Figure 3. Firstly, the modified angle of the profile of orbiting scroll tip was considered as the input parameter of the optimized model through the modelling software, which ranged from 65 to 75°. By modifying this parameter, the original model can be modified and saved as an STL file. Secondly, the CFD software Pumplinx was integrated into the ISIGHT platform to import the STL file and create the rotor fluid domain grid which was added to the case file of the original model. The inlet and outlet fluid domain grids were not modified, and the calculation of the new case file was controlled through the batch file. Afterwards, the CFD calculation result file was imported into EXCEL to obtain the parameters required to calculate the objective function value. Then, the optimization algorithm determines the input parameter value for the next calculation, and the iteration started again. Until the best isentropic efficiency was obtained, the iteration stopped and the optimum results were recorded.

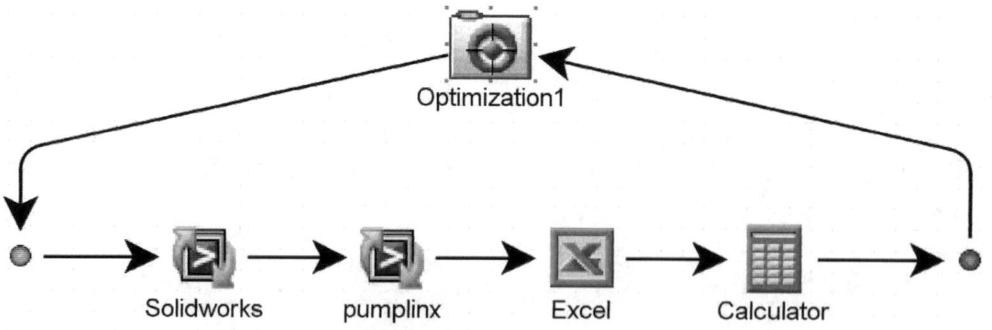

Figure 3. Schematic diagram of optimization procedure

3. Results and Discussion

In this optimization, the modification angle of the profile of orbiting scroll tip ranged from 65° to 75°. The initial correction angel was 75°. The isentropic efficiency was the target of optimization. After 9 iterations of calculation, the optimal modification angle was determined to be 66.3°. Table2 shows the time-average performance parameters of the compressor before and after optimization. The volumetric efficiency of the optimized model was improved, and the exhaust temperature is reduced by 6.57K. The adiabatic efficiency of the optimized model is 60.43%, which is 4.91% higher than the original model. This is because the optimized model changed the under-compression state of the prototype compressor's working chamber and improved the compressor's volumetric efficiency and adiabatic efficiency.

Table 2. Performance of two compressor

	suction mass flow rate[kg·s^{-1}]	volumetric efficiency[%]		isentropic efficiency[%]		discharge temperature[K]
		Calculated value	Increased value	Calculated value	Increased value	
Original model(75°)	0.08854	92.35		55.52		384.36
Optimized model(66.3°)	0.09354	96.56	4.11	60.43	4.91	377.79

Figure 4 shows the variation of discharge mass flow rate of the original model and the optimized model in one period. The 0 degree was defined when the suction process in Suc_A was finished. Both diagrams fluctuated with the crank angle and reach the minimum value at the crank angle of 54° and 46°, respectively. As the profile of orbiting scroll was cut, the outer working chamber Com_2B(Figure 5) firstly connected to the discharge chamber. At this moment, as the pressure in working chamber was lower than discharge pressure, some gas flowed back into the working chamber, which led to the decreases of discharge mass flow rate. In the original model, the mass flow rate became minus at the crank angle of 54°, which meant that the backflow was severe and led to large flow loss. After optimization, the minimum mass flow rate happened at about 46° and kept positive. Therefore, the discharge mass flow rate of the optimized model was larger than the original one, which led to the improvement of the volumetric efficiency.

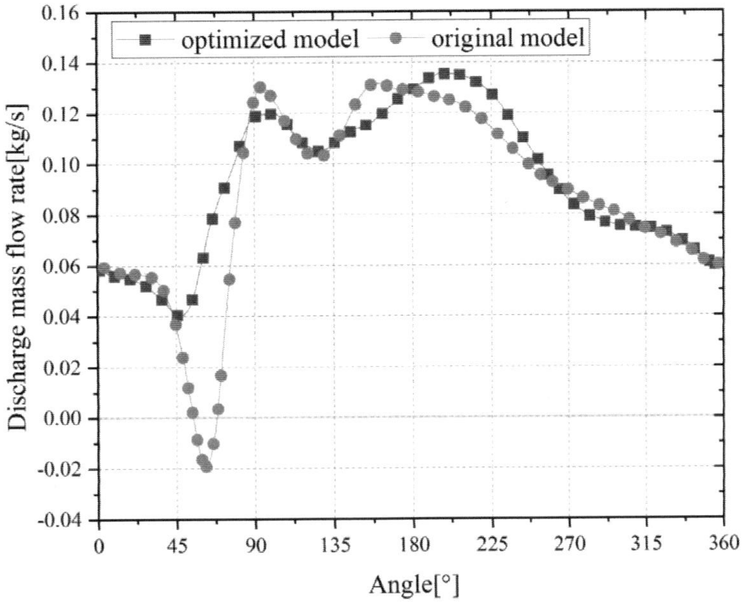

Figure 4. Discharge mass flow rate of different models

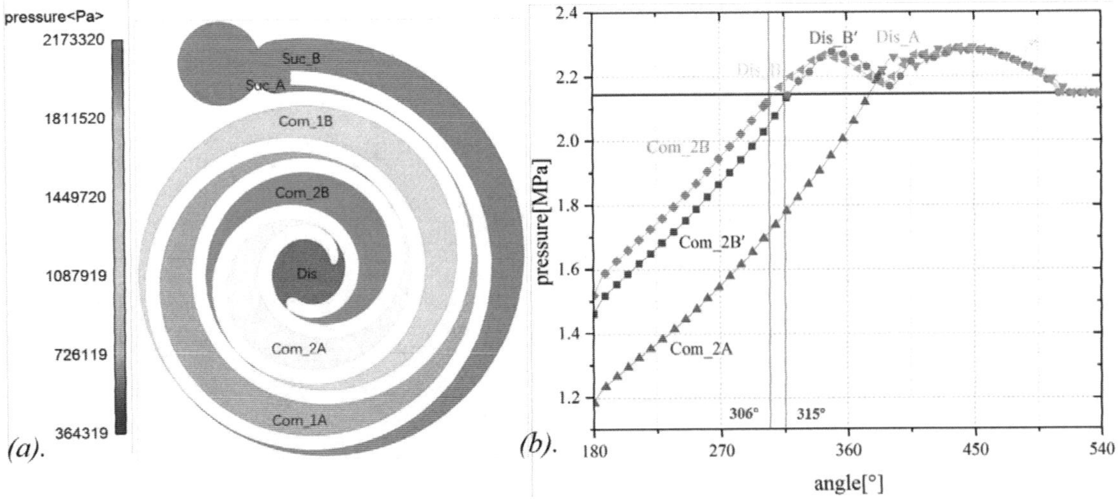

Figure 5. Pressure distribution *(a)* and p-θ diagrams with crank angle (b)in different compressor

Figure 5 shows the pressure distribution of the optimized model and the *p-θ* diagrams for both models during discharge process. The simulation results show that the pressure in one working chamber is almost uniform, but it is different from the other corresponding chambers, such as the Com_1A and Com_1B. This is caused by the asymmetric suction chamber Suc_B which is 180° earlier than the Suc_A; the gas in Suc_B and Com_B was compressed 180° earlier and their pressure was higher. In Figure5(b), the 0 degree was defined when the suction process in Suc_A was finished. Because the process in Suc_B was 180 degree earlier than that in Suc_A, the compression chamber Com_1B become Com_2B at 180degree in Figure5(b). Hence, the Figure5(b) actually shows the p-θ diagrams in the working chambers Com_2A, Com_2B and Com_2B'(optimized model) in a period. In the optimized model, the modification angle changed from 75° to 66.3°, which led to the delay of the discharge angle of Com_2B'. As shown in Figure5(b), the discharge angle of Com_2B' was 315° and the pressure in Com_2B' reached the discharge pressure at the crank angle. However, the discharge angle of Com_2B(the original model) was 306° when the pressure in Com_2B was lower than the discharge pressure, which resulted in the under-compression and the backflow. And the discharge loss of the optimized model was lower than that of the original model. The pressure in Com_2B was higher than that of Com_2B', so the leakage mass flow rate from the Com_2B was larger than that in Com_2B', and the volumetric efficiency of the original model was lower.

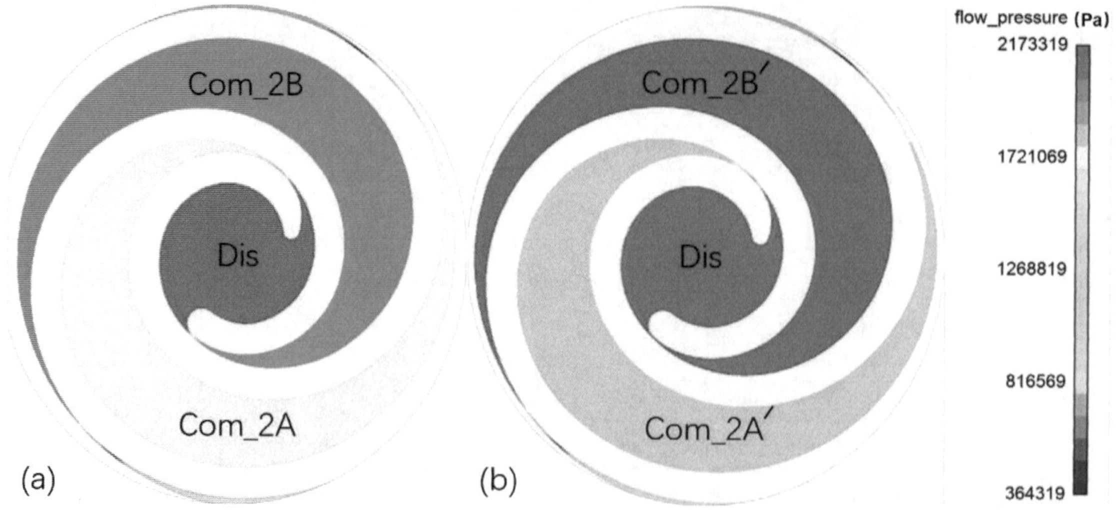

Figure 6. Pressure distribution of discharge and compression chamber in different model

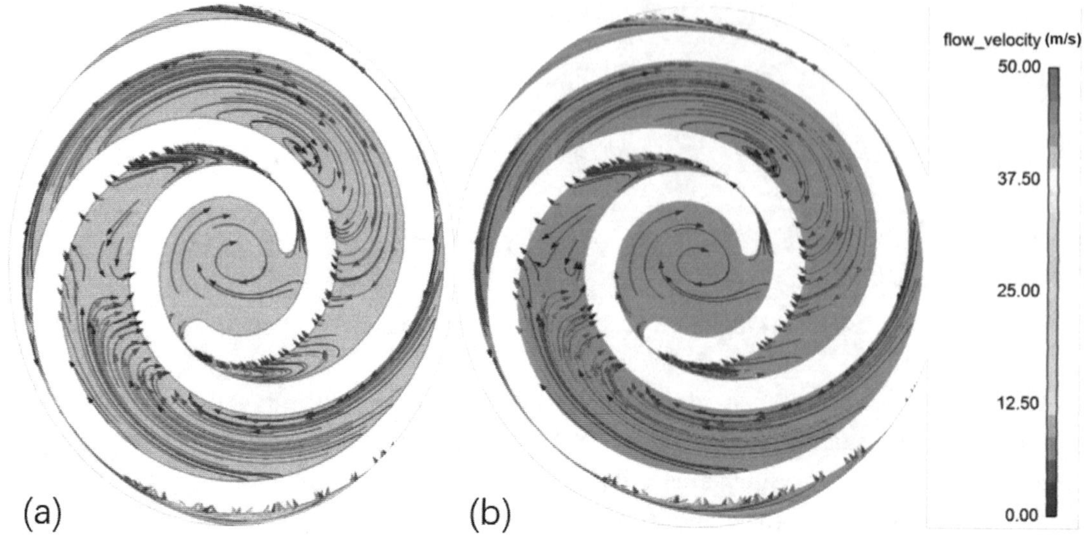

Figure 7. Velocity field of discharge chambers(a) original model (b) optimized model

Figure 6 shows the pressure distribution in the discharge chamber of the original model(a) and the optimized model(b) When the crank angle is 126°, the compression chamber Com_2B of the original model begins to connect with the discharge chamber. The compression process of the compression chamber Com_2A has not yet finished and its pressure is far lower than the discharge pressure. After optimization, the pressure in Com_2B' increases due to longer compression time.
Figure 7 shows the velocity field of discharge chamber of the two models at the beginning of the discharge process. Because the pressure in the discharge chamber is is higher than the compression chamber, the working fluid flows back into the compression chamber in the form of gap jets. The backflow has high velocity and interacts with the main flow in the working chamber to form a complex secondary vortex. As shown in Figure 7(b), the velocity of backflow in the optimized compressor is lower than that of the original flow, and the flow loss is also lower.

4. Conclusions

In order to reduce the discharge loss of scroll compressor with the asymmetric suction chamber and improve its isentropic efficiency, this paper established the optimization model based on the CFD method using the ISIGHT software, and optimized the orbiting scroll tip profile by modifying the initial angle. The best modification angle was obtained and the results were analysed and compared. The following conclusions were drawn:
(1) After optimization, the modification angle of the tip profile of orbiting scroll was reduced from 75° to 66.3°, the volumetric efficiency of the compressor increased by 4.11%; and the isentropic efficiency increased by 4.91%; the optimization effect was obvious.
(2) After optimization, the pulsation of discharge mass flow was significantly reduced, and the backflow in discharge pipe was reduced.
(3) In optimized compressor, the pressure in compression chamber Com_2B' at the beginning of discharge process was basically equal to the discharge pressure, which lead to the lower backflow rate and discharge loss.

Acknowledgments

This work was supported by the National Natural Science Foundation of China (51839010), Shaanxi Natural Science Basic Research Program-Yin Han Ji Wei Joint Fund(2021JLM-55), Xi 'an Science and Technology Project (2019217814GXRC014CG015-GXYD14.14), The Scientific Research Program of Engineering Research Centre of Clean Energy and Eco-hydraulics in Shaanxi Province(QNZX-2019-05, QNZX-2019-06) and The Youth Innovation Team of Shaanxi Universities(2020-29). All supports are acknowledged.

References

[1] Sun S, Wu K, Guo P and Yan J 2017, Analysis of the three-dimensional transient flow in a scroll refrigeration compressor [J]. *Applied Thermal Engineering*, 127: 1086-1094.
[2] Shuaihui SUN, Xianwen WANG, Pengcheng GUO, Xiaobo ZHENG, Jianguo YAN 2019, Optimization of Suction Chamber Structure in a Scroll Refrigeration Compressor [J]. *IOP Conference Series: Materials Science and Engineering,* 604: 012080.
[3] Suefuji K.S, Sendhu T.S, Arata T.S, 1990, Scroll-type fluid machine with different terminal end wrap angles[P]. *USA*.

[4] Qiao Z, Xu B, Xiong Z, et al. 1994,Tentative study of asymmetry wraps scroll compressor. *International Compressor Engineering Conference*, Purdue University, 983.

[5] Shaffer B.R, Groll E.A, 2013,Variable wall thickness scroll geometry modeling with use of a control volume approach. *International Journal of Refrigeration*, 36(7): 1809-1820.

[6] Sun S, Wang X, Guo P, et al. 2020, Investigation on the modifications of the suction flow passage in a scroll refrigeration compressor[J]. *Applied Thermal Engineering*, 170,115031.

[7] Cao C, Cui S, Wang J, et al 2018, Asymmetric Modified Method of the Scroll Wrap and the Research on Performance of a Novel Asynchronous Suction Scroll Compressor[J] *Compressor technology*, (06):6-11.

[8] Wang K, Liu H, Yuan S, et al. 2011, Automatic optimization of impeller meridional shape for centrifugal pumps based on CFD. *Transactions of the Chinese Society of Agricultural Engineering*; volume 27: 39-43(35).

[9] Zuo H, Liu C, Yang H, et al. 2016, Numerical Study on the Effect of Swept Blade on the Aerodynamic Performance of Wind Turbine at High Tip Speed Ratio. *Journal of Physics Conference Series*, 753: 102010.

[10] G. Cavazzini et al. 2020, CFD-based optimization of scroll compressor design and uncertainty quantification of the performance under geometrical variations[J]. *Energy*, 209

[11] Wang X. 2020, Numerical SimulationI OF Transient Flow and Investiogation on the Modifications of the Scution Folw Passage in a Scroll Refrigeration Compressor [D]. *Xi'an University of Technology*.

[12] National Institute of Standards and Technology(NIST), 2021, Data from NIST Standard ReferenceDatabase 69: NIST Chemistry WebBook, https://webbook.nist.gov/chemistry/fluid/.

[13] Lai Y, 2012. Isight parameter optimization theory and detailed examples[M]. *Beihang University Press*.

IOP Conf. Series: Materials Science and Engineering 1180 (2021) 012015 doi:10.1088/1757-899X/1180/1/012015

Flow computation inside a scroll compressor based on open-source code

E Fadiga[1], N Casari[1], B Angel[2], A Picavet[2]

[1] Department of Engineering, Università di Ferrara, Via Saragat 1, 44122 Ferrara, Italy
[2] Danfoss Commercial Compressors, Parc TECHNOLAND, 2 Allée Toscane, Zi du Champ Dolin, 69800 Saint-Priest, Francia

E-mail: ettore.fadiga@unife.it

Abstract. Computational fluid dynamics (CFD) is an essential tool for improving the performance of positive displacement machines, which are commonly adopted in vapour compression applications. The numerical analysis of such machines is characterized by several modelling challenges. Firstly, the relative motion of the wraps and the presence of particularly small gap sizes need to be considered during the implementation of a reliable meshing strategy. In addition to that, the operating conditions are close to the saturated-vapour line, requiring a real gas equation of state for the determination of fluid thermophysical properties.

In this work, a commercial scroll compressor has been simulated at three different operating conditions. The geometrical modelling, meshing strategy and compression process have been obtained thanks to a set of libraries built in OpenFOAM-v2006. The results have been compared with experimental data of the pressure evolution in the compression chambers. Furthermore, a complete fluid flow analysis is presented to assess the capabilities of the developed tool when applied to a compressor operating with R410a.

1. Introduction

Among positive displacement (PD) machines, the scroll compressor is a device largely employed in refrigeration systems and air conditioners since the 1980s. These machines have become one of the most commonly adopted compressors in such sectors, so it is important to study their dynamic behaviour to improve their overall performance.

The most common strategies to analyze scroll compressors are experimental campaigns [1–3], aided by theoretical studies [4, 5], semi-empirical modelling [6] and thermodynamic analyses [7, 8]. However, the flow inside positive displacement machines has a complex three-dimensional nature, especially in suction and discharge regions. Unfortunately, such flow features are either impossible or extremely costly to determine experimentally. Hence, Computational Fluid Dynamics (CFD) represents a suitable method to perform a complete analysis of the dynamic behaviour of positive displacement PD machines.

The majority of the works presented in the literature is based on the definition of a structured grid (in the scroll set), and the operation is simulated by means of pre-defined motion of the nodes. This method allows obtaining good accuracy reducing conservation errors, as extensively described in section 2. However, a number of works have been performed with re-meshing techniques [9, 10] or cartesian cut-cells methods [11, 12].

The analyses based on pre-defined mesh generation available in the literature are mainly realized using commercial CFD software. These software are usually coupled with mesh generators focused on positive displacement machines. In particular, the mesh generator *Twin Mesh* has been used by Hesse et al. [13] to simulate a scroll vacuum pump considering thermal deformation with ANSYS CFX. The same computational workflow has been applied by Picavet et al. [14, 15] to simulate the same compressors studied in this work, respectively with and without Intermediate Discharge Valves (IDVs). The mesh generator and CFD software SimericsMP+ has been used by Cavazzini et al. [16] to perform a CFD-based optimization and uncertainty quantification of a scroll compressor.

In this work, the authors have performed a full three dimensional CFD simulation of a commercial scroll compressor produced by Danfoss using open-source software. The numerical results, based on the framework developed by Fadiga et al. [17], have been experimentally validated using the data obtained by Picavet et al. [3]. In particular, the effects of the discharge region geometry on the convergence of the simulation have been investigated, as well as the gap size impact on leakage flows. The analyses conducted by Fadiga et al. in [17] and [18] along with the present investigation, represent a series of studies that have been carried out with a set of utilities based on Open-FOAM. To the authors' knowledge, these are the only CFD simulations of scroll machines performed with open-source software. The results testify the reliability that open-source software have gained over the last years.

2. CFD simulations of scroll compressors

The CFD simulations presented in this article have been performed within the OpenFOAM-v2006 framework. This open-source software is constituted by a set of C++ libraries, including the implementation of different solvers suitable for the study of PD machines, thanks to the availability of compressible flow solver, Arbitrarily Coupled Mesh Interfaces (ACMIs) [19] and dynamic mesh treatment.

Fadiga et al. [17] extended the capability of the software, by developing a series of utilities to generate structured meshes for scroll-type machines and to perform moving mesh simulations based on user-defined nodal displacements. These methodologies, combined with the built-in OpenFOAM mesh generator *snappyHexMesh*, have been employed in this work for creating, manipulating and moving the computational mesh. The OpenFOAM solver adopted by the authors is *rhoPimpleFoam*, a pressure-based segregated solver for transient compressible flows.

2.1. Deforming geometry and mesh motion

The orbital movement of the rotor leads to a change in the solution domain due to the displacement of the solid boundaries. Consequently, the grid nodes have to move to accommodate such displacement, resulting in the introduction of the relative velocity in the convection terms of the Navier-Stokes equations. As an example, this modification is reported for the integral form governing equation of a general tensorial property, Φ:

$$\frac{d}{dt}\int_V \rho\Phi\, dV + \int_S \rho\Phi(\mathbf{u}-\mathbf{u}_S)\cdot\mathbf{n}\, dS - \int_S \Gamma\nabla\Phi\cdot\mathbf{n}\, dS = \int_V q_\Phi\, dV \tag{1}$$

where V is an arbitrary moving volume, S is the surface bounding V, ρ is the density, $\mathbf{n}$ is the outward pointing unit normal vector, $\mathbf{u}$ is the fluid velocity, $\mathbf{u}_S$ is the velocity of the boundary surface, Γ is the diffusivity and q_Φ is the volume source of Φ. Demirdžić and Perić [20] suggest that in order the mesh deformation to respect continuity, the Space Conservation Law (SCL), which describes the conservation of space in case of moving or deforming cells, should be strictly respected. This law is based on the principle that the sum of volume fluxes through cell faces must correspond to the volume rate of change:

$$\frac{d}{dt}\int_V dV + \int_S \mathbf{u}_S \cdot \mathbf{n}\, dS = 0 \tag{2}$$

Ferziger and Perić [21] illustrate one of the most common methods to satisfy the SCL and ensure mass conservation, which is also implemented in OpenFOAM. This method is based on the calculation of the volumes swept by each cell face over a single time step, which are employed to evaluate the volume fluxes avoiding to explicitly define the grid lines velocities (using these last quantities can easily lead to errors).

In this work, the orbiting motion of the mobile scroll is handled using a user-defined nodal displacement technique. This method is based on the generation of a set of structured grids corresponding to a number of angular positions predefined by the user. The nodal positions are stored and serve as guidelines for the mesh nodes during the scroll revolution, with the option of linearly interpolate between successive grids to adopt arbitrary time steps, which are constrained in accordance with a user-defined threshold of Courant-Friedrichs-Lewy (CFL) number calculated at the end of every time step. This procedure, extensively described in [17], is in accordance with the theoretical basis described in the previous paragraphs of this section.

On the other hand, the key-frame re-meshing method includes a mesh replacement for each time step, leading to possible infringements of the SCL, as reported by Rane et al. [22]. Another method that does not involve motion of the mesh, but is suitable for the correct replication of the scroll operation is the Cut-Cell method, as presented in [11]. Such an approach involves the generation of a background Cartesian grid in which the geometry to be investigated is contained. In correspondence of the walls, the mesh is adapted to represent the boundary, and the cells falling into the "solid" part of the scroll are discarded. In the framework of the cut-cell method, the space conservation law is not automatically respected, as reported in [23]. A non-trivial treatment is required to enforce the mass conservation that follows from such volume mismatch.

2.2. Compressible flow and real gas properties

In PD machines, the fluid density is strongly influenced by its pressure. Density variations, which must be considered in the numerical analysis, are triggered by two different phenomena. Firstly, supersonic or transonic leakage flows develop in gap regions, which can be thought of as convergent-divergent nozzles. Additionally, chamber volumes are subjected to considerable variations to achieve compression or expansion of the operating fluid.

In the present analysis, the large part of the domain presents little compressibility, hence the authors have decided to adopt a pressure-based solver, trying to reach higher accuracy in low-Mach regions. The thermophysical properties have been evaluated using the native OpenFOAM implementation of the Peng-Robinson EOS, associated with polynomials models for transport properties. This decision is supported by the findings of Papes et al. [24] and Fadiga et al. [25, 26], who have investigated the real gas modelling in CFD simulations with refrigerant fluids.

2.3. Leakages and sealing

Internal leakages represent a phenomenon that is inherently included in scroll compressors. Considering that the gap sizes are usually below 100 µm, even small variations can produce non-negligible effects on the volumetric efficiency of the compressors. Such variations can be caused by heating and cooling phenomena linked to the compression process, as well as mechanical loading. An additional aspect to be considered is the presence of tip sealing components, which reduce axial leakages. The sealing capacity of tip seals is not constant, since friction and wear influence the mass flow rate of refrigerant. Modelling all these aspects would require a wide range of complex computational methodologies, ranging from Fluid Structure Interaction (FSI) to special treatments for thermal deformation and tip seals efficiency. However, it is possible to reach good levels of accuracy by setting a constant gap size, which can be tuned with the help of experimental data. When structured grids are adopted that gap modelling is naturally included as the mesh is squeezed with little quality degradation. In this work, the theoretical orbiting radius of the mobile scroll has been slightly reduced to generate a constant flank gap, combined with the assumption of perfect axial sealing.

3. Computational domain and numerical set-up

The scroll compressor analyzed in this work is one of the machines of the *Danfoss SH* compressor family.

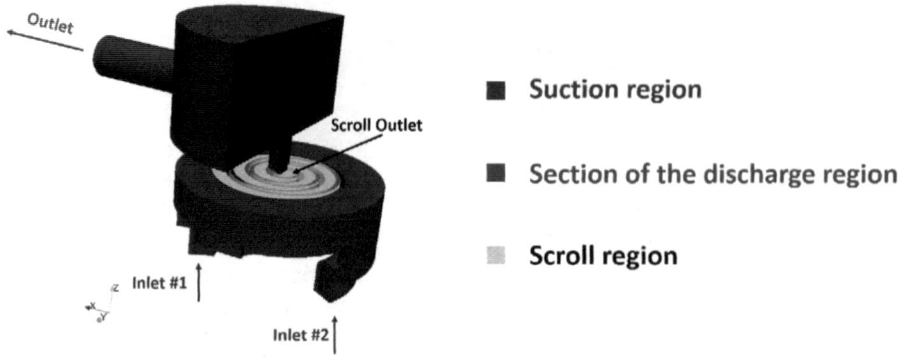

Figure 1. Computational domain of the scroll

The simulations have all been performed in oil-free conditions, although the original compressor works as an oil-flooded machine. The sealing action of the oil has been considered by acting on the flank gap size, while the cooling effects have not been considered. Figure 1 reports the computational domain, consisting in a fixed suction port (blue), a moving mesh region in correspondence of the scroll wraps (grey) and a large reservoir included in the fixed discharge volume (red). While the moving region has been discretized with a structured mesh, in accordance with the methodology reported in section 2, the stationary parts have been meshed with the unstructured mesh generator *snappyHexMesh*. The patch indicated with the name *Scroll Outlet* is an ACMI interface between the moving part and the discharge region. This patch has been used to evaluate the mass flow rate at the machine outlet. The resulting structured mesh is aligned with the software requirements in terms of mesh quality, as confirmed by the OpenFOAM native utility *checkMesh*. More particularly, the gap regions are characterized by regular stretched cells with an aspect ratio of approximately 220, as shown in figure 2.

The representation of the structured grid at an orbiting angle of 0° is reported in figure 2, where the region in correspondence of the end of the mobile spiral has been highlighted. This important area, which is in the proximity of the ACMI interface between the moving domain and the suction port, has been meshed with the aid of a quadratic grading function that reduces the size of the cells in accordance with the local width of the flank gap. The grading function has been applied to the edge reported as *Smoothed edge* in figure 2. This adaptive method has been implemented to improve the accuracy in the computation of the flow in the admission area, avoiding significant differences in the size of adjacent cells. The simulations have run fully parallel on a 40 cores machine, with a computational time of more than four scroll revolutions per 24 computational hours on a mesh with more than 2.5E+06 cells (corresponding to the case with the reservoir). Approximately, 600k and 800k cells have been used to discretize respectively

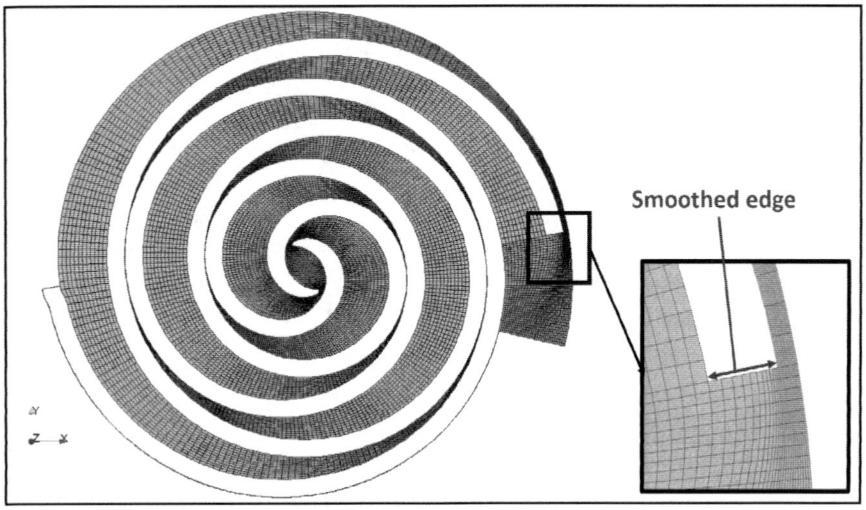

Figure 2. Structured mesh of the moving domain. Highlighted area: adaptive grading of the additional block cells.

the moving and the suction domains. The case with no reservoir, with more than 1.5E+06 cells, runs more than five scroll revolutions per 24 computational hours. The average values of Y-plus calculated at the domain boundaries range from approximately 20 to 400. The highest values have been evaluated at the reservoir walls, where particular accuracy is not required considering the purpose of this work. A grid independence study has not been included in this work, because the mesh size has been chosen according to the authors' experience in similar computational studies performed on scroll machines.

3.1. Simulation campaign

The numerical campaign is characterized by three main analyses, which have been performed on the full computational domain, inclusive of the outlet reservoir. The pressure ratio has been varied to inspect under and over-compression, while keeping the flank gap size at a constant value of 60 µm. The operating point corresponding to the Air-Conditioning and Refrigeration Institute (ARI) standard has been taken as reference, with a rotational speed of 2900 rpm. The pressure ratio of the ARI point is multiplied by 1.83 to reach the high PR point, while the low PR condition is set to a pressure ratio corresponding to 51 % of the reference one. The experimental study proposed by Picavet et al. [3] has been employed as validation for the three analyses reported above.

From the first set of analyses, it has been noticed that the presence of a large reservoir in the discharge region produces a low-frequency pressure wave, much lower than the scroll revolution frequency. This phenomenon is the cause of a lack of convergence at the outlet of the computational domain, as highlighted in section 4. Thus, the authors have decided to perform an additional simulation removing the reservoir and adding a non-reflective boundary condition, in order to isolate the reservoir effects. The last analysis regards the flank gap size: starting from the simulation with no reservoir, the radial distance between the two spirals has been increased from 60 µm to 100 µm.

3.2. Numerical schemes and boundary conditions

Fully turbulent adiabatic simulations employing Menter's k-ω SST model with updated coefficients have been performed. A second-order discretization scheme has been applied to spatial and time-dependent terms, with the exception of turbulent quantities. Furthermore, gradient limiters have been necessary to ensure the stability of the analyses.

The boundary conditions imposed for the full domain cases include fixed static pressure at inlet and outlet and inflow conditions based on specified mixing length for turbulence specific dissipation. The turbulent kinetic energy is evaluated at inlet and outlet, in case of inflow, according to a user-supplied turbulence intensity value. Temperature values have been imposed in case of flows entering the domain, while zero gradient conditions have been applied on outflows. It is particularly important to take care of reverse flow both at inlet and outlet because it is an inherent phenomenon in positive displacement machines, especially in the case of under-over compression. Accordingly, the velocity boundary conditions have been set as follows: for the flow leaving the domain, a zero gradient condition has been applied. On the other hand, for the flow entering in the domain a velocity value based on the flux in the patch-normal direction has been imposed. Lastly, for the simulations performed with the modified discharge port, with no reservoir, the outlet boundary conditions on velocity and pressure have been replaced with the OpenFOAM native implementation of non-reflective boundary conditions.

4. Results

Table 1 contains values of power and mass flow rate averaged over a complete revolution of the compressor. It is interesting to notice how the presence of the reservoir generates an imbalance between the inlet and outlet mass flow rate. The reason is clearly represented in figure 3, where the different periodicity of the mass flow rate trend at the domain outlet (*Outlet*) is highlighted. The pressure waves that produce this mass flow rate trend is influencing also the region in proximity of the scroll outlet, increasing the time needed to reach an acceptable level of convergence. These issues are the main reasons behind the decision of performing an additional simulation with no reservoir.

	$\dot{m}_{in}/\dot{m}_{in,ARI}$	$\dot{m}_{out}/\dot{m}_{in,ARI}$	$\dot{m}_{out}/\dot{m}_{in}$	VE
ARI	1.0	1.017	1.017	0.967
HP	0.381	0.396	1.039	0.941
LP	1.214	1.213	0.9991	0.975
ARI NR	1.002	1.005	1.003	0.969
ARI NR 100 µm	0.941	0.948	1.007	0.910

Table 1. Revolution-averaged mass flow rate, mass imbalance and volumetric efficiency. (NR = No Reservoir, $\dot{m}$ = Mass Flow Rate, VE = Volumetric Efficiency).

The pressure trend during the compression process is very well captured for all the operating conditions. In this work, the *direct* chamber is the compression chamber which firstly opens towards the discharge port, while *indirect* refers to the other. In figures 4, 5, 6 and 7, *pDirect* and *pIndirect* refer to the pressure recorded in the direct and indirect chambers. *pSuction* is the pressure recorded by a probe in the suction region, while *pDischarge* has been evaluated in the discharge region. Lastly, *pExDir* and *pExInd* are the experimental trends collected by Picavet et al. [3], respectively for the direct and indirect chamber.

ARI: the pressure trend (figure 4) is accurately reproduced, especially in the second half of the compression process. There is a small overestimation of the two peaks that occur in the discharge phase. This difference is probably due to the absence of tip seals in the final part of the scroll wrap (not modelled in this work), as reported in [3].

High pressure ratio: The under-compression generates reverse flow as the discharge port opens (figure 5). The consequent pressure peaks are very well captured by the numerical simulation. This validation is important as it is not highly influenced by the gap sizes. On

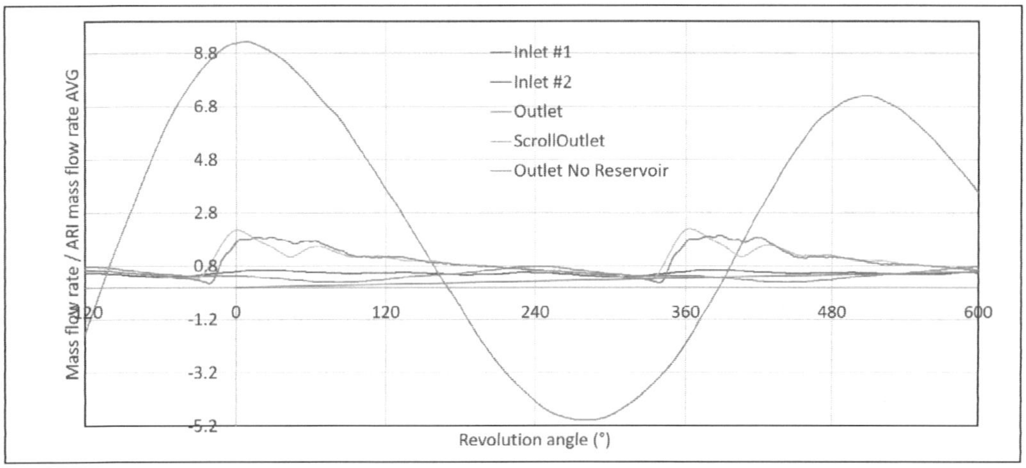

Figure 3. Mass flow rate as a function of the revolution angle. ARI point. Mass flow rate values have been divided by the revolution average of the inlet mass flow rate at ARI conditions (with reservoir).

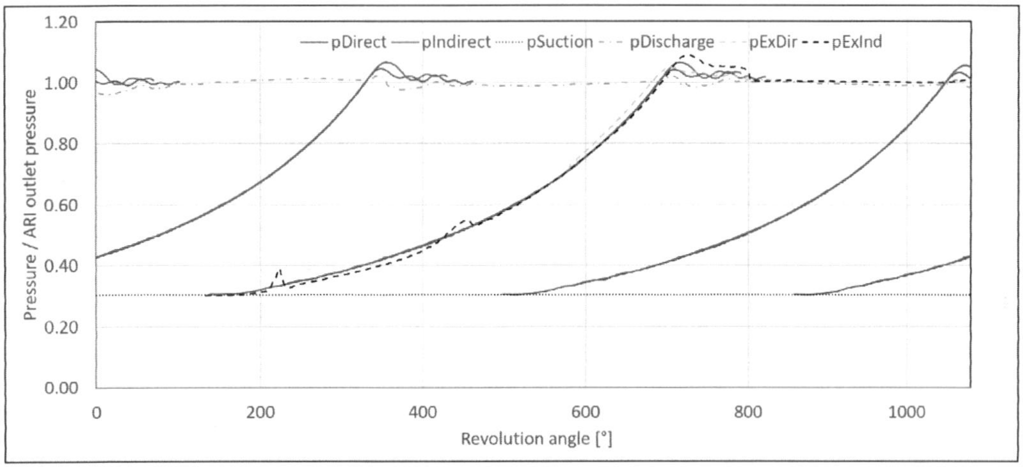

Figure 4. Pressure trends as functions of the revolution angle. ARI. Pressure values have been divided by the outlet pressure of the ARI point.

the other hand, underestimation of the pressure values can be seen in the final part of the compression. The higher pressure ratio amplifies the effects of the modelling assumptions described in section 2.3.

Low pressure ratio: The most interesting aspect of the low pressure ratio trend (figure 6) is related to the expansion process corresponding to the opening of the compression chamber. The discharge pressure experiences a fast decrease before stabilizing again at the outlet pressure levels. In this case, the peaks of the simulated trends are higher than the experimental data.

4.1. Impact of the reservoir and gap size effects

Regarding the pressure values in the compression chambers, the absence of the reservoir does not produce significant variations. Conversely, the removal of the reservoir has a beneficial effect on the global conservation of the mass flow rate. Table 1 highlights a consistent reduction (from

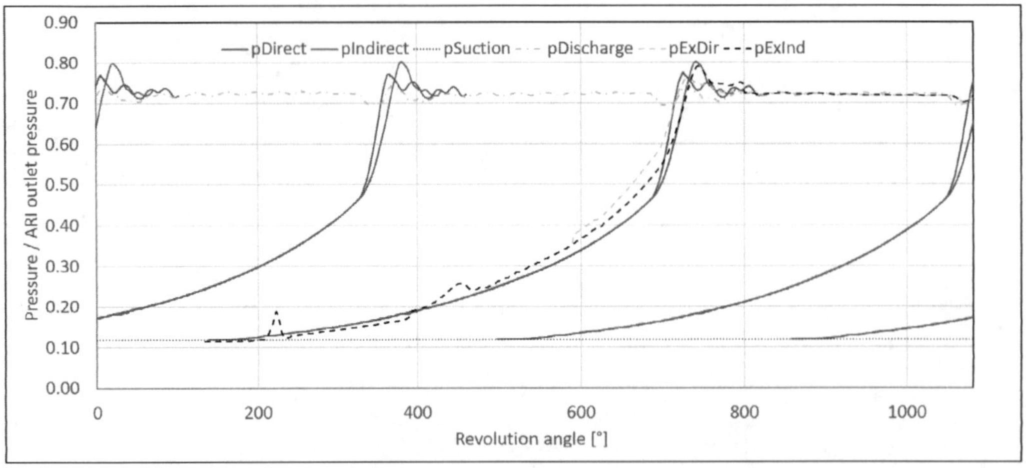

Figure 5. Pressure trends as functions of the revolution angle. High pressure ratio. Pressure values have been divided by the outlet pressure of the ARI point.

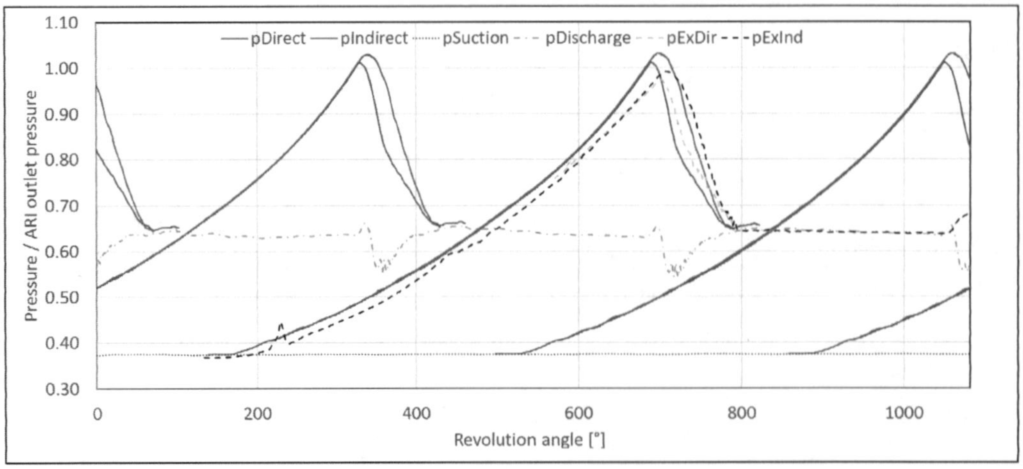

Figure 6. Pressure trends as functions of the revolution angle. Low pressure ratio. Pressure values have been divided by the outlet pressure of the ARI point.

the 1.6 % to the 0.3 %) in the difference between the suctioned and discharged mass flow rate over a complete revolution of the machine. The comparison between the *Outlet* and *Outlet No Reservoir* cases are represented in figure 3. The low frequency oscillations in the mass flow rate trend recorded at the domain outlet have been completely removed by eliminating the reservoir.

The determination of the flank gap size leading to the most accurate solution is certainly a challenging task. The comparison between figure 7 and figure 4 shows that the value of 60 µm produces a solution that is closer to the experimental reference. However, the low pressure ratio and high pressure ratio trends show that there is still room for improvement, probably in the area of the tip region modelling. It is important to underline that the introduction of axial gap modelling requires additional interfaces, which would be placed in regions where cells with high aspect-ratio are needed. This is definitely a topic that should be covered in future developments of the open-source framework for scroll machines presented in this article.

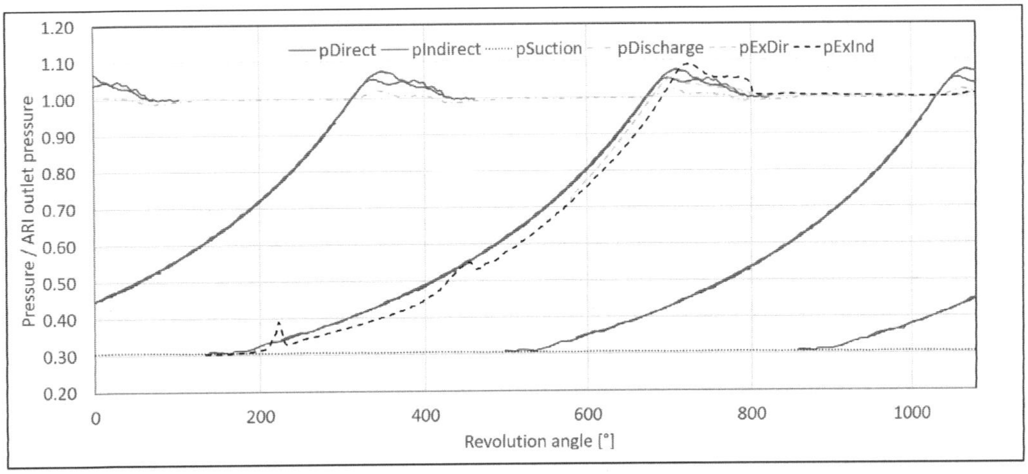

Figure 7. Pressure trends as functions of the revolution angle. ARI point, no reservoir, 100 µm. Pressure values have been divided by the outlet pressure of the ARI point.

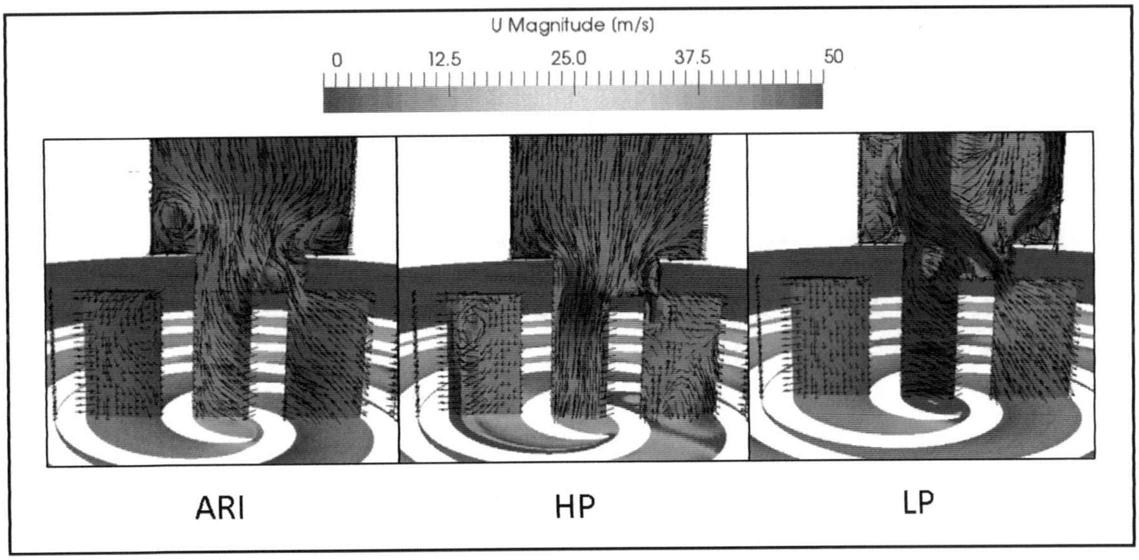

Figure 8. Contour of the velocity magnitude.

4.2. Under and over-compression

The discharge process is represented in the contour of the velocity magnitude (figure 8). There are several fluid dynamics phenomena to be highlighted: in the ARI case, the separation zone downstream the main outlet duct is clearly visible on the left. This phenomenon is strongly amplified when a lower pressure ratio is imposed between outlet and inlet, leading to a jet flow that occurs when the compression chambers open. Lastly, the high pressure ratio case is characterized by a large amount of reverse flow, which enters in the scroll domain influencing also the upstream chambers. The effects of under and over-compression are also reflected in the volumetric efficiency of the compressor, as reported in table 1. The HP case is characterized by a reduction of the volumetric efficiency close to 2.5 %, due to the increased leakages mass flow rate. Overall, the compressor shows good volumetric performance, even in presence of strong variations in the pressure ratio.

5. Conclusions

The authors have presented the first open-source simulation of a commercial scroll compressor, complete with experimental validation. This achievement could represent a starting point for including new modelling options in the field of positive displacement machines CFD. In fact, open-source frameworks are particularly suitable for the implementation of multiphase models, studies on heat transfer or optimization procedures.

The post-processing highlights the importance of CFD in the analysis of off-design conditions characterized by over and under-compression phenomena. The fluid dynamic behaviour of the compressor during the discharge processes can be analyzed at different operating points, with the possibility of extracting loss coefficients that can be adopted in reduced-order models. Additionally, the flank gap size has been the object of a preliminary analysis, with the aim of evaluating the variations of the pressure trends in the compression chambers, as well as the reduction in the volumetric efficiency.

References

[1] Ramaraj S, Yang B, Braun J, Groll E and Horton W 2014 *International Journal of Refrigeration* **46** 185–195
[2] Bell I, Lemort V, Groll E, Braun J, King G and Horton W 2012 *International Journal of Refrigeration* **35** 1890–1900
[3] Picavet A and Ginies P 2014 Experimental pressure-volume diagrams of scroll compressors *Proc. 22nd Int. Compressor Eng. Conference at Purdue*
[4] Bell I, Groll E, Braun J, Horton W and Lemort V 2014 *International Journal of Refrigeration* **45** 223–242
[5] Gravesen J and Henriksen C 2001 *Society for Industrial and Applied Mathematics* **43** 113–126
[6] Tello-Oquendo F M, Navarro-Peris E, Barceló-Ruescas F and Gonzálvez-Maciá J 2019 *International Journal of Refrigeration* **106** 308–326
[7] Park Y C, Kim Y and Cho H 2002 *International Journal of Refrigeration* **25** 1072–1082
[8] Sung J P, Boo J H and Jung E G 2020 *Energies* **13**
[9] Emhardt S, Tian G, Song P, Chew J and Wei M 2020 *Energy* **199** 117399
[10] Morini M, Pavan C, Pinelli M, Romito E and Suman A 2015 *Applied Thermal Engineering* **80** 132–140
[11] Pham H D, Brandt T and Rowinski D H 2018
[12] Rowinski D H, Li Y and Bansal K 2018
[13] Hesse J, Spille-Kohoff A, Andres R and Hetze F 2018 Cfd simulation of scroll compressors with axial and radial clearances and thermal deformation *18. ISS* pp 123–137
[14] Picavet A and Angel B 2016 Numerical simulation of the flow inside a scroll compressor equipped with intermediate discharge valves *Proc. 23th Int. Compressor Eng. Conference at Purdue*
[15] Picavet A and Genevois D 2018 Three-dimensional navier-stokes simulations of working of scroll compressors *Proc. 24th Int. Compressor Eng. Conference at Purdue*
[16] Cavazzini G, Giacomel F, Ardizzon G, Casari N, Fadiga E, Pinelli M, Suman A and Montomoli F 2020 *Energy* **209** 118382
[17] Fadiga E, Casari N, Suman A and Pinelli M 2020 *Energies* **13(3)** 666
[18] Fadiga E, Casari N, Suman A and Pinelli M 2020 Multi-component numerical investigation of a micro organic rankine cycle *Rankine 2020*
[19] Farrell P and Maddison J 2011 *Computer Methods in Applied Mechanics and Engineering* **200** 89–100
[20] Demirdžić I and Perić M **8** 1037–1050
[21] Ferziger J H and Perić M 1999 *Computational Methods for Fluid Dynamics* 2nd ed (Berlin: Springer)
[22] Rane S, Kovacevic A, Stosic N and Kethidi M **36** 1883 – 1893
[23] Bennett W, Nikiforakis N and Klein R 2018 *Journal of Computational Physics* **368** 333–358
[24] Papes I, Abdelli L, Degroote J and Vierendeels J 2015 3d cfd analysis of a twin screw expander with different real gas models for r245fa *ASME 2015 Int. Mechanical Engineering Congr. and Exp.* p V07AT09A045
[25] Fadiga E, Casari N, Suman A and Pinelli M 2020 *Computer Physics Communications* **250** 107047
[26] Fadiga E, Casari N, Suman A and Pinelli M 2020 Fluid thermophysical properties modelling in an opensource platform: Coolfoam *ECOS 2020*

Implementing the principles of operating processes schematization and of performance loss distribution when designing long-stroke reciprocating compressor stages

Vladimir L. Yusha[1], Sergey S. Busarov[1], Nikolay Yu. Filkin[1], Maria A. Fedorova[1], Vladimir V. Denisenko[2], Alexander A. Goncharenko[2], Vyacheslav B. Shipov[2] and Oleg G. Bessonov[2]

[1]Omsk State Technical University, Russian Federation
[2] JSC "RUMO", Nizhny Novgorod, Russian Federation

sidorova_ma79@mail.ru, yusha@omgtu.ru

Abstract. To compress gases up to 10.0 MPa and higher, diaphragm and multistage reciprocating compressor units are used. They are heavy, massive and complicated. The results of analyzing laboratory examples of slow-speed long-stroke stages of reciprocating compressors while gas pressurizing up to such pressure have shown the possibility of solving the problems mentioned. Such stages differ from others, that is why we cannot use common design methods in this case. The performance design method for slow-speed long-stroke compressor stages with a discharge pressure of up to 10.0 MPa is based on the well-known principles of schematization of operating processes and distribution of performance losses. With regard to the experimental studies results (indicator and temperature diagrams, integral characteristics) we have found: the volumetric efficiency, which characterizes the ratio between real and theoretical compressor performance, its main components: the volume coefficient, the pressure coefficient, the temperature coefficient, and the leakage coefficient. Dependences are obtained for calculating these coefficients, which differ significantly from the known dependences for calculating high-speed compressor stages. The analysis of the influence of various factors on the performance losses of low-capacity long-stroke compressor stages of medium and high pressure revealed that the gaps in the working chamber have the maximum impact on the performance losses, the leakage coefficient significantly depending on the discharge pressure change. The volume coefficient has much less impact due to the increased stroke-to-diameter ratio of the piston, which provides a small amount of relative dead volume. The developed method can be used for preliminary calculation of compressor stages of the considered type.

1. Introduction

The study results of laboratory samples of reciprocating compressors slow-speed long-stroke stages, performed at the Department of Refrigeration and Compressor Equipment and Technology of the Omsk State Technical University, proved the possibility of compressing air in one stage from atmospheric pressure to pressures of at least 10.0 MPa at a temperature of the injected gas that does not exceed the regulatory requirements [1-3]. In accordance with their specific design and operational differences, such stages can be called long-stroke (since parameter $Sp/Dc > 10$) low-speed (since the working cycle time $\tau > 1$s) [2, 4]. These parameters play a key role in the organization of the working process, providing a mode of operation close to isothermal. It should be noted that in the high-pressure stages of multi-stage compressors, the parameter *Sp/Dc* can also be close to 10. However, in such stages, the ratio of the discharge pressure to the suction pressure usually does not exceed 3...5, while the working cycle time is 0.2...0.02 s [5, 6].

The expected advantages of single-stage compressors of this type over multi-stage piston and diaphragm compressors of a similar purpose are: reducing the number of compression stages and interstage components (heat exchangers, pipelines, oil and condensate separators, filters, safety valves, etc.), increasing the degree of unification of the unit, as well as reducing the weight and overall dimensions [5]. In addition, such units do not have normal forces in the cylinder-piston group, which positively affects their performance characteristics [5].

When designing stages of reciprocating compressors, their main dimensions and parameters are often calculated based on the method of determining performance losses, including the calculation of the volumetric efficiency (Frenkel M. I., Plastinin P. I., Minta M., Barclay M., Davies R., John F. McLaren, Corberan, J. M., Hafner, J., Gaspersic, B., Luszczycki, M., Prakash, R., and others) [6-9]. One of the options for implementing this method is to determine the volumetric efficiency according to the following dependence [8]:

$$\lambda = \lambda_0 \cdot \lambda_p \cdot \lambda_T \cdot \lambda_h \cdot \lambda_L , \qquad (1)$$

where λ_0 is the volume coefficient, λ_p is the pressure coefficient, λ_T is the temperature coefficient, λ_h is the humidity coefficient and λ_L is the leakage coefficient.

At the same time, each of the coefficients included in equation (1) is determined according to the known ratios or recommendations that are valid for existing high-speed compressor stages [8]. It can be assumed with sufficient probability that the significant differences between the long-stroke slow-speed compressor stages considered in this work and the high-speed analogues do not allow us to correctly calculate the volumetric efficiency of such a stage according to the known method. Therefore, the task of this study is to obtain dependencies and recommendations for determining each of the above-mentioned components in formula (1).

2. Object of the study

The object of the study is the stage of a low-capacity compressor (the described volume is about 2.0...3.0 m^3/h) long-stroke slow-speed piston compressor without feeding lubricant to the flow part. The design of the stage is described in detail in [10, 11]; the lip piston seals are made of self-lubricating material; the automatic suction and discharge valves are of the poppet type. The experimental stage has the following main dimensions and parameters: cylinder diameter is 0.05 m, piston stroke is 0.5 m; cycle time is 2...3 s; relative dead volume is 0.026 %. External cylinder cooling is water; water temperature is 290 K; compressed gas is air; intake gas temperature is 290 K; suction pressure is 0.1 MPa; discharge pressure is 0.5…12.0 MPa.

3. Experimental research methodology

The scheme of the experimental stage is shown in (Fig. 1). Piston 1 is driven through rod 2 from the rod of the hydraulic cylinder, which in turn is driven from the hydraulic accumulator station. The suction and discharge valves are located in the valve plate and are not shown conditionally. An experimental unit with a linear (hydraulic) drive has been developed for experimental studies [12]. Experimental studies of the working processes of the stage include measurement of the instantaneous parameters of the state of the working gas in the cylinder and its integral characteristics. The pressure sensor 9 and the temperature sensor 8 allow us to measure the corresponding instantaneous parameters in the working compression chamber. The flow sensor 12 allows us to determine the actual compressor performance of the experimental stage. To measure the rapidly changing gas pressure in the working chamber of the stage, silicon pressure sensors 9 (Fig. 1) of the D16 type were used [12-14]. The temperature measurement was performed by thermistors 8, 11 (Fig.1) with a negative temperature coefficient of resistance [15-17].

The actual gas flow rate at the stage discharge was determined by the flow sensor 12 (Fig.1) of the AWM720P1 type. Data from the temperature, the pressure, and the flow sensors is transmitted to the digital oscilloscope 14 via amplifier 13 and personal computer 15.

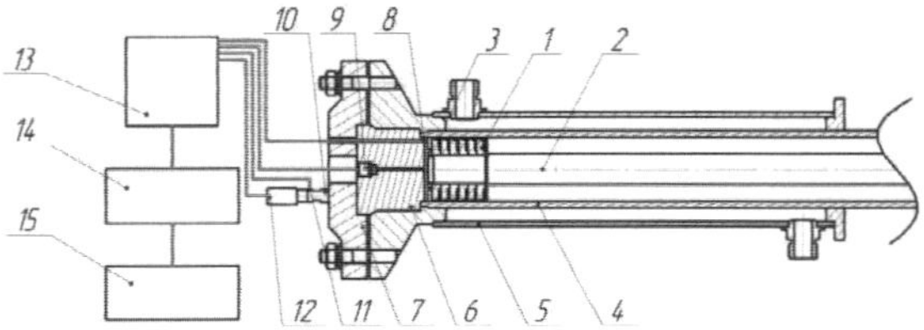

Figure 1. Scheme of the experimental stage

1-piston; 2-rod; 3-lip seals; 4-cylinder; 5-cooling jacket; 6-plate; 7-cover-flange; 8,11-temperature sensor; 9-pressure sensor; 10-discharge nozzle; 12-flow sensor; 13-signal amplifier; 14-digital storage oscilloscope; 15-personal computer

Let us estimate the error of measuring temperature, pressure, and flow. The error of the pressure sensor is determined by the formula [13, 14, 15, 16]:

$$\delta_{ps} = \sqrt{\delta_p^{\ 2} + \delta_{PG}^{\ 2} + \delta_0^{\ 2}} \tag{2}$$

where δ_p is the relative error of the pressure sensor,%; δ_{PG} is the relative error of the sample pressure gauge,%; δ_0 is the relative error of the oscilloscope,%.

The relative error of the pressure sensor according to the passport $\delta_p = 1.4\ \%$. The relative error of the sample pressure gauge δ_{PG}=1 .5 %. The relative error for the oscilloscope is determined according to the passport $\delta_0 = 3\ \%$. Then the total error of the pressure sensor is equal to:

$$\delta_{ps} = \sqrt{1.4^2 + 1.5^2 + 3.0^2} = 3.63\ \%.$$

The air flow rate is measured by the AWM720P1 thermal flow sensor with an intrinsic accuracy δ_{VD}= 0.3%. Then the total error of the flow measurement is determined by the formula:

$$\delta_V = \sqrt{\delta_0^{\ 2} + \delta_{VD}^{\ 2}} = \sqrt{3^2 + 0.3^2} = 3.02\ \%.$$

To measure the instantaneous air temperature in the working chamber of the stage, sensors based on a bead thermistor of st1-18A type were used [16, 17, 25]. The data from the temperature sensor is transmitted to the digital oscilloscope via an amplifier. We will find the total accuracy of the temperature sensor based on a bead thermistor [19-22]:

$$\delta_T = \sqrt{\delta_M^{\ 2} + \delta_t^{\ 2} + \delta_V^{\ 2} + \delta_F^{\ 2}}, \tag{3}$$

where δ_M is the relative accuracy of the oscilloscope, 0.05 %; δ_t is the is the thermometer accuracy, determined by the accuracy of the device, 0.1 %; δ_V is the voltmeter accuracy, 0.3 %; δ_F is the accuracy taking into account non-linear correlation between voltage and temperature, 1.5%.

Thus, the accuracy in measuring the instantaneous air temperature in the working chamber of the experimental stage will be:

$$\delta_T = \sqrt{0.05^2 + 0.1^2 + 0.3^2 + 1.5^2} = 1.53\ \%.$$

The bead design of the temperature sensors ensures a minimum dead volume in the working chamber of the stage, which is crucial for high values of the ratio of the discharge pressure to the suction pressure. Note that this measurement method does not allow temperature measurements in high-speed stages due to the significant inertia of the bead sensing elements; however, in low-speed stages of the type under consideration, the working cycle time of which is one or two orders of magnitude longer, the use of bead thermistors is well-justified.

4. Method for estimation the components of the volumetric efficiency

It is known that the volumetric efficiency can be represented as [24]:

$$\lambda = \lambda_{\text{ind}} \cdot \lambda_{hu} \tag{4}$$

where $\lambda_{\text{ind}} = \lambda_0 \cdot \lambda_p$ is the coefficient that takes into account the indicator compressor performance loss; $\lambda_{\text{hu}} = \lambda_T \cdot \lambda_L \cdot \lambda_{\text{h}}$ is the coefficient that takes into account the so-called “hidden” compressor performance loss.

We will use the results of our experimental studies to determine the components of the volumetric efficiency listed above

5. Estimation of indicator compressor performance loss

To calculate the volume coefficient, we apply the existing calculation method [23]:

$$\lambda_0 = 1 - a_m((\frac{P_{dis}}{P_{suc}})^{1/m} - 1), \tag{5}$$

where P_{suc} is the suction pressure, Pa; P_{dis} is the discharge pressure, Pa; m is the “polytropy index of the boundary parameters”; a_m is the value of the relative dead volume.

In formula (5), all values except m are set according to the design and operating mode of the compressor stage. Let us find the polytropy index of the final parameters m using the obtained experimental indicator diagrams. For this purpose, let us apply the well-known principle of schematization of the process of reverse expansion by replacing the indicator diagram of a valid reverse process of expansion occurring with varying polytropic exponent, on a curve with a relatively constant polytropic inverse exponent m, with both curves start and end points of the expansion process must be the same [23]. In this case, the schematized process allows us to quantify, with an acceptable margin of error, the actual performance losses caused by the expansion of the gas from the dead volume (Fig. 2).

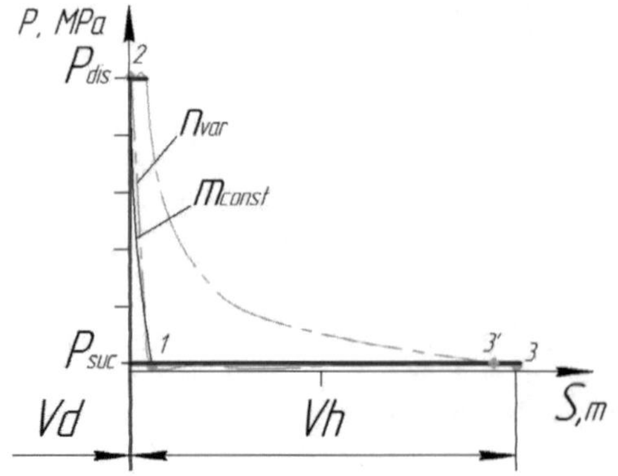

Figure 2. An example of schematization of the final parameters by polytropic at P_{suc} = 0.1 MPa; P_{dis} =10.0 MPa; Dc = 0.05m; Sp = 0.5m; τ = 3 s

According to the well-known recommendations, the formula (6) can be used to calculate the “polytropy index of the final parameters” [22, 23]:

$$m = 1 + A(k - 1), \tag{6}$$

where A is the empirical coefficient.

The analysis of the obtained experimental indicator diagrams (Fig. 3...6, Table 1) allows for the considered range of operating modes of the air slow-speed long-stroke stage to preliminarily accept the following recommendations: for $Pdis \leq 5.0$ MPa – A= 0.125 (when $m \approx 1.05$); for 5.0 MPa < $Pdis \leq 12.0$ MPa – A= 0.25 (when $m \approx 1.1$). The volume coefficient in this case varies from 0.98 at a pressure of 5.0 MPa to 0.95 at a pressure of 12.0 MPa.

Note that for high-speed air stages, the value of *m* at *Pdis*<3.0 MPa lies in the range of 1.2...1.4, at *Pdis*>3.0 MPa is assumed to be 1.4 [26].

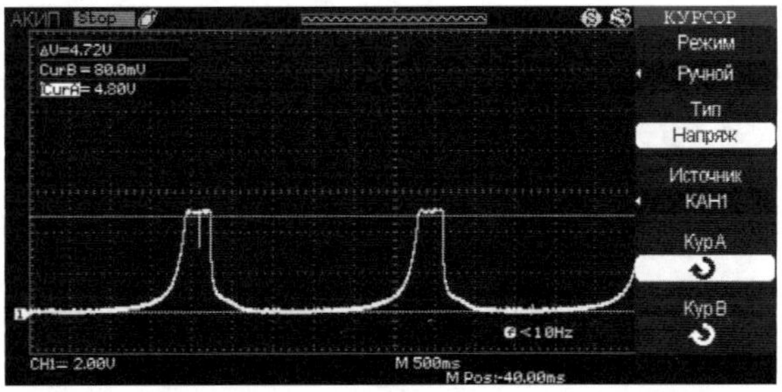

Figure 3. Experimental indicator diagram at *Pdis*=5.0 MPa, τ=3 s

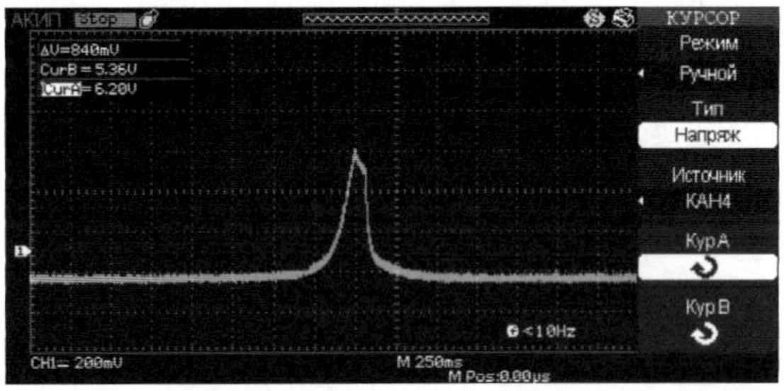

Figure 4. Experimental indicator diagram at *Pdis*=10.0 MPa, τ=3 s

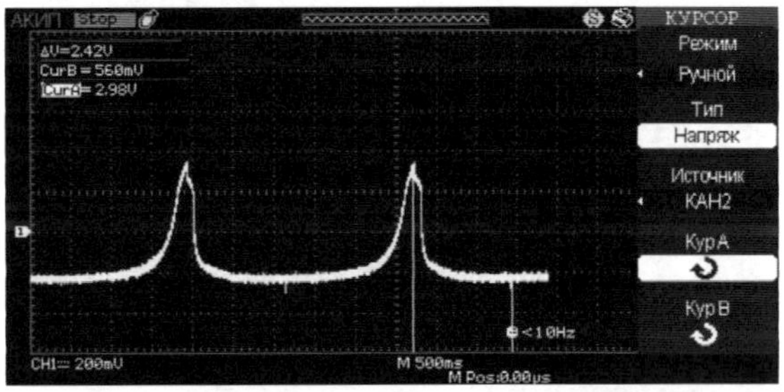

Figure 5. Experimental indicator diagram at *Pdis*=12.0 MPa, τ=3 s

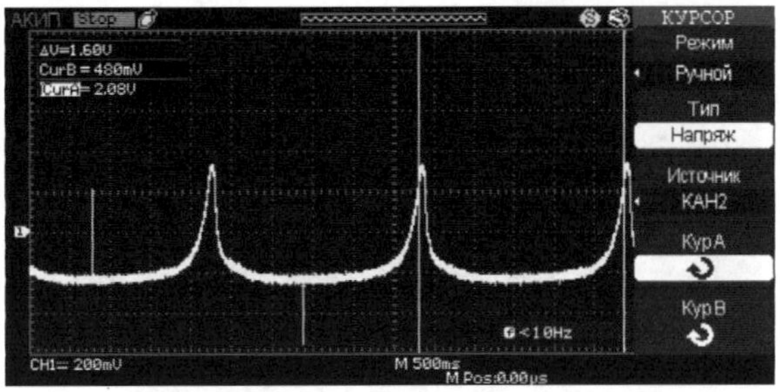

Figure 6. Experimental indicator diagram at *Pdis*=12.0 MPa, τ=2 s

Table 1. Indicator volumetric efficiency and its components

Experiment No.	***Pdis*, MPa**	**Cycle time, s**	λ_{ind}	λ_0	λ_p
1	5	3	0.97	0.98	0.99
2	10	3	0.96	0.97	0.985
3	12	3	0.93	0.95	0.984
4	12	2	0.93	0.95	0.981

Throttle compressor performance losses are determined on the indicator diagram (Fig. 2) by the segment 3 – 3' [23], because during this period of the working process, the volume of the working chamber decreases when the suction valve is open. The values of the pressure coefficient, with an error are in the range of 0.98...0.99.

6. Estimation of hidden compressor performance loss

The temperature coefficient characterizes the reduced mass of the new portion of the intake air that fills the working chamber of the piston during the process of absorption, due to its heated surfaces forming the flow path of a stage. It is defined as the ratio of the air temperature in the standard point of suction to the air temperature in the stage working chamber at the end of the suction process [23]:

$$\lambda_T = \frac{T_{suc}}{T_{suc3}}, \qquad (7)$$

where T_{suc} is the air temperature at the standard suction point, T_{suc3} is the air temperature in the working chamber at the end of the suction process.

This component of the volumetric efficiency should be estimated based on the results of the experimental studies, which resulted in temperature diagrams that allow us to estimate the change in air temperature in the working chamber of the stage during the working cycle of the compressor stage (Fig. 7 – 9). Table 2 shows the results of the obtained experimental data processing.

The experimental data obtained made it possible to determine the gas temperature at the end of the suction process and calculate the heating coefficient for the considered range of the operating parameters using formula (7).

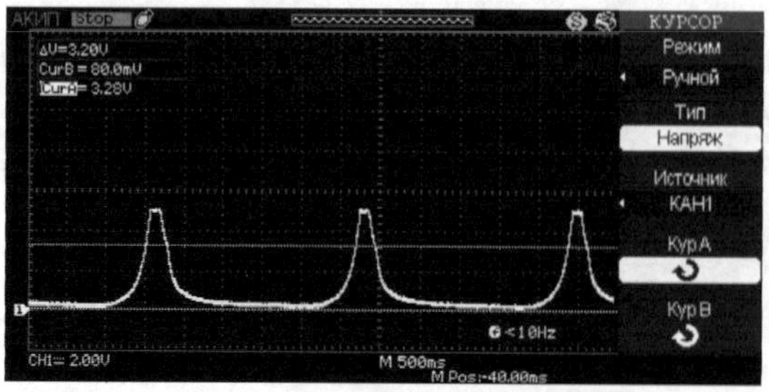

Figure 7. Experimental temperature diagram at a discharge pressure of 3.0 MPa, cycle time is 3 s

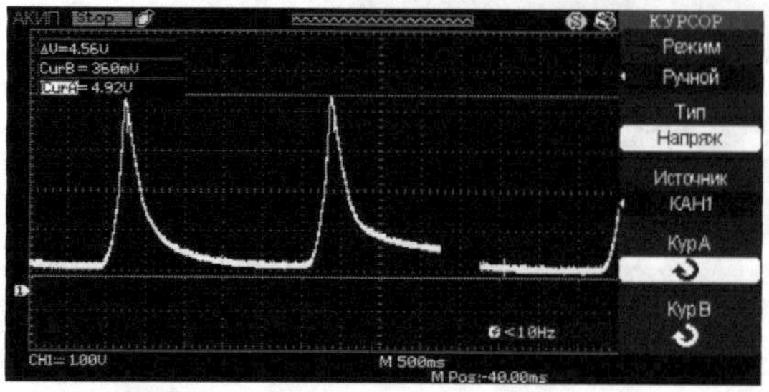

Figure 8. Experimental temperature diagram at a discharge pressure of 12.0 MPa, cycle time is 3 s

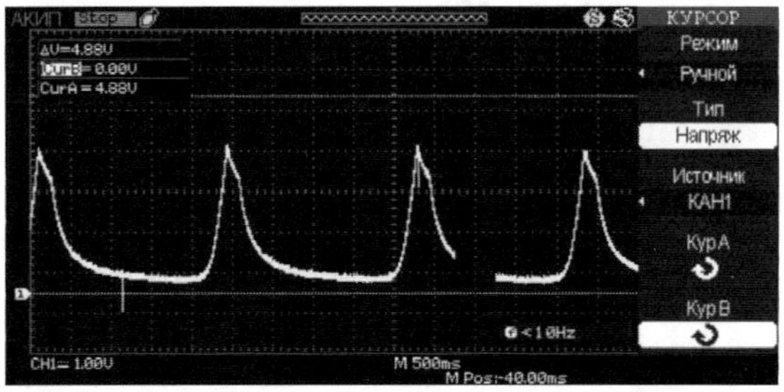

Figure 9. Experimental temperature diagram at a discharge pressure of 10.0 MPa, cycle time of 2 s

Table 2. Experimental data for calculating of the temperature coefficient

Experiment No.	P_{dis}, MPa	Cycle time, s	T_{suc}	T_{suc3}	λ_T
1	0.5	3	293	294.5	0.996
2	3	3	293	300	0.975
3	6	3	293	307.5	0.953
4	10	3	293	321	0.91
5	12	3	293	329	0.89
6	0.5	2	293	296	0.99
7	3	2	293	305	0.96
8	6	2	293	310	0.945
9	10	2	293	326	0.9
10	12	2	293	333	0.88

In the engineering method of calculating high-speed reciprocating compressor stage, the formula for calculating the heating coefficient is as follows [23]:

$$\lambda_T = 1 - B(\varepsilon_c - 1) \tag{8}$$

where ε_c is the ratio of the pressure value at the end of the compression process to the pressure value at the beginning of the compression process; B is the empirical coefficient, the value of which is approximately 0.01 when calculating high-speed stages [23]. According to the results of processing the experimental data presented in Table 2, in relation to slow-speed long-stroke stages, it is preliminarily recommended to assign B=0.001.

Unlike the volume coefficient, the pressure coefficient and the temperature coefficient, the leakage coefficient cannot be estimated from experimental indicator and temperature diagrams. One of the ways to solve this problem is to determine the leakage coefficient from the known values of the volumetric efficiency [23, 24] and its four components:

$$\lambda_L = \frac{\lambda}{\lambda_0 \cdot \lambda_p \cdot \lambda_T \cdot \lambda_h} \tag{9}$$

In this case, the total volumetric efficiency (which is in the numerator) can be defined as the ratio of the measured actual productivity to the value of the theoretical productivity calculated from the known basic dimensions and parameters of the stage [23]. In the denominator, the three coefficients are defined above, and the humidity coefficient does not depend on the design features of the compressor stage and is determined in all cases by the known thermodynamic dependences [22-24].

Some results of processing the experimental data obtained for the considered range of the design and operating parameters of the air slow-speed long-stroke compressor stage are presented in Tables 3 and 4.

Table 3. The base for calculation the leakage coefficient at cycle time τ=2

	ε=5	ε=40	ε=60	ε=100
λ_0	0.996	0.96	0.94	0.92
λ_p	0.99	0.99	0.98	0.98
λ_T	0.996	0.94	0.92	0.91
λ_h	0.99	0.99	0.99	0.99
λ	0.95	0.75	0.52	0.15
$\lambda_L = \lambda/(\lambda_0\, \lambda_p\, \lambda_T\, \lambda_h)$	**0.98**	**0.85**	**0.63**	**0.184**

Table 4. The base for calculation the leakage coefficient at cycle time τ=3

	ε=5	ε=40	ε=60	ε=100
λ_0	0.98	0.96	0.94	0.92
λ_p	0.99	0.98	0.97	0.96
λ_T	0.98	0.92	0.9	0.88
λ_h	0.99	0.99	0.99	0.99
λ	0.93	0.69	0.52	0.15
$\lambda_L = \lambda/(\lambda_0\, \lambda_p\, \lambda_T\, \lambda_h)$	**0.97**	**0.8**	**0.47**	**0.1**

The analysis of these results showed that it is convenient to represent the leakage coefficient in the form of the following expression to estimate the productivity losses, taking into account the effects of air flows through the gaps in the working chamber:

$$\lambda_L = 1 - x \cdot \varepsilon, \qquad (10)$$

where x is the empirical coefficient (for the considered range of design and operating parameters x=**0.008);** ε is the ratio of the discharge pressure to the suction pressure.

The method of calculating the actual performance slow-speed long-stroke compressor stage of, based on the definition of the flow coefficient as a set of factors reflecting the influence of various factors on the loss of productivity allows us to estimate the influence of these factors on the performance loss of the object. Figure 10 shows data on the ratio of individual components of performance losses depending on the degree of pressure increase for the considered slow-speed long-stroke stage (Fig. 10a) and for the known high-speed stages (Fig. 10b); in relation to them, the presented comparative results show essential difference in the influence of individual factors on the compressor performance losses.

It is known, in high-speed reciprocating compressor stages, having satisfactory technical condition, the determining factor is the value of the relative dead volume, the effect of which on the compressor performance losses is reflected (presented) by the volume coefficient [6, 8, 23, 24], which is confirmed by the calculation results presented in Figure 10b. In the considered slow-speed stages, this factor is minimized by increasing the ratio of the stroke of the piston to its diameter (in this case, the relative dead volume becomes less than 0.1%); but the long working cycle time and, accordingly, long time of the processes of air flow through the gaps, leads to the fact that the leakage coefficient becomes a key one (Fig. 10a).

7. Conclusion

The analysis and processing of the results of the experimental studies made it possible to develop a method for calculating the actual volumetric efficiency of slow-speed long-stroke air compressor stages, based on the principles of schematization of working processes and separation of losses. It is shown that for the object under consideration, in contrast to the high-speed reciprocating compressors the most important factor in determining the loss of productivity is the leakage coefficient, reflecting the growing influence of mass transfer through the gaps in the working chamber with increasing time cycle stage.

The results obtained determine the current directions for improving the design of slow-speed reciprocating compressor stages, influencing the development of promising sealed self-acting valves and cylinder-piston seals.

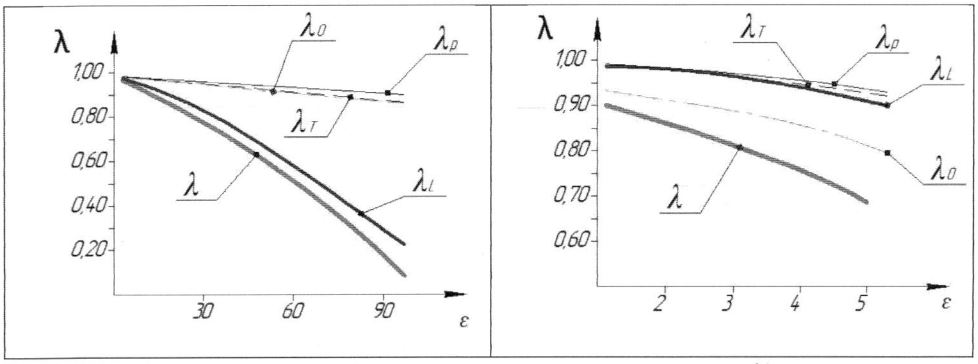

a) b)

Figure 10. Dependence of the components of the flow coefficient on the values of the discharge pressure to the suction pressure a) slow-speed compressor stage (DC=0.05 m; Sp=0.5 m; τ=3 c; am=0.00026); b) high-speed compressor stage (DC=0.05 m; Sp=0.05 m; τ=0.02 c; am=0.05)

Obviously, the described results is only the first step in improving the existing methodology for calculating piston compressor stages in relation to slow-speed long-stroke piston compressor stages. Further experimental and theoretical studies will be aimed at clarifying the total volumetric efficiency and its components, including when compressing gases with different properties.

List of symbols

λ_0	volume coefficient
λ_p	pressure coefficient
λ_T	temperature coefficient
λ_h	humidity coefficient
λ_L	leakage coefficient
λ_{ind}	indicator volumetric capacity
λ_{hi}	"hidden" volumetric capacity
P_{suc}	suction pressure
P_{dis}	discharge pressure
k	ratio of specific heats
a_m	relative dead volume
Sp	stroke of the piston
Dc	diameter of cylinder
δ_p	relative error of the pressure sensor
δ	accuracy
m	polytropy index of the ultimate, boundary parameters
τ	working cycle time
T_{suc}	air temperature at the standard suction point
T_{suc3}	air temperature in the working chamber at the end of the suction process
ε_c	ratio of the pressure value at the end of the compression process to the pressure value at the beginning of the compression process
ε	ratio of the discharge pressure to the suction pressure

References

[1] Yusha V. L., Busarov S. S., R.Yu. Goshlya, A. V. Nedovenchanyi, B.S. Sazhin, M.A. Chizhikov, I.S. Busarov, The experimental research of the thermal conditions in slow speed stage of air reciprocating compressor, International Conference on Oil and Gas Engineering, OGE-2016. - pp. 297-302.

[2] Yusha V L, Busarov S S, Gromov A Yu Assessment of the prospects of development of medium-pressure single-stage piston compressor units *Chemical and Petroleum Engineering* 2017 **53(3)** DOI: 10.1007/s10556-017-0362-2

[3] Yusha V L, Busarov S S Prospects for creating slow-speed compressor units of medium and high pressure on the basis of unified slow-speed long-stroke stages *Saint Petersburg Technical University Journal. Engineering science and technology.* 2018 **24 (4)** 80-89 DOI: 10.18721/JEST. 24408. (in Russian)

[4] Yusha V. L., Busarov S. S., Nedovenchanyi V. A. Analysis of the operating cycle efficiency of the long-stroke slow stage under thechanging ratio of the piston forward and backward stroke time / AIP Conf. Proc. 2018. Vol. 2007. P.030058-1–030058-5.

[5] Vetter H. The Sulzer Oil-Free Labyrinth Piston Compressor *Int Compressor Eng Conf*, Purdue University 1972 221–228

[6] Plastinin P, Fedorenko S Simulation of Transient Gas-Temperatures in Cylinders of Reciprocating Compressors Using Identification Techniques with a Mathematical Model *Int Compressor Eng Conf*, 1978 **295** http://docs.lib.purdue.edu/icec/295

[7] Davies R, Bell A Mathematical modeling of reciprocation air compressors *Mining Technol* 1987 **69, 795** 13-14

[8] Chrustalev B S, Zdalinsky V B, Bulanov V P Mathematical Model of Reciprocating Compressor with One or Several Stages for the Real Gases" *Int Compressor Eng Conf*, 1996 **1108**. http://docs.lib.purdue.edu/icec/1108

[9] Luszczycki M, Cyklis P, Zelasko J Developed Mathematical Model of the Self-Acting Valves of the Reciprocating Compressor and Its Application for Tongue Valves. *Int Compressor Eng Conf* 1992 918 http://docs.lib.purdue.edu/icec/918

[10] Yusha V L, Busarov S S, Goshlya R Yu., Vasiliev V K The Experimental Research of the Operating Processes in Slow Speed Stages of Air Reciprocating Medium-pressure in Int Oil and Gas Eng Conf (OGE-2017) *AIP Conf Proc* **1876(1)** 020039 DOI:10.1063/1.4998859

[11] Yusha V L, Busarov S S, Goshlya R Yu, Nedovenchanyi A V Experimental research of the operating processes of slow-speed long-stroke oil-free reciprocating compressor stages at discharge pressures up to 10 MPa *AIP Conf Proc* **2007** 030057 (2018); p. 030057-1 – 030057-6. Available at: https:// doi.org/10.1063/1.5051918.

[12] Nedovenchanyi A V, Yusha V L, Busarov S S Experimental evaluation of the efficiency of long-stroke, low-speed reciprocating compressor stages in compression of different gases *Chem and Petr Eng* 2018 **54** 593-597

[13] Moril S Dry gas seals point the way to oil-free compressors *Chem Eng* 1988 **451** 21

[14] Tesfamichael T, Ahsan M, Notarianni M et al. Gas sensing of ruthenium implanted tungsten oxide thin films *Thin Solid Films* 2014 **558** 416–422

[15] Haugen J-E, Tomic O, Kvaal K A calibration method for handling the temporal drift of solid state gas-sensors *Analytica Chimica acta* 2000 **407** 23-39

[16] Helwig A, Müller G, Garrido J A, Eickhoff M Gas sensing properties of hydrogen-terminated diamond *Sens Actuators, B* 2008 **133** 156-165

[17] Kester W, Briant J, Jung W Practical design techniques for sensor signal conditioning. *Cham. 7 Circuit desig.* 2000 **3** 366

[18] Lewandowski D, Awrejcewicz J Dynamical simulation of a nonlinear stepper motorsystem 2014 **5**. *Int. J. Dynam. Control* 31-35

[19] Bland J M, Altman D G Statistics notes: measurement error *Bmj* 1996 **313 (7059)** 744

[20] Cochran W G Errors of Measurement in Statistics, *Techno metrics* 1968 **10** (4) 637–666

Available at: http://www.jstor.org/stable/1267450

[21] Krstev I, Helwig A, Müller G et al Detection of random vapour concentrations using an integrating diamond gas sensor *Sensors and Actuators B: Chemical* 2014 **195** 603

[22] Diamantopoulos A Export performance measurement: Reflective vs formative indicators *Intl Markt Rev* 1999 **16(6)** 444–457

[23] Plastinin P I *Reciprocating compressors* In 2 vols. Vol. 1. Theory and calculation. 3rd ed., reprint. and add. Moscow: Kolos, 2006 (in Russian)

[24] Frenkel M I *Piston compressors* Kolbenverdichter, Theorie, Konstruktion Und Projektierung Berlin: Technik 1969 (in German)

[25] Nedovenchanyi A V, Busarov S S, Gromov A Yu, Goshlya R Yu, Busarov I S The application of compact thermistors for the temperature conditions analysis of small-sized long-stroke low-speed stages of piston compressors *Proc Eng* 2016 **152** 281-287

[26] Yusha V L, Busarov S S, Nedovenchannyi A V Experimental evaluation of the efficiency of working processes of low-speed long-stroke piston compressor stages in the compression of various gases. *Chem and Oil and Gas Eng* 2018 **8** 27-29

IOP Conf. Series: Materials Science and Engineering 1180 (2021) 012017 doi:10.1088/1757-899X/1180/1/012017

Numerical-experimental investigation of liquid slugging in the suction muffler of a hermetic reciprocating compressor

M Bianchi[1], C J Deschamps[1], T T Rodrigues[2] and E E Paladino[3]

[1] POLO Research Labs, UFSC, Florianopolis, Brazil
[2] R&D, Nidec-GA, Joinville, Brazil
[3] SINMEC Computational Fluid Dynamics Lab, UFSC, Florianopolis, Brazil

deschamps@polo.ufsc.br

Abstract. Refrigerant in the liquid phase and lubricating oil that circulate in refrigeration systems may reach the compression chamber of compressors. This phenomenon of liquid slugging is detrimental to reciprocating compressors because it can damage valves and other components. A three-dimensional Eulerian multiphase model was developed using the software ANSYS CFX to predict the gas-liquid flow in a simplified suction muffler geometry. In addition, an experimental bench was designed to allow the injection of specified volumes of lubricant oil into the suction muffler and measurement of the volume that reaches the muffler outlet. The liquid slugging was observed through glass windows positioned in the test section and recorded with a high-speed camera. Measurements showed that the phenomenon of liquid slugging is random and that the volume of liquid reaching the muffler outlet increases with the oil injection pressure. The numerical model was able to predict the oil volume that reaches the muffler outlet in agreement with measurements in the case of high injection pressures, but it overpredicted the volume when the injection pressure was decreased. The visualization showed changes in the liquid phase morphology that are difficult to predict since closure models for a two-fluid model are typically developed for dispersed morphology.

1. Introduction

Under normal operating conditions, superheated vapor refrigerant is admitted into the cylinder of reciprocating compressors. However, liquid refrigerant and lubricating oil circulate in the refrigeration system to which the compressor is connected. Under some operating conditions, liquid can enter the suction system and reach the compression chamber. According to Singh et al. [1], in these situations the pressure inside the compression chamber can reach a value 10 times higher than the maximum pressure under normal operating conditions, due to the incompressibility of the liquid. This event, called liquid slugging, can damage valves and other components, reducing the reliability of the compressor.

In the suction of reciprocating compressors, an acoustic muffler is installed in order to attenuate the noise generated by the pressure pulsation in the flow, induced by the reciprocal movement of the piston. Due to the complexity of the flow inside the muffler, the effect of this acoustic muffler on the amount of liquid that reaches the compression chamber during a liquid slugging event is not well understood. The main objective of this study is to investigate the gas-liquid flow in a simplified geometry of the suction muffler of a small reciprocating compressor, aiming to understand the effect of the muffler geometry on the liquid volume that reaches the compression chamber after a liquid slugging event.

IOP Conf. Series: Materials Science and Engineering 1180 (2021) 012017 doi:10.1088/1757-899X/1180/1/012017

The majority of studies reported in the literature investigate the return of liquid refrigerant to the compressor, and its effects when it reaches the compression chamber [1-5]. Regarding lubricating oil, there are a variety of studies that analyze the effects of oil circulation in the compressor and the system, as shown by Youbi-Idrissi et al. [6]. However, no analysis of oil circulation related to the occurrence of liquid slugging, and subsequent damage to the compressor, could be found in the literature. An experimental investigation of gas-liquid flow in suction mufflers of reciprocating compressors also appears to be lacking. Regarding numerical modeling, several studies have been aimed at analyzing the flow features, heat transfer and noise attenuation in mufflers through CFD [7-10]. However, these studies consider the flow of pure gas refrigerant.

Rodrigues [11] adopted a two-dimensional formulation to numerically solve the gas-liquid flow in simplified geometries of suction mufflers, using the commercial software ANSYS CFX. The author sought to assess the influence of geometric parameters on the amount of liquid that reaches the muffler outlet from different liquid pulses of known volume. Rodrigues [11] observed that the geometry affects considerably the amount of liquid that reaches the muffler outlet.

This paper reports the results of a three-dimensional simulation model for gas-liquid flow in a simplified geometry representing a suction muffler. Furthermore, an experimental bench was also designed to measure the volume of oil that reaches the muffler outlet after the injection of an oil pulse of known volume, and to allow the visualization of the flow inside the component. For simplicity, lubricating oil was adopted for the liquid phase. The refrigerant R134a was selected for the gas phase as it is associated with a high probability of the occurrence of liquid slugging [11].

2. Numerical model

2.1. *Flow geometry*

The actual geometry of the suction muffler (figure 1) is modeled as a three-dimensional rectangular muffler with square inlet and outlet tubes, as well as a domain extended around the muffler to represent the internal environment of the compressor housing and the housing walls, as shown in figure 2. The suction duct is connected to the housing, corresponding to a semi-direct configuration. A storage reservoir is placed at the muffler outlet to monitor the oil volume, which is later used for comparison with the experimentally obtained volumes.

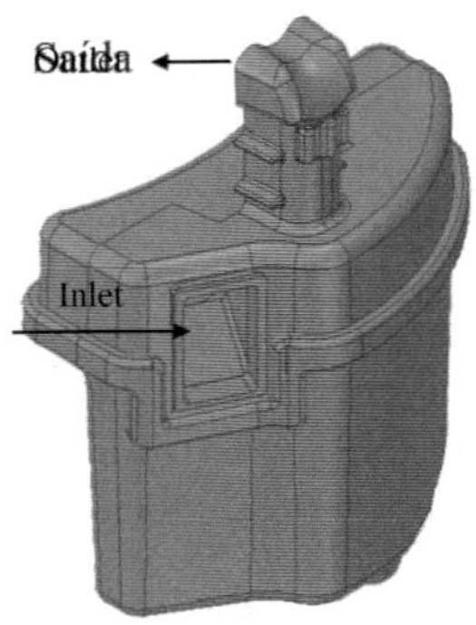

Figure 1. Actual suction muffler.

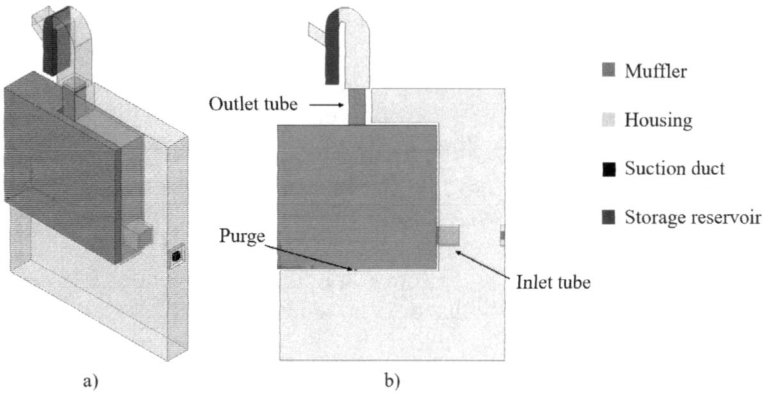

Figure 2. Solution domain. a) three-dimensional view and b) side view.

The internal volume of the muffler and the dimensions of the tubes were approximated to the dimensions of an actual reference muffler. An important difference of this 3D geometry, in relation to the 2D geometry of Rodrigues [11], is the presence of a small orifice (purge) under the muffler to assist in the removal of the liquid that enters the component. A structured mesh with hexagonal elements was adopted to reduce truncation errors and improve the convergence of the solution procedure. Mesh refinement analysis was also performed, with the mesh being refined until the liquid mass reaching the outlet showed an asymptotic value.

2.2. Governing equations

For the modeling of the gas-liquid flow in the suction muffler, mass and momentum conservation equations for each phase were numerically solved, in the framework of the two-fluid model, enabling the evaluation of velocity, pressure and volume fraction fields for liquid and gas phases inside the muffler. The adopted Eulerian heterogeneous model (two-fluid model) solves the governing equations for each phase, with interfacial transfer terms, which are responsible for modeling the interaction between phases. The liquid volume fraction is equal to 1.0 at the entrance of the solution domain during the oil injection. After that, the liquid distribution is variable within the domain and will result from the solution of the equation of mass conservation for the liquid phase. The set of governing equations was solved using the finite volume method available in the commercial software ANSYS CFX.

The turbulent flow was considered isothermal, compressible for the gas phase (R134a) and incompressible for the liquid phase (oil). The transient problem is characterized by an oil pulse prescribed at the suction line for a short period, after which the duct is closed. The R134a fluid properties are obtained using the Redlich-Kwong equation of state, while the properties of the liquid phase (oil) were prescribed (density $\rho = 926$ kg/m^3, viscosity $\mu = 0.01389$ Pa.s, interfacial tension $\gamma_{LG} = 12$ mN/m). The liquid phase was set as dispersed and the gas phase as continuous. In fact, although the liquid phase enters as a continuous morphology in solution domain, it breaks-up within the suction muffler changing its morphology to dispersed droplets with different sizes. This represents a challenge from the modelling point of view, as model closures are usually developed for a given phase morphology.

The oil droplet diameter, which is an input parameter for the interfacial transfer closure model, was used as a fitting parameter of the model. Although this set-up was able to reproduce qualitatively the experimentally observed flow, further investigation on the numerical model is necessary to capture adequately the phase configurations and morphological changes within the muffler and obtain more accurate quantitative results.

Turbulence modelling in gas-liquid two-phase flow is another challenging aspect of the numerical simulation, especially when in the presence of changes in the phase morphology. Adequate phase-dependent turbulence closure requires complex models and are, in general, specific closure models. Since there is no detailed information about the turbulence interactions between phases, we have used the SST turbulence model [11] with both phases sharing the solved turbulence field.

2.3. Boundary and initial conditions

The inlet liquid velocity at the suction duct during the pulse period was estimated by tracking the jet front via camera frame counting for different injection pressures (3.2 m/s for 0.7 bar, 3.9 m/s for 1.0 bar, 5.0 m/s for 1.5 bar and 5.5 m/s for 2.0 bar). A turbulence intensity of 5% was assumed as inlet boundary condition. The constant average pressure measured in the experimental apparatus was prescribed at the outlet of the storage reservoir. The boundaries representing the internal housing regions were represented by the suction pressure conditions (front, rear and bottom faces), allowing the gas to enter the muffler inlet and the oil that does not enter the muffler to be removed from the solution domain. The initial conditions assumed were that the solution domain was full of gas refrigerant (zero liquid fraction) under stagnant conditions and the pressure was equal to the saturation pressure for a temperature of -23.3 ° C.

2.4. Spatial and temporal discretization

The grid convergence index (GCI) derived from the theory of generalized Richardson extrapolation [12] was adopted to verify convergence regarding spatial and temporal discretization, i.e., to identify the point from which further refinement in these discretizations did not produce a variation greater than 3% in the volume of oil collected in the storage reservoir. The adopted computational grid had 965,324 cells with very small sizes near solid walls, whereas the time step was set to 0.0025 s.

IOP Conf. Series: Materials Science and Engineering 1180 (2021) 012017 doi:10.1088/1757-899X/1180/1/012017

3. Experimental setup and procedure

The experimental test bench is composed of three main components: calorimeter, oil injection system and test section (figure 3). It should be noted that the superheated gas cycle calorimeter was just used to submit the muffler (which is inserted in the test section) to different operating conditions and to provide the mass flow of R-134a through the test section. The compressor discharge pressure was regulated via a hand-operated needle valve. An oil separator was installed in the discharge line followed by an accumulator to dampen pressure oscillations. Downstream of it, another hand-regulated needle valve was operated to expand the gas from the intermediate pressure down to the desired suction line pressure.

The oil injection system (figure 3) consists of an oil reservoir under controlled pressure, supply line and a solenoid valve that controls the injections in the test section. The supply line is connected to the test section and the distance between its outlet and the muffler inlet is similar to that found in the reference compressor, corresponding to a semi-direct suction system.

The test section, illustrated in figure 4, consists of a container with a sight glass where the transparent muffler made of acrylic is inserted. The test section is positioned in the calorimeter circuit before the suction duct of the compressor, so that the whole container is under the selected suction pressure conditions, simulating operation inside the hermetic compressor.

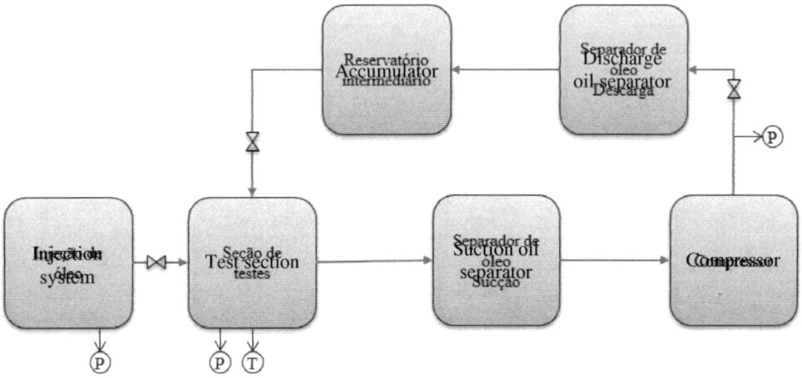

Figure 3. Schematic representation of the test bench.

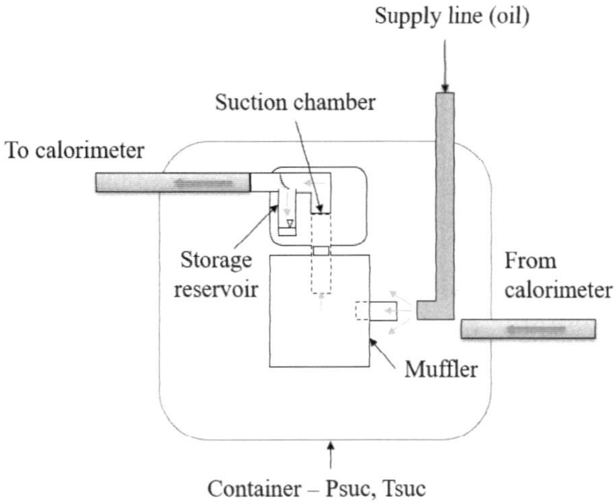

Figure 4. Schematic representation of the test section.

A high-speed camera is placed in front of the sight glass, to allow observation of the flow through the muffler. The muffler outlet is connected to the suction chamber, storage reservoir and calorimeter line. The storage reservoir is also made of acrylic with dimensions that allow the level to be observed and measured by image processing. After establishing the suction and discharge pressures in the calorimeter and in the test section, the oil reservoir is loaded with the desired pulse volume. The oil reservoir is then pressurized with gas refrigerant, as the flow in the test section is promoted by the pressure gradient. The solenoid valve is opened with the high-speed camera being activated simultaneously, capturing the oil jet at the entrance and the flow inside the muffler. Finally, the compressor is turned off and an image of the storage reservoir is used to measure the volume of oil collected.

4. Results

4.1. Quantitative analysis

The experimental tests require the definition of three main parameters: injection pressure in the oil reservoir, oil pulse volume and mass flow rate through the system, the latter being determined mainly from the suction and discharge pressures in the compressor. The aim was to analyze the effect of the injection pressure, keeping the pulse volume and mass flow rate at fixed values of 16 mL and 10 kg/h, respectively. The injection pressures were 0.7, 1.0, 1.5 and 2.0 bar above the suction line pressure of 1.15 bar corresponding to the evaporating temperature of -23.3 °C, with a total of 9 injections being performed for each injection pressure in order to obtain an average value for the volume of oil collected at the muffler outlet. In the numerical model, varying the injection pressure resulted in different velocities at the suction duct entrance.

The measurements in figure 5 indicate that the volume of oil collected increases with the injection pressure, but the relationship is not linear. This suggests that the increase in the jet velocity gives rise to flow patterns that are more likely to direct the oil towards the muffler outlet. Hence, the higher the velocity, the higher the oil dispersion. There is a considerable increase in the volume of oil collected when the pressure is raised from 1.0 to 1.5 bar and a tendency for the volume to stabilize when the pressure is raised from 1.5 to 2.0 bar. As also shown in figure 5, no oil was detected at the muffler outlet for the injection pressure of 0.7 bar.

Figure 6 shows the experimental and numerical results for the volume of oil collected in the storage reservoir. The dashed lines correspond to the average value obtained in the experiments at each injection pressure, while the solid lines correspond to the numerical prediction of the oil reaching the storage reservoir as a function of time. Table 1 shows measurements of collected volumes compared with the maximum collected volumes obtained from the model over a period of 1 s. As can be seen, the model is able to capture the trend of the volume of oil collected increasing with the injection pressure. It is also observed that the model estimates for the volume of oil collected are in good agreement with the experimental measurements for the higher injection pressures (1.5 bar and 2.0 bar), but the model overestimates the volume for the lower pressures (1.0 bar and 0.7 bar). The explanation for this could be related to the magnitude of the jet velocity at the entrance of the solution domain. Cases with higher injection pressure have higher velocities, with inertial forces, better captured by the model, being dominant over interfacial forces, which rely on less accurate closure models, particularly for this case in which phase morphologies and their changes along the fluid domain are not properly resolved.

The numerical results showed that a small amount of oil is dragged out of the solution domain through the exit section and is thus not captured by the storage reservoir. In addition, the volume of oil collected in the storage reservoir varies over time, which is associated with the predicted gas flow in that region. The adoption of a dispersed configuration for the liquid phase may contribute to this phenomenon, with the oil level in the storage reservoir being disturbed due to oil being dragged out of the domain. This disturbance is more visible for the case with an injection pressure of 2.0 bar. This case results in a higher volume of oil being collected and, consequently, a greater height of the oil level in the storage reservoir becomes exposed to the gas flow. It should be noted that it was not possible to quantify the amount of oil dragged out in the experimental bench system.

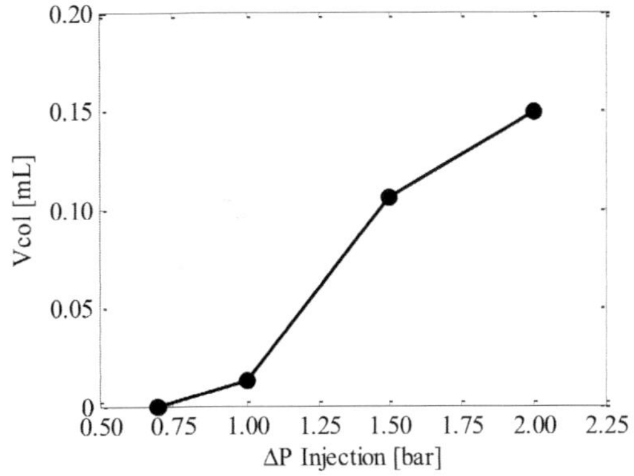

Figure 5. Measured volume of oil collected as a function of injection pressure.

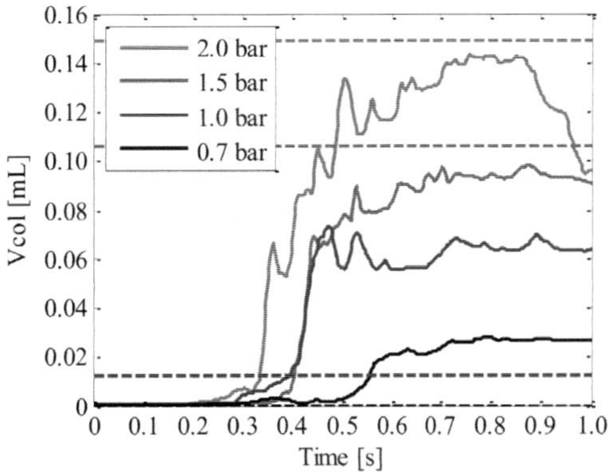

Figure 6. Experimental and numerical results for volumes of oil collected.

Table 1. Experimental and numerical results for the volumes of oil collected.

Injection Pressure [bar]	Volume [mL] Exp.	Volume [mL] Num.	Δ [mL]
0.7	0.0000	0.0281	-0.0281
1.0	0.0124	0.0732	-0.0608
1.5	0.1061	0.0978	0.0083
2.0	0.1493	0.1436	0.0057

The time of the experiment is 4 s from the start of the oil pulse to the compressor shutdown. If the total simulation time was also 4 s, it is possible that the volume of oil collected would still vary. On the other hand, the volume fraction iso-surfaces for the model and the high-speed camera videos indicate that after approximately 0.6 s there is no more liquid entering the muffler outlet tube. Hence, the oil directed to the outlet had already left, either being collected in the storage reservoir or having left the solution domain. Taking this into consideration and the fact that the computational processing time is high (an average 60 h/run of 1 s), the simulation was limited to a period of 1 s.

Some variability was found in the measurement of the volume of oil collected for each injection, as previously indicated by Singh et al. [1], which can be attributed to the chaotic characteristic of the injection process. figure 7 shows snapshots of the oil flow inside the muffler at the same time for three distinct injections with 2.0 bar. These images show that the jet breaks when it collides with the muffler walls, assuming very different flow patterns for each injection.

The results for the volume of oil collected in the nine experiments with injection at 2.0 bar are shown in figure 8, where the dashed line represents the average value. Table 2 shows the average value after three, six and nine injections, indicating that it reaches a well-defined value as the number of injections increases, despite the variability of the phenomenon.

4.2. Qualitative analysis

In order to visualize the oil flow from the numerical results, iso-surfaces were constructed with volume fractions ranging from 0.1 to 1.0, with increments of 0.1, totalizing ten iso-surfaces. These iso-surfaces were plotted simultaneously every 0.1 s and compared with snapshots from the camera video frame at the same time, allowing a qualitative comparison of the flow configuration inside the muffler. The comparisons are presented herein for the first intervals (0.1 to 0.4 s) with an injection pressure of 2.0 bar. However, the qualitative observations are similar for the other injection pressures.

Figure 7. Flow pattern for t = 0.2 s for three distinct injections at 2.0 bar.

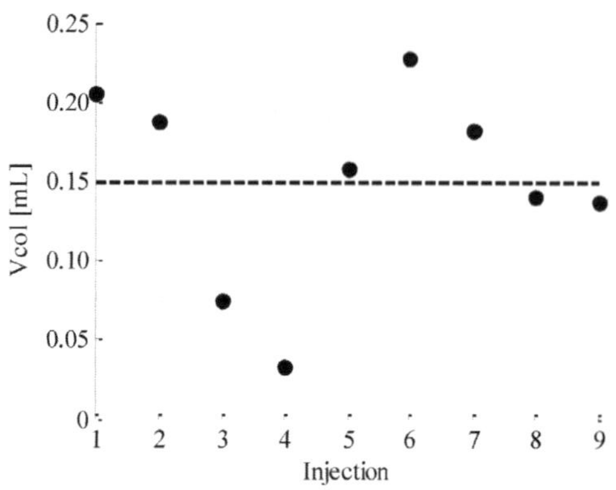

Figure 8. Volume of oil collected for different injections at 2.0 bar.

Table 2: Volume of oil collected per injection at 2.0 bar

Injection	Vcol [mL]	Mean 3-6-9
1	0.2054	
2	0.1880	0.1561
3	0.0750	
4	0.0324	
5	0.1574	0.1476
6	0.2276	
7	0.1815	
8	0.1395	0.1493
9	0.1371	

Figure 9 was prepared for t = 0.1 s and shows the numerical model predicted higher vertical oil dispersion, which may be related to the adoption of a dispersed model configuration for the liquid phase. A slight downward slope is observed in the jet due to the effect of gravity. We found this jet inclination is even more prominent in cases with lower injection pressure, a consequence of the lower momentum of the jet at the entrance. The experiments showed that the flow presents a continuous-continuous pattern until approximately the end of the jet injection, with a well-defined interface between the oil and the gas. The numerical and experimental results of the oil film advancing on the bottom surface of the muffler are in good agreement and no oil is present in the outlet tube at this time.

As shown in figure 10 for 0.2 s, the numerical and experimental results for the oil level on the bottom surface of the muffler seem to be in good agreement, with oil present at the outlet tube walls. For a 16 mL pulse of oil, the injection time in the model is 0.1786 s. In the experimental test section, the oil is injected via a pressure load applied to the reservoir. Therefore, the end of the jet pulse is characterized by a mixture of gas and liquid, with the jet taking longer to cease when compared to predictions. This also gives rise to higher dispersion of the oil jet inside the muffler, which is well suited for predictions with the dispersed model adopted in the simulations. Part of the discrepancy between numerical and experimental results can be attributed to the measurement of the jet velocity, which is estimated by tracking the jet front via camera frame counting.

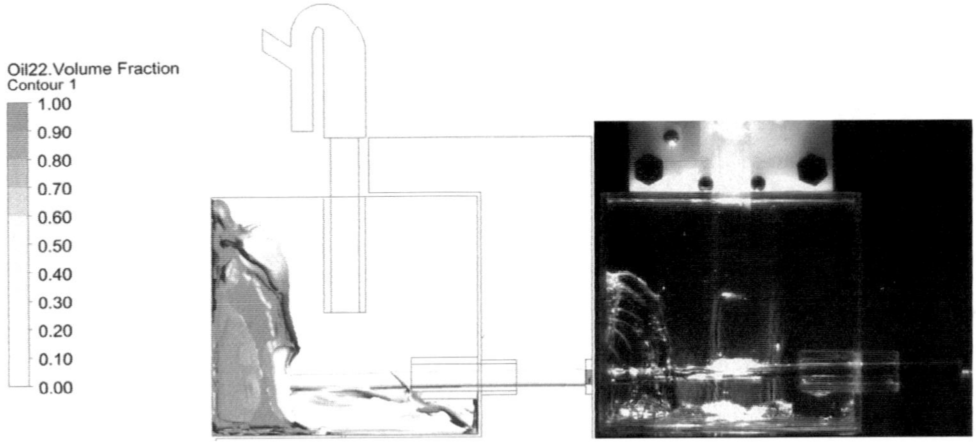

Figure 9. Qualitative comparison for t = 0.1 s.

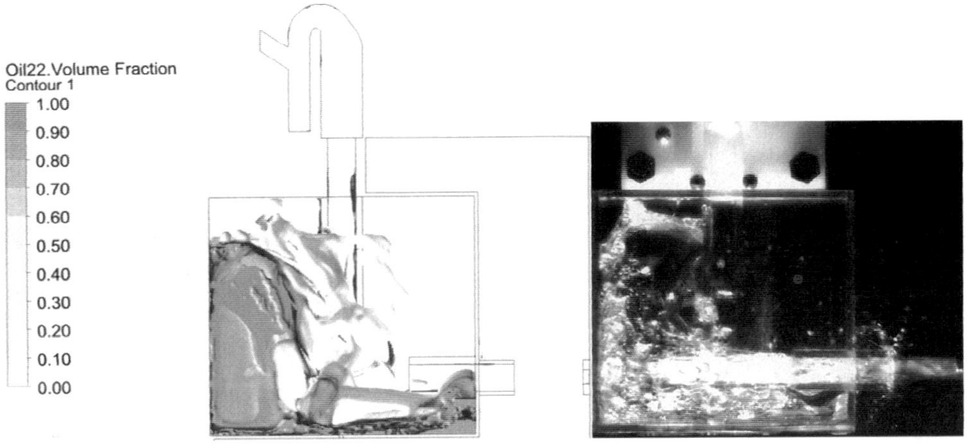

Figure 10. Qualitative comparison for t = 0.2 s.

Figure 11 shows some similarity between the numerical and experimental results (t = 0.3 s) for the flow configuration at the left wall of the muffler. In addition, the oil level is seen to increase at the lower right corner of the muffler, both in the model and in the experiment. In fact, this increase was observed at 0.2 s (figure 10), but it was intensified at 0.3 s due to the phenomenon of "sloshing", in which the oil slides from left to right. This sliding and the presence of oil in the outlet section become more evident in the subsequent figures as time progresses. At t = 0.4 s (figure 12), oil dispersed in the muffler central region is directed towards the outlet. It is also possible to observe the level of the oil settling at the bottom. The numerical result shows that the wave formed by the sloshing phenomenon begins to move towards the left side of the muffler.

As time continues to progress (figure 13), basically three phenomena are observed: (i) a reduction in the dispersion morphology in the muffler central region; (ii) stabilization of the oil level at the bottom of the muffler; (iii) a variation in the oil level in the storage reservoir, which is observed exclusively in the numerical model. The high-speed camera videos indicated that the oil reaches the outlet via two mechanisms: (i) dragging of dispersed oil particles that approach the entrance of the outlet tube; and (ii) suction of oil film that flows downwards on the external surface of the outlet tube, which is dragged as soon as it reaches the entrance of the outlet tube.

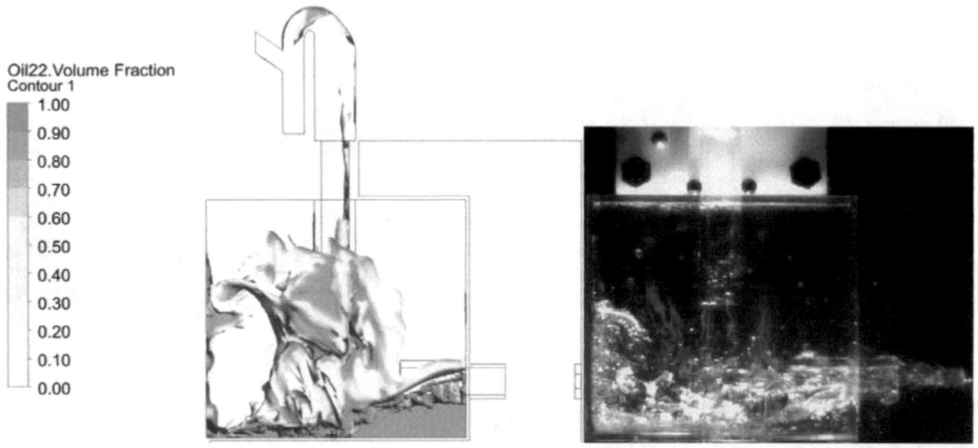

Figure 11. Qualitative comparison for t = 0.3 s.

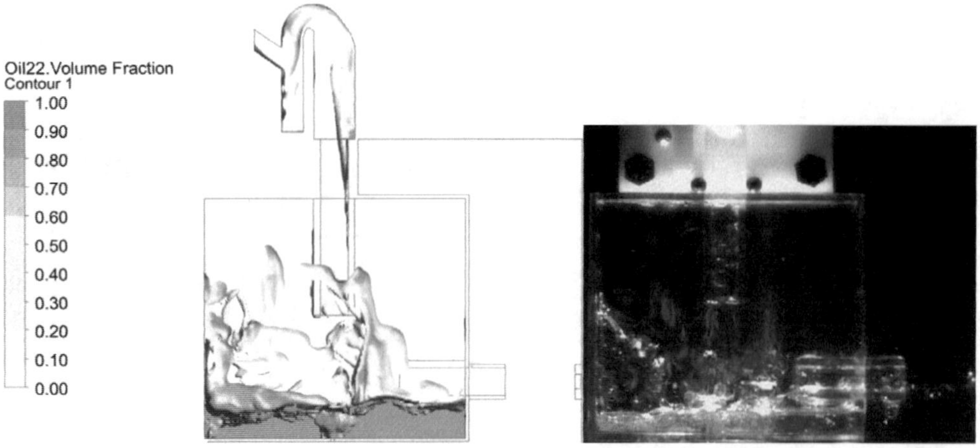

Figure 12. Qualitative comparison for t = 0.4 s.

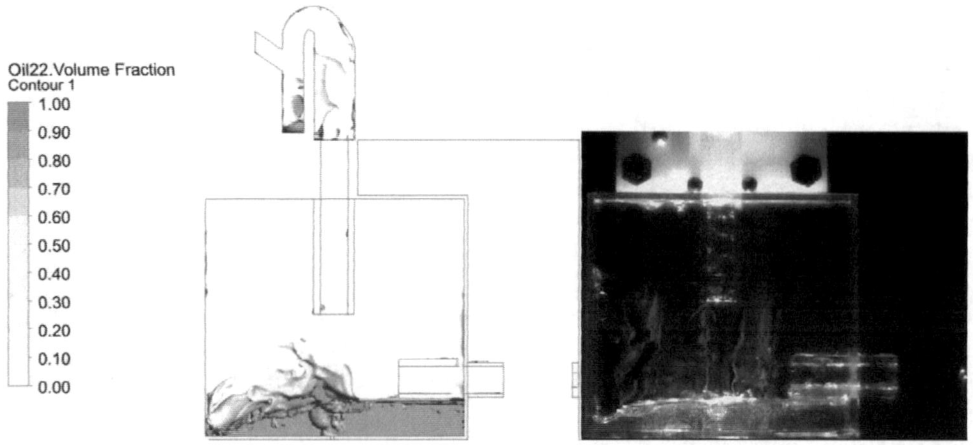

Figure 13. Qualitative comparison for t = 1.0 s.

5. Conclusions

A numerical-experimental investigation of gas-liquid flow in a simplified suction muffler was conducted, aimed at deepening our understanding of liquid slugging in reciprocating compressors of small refrigeration systems. Measurements were carried out in an innovative experimental apparatus and showed that the volume of liquid reaching the muffler outlet increases with the oil injection pressure at the compressor entrance. In addition, we found that the phenomenon of liquid slugging is random, requiring mean values to quantify the liquid volume that reaches the muffler outlet. A three-dimensional Eulerian multiphase model was developed to predict the gas-liquid flow. The model was able to predict satisfactorily the main flow features and the volume of oil that reaches the muffler outlet under high injection pressures. The visualization clearly showed that the liquid phase morphology changes from continuous to dispersed in some regions of the muffler, and to continuous again when the oil injection ceases. These morphological changes are difficult to predict since closure models for a two-fluid model are typically developed for dispersed morphology while interface-capture methods cannot deal with changes from continuous to dispersed morphology. This aspect of lquid slugging in reciprocating compressors merits extensive investigation in future studies.

Acknowledgements

The present study is part of a technical-scientific cooperation program between the Federal University of Santa Catarina (UFSC) and Nidec-GA (Embraco). Financial support provided by the National Institutes of Science and Technology (INCT) Program (CNPq Grant No. 443696/2014-4; FAPESC Grant No. 2018TR1576) and EMBRAPII Unit POLO/UFSC and CAPES is duly acknowledged.

References

[1] Singh R, Nieter J J and Prater G 1986 Prediction of Slugging-Induced Cylinder Overpressures *Proc. Int. Compressor Eng. Conf. at Purdue (West Lafayette, USA)*

[2] Simpson F and Lis G 1988 Liquid Slugging Measurements in Reciprocating Compressors *Proc. Int. Compressor Eng. Conf. at Purdue (West Lafayette, USA)*

[3] Shiva-Prasad B G 2002 Effect of Liquid on a Reciprocating Compressor. *J. Energy Resources Technology*, **124** 187-190

[4] Liu Z and Soedel W 1994 An Investigation of Compressor Slugging Problems *Proc. Int. Compressor Eng. Conf. at Purdue (West Lafayette, USA)*

[5] Laughman C R, Armstrong P R, Norford L K and Leeb S B 2006 The Detection of Liquid Slugging Phenomena in Reciprocating Compressors via Power Measurements *Proc. Int. Compressor Eng. Conf. at Purdue (West Lafayette, USA)*

[6] Youbi-Idrissi M and Bonjour J 2008 The effect of oil in refrigeration: Current research issues and critical review of thermodynamic aspects. *Int. J. Refrigeration*, **31** 165-179

[7] Nakano A and Kinjo K 2008 CFD Applications for Development of Reciprocating Compressor. *Proc. Int. Compressor Eng. Conf. at Purdue (West Lafayette, USA)*

[8] Morriesen A, Deschamps C J, Pereira E L L, Dutra T and Ribas Jr F A 2009 Numerical Prediction of Superheating in the Suction of a Hermetic Reciprocating Compressor *Proc. 20th Int. Congress of Mech. Eng. (Gramado, Brazil)*

[9] Sarioglu K, Ozdemir A R, Oguz E and Kaya A 2012 An Experimental and Numerical Analysis of Refrigerant Flow Inside the Suction Muffler of Hermetic Reciprocating Compressor *Proc. Int. Compressor Eng. Conf. at Purdue (West Lafayette, USA)*

[10] Coskun U C, Gunes H and Sarioglu K 2020 Numerical Investigation of Suction Muffler in Household Refrigerator Compressor. *Int. J. Geology*, **14** 27-31.

[11] Rodrigues T T 2018 Refrigerant Liquid Slugging in the Suction System of Reciprocating Compressor. *Proc. Int. Compressor Eng. Conf. at Purdue (West Lafayette, USA)*

[12] Roache P J 1998 *Verification and Validation in Computational Science and Engineering*, (Albuquerque: Hermosa Publishers)

IOP Conf. Series: Materials Science and Engineering 1180 (2021) 012018 doi:10.1088/1757-899X/1180/1/012018

Mechanical and deformation characteristics of rotary compressor discharge valve

Y S Hu[1,3], J Xu[1,2,3], O X Yang[1,2,3], J Gao[3], X Y Zhang[3] and Y P Hu[1,2,3*]

[1]State Key Laboratory of Air-conditioning Equipment and System Energy Conservation, Zhuhai, Guangdong 519070, China
[2]Guangdong Key Laboratory of Refrigeration Equipment and Energy Conservation Technology, Zhuhai, Guangdong 519070, China
[3]Gree Electric Appliances, Inc., Zhuhai, Guangdong 519070, China
[*]Correspondence: huyuanpei@lnm.imech.ac.cn

Abstract. The discharge valve plays an important role in the rotary compressor. For better understanding the mechanical characteristics of discharge valve in the process of movement, a mechanical model considering the coupling effect with limiter is established. Some major parameters, i.e., displacement, stress, stiffness and elastic force, are obtained to characterize the mechanical properties of the discharge valve. Two cases of valve component are calculated by the mechanical model, and the results indicated that the profile of limiter has an important influence on the movement of discharge valve. In order to verify the accuracy of the model, finite element method and high-speed photography are used. The comparison results show the calculation error of valve stiffness obtained by the model is within 3.9% compared with the finite element method results, and the valve motion is consistent with the high-speed photography results. Based on the above analysis, the profile of limiter in one valve component is optimized. The initial winding displacement of the valve is advanced from 2.115 mm to 1.453 mm after the optimization of limiter profile, which exhibits better limiting effect.

1. Introduction

As the heart of air conditioning, compressor plays an important role in refrigeration cycle, and the discharge valve is the core component to achieve the normal working of the compressor. A large number of research results show that the discharge valve component is important to the reliability, performance and noise level of compressor [1-5]. When the discharge valve slaps the limiter and the valve seat repeatedly, the bending fracture and the impact fracture often occur. In order to solve this problem, the bending stress and impact stress of the discharge valve are usually reduced by adjusting the lift and stiffness of the discharge valve, respectively. However, this method often leads to the sacrifice of compressor performance [6].

For better improving the reliability and performance of compressor, the motion characteristics of discharge valve under the coupling action with the limiter should be better understood. The limiter is main to limit the lift of discharge valve, avoiding fracture of discharge valve due to excessive bending stress. Kim et al. [7] provided a novel method for structural stability of limiter by calculating the maximum contact stress of limiter using Adina simulation. Zhang et al. [8] found the impact speed of valve mainly depends on the lift of limiter, and gave the reasonable range of lift and impact speed. Kwon et al. [9] found that the stability of compressor is better with the decrease of limiter length, and the performance of compressor is better with the increase of limiter length.

The above researches are mostly aim at the reliability and performance of compressor, but not consider the effect of limiter profile. The limiter profile mainly includes linear, double linear, single curvature, single curvature linear, double curvature and so on, and the limiter profile has a great influence on the valve movement. Zheng et al. [10] analyzed the stress distribution of three limiter profiles, and indicated that valve has better motion characteristics with single curvature type. Wang et al. [11] found that single curvature and single curvature linear limiter have no effect on the design of small compressor, but single curvature linear limiter has larger effective flow area and smaller characteristic lift. Ho et al. [12] verified the maximum stress of valve under two kinds of limiter, the results shown the maximum stress of valve under limiter A is higher 26% than it under limiter B. Mu et al. [13] proposed a valve model considering the coupling effect with limiter, and established the relationship between the effective working length of valve and the elastic force of valve with lift. Nevertheless, the above researches have not carried out a detailed study on the mechanical deformation characteristics of valve under the restriction of limiter, and it is hard to understand the actual movement and stress state of valve in the process of movement.

In this paper, a mechanical model of discharge valve considering the coupling effect with limiter is established, and some major parameters of the discharge valve are obtained. This model is used to analyze the valve movement under the two limiter profiles, and the accuracy of the model is verified by simulation and experiment. Furthermore, the profile of limiter B is optimized.

2. Theoretical model

In the process of valve opening and closing, the root of the valve is constrained by the limiter and rivet. The parallel part of the limiter can be considered as the fixed end constraint. The head of the valve is covered on the flange vent hole and bears the gas force, as shown in Figure 1a. Based on the above structure and mechanical characteristics, the valve can be simplified as a cantilever beam. One end is fixed, and the other end is under uniform load. In order to facilitate the calculation, the load is equivalent by using the superposition principle, as shown in Figure 1b. It should be noted that the model does not consider the variable cross-section of the valve head.

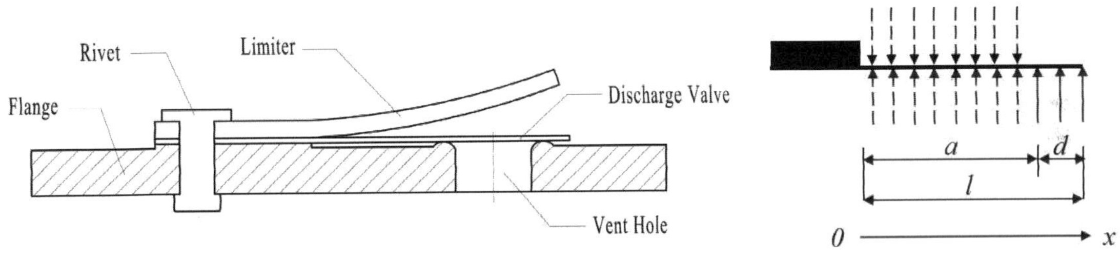

(a) Diagram of discharge valve component (b) Cantilever beam model

Figure 1. Diagram of discharge valve component and cantilever beam model.

Because the equivalent cantilever beam bears two different loads, it is necessary to carry out sectional analysis, and the moment M of valve is

$$M(x)=\begin{cases} q(l-a)x-\frac{q}{2}\left(l^2-a^2\right) & 0\le x\le a \\ -\frac{q}{2}(l-x)^2 & a\le x\le l \end{cases} \tag{1}$$

where q is uniform load. According to the bending force and deformation formula of material mechanics, there is

$$y''=\frac{\mathrm{d}^2w}{dx^2}=\frac{M(x)}{EI_0}$$

$$y'=\frac{\mathrm{d}w}{dx}=\int\frac{M(x)}{EI_0}dx+C$$

$$y = w = \int \left[\int \frac{M(x)}{EI_0} dx \right] dx + Cx + D \tag{2}$$

where w is the lift of valve head center, E is Elastic modulus, I_0 is moment of inertia, C and D are undetermined coefficient.

According to geometric boundary conditions, there is

$$\begin{gathered} y_1\big|_{x=0} = 0,\ y_1'\big|_{x=0} = 0 \\ y_1\big|_{x=a} = y_2\big|_{x=a},\ y_1'\big|_{x=a} = y_2'\big|_{x=a} \end{gathered} \tag{3}$$

Where y_1 and y_2 are displacements for $0<x\leq a$ and $a\leq x<l$, respectively. Combining equations (1), (2) and (3), we can achieve the deflection curve and moment of valve, that is

$$y(x) = \begin{cases} \dfrac{96(l^2-a^2)x^2 - 64(l-a)x^3}{17d^4 + 32ad(3l^2+a^2)} w & 0 \le x \le a \\ -\dfrac{384w\left[-16(l-x)^4 - (192ald + 64d^3)(x-a) + 16d(-6a^2l - 2a^3 + d^3)\right]}{17d^4 + 32ad(3l^2+a^2)} & a \le x \le l \end{cases} \tag{4}$$

$$M(x) = \begin{cases} -\dfrac{384EI_0w}{17d^4 + 32ad(3l^2+a^2)}\left[(l-a)x - \dfrac{1}{2}(l^2-a^2)\right] & 0 \le x \le a \\ \dfrac{192EI_0w}{17d^4 + 32ad(3l^2+a^2)}(l-x)^2 & a \le x \le l \end{cases} \tag{5}$$

Then, the stress and stiffness can be achieved, namely

$$\sigma(x) = \begin{cases} -\dfrac{384EI_0w}{17d^4 + 32ad(3l^2+a^2)}\left[(l-a)x - \dfrac{1}{2}(l^2-a^2)\right]\dfrac{t}{2I_0} & 0 \le x \le a \\ \dfrac{192EI_0w(l-x)^2}{17d^4 + 32ad(3l^2+a^2)}\dfrac{t}{2I_0} & a \le x \le l \end{cases} \tag{6}$$

where t is thickness of valve.

$$k = \frac{qd}{w} = \frac{384EI_0d}{17d^4 + 32ad(3l^2+a^2)} \tag{7}$$

The above theoretical process only considers the mechanical and deformation of the valve under the action of gas force on the head with the root restraint, and does not consider the coupling effect between the valve and the limiter. With the gradual opening of the valve, the radius of curvature gradually decreases. When the radius of curvature of the valve decreases to equal to the radius of curvature of the limiter, the two start to wind. Firstly, the initial contact position of valve and limiter (Initial winding point) can be determined based on the deflection curve of the valve and the profile line of the limiter. The limiter profile has two common types: continuous type and discontinuous type, as shown in Figure 2.

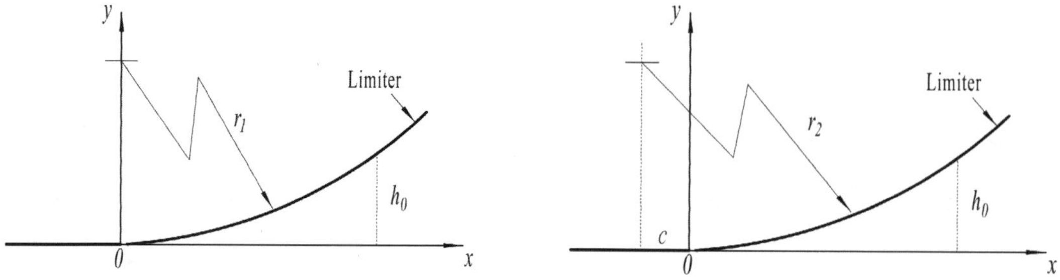

(a) Continuous type (b) Discontinuous type

Figure 2. Diagram of two limiter profiles.

For continuous type, the center of limiter profile is just above the starting point, while the center of limiter profile is on the starting point left for discontinuous type. The continuous type limiter profile is

$$y = r_1 - \sqrt{r_1^2 - x^2} \tag{8}$$

and the discontinuous type limiter profile is

$$y = \sqrt{r_2^2 - c^2} - \sqrt{r_2^2 - (x+c)^2} \tag{9}$$

Because the analysis process of two limiters is same, taking the continuous type limiter as an example, the mechanical and deformation of valve is studied with the coupling effect. Combining the equations (4) and (8), the initial winding displacement w_0 at the center of valve top (Valve displacement) can be achieved. For the continuous type limiter, the initial winding point is always at the point $x = 0$.

Based on equation (7), it can be seen that the valve stiffness is mainly related to the material parameters and structural parameters of the valve. When the valve displacement is less than w_0, the valve and the limiter have not yet been winded. At this time, the valve stiffness is a constant value. When the valve displacement is greater than w_0, the valve and the limiter begin to wind, and the effective working length l' gradually decreases, which is not the deformation length l any more. The valve stiffness analysis is divided into two stages, as shown in Figure 3. Stage I is before that the valve winding point does not reach the uniform load position, that is $l' \geq l - a$. Stage II is when the valve winding point has reached the uniform load position, that is $l' < l - a$. Therefore, the two stages have different stress forms.

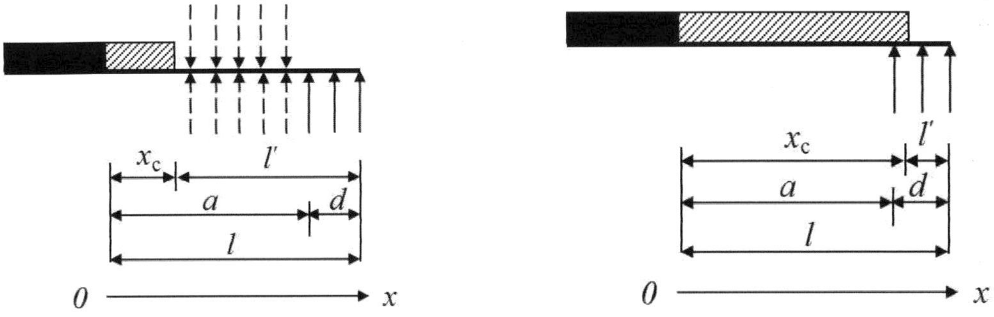

(a) Mechanical model in stage I (b) Mechanical model in stage II

Figure 3. Mechanical model of valve under two stages.

The stiffness of valve in two stages can be calculated, namely

when $l' \geq l - a$,

$$k_1 = \frac{qd}{w} = \frac{384EI_0 d}{17d^4 + 32d(a - x_c)[3(l - x_c)^2 + a^2]} \tag{10}$$

when $l' < l - a$,

$$k_2 = \frac{ql'}{w} = \frac{24EI_0(l - x_c)}{(l - \frac{d}{2} - x_c)^2[(l - \frac{d}{2} - x_c)^2 - 4(l - x_c)(l - \frac{d}{2} - x_c) + 6(l - x_c)^2]} \tag{11}$$

The relationship between the valve stiffness and the length of deformation section can be established by equations (7), (10) and (11). In order to get the relationship between the stiffness and valve displacement, it is necessary to establish the relationship between the displacement and the length of the valve. During calculating the valve displacement, it can be divided into three parts, as shown in Figure 4. Part one y_1 is the height of limiter at the winding point, part two y_2 is the height of valve center due to the angle α at the winding point, part three y_3 is the valve displacement because of the deformation of the rest length.

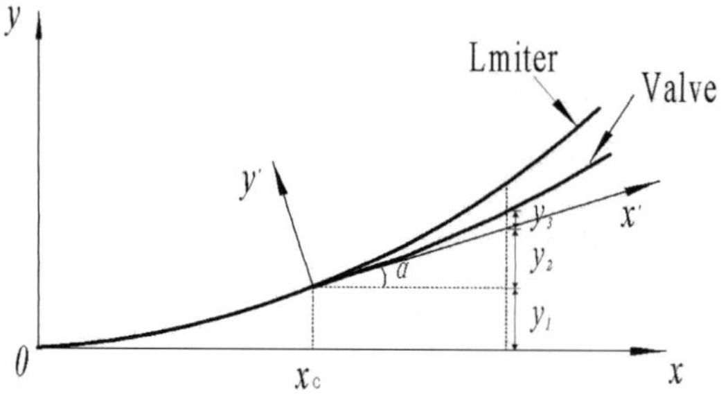

Figure 4. Three parts of valve displacement.

Based on the limiter profile, y_1 and y_2 can be easily obtained.

$$y_1 = r_1 - \sqrt{r_1^2 - x_c^2} \tag{12}$$

$$y_2 = (l - x_c - \frac{d}{2})\frac{x_c}{\sqrt{r_1^2 - x_c^2}} \tag{13}$$

Based on the mechanical model in Figure 3, y_3 can be obtained.

$$y_3 = \begin{cases} -\dfrac{17qd^4 + 96qd(a-x_c)(l-x_c)^2 + 32qd(a-x_c)^3}{384EI_0} & 0 \le x_c \le a \\ \dfrac{q(l-\frac{d}{2}-x_c)^2}{24EI_0}\left[(l-\frac{d}{2}-x_c)^2 - 4(l-x_c)(l-\frac{d}{2}-x_c) + 6(l-x_c)^2\right] & a \le x_c \le l \end{cases} \tag{14}$$

In equation (14), the load q is uncertain and need to be obtained. According to the define of bending moment,

$$M(x_c) = \frac{EI_0}{r_1} \tag{15}$$

where r_1 is the curvature radius of limiter. Then the load q will be

$$q(x_c) = \begin{cases} \dfrac{EI_0}{r_1\left[(l-a)x_c - \dfrac{(l^2-a^2)}{2}\right]} & 0 \le x_c \le a \\ -\dfrac{2EI_0}{r_1(l-x_c)^2} & a \le x_c \le l \end{cases} \tag{16}$$

Combining equations (7), (10), (11), (12), (13), (14) and (16), we can achieve the relation of valve stiffness and valve displacement. So the elastic force is easily achieved by equation (17).

$$F_{sp} = k \cdot w \tag{17}$$

The calculation of the maximum bending stress of the valve is divided into two processes, that is, the maximum bending stress of the valve occurs at the root of the valve before winding, as shown in equation (18), and the maximum bending stress of the valve occurs at the winding position after winding, and changes with the change of the winding position, as shown in equation (19).

$$\sigma_{\max} = \frac{96EI_0w(l+a)}{17d^3 + 32a(3l^2 + a^2)}\frac{t}{I_0} \tag{18}$$

$$\sigma_{\max} = \frac{M}{W_z} = \frac{EI_0}{r_1W_z} = \frac{Et}{2r_1} \tag{19}$$

According to equation (18) and (19), we can get the change of the maximum bending stress of the valve with the opening process of the valve. It is worth noting that the maximum bending stress of the valve is a constant value when the valve is winded on the limiter, and it occurs at the real-time position of the winding point.

3. Results and Discussion

3.1 Calculated Results

Table 1 shows the parameters of the two valve components used in this calculation. The limiter in valve component A is a continuous limiter, and the limiter in valve component B is a discontinuous limiter. Based on the above theoretical principle, the results of two valve components can be obtained.

Table 1. The structural and material parameters of two valve components.

Parameters	Valve component A	Valve component B
Deformation length of valve/mm	23.05	27
Valve width/mm	4.2	6
Valve thickness/mm	0.254	0.381
Valve head diameter/mm	12	13
Elastic modulus/GPa	210	210
Limiter lift/ mm	2	2.2
Curvature radius of limiter/mm	73.57	130
Center offset of limiter/mm	0	3.55

As shown in Figure 5, the initial winding point of the valve component A under the constraint of continuous limiter occurs at $x = 0$ when the valve displacement reaches 1.321 mm. For valve component B, when the valve displacement reaches 2.115 mm, the initial winding point of the valve is located at the position of $x = 14.8$ mm. According to the two groups of results, it can be seen that the valve is gradually winding and opening under the continuous limiter profile, while the valve is directly hitting the limiter at a certain distance from the root under the discontinuous limiter profile.

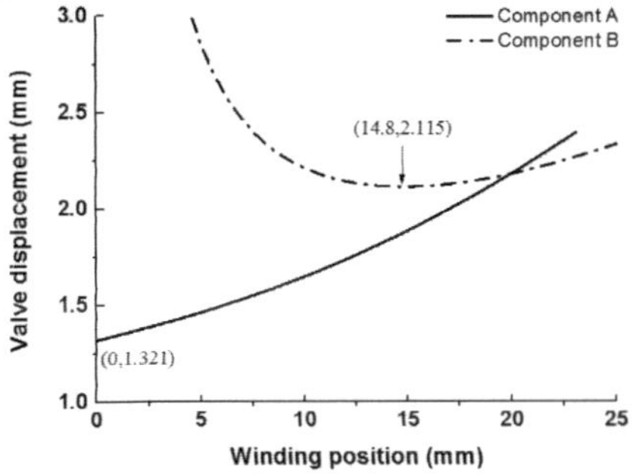

Figure 5. The initial winding position of two valve components.

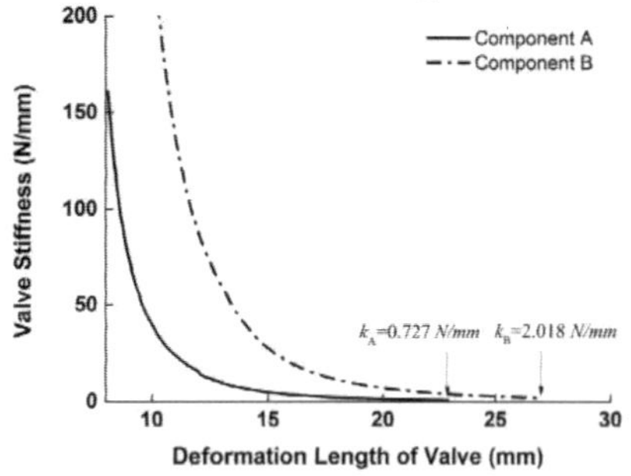

Figure 6. The change of valve stiffness with the deformation length of valve for two components.

Figure 6 shows the variation of valve stiffness of the two valve components with the effective working length. When the effective working length is the initial length, the initial stiffness of valve component A is 0.727 N/mm, and that of valve component B is 2.018 N/mm. The results show that the effective working length of the valve will decrease when the valve is winded or impacted on the limiter, and the valve stiffness will increase with the decrease of effective working length.

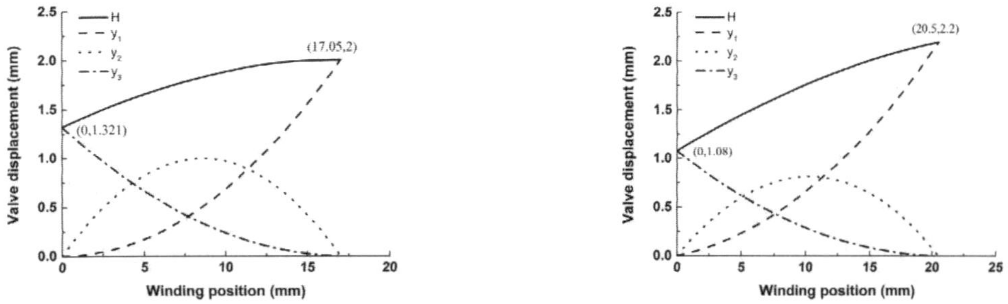

(a) Component A (b) Component B

Figure 7. The relation of valve displacement and effective working length.

In order to establish the relationship between the valve stiffness and the valve displacement, the relationship between the valve displacement and the effective working length is obtained, as shown in Figure 7. The results show that the total displacement H increases with the decrease of effective working length for both valve components, where y_1 increases gradually, y_2 increases first and then decreases, and y_3 decreases gradually.

The relationship between valve stiffness and displacement is shown in Figure 8a. The valve stiffness is constant when the displacement is less than the initial winding displacement, and increases sharply after the displacement exceeds the initial winding displacement. As the valve component A adopts continuous limiter, its initial winding displacement 1.321 mm is more forward than that of valve component B 2.115 mm with discontinuous limiter. The initial valve stiffness in valve component A is 0.727 N/mm, and that of valve B is 2.018 N/mm. Since the initial winding displacement of valve component B is 2.115 mm, which is close to the lift of 2.2 mm, the valve is basically in a constant state. According to the relationship between the valve stiffness and displacement, the relationship between the elastic force and displacement in Figure 8b can be calculated. It can be seen from the figure that the elastic force in valve component A first increases linearly and then increases sharply, while the elastic force in valve component B almost always increases linearly. According to the relationship between the elastic force and the displacement, the work done by the elastic force in the process of movement can be obtained. The work of valve from 0 to 2 mm in valve component A and B are 12.13 mJ and 4.04 mJ, respectively. Although the initial valve stiffness of valve component B is higher than that of valve component A, the work of valve component B is lower than that of valve component A because of the difference of limiter profile. In addition, the maximum bending stress is calculated. The results indicated that the maximum bending stress of valve component A increases linearly before the displacement of 1.321 mm, and keeps constant at 363 MPa when it reaches the winding point. However, the maximum bending stress of valve component B increases linearly all the time, and the maximum value reaches 601 MPa, which is higher 238 MPa than that of valve component A.

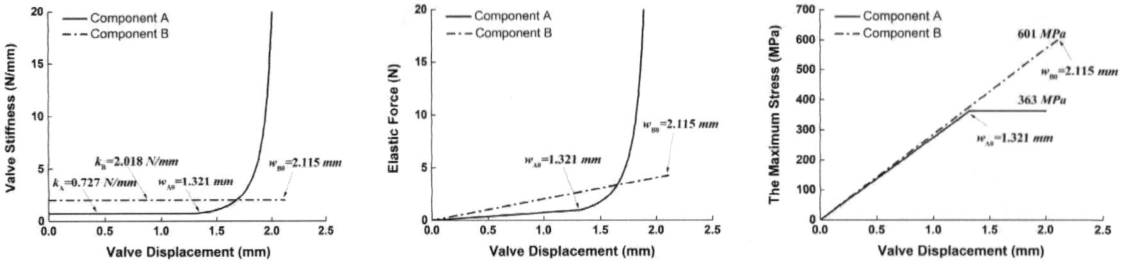

(a) Valve stiffness (b) Elastic force (c) Maximum stress

Figure 8. The relationship between the stiffness, elastic force and maximum stress of valve and the valve displacement.

It can be seen that the limiter profile has an importance on the winding displacement, valve stiffness, elastic force, work and maximum stress. For continuous limiter, the valve can show variable stiffness characteristics in the process of movement, while it is always in the state of equal stiffness for discontinuous limiter.

3.2 Simulation Verification

In order to verify the accuracy of the above theoretical methods and results, finite element method (FEM) is used to simulate the mechanical and deformation of the continuous limiter and discontinuous limiter coupling valve. By applying uniform load step from small to large on the valve head, the valve displacement under different loads is calculated, and then the valve stiffness is calculated and compared with the above theoretical results.

Figure 9 shows the present investigation and FEM results of the two valve components. For valve component A, the initial valve stiffness obtained by the FEM is 0.725 N/mm, which is slightly lower than the present valve 0.727 N/mm, and the calculation error is 0.28% based on the FEM data. For valve component B, the initial stiffness of the valve obtained by the FEM is 2.100 N/mm, which is slightly lower than the present valve 2.018 N/mm, and the calculation error is 3.90% based on the FEM data. It can be seen that the valve stiffness value obtained by theoretical method is relatively accurate. In Figure 9, the elastic force obtained by the FEM also presents a process of linear increase first and then sharply increase, which is consistent with the present investigation. However, the results obtained by the two methods have some errors, and the FEM has greater deformation under the same elastic force after valve reaching the winding displacement, the maximum difference is less than 0.2 mm. This is because the elastic deformation of the limiter is considered in the FEM, while the theoretical calculation method treats the limiter as a rigid body, so it is normal to have this error.

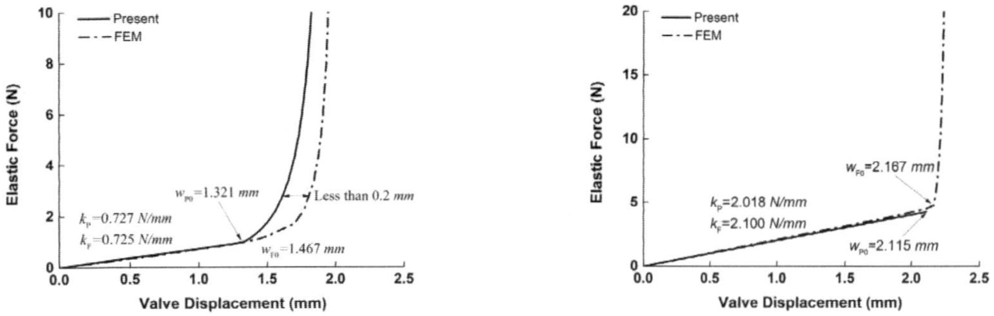

(a) Component A (b) Component B

Figure 9. The comparison of present investigation and FEM results.

3.3 Experimental Verification

In order to further verify the accuracy of the mechanical model in this paper, the high-speed photography (HSP) experiment of valve movement was carried out to observe the change trend of valve under the action of continuous limiter and discontinuous limiter, as shown in Figure 10. During the experiment, a thin rod was used to manually open the valve head from the flange hole. A high-speed camera was used to observe the valve component from the side. The sampling rate of the high-speed camera was set at 3000 s^{-1}, and the resolution was 1024 × 768. It can be seen from the results that the opening process of the valve under the continuous limiter is gradually around the limiter, while under the discontinuous limiter, the valve directly impacts with the limiter at the neck. Figure 10c shows the image of valve and limiter for valve component B at the winding moment. The winding position is at 15.8 mm away the starting point, which is very close to the theoretical value 14.8 mm. When the valve impacts on the limiter at the neck, clearly clearance can be found between the valve and limiter at the waist. This phenomenon is consistent with the theoretical analysis process, which proves the accuracy of the mechanical model in this paper again.

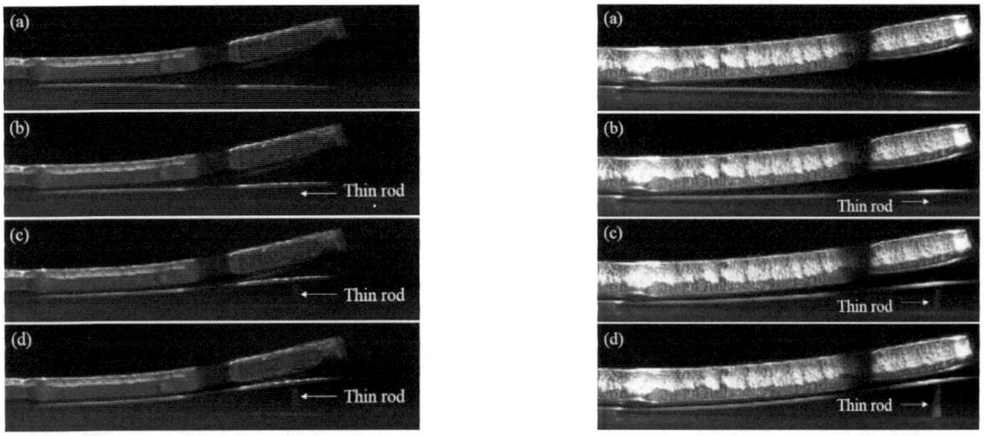

(a) Component A (b) Component B

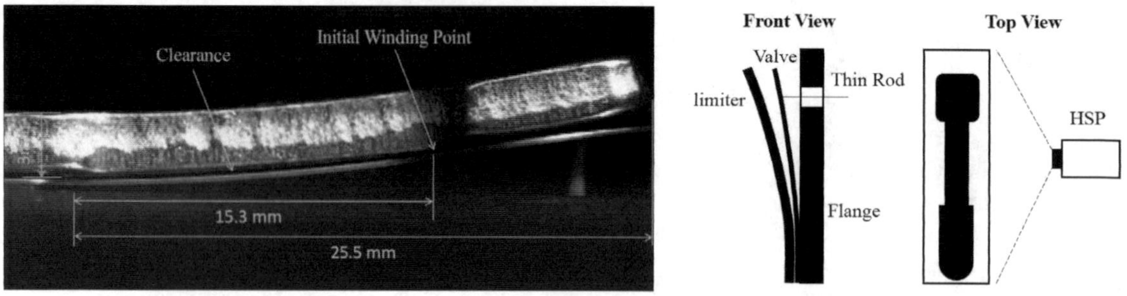

(c) The initial winding point of component B (d) The sketch of the experiment set-up

Figure 10. High speed photography of valve opening process.

3.4 Structural Optimization

Due to the importance of limiter profile, the discontinuous limiter profile in valve component B is optimized. When the center distance and lift are kept unchanged, the limiter profile is adjusted to continuous type with the limiter curvature radius changed from 130 mm to 96.6 mm, as shown in Figure 11.

The initial winding point was obtained in Figure 11. Compared with original structure, the optimized structure changes the initial winding position from x = 14.8 mm to x = 0, and change the winding displacement from 2.115 mm to 1.453 mm. The optimization greatly advances the winding effect of valve and limiter, and avoids the impact of the valve on limiter.

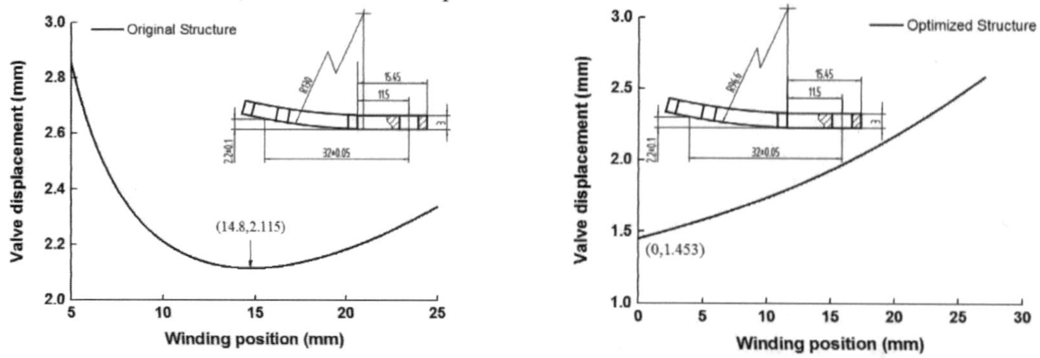

(a) Original structure (b) Optimized structure

Figure 11. Optimization of limiter profile and the initial winding position of valves.

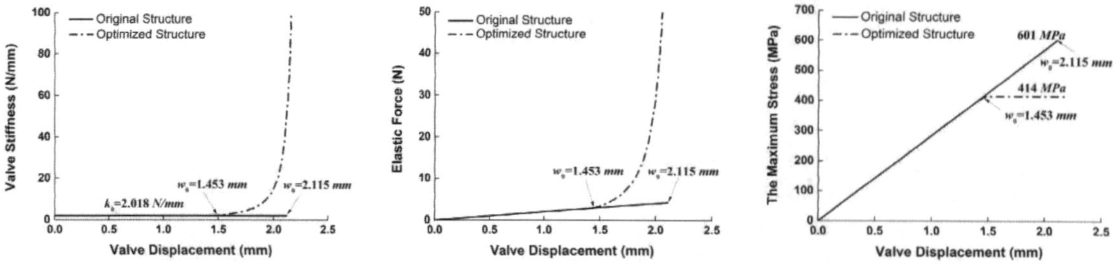

(a) Valve stiffness (b) Elastic force (c) Maximum stress

Figure 12. The relationship between the stiffness, elastic force and maximum stress of valve with the valve displacement.

Figure 12 shows the change relationship between the stiffness, elastic force and maximum stress of valve with the valve displacement before and after the limiter optimization. It can be seen that the stiffness and elastic force begin to change sharply after 1.453 mm, which makes the limiter have better restraint effect. The work of valve from 0 to 2 mm for original and optimized structure are 4.04 mJ and 7.42 mJ, respectively, and the work of valve in the optimized structure increased 83.7% than it in the original structure. After optimization, the maximum bending stress of valve is reduced to 414 MPa, which is reduced by 187 MPa and 31.1% compared with 601 MPa before optimization.

4. Conclusions

In this paper, the mechanical and deformation characteristics of the discharge valve under the restriction of limiter are studied by means of theory, simulation and experiment. The main conclusions are as follows:

(1) The mechanical model of discharge valve considering the coupling effect of limiter is established, and the important parameters such as displacement, stress, stiffness, elastic force and work of valve are obtained.

(2) The results show that the limiter profile has a significant effect on the movement of valve. For continuous limiter, the valve is gradually winded with limiter and shows variable stiffness characteristic, while the valve directly impacts the limiter and always in the state of equal stiffness for discontinuous limiter.

(3) The error between the present investigation and the FEM is less than 3.9%, and the HSP results show that the motions of valve are consistent with the model results.

(4) The limiter profile in component B is optimized. After optimization, the initial position of the valve is advanced from 2.115 mm to 1.453 mm, which shows a better constraint effect.

Acknowledgments

The authors gratefully acknowledge the support of the Guangdong Basic and Applied Basic Research Foundation (2019A1515010787).

References

[1] Dusil R and Appell B 1976 Fatigue and Fracture Mechanics Properties of Valve Steels *International Compressor Engineering Conference* **174** 82–90

[2] Dusil R and Johnansson B 1978 Fatigue Fracture Behaviour of Impact Loaded Compressor Valves *International Compressor Engineering Conference* **255** 124–128

[3] Glaeser W A 1999 Failure Mechanisms of Reed Valves in Refrigeration Compressors *Wear* **225– 229** 918–924

[4] Drews R E 1986 The Development of a Longer Life and Superior Performance Compressor Plate Valve *International Compressor Engineering Conference* **538** 362–364

[5] He Z L, Chen Q, Li D T, Wang J, Xia P and Wang T 2017 Research on Noise Reduction of Reed Valves of a Hermetic Refrigerator Compressor *IOP Conference Series: Materials Science and Engineering* **232** 012014

[6] Hareland M, Hoel A, Jonsson S, Liang D, and Chai G C 2014 Selection of Flapper Valve Steel for High Efficient Compressor *International Compressor Engineering Conference* **2330** 1408

[7] Kim J Y， Park S G， Oh I K 2009 Fluid Structure Interaction Analysis of a Reciprocation Compressor *International Compressor Engineering Conference* **1936** 1112

[8] Zhang C N 1998 Influence of The Lift of Reed Valve on Its Service Life *Fluid Mechinery* **26** 4

[9] Kwon Y K, Lee G H and Lee T J 2004 The Design of Compressor Valve to Consider the Flexibility and Reliability *International Compressor Engineering Conference* **1705** C127

[10] Zheng X P, Wang C Z and Guo X P 1999 Research on Profile of Lift Limiter for Reed Valve *Compressor technology* **6** 7–10

[11] Wang F 2011 Study on the Reed Valve of Piston Refrigeration Compressor *Fluid Mechinery* **39** 42– 45

[12] Ho J S and Chen R Y 1992 The Effect of Profile of Backing Plate Upon the Fatigue Life of a Cantilever Discharge Valve Reed *International Compressor Engineering Conference* **886** 927–933

[13] Mu G Y, Wang F and Mi X 2017 Comparison of Motion Models of Reciprocating Compressor Discharge Reed Valve and Analysis for Factors Affecting Dynamic Characteristics *Journal of Xi'An JiaoTong University* **51** 69–76

Force assisted discharge valve for piston compressors

L Dür[1], A Egger[1], M Lang[1], R Almbauer[1]

[1]Graz University of Technology, 8010 Graz, Austria

duer@ivt.tugraz.at

Abstract. Due to the enormous number of domestic refrigeration appliances worldwide and the associated high overall energy demand of the cooling sector, manufacturers are required to develop more efficient cooling devices. For this reason, variable speed compressors are increasingly being used in order to better match refrigeration demand and supplied cooling capacity. However, variable speed operation is associated with challenges concerning the valve dynamics. Valve oscillation as well as opening and closing delays reduce the energy efficiency (COP) of the compressor.

The underlying study is about a simulation based potential analysis of a novel force assisted discharge valve to improve the valve dynamics of a small hermetic reciprocating compressor for domestic refrigeration. Discharge valve and valve support mechanism are modelled by a 1d-spring-damper-mass multi-body system and implemented in an existing in-house compressor simulation model.

Systematic methods of statistical design of experiments (DOE) are applied not only to improve the COP but also to generate a robust valve behaviour which is largely independent of the operating conditions of the compressor. The findings of the force assisted discharge valve are basically transferable to any other piston compressor with flutter valves.

1. Introduction

In reciprocating compressors for domestic refrigeration appliances reed valves are widely used to control the in- and outflow of the cylinder. The advantages of such valves are their simple and inexpensive design as well as their capability to automatically adapt to different operating conditions as the actuation is caused by the pressure difference between cylinder and suction or discharge line. However, reed valves are crucial components in terms of efficiency, cooling capacity, acoustics and reliability of the compressor.

The oscillation frequency of the valve must be matched to the compressor speed to ensure correct timing of the opening and closing process. This is well manageable for fixed-speed compressors since the oscillation frequency of conventional reed valves mainly depends on the effective moving mass, the stiffness and the preload and is therefore a system constant. In variable-speed compressors, however, correct valve timing is far more challenging as the duration of the suction and discharge phase varies greatly. Variable-speed compressors are increasingly used in domestic refrigeration as they allow for a better match of refrigeration demand and supplied cooling capacity of the refrigeration unit [7]. A further improvement of the valve dynamics of variable-speed compressors requires the consideration of new valve concepts.

The connection between valve closing delays and gas back-flow is investigated in [9], where the authors measured the valve motion as well as the pV-diagram of a hermetic reciprocating compressor. [8] measured the suction valve motion of a reciprocating compressor for varying compressor speeds. It was found that the volumetric efficiency varies in relation to the compressor speed. Reed valves open and close several times during the suction phase and the discharge phase, which is commonly referred to as valve flutter. Since this flutter frequency is a system constant, different compressor speeds lead to valve closing delays that reduce the volumetric efficiency. The importance of reducing the valve flutter to increase the efficiency is also mentioned in [2]. The authors highlighted, that the COP improvement potential of a theoretical suction valve without valve flutter is 2%.

Recently, a few studies [3–5] were published which focus on practical approaches to reduce valve flutter and increase the valve dynamics of suction reed valves in reciprocating compressors for domestic refrigeration. The authors investigated electromagnetic and mechanical actuations to influence the valve dynamics of conventional suction reed valves.

The focus of this study is on the valve dynamics of the discharge valve in a hermetic reciprocating compressor for domestic refrigeration. A novel concept of a force assisted discharge valve based on a valve support mechanism is investigated. For this purpose, an in-house 0d/1d compressor simulation model is adopted to allow a simulation based potential analysis of the force assisted discharge valve.

Methods of statistical design of experiments (DOE) are applied to find optimal parameters for the discharge valve support mechanism.

2. Methodology

The first part of this chapter gives a short overview of the discharge valve support mechanism and the overall simulation strategy that was chosen for the simulation based potential analysis. Subsequently, the selected compressor simulation model and the modelling approach of the force assisted discharge valve are introduced. The last part of this chapter outlines the applied design of experiments approach of the simulation based potential analysis.

2.1. Discharge valve support mechanism

The principle of the discharge valve support mechanism can be seen in Figure 1 a.) and b.). The mechanism basically consists of a spring element and a rotating lever which transfers the supporting force to the discharge valve. This additional supporting force should lead to faster valve opening, reduced valve flutter as well as timely valve closing. Shortly before the piston reaches top dead center, the lever is moved back to allow the valve to close properly in time.

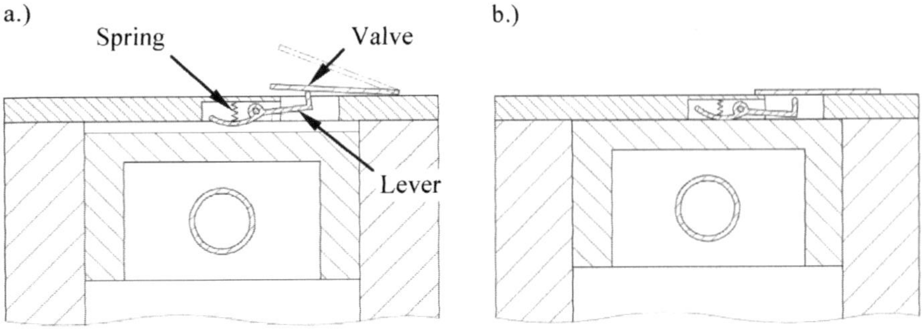

Figure 1. Force assisted discharge valve

2.2. Overall simulation strategy

The underlying study is about a first potential analysis of a novel discharge valve system. Due to the novelty of the system and the associated lack of system knowledge, a large number of potentially relevant system factors were identified. Many simulations are therefore required to determine the improvement potential of this new system, which is why a computationally intensive 3d simulation approach is not considered appropriate. Instead, an already existing in-house 0d/1d compressor simulation model based on [6] is adapted to comply with the compressor under consideration. It resolves the most important physical effects with sufficient accuracy and is therefore particularly suitable for a potential analysis.

The simulation based potential analysis focuses on two different operating conditions of the ASHRAE process [1], which only differ with regard to the condensing temperature level. In the following, these two operating conditions are labeled as -23.3°C/55°C and -23.3°C/45°C, referring to the evaporating and condensing temperature level. Since the compressor is operated by the refrigerant R600a, the corresponding pressure levels are 0.64 bar/7.73 bar (-23.3°C/55°C) and 0.64 bar/6.04 bar (-23.3°C/45 °C).

2.3. The 0d/1d compressor simulation model

The schematic gasline structure of the compressor simulation model is illustrated in Figure 2. It mainly consists of volumes connected by pipes. The cylinder is modelled based on the first law of thermodynamics according to equation (1).

$$-p * \frac{dV}{dt} + \delta\dot{Q}_a + \sum_i \left(\frac{dm_i}{dt} * h_i\right) = \frac{d(m * u)}{dt} \tag{1}$$

The volumes of the compressor simulation model are calculated similarly to the cylinder. However, compared to equation (1), the first term corresponding to the work associated with a change in volume is not present. In contrast to the cylinder and the volumes, which are modelled as 0d components, the pipes are modelled as 1d components. The flow calculation inside the pipes is carried out numerically based on a 1d-CFD solver which is implemented in the compressor simulation model. The mass flows through the valves are calculated through equation (2), where μ is a flow coefficient capturing all flow losses and $\dot{m}_{th}$ is the theoretical mass flow of an isentropic flow assuming perfect gas behaviour. The flow coefficient μ is a function of the valve lift and has to be calculated through 3d-CFD simulations or, alternatively, can be determined by measurements. The valve movement of the compressor simulation model is modelled using a 1d spring-damper-mass system. A more detailed description concerning the modelling of the force assisted discharge valve is given in the next section. The heat transfer is represented by a thermal network which is included in the compressor simulation model. For more information regarding the modelling approach of the compressor simulation model, see [6].

$$\dot{m} = \mu * \dot{m}_{th} \tag{2}$$

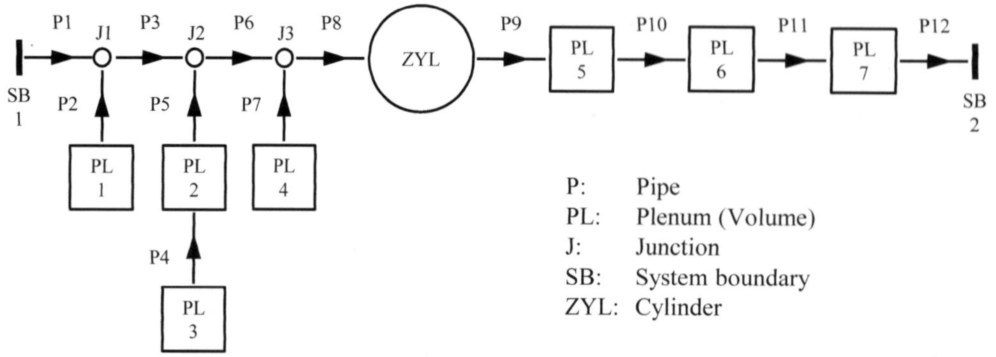

Figure 2. Schematic gasline structure of the in-house 0d/1d compressor simulation model [6]

2.4. Modelling of the force assisted discharge valve

Figure 3 shows the standard discharge valve system of the considered compressor, a $9.5cm^3$ hermetic reciprocating compressor for domestic refrigeration appliances. The valve has a three-part structure consisting of the valve lamella, a damper and the clamp that fixes damper and lamella to the valve plate. The damper tensioned between the discharge valve and the clamp increases the stiffness and limits the lift of the lamella.

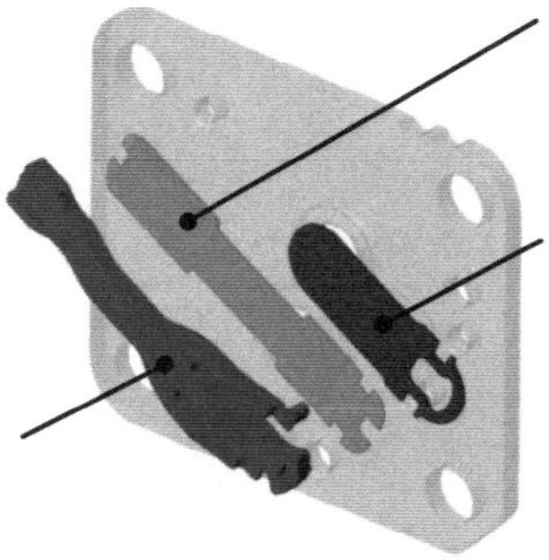

Figure 3. Discharge valve system

Due to the symmetrical design of the discharge valve, the system can be reduced from 3d to 2d. Since only the valve lift influences the mass flow rate, a description of the valve system in 1d is sufficient. Therefore, simple spring-damper-mass systems for moving parts like the valve lamella and the damper were introduced.

$$\omega_{0_1} = \sqrt{\frac{c_1}{m_1}} \tag{3}$$

In addition to the parts of the standard discharge valve system, the valve support mechanism is included in the force assisted discharge valve model, forming a 1d multi-body-system as illustrated in Figure 4.

The point mass m_1 which corresponds to the valve lamella is modelled according equation (3) where c_1 is the stiffness of the lamella in valve opening direction and ω_{0_1} is the first undamped natural angular frequency of the lamella. Both system constants, ω_{0_1} and c_1, were obtained from a 3d-FEM calculation. The second point mass m_2 which corresponds to the damper element was modelled according to the same principle as the lamella, with m_2 located at the contact point between the damper and the valve lamella. The discharge valve support mechanism is modelled as a rigid rotary oscillator. The piston, clamp and valve plate are also modelled as rigid bodies, whereby in contrast to the piston, the clamp and valve plate are fixed in position.

All contacts are modelled based on 1d force elements, except for the contact between valve lamella and valve plate, where a rigid body contact model is used. Force elements basically consist of a spring and a damper element. A trigger function, which is influenced by displacement and velocity differences of the contact points, ensures that only compressive contact forces are transmitted. The rigid body contact between valve lamella and valve plate is described as an impact without temporal expansion. A so-called restitution factor ε is used to calculate the velocity of the valve lamella after the impact on the valve plate.

The valve motion is influenced by the conventional valve force driven by the pressure difference between upstream and downstream side of the discharge valve, as well as the additional supporting force generated by the valve support mechanism. This supporting force is influenced by the piston position, the torsional spring stiffness, the preload angle and the moment of inertia of the lever.

The multi-body-system of the force assisted discharge valve is mathematically described by an inhomogeneous, non-linear differential equation system of first order which is solved through a fourth-order Runge-Kutta approach.

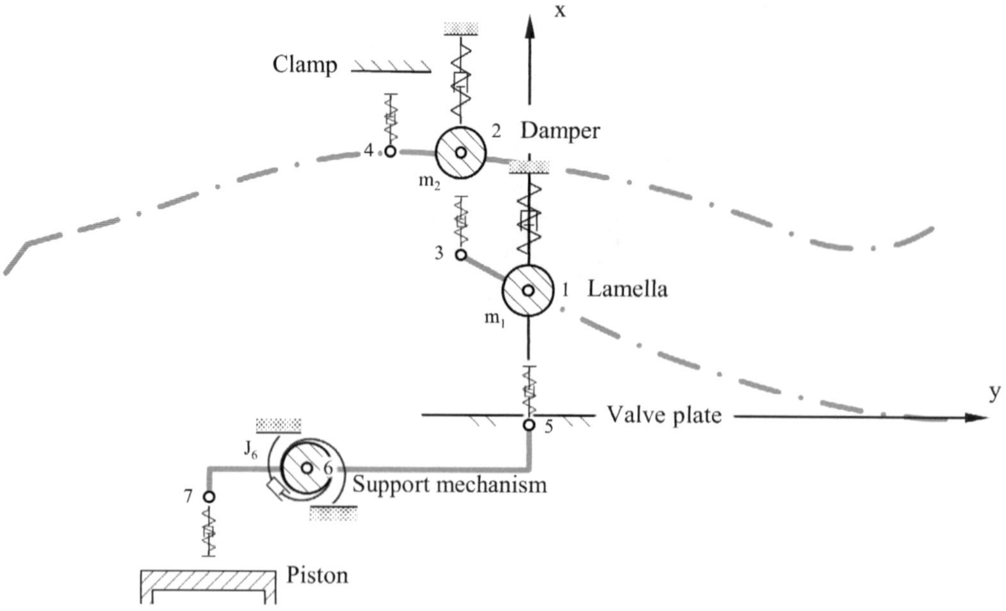

Figure 4. 1d multi-body-system of the force assisted discharge valve

2.5. *Application of the design of experiments approach*

The potential analysis of the force assisted discharge valve is carried out using a design of experiments (DOE) approach. The DOE approach requires a clear definition of the system und consideration with all its potential influences. For this purpose, a system boundary must first be defined. The parameter diagram shown in Figure 5 systematically summarises all system variables related to the force assisted discharge valve system.

The input parameters are divided into design and signal parameters. The design parameter pool includes the geometrical shape of the support mechanism as well as the spring stiffness, the preload angle and the density of the material, while the compressor speed and the condensing pressure level are defined as signal parameters. Factors represent the subset of the parameters whose values (levels) will be varied within the DOE approach to determine the effect on the responses. These responses represent measurable output variables which provide a meaningful information about the system under study. In this case, the discharge mass and the discharge work are chosen as responses, as both are directly influenced by the discharge valve dynamics. It should be noted that depending on the final design of the discharge valve support mechanism, the cylinder clearance volume might be influenced. However, this influence is not yet considered in this potential analysis.

Once the factors and the responses are defined, an adequate experimental design must be selected. Effect diagrams, i.e. main effect and interaction effect plots which involve simple linear models, are chosen as primary tool to improve the design parameters of the force assisted discharge valve. Thus, mainly two-level fractional factorial experimental designs are selected as experimental designs for the simulation based DOE.

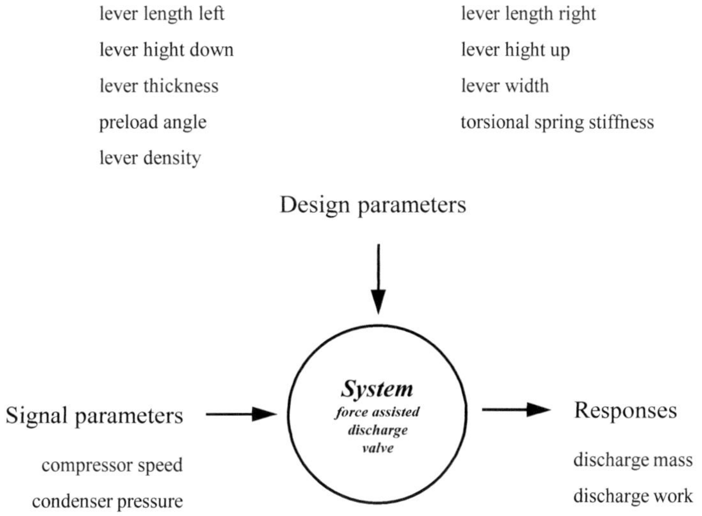

Figure 5. Parameter diagram of the force assisted discharge valve

Several design parameter adjustment loops resulted in a set of design parameters (see Table 1) that lead to a better overall performance of the compressor and its responses.

Table 1. final set of the design parameters

nr.	parameter	unit	value
1	lever length left	[mm]	2.0
2	lever length right	[mm]	13.5
3	lever hight down	[mm]	0.5
4	lever hight up	[mm]	1.8
5	lever thickness	[mm]	0.9
6	lever width	[mm]	6.0
7	preload angle	[°]	-4.5
8	torsional spring stiffness	[N*m/rad]	0.135
9	lever density	[kg/m^3]	7900

3. Results

In this chapter, the final results of the potential analysis are presented. Figure 6 a.) and b.) show the discharge valve motion over the crank angel and the corresponding pV-diagram at -23.3°C/45°C and a compressor speed of 3000 rpm. The black solid line in Figure 6 a.) refers to the displacement of the discharge valve lamella at point 1, while the black dashed line shows the displacement of the valve support mechanism at point 5 (see Figure 4). The grey solid line represents the standard discharge valve lamella motion, having no valve support mechanism. It can be seen that the discharge valve support mechanism eliminates the intermediate oscillation (valve flutter) of the valve lamella, resulting in a smoother valve motion and a reduced valve closing delay. As a consequence of the smoother valve motion, the pressure fluctuations during the discharge phase are reduced, see Figure 6 b.).

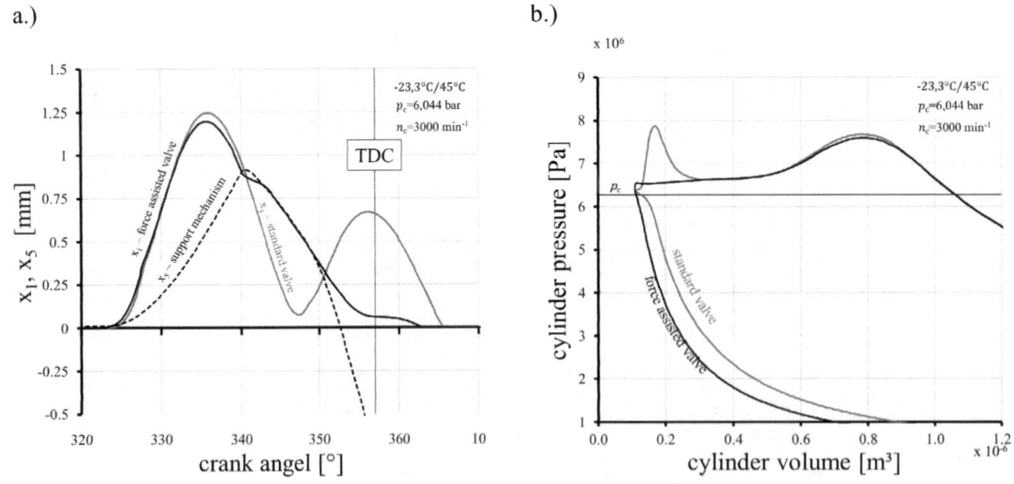

Figure 6. Comparison of standard valve and force assisted discharge valve

Figure 7 a.) and b.) show the responses discharge mass and discharge work over a wide compressor speed range. The force assisted discharge valve leads to a smoother relationship between discharge mass and compressor speed. Thus, between 3000 rpm and 4500 rpm a significant improvement of the discharge mass can be observed. Furthermore, the difference between the two operating conditions -23.3°C/45°C and -23.3°C/55°C is smaller, indicating that the delivered discharge mass is less dependent on the operating conditions if a force assisted discharge valve is used.

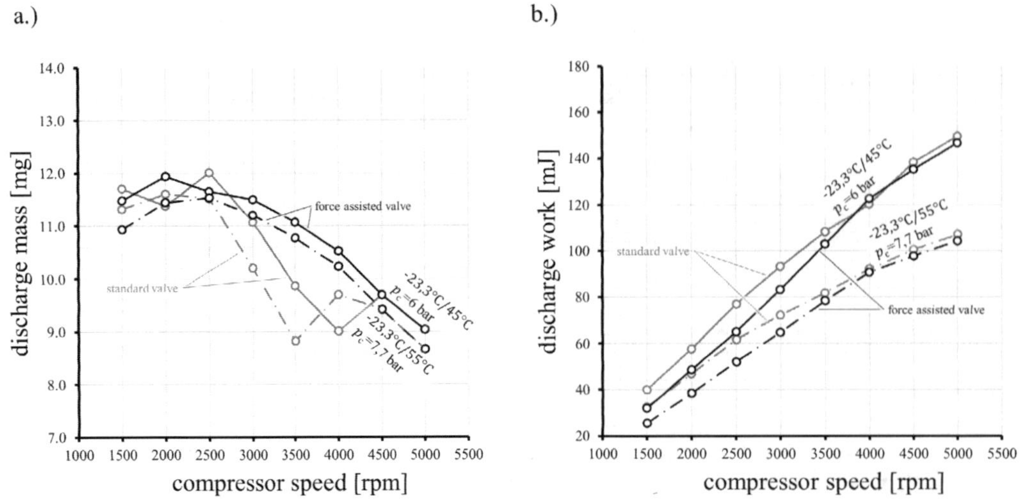

Figure 7. Reponses of the force assisted discharge valve in comparison with the standard valve

Figure 7 b.) shows the corresponding curves of the discharge work over the compressor speed. Although the qualitative behaviour of the force assisted discharge valve is similar to that of the standard valve, some improvements of the discharge work are obtained especially at lower compressor speeds. At low compressor speeds, valve flutter is more intensive than at high compressor speeds, leading to significant cylinder pressure oscillations during the discharge phase which increase the discharge work. The force assisted discharge valve reduces these pressure oscillations and thus reduces the required discharge work.

Table 2. Results of the force assisted discharge valve and comparison with the standard valve

T_0/T_c [°C]	speed [rpm]	discharge mass [mg]	Δ [%]	discharge work [mJ]	Δ [%]	ΔCOP [%]
-23.3/45	1500	11.480	-1.9	31.966	-19.8	0.4
-23.3/55	1500	10.926	-3.4	25.501	-21.5	0.2
-23.3/45	2000	11.933	4.9	48.550	-15.8	0.5
-23.3/55	2000	11.441	-1.4	38.357	-18.0	0.4
-23.3/45	2500	11.649	-3.0	65.137	-15.4	0.6
-23.3/55	2500	11.529	0.1	51.982	-15.8	0.6
-23.3/45	3000	11.497	3.9	83.277	-10.7	0.9
-23.3/55	3000	11.199	9.8	64.880	-10.2	1.0
-23.3/45	3500	11.064	12.2	102.826	-5.0	1.2
-23.3/55	3500	10.768	22.1	78.524	-4.0	1.8
-23.3/45	4000	10.518	16.8	122.703	2.0	1.7
-23.3/55	4000	10.226	5.5	90.707	-1.6	0.4
-23.3/45	4500	9.688	0.0	135.217	-2.3	-0.2
-23.3/55	4500	9.411	0.0	97.767	-2.6	0.1
-23.3/45	5000	9.034	0.1	146.680	-1.9	0.0
-23.3/55	5000	8.661	0.0	104.188	-2.7	0.2
	mean		4.1		-9.1	0.6

The quantitative values of the responses as well as their relative change compared to the standard valve are shown in Table 2. On average, the force assisted discharge valve increases the discharge mass by 4.1% and reduces the discharge work by 9.1%, considering two different operating conditions and a wide compressor speed range. The last column in Table 2 represents the theoretical COP improvement neglecting changes in the mechanical and the electrical efficiency of the compressor. Although significant improvements of discharge mass and discharge work are achieved, the theoretical COP improvement is only moderate.

4. Conclusion

The underlying study includes a potential analysis of a novel force assisted discharge valve system. The results of the potential analysis show that the valve support mechanism significantly improves the valve dynamics of the discharge valve. In general, the valve dynamics is more robust in relation to different operating conditions and compressor speeds. Especially in variable speed operation, reduced valve oscillation and reduced valve closing delays lead to improvements of the discharge mass and the discharge work over a wide compressor speed range. Local break-ins in the discharge mass as a result of an unfavorable valve timing can be avoided. Depending on the operating condition and the compressor speed, the discharge mass increases up to 22.1 % and discharge work reductions up to 21.5% were obtained. Furthermore. the analysis indicated a theoretical COP improvement potential up to 1.8%. However, such high improvements can only be achieved locally rather than over the entire compressor speed range. The average COP improvement potential of the force assisted discharge valve over two different operating conditions and a wide compressor speed range was determined to be 0.6%.

The potential analysis of the force assisted discharge valve shows promising results, however, the practical implementation of the valve support mechanism still faces some challenges regarding the influence on reliability and cylinder clearance volume.

Finally, it should be mentioned that the investigated concept of a force assisted discharge valve is not limited to hermetic reciprocating compressors for domestic refrigeration. It can be applied to any piston compressor equipped with reed valves.

IOP Conf. Series: Materials Science and Engineering 1180 (2021) 012019 doi:10.1088/1757-899X/1180/1/012019

Nomenclature

c_1	spring stiffness valve lamella
h	enthalpy
J_6	moment of inertia support mechanism
m	mass
m_1	effective mass valve lamella
m_2	effective mass valve damper
$\dot{m}$	mass flow
$\dot{m}_{\mathrm{th}}$	theoretical mass flow (isentropic flow)
n_c	compressor speed
p	pressure
p_c	condenser pressure
$\dot{Q}_a$	external heatflux
u	internal energy
V	volume
x_1	displacement valve lamella – centre
x_5	displacement support mechanism – contact point to valve lamella
ε	restitution factor
μ	flow coefficient
ω_{0_1}	first undamped angular frequency valve lamella

Abbreviations

0d	zero-dimensional
1d	one-dimensional
2d	two-dimensional
3d	three-dimensional
ASHRAE	American Society of Heating, Refrigerating and Air-conditioning Engineers
CFD	Computational Fluid Dynamics
COP	Coefficient Of Performance
DOE	Design Of Experiments
FEM	Finite Element Method
rpm	revolutions per minute
TDC	Top Dead Centre

References

[1] American Society of Heating, Refrigerating and Air-Conditioning Engineers. 2010. ANSI/ASHRAE Standard 23.1-2010. Methods of Testing for Rating the Performance of Positive Displacement Refrigerant Compressors and Condensing Units that Operate at Subcritical Temperatures of the Refrigerant, Atlanta, Ga.

[2] Burgstaller, A., Nagy, D., Almbauer, R., and Lang, W. 2008. Influence of the Main Parameters of the Suction Valve on the Overall Performance of a Small Hermetic Reciprocating Compressor. *International Compressor Engineering Conference*.

[3] Egger, A., Almbauer, R., Dür, L., Hopfgartner, J., and Lang, M. 2020. Multi-Response optimization applied to a mechanically assisted reed valve of a hermetic reciprocating compressor. *International Journal of Refrigeration* 119, 119–130.

[4] Egger, A., Almbauer, R., Dür, L., Hopfgartner, J., Zuber, B., and Stangl, S. 2019. Experimental Investigation of a Mechanically Assisted Suction Reed Valve in a Small Hermetic Reciprocating Compressor. *IOP Conference Series: Materials Science and Engineering* 604, 12018.

[5] Hopfgartner, J., Posch, S., Zuber, B., Almbauer, R., Krischan, K., and Stangl, S. 2017. Reduction of the suction losses through reed valves in hermetic reciprocating compressors using a magnet coil. *IOP Conference Series: Materials Science and Engineering* 232, 12034.

[6] Lang, W. 2010. *Erstellung von Simulationsmodellen für ausgewählte Problemstellungen bei der Entwicklung von hermetischen Kolbenkompressoren*. Dissertation.

[7] Letschert, V., Desroches, L.-B., Ke, J., and McNeil, M. 2013. Energy efficiency – How far can we raise the bar? Revealing the potential of best available technologies. *Energy* 59, 72–82.

[8] Nagata, S., Nozaki, T., and Akizawa, T. 2010. Analysis of Dynamic Behavior of Suction Valve Using Strain Gauge in Reciprocating Compressor. *International Compressor Engineering Conference*.

[9] Real, M. A. and Gomes Pereira, E. A. 2010. Using PV Diagram Synchronized With the Valve Functioning to Increase the Efficiency on the Reciprocating Hermetic Compressors. *International Compressor Engineering Conference*.

Experimental analysis of reed valve movement for different reed valve designs tested in an impact fatigue test system

Muhammad Waqas Tofique[1*], Alexander Löf[1], Chris Millward[1]
[1]voestalpine Precision Strip AB, Uddeholmsvägen 20, Munkfors, Sweden

waqas.tofique@voestalpine.com

Abstract. Impact fatigue is a phenomenon that is the main cause of failure of reed valves in compressor systems. This phenomenon occurs in compressors due to repeated opening and closing of reed valves during suction and exhaust cycles of compressors. When the valve opens it creates bending fatigue stresses in the body of reed valve; when the valve closes it strikes against the valve plate creating impact stresses. This reed valve movement and impact is repeated billions of times. This cyclic movement influences the impact fatigue life of the reed valve and, hence, the performance of a compressor. The valve movement can be defined in terms of valve frequency, valve lift, valve velocity and impact velocity. Inside a reciprocating compressor, a number of parameters including the valve design, valve material, compressor operating frequency and suction/exhaust pressure influence the reed valve movement.

In this paper, we studied the valve movement parameters for three different suction reed valve designs manufactured from Flap-X – a martensitic stainless steel grade developed for reed valves – tested in a custom-built impact fatigue test system. The valves were excited into movement using compressed air pulses varying in frequency up to 300 Hz and varying pulse width. The valve displacement and frequency was recorded by a laser sensor at 10 000 frames per second. Before starting each test, the operating conditions such as the operating frequency and the air pulse width were pre-set in the impact fatigue test system's control software while the applied pressure was kept constant during the test. The valve response was measured to be different for the different valve designs tested in this study. The reed valve design influenced the important reed movement parameters such as the valve lift, valve velocity, impact velocity and the frequency of valve vibrations. The valves were not tested to failure as the focus of this study was to collect the dynamic data of valve movement. However, for a practical impact fatigue testing of the reed valves intended to achieve fracture of valves, the testing should be conducted at higher frequencies to reduce test times as well as for a stable amplitude of vibration and impact velocity as shown in this study. The information obtained from this study shows the applicability of the presented impact fatigue test system to study the reed valve movement as well as their impact fatigue characteristics.

1. Introduction

In compressor systems, the reed valve is a crucial component as its smooth operation allows the reliable operation of a compressor and, hence, the operation of any appliance that the compressor is installed in.

Such compressors are usually installed in the household or vehicle air conditioning systems, refrigeration systems and other cooling/ventilation systems.

Importance of the reed valve movement can be judged by its effect on the impact fatigue and bending fatigue life of the reed valve. Long impact fatigue and bending fatigue lives of the reed valves and, hence, their compressor systems are desired. The valve movement parameters, such as valve lift, affect the coefficient of performance (COP) of a compressor. Similarly, higher valve lift in a compressor would mean higher reed velocity due to valve's elasticity and, hence, higher impact velocity when the valve shuts down and strikes against the valve seat. The higher impact velocity means higher impact fatigue stresses are established inside the reed valve body after the impact. Similarly, higher bending fatigue stresses arise in the reed valve body when the valve lift is higher.

The problem of impact of reed valves has been studied by several researchers using various experimental as well as numerical modelling techniques. Svenzon [1] and Futakawa et al. [2] conducted the impact fatigue experiments using their custom-built impact fatigue equipment and analysed the fractured valve surfaces. Altunlu et al. [3] have also studied the effect of different tumbling times as well as the different impact velocity levels on the life length of the tested reed valves in their own custom-built impact fatigue test system. Tofique et al. [4] have studied the impact fatigue phenomenon of reed valves when tested against impact plates of different materials. Their study focussed on understanding the influence of the material properties on the impact fatigue life of reed valves. Some other researchers have contributed towards understanding of the impact fatigue phenomenon using different numerical modelling techniques. For instance, Böswirth [5], Soedel [6] and Rigola [7] have proposed different numerical models to deepen the understanding of the reed valve movement behaviour incorporating both the fluid flow parameters as well as the structural properties of the reed valve. These modelling techniques have allowed researchers to study different geometrical aspects of the reed valve as well as to simulate the valve movement behaviour under compressors' operating conditions. Other researchers have carried out validation of the numerical modelling techniques with their experiments. For instance, Mayer et al. [8] and Tao (2018) carried out a validation of Fluid-Structure Interaction (FSI) computational methods by reproducing experimental results of reed valves. There have also been numerous theoretical studies in the past such as by Böswirth [9], Pandeya et al. [10] and more recently by Yu et al. [11] that discussed the dynamic behavior of a generic geometry of reed valves and the impact fatigue stress state generated due to their impact against the impact seat during compressor operation.

Another important piece in solving this puzzle is the effect of the reed valve's design. The design of the reed valves plays an important role in determining not only the magnitude of the valve movement parameters but also the life length of the reed valves. However, in the compressor market there are several types of valve designs that give varying amounts of advantages and disadvantages during the reed valve operation. Therefore, in this experimental study, different types of valve designs were selected to study their valve movement behavior.

The aim of this study was to analyze the influence of different valve designs on the reed valve movement parameters when tested in a custom-built impact fatigue test system. Based on this information, the valve parameters suitable for testing the impact fatigue life of reed valves were proposed.

2. Experimental

Reed valve movement experiments were conducted on a custom-built impact fatigue test rig that, in normal operation, uses air pulses to produce movement of the valves at a range of frequencies (Hz) and pulse widths (milliseconds). The frequency, and pulse-width, is provided as input to the control software on the connected computer.

A schematic sketch of this custom-built impact fatigue test rig is shown in Figure 1. A dedicated compressor provides compressed air at up to 13 bars pressure. This compressed air gets stored in a storage tank. The compressed air is transferred through tubes of 4 mm inner diameter and passes through a flow meter that measures the airflow rate in litres per minute. The air pressure regulator regulates the magnitude of the compressed air pressure before it passes through a high frequency solenoid valve. There is a possibility to increase or decrease the magnitude of pressure supplied to the solenoid valve, and hence the airflow rate, through the pressure regulator. The compressed air pressure reported in this paper is measured at the solenoid valve opening.

The reed valve was made to strike the valve plate repeatedly inducing impact fatigue stresses in them, hence, mimicking the movement of a reed valve in an actual compressor. A laser sensor measures the reed valve's displacement, and thus velocity, and its operating frequency at a sampling rate of 10000 per second. This data is displayed and stored on the control software.

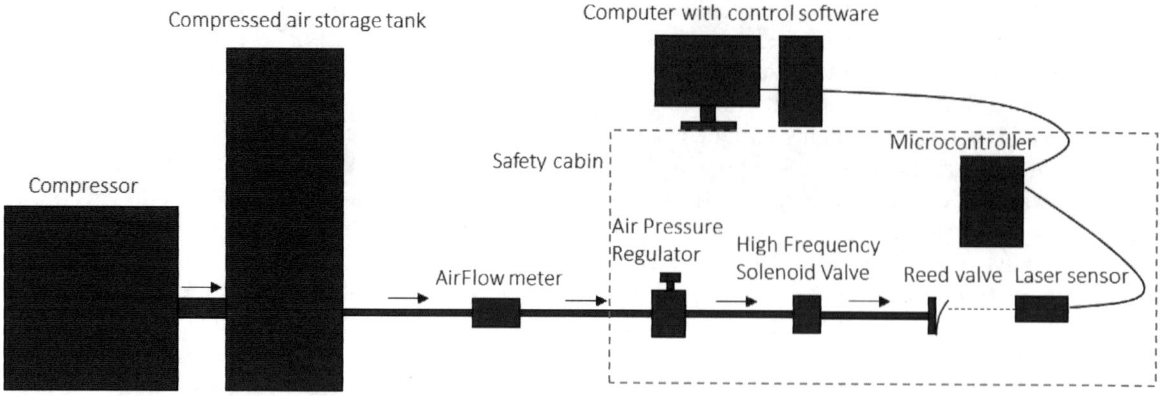

Figure 1. Schematic drawing of the working of the custom-built impact fatigue test system available at voestalpine Precision Strip AB, Sweden that was used to test the reed valve (test specimen) movement in this study.

In the custom-built impact fatigue test system, one important component that influences the reed valve movement is the high frequency solenoid valve. It is a two-way, in-line valve that is electrically actuated with moulded-in cable. Installed in the impact fatigue test system, it receives compressed air after the pressure regulator and through its fast switching action converts it into pressurized air pulses of fixed frequency and pulse-width. These pressurized air pulses are then supplied to a reed valve/test specimen that moves in response. A schematic representation of the switching behavior of the solenoid valve for a trigger pulse-width of 5 milliseconds is shown in Figure 2. It shows that the solenoid valve takes about 1 millisecond to switch on as the electric current flows through the moulded-in wire and into the solenoid coil. As the solenoid valve opens the compressed air pushes through a small opening as the pressure starts to rise to reach its maximum value of 6 bars; the air pressure then stabilizes for approx. 3 milliseconds before the valve switches off; the valve switching off time delay is approx. 0.6 milliseconds; as the solenoid valve starts to close the pressure starts to reduce to its minimum value. This pulsing cycle is repeated a few thousand times as the reed valve movement data is recorded on the control software.

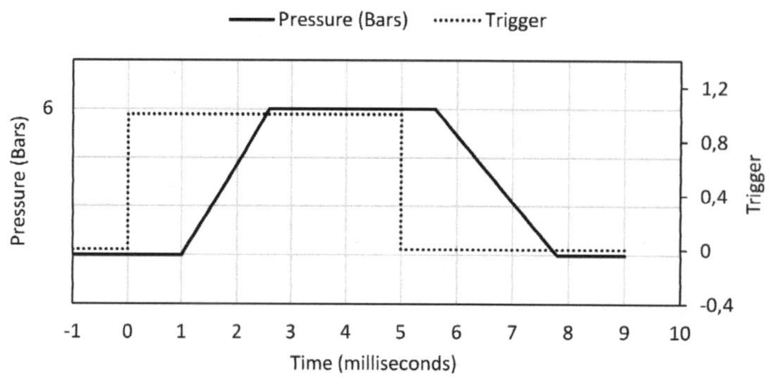

Figure 2. Schematic representation of the switching behavior of the high frequency solenoid valve provided by its manufacturer for compressed air pressure of 6 bars.

2.1. *Test specimen designs & material*

The geometry of the reed valves tested in this work is presented in Figure 3. The valve designs were chosen as typical examples of the suction valves found in the commercially available compressors. The suction valves, typically, are not limited by a valve stopper or space limiter in compressors. Therefore, only suction valve designs were chosen to be studied here. The reed valves were produced by photo-chemical etching process along the rolling direction of the 0.203 mm thick Flap-X material. The reed valves were subsequently tumbled as is common practice in the compressor industry.

The reed valve design 1 consists of a neck and head with a gradual curvature connecting these two components of the valve, see Figure 3a. The reed valve design 2 is like a cantilever beam with the head diameter equal to that of the valve design 1, see Figure 3b. The valve design 3 is a two-legged cantilever beam type with an empty groove in the middle separating the two legs, see Figure 3c.

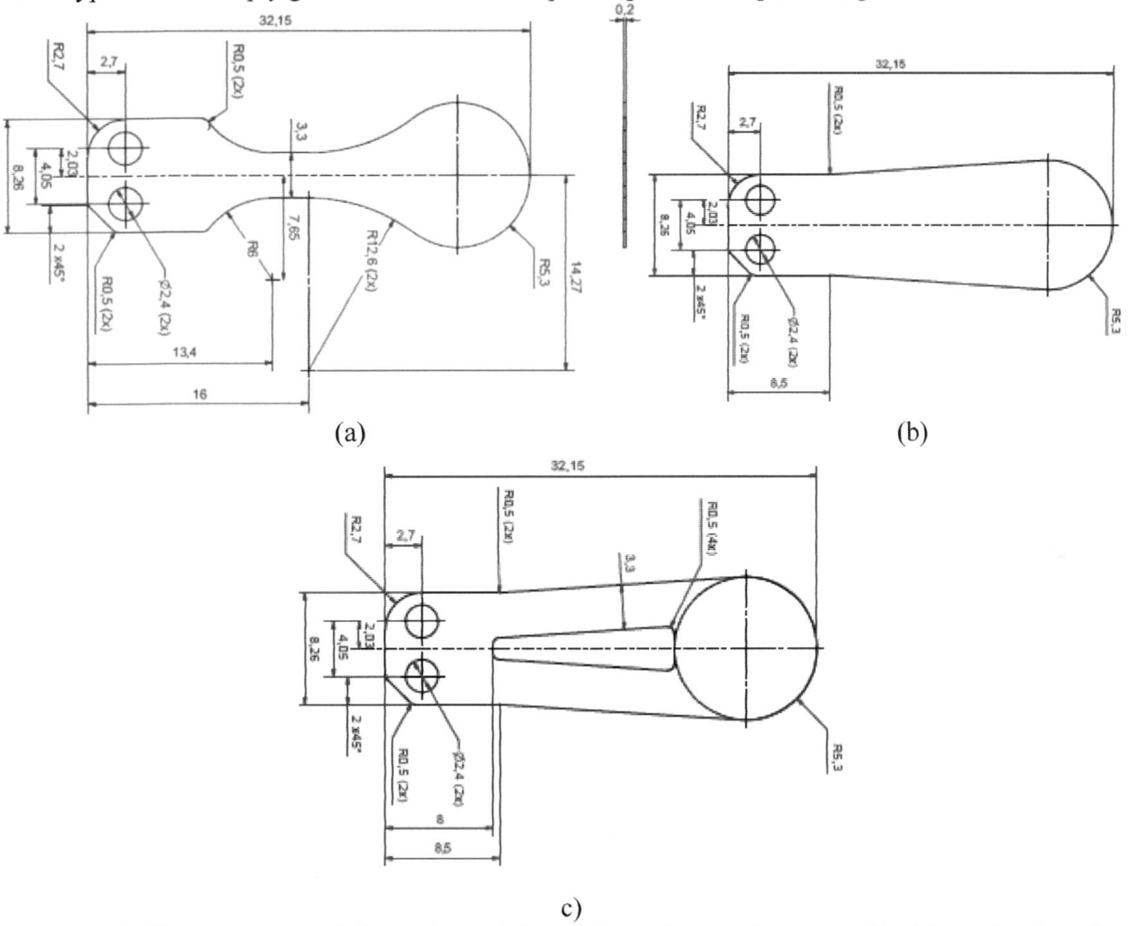

Figure 3. The geometry and dimensions of the suction valve specimens used in this study: a) reed valve design 1; b) reed valve design 2; c) reed valve design 3.

In the impact fatigue test system, all the valve specimens tested in this study were mounted through two pins attached to a steel block passing through two holes in the valves. This steel block covered the bottom end of the valves over a length of 8.5 mm, thus, clamping the valve's movement in any direction in that area, see Figure 4. The reed valves were tested against the tool steel valve plates. No valve limiter/stopper was used instead full free movement of the suction valve was allowed in the outward direction.

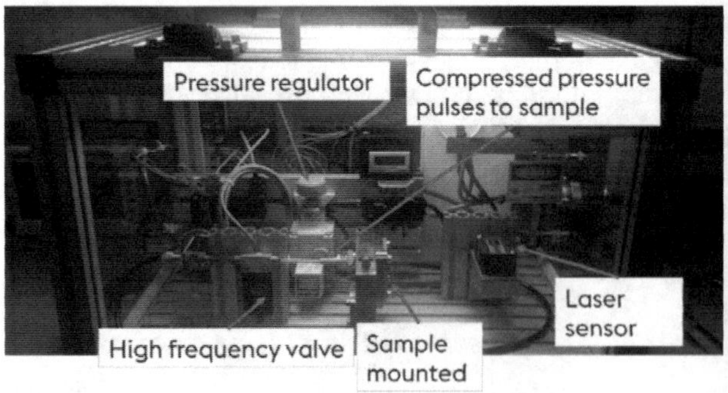

a) b)

Figure 4. In the impact fatigue test system, the reed valve was mounted using a steel block that covered a part of its length. The measuring point of the laser detecting valve movement detecting valve movement data was 2.5 mm from the top edge of the valve head along the centerline.

The reed valves were manufactured from 0.203 mm thick strip of Flap-X steel grade. Flap-X is a martensitic stainless steel that is hardened and tempered to achieve a combination of high tensile strength, ductility and high cleanness that are vital for good impact fatigue and bending fatigue properties. The nominal chemical composition of Flap-X is shown in Table 1

Table 1. Nominal chemical composition (wt. %) of the Flap-X steel grade used for the reed valve specimens

Steel grade	C	Si	Mn	Cr	Mo	P	S
Flap-X	0.38	0.45	0.55	13.5	1.00	≤0.025	≤0.015

3. Results and Discussion

3.1. Reed valve frequency

The reed valve frequency data of the reed valve design 1 as recorded by the laser sensor is plotted in Figure 5. The frequencies plotted here are the response frequency of the reed valves when they are subjected to air pulses of certain frequencies and pulse-widths. The main feature of the plots in Figure 5a and 5b is that the response frequencies of the reed valves as indicated by the peaks are multiples (0.5, 1, 1.5, 2, 2.5, 3….) of the frequency of the input air pulse. The response plot in Figure 5c shows that although one of the peaks matches the input frequency of the applied air pulses of 305 Hz but there are a few other peaks that are not multiples of the input frequency. This could be due to the fact that the pulse width of 2.49 ms is 3/4th of the time period compared to ½ of the time period for lower frequencies of 50 Hz and 110 Hz. The longer pulse-width of 2.49 ms was applied because half of the time period of the input pulse of air would be too little time for the high frequency solenoid valve to be able to open and close to generate pulses (see Figure 2). In terms of the frequency data, the response of the other two valve designs was also similar with minute differences.

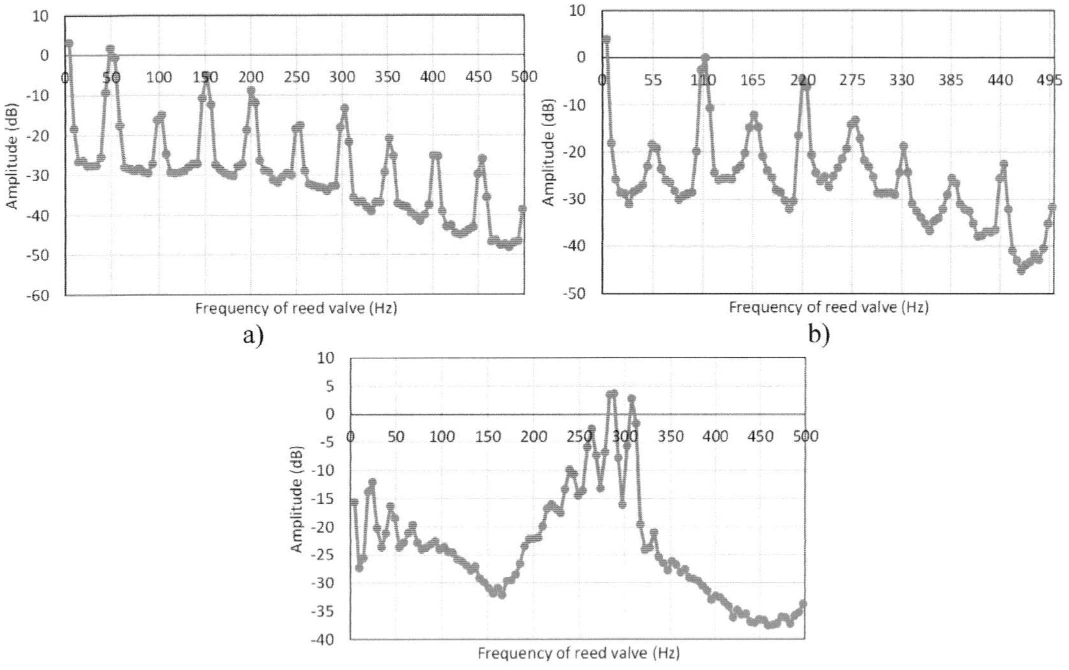

c)

Figure 5. Reed valve displacement plotted as a function of time for the valve design 1 specified to the solenoid high frequency valve: a) test frequency = 50 Hz, pulse-width = 10 ms, airflow rate = 90 litres/minute; b) test frequency = 110 Hz, pulse-width = 4.54 ms, airflow rate = 76 litres/minute; c) test frequency = 305 Hz, pulse-width = 2.47 ms, airflow rate = 131 litres/minute.

3.2. Valve lift measurements

The valve lift measurements were done by the laser sensor installed in the impact fatigue test system. The plots in the following sections show the response displacement of the valve recorded at a sampling rate of 10 000 Hz by the laser sensor i.e. each data point on the curves shows position of the valve recorded after an interval of 0.0001 seconds.

3.2.1. Test frequency = 50 Hz, Pulse width = 10 ms: The reed displacement response is plotted for the three specimen designs at input frequency of 50 Hz and pulse-width of 10 ms in Figure 6. It can be observed that the reed valve opens with a slight delay after the trigger voltage is applied to the high frequency solenoid valve. After the valve is pushed open to the maximum amplitude, the valve flutters without striking against the valve plate that is positioned at approx. -4 mm. After the air pulse is finished – observe the trigger voltage pulse in Figure 6 – the reed valves strike against the valve plate. There are at least three bounces of the reed valve observed for all the valve designs before the reed valve is opened by another incoming air pulse. One can observe that the highest reed valve opening amplitude is the highest for reed valve design1 and the lowest for the design 3. Furthermore, some variations in the shapes of the response displacement curves can also be observed between the different reed valve designs. These differences are mainly due to the different designs of these valves as all the other input test conditions such as the air pressure (approx. 7 bars) and airflow rate (90 litres per minute) were kept the same.

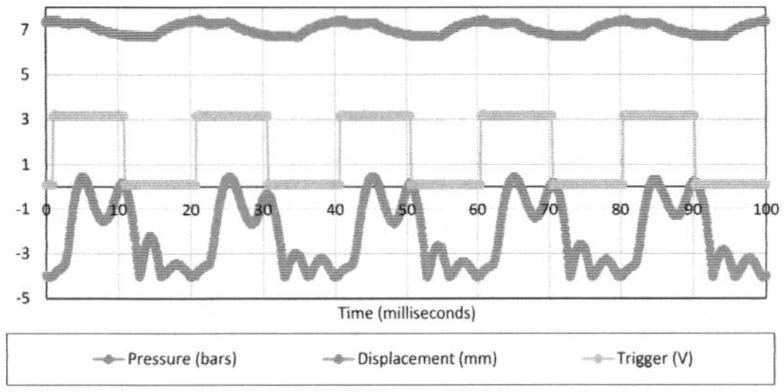

a)

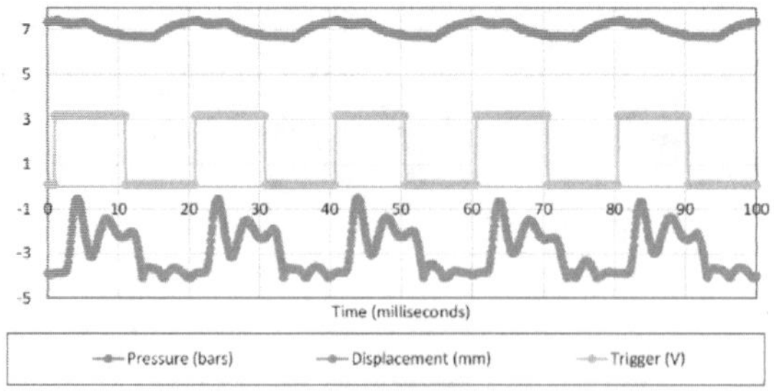

b)

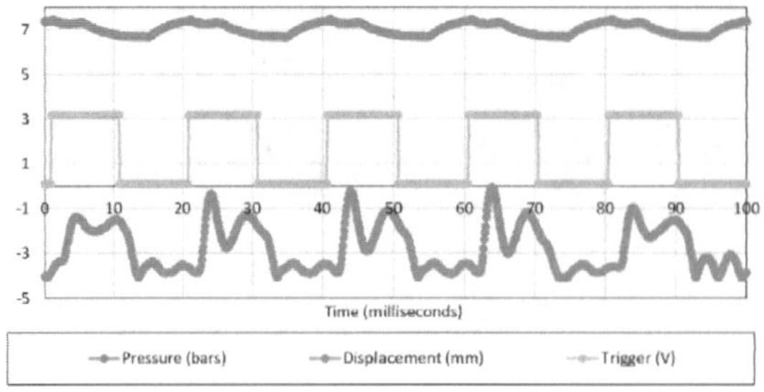

c)

Figure 6. Reed valve displacement plotted as a function of time for input frequency of 50 Hz and pulse-width of 10 milliseconds specified to the solenoid high frequency valve: a) specimen design 1; b) specimen design 2; c) specimen design 3.

3.2.2. Test frequency = 110 Hz, Pulse-width = 4.54 ms: The response displacement is plotted for three different valve designs for higher frequency of 110 Hz and smaller pulse-width of 4.54 ms in Figure 7. Here, it can be seen that the valves of design 1 bounced two times against the valve plates for each applied pulse of compressed air. However, no fluttering of the reed valves was observed for this valve design, see Figure 7a. On the other hand, the shape of reed valve's response displacement curves for the design 2 varies from one applied pulse to another even though cyclic patterns can be observed, see Figure 7b. The maximum opening amplitude varies considerably for different cycles of movement. In addition, fluttering of reed inter-mixed with multiple bounces against the valve plate were observed. Similarly, for the valve design 3 the valve lift curve shows considerable variations in response to successive pulses of applied compressed air, see Figure 7c. The results from this testing shows that the valve design 1 is not only more efficient as it shows higher displacement amplitude for same input test conditions but it shows the most steady movement as well.

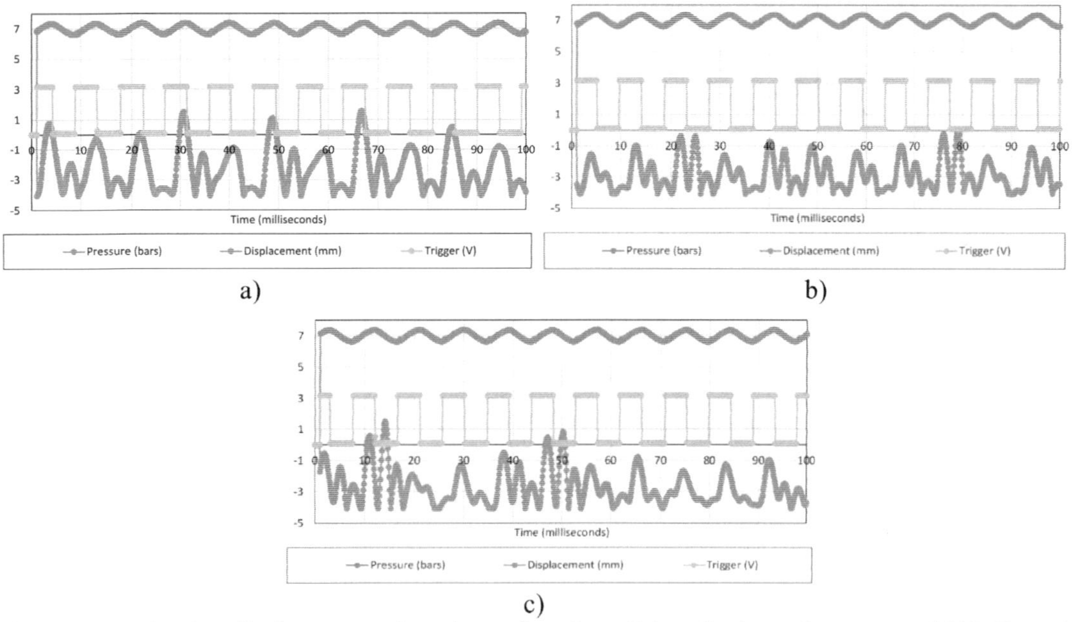

Figure 7. Reed valve displacement plotted as a function of time for input frequency of 110 Hz and pulse-width of 4.54 milliseconds specified to the solenoid high frequency valve: a) specimen design 1; b) specimen design 2; c) specimen design 3.

3.2.3. Test frequency = 300 Hz, Pulse-width = 2.49 ms: In this test condition, the valves were subjected to a much higher frequency of 300 Hz and pulse-width of 2.49 ms that makes 3/4th of the total time period of the applied frequency, see Figure 8 for the displacement response. Since, the pulse width is quite small there is very little variation in the applied pressure magnitude as shown by the almost flat line at the top. This is due to the switching behaviour of the high frequency solenoid valve, its inherent time delay and the limitation of its response time, see Figure 2. The small pulse-width also means that there is only one clean vibration of the reed valve per pulse-width without any valve fluttering or multiple bounces. Moreover, such a valve movement means more stable and constant amplitude of displacement over many cycles which allows more accurate evaluation of the average valve impact velocity as the loading parameter. It can also be seen that the valve lift or the highest opening amplitude of the reed valve is the highest for design 1 and the lowest for the design 3. Therefore, the valve design 1 has once again proven itself to be more efficient at higher frequency too.

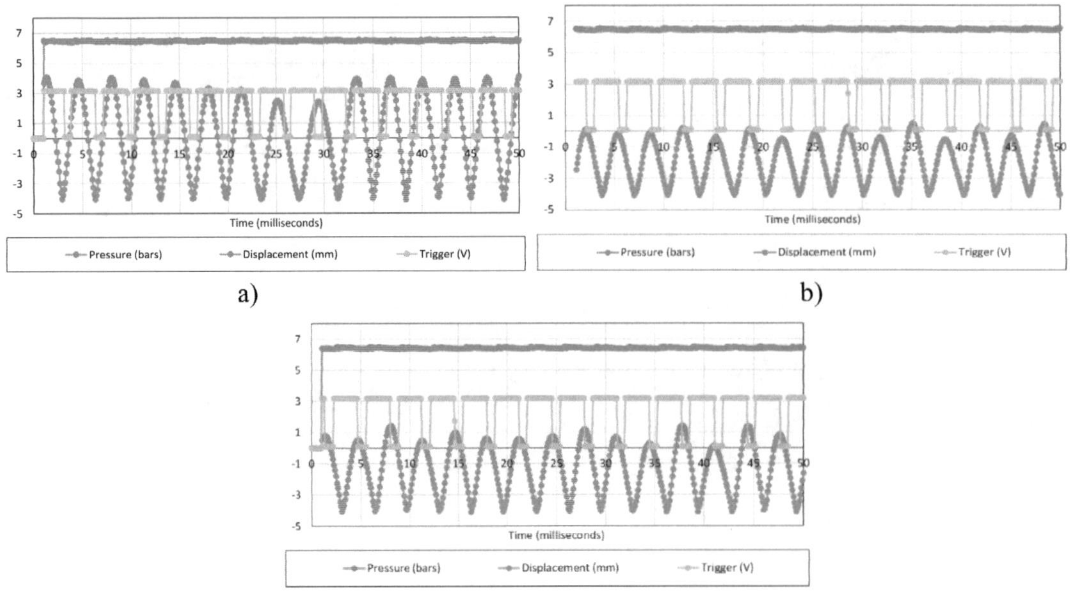

a) b) c)

Figure 8. Reed valve displacement plotted as a function of time for input frequency of 300 Hz and pulse-width of 2.49 milliseconds specified to the solenoid high frequency valve: a) specimen design 1; b) specimen design 2; c) specimen design 3.

3.3. Reed velocity measurements

The reed valve's velocity is one of the most important parameters as it helps determine the intensity of its impact against the valve plate. The reed valve velocity was determined by subtracting the successive points of displacement, say x_1 and x_2, recorded by the laser sensor at successive time points, say t_1 and t_2, as shown in the following formula:

$$v_1 = \frac{x_2 - x_1}{t_2 - t_1} \quad (1)$$

The intensity of the impact is determined by the kinetic energy transferred by the valve to the valve plate. This kinetic energy depends on the velocity of the valve, thus termed the impact velocity of the valve, just before it strikes against the impact plate.

3.3.1. Test frequency = 50 Hz, Pulse-width = 10 ms: The reed velocity of the different valve designs studied in this paper are shown in Figure 9 for test frequency of 50 Hz and pulse-width of 10 ms. The positive side of the scale showing reed velocity is for the valve opening phase whereas the negative for the closing. In this study, we are more interested when the valve is in the closing phase, i.e. the negative velocity values in Figure 9. For each pulse-width the highest negative are the ones when the valve is about to strike against the valve plate, thus, its impact velocity. It can be observed that the highest impact velocity was attained by the valves of design 1 and the lowest by the design 3.

IOP Conf. Series: Materials Science and Engineering 1180 (2021) 012020 doi:10.1088/1757-899X/1180/1/012020

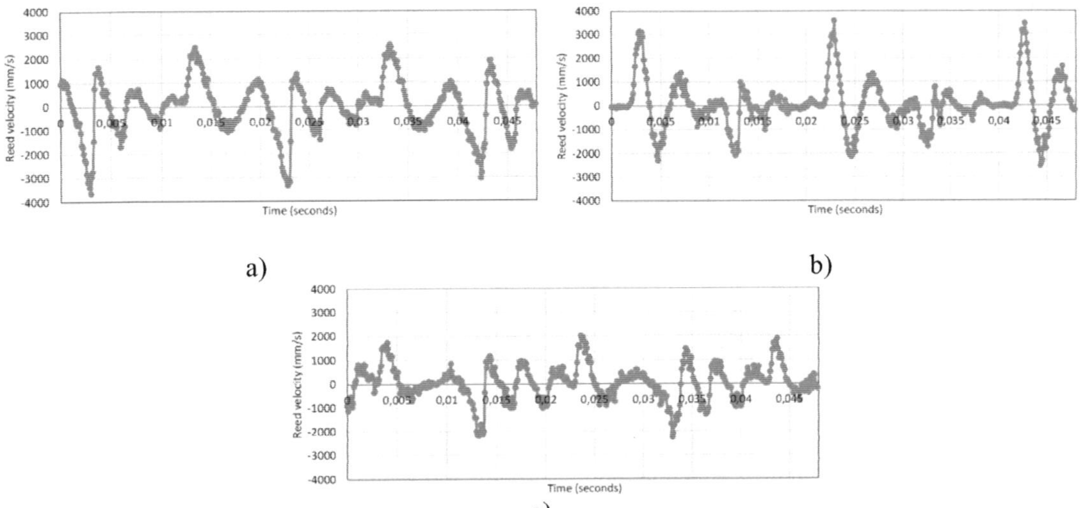

Figure 9. Reed valve velocity plotted as a function of time for input frequency of 50 Hz and pulse-width of 10 milliseconds specified to the solenoid high frequency valve: a) specimen design 1; b) specimen design 2; c) specimen design 3.

3.3.2. Test frequency = 300 Hz, Pulse-width = 2.49 ms: Similarly, reed velocity data for the three different valve designs is shown in Figure 10 for a much higher test frequency of 300 Hz and smaller pulse pulse-width of 2.49 ms. Testing these valve designs at this frequency provides too advantages: 1) accelerated impact fatigue testing that reduces test times to practical limits, 2) the loading parameter of the impact fatigue tests i.e. impact velocity is much more stable in magnitude. It can be observed from Figure 10 that the impact velocity – the negative peaks of the reed velocity curve – is the highest for the valve design 1. The higher the impact velocity of a reed valve the higher the impact fatigue stresses generated in it. Thus, from this point of view, the highest impact fatigue stresses would be generated in the valve design 1 due to its higher valve lift and velocity. However, other design parameters of valves also dictate the concentration and magnitude of impact fatigue stresses in them [11].

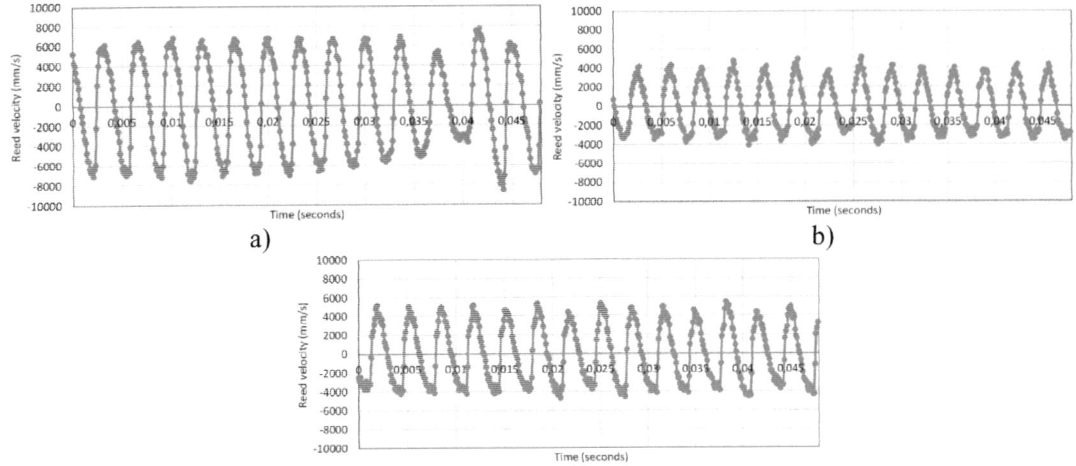

Figure 10. Reed valve velocity plotted as a function of time for input frequency of 300 Hz and pulse-width of 2.49 milliseconds specified to the solenoid high frequency valve: a) specimen design 1; b) specimen design 2; c) specimen design 3.

In order to conduct the fatigue testing of a material, at least 15 specimens should be tested for indicative testing of the fatigue strength [12]. Following the same standard and applying it to the impact fatigue testing would mean that the test times could be very long at the lower frequencies such as 50 Hz or 110 Hz. Testing at higher frequencies, such as 300 Hz for the valve designs tested in this study, would mean reducing the testing times to more practical times since the reed valves undergo several hundred millions of cycles during their operation. Moreover, the compressor manufacturers are interested in knowing the critical impact velocity values at which the reed valves suffer impact failure. Therefore, it is advisable to test the reed valves at frequencies where there is only one impact per pulse-width of compressed air and the scatter of impact velocities is as low as possible. Depending on the valve design, this optimal frequency could be different for different reed valves. Hence, an initial testing of a specific reed valve is suggested to determine the frequency and pulse-width at which the scatter reed valve displacement and velocity is minimum. This study further provides a template to test the bending fatigue and impact fatigue characteristics using the custom-built impact fatigue test system used here.

4. Conclusions

Following are the conclusions drawn from the suction valve movement experiments conducted using the impact fatigue test system in this study:

1. The reed valve design influences the important reed movement parameters such as the valve lift, valve velocity, impact velocity and the frequency of valve vibrations.
2. The valve design 1 was found to be more efficient as it produced higher valve lift while all the input parameters were kept the same.
3. The impact velocity of the valve design 1 was also found to be the highest, thereby, exposing the valve to possibility of higher impact fatigue stresses.
4. For a practical impact fatigue testing of the reed valves, the testing should be conducted at higher frequencies to reduce test times as well as for a stable amplitude of vibration and impact velocity.
5. The custom-built impact fatigue system used in this study shows prospects to conduct further research on the reed valves.

References

[1] M. Svenzon, "Impact fatigue of valve steel," i *International Compressor Engineering Conference*, Lafayette, Indiana, USA, 1976.

[2] A. Futakawa och K. Namura, "Impact fatigue behaviour of flapper valve steel," i *International Compressor Engineering Conference*, Lafayette, Indiana, USA, 1982.

[3] A. C. Altunlu, I. Lazoglu, E. Oguz och S. Kara, "An investigation on the impact fatigue characteristics of valve leaves for small hermetic reciprocating compressors in a new automated test system," *Fatigue & Fracture of Engineering Materials & Structures,* pp. 826-841, 2012.

[4] M. W. Tofique, C. Millward, A. Löf och A. Nawaz, "Investigation of valve reed-impact plate interaction with respect to their impact fatigue properties," i *24th International Compressor Engineering Conference*, Lafayette, Indiana, USA, 2018.

[5] L. Böswirth, "A new valve simulation program and its use for the design of valves," i *International Compressor Engineering Conference*, Lafayette, India, USA, 1996.

[6] W. Soedel, Sound and vibration of positive displacement compressors, ISBN: 0-8493-7094-3, Taylor & Francis Group, 1st Ed., 2007.

[7] J. Rigola, D. Aljure, O. Lehmkuhl, C. D. Perez-Segarra och A. Oliva, "Numerical analysis of the turbulent fluid flow through valves. Geometrical aspects influence at different positions," i *9th International Conference on Compressors and their Systems*, London, 2015.

[8] J. Mayer, P. Bjerre och F. Brune, "A comparative study of different numerical model for flapper valve motion," i *International Compressor Engineering Conference*, Lafayette, Indiana, USA, 2014.

[9] L. Böswirth, "Hypothesis on the failure of spring loaded compressor valve plates," i *Proceedings of the International compressor engineering conference at Purdue, Purdue University (198-206), Paper 332.*, Indiana, USA, 1980.

[10] P. N. S. W. Pandeya, "Analysis of the influence of seat-plating or cushioning on valve impact stresses in high speed compressors," i *Proceedings of the International compressor engineering conference at Purdue, Purdue University*, Indiana, USA, 1978.

[11] X. Yu, Q. Tan, Y. Ren, X. Jia och L. Jin, "Numerical study of the reed valve impact in the rotary compressor by FSI model," *Energy Procedia 105,* pp. 4890-4897, 2017.

[12] Metallic materials - Fatigue testing - Statistical planning and analysis of data (ISO 12107: 2012, IDT), 2012.

Simulation and Experiment Study on Dynamic Characteristic of Reed Valve of Reciprocating Compressor Based on Fluid-Structure Interaction Model

Yanbin Wang, Bei Guo, Kai Ma, Yuhang Zhou,
School of Energy and Power Engineering, Xi'an Jiaotong University, Xi'an 710049, China;
Email: wangyanbin@stu.xjtu.edu.cn

Abstract. The working process of the suction valve with complex shape called "boqi" and the discharge valve with four arms were simulated by the three-dimensional fluid-structure interaction method in this paper. By setting the gap boundary condition between the valve and the valve plate, hanged the connection between the valve and the valve plate in the model, and the problem of flow area connectivity caused by the opening and closing of the valve was solved. It solved the mesh overlapped and distorted problem caused by the valve motion. According to the simulation, the results of valve's dynamic characteristic and gas pressure were obtained. A valve displacement test bench of reciprocating compressor was built, and the valve motions under different rotational speed and different pressure ratio was tested. Comparing the simulation results with the experimental results, it shows that the three-dimensional fluid-structure interaction model can be used to simulate the interaction between the valve with complex shape and the compressor fluid field. From the simulation results, the pressure cloud map of the valve under different conditions was extracted. Finally, the method proposed in this paper was used to analyze the valve's displacement, speed and stress.

1. Introduction

Reciprocating compressors are widely used in various fields, which are the core components of central air conditioners and commercial refrigerators. Reed valve is mainly used in small and medium-sized refrigeration compressors and air compressors. Valves are easy to fail, affecting economic benefits, so the research on valves is very important.

Early studies mostly used lumped parameter models to solve the valve motion problem. Tuhovcak, Hejcik and Jicha [1] analyze the heat transfer of the compressor working cavity, while Dutra and Deschamps [2] establishe a mathematical model of the compressor working process. The similarity between their models is that they both use a single degree of freedom model to predict the valve position. Papastergiou [3] uses the finite element method to solve the displacement of the reed valve, and the simulation and experiment results are consistent.

Mistry [4] and Aigner [5-6] use the same method to model the flow field and valve separately, and obtain the pressure field of the working cavity and the movement of the suction and exhaust valves.

At present, scholars have realized the interaction of the valve and the flow field through the three-dimensional fluid-solid interaction method, and obtained the dynamic characteristics of the valve and the flow field distribution of the cylinder. Wang, Guo, and He [7] used a three-dimensional fluid-solid interaction model to study the effects of speed and valve parameters of a refrigerator compressor on the delayed closing and flutter of the valve, and got the characteristic speed of the reed valve. Zhao, Jia, and Wen [8] studied the double-acting reciprocating compressor, and accurately obtained the relationship between pressure pulsation and valve movement. Silva and Deschamps [9] predicted the gas leakage of the reed valve of a small reciprocating compressor used in household refrigeration when the seal is not complete. This model takes into account the viscous friction, viscous flow state and compressibility of the refrigerant. Song, Wu and Li [10] studied the failure of the whole machine due to the damage of the suction valve and exhaust valve in the automobile air-conditioning compressor. Tan, Pan, Feng [11] and others have studied rolling piston compressors. The p-V diagram measured by experiments verifies the accuracy of the fluid-structure interaction model. Yu, Tan, and Ren [12] tested the p-θ diagram in the experiment to verify the accuracy of the model. Ba [13] used ADINA software to perform fluid-solid interaction simulations on the compressor valve, analyzed the dynamic characteristics of the valve and the thermal process of the compressor, and studied the influence of the thickness of the valve and the crankshaft speed on the movement of the valve;

This paper uses ADINA software to numerically simulate the working process of the complicated type Boqi suction valve and the four-arm strip exhaust valve in the reciprocating compressor based on the three-dimensional fluid-structure interaction method, and build an experimental bench for real-time acquisition of suction valve displacement to verify the simulation results. Obtain the dynamic characteristic curve of displacement, speed and stress of the suction valve.

2. Physical model and meshing

The research object of this article is a reciprocating air compressor, the two cylinders of the compressor are distributed in a "V" shape. Only one of the cylinders is selected for research. The main dimensions of this cylinder and the intake and exhaust valve groups are shown in Table 1.

Table 1. Main dimensions of compressor model

Item	Value	Item	Value
Cylinder diameter /mm	90	Piston stroke /mm	60
Link length /mm	170	Crankshaft rotation radius /mm	30
Suction valve thickness /mm	0.4	Exhaust valve thickness /mm	0.4
Suction orifice diameter /mm	5	Exhaust orifice diameter /mm	5

The suction valve is a Boqi type reed valve and the exhaust valve is a four arm strip reed valve. The root of the suction valve is pressed between the valve plate and the cylinder, which is simplified as a fixed part. The suction and exhaust valve lift limiter is shown in Figure 1 .The suction valve lift limiter is a groove in the cylinder. Use a crescent-shaped plane instead in the model. Both the suction valve and the exhaust valve are made of 60CrMnA.

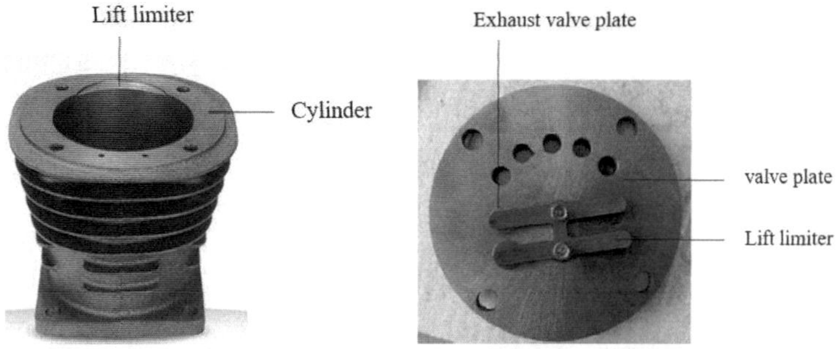

(a) Actual suction lift limiter (b) Actual exhaust lift limiter

Figure 1. Suction and exhaust valve lift limiter diagram

When dividing the grid, first divide into blocks, then set the nodes, and finally generate the grid. In order to ensure that the mesh can maintain high quality after deformation, the flow field model is divided into 3D-8 node element meshes. The rigid surface of the suction valve plate, suction lift limiter, exhaust valve plate and exhaust lift limiter adopt 2D-4 nodes. The suction valve and the exhaust valve adopt a 3D-27 node grid. The grid division diagram is shown in Figure 2.

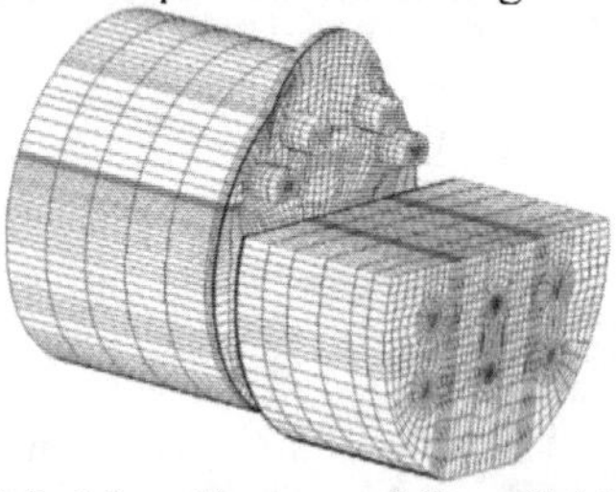

(a) Meshing diagram of flow field

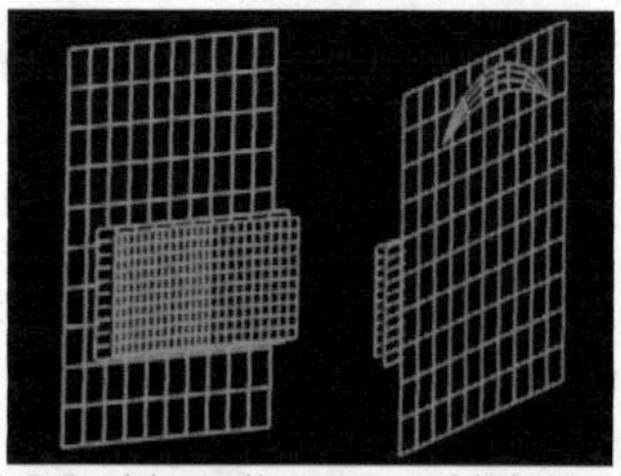

(b) Meshing diagram of lift limiter and valve plate

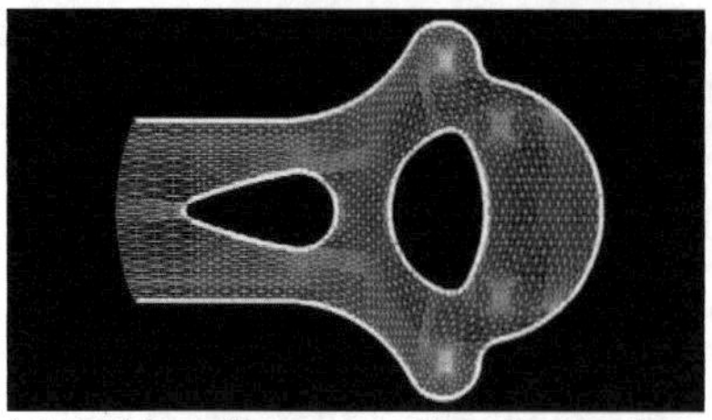

(c) Meshing diagram of suction valve

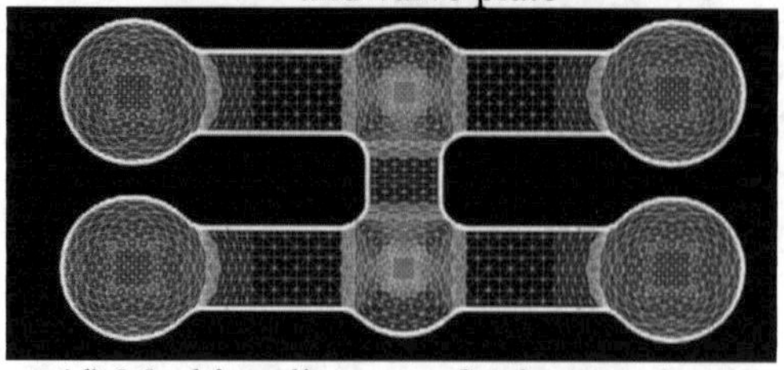

(d) Meshing diagram of exhaust valve

Figure 2. Finite element model

3. Solve the governing equations

3.1 Flow field governing equation

Mass conservation equation:

$$\frac{\partial \rho}{\partial t}+div(\rho\mathbf{u})=0 \tag{1}$$

Where ρ is Fluid density, t is time, $\mathbf{u}$ is Fluid velocity vector, $\mathbf{u}=(u_1,u_2,u_3)$.

Momentum conservation equation:

$$\frac{\partial}{\partial t}(\rho\mathbf{u})+div(\rho\mathbf{u}\otimes\mathbf{u}+pI)=\rho\mathbf{F} \tag{2}$$

Energy conservation equation:

$$\frac{\partial}{\partial t}(\rho e+\frac{1}{2}\rho\mathbf{u}^2)+div(\rho e+\frac{1}{2}\rho\mathbf{u}^2+\rho)\mathbf{u}=\rho\mathbf{F}\cdot\mathbf{u} \tag{3}$$

3.2 Structural control equations

The governing equation expression of the structure is:

$$M\ddot{\mathbf{U}}+C\dot{\mathbf{U}}+K\mathbf{U}=\mathbf{R}-\mathbf{F} \tag{4}$$

Where M is Quality matrix, $\ddot{\mathbf{U}}$ is Acceleration vector , C is Damping matrix, $\dot{\mathbf{U}}$ is Velocity vector, K is Stiffness matrix, $\mathbf{U}$ is Displacement matrix , $\mathbf{R}$ is External force vector, $\mathbf{F}$ is Unit internal force.

When simulating the process of the gas valve colliding with the lift limiter and the valve plate, the contact boundary conditions are set on each contact surface, and the contact boundary is expressed as:

$$g \geq 0, \quad \xi \geq 0, \quad g\xi = 0 \tag{5}$$

Where g is Distance between two contact surfaces, ξ is Normal contact pressure.

3.3 Fluid-solid interaction interface equation

The Euler (EULER) description method and the Lagrange (Lagrange) description method were used respectively in the flow field and the structure field. In the fluid-solid interaction calculation, if the structure field produces a large displacement, the difference between the two description methods will cause the fluid field and the structure field to separate at the originally coincident node on the fluid-structure interaction interface. In order to solve this problem, in the fluid-structure interaction calculation, the flow field adopts the Arbitrary Lagrangian-Eulerian coordinate system (Arbitrary Lagrangian-Eulerian, ALE).

Displacement coordination and force balance can be expressed as:

$$\mathbf{d}_f = \mathbf{d}_s \tag{6}$$

$$\mathbf{n} \cdot \boldsymbol{\tau}_f = \mathbf{n} \cdot \boldsymbol{\tau}_s \tag{7}$$

Where $\mathbf{d}_f$ is Fluid displacement vector, $\mathbf{d}_s$ is Displacement vector of structure, $\boldsymbol{\tau}_f$ is Fluid stress vector, $\boldsymbol{\tau}_s$ is Structural stress vector.

4. Boundary conditions

4.1 Gap boundary

In the expansion process and compression process, the compressor cylinder and other parts are not connected, but in the suction process and the exhaust process are connected, so in a working process of the compressor, the topological structure of the flow field will constantly change. Gap boundary conditions can effectively solve this problem. When using this boundary, a small gap needs to be reserved, and the Gap boundary is set in this gap flow field. In this paper, the capitalized Gap represents a special boundary condition in the model.

There are three ways to control the opening and closing of Gap: average pressure difference, time function and size of Gap. The third one is similar to the working process of the compressor valve, so this article adopts the third method: the Gap is controlled by the size of the grid. Extend the suction orifice to form a cylindrical surface, and the part in the clearance between the valve and the valve plate is the Gap boundary; In the same way, there is also a Gap boundary around the exhaust orifice.

4.2 Import and export boundary conditions and initial conditions

The boundary conditions imposed on the fluid field in this paper are shown in Figure 3. It is necessary to set boundary conditions at the inlet of the intake cavity and the outlet of the exhaust cavity.

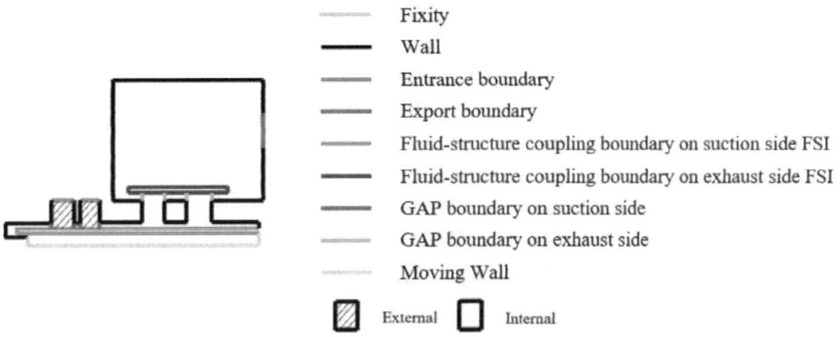

Figure 3. Boundary condition setting of flow field model

In the calculation, the time when the piston reaches the top dead center position is zero, and the suction valve is closed at this time. As the gas expands, the suction valve will gradually open. Therefore, with the Gap on the suction side as the boundary, two relatively independent fluid regions are formed respectively, the area on the side of the suction port is called EXTERNAL , and the other part is called INTERNAL. Set the initial conditions respectively.

4.3 Setting of boundary conditions of solid field
The boundary condition setting of the solid field includes fluid-structure interaction boundary, displacement fully constrained boundary and contact boundary, the fluid-structure interaction boundary corresponds to the fluid-structure interaction boundary in the flow field. The full restraint of the suction valve is set at the root of the valve where the actual valve installation screw. The contact boundary includes two contact pairs, as shown in Figure 4. The exhaust valve is similar to the suction valve.

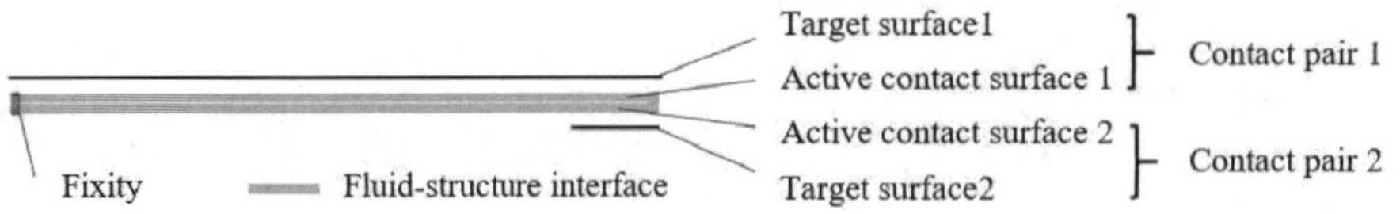

Figure 4. Boundary condition setting of suction valve

5. Reciprocating compressor suction valve displacement test experiment

5.1 Experimental setup
The entire system is mainly composed of a reciprocating air compressor, an air storage tank, a frequency converter and a test system. The test system consists of eddy current displacement sensor, Hall sensor and data acquisition system.The experimental compressor is V-0.67/8 reciprocating air compressor . The motor's rated speed is 980 /r·min-1 and rated power is 5.5 kW. Intake pressure of the compressor is 0.1 MPa and rated exhaust pressure is 0.8 MPa.
Change the degree of valve opening to adjust the pressure of the gas tank, and start data collection when the pressure stabilizes at the set working condition. The eddy current displacement sensor is used to measure the valve displacement, and the installation position is shown in Figure 5.

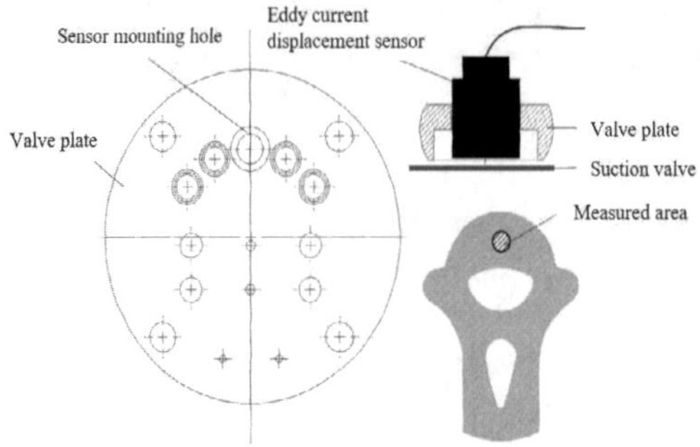

Figure 5. Schematic diagram of the installation position of the eddy current displacement sensor

5.2 Verification of fluid-structure coupling model
This paper sets up 9 working conditions: the pressure ratio is 3, 4 and 5; the speed is 408r·min^{-1}, 603r·min^{-1} and 980r·min^{-1}. The trend of valve movement curve in simulation and experiment is basically the same. The maximum displacement, opening angle and closing angle of the valve are shown in Table 2. The maximum error of the maximum displacement, opening angle and closing angle is within the allowable range.

Table 2. Main thermal parameters of compressor

Pressure ratio	Rotating speed/ $r \cdot min^{-1}$	Maximum displacement /mm		error /mm	Opening angle /°		error /°	Close corner /°		error /°
		simulation	experiment		simulation	experiment		simulation	experiment	
3	408	2.00	2.00	0.00	29.4	29.6	-0.2	181.2	184.3	-3.1
	603	2.11	2.00	0.11	30.4	27.3	3.1	193.7	194.5	-0.8
	980	2.10	2.04	0.06	32.3	29.5	2.8	201.7	205.6	-3.9
4	408	2.03	1.98	0.05	34.9	37.9	-3.0	180.5	184.7	-4.2
	603	2.10	1.96	0.14	37.5	38.2	-0.7	192.1	193.6	-2.5
	980	2.1	1.98	0.12	38.2	41.3	-3.1	204.0	200.7	3.3
5	408	2.05	1.99	0.09	39.8	40.8	-1.0	180.5	184.8	-4.3
	603	2.10	1.96	0.14	42.3	42.6	-0.2	192.7	193.9	-1.2
	980	2.10	2.19	-0.09	44.0	41.3	2.7	209.3	210.5	-1.1

In summary, the three-dimensional fluid-solid interaction model established in this paper is effective, and it is suitable for the interaction analysis of the complex shape valve and the surrounding flow field in the compressor.

6. Research on Dynamic Characteristics of Reed Valve of Reciprocating Compressor Based on Fluid-Structure Interaction

6.1 Pressure analysis of the suction valve surface

The pressure difference controls the movement of the valve, so the pressure distribution on the upper and lower surfaces of the valve is extracted from the simulation results. The cylinder side is regarded as the lower surface, and the cylinder head side is regarded as the upper surface.Here only shows the pressure distribution when the speed is $980 r \cdot min^{-1}$ and the pressure ratio is 3. Five stages are taken for analysis, the result shown in Figure 6.

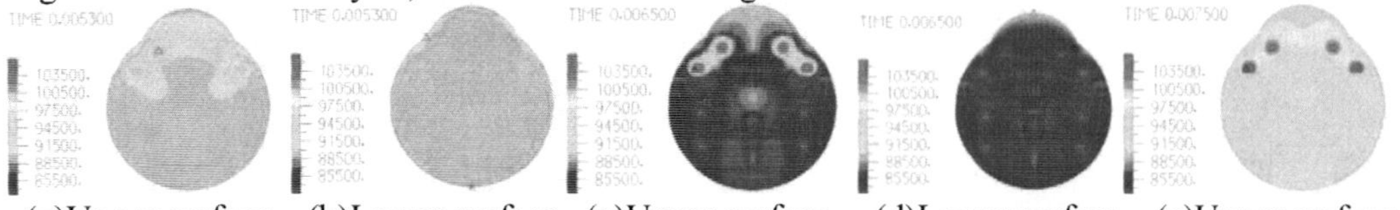

(a)Upper surface just opened (b)Lower surface just opened (c)Upper surface half-opened (d)Lower surface half-opened (e)Upper surface fully opened

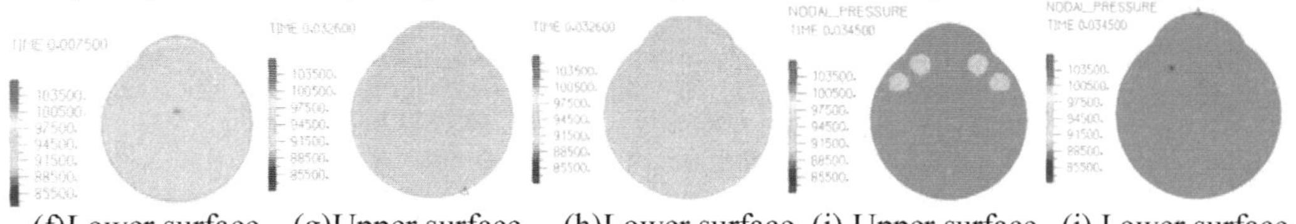

(f)Lower surface fully opened (g)Upper surface half-closed (h)Lower surface half-closed (i) Upper surface fully closed (j) Lower surface fully closed

Figure 6. Cloud diagram of pressure distribution on the upper and lower surfaces of the valve

It can be seen that the pressure distribution of the suction valve lower surface is very uniform in all time periods. During the opening phase, an obvious low pressure area appears on the upper surface of the valve, of which the pressure is less than the intake pressure. This is because the clearance on the upper surface of the valve is narrow and the air flow is complicated, so the pressure distribution is uneven.

In order to better understand the relationship between the pressure on the upper surface of the valve and the speed, the pressure distribution along the width and length of the valve when the valve is half-open at different speeds is studied. Set up three pressure monitoring coordinate axes and six displacement monitoring points as shown in Figure 7. Points 3,4,5, and 6 are located in the center of the suction valve holes. Point 2 is located at the test area of the eddy current displacement sensor above. Point 1 is located at the head of valve. Axis X1 and X2 are horizontal axis, passing through points 3,4 and 5,6. Axis Y is the vertical axis, which is the symmetry axis of the valve.

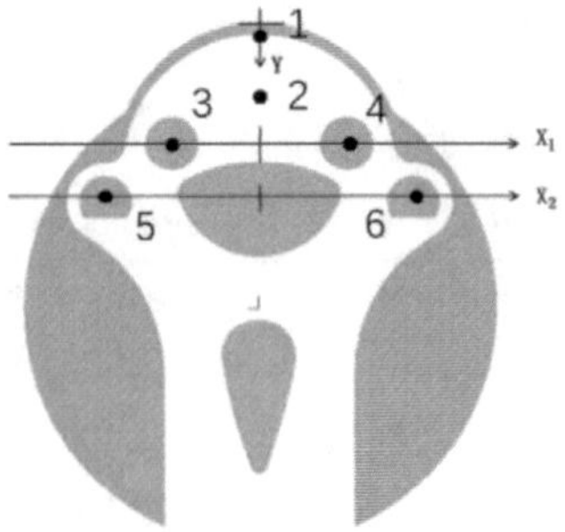

Figure 7. Selection of monitoring coordinate axes and points of suction valve

Figure 8 shows the pressure distribution in the X and Y direction of the valve. Due to similar trends, only a few cases are selected as examples. It can be seen that with the increase of speed, the pressure distribution in different areas becomes more uneven, and the pressure valley value is obviously reduced, while the pressure peak value do not have much difference. With the increase of the speed, the pressure of the valve along the length direction becomes smaller.

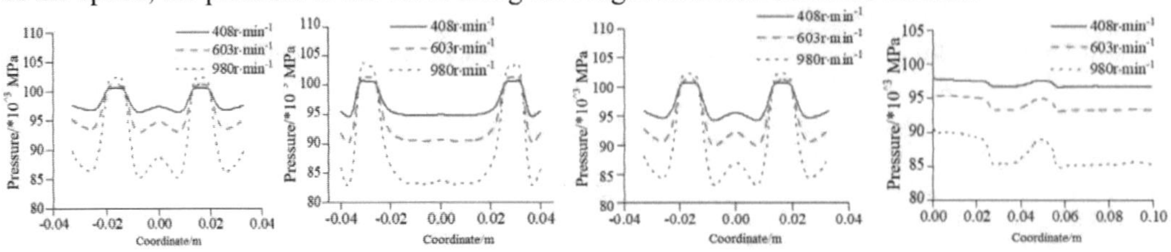

(a) X1 pressure ratio 3 (b) X2 pressure ratio 4 (c) X1 pressure ratio 5 (d) Y pressure ratio 3

Figure 8. Pressure graph at different speeds along the coordinate direction

6.2 Analysis of dynamic characterists of suction valve

6.2.1Displacement

The valve motion curve is an important factor to analyze whether the valve structure is reasonable.Through the displacement curve, the valve opening angle, closing angle and other information can be obtained. Therefore, this article analyzes the movement law of the suction valve at different speeds and pressure ratios.

By monitoring the points 3,4,5, and 6, the motion curves of the symmetric points are coincident. It shows that there is no side deviation of the suction valve during the working process. Therefore, only monitoring points 1 and 2 are analyzed.

The deformation curve is shown in Figure 9. It can be seen that the valve reaches the maximum lift in a very short time after opening, and then the head of the valve is close to the lift limiter. With the increase of speed, the opening angle does not change much, the closing angle increases obviously, the valve closes later, and the working cycle becomes longer. Because the intake pressure loss increases with the increase of the speed, the pressure difference between the two sides of the valve is greater when the speed is high, and the cylinder pressure needs to be further increased, so the valve closes later.

In order to understand the movement of other parts after the valve head is close to the lift limiter, extract the displacement curve of point 2 at different pressure ratios and speeds. After the valve hits the limiter and closes tightly, the monitoring point 2 begins to vibrate, and as the speed increases, the frequency of reciprocating motion becomes smaller. And as the speed increases, the opening angle and closing angle increase, the oscillation frequency decreases.

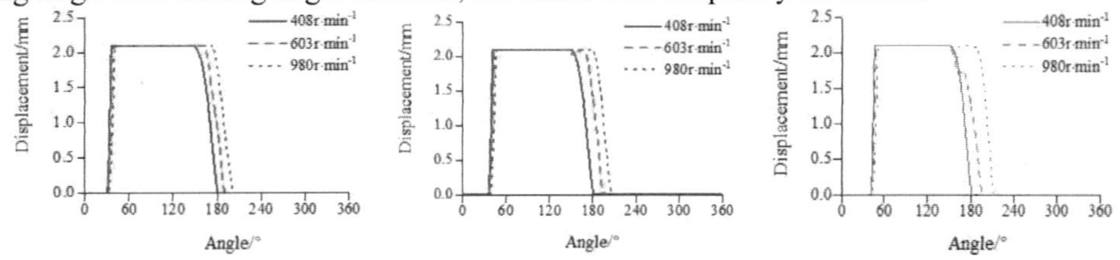

(a) Pressure ratio 3 (b) Pressure ratio 4 (c) Pressure ratio 5

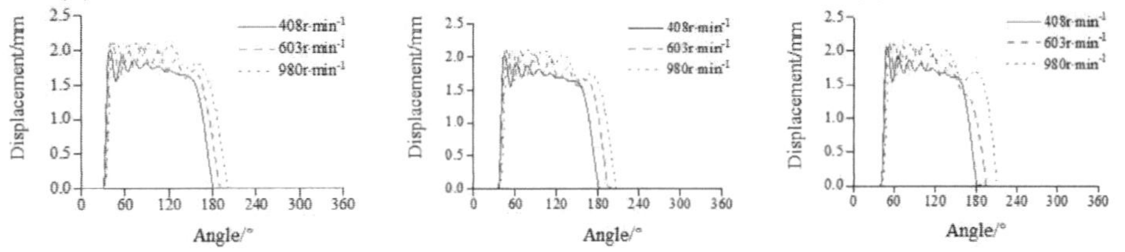

(d) Pressure ratio 3 (e) Pressure ratio 4 (f) Pressure ratio 5

Figure 9. The movement of monitoring point 1 and 2 of the suction valve changes with the speed in same pressure ratio

6.2.2 Velocity

The relationship between the movement speed and the angle of rotation at different speeds of detection point 1 and 2 is shown in Figure 10. It can be seen that in the figure (a) the speed of the suction valve suddenly increases and then drops to zero quickly. Until the valve falls back, there will be a rebound, and the rebound speed is less than $0.5 m \cdot s^{-1}$, so when the valve is closed, the impulse is not large, and then the speed becomes 0 again. This is a working cycle of the suction valve.

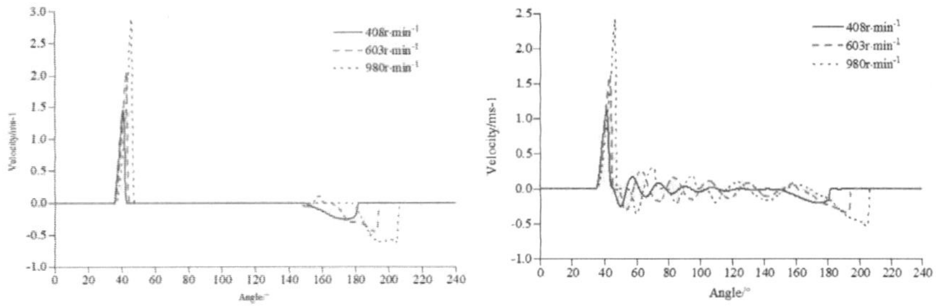

(a) Point 1 (b) Point 2

Figure 10. Speed change graph of monitoring point 1 and point 2

According to the above analysis of the pressure distribution on the surface of the suction valve, it can be seen that during the opening phase of the suction valve, a low pressure area is generated on the upper surface, and the pressure in the low pressure area is less than the inlet pressure of the valve port. With the change of speed, the minimum pressure of the low pressure zone changes, and the area of the low pressure zone changes, so the acceleration of the valve movement changes. It can be seen that in the figure (b) the speed of monitoring point 2 initially increases suddenly, then decreases and starts to increase in the reverse direction, but the maximum reverse speed is small. The change of speed then shows a simple harmonic vibration with a small amplitude. In a cycle, the valve swings up and down many times, indicating that the alternating stress at the root of the intake valve changes rapidly.

6.2.3 Stress

From the above analysis, it can be seen that the valve vibrates during operation, it will generate cyclic stress and cause fatigue damage, so it is very necessary to perform stress analysis on the valve. In this paper, the movement process of four working periods are selected for stress analysis when the pressure ratio is 3 and the speed is 480r·min^{-1}, and the stress distribution cloud diagram as shown in Figure 11.

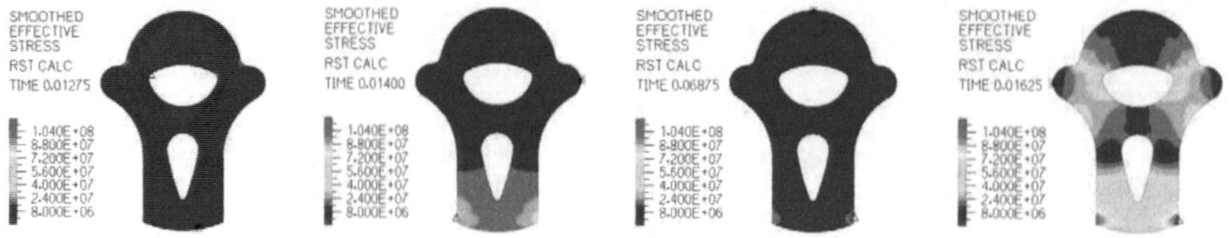

(a) Just opened (b) Half opened (c) Fully open (d) Half closed

Figure 11. Stress distribution diagram of suction valve

It can be seen from Figure 11 that the maximum stress of the valve is mainly concentrated at the root of the valve. Figure 12 shows the change of the maximum stress at the root of the suction valve with the crankshaft angle. It can be seen that the maximum stress occurs at an angle of 40 degrees, which has the same trend as the displacement diagram.The maximum stress at the root changes continuously with the vibration of the monitoring point 2. When the displacement of monitoring point 2 is the maximum, the root stress is the maximum.

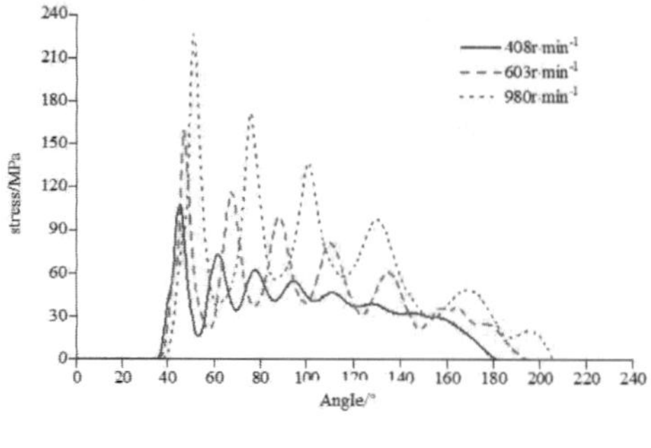

Figure 12. The maximum stress at the root of the suction valve varies with the angle of rotation

7. Conclusions

In this paper, ADINA software is used to simulate the reciprocating compressor air valve working process by fluid-structure interaction method, and obtain the motion curve. Compared with the experiment results, and get the following conclusions:

1) In three-dimensional fluid-solid interaction problems, changing the properties of the fluid connection surface dynamically by setting the gap boundary conditions can solve the the flow area connection problem caused by the opening and closing of the valve. It can control the coupling boundary and the flow field grid near it to solve the problem of excessive deformation and overlap of the flow field grid caused by the movement of the valve;

2) The movement law curve of the suction valve obtained by the three-dimensional fluid-solid interaction model is basically consistent with the movement law of the suction valve measured by the experiment. It shows that the fluid-solid interaction model can accurately simulate the interaction law between the reed valve and the compressor flow field, and is suitable for valve models with complex shapes;
3) In this paper the dynamic characteristics of the compressor valve is analyzed. The results show that the head speed of the suction valve increases suddenly after the valve is opened, and then it will keep close to the lift limiter, the rest of the part will vibrate. With the increase of speed, the vibration frequency and opening angle change little, but the closing angle increases; As the pressure ratio increases, the valve opening angle increases significantly. In the working process of the suction valve, the maximum stress is concentrated at the root of the valve, and the change trend of the maximum stress and displacement of the root is basically the same.

References

[1] Tuhovcak J, Hejcik J, Jicha M. Comparison of heat transfer models for reciprocating compressor[J]. Applied Thermal Engineering, 2016, 103:607-615.
[2] Dutra T, Deschamps CJ. A simulation approach for hermetic reciprocating compressors including electrical motor modeling[J]. International Journal of Refrigeration, 2015, (59):168-181.
[3] Papastergiou S. The dynamic behaviour of valve reeds in reciprocating gas compressors analytical study[C]. International Compressor Engineering Conference: Purdue University, W lafayette, IN, USA, 1980:263-272.
[4] Mistry H, Bhakta A, Dhar S. Capturing valve dynamics in reciprocating compressors through computational fluid dynamics[C]. International Compressor Engineering Conference: Purdue University, W lafayette, IN, USA, 2012:1210.
[5] Aigner R, Meyer G, Steinruck H. Valve dynamics and internal waves in a reciprocating compressor[C]. 4th Conference of the EFRC. Antwerp, USA, 2005.
[6] Aigner R, Steinruck H. Modelling fluid dynamics, heat transfer and valve dynamics in a reciprocating compressor[C]. 5th Conference of the EFRC, Prague, Czech Republic,2007:171-180.
[7] Wang T, Guo Y, He ZL, et al. Investigation on the delayed closure of the suction valve in the refrigerator compressor by FSI modeling[J]. International Journal of Refrigeration, 2018, 91:111-121.
[8] Zhao B, Jia X, Wen J. FSI model of valve motion and pressure pulsation for investigating thermodynamic process and internal flow inside a reciprocating compressor[J]. Applied Thermal Engineering, 2018, 131:998-1007.
[9] Silva L R, Deschamps C J. Modeling of gas leakage through compressor valves[J]. International Journal of Refrigeration, 2015, 53:195-205.
[10] Song MY, Wu WF, Li Z. Simulation of the movement law of automobile air-conditioning compressor valve ［J］. Journal of Xi'an Jiaotong University, 2015, 49(12): 149-155.
[11] Tan, Q, Pan, S. L, Feng, Q, et al. Fluid-structure interaction model of dynamic behavior of the discharge valve in a rotary compressor[J]. Proceedings of the Institution of Mechanical Engineers Part E Journal of Process Mechanical Engineering, 229(4):280-289.
[12] Yu X, Tan Q, Ren Y, et al. Numerical study of the reed valve impact in the rotary compressor by FSI model[J]. Energy Procedia, 2017, 105:4890-4897.
[13] Bai CC. Numerical simulation analysis of reed valve of reciprocating compressor based on fluid-structure coupling [D]. Dalian: Dalian University of Technology, 2017.

IOP Conf. Series: Materials Science and Engineering 1180 (2021) 012022 doi:10.1088/1757-899X/1180/1/012022

Development of the Universal Modelling Method mathematical model and the practice of its application

Y.B. Galerkin[1], O.A. Solovyeva[2], A.A. Drozdov[3] and V.B. Semenovskiy[3]

1. Leading research center "Digital Design and Modeling (Smart Design)" Peter the Great St.Petersburg Polytechnic University, Polytechnical st. 29, St.Petersburg, Russia
2. Higher School of Hydraulic and Power Engineering Construction Engineering Institute Peter the Great St.Petersburg Polytechnic University, Polytechnical st. 29, St.Petersburg, Russia
3. Higher School of Power Engineering Institute of Power Engineering Peter the Great St.Petersburg Polytechnic University, Polytechnical st. 29, St.Petersburg, Russia

Email: A_drozdi@mail.ru.

Abstract. We present the latest improvements to the Universal Modelling Method for optimal gas-dynamic design of centrifugal compressors. The calculation of the gas velocity at the inlet to the impeller blade cascade was corrected. The flow restructuring scheme at the inlet to the impeller blade cascade has been refined. In accordance with this, the mathematical model of the non-incidence gas inlet to an impeller has been improved. A new approach to modelling the velocity diagram on the blade surfaces is proposed, based on the flow parameters corresponding to the non-incidence flow inlet. A mathematical model with quasi-three-dimensional approach to calculating losses in 3D impellers has been developed. The method of gas-dynamic functions is applied to the calculation of flow parameters in control sections. The new mathematical model was identified based on the test results of model stages and compressors. The mathematical model showed a 0.525% error in modelling efficiency of stages with 2D impellers for the design flow rate. For stages with 3D impellers, the modelling error is 0.889%. To verify the new model, the characteristics of 11 low-flow-rate model stages were calculated. The efficiency modelling error for the design flow rate is 1.08%. A series of model stages with a range of design flow rates coefficient 0.15–0.015 and design loading factor of 0.5 was designed. Four of these model stages were tested on the Gas dynamics of turbomachines laboratory ECC-55 test rig. The measured gas-dynamic characteristics confirmed those declared in the design.

1. Introduction

Centrifugal compressors are used in almost all key industries. The beginning of their use dates back to the end of the 19th century. Initially, one of the main areas of application was ferrous and then non-ferrous metallurgy. Further industrial development has led to the emergence of areas requiring the use of centrifugal compressors, such as chemical production, pneumatics, refrigeration, etc. [1].

Economic development leads to an increase in demand for compressor equipment and a corresponding increase in their production volume. This is associated both with the construction of new industrial enterprises utilizing compressor equipment or with construction of new gas pipelines and with the need to renew the existing compressor park due to wear or changes in their operating conditions. At the same time, the unit capacity of individual compressors also increases (up to 64 MW inclusive). The global market for centrifugal refrigeration compressors is estimated to have reached 18,000 units in 2017, up 3.5% compared to the same period in 2016 due to the rapid development of the magnetic bearing market [2]. The compressor drive consume up to 20% of the generated electricity [3], [4]. In total, more than 100 million compressors are produced annually in the world [5]. Although centrifugal compressors are not the most numerous, their total capacity is very high. In this regard, an important task is to ensure the maximum energy efficiency of centrifugal compressors.

In general, the industry requires design methods that will ensure high compressor efficiency while obtaining the specified design parameters (mass flow rate and pressure ratio) and at the same time do not require experimental verification of the designed compressor or its stages. For this, both commercial CFD programs and proprietary software systems can be used. However, the experience of using CFD calculations shows that CFD programs do not allow simulating the gas-dynamic characteristics of a centrifugal compressor with the required accuracy. Thus, it is necessary to create specialized engineering programs.

For this, compressor equipment manufacturers develop their own design methods and mathematical models based on them. In Russia, such work and research is presented in [6] - [10]. Similar researches are carried out by foreign experts [11] - [18].

A well-known foreign model for calculating centrifugal compressors is used in the Concepts NREC software package, authored by Dr. D. Japikse [12], [13]. The calculation of head losses in the mathematical model uses the techniques of the boundary layer theory and is based on a two-zone scheme: friction loss and mixing loss. When designing a centrifugal stage, one-dimensional optimization of the stage dimensions is initially performed, then an inviscid quasi-three-dimensional calculation of the impeller is performed, and at the final step, gas-dynamic characteristics are calculated using the CFD program. The mathematical model contains empirical formulas for calculating losses in individual elements of the flow path of the stage, which implies identification.

In paper [16], the authors compare the experimental gas-dynamic characteristics of an internal combustion engine turbocharger at three speeds with calculations using a proprietary computer program Compressor Off-design Radial Analysis (CORA). The CORA program is based on mathematical models for calculating flow parameters previously obtained by various authors.

Three different models were used to calculate the losses in the CORA program in the impeller: Auinger [14], Oh [19], and Galvas [20]. These three models take into account a different number of loss components in the impeller, so the Auinger model is the most detailed and takes into account most of the loss components, and the Galvas model is the least detailed (it does not take into account mixing losses, leakage losses, etc.). The model for calculating losses in the vaneless diffuser is taken from Herbert [21], and the model of losses in a scroll from Weber and Koronovski [22]. According to the authors, a combination of the Galvas and Auinger loss models for the impeller matches experimental data well. The authors emphasize the relevance of one-dimensional models for calculating the gas-dynamic characteristics of centrifugal compressors and the importance of their further development.

The authors of papers [23], [24] compared the gas-dynamic characteristics of a stage with a 3D impeller, a vaneless diffuser and a return channel using the one-dimensional calculation program of their own production KSTK and the CFD program NUMECA. The authors used the one-dimensional KSTK model to design a series of centrifugal compressor stages. This demonstrates the wide practical application of one-dimensional models.

The paper [25] presents the identification and verification of a one-dimensional model for calculating turbocharger centrifugal compressors, a long-term project of the Dept. of Industrial Engineering of the Naples University. The creators of this mathematical model pay special attention to the system for specifying the shape of the flow path of the stage, and in the process of verification, to the exact repetition of the 3D impeller shape. The mathematical model takes into account the external heat exchange with the casing. Losses in the impeller are divided into incidence losses in the flow around the leading edges of the blades, losses in shock waves, friction losses on surfaces, and recirculation losses. Scroll losses are divided into tangential and radial losses. The authors used the results of gas dynamic tests for 5 different compressors in the model identification process.

In the paper [26], the application of the quasi-three-dimensional method to the calculation of a turbocharger centrifugal compressor is presented. The applied model is based on the equations of conservation of mass, momentum, and energy for unsteady and compressible flows. The foundations of the model are taken from the Navier-Stokes equations, but it uses the inviscid flow assumption, neglecting the terms of the second-order partial differential equation. Also, the effect of thermal diffusivity was not taken into account. The model allows calculating the parameters of a single-stage centrifugal compressor consisting of 3D impeller, vaneless diffuser and scroll.

The mathematical model was verified based on the results of turbocharger tests carried out at the University of Genoa. The comparison was made for tests with four different Mach numbers. The results showed good agreement between the calculated and experimental data and the importance of applying quasi-three-dimensional approaches to modeling centrifugal compressors.

For decades, research on centrifugal compressors has been carried out at LPI-SPbPU [27]. Analysis and generalization of the results obtained allowed the prof. Yu.B. Galerkin to create a method for gas-dynamic design of centrifugal compressors and a mathematical model based on it. It allowed to determine the characteristics of the compressor by dimensions of its flow path and similarity criteria (isentropic exponent, conventional Mach and Reynolds numbers). The mathematical model was implemented in the form of computer programs called the Universal Modeling Method (further referred to as the Method) [28] - [31].

Mathematical models of the Universal Modeling Method use simplified one-dimensional, two-dimensional and quasi-three-dimensional models, the calculation of which implies the use of empirical equations and the presence of simplifying assumptions. The values of empirical coefficients are determined in the process of identifying a mathematical model based on the results of experimental researches. The empirical equations themselves are implemented in the form of iterative processes. All this imposes a number of restrictions on the capabilities of programs.

The accumulation of new calculated and experimental data made it possible to make the mathematical models of the 9th version of the Universal Modeling Method more detailed and complex, to move away from the organization of iterative processes to the use of gas-dynamic functions. To transfer information to designers and carry out checks using CFD calculations, the Universal Modeling Method was supplemented with a program for constructing meridional and radial schematic images of the flow path and a program for constructing a three-dimensional impeller model. The main improvements of the 9th version of the Universal Modeling Method mathematical model, its identification and application practice are shown below.

2. Innovations of the mathematical model

Impeller inlet velocity. 1-D approach to calculate inlet velocity rate in the 8th and earlier versions of the Method [32]:

$$\bar{w}_1'' = \frac{\Phi_0}{4\tau_1 \bar{D}_1 \bar{b}_{1m} \sin\beta_{bl1}\varepsilon_1''}, \quad (1)$$

where τ_1 is blade blockade factor at the inlet to the impeller; ε_1'' is coefficient of compressibility at the inlet; Φ_0 is flow rate coefficient of an impeller; β_{bl1} is inlet blade angle; $\bar{D}_1$ is inlet diameter of the impeller relative to D_2; $\bar{b}_{1m}$ is height of impeller blades at the inlet relative to D_2.

Correct value of $\bar{w}_1''$ is important to calculate loss efficiency:

$$\Delta\eta_{imp} = 0.5\frac{\zeta_{imp}}{\psi_T}\bar{w}_1''^2, \quad (2)$$

where ζ_{imp} is impeller loss coefficient, ψ_T is loading factor.

IOP Conf. Series: Materials Science and Engineering 1180 (2021) 012022 doi:10.1088/1757-899X/1180/1/012022

The problem of impeller efficiency simulation of stages in a wide range of Φ_{des} and ψ_{Tdes} stressed the necessity to investigate inlet velocity calculation. The PC program 3DM-023 [28] executed inviscid Q3D calculations in impellers with flow rate coefficients Φ_{des}= 0.015–0.15, loading factors $\psi_{T\,des}$ = 0.5–0.65, relative hub ratios $\bar{D}_{hub}$ = 0.25–0.45. Measurements inside impellers [28] demonstrated good simulations of inlet velocity by Q3D inviscid calculations. Figure 1 demonstrates a sample of calculation.

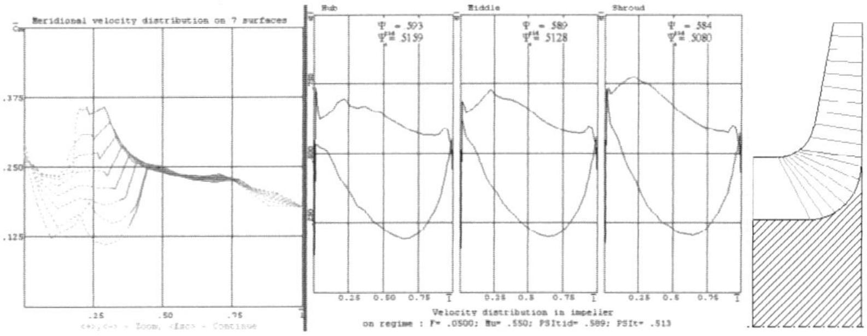

Figure 1. Sample of Q3D inviscid flow calculation. Left, meridian velocities on eight blade-to-blade surfaces; middle, velocity diagrams on three blade-to-blade surfaces; right, 2-D impeller meridian projection.

Mean Q3D velocities are less than the velocity calculated using the eq. (1). The coefficient $K_{w_1''} = \frac{w_{1MM}}{w_{1Q3D}} > 1$ is inserted into the eq. (1) in the 9th version of the mathematical model:

$$\bar{w}_1'' = \frac{\Phi_0}{K_{w_1''} 4\tau_1 \bar{D}_1 \bar{b}_{1m} \sin\beta_{bl1} \varepsilon_1''}, \tag{3}$$

Values of $K_{w_1''}$ lie in range 1.2–1.5 for different impellers. Formula (3) is a satisfactory approximation of the calculation experiment:

$$K_{w_1''} = 2.3226 \cdot 10^{-6} \beta_{bl1}^4 - 0.000273\beta_{bl1}^3 + 0.012052\beta_{bl1}^2 - 0.24047\beta_{bl1} + 2.9458. \tag{4}$$

A graphic representation on Figure 2 demonstrates accuracy of this approximation. The introduction of the empirical coefficient made it possible to increase the accuracy of calculations for the 9th version of the mathematical model compared to the 8th version.

Inlet flow conditions. The Method calculates characteristics as 110–150 points at different flow rates $\bar{m} < \bar{m}_{des} < \bar{m}$. For correct characteristics' simulation it is important to start with properly defined design flow rate. The design flow rate corresponds to the impeller non-incidence inlet condition, Figure 3. The flow pattern was obtained as a result of calculations using the NUMECA Fine/Turbo program.

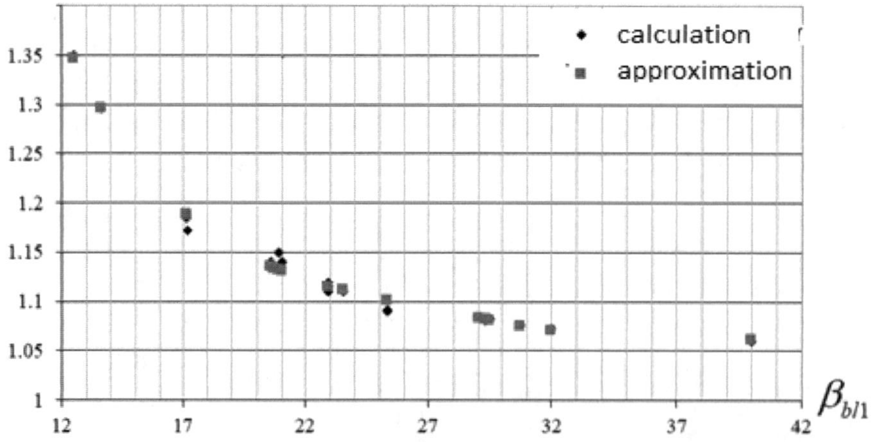

Figure 2. Calculated ratio $\frac{w_{1MM}}{w_{1Q3D}}$ and approximation by the formula (4)

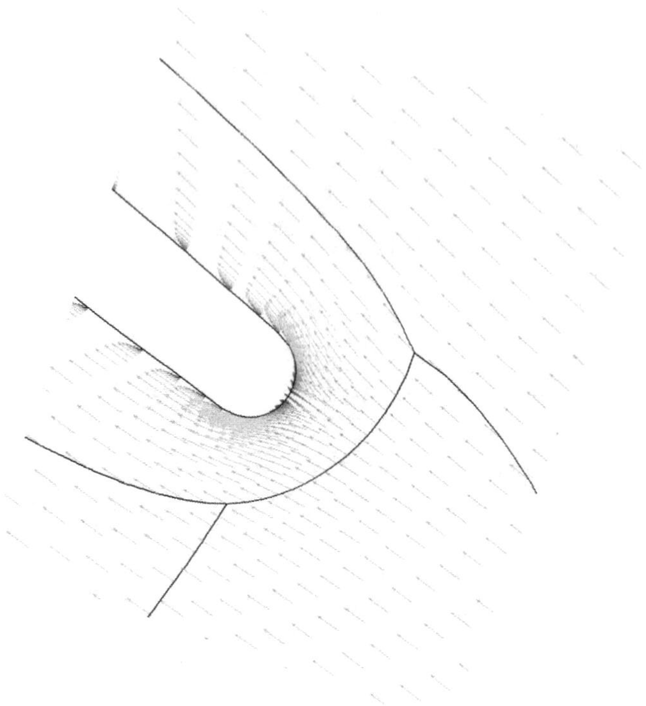

Figure 3. Velocity vectors around a blade leading edge. Non-incidence inlet.

The condition of a non-incidence inlet: stagnation point is located at the center of the leading edge. Blade load influences critical streamline direction. The scheme in Figure 4 (where $\overline{w}'_{1ni}$ is relative flow velocity at the inlet to the impeller, taking into account constraint in non-incidence mode) replaces blade load by a vortex with the same circulation.

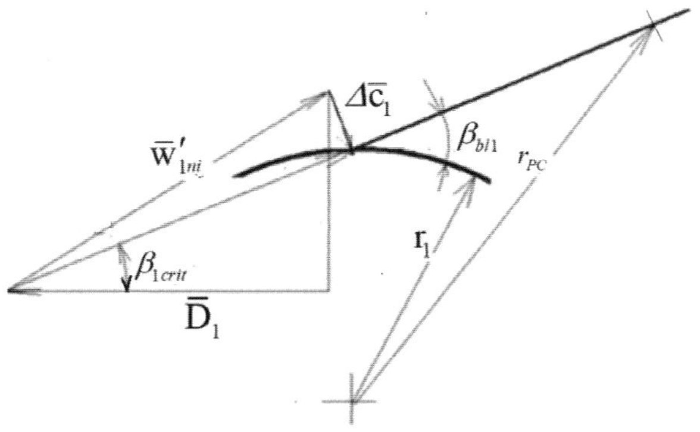

Figure 4. Scheme of the effect of a vortex (non-incidence inlet flow rate)

The vortex position is r_{pc}. Non-incidence condition: a critical streamline angle is equal to the inlet blade angle $\beta_{1crit} = \beta_{bl1}$. Average relative inlet flow velocity $\bar{w}'_{1ni} = w'_{1ni} / u_2$ differs from the velocity of a critical streamline by a component $\Delta\bar{c}_{1crit} = \Delta c_{1crit} / u_2$. The vortex creates this component:

$$\Delta\bar{c}_1 = X\frac{\psi_T}{z}\frac{\sin\beta_{bl1}}{(1-K_{PC})(1-\bar{D}_1)}, \quad (5)$$

where z is number of blades, $K_{PC} = \frac{r_{pc} - r_1}{r_2 - r_1}$ is the coefficient that represents blade load distribution along the blade radius, X is the empirical coefficient to be identified. The viscosity of the fluid is taken into account when calculating the loading factor using the empirical coefficient K_μ.

The value of the velocity $\Delta\bar{c}_1$ that determines the restructuring of the flow is involved in the calculation of incidence losses. The formulas for calculating incidence losses presented in [31] include empirical coefficients determined in the process of identifying a mathematical model.

Twenty impellers with different design parameters were the object of the analysis. Non-incidence flow rates were obtained by iterative calculations of inviscid Q3D flow. Inlet velocity triangles demonstrated non-incidence inlet with different X values for different impellers. The geometric parameter, relative throat width $\bar{a}_1 = a_1 / D_2$, influences the coefficient X the most. Approximating formulae for different types of impellers are:

- 3D impellers (two different series):

$$X = 1 + 9.75\Delta\bar{w} - 31.9\bar{a}_1; \quad (6)$$

- 2D impeller with profiled blades (controlled blade load distribution along radius):

$$X = 1 + 5\Delta\bar{w} - 24.9\bar{a}_1. \quad (7)$$

- 2D impellers with arc blades:

$$X = 1.37 = \text{const}; \quad (8)$$

Figure 5 demonstrates the accuracy of the formulae (6–8). Non-incidence flow rate coefficient Φ_{niMM} obtained using the formulae (6–8) is related to the Φ_{ni3DM} calculated by the 3DM-023 PC program.

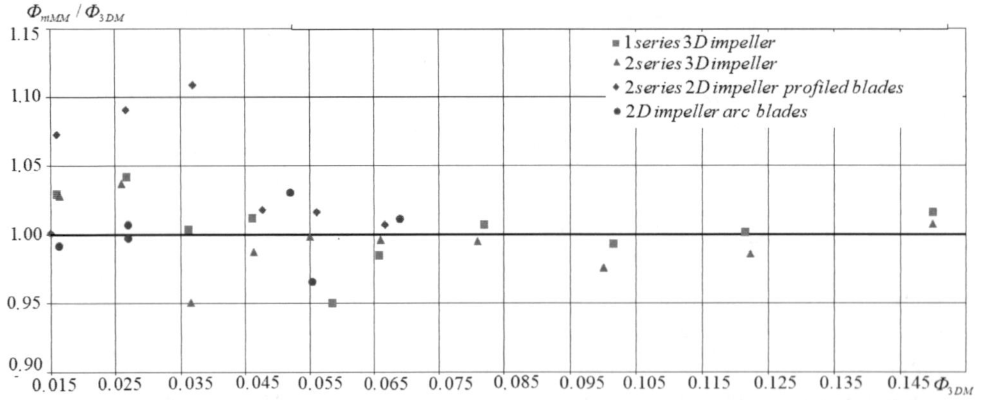

Figure 5. Comparison of non-incidence flow rates. Dependence $\Phi_{\mathrm{niMM}} / \Phi_{\mathrm{ni3DM}} = \mathrm{f}\left(\Phi_{\mathrm{ni3DM}}\right)$ for stages with 3D and 2D impellers with arc and profiled blades when calculating new formulas for calculating flow restructuring

The accuracy of formulae (6–8) is sufficient for a preliminary design. To calculate characteristics of an existing compressor, better simulation is necessary. New test results will be approximated in the future.

Q3D approach for 3D impellers. A significant non-uniformity of the flow along the height of the 3D impeller blades leads to the fact that using the one-dimensional approach used in the 8th version of the mathematical model is not the best solution. In the 9th version of the Method, quasi-three-dimensional approach is applied. The flow in the blade cascade is divided by seven blade-to-blade surfaces. The mathematical model calculates flow parameters in eight elementary blade cascades. Friction losses on the hub and shroud surfaces occur on the 1st and 8th blade elements. Mixing losses in the case when the VLD width is higher than the height of the impeller blades at the outlet $\zeta_{mixb3/b2}$ are calculated using the 1-D formulation.

At the entrance to the blade cascade, it is assumed that the flow coefficient $\varphi'_{1i} = c_{1i} / u_2$ is the same on each streamline. Relative velocity at the entrance to the blade cascade:

$$\bar{w}_{1i} = \sqrt{\bar{D}_{1i}^2 + \varphi_{1i}'^2}\,. \tag{9}$$

After determining all the components of the head losses in the blade cascade, the polytrophic coefficient of the impeller is calculated. Its value determines the polytrophic process: $\mathrm{pv}^{\mathrm{n}} = \mathrm{const}$. The sum of losses of all blade elements is in the square brackets:

$$\frac{n_{1-2}}{n_{1-2}-1} = \frac{k}{k-1} - \lambda_{u0}^2 \frac{k}{k+1} \left[\begin{array}{l} \sum_{i=1}^{8}\left(\zeta_{frbli} + \zeta_{mixi}\right)\bar{w}_1''^2 + \zeta_{mixb3/b2}\bar{w}_1''^2 + \zeta_{\lim i=1}\bar{w}_{1i=1}''^2 + \\ +\zeta_{\lim i=8}\bar{w}_{1i=8}''^2 + \sum_{i=1}^{8}\left(\zeta_{inci}\bar{w}_{1i}''^2\right) + 2\psi_T\left(\beta_{leak} + \beta_{fr}\right) \end{array} \right] \Bigg/ \left[\frac{T_1'}{T_0^*}\left(\frac{T_2}{T_1'} - 1\right) \right]. \tag{10}$$

where β_{leak} is labyrinth seal leakage coefficient; β_{fr} is disc friction coefficient; k is isentropic coefficient; ζ_{frbli} is coefficient of friction loss on blade surfaces; ζ_{mixi} is coefficient of mixing losses; ζ_{inci} is coefficient of incidence losses; T_1' is static temperature at the impeller inlet; T_0^* is total temperature at the impeller inlet; T_2 is static temperature at the impeller outlet; λ_{u0} is velocity coefficient at the impeller inlet.

Q3D approach more precisely calculates blade surface area. It is important for the friction loss coefficient calculation. The PC program 3DM.023 calculates blade area quite precisely. However, these calculations are too time-consuming to be directly used in the Method. Figure 6 shows a comparison of the blade areas of the 3D impellers, calculated by the 3DM.023 program with 1-D approach of the 8th and Q3D approach of the 9th Method versions.

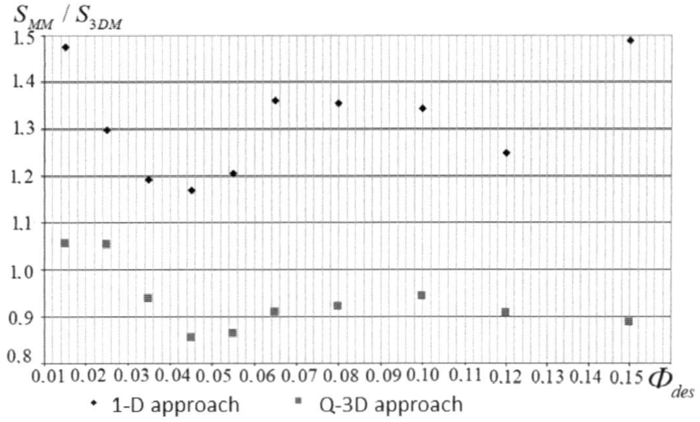

Figure 6. The blade area ratio S_{MM} / S_{3DM} of 3D impellers with different Φ_{des}

The average error is 31% in the 1-D approach and 7% in the Q3D approach.

CFD mathematical model for vaneless diffusers. The CFD model substitutes a rather precise but complicated model that calculates friction losses with a number of empirical coefficients. The ANSYS CFX program was used to calculate efficiency characteristics and outlet parameters of diffusers with their relative width b / D_2 = 0.006–0.10 and with radial length D_4 / D_2 = 1.6–2.0 in the range of velocity coefficients λ_{c2} = 0.39–0.82 and Reynolds numbers Re_{b2} =36 800–73 600, inlet flow angles α_2 =10°–90°. The 8th version of the mathematical model guaranteed the adequacy of the application for the relative width b / D_2=0.014–0.10. Loss coefficient and flow outlet angle are functions of geometry and gas dynamic parameters:

$$\zeta_{dif}, \alpha_4 = f(\bar{b}_2, \bar{D}_4, \alpha_2, \lambda_{c2}, \mathrm{Re}, \bar{k}_{Rz}), \tag{11}$$

where $\bar{b}_2$ is relative width, $\bar{D}_4$ is relative length, α_2 is inlet flow angle, λ_{c2} is inlet velocity coefficient (compressibility criteria), Re_{b2} is Reynolds number, $\bar{k}_{Rz}$ is relative roughness.

The data of calculations were approximated by sets of algebraic equations. A sample for a loss coefficient:

$$\zeta_{dif} = \frac{A \cdot \alpha^{B} \cdot K_{D4}}{K_{\mathrm{Re}\bar{k}_{Rz}}}, \tag{12}$$

where $\mathrm{A}, \mathrm{B}, \alpha$, K_{D4}, $K_{\mathrm{Re}\bar{k}_{Rz}}$ are sets of algebraic equations.

Mathematical model of tangential exit nozzles. The model establishes nozzle dimensions and calculates loss coefficient. The continuity equation for a scroll:

$$\bar{m}_{\Theta} = \frac{\Theta}{360^0} \bar{m}_{des} \approx \rho c_{u4} r_4 \int_{r_4}^{r_{out}} b \frac{dr}{r}. \tag{13}$$

An average flow velocity at the outlet and an area of the exit follow from eq. (13):

$$\bar{c}_{des360} = \frac{1}{K_s}\bar{c}_{u4des}\frac{\bar{D}_4}{\bar{D}_{360avr}}, \tag{14}$$

where the empirical coefficient $K_s = 1.25$ [33]. Equations to calculate sizes of scrolls in Figure 7 and others are formulated assuming proper functions $b = f(r)$.

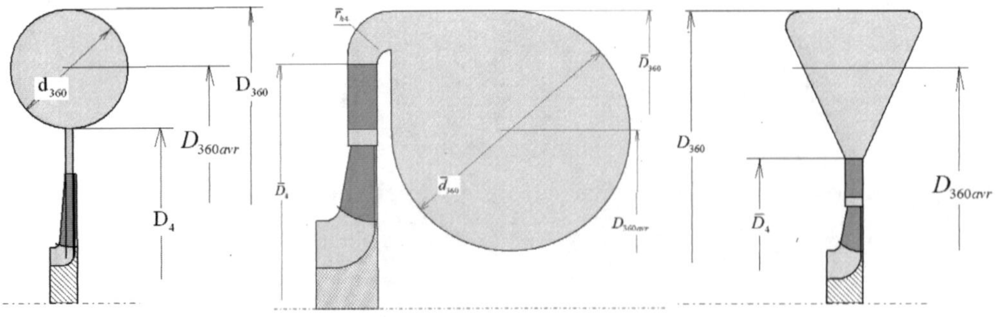

Figure 7. Dimensions of scrolls of different types

The model of a loss coefficient summarizes losses in different parts of a scroll and losses of different nature: friction, separation, secondary, and incidence. The new mathematical model contains 10 empirical coefficients, which is 1.5 times less than in the previous model.

Flow parameters in control planes calculation using the velocity coefficient λ. The engineering 1-D model calculates static and total pressures and temperatures, density and velocity vectors in control planes throughout the flow path: from its inlet to the outlet. The continuity equation $\bar{m} = cf\frac{p}{RT} = const$ solves the problem iteratively. It creates numeric problems in some special cases. To avoid it, the continuity equation is presented in the form that does not require iterations:

$$\frac{\lambda_{i+1}}{\lambda_i}\sqrt{\frac{T^*_{i+1}}{T^*_i}}\left(\frac{1-\frac{k-1}{k+1}\lambda^2_{i+1}}{1-\frac{k-1}{k+1}\lambda^2_i}\right)^{\frac{n}{n-1}} = \frac{\Phi_i}{\Phi_{i+1}}\frac{\bar{f}_i}{\tilde{f}_{i+1}}, \tag{15}$$

and

$$\Phi = \left(\frac{k+1}{2}\right)^{\frac{1}{k-1}}\left(1-\frac{k-1}{k+1}\right)^{\frac{1}{k-1}}\frac{\bar{f}}{\lambda_u}\lambda\left(1-\lambda^2\right)^{\frac{1}{k-1}} \tag{16}$$

where: $\lambda = \frac{c}{\sqrt{\frac{2k}{k+1}RT^*}}$, $\lambda_u = \frac{u}{\sqrt{\frac{2k}{k+1}RT^*_{inl}}}$, $\Phi = \frac{4\bar{m}}{\rho\pi D_2^2}$, n is polytrophic coefficient.

3. Identification and Verification of the 9th Version of the mathematical model

The main objects of modeling are loss coefficients in the eq. (10) and in similar equations for other elements of the flow path. The model treats the loss coefficient as a sum of coefficients of friction (secondary flow is included), mixing (separation) and incidence (for blades and vanes only) losses. Calculation of losses friction loss coefficient ζ_s of a blade suction side is shown as an example:

$$\zeta_s = c_{ws} z \frac{4\bar{b}_{avr} \bar{l}\, \varepsilon_{avr} \bar{w}_{s\,avr}}{\pi \Phi} \left(\frac{w_{s\,avr}}{w_1} \right)^2 , \tag{17}$$

where $\bar{b}_{avr} = b_{avr} / u_2$ is average blade height, $\bar{l} = l / D_2$ is blade length, $\bar{w}_{avr} = w_{avr} / u_2$.

In the eq. (17), the mathematical model associates the coefficient of friction force of a blade c_w with the coefficient of a thin plate friction force $c_f = 0.0306\,\mathrm{Re}^{-1/7}$ [34]:

$$c_{ws} = c_f \left(1 + X_i \left(1 - \frac{w_{s2}}{w_{s1}} \right)^{X_{i+1}} \right) \left(1 + X_{i+2} Ro'^{X_{i+3}} \right). \tag{18}$$

In the eq. (18), the empirical coefficient X of the mathematical model needs to be identified. The expression in the first bracket represents the influence of a velocity gradient along a blade surface. The expression in the second bracket represents the influence of a velocity gradient normal to a blade surface. The normalized normal gradient for an impeller is $Ro' = 4 - \bar{w}_{avr} / \bar{R}_{bl\,avr}$ ($\bar{R}_{bl\,avr} = R_{bl\,avr} / D_2$ is the average radius of a blade curvature).

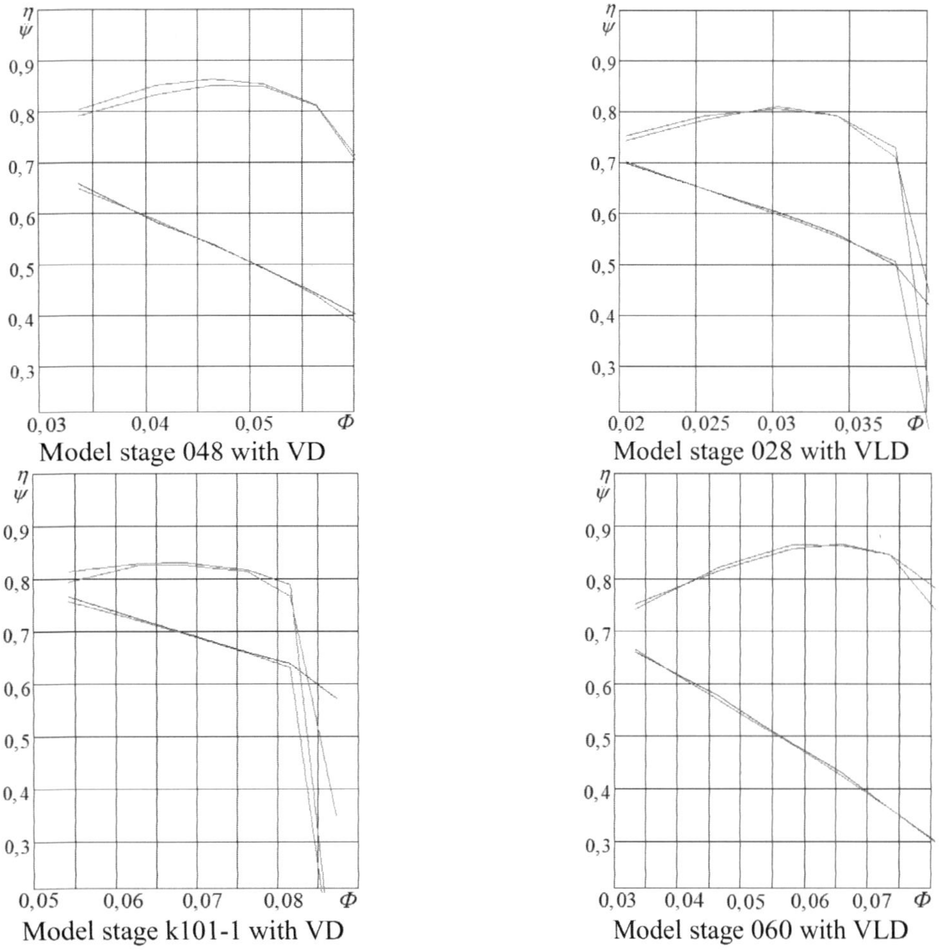

Model stage 048 with VD

Model stage 028 with VLD

Model stage k101-1 with VD

Model stage 060 with VLD

Figure 8. Identification results. Comparison of calculated (red lines) and experimental (blue and green lines) gas dynamic characteristics

The identification of the mathematical model is carried out in the specialized program IDENT [32]. To identify the loss model, the characteristics of the model stages [27] were used. In total, 70 stage characteristics were used with 6 values of the flow coefficient for each test. Figure 8 demonstrates comparison of measured and calculated characteristics.

Average inaccuracy of efficiency calculation after the identification is ±1.0 % [35].

To verify the mathematical model, the IDENT program calculated characteristics of the stages that did not participate in identification. Low flow rate model intermediate type stages in Figure 9 (left) have very short VLD and 2D impellers with blades of constant height. The model 1:2 of a single-stage 32 MW pipeline compressor in Figure 9 (right) has the axial inlet nozzle, long VLD, and a scroll of design according to the part 2 of this text [36].

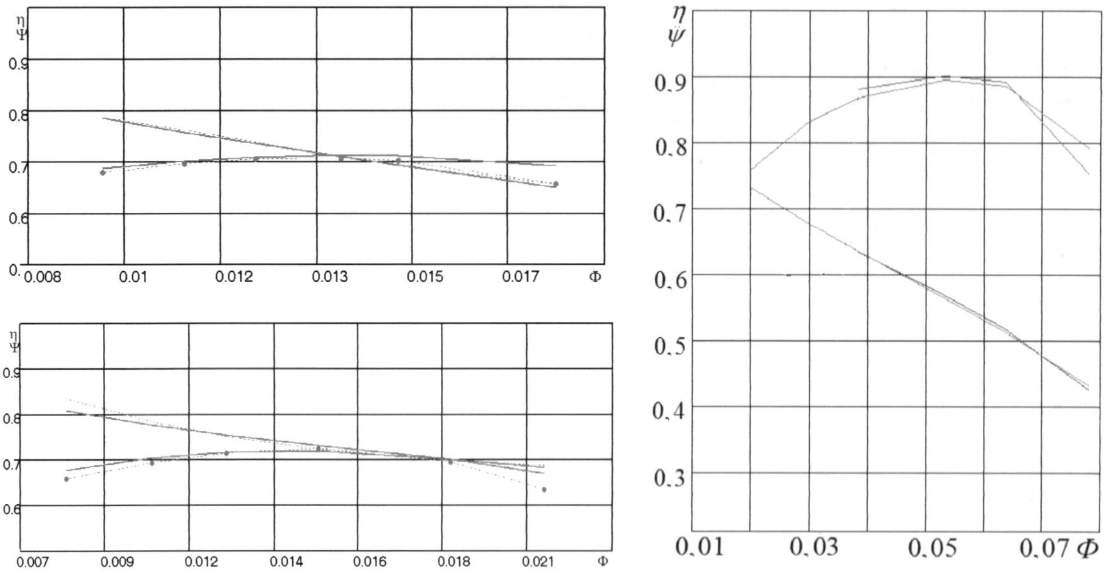

Figure 9. Verification results. Left, intermediate type low flow rate model stages: measured (dashed lines) and calculated (solid lines); right, the model of a single-stage 32 MW pipeline compressor: experiment (blue and green lines) and calculation (red lines).

The accuracy of verification calculations is the same as in identification.

4. The 9th Version Mathematical Model design results

The programs of the 9th version of the Universal Modeling Method in varying degrees of readiness have been used since 2014 in the design practice of the Laboratory "Gas Dynamics of Turbomachines".

Twenty designed model stages have the following design parameters: flow rate coefficient Φ_{des} = 0.015–0.15 and impeller diameter of 0.35 m. Two stages with Φ_{des} = 0.025 and 0.035 were tested on the test rig ECC-55 of the Laboratory [37]. Design and test characteristics are in Figure 10.

The mathematical model accurately determines the flow rate range of both stages. Near the design flow rate, the mathematical model overestimates the efficiency by 0.8–1.3%. In case of the stage with Φ_{des} = 0.025 at maximum flow rate the measured efficiency is even higher. The measured characteristic of the stage with Φ_{des} = 0.035 shifts to the left in comparison with the calculated one. The test data of all new model stages will be the material for future improvements of the model stage.

The turbocharger compressor was jointly designed by the Laboratory and NPO Turbotekhnika JSC. The industrial partner manufactured and tested the compressor at his test rig [38]. The turbocharger is intended for a low speed diesel converted for natural gas. The pressure ratio is not high. Gas-dynamic characteristics of the TKR 140E compressor are presented on Figure 11.

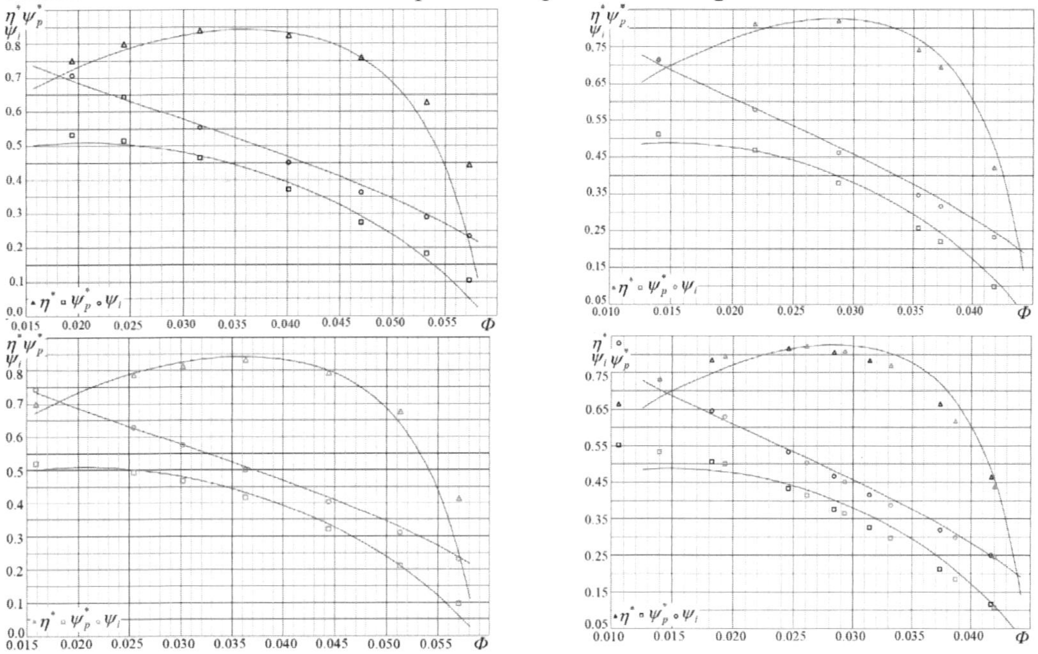

Figure 10. Gas dynamic characteristics of a stage with $\Phi_{des} = 0.035$ at 8700 rpm, $M_u = 0.464$ (left) and stage with $\Phi_{des} = 0.025$ at 5300 rpm, $M_u = 0.282$ (right): calculation (solid line) and experiment (dots)

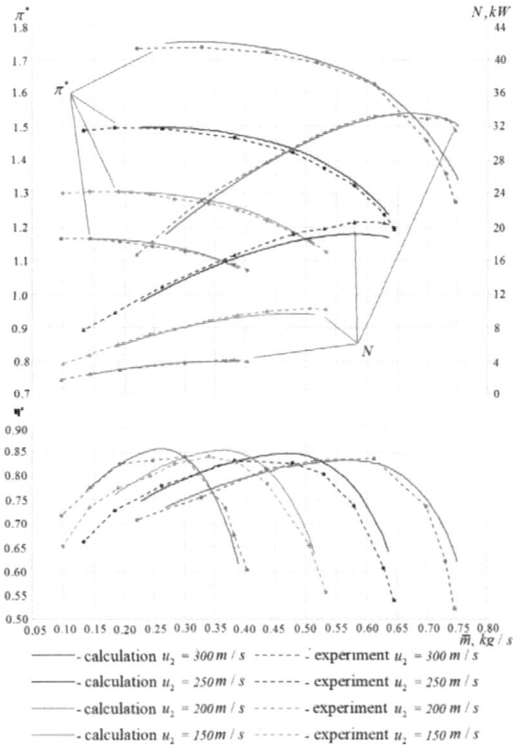

Figure 11. Gas-dynamic characteristics of the TKR 140E compressor, design and test

The industrial partner is satisfied with the results of cooperation.

Conclusion

Starting from the 4th version (1990s), all versions of the Universal Modeling Method execute optimal design and calculate gas-dynamic characteristics of centrifugal compressors. However, in the 4th version, it was necessary to use several sets of empirical coefficients X for stages with different design parameters. Due to the significant detailing of the mathematical model, the 5th version (2010s) correctly calculates the efficiency at the design point. To calculate the entire range of characteristics, individual sets of empirical coefficients are required. Further improvements allowed the 6-8th versions to operate with single sets of X for all stages to calculate the entire characteristics. In the 9th version presented above, the characteristics of very low flow rate stages are correctly modeled by introducing the vaneless diffuser model based on CFD calculations. To calculate high-flow rate 3D impellers, a quasi-three-dimensional calculation is introduced. The new model of exit nozzles has increased the accuracy of calculating of related stages. The Method, like other engineering techniques, cannot guarantee 100% accuracy in calculating the characteristics of any compressor. There are directions for further development. The introduction of a quasi-three-dimensional inviscid calculation into the mathematical model is very promising. This calculation will provide the most realistic information about the local flow velocities and dimensions of the flow path. It is necessary to bear in mind the opposite trend in the development of computational gas dynamics. As CFD calculations of centrifugal compressor performance become reliable, engineering techniques will play a modest role in preliminary design. The future will show for how long the further development of the Method will be relevant.

Acknowledgments

The research was performed by a Grant of the President of the Russian Federation for young PhD MK-1893.2020.8.

References

[1] Khisameev, I.G., Maximov, V.A., Batkis, G.S., Guzelbaev, Ya.Z. *Design and operation of industrial centrifugal compressors*. Kazan. P. 671. 2012 [in Russian]

[2] https://www.c-o-k.ru/market_news/mirovoy-rynok-kompressorov-2017 access 11.11.2020 [Electronic resource].

[3] Maksimov, V.A., Miftakhov, A.A., Khisameev, I.G. Compressor and refrigeration engineering at the present stage. *Bulletin of KSTU*. **1**. p. 104-113. 1998 [in Russian]

[4] Safin, A.Kh. Trends in the technical and economic structure of production and development of compressor equipment. *Compressors and pneumatic*. **2**. p.4-9. 2002 [in Russian]

[5] Maximov, T.V., Maximov, V.A., Egorov, A.G. Trends in the development and production of compressor technology at the present stage. *Bulletin of Kazan Technological University*. **5**. p. 176-179. 2013 [in Russian]

[6] Afanasyev, B.V., Lunev, A.T., Mustafin, N.G., Potasheva, E.V. Designing a compressor impeller using an inverse problem for a rotating lattice of profiles on an axisymmetric current surface. *Compressors and pneumatic*. **1-2**. p. 33-37. 1996 [in Russian]

[7] Den, G.D. *Design of the flow path of centrifugal compressors*. Publishing house "Mechanical engineering". - Moscow. p. 232. 1980 [in Russian]

[8] Lyubimov, A.N. Improvement of methods for calculating the gas-dynamic characteristics of the flow path of stationary centrifugal compressors: *PhD Thesys*. Saint Petersburg, 2016. p. 138. [in Russian]

[9] Selyanskaya, E.L., Zhenikhov, S.V., Goldobin, A.S. Compressors NPO Iskra for underground gas storage stations. *Compressors and pneumatic*. **4**. p. 2-7. 2015 [in Russian]

[10] Riess, V.F. *Centrifugal compressor machines*. L .: Mechanical engineering. p. 336. 1964 [in Russian]

[11] Guidotti, G.E. Towards Centrifugal Compressor Stages Virtual Testing. *Ph.D. thesis* Università degli Studi di Bologna, 2014.

[12] Japikse, D. Agile design system in the age of concurrent engineering. *JANNAF Conference*. Albuquerque. 1996.

[13] Japikse, D. Design system development for turbomachinery (turbopump) designs - 1998 and a decade beyond. *JANNAF Conference*. Cleveland. Ohio. 1998.

[14] Aungier, R.H. *Centrifugal Compressors: A Strategy for Aerodynamic Design and Analysis*. ASME Press, ISBN 0-7918-0093-8, USA. 2000

[15] Brebenel, M. An analytical method for simulation of centrifugal compressors. *INCAS BULLETIN*, **12**, Issue 1 2020, p. 35 – 49.

[16] Harley, P., Spence, S., Filsinger, D., Dietrich, M., Early, J. Assessing 1D loss models for the off-design performance prediction of automotive turbocharger compressors. *Proceedings of ASME Turbo Expo 2013: Turbine Technical Conference and Exposition GT2013* June 3-7, 2013, San Antonio, Texas, USA GT2013-94262

[17] Monje, B. 1D and 3D Tools to Design Supercritical CO2 Radial Compressors: A Comparison. *Proceedings of the 2nd International Seminar on ORC Power Systems*, Rotterdam, 2013.

[18] Swain, E. Improving a one-dimensional centrifugal compressor performance prediction method. *Part A: J. Power and Energy IMechE.* **219.** p. 653-659. 2005

[19] Oh, H.W., Yoon, E.S., Chung, M.K. An optimum set of loss models for performance prediction of centrifugal compressors. *Proceedings of the Institution of Mechanical Engineers, Part A: Journal of Power and Energy* 1997, **211**, London, England: 331. DOI: 10.1243/0957650971537231.

[20] Galvas, M.R., "Fortran Program for Predicting Offdesign Performance of Centrifugal Compressors". 1974 *NASA Lewis Research Cente*r, Cleveland, Ohio, USA, NASATN-D-7487.

[21] Herbert, M.V. "A Method of performance prediction for centrifugal compressors". *ARC R&M No. 3843*, 1980, H.M. Stationery Office, London, England.

[22] Weber, C.R., Koronowski, M.E. "Meanline Performance Prediction of Volutes in Centrifugal Compressors". *ASME Conference Proceedings*, New York, USA, 1986, 86-GT-216.

[23] Syka, T., Luňáček O. Numerical simulation of radial compressor stage. *EPJ Web of Conferences* **45**, 01088 (2013) DOI: 10.1051/epjconf20134501088.

[24] Syka, T., Luňaček, O., Kňourek J. Numerical simulation of radial compressor stages with seals and technological holes. *EPJ Web of Conferences* **67**, 02115 (2014) DOI: 10.1051/ epjconf/20146702115.

[25] De Bellis, F., Bozza, F., Bevilacqua, M., Bonamassa, G., et al. "Validation of a 1D compressor model for performance prediction". *SAE Int. J. Engines* **6(3)**:2013, doi:10.4271/2013-24-0120.

[26] Montenegro, G. A., Tamborski M., Marinoni A., Della A. Torre, A.Onorati, S.Marelli Quasi 3D approach for the modelling of an automotive turbocharger's compressor. *AIP Conference Proceedings* **2191**, 020113 (2019); https://doi.org/10.1063/1.5138846.

[27] *Proceedings of the scientific school of compressor engineering SPbSPU.* / Edited by Galerkin Yu.B. // - M .: Ed. SPbSPU - 2010. [in Russian]

[28] Galerkin, Yu.B., Danilov, K.A., Mitrofanov, V.P., Popova, E.Yu. To the use of numerical methods in the design of the flow path of centrifugal compressors. SPb. *SPbSTU*. p. 68. 1996 [in Russian]

[29] Galerkin, Yu.B., Danilov, K.A., Popova, E.Yu. Development of the method of universal modeling of the workflow of the central control center, software systems of the first level (third generation), experience in the development and practical use of the complex of the third level. SPb. *SPbSTU*. 1995. [in Russian]

[30] Galerkin, Yu.B., Danilov, K.A., Popova, E.Yu. Numerical modeling of centrifugal compressor stages (physical foundations, state of the art). *Compressors and pneumatic*. **2**. 1993 [in Russian]
[31] Danilov, K.A. Creation of a mathematical model and software systems for optimal gas-dynamic design of refrigeration centrifugal compressors. *PhD Thesys*; SPbSTU. SPb., 1999 p. 176 [in Russian]
[32] Galerkin, Yu.B., Soldatova, K.V., *Modeling the workflow of industrial centrifugal compressors. Scientific foundations, stages of development, current state*. Monograph. Publishing House of the Polytechnic University, 2011. - p. 327. [in Russian]
[33] Miftakhov A.A., Zykov V.I. *Inlet and outlet nozzles of centrifugal compressors*. - "Fan". - Kazan. - 1996. [in Russian]
[34] Loytsansky, L. G. *Mechanics of liquid and gas*. M. Nauka., 1978 .- P. 736.
[35] Galerkin, Yu.B., Rekstin, A.F. *Research methods of centrifugal compressor machines*. L .: Mechanical Engineering. p. 303. 1969 [in Russian]
[36] Galerkin, Yu.B., Rekstin, A.F., Soldatova, K.V., Drozdov, A.A. Highly efficient single-stage full-pressure compressor for gas compressor units (gas-dynamic design, result of model tests). *Compressors and pneumatic*. **8**. p. 19-25. 2014 [in Russian]
[37] Borovkov, A.I., Galerkin, Yu.B., Drozdov, A.A., Rekstin, A.F., Semenovsky, V.B., Yadykin V.K. Stand ECC-55 with direct high-frequency drive for gas-dynamic research of industrial centrifugal compressors. *Compressors and pneumatic*. **4**. p. 32-43. 2020 [in Russian]
[38] Galerkin, Yu.B., Rekstin, A.F., Drozdov, A.A., Kaminsky, R.V., Sibiryakov, S.V., Turgenev, T.I., Usenko, A.E. Experience in creating a low-pressure turbocharger for pressurizing an internal combustion engine using a modern version of the Universal Modeling Method. *Compressors and pneumatic*. **2**. p. 2-10. 2019 [in Russian]

Continuous Cooling Compressor for water refrigerant heat pump

T. Shoyama[1,2], H. Sun[1], B. Kawano[1] and M. Matsui[1]

[1]Appliances Company, Panasonic Corporation, Moriguchi, Osaka, 570-8501, Japan
[2]shoyama.tadayoshi@jp.panasonic.com

Abstract. A turbo compressor was developed for water refrigerant heat pumps equipped with continuous cooling compression, in which the vapour is continuously cooled during the compression process by the latent heat of evaporation of the liquid water refrigerant sprayed into the impeller channels. One of the disadvantages of water refrigerant is that the discharge temperature during vapour compression is higher than that of fluorocarbon refrigerant. Multi-stage compression with intercooling are one of the solutions, the system becomes large. In continuous cooling compression, fine particles of water are sprayed from a nozzle embedded inside the impeller, and the water vapour in compression process is directly cooled to the saturation temperature. Therefore, if sufficient amount of spray is present, the compression process follows the saturation line in p-h diagram. Water is supplied to the spray nozzle after pressurized to 10 MPa or more using the centrifugal effect of rotor rotation. A small nozzle was developed that suppresses rotor imbalance and can withstand the load of centrifugal force. A followability index was defined as a reference for the droplets to follow the vapour flow and evaporate before colliding with the blade surface. A water-refrigerant heat pump system was constructed, and the degree of superheat of the discharged vapour was evaluated while changing the spray amount. Increasing the amount of spray decreased the superheat but COP also decreased. Saturated compression was demonstrated above a certain amount of spray, but the impeller flow was choked due to the droplet evaporation. It is desirable to assume a sufficient choke margin, or to develop a special CFD code that can take into account the droplet evaporation.

1. Introduction

Water refrigerant heat pump has high theoretical COP and environmental friendliness [1]. Its applications are not limited to chillers, but also include ice generation [2] and high-temperature heat pumps [3]. Commercial products have been released such as Efficient Energy's eChiller and Kawasaki Heavy Industries' MizTurbo. However, due to the high specific volume of water refrigerant, the equipment dimension tends to be large, and the cost of the compressor becomes high [4]. Therefore, a low-cost and high-performance compressor is necessary. We have developed a water-refrigerant system with a two-stage turbo compressor that uses water as both a refrigerant and a lubricant, and achieved COP = 5.0 [5]. One of the difficulties in a water-refrigerant turbo compressor is a vibration of the rotating system and bearings. Thus it is preferable to use water-lubricated bearings, which are simple, low cost, and have a long life [6,7].

Another difficulty is a high degree of superheat during steam compression. Fluorocarbon refrigerants, being designed for the refrigeration cycle, has an adiabatic compression line increasing along the saturation line in a p-h diagram, reducing the theoretical compression energy. On the other hand, water has an adiabatic compression line increasing with a low gradient. Therefore, lowering the discharge temperature by intercooling by mist spraying leads to low theoretical compression energy. This has been theoretically shown [4] and demonstrated also in our previous study [5]. In the actual vapour compression process, droplets mixed in the suction flow of the compressor causes erosion of the impeller and a decrease in surge margin, whereas the compressed hot vapour is cooled by the latent heat of vaporization of the droplets, resulting in a compression process close to saturation, and the compression power is lowered [8]. Wu et al. [9] showed that an initial droplet size must be less than 10 μm for complete evaporation of the droplet and is independent on the relative spray velocity. However, for atomization to the droplet size of 10 μm or less, water must be supplied at a high-pressure exceeding 10 MPa to the nozzle. A pump capable of such high pressure is too large and too energy-consuming to be equipped with separately. In this paper, continuous cooling compression (CCC) is proposed, in which sprayed mist with an average particle size of 5.8 μm was achieved using the centrifugal effect of the impeller. The fine mists are sprayed from the impeller, with low relative velocity to the blade, suppressing the erosion. The theory and the experimental results are reported.

2. Theory of continuous cooling compression

2.1. Thermodynamics

Here CCC is defined as a thermodynamic process. Nozzles are embedded in rotating impellers. They spray water droplets that is fine enough to accompany the compressed vapour flow between the blades. The vapour is cooled during the compression process by the latent heat of vaporization of the droplets. The thermodynamic process of CCC is depicted in figure 1 as p-h and p-v diagrams.

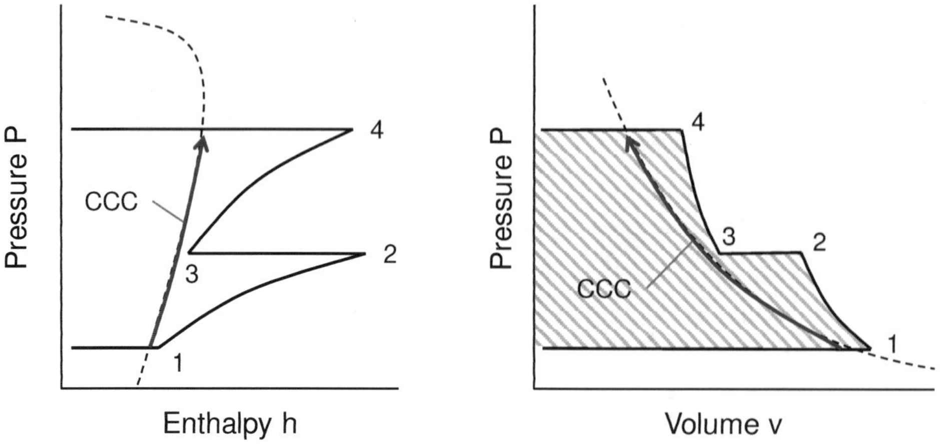

Figure 1. Thermodynamic diagrams of CCC

In the case of a two-stage compression process with intercooling (IC), compressed vapour is cooled close to the saturation temperature in the isobaric cooling process (2-3) between the first-stage adiabatic compression (1-2) and the second-stage adiabatic compression (3-4). The required cooling heat is $c_2 = h_3 - h_2$, and the total theoretical compression power is expressed by the following equation, ignoring the change in entropy.

$$\Delta h_2 = (h_2 - h_1) + (h_4 - h_3) = \int_{P_1}^{P_2} v_{\mathrm{ad}} dP + \int_{P_3}^{P_4} v_{\mathrm{ad}} dP \quad (1)$$

where v_{ad} is the specific volume in adiabatic compression, and Δh_2 is equal to the hatched area on the p-v diagram. It should be noted that the subscript 2 of c_2 and Δh_2 is the number of stages. When it is increased to n, the theoretical compression energy Δh_n and the intercooling heat c_n are expressed by the following equations.

$$\Delta h_n = \sum_{i=1}^{n} (h_{2i} - h_{2i-1}) = \sum_{i=1}^{n} \int_{P_{2i-1}}^{P_{2i}} v_{\mathrm{ad}} dP \tag{2}$$

$$c_n = \sum_{i=1}^{n-1} (h_{2i} - h_{2i+1}) \tag{3}$$

CCC process is the limit of $n \to \infty$. As shown in figure 1, the process is along the saturation line. The theoretical compression energy Δh_∞ is the limit of Eq. (2).

$$\Delta h_\infty = \sum_{i=1}^{\infty} \int_{P_{2i-1}}^{P_{2i}} v_{\mathrm{ad}} dP = \int_{P_1}^{\gamma P_1} v_{\mathrm{sat}} dP \tag{4}$$

where v_{sat} is the specific volume in saturation, γ is the pressure ratio. Just like $\Delta h_2 \neq h_2 - h_1$ in two-stage compressor, $\Delta h_\infty \neq h_{\gamma P_1} - h_1$ in CCC. Therefore, the theoretical compression power of the CCC process must be calculated from the saturation line on the P-v diagram, or the calculation of Eq. (2) must be performed for a sufficiently large number of n. When the calculation was performed, it converged in 100 steps. The area between the adiabatic line and the saturation line shown in the P-v diagram is the theoretical limit of the power reduction effect by CCC, which is expressed as follows.

$$\int_{P_1}^{\gamma P_1} v_{\mathrm{ad}} dP - \int_{P_1}^{\gamma P_1} v_{\mathrm{sat}} dP = \int_{P_1}^{\gamma P_1} (v_{\mathrm{ad}} - v_{\mathrm{sat}}) dP \tag{5}$$

Thus, theoretical work in CCC is always lower than those in the intercooled compression. In the present case, 5% reduction of theoretical work was estimated.

2.2. Centrifugal pressure spraying

There are two types of method to spray droplets for CCC: external spraying and centrifugal pressure spraying. In the former, mist is generated outside the compressor and sprayed into the compressor suction path together with the vapour. This is used in wet compression technology [9] and advanced humid air turbine (AHAT) systems [10]. This method is simple and has similar performance gain, but has two problems. The first is erosion at the leading edge of the blade due to the droplet collision because the relative velocity of the droplet to the blade is high. The second is necessity of a high-pressure pump of over 10 MPa that allow to generate sufficiently fine droplets. Centrifugal pressure spraying solves these problems at the same time by spraying from a nozzle embedded inside a rotating impeller as shown in figure 2. Water is supplied from an axial flow path through the centre of the axis connected radial flow paths in the impeller. Nozzles are attached to the tip of the radial flow path, to which the water is supplied and pressurized by centrifugal effect of the rotation. This axial flow path is shared with the centrifugal pressurization and supplying to the water-lubricated bearing. The low relative velocity between the droplet and the impeller, and the high-speed rotation of the impeller provides both of sufficiently high pressure and no erosion. Despite a problem of vibration due to increased imbalance due to the existence of flow paths and nozzles in the rotating body, the centrifugal pressure spraying was adopted in this study, by solving them by the symmetric arrangement conditions described later.

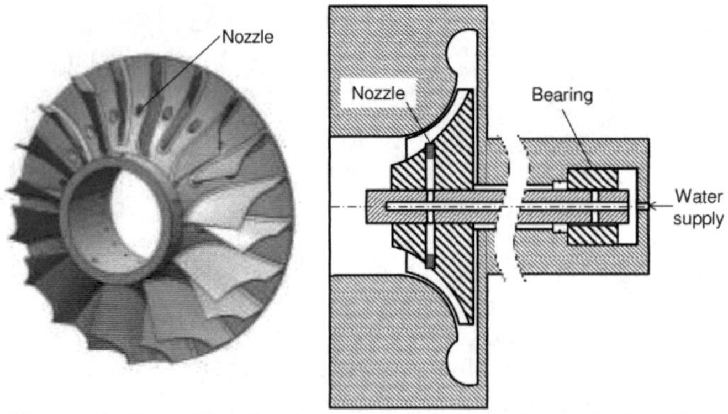

Figure 2. Centrifugal spray mechanism and nozzle location.

In order to avoid cavitation inside the axial flow path, it is necessary to supply water with positive pressure P_s from the outside. Assuming that the rotational angular velocity of the rotor is ω, the radius position of the centrifugal spray nozzle is r_n, the radius of the axial flow path is r_0, and the density of water is ρ, the centrifugal pressure applied to the nozzle is given by the following equation.

$$P_n = P_s + \frac{1}{2}\rho\omega^2({r_n}^2 - {r_0}^2) \approx P_s + \frac{1}{2}\rho(\omega r_n)^2 \tag{6}$$

Here, the approximation holds because when $r_0 \ll r_n$. The pressure loss in the flow path is negligible compared to P_n. Assuming that the spray volume flow rate is Q_n, the centrifugal spray power is calculated by the following equation.

$$W_n = (P_n - P_s)Q_n = \frac{1}{2}Q_n\rho(\omega r_n)^2 \tag{7}$$

2.3. Particle traceability

If the particle size is too large to follow the vapour flow, it will collide with blades adjacent to the nozzle or shrouds. The colliding droplets don't effectively contribute to the continuous cooling. Here we consider the relation between particle size and its followability to the vapour flow.

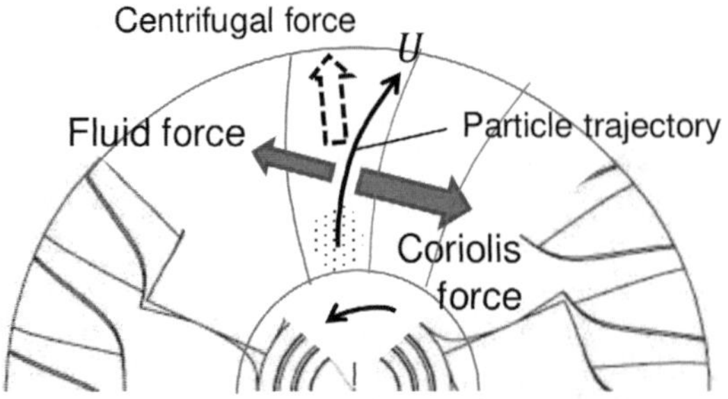

Figure 3. Dynamics of particles in rotating coordinate system.

Figure 3 shows the motion of droplets sprayed from a nozzle at the hub surface of a rotating impeller in a rotating coordinate system. It is assumed that the droplets are sprayed into the water vapour flowing through between the blades at a relative velocity U. Centrifugal force accelerates the particles in almost the same direction as the steam flow, so the effect on impeller collision is ignored. The Coriolis force accompanied by the radial movement of the particles causes the droplets to bend their trajectory to the opposite direction of rotation, which can cause a collision to the subsequent blades. If the drag force that prevents this tangential motion is dominantly larger than the Coriolis force, it exits the impeller with vapour before colliding with subsequent blades. The ratio of these two opposing forces is defined as a dimensionless traceability factor:

$$\mathrm{Tr} = \frac{\text{Fluid force}}{\text{Coriolis force}} = \frac{0.5\rho_g U^2 C_D S_p}{\rho_p V_p (2U\omega)} = \frac{9\mu}{\omega \rho_p {D_p}^2} \tag{8}$$

where S_p and V_p are the projected areas and volumes of droplets with a diameter of D_p. The relative velocity of particles to vapour is small, Stokes equation was used for the drag coefficient.

$$C_D = \frac{24}{\mathrm{Re}} = \frac{24\mu}{\rho_g U D_p} \tag{9}$$

It is necessary to atomize the droplet to at least D_p or less so that $\mathrm{Tr} > 1$ in order to avoid collision with the subsequent blade. Figure 4 shows the relationship between the followability Tr and the droplet diameter D_p, and the thick blue line is the designed rotational speed of the compressor in this paper. $\mathrm{Tr} > 1$ is satisfied when $D_p < 8$ μm. The trajectories of droplets of various diameters were calculated using a three-dimensional CFD with an actual rotating impeller shape as shown in figure 5. The analysis was performed with ANSYS Fluent with discrete phase model (DPM) for droplets of various diameters. Tetrahedral mesh of 2,684,072 cells and SST k-ω turbulence model were used. Fluid-particle interaction was only considered in the one-way mode. The particles that has entered from the inlet, did not collide to the blade if the particle size was less than 5 μm, where $\mathrm{Tr} = 3$ as obtained from figure 4. This value can be used as a lower limit of followability to avoid the collision. The condition required for complete evaporation of the droplet is $D_p < 10$ μm [8], which is also satisfied in this case. Figure 4 also shows the case of other rotation speeds. For example, when a rotational speed is 100,000 rpm and the particle size of $D_p > 2.4$ μm, there is a high possibility of collision because the Coriolis force becomes dominant.

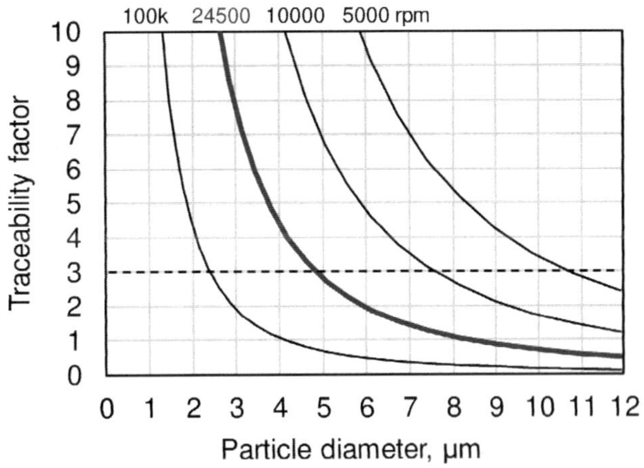

Figure 4. Traceability factor and rotational speed

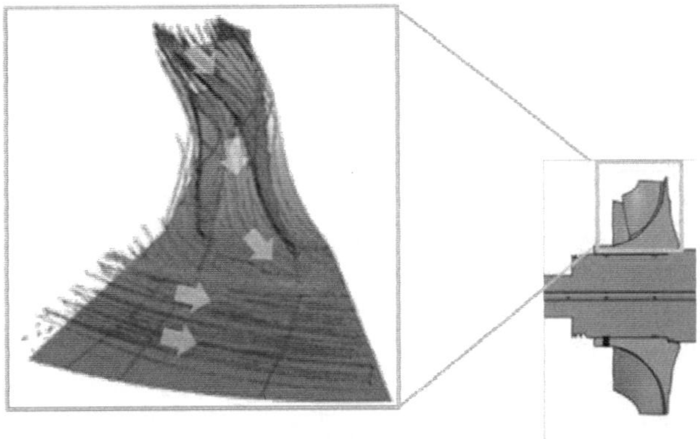

Figure 5. Particle trajectory analysis in CFD.

3. Continuous Cooling Compressor

Table 1 shows the specifications of the turbo compressor to demonstrate CCC process. Compared with the previous study [5], the operating conditions and capacity are the same and the stage number and specific speed were increased. The cross-sectional view of the compressor is shown in figure 6.

Table 1. The design condition of turbo compressor

Refrigerant	Tap water
Capacity, kW	100
Pressure ratio	5.3
Suction pressure, kPa	0.9
Suction temp.	5.4 °C
Design speed	24500 rpm

Stage	1st	2nd	3rd
Pressure ratio	2.0	1.9	1.6
Specific speed	0.8	0.7	0.7
Volumetric flow, m^3/s	5.7	3.6	2.4
Reynolds number	2.3×10^4	4.6×10^4	7.6×10^4
Isentropic efficiency	76%	76%	78%

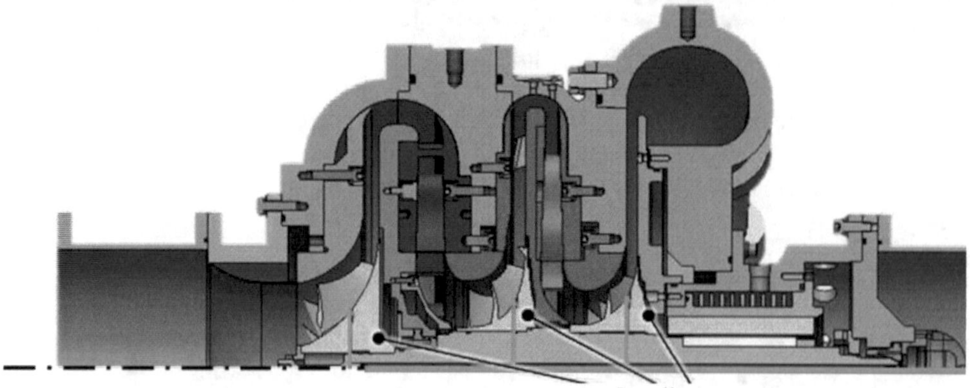

Figure 6. Structure of the continuous cooling compressor.

3.1. Aerodynamics

Aiming to reduce the overall size of the compressor, specific speed of each impeller was increased by using three-stage impellers of low pressure ratios. As a result, the maximum tip speed decreased to 384 m/s from 530 m/s. As the maximum stress was also decreased, A2024-T6 material was used for the 2nd and 3rd impellers. Since the volumetric flow rate is lower for downstream stage, the pressure ratio was designed to be the highest in the first stage in order to keep the specific speed below 0.8, which is desirable value for high adiabatic efficiency. As a characteristic of water refrigerant, the Reynolds number is the order of 10^4. This value is very low for a general turbo compressor, so care must be taken to suppress loss due to the secondary flow vortex. As a result, the blade shape with an adiabatic efficiency of 76% was obtained.

3.2. Rotordynamics and bearing

The shaft was designed as a rigid shaft, supported by saturated water lubricated bearings [11], to which liquid water was supplied as a lubricant. Since the three impellers are facing the same direction, the thrust forces of do not cancel each other. Therefore, an integrated bearing shown in figure 7 was developed, which has a thrust bearing surface adjacent to the end of a radial bearing. It is located between the 1st-stage and the 2nd-stage impeller as shown in figure 6. The stepped portion between the radial bearing part with a diameter of 50 mm and the cylindrical part to shrink-fit the 2nd impeller with a diameter of 100 mm was used as a thrust support surface, instead it does not have a thrust collar, which is generally used in turbo compressors for high-pressure working fluids. Such a configuration was possible because the diameter of the central part of the shaft is large to increase the critical speed of the shaft. The bearing loss was estimated to be 2 kW from Petroff's equation, which was twice of the tapered bearing in the previous reports, being a disadvantage of the three-stage configuration.

As shown in figure 2, centrifugal spray nozzles are located at equal intervals between all blades. The spray flow rate is changed with the number of nozzles, which is adjusted by selecting whether a nozzle or a plug is attached to the tip of the flow path. In order to reduce the rotor unbalances, a symmetric arrangement of nozzle is necessary. Therefore, there are restrictions on the number of nozzles and blades. Given that the number of blades is B, the number of nozzles must be "a divisor of B" or "B minus a divisor of B". For example, B = 15, then the allowable numbers are 0, 3, 5, 10, 12, and 15. Thus, the number of blades should not be a prime numbers, in which the nozzles must be located between all the blades or nothing. The impeller and assembled rotor were dynamically balanced each time when the nozzle was replaced to meet the balance grade G2.5.

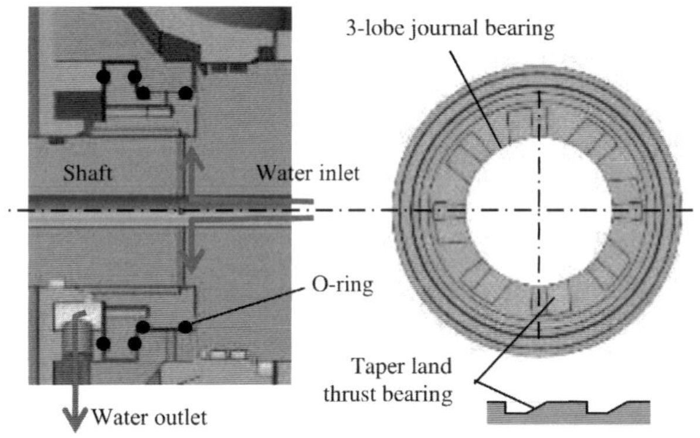

Figure 7. Structure of the water-lubricated integrated bearing.

3.3. *Spraying nozzle*

Compactness and strength are required for a nozzle that is embedded into an impeller and rotates together with the impeller, as well as atomization performance. Figure 8 shows the structure of the centrifugal pressure nozzle. The fastening thread portion has sufficient strength to withstand the pulling force acting to the nozzle due to the centrifugal effect of high speed rotation. The main functional components are the swirl chamber that produces the swirl flow and the orifice in its centre. The water pressurised by the centrifugal force flows through the filter into the chamber along the tangent line of the spiral chamber. Finally, it flows into the orifice in the centre while increasing the swirling speed. Inside the orifice, a thin swirling liquid film is formed on the outer wall of the hole, and injected to the flow channel of the impeller. The generated liquid film further divides to produce fine particles.

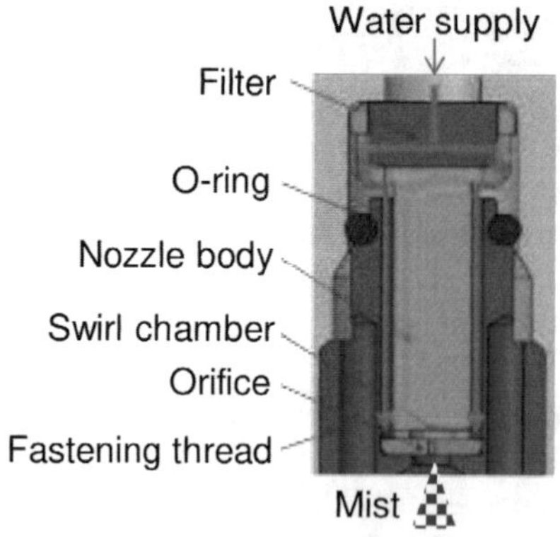

Figure 8. Centrifugal nozzle embedded in the impellers

4. Experiment

4.1. Test rig

To demonstrate CCC, a water-refrigerant heat pump system, which was comprised of a compressor, a spray-type evaporator, a condensing ejector, and an economiser tank, was constructed. Figure 9 shows a photograph of the demonstration device and a cycle diagram. It was a three-stage compression three-stage expansion economizer cycle. When it was operated at 24,500 rpm, the pressure ratios of 1st, 2nd, and 3rd stage were obtained as 2.07, 1.85, 1.63, respectively, and 5.6 in total. Those values agreed with the designed conditions as shown in Table 1.

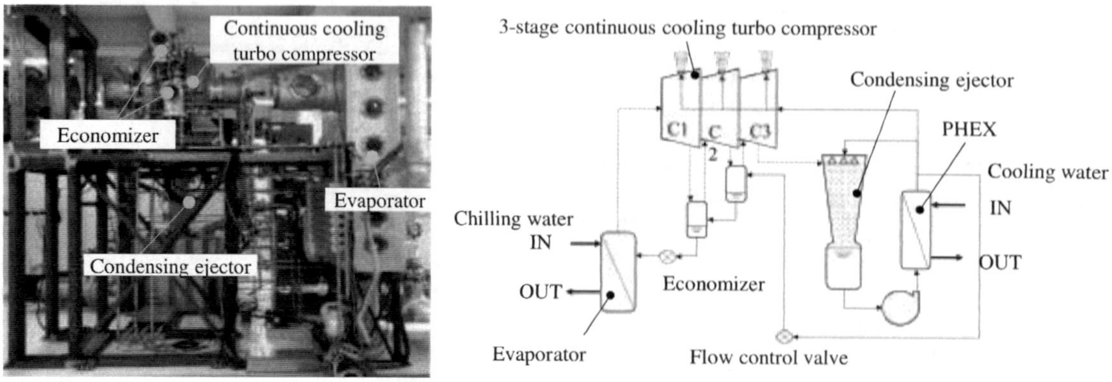

Figure 9. Demonstration test rig.

4.2. Discharge temperature

The discharge superheat reaches 200 K when compressed without any nozzles and without cooling, assuming the impeller efficiency of 76%. In order to investigate the continuous cooling effect of centrifugal spraying, the amount of spraying was varied by changing the number of nozzles attached to the impeller. Figure 10 shows the relation between the total spray flow rate of all stages and the discharge superheat of 3rd stage. Initially, it was estimated that the discharged vapour could be cooled to the saturation, i.e., discharge superheat of 0 K with spray flow rate of 4.4 g/s; however, the superheat of 55 K was left uncooled in the experiment. The cooling effect of 145K was confirmed, but it was insufficient. It is considered that this is because droplets larger than the Sauter average diameter do not contribute to cooling, and the adiabatic efficiency of the impeller is decreased from the analytical value. Therefore, additional experiments were performed with increased spray flow rate by increasing the number of nozzles. The discharge superheat decreased as the spray flow rate was increased, and the superheat became zero when it was 9.5 g/s.

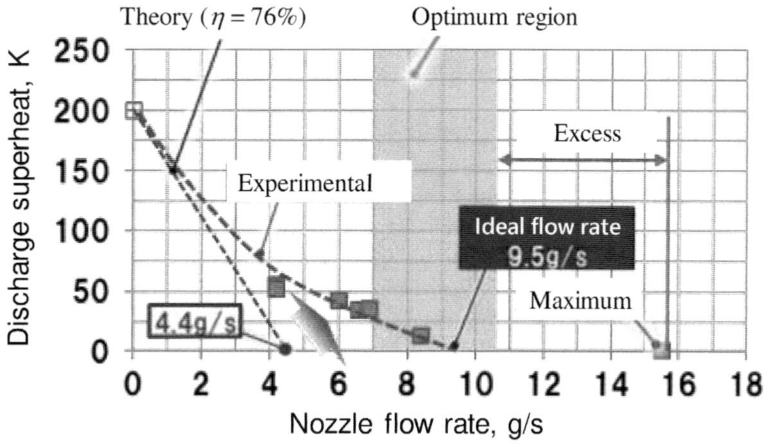

Figure 10. Discharge cooling effect of spray from the nozzles.

4.3. *Performance*

When the spray flow rate was 4.4 g/s with the discharge superheat of 55 K, the COP had a value of 4.5. When the superheat was reduced by increasing the number of nozzle, the COP increased, reaching a peak COP value of 4.86 when the discharge superheat was 43 K. This value is lower than 5.0 that was obtained with the 2-stage intercooled compressor in the previous study; however, the increase in the efficiency by decreasing the discharge superheat with CCC was demonstrated.

When the spray flow rate was increased further, the discharge superheat decreased but the COP also decreased. In this case, the increase in the cooling capacity and pressure ratio was very small even if the compressor speed was increased. These results inferred that the impeller was choking, due to the increase in the flow rate by the existence of sprayed mist and its evaporated vapour. Although the mist was sprayed downstream of the throat, when the flow blockage due to the droplets and increase in volumetric flow is significant, the actual location of throat, at which the area is the smallest for the flow rate, would exist at downstream of the geometrical throat. From these results, it was found that there is an optimum point of CCC, and it is important to take a sufficient choke margin by increasing the exit area of the impeller of the continuous cooling compressor.

4.4. *Durability*

Erosion was observed on the leading edge of the blade of the impeller. This is due to a minor problem at the early stage of the experiments. It did not progress further after the problem was solved. Furthermore, no erosion was observed on the blade located downstream from the spray nozzle. Thus, the anti-erosion effect of the small relative velocity between the droplet and the blade by embedding the nozzle in the impeller was confirmed. As for the water-lubricated bearings, progressive wear was not observed, and long-term reliability owing to non-contact lubrication was confirmed.

5. Conclusion

A continuous cooling turbo compressor was proposed that sprays water droplets from the nozzle embedded in the impeller in order to reduce the superheat of discharged refrigerant water. The effect was demonstrated by experiments, and the conclusions are as follows:

1. Droplets sprayed between the blades of the impeller cool the vapour reducing the discharge superheat. As the spray flow rate was increased, the discharge superheat of the vapour decreased to the saturation. The COP was improved, given the spray flow rate was appropriate.

2. No erosion was observed at the blades due to collision of centrifugal spray droplets.
3. There is an optimum spray flow rate, at which the COP took its maximum value. When the spray flow rate is increased further, a choking at the impeller decreases the COP. This is due to the increase in the volumetric flow rate by the evaporated droplets. It is desirable to assume a sufficient choke margin or to estimate the increase in the volumetric flow rate using a special CFD code that can consider the divergence due to droplet evaporation.

References

1. Kilicarslan A, Müller N. A comparative study of water as a refrigerant with some current refrigerants. *Int J Energy Res* 2005; 29: 947–959.
2. Honke M, Safarik M, Herzog R. R718 turbo chillers and vacuum ice generation - Two applications of a new generation of high speed, high capacity R718 centrifugal compressors. *Refrig Sci Technol* 2015; 1534–1541.
3. Madsboell H, Weel M, Kolstrup A. Development of a water vapor compressor for high temperature heat pump applications. *Refrig Sci Technol* 2015; 1613–1620.
4. Lachner BF, Nellis GF, Reindl DT. The commercial feasibility of the use of water vapor as a refrigerant. *Int J Refrig* 2007; 30: 699–708.
5. Shoyama T, Kawano B, Ogata T, et al. Novel Turbo Compressor for Heat Pump Using Water as Refrigerant and Lubricant. In: *IOP Conference Series: Materials Science and Engineering*. Institute of Physics Publishing, 2019. Epub ahead of print 3 September 2019. DOI: 10.1088/1757-899X/604/1/011010.
6. Shoyama T, Kouda K, Ogata T. Saturated Water Journal Bearings of a Turbo Compressor. *Proc ASME Turbo Expo, GT2016-58131.*
7. Shoyama T. Nonlinear Vibration of Saturated Water Journal Bearing and Bifurcation Analysis. *J Vib Acoust*; 141. Epub ahead of print 2019. DOI: 10.1115/1.4042041.
8. Wu H, Yin H, Li Y, et al. Effect of droplets on water vapor compression performance. *Desalination* 2019; 464: 33–47.
9. Sun L, Zheng Q, Luo M, et al. On the behavior of water droplets when moving onto blade surface in a wet compression transonic compressor. *J Eng Gas Turbines Power*; 133. Epub ahead of print 2011. DOI: 10.1115/1.4002822.
10. Higuchi S, Hatamiya S. A Study of Performance on Advanced Humid Air Turbine Systems. *Proc Int Gas Turbine Congr 2003 Tokyo* 2003; 1–7.
11. Shoyama T, Kouda K, Ogata T. Saturated water journal bearings of a turbo compressor. In: *Proceedings of the ASME Turbo Expo*. 2016. Epub ahead of print 2016. DOI: 10.1115/GT2016-58131.

Modeling the loading factor characteristic of an industrial centrifugal compressor impeller

Yu Galerkin[1], O Solovyeva[2], K Soldatova[3] and A Nikiforov[4]

[1]Leading research center "Digital Design and Modeling (Smart Design)" Peter the Great St.Petersburg Polytechnic University, Polytechnical st. 29, St.Petersburg, Russia
[2]Higher School of Hydraulic and Power Engineering Construction Engineering Institute Peter the Great St.Petersburg Polytechnic University, Polytechnical st. 29, St.Petersburg, Russia
[3]R&D Laboratory "Gas dynamics of turbo machines" Peter the Great St.Petersburg Polytechnic University, Polytechnical st. 29, St.Petersburg, Russia
[4]Department of Mechanization, Smolensk State Agricultural Academy, 214000, Bolshaja Sovetskaja, 10/2, Smolensk, Russia

Email: solovyeva.oa@yandex.ru

Abstract. Test data on compressor stages demonstrate that loading factor versus flow coefficient at an impeller exit is of linear character independent of compressibility. Mathematical models of the Universal Modeling Method use this pattern at their core. In the first versions of the Method, two points determined the loading factor characteristic: loading factor at the design point, and at zero flow rate. To calculate these parameters, the user had to define values of two empirical coefficients. The choice of the coefficient is based on experience. An attempt to obtain approximating formulas for these coefficients did not lead to a result with the required accuracy.

In the modern version of the Method, the loading factor characteristic is defined by the angle of its inclination to the ordinate axis and by the loading factor at zero flow rate. The authors have researched test performances of model stages. Simple and definite equations with four geometric parameters were proposed for loading factor characteristics. The calculation error is 1%.

A trailing edge shape influences the loading factor. Blunt and symmetrically tapered blades have the same loading factor. Non-symmetric tapering changes the loading factor by up to plus/minus 4%. The results of impeller tests with different blade trailing edges and with vane and vaneless diffusers were generalized. The influence of these factors was taken into account as correcting coefficients in the formulas. The alternative model of loading factor characteristic modeling is included in the Universal Modeling Method and demonstrates good results in design practice.

1. Introduction

Research on centrifugal compressors has been carried out at LPI-SPbPU for decades [1],[2]. Dr-Ing. hab., Prof. Yu.B. Galerkin created a method for gas-dynamic design of centrifugal compressors and a mathematical model based on experiments and experience. Dimensions of a compressor flow path and similarity criteria (isentropic coefficient, Mach and Reynolds numbers) determine its gas dynamic characteristics. The mathematical model was implemented in the form of computer programs called the Universal Modeling Method [3], [4], [5], [6], [7]. The method is constantly being developed and improved at the Research Laboratory "Gas Dynamics of Turbomachines".

A centrifugal compressor does not always operate at design flow rate. When discharge pressure changes in a pipeline, the gas flow rate changes too. The discharge pressure versus gas flow rate characteristic is important for assessing the compressor-pipeline interaction.

The mathematical model of the Method involves calculation and provision of gas-dynamic characteristics in a dimensionless form. Head according to Euler's formula, is $h_T = c_{u2}u_2 - c_{u1}u_1$ (where c_u is the tangential component of absolute flow velocity c, u is blade velocity, subscripts 1 and 2 indicate control sections). For centrifugal process compressors c_{u1}=0. The dimensionless loading factor is $\psi_T = c_{u2}/u_2$. Figure 1 presents velocity triangles at the outlet of a centrifugal impeller with infinite and finite number of blades (w is relative velocity, c_r is the radial component of absolute velocity, β_{bl} is the blade angle).

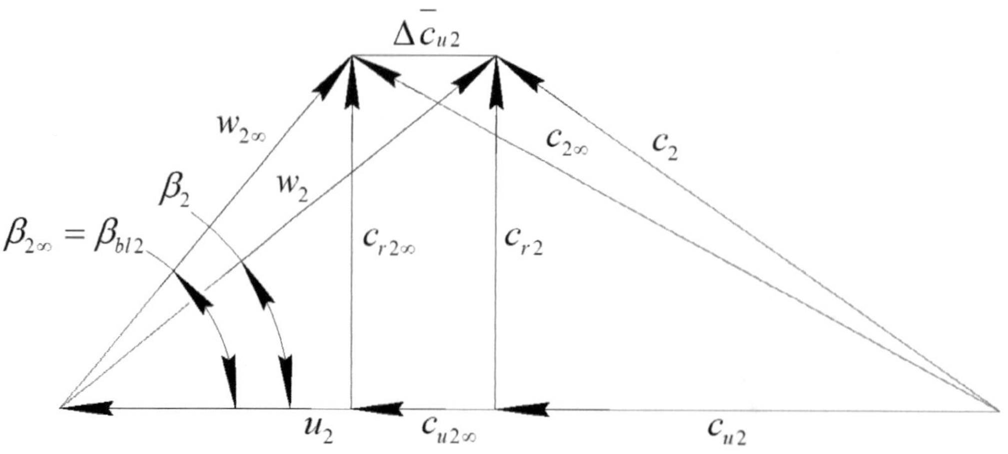

Figure 1. Velocity triangles at the outlet of a centrifugal impeller with infinite and finite number of blades

For an impeller with an infinite number of blades, the loading factor:

$$\psi_{T\infty} = 1 - \varphi_{2des}\, ctg\beta_{bl2}, \tag{1}$$

where φ_{2des} is the design flow rate coefficient at the impeller outlet.

For a real impeller with a finite number of blades, the flow angle β_2 differs from the blade angle due to the flow lag. The value of lag in engineering formulas is usually determined by $\Delta\bar{c}_{u2}$, Figure 1. For industrial compressors in the middle of the last century, the well-known formula of A. Stodola was used. This formula was introduced into Russian practice by the author of [8]:

$$\psi_T = 1 - \varphi_{2des}\, ctg\beta_{bl2} - \frac{\pi}{z_{imp}}\sin\beta_{bl2}, \tag{2}$$

where z_{imp} is the number of blades.

The formula is based on the representation of the relative velocities field in the impeller channel as the sum of the transit flow and the axial vortex. Stodola's formula does not take into account the influence of viscosity, blade dimensions and shape, etc. The accuracy of the formula is acceptable for a limited class of impellers.

In [9], B. Eckert analyzed equilibrium and continuity equations and presented the loading factor formula without taking into account the effect of viscosity:

$$\psi_{\text{T}nonvisc} = \frac{\psi_{\text{T}\infty}}{1 + \dfrac{\pi D_2^{\,2} b_2}{8 z_{imp} S} \sin \beta_{bl2}}, \tag{3}$$

where D_2 is the impeller diameter; b_2 is the outlet blade height; $S = \int_{r_1}^{r_2} br\,dr$.

The book [9] presents the integral S for "the most common case br = const". To take into account the effect of viscosity, the ratio h'/h is introduced. There h is the breadth of the channel, h' is the width of the active part of the flow ("jet"). Then the formula for an impeller with br = const will be:

$$\psi_{\text{T}} = \frac{\psi_{\text{T}\infty}}{1 + \dfrac{h'}{h} \dfrac{\pi}{2 z_{imp} \left(1 - r_1 / r_2\right)} \sin \beta_{bl2}}. \tag{4}$$

B. Eckert [9] recommended h'/h = 0.7–0.8 (a smaller value for blades with β_{bl2} < 90°). Referring to him, we can write the following empirical formula:

$$\frac{h'}{h} \pi \sin \beta_2 = 1.5 + 1.1 \frac{\beta_2}{90^\circ}, \tag{5}$$

The formula (4) for impellers with an arbitrary dependence $\mathrm{b} = \mathrm{f}(\mathrm{r})$ taking into account the viscosity becomes:

$$\psi_{\text{T}} = \frac{\psi_{\text{T}\infty}}{1 + \dfrac{\pi D_2^{\,2} b_2}{8 z_{imp} S} \dfrac{h'}{h} \sin \beta_{bl2}}. \tag{6}$$

Experiments of the LPI Compressor Engineering Problem Laboratory back in the 1980s showed [10] that for impellers with 2D blades, the most advantageous meridional shape is a linear function $\mathrm{b} = \mathrm{f}(\mathrm{r})$. In this case, $S = \int_{r_1}^{r_2} brdr = r_2^2 b_2 \left(1 - \dfrac{r_1}{r_2}\right)$ and the loading factor (linear function $\mathrm{b} = \mathrm{f}(\mathrm{r})$, blade load along r is constant) is:

$$\psi_{\text{T}} = \frac{\psi_{\text{T}\infty}}{1 + \dfrac{h'}{h} \dfrac{\pi \sin \beta_{bl2}}{2 z_{imp} \left(1 - \bar{D}_1\right)}} \tag{7}$$

where $\bar{D}_1$ is blade inlet relative diameter.

B. Eckert's approach seems to be well grounded. K.I. Strakhovich [11] introduced this approach in Russian practice. However, the samples of velocity diagrams calculated by B. Eckert in [9] show that blade load is far from being constant along r. All others formulas of this kind do not take into account actual blade load distribution too.

The flow lag at the impeller outlet is the result of the pressure field created by the blade load. In the scheme proposed by Yu. Galerkin [12], [13], blade load is replaced by a vortex with circulation $\Gamma_{bl} = c_{u2} 2\pi r_2 / z_{imp}$. The vortex is located on the radius r_p. If the load is uniform, the vortex is located in the middle between r_1 and r_2. If the shape of the blade is such that the load is greater at the beginning of the blade, r_p is closer to r_1, and vice versa. In accordance with such a scheme, the vortex induces the following velocity component at the impeller outlet:

$$\Delta c_{u2} = \frac{\Gamma_{bl} \sin \beta_{bl\,av}}{2\pi (r_2 - r_p)}. \quad (8)$$

The radius of the vortex position is set by the empirical coefficient:

$$K_{pc} = \frac{r_2 - r_p}{r_2 - r_1}. \quad (9)$$

Formula (8) in dimensionless form becomes:

$$\Delta \bar{c}_{u2} = \frac{\psi_T}{2\, z_{imp}\, \bar{l}_{bl}\, K_{pc}}, \quad (10)$$

where $\bar{l}_{bl}$ is dimensionless length of a blade.

The load distribution along the radius depends on the shape of the blade mean line and the meridional shape of the blade channel. To take into account the real nature of the flow, an empirical coefficient K_μ is introduced. The coefficient takes into account the increase in flow lag due to viscosity. The formula for the loading factor at the design flow rate is:

$$\psi_{T\,des} = \frac{1 - \varphi'_{2des} ctg \beta_{bl2}}{1 + K_\mu \dfrac{1}{z(1 - \bar{D}_1) K_{pc\,des}} \sin \dfrac{\beta_{bl1} + \beta_{bl2}}{2}}, \quad (11)$$

where φ'_2 is the flow rate coefficient taking into account the blade blockade factor.

The accuracy of modeling the loading factor characteristics of a compressor affects the efficiency of its gas-dynamic design results, as well as the results of calculating the stage efficiency. This paper is devoted to the choice of the most correct and user-friendly method for modeling the loading factor characteristics of a centrifugal compressor stage.

2. Research objects

The equations presented below for calculating the loading factor characteristic $\psi_T = f(\varphi_2)$ are based on gas dynamic characteristics of the 20CE family of model stages [14]. The stages consist of 2D impellers (the blade midline shape does not change along its height), vaneless or vane diffusers and return channels. The impeller blades have the mean line as a circular arc, or had a "profiled" mean line, to provide one or another velocity diagram, load distribution.

The gas dynamic and geometric parameters of the 20CE stages vary as follows: design flow rate coefficient Φ_{des} =0.028–0.080, design loading factor $\psi_{T\,des}$ = 0.45–0.65, hub relative diameter $\bar{D}_{hub}$ = 0.25–0.373, diffuser outlet diameter $\bar{D}_4$ = 1.428–1.615, impeller Mach number M_u = 0.60–0.86, impeller diameter Reynolds number $Re_u = 4.8 \cdot 10^6 - 6.9 \cdot 10^6$. The experiments were carried out on a test rig with an open circuit. The relative uncertainty in determining the gas-dynamic characteristics of the model stages ± 1,0 % [15].

In experiments with model stages, characteristics are measured in the form $\psi_i = f(\Phi)$, where the work coefficient $\psi_i = \frac{c_p (T^*_{out} - T^*_{in})}{u_2^2}$ (c_p is heat capacity at constant pressure; T^*_{out} is total temperature at the stage outlet; T^*_{in} is total temperature at the stage inlet). To obtain empirical equations $\psi_T = f(\varphi_2)$, the flow rate coefficient at the outlet from the impeller is calculated using the Universal Modeling Method. To calculate the loading factor, the proposed by V.F. Ries [8] schematization of the power transmission process can be applied:

$$\psi_i = \psi_T\left(1+\beta_{fr}+\beta_{leak}\right), \tag{12}$$

where β_{leak} is the labyrinth seal leakage coefficient; β_{fr} is the disk friction coefficient. The Universal Modeling Method also performs the calculation of the coefficients of disk friction and leakage, calculation formulas are presented in [12].

3. Two ways of presenting the loading factor characteristic

The simulation of the loading factor characteristic is based on the linear dependence of the loading factor versus the flow rate at the outlet of the impeller φ_2 [16]. The first versions of the Universal Modeling Method contained formulas for calculating the loading factor characteristic based on the loading factor at the design flow rate and the loading factor at zero flow rate. To determine these two, the user chose two empirical coefficients: a coefficient for a design flow rate K_μ (equation (11)), and a coefficient $X_{\psi_{T0}}$ that determines the loading factor at zero flow rate:

$$\psi_{T0} = 1 - X_{\psi_{T0}}\frac{\Delta\bar{c}_{u2des}}{2}. \tag{13}$$

A loading factor $\psi_T = f(\varphi_2')$:

$$\psi_T = \psi_{T0} - \frac{\psi_{T0}-\psi_{Tdes}}{\varphi_{2des}'}\varphi_2'. \tag{14}$$

The wide range of recommended empirical coefficients K_μ and $X_{\psi_{T0}}$ values creates certain difficulties. In [17] and [18], the authors proposed to model the loading factor characteristic using the angle of inclination β_T and the coefficient of loading factor at zero flow rate ψ_{T0}, Figure 2.

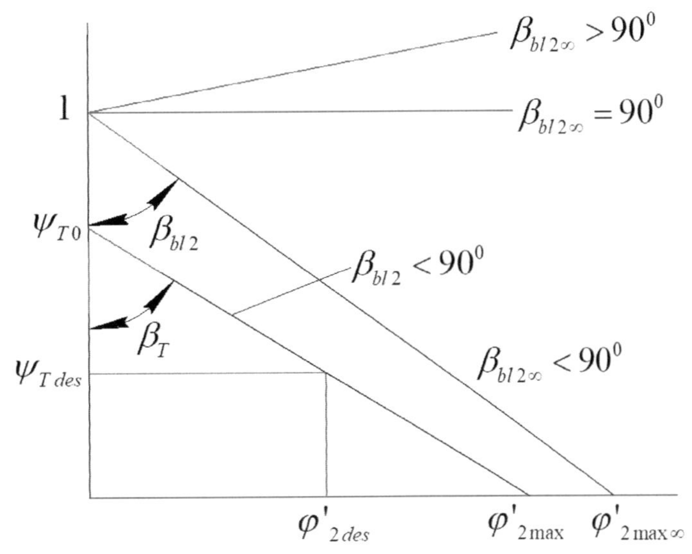

Figure 2. The loading factor characteristics of ideal impellers with an infinite number of blades and the linear loading factor characteristic of the impeller with $\beta_{bl2} < 90^0$

The alternative description of the loading factor characteristic is:

$$\psi_T = \psi_{T0} - \varphi_2 \mathrm{ctg}\beta_T . \tag{15}$$

3.1. First way of modeling. Approximation formulas for coefficients K_μ and $X_{\psi T0}$

Model stages of the 20CE family and multistage compressors tests from [14] were reduced for K_μ and $X_{\psi T0}$ approximation formulas. The authors investigated several variants of formulas with different arguments and for different groups of impellers. The simple formulas demonstrated better results:

$$X_{\psi T0} = 2.588 - 0.000576\beta_{bl2} + 3.655 \cdot 10^{-8} \beta_{bl2}^2 ; \tag{16}$$

$$K_\mu = 1.545 - 3.141 \cdot 10^{-5} \beta_{bl2} . \tag{17}$$

Figure 3 demonstrates the quality of approximation.

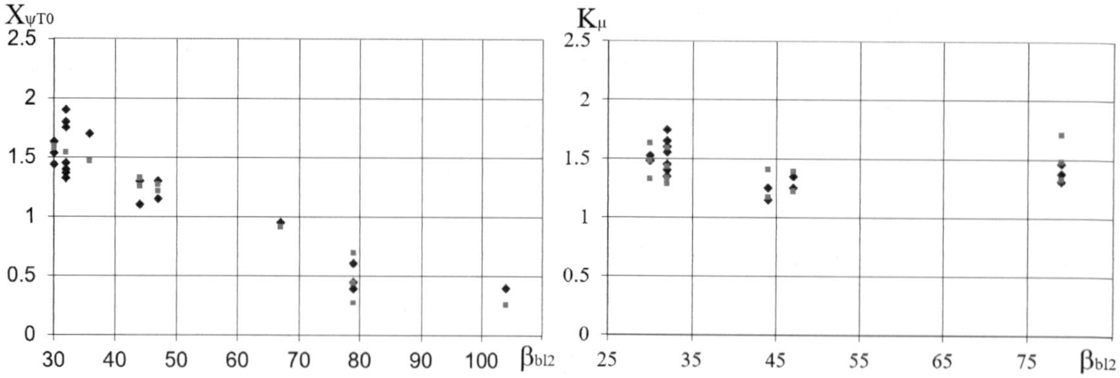

Figure 3. Empirical coefficients in equations (11, 13) versus blade exit angle. Blue is test, red is approximation

The authors accept validity of the formulas (16, 17) for preliminary design. Q3-D PC program 3DM-023 [12] calculates head characteristics of final design quite precisely. However, for calculating head characteristics of a compressor of unknown design we need a solution that is more precise.

3.2. Second way of modeling. Approximation formulas for coefficients β_T and ψ_{T0}

The authors of [17] investigated the influence of the main geometric parameters of impellers (the outlet blade angle, the thickness of the blades, the number of blades, the ratio b_2 / b_1, the shape of the blade mean line, etc.) on head characteristics. In addition to developing formulas for calculating parameters β_T and ψ_{T0}, it is also necessary to take into account the shape of the impellers trailing edge and the type of diffuser. Both of these parameters affect the shape of the stage head characteristic.

The tests of model stages with different shapes of blade trailing edges and types of diffusers demonstrated different head characteristics [12]. Figure 4 shows studied trailing edges configurations.

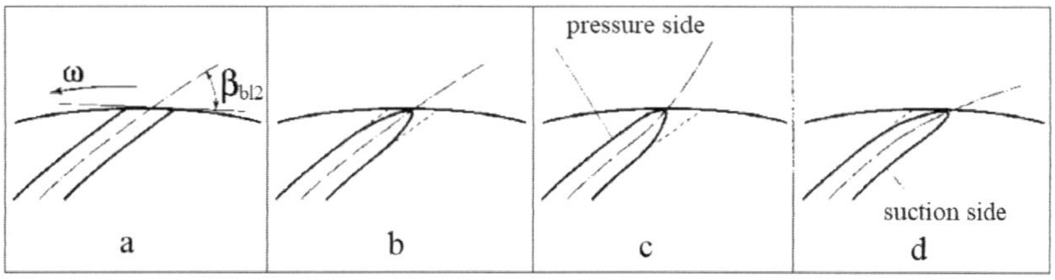

Figure 4. Experimentally studied blade trailing edge shape [12]: (a) blunt; (b) symmetrically tapered; (c) tapered to the pressure side; (d) tapered to the suction side

Welded impellers usually have trailing edge (a). Recommended configuration for an impeller with a riveted shroud is (b). Loading factor is the same. Efficiency is a bit better with (b). The configuration (d) ensures highest efficiency at the cost of up to –4% of the loading factor, and with (c) it is the other way around. To take into account the effect of the blade edge type on a loading factor characteristic, correction coefficients $K_{\beta\tau ed}$ and $K_{\psi\tau 0\,ed}$ are introduced into approximating formulas. Vanes of a diffuser influence flow conditions in an impeller. Loading factor of an impeller coupled with a vane diffuser is higher. To take into account the influence of the vane diffuser, correction coefficients $K_{\beta\tau dif}$ and $K_{\psi\tau 0\,dif}$ were introduced.

The formula for the inclination angle β^0_T of the loading factor characteristic provides accuracy acceptable for design practice:

$$\beta^0_T = \left(0.00478\cdot\beta^{0\ 2}_{bl2} + 17.4802\left(\frac{1}{t}\right)^{0.4} + 18.22\bar{b}_2^{0.5} - 5.702\left(\frac{b_2}{b_1}\right)^{0.8}\right)\cdot K_{\beta T\,ed}\cdot K_{\beta T\,dif}\,. \tag{18}$$

This formula (18) was obtained as a result of processing the results of experiments with blunt and symmetrical trailing edges. For tapered trailing edges, the values of the empirical coefficients $K_{\beta\tau ed}$ were selected. The results showed that for the trailing edge tapered from the pressure side, the coefficient $K_{\beta\tau ed}$ is constant and does not depend on $\psi_{\tau des}$. For the trailing edge tapered from the suction side, a graph was built (Figure 5), the approximation the data of which made it possible to obtain a linear dependence $K_{\beta\tau ed}$ on $\psi_{\tau des}$. Thereby:

- for a blunt and symmetrical trailing edges:

$$K_{\beta\tau ed} = 1\,; \tag{19}$$

- for the trailing edge tapered to the pressure side:

$$K_{\beta\tau ed} = 0.97\,; \tag{20}$$

- for the trailing edge tapered to the suction side:

$$K_{\beta\tau ed} = -1.2457\psi_{\tau des} + 1.6218\,. \tag{21}$$

Accuracy of approximation is demonstrated in figures below.

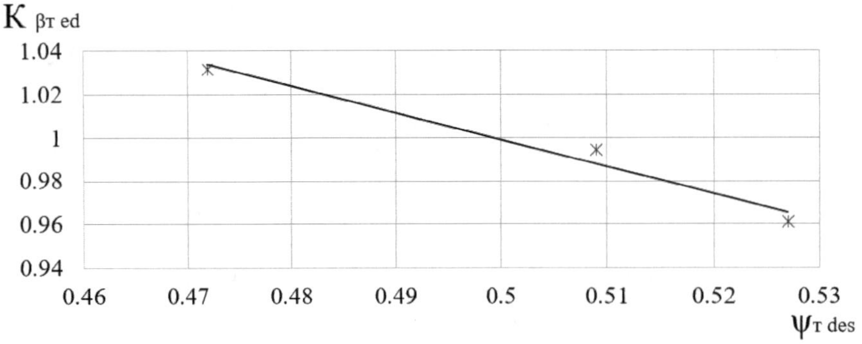

Figure 5. Coefficient $K_{\beta T ed}$ for the trailing edge tapered to the suction side versus a design loading factor. ✱ is test data, line is approximation

Similar researches have been done to account for the influence of the diffuser type. When developing formula (18), the experimental results of model stages with VLD were used. The experimental data were approximated using a linear dependence presented in formula (23). It shows insufficient accuracy of the approximation. However, due to the current lack of experimental results for model stages with different values $\psi_{T des}$, it is not possible to refine formula (23). Correction coefficients for a diffuser type:

- for a stage with a vaneless diffuser:

$$K_{\beta T\, dif} = 1; \quad (22)$$

- for a stage with a vane diffuser:

$$K_{\beta T\, dif} = -0.3413 \cdot \psi_{T des} + 1.1375. \quad (23)$$

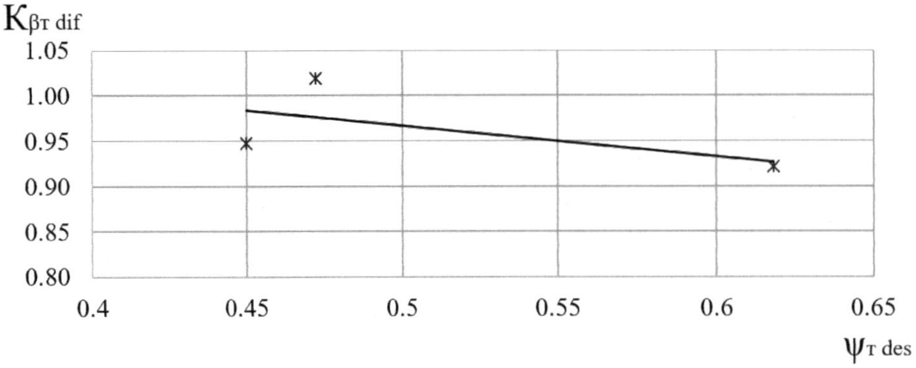

Figure 6. Coefficient $K_{\beta T\, dif}$ for stages with vane diffuser versus design loading factor. ✱ is test data, line is approximation

The accuracy of approximation by formula (18) is 98.96%. Comparison of the calculated and experimental data is shown in Figure 7.

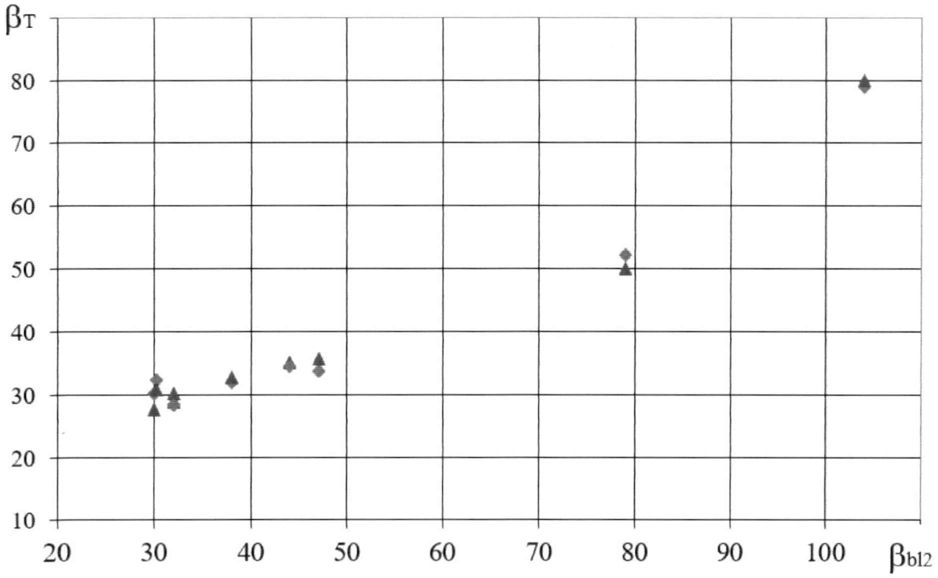

Figure 7. Comparison of the inclination angle of the loading factor characteristic calculated by formula (18) and the experimental data. Blue is test, red is approximation

The formula for the loading factor at zero flow rate:

$$\psi_{T0} = \left(1 - 0.0479\left(\frac{1}{t}\right)^{0.9} - 0.0025\left(\frac{b_2}{b_1}\right)^2 + 0.0255\left(\frac{\beta_{bl2}}{40}\right)^{0.7} - 3.7462\bar{b}_2^{1.5}\right) \cdot K_{\psi T0\,ed} \cdot K_{\psi T0\,dif}, \quad (24)$$

where

- for a blunt and symmetrical trailing edge of the blade:
$$K_{\psi T0\,ed} = 1; \quad (25)$$
- for the trailing edge of the blade tapered to the pressure side:
$$K_{\psi T0\,ed} = 1.04; \quad (26)$$
- for the trailing edge of the blade tapered to the suction side:
$$K_{\psi T0\,ed} = 21.349 \cdot \psi_{T\,des}^2 - 20.673 \cdot \psi_{T\,des} + 5.9484; \quad (27)$$

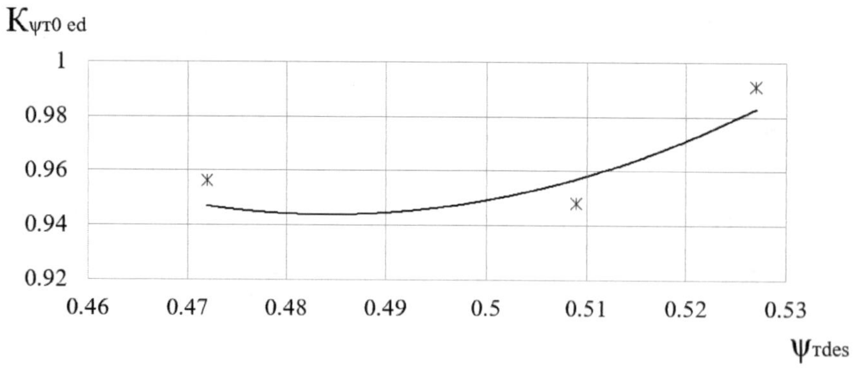

Figure 8. Coefficient $K_{\psi T0\ ed}$ for the trailing edge of the blade tapered to the suction side on the design loading factor. ✱ is test data, line is approximation

- for a stage with a vaneless diffuser:

$$K_{\psi T0\ dif} = 1\,; \tag{28}$$

- for a stage with a vane diffuser (the approximating equation is shown in Figure 9):

$$K_{\psi T0\ dif} = 0.0979 \cdot \psi_{T\,des} + 0.9747\,. \tag{29}$$

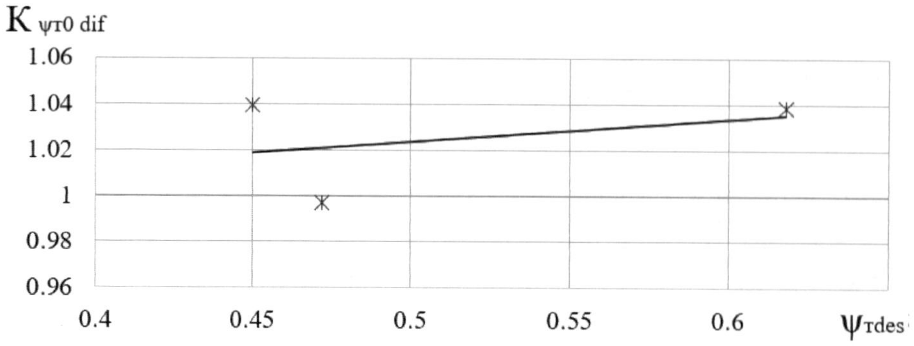

Figure 9. Coefficient $K_{\psi T0\ dif}$ for stages with vane diffusers. ✱ is test data, line is approximation

Comparison of the calculated and experimental data is shown in Figure 10.

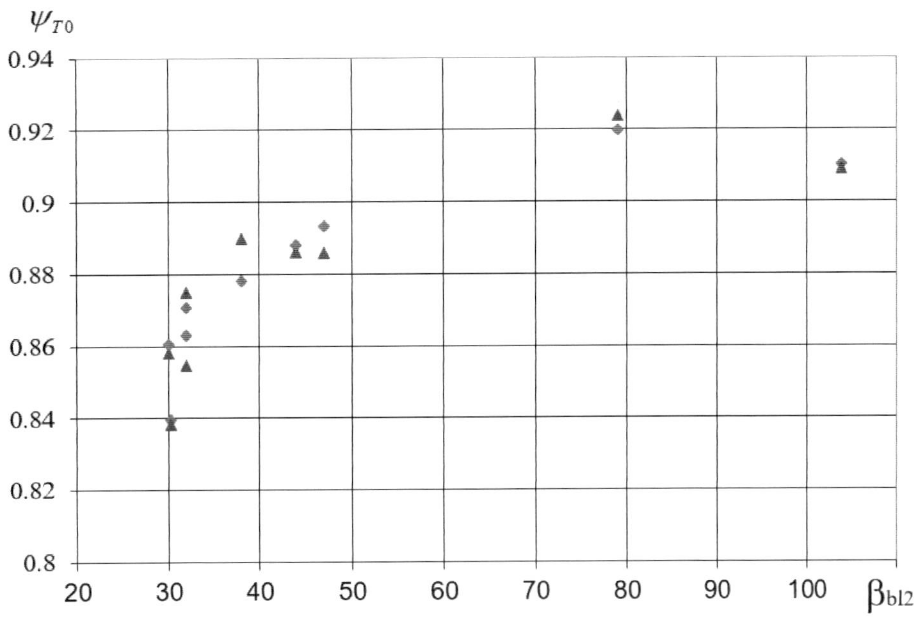

Figure 10. Comparison of the loading factor at zero flow rate calculated by the formula (24) and the experimental data. Blue is test, red is approximation

The approximation accuracy is 88.62%.

4. Conclusion

Any model of the loading factor characteristic is based on its linear dependence on the flow rate coefficient. The model determining the characteristic by an inclination angle and the loading factor at zero flow rate demonstrates its advantages. The algebraic formulas use blade exit angle and blade relative width for calculating characteristics. The correcting coefficients take into account the shape of blades trailing edge and the diffuser type. The application of this model in the Universal Modeling Method shows that the model accuracy satisfies the preliminary design demand, however, calculations of performance of arbitrary compressors are less reliable. The authors will continue to correct the approximating formulas as the experimental data accumulate on the new test rig ECC-55 [19] that was built for the compressor group of the Laboratory "Technological Processes Modeling and Design of Power Equipment". The proposed approaches to modeling the loading factor characteristics of centrifugal compressors impellers are based on processing the results of experimental researches of single-row radial impellers without splitter blades for a shroud without cutouts and holes in it. The applicability of the proposed model for other impellers will be tested separately when comparing the calculated and experimental data.

Acknowledgments
The research was performed by a Grant of the President of the Russian Federation for young PhD MK-1893.2020.8.

References
[1] Edited by Galerkin YuB 2005 *Proceedings of the scientific school of compressor engineering SPbSPU* (M: Ed. KHT)
[2] Edited by Galerkin YuB 2010 *Proceedings of the scientific school of compressor engineering*

SPbSPU (M: Ed. SPbSPU)

[3] Galerkin Yu B 2000 Formation of views on working processes and the current state of gas-dynamic methods for designing industrial centrifugal compressors. *Compressors and pneumatics* **2** 9-14.

[4] Galerkin Yu B, Danilov K A, Mitrofanov V P and Popova E Yu 1996 *On the use of numerical methods in the design of the centrifugal compressors flow path* (SPb. – SPbSTU) p. 68

[5] Galerkin Yu B, Danilov K A and Popova E Yu 1995 *Development of the method of universal modeling of the centrifugal compressors working process - software complexes of the first level (third generation), experience in the development and practical use of the third level complex* (SPb)

[6] Galerkin Yu B, Danilov K A and Popova E Yu 1993 Numerical modeling of centrifugal compressor stages (physical bases, current state) *Compressors and pneumatics* **2**

[7] Danilov K A 1999 *Creation of a mathematical model and software systems for optimal gas-dynamic design of refrigerated centrifugal compressors* PhD thesis. SPb

[8] Ries V 1951 *Centrifugal compressors* (L: Mechanical Engineering) p. 245

[9] Eckert B 1959 Axial and centrifugal compressors. Application, theory, calculation (M) p. 680

[10] Edited by Galerkin YuB 2000 *Proceedings of the scientific school of compressor engineering SPbSPU (* M: Ed. KHT)

[11] Strahovich K I et al. 1961 *Compressor machines* (M.) p. 600

[12] Galerkin Yu 2010 *Turbocompressors* (LTD information and publishing center KHT, Moscow) p. 596

[13] Seleznev K and Galerkin Y 1982 *Centrifugal compressors* (L.: mechanical engineering) p. 271

[14] Galerkin Yu B and Soldatova K V 2011 *Modeling the working process of industrial centrifugal compressors. Scientific foundations, stages of development, current status* (SPb .: Publishing house of SPbPU) p. 327

[15] Galerkin, Yu.B., Rekstin, A.F. Research methods of centrifugal compressor machines. L .: Mechanical Engineering, 1969. - p. 303.

[16] Karpov A N 2011 *Methodology for modeling the pressure characteristics of a centrifugal compressor wheel based on the test results of model stages* PhD thesis. SPb p. 137

[17] Soldatova K V and Galerkin Yu B Loading factor performance of a centrifugal compressor impeller. Specific features and way of modeling 2016 *ICMSE 2016: 18th International Conference on Mathematics and Statistical Engineering* (Australia. Sydney) **3** (12) 615–624

[18] Galerkin Yu B, Drozdov A A, Rekstin A F and Soldatova K V 2017 An alternative method of centrifugal compressor loading factor modeling *International Conference on Compressors and their Systems* (London. City University. UK)

[19] Borovkov A I , Galerkin Yu B, Drozdov A A, Rekstin A F, Semenovsky V B, Yadykin V K 2020 Stand ECC-55 with direct high-frequency drive for gas-dynamic research of industrial centrifugal compressors *Compressors and pneumatic* **4** 32-43

IOP Conf. Series: Materials Science and Engineering 1180 (2021) 012025 doi:10.1088/1757-899X/1180/1/012025

Theory and practice of neural networks application to building mathematical model of centrifugal compressor vane diffusers

Aleksandr Nikiforov[1], Andrei Rekovetc[1], Yuri Galerkin[2], Evgeniy Petukhov[2], Aleksey Rekstin[2], Vasiliy Semenovsky[2], Olga Solovyeva[2]

[1] Smolensk State Agricultural Academy, Bolshaja Sovetskaja 10/2, Smolensk, 214000 Russian Federation

[2] Peter the Great St. Petersburg Polytechnic University, Polytechnicheskaya 29, Saint Petersburg, 195251 Russian Federation

Email: eugene@lamm.spbstu.ru

Abstract. Optimal design of centrifugal compressor stages needs special computational and experimental methods, both of them could be costly enough. So new advanced design methods which can provide optimal solution faster are needed. Authors developed the set of mathematical models – Universal modeling method - for describing compressor stages characteristics. Its models are being widened and improved. One the most advanced approaches of model building is based on machine learning. A neural network based method for predicting centrifugal compressor vane diffuser characteristics was developed. Input data for network training was obtained from CFD simulations. The resulting model for diffuser loss coefficient shows good approximation quality and can be used for improvement of VD model in Universal modeling method.

Introduction

The use of vane diffusers (VD) in the stages of centrifugal compressors is characterized by the possibility of higher efficiency factor in comparison with vaneless diffusers, but at the expense of the width of the operating area. With the help of vanes the higher slowdown of the flow velocity and pressure build-up is achieved. In these conditions the flow structure in VD after the impeller remains quite complicated with often detachable character. In order to accomplish an objective of optimal design it is necessary to develop a special approach that can take into account peculiarities of every element of centrifugal compressor's flow channel. At the Research Laboratory Gas Dynamics of Turbomachines at the St. Petersburg Polytechnic University, under the guidance of Professor Yu.B. Galerkin the Universal modeling method (UMM) [1-7] is developed, based on mathematical models that describe diffusers characteristics adequately. The method is being constantly upgraded, according to new research results the models are being specified and completed.

The trials of model stages as well as numerical results from CFD simulations are used for identification. Commercial CFD software seems to be suitable tool for designed compressor characteristics estimation [8-11] but it cannot provide acceptable fidelity for design practices [12-16].

Nevertheless, in case of stationary elements simulations CFD methods provide correct solutions [10, 17, 18]. In 8th version of UMM the model of vaneless diffuser was developed [4, 17-20] based on CFD results summary. Similarly, a methodology for vane diffuser CFD simulation was developed and tested for UMM VD model enhancement [21-23]. Also method for processing simulations data with neural networks (NN) was used, the possibility of its application was shown before [3, 24-26].

In [24] the performances in terms of efficiency, pressure ratio and work coefficient of 39 model stages were modeled with mean squared error 1.5 %. In addition, the loss and friction coefficients of vaneless diffusers of relative widths 0.014-0.10 are modeled with mean squared error 2.45 %.

The paper [25] describes the technology for constructing neural network models, which includes preparing a sample of input data and determining the optimal structure of the neural network. Based on the obtained mathematical models, a computational experiment was carried out in order to determine the influence of the main geometric and gas-dynamic parameters on the efficiency of vane diffusers.

A simulation experiment based on neural models for studying the influence of dimensions, diffuser shape, and similarity criteria in [26] made it possible to check the physical adequacy of mathematical models, to obtain new data on energy conversion processes and to establish a number of recommendations on the optimal design of vaneless diffusers.

To improve the model of VD the authors introduce the calculation methods based on data processing of hydrodynamic calculations using neural networks (NN). Based on results of multivariant CFD simulations of VD their characteristics are calculated and together with geometric parameters are used as input data for building NN.

Materials and Methods

The aim of the work is to build mathematical models of the gas-dynamic characteristics of the vane diffuser at the intermediate stage of a centrifugal compressor, which allow performing a computational study and searching for the optimal design of the VD for the given parameters in the process of developing new designs of centrifugal compressors.

In order to achieve the aim of work, the most modern method of processing the input data, which is the neural modeling, was used. The main task is the methodology of training a neural network on a sample of input data. As separate stages of training, it is envisaged to provide analysis and selection of types of neural networks and activation functions; creation of several test neural networks of different architecture; analysis of main coefficients of input neurons; perturbation of input parameter values and analysis of the response of the neural network to these perturbations; sequential exclusion of input neurons and observation of a network generalization error; testing trial neural networks of different architectures; the choice of the ultimate neural network architecture. Also, one of the essential tasks for achieving the aim of the work is to develop methods and to implement them by preliminary preparing a sample of initial data for training a neural model of the energy characteristics of a vane diffuser, which includes determining the vector of output and input parameters of the model; identification of conflicting examples; determination of the required minimum sample size to create a neural network; sample conversion in order to improve the quality of training of neural networks with insufficient sample size (multiple cross-validation, multiple sampling and changing the order of training examples); identification of outliers in the input data; excluding outliers from the training set; rationing of input data; adding noise to training samples.

Base object

Vane diffusers, to a large extent, determine the overall dimensions and energy characteristics of the centrifugal compressor as a whole [27-30]. The use of vane diffusers in centrifugal compressors allows to obtain a greater deceleration of the gas flow and, accordingly, to reduce losses in the rotary elbow and the return guide apparatus. The coordination of the optimal modes of the impeller and of the vane diffuser is achieved due to the special installation of the diffuser vanes, which allows to increase the efficiency factor by 2-4% in the design mode with a certain narrowing of the stage operation zone. Along with this, we know designs of stages with vaneless diffusers providing high efficiency in a wide operating area. On the other hand, the use of vane diffusers in low-flow stages did not lead to a significant narrowing of the stage operation zone. Thus, the choice of the type of diffuser and the determination of its geometric parameters requires special analysis.

The main structural elements of the vane diffuser are shown in Figure 1. As a rule, the height of the vanes is taken to be constant along the length of the diffuser $b_3 = b_4$, in order to simplify the design of the diffuser; the midline of the vane profile is made along an arc. In this case, the task of determining the optimal design of the vane diffuser is to search for the main structural dimensions:

- the radii of the input r_3 and of the outlet r_4 from the diffuser or diameters D_3 and D_4 respectively;
- the inlet vane angle α_{v3};
- the outlet vane angle α_{v4} or the deviation angle $\Delta\alpha_v = \alpha_{v4} - \alpha_{v3}$;
- the number of vanes *z* or row density *l/t;*
- the height of the vanes *b*.

All linear dimensions are assumed as ratio to the impeller diameter D_2. Figure 1 shows a diagram of the geometry of the investigated vane diffuser, the vane is limited by sections "3" and "4".

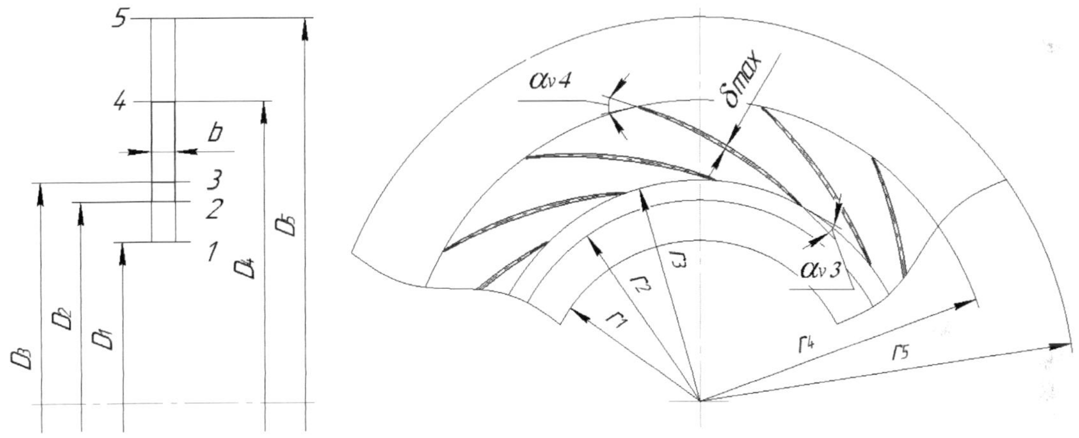

Figure 1. Scheme of the geometry of the vane diffuser of a centrifugal compressor, (a) meridional section, (b) cross-section.

Previously the influence of objective setting (2D-3D), the net dimensions, the intervane channels and turbulence model on the VD characteristics was investigated. The obtained results of modeling in Ansys CFX were used to analyze the flow structure and to calculate the VD characteristics versus incidence angle (which controls mass flow rate). The following characteristics were introduced to assess efficiency:

- polytropic efficiency factor $\eta = \frac{h_p}{-h_d} = \frac{\ln(p_4/p_2)}{\frac{k}{k-1}\ln(T_4/T_2)}$;
- loss coefficient $\zeta = \frac{h_w}{0.5c_2^2} = (1-\eta)\left(1-\frac{c_4^2}{c_2^2}\right)$;
- recovery coefficient $\xi = \frac{h_p}{0.5c_2^2} = 1 - \frac{c_4^2}{c_2^2} - \zeta$;
- velocity ratio (flow slowdown) c_4/c_2;
- outlet VD flow deviation angle $\Delta\alpha = \alpha_{v4} - \alpha_4$.

For calculations the flow parameters in section "5" were used. In this section the process of mixture of high energy particles and low energy particles is finished and the flow is quite steady. As in the vaneless space between sections "4" and "5" while objective setting the gas sticking to the walls is exluded and the flow is practically incompressible, it is correct to calculate the flow parameters by the ratios as follows:

- VD outlet velocity $c_4 = \frac{c_5}{D_4/D_5}$;
- outlet VD flow angle $\alpha_4 = \alpha_5$;
- temperature $T_4 = T_{inl}^* - \frac{c_4^2}{2c_p}$;
- temperature $T_5 = T_{inl}^* - \frac{c_5^2}{2c_p}$;
- VD outlet static pressure $p_4 = p_5(T_5/T_4)^{-3.5}$.

Thus, the VD with constant parameters of relative diameters of the beginning (1.1) and the end (1.5) of the vanes, the number of vanes (according to the recommended row density 2 [30]). The investigated parameters and the data ranges are shown in Table 1. The total number of the cases is 522 (for $\alpha_{v3} = 15°$ are also analyzed $\alpha_{v3} = 20, 30°$, in some cases i=-6°..+6°).

Table 1. The data ranges

Parameter	Values
b	0.025, 0.045, 0.06, 0.08
α_{v3}	15, 20, 25, 30°
$\Delta\alpha_v$	10, 15, 20°
i	-10..+10° at 2° interval

Neural network data processing approach

Neural networks as a universal approximator allow to build general models based on significant volume of data. Main ideas, peculiarities and advantages of neural network approach in modeling of characteristics of centrifugal compressors are stated in the papers by A.Nikiforov et al [31-33].

The full sample consisting of 522 controlling cases is analyzed. The neural network is built for Loss coefficient $\zeta = f(b; a_{v3}; a_{v4}; l/t; i; z)$, and the conflicting samples are ignored. The minimum and maximum values of the analyzed samples are shown in Table 2.

Table 2. Minimum and maximum parameters

Parameter	b	α_{v3}	α_{v4}	z	l/t	i	ζ
Minimum value	0,025	15	25	14	1,96	-10	0,06
Maximum value	0,08	30	50	26	2,05	10	0,99

In order to analyze the sample, a frequency analysis was performed for each of the parameters. For repeated parameters the frequency of coefficients was calculated in the sample for every value. For the parameters with different values the frequency of coefficients was calculated for the ranges. Frequency analysis of the sample allows us to see in the diagrams which ranges of the changed or other input parameters are covered most fully by the values, and which areas in the sample were represented to a lesser extent. It guarantees the most accurate modeling of the areas described in the initial sample by neural networks.

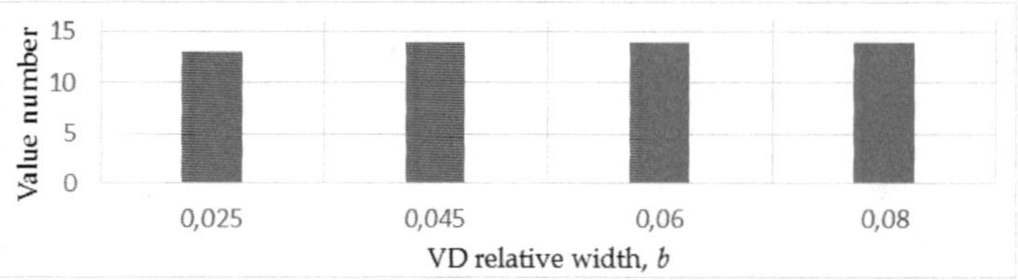

Figure 2. Frequency diagram of values of the relative width of VD diffuser.

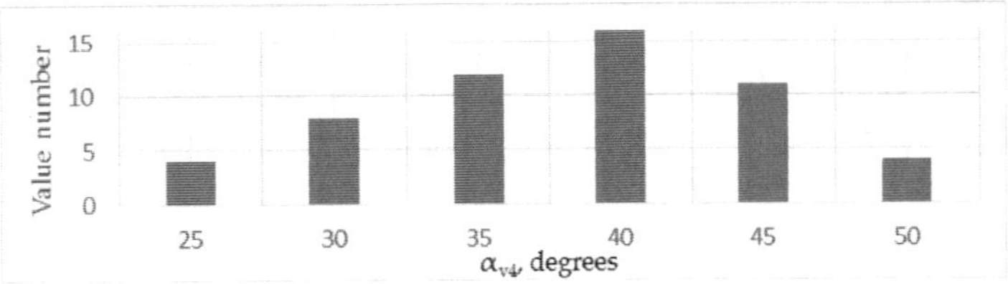

Figure 3. Frequency diagram of the angle value of output vane setting.

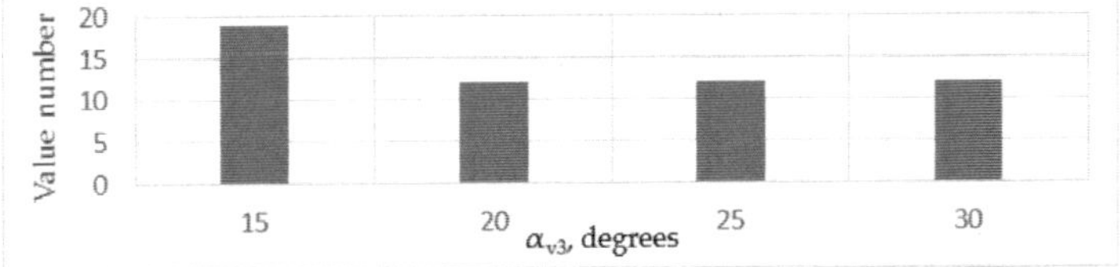

Figure 4. Frequency diagram of the angle value of input vane setting.

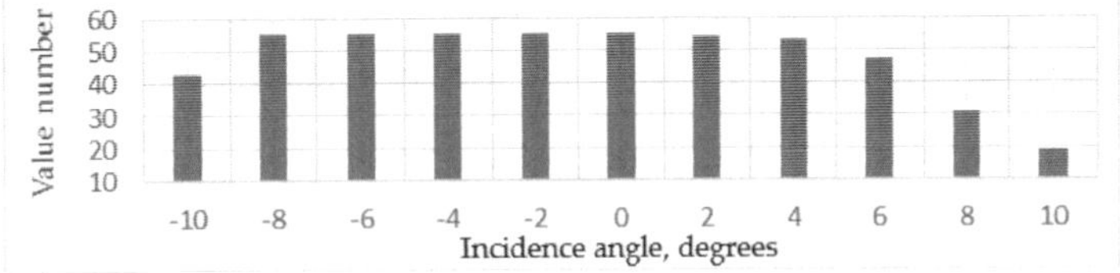

Figure 5. Frequency diagram of the angle of attack value.

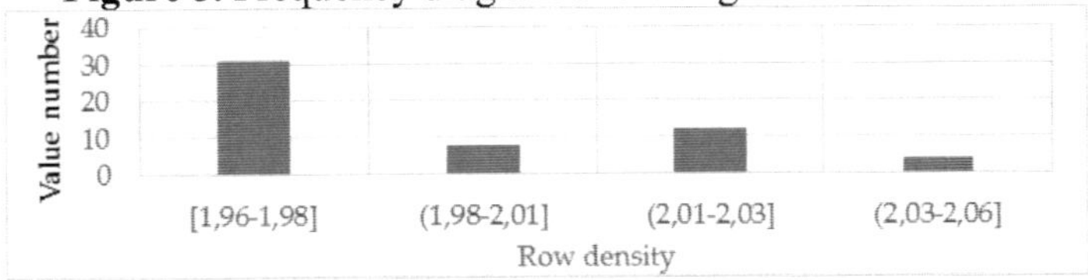

Figure 6. Frequency diagram of the VD cascade solidity.

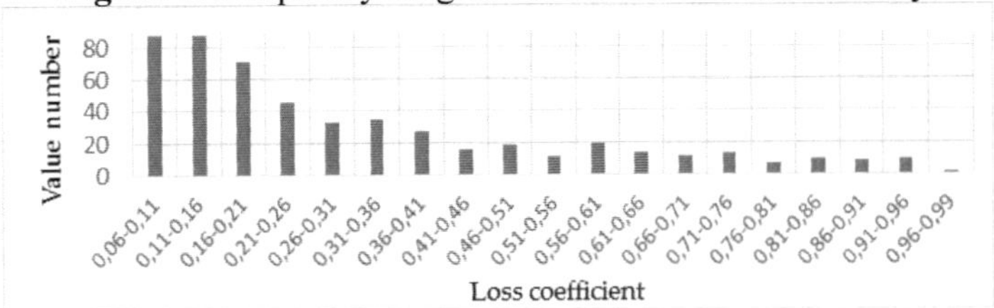

Figure 7. Frequency diagram of the VD loss coefficient.

Based on the previous neural network modeling experience [3] two-layer neural networks with 25 neurons in the first (hidden) layer were created. The simulation accuracy was checked when changing the training functions (trainfunction) as follows: with the BFGS quasi-Newtonian method (trainbfgs), with the Levenberg-Marquardt optimization method (trainlm) and Levenberg-Marquardt optimization with Bayesian regularization (trainbr). In the first case hyperbolic tangent function (tansig) of activation in both layers was used. In the second case neural networks with a logical sigmoid (logsig) and a linear activation function (purelin) for each learning function were used.

As a result of the training the root-mean-square inaccuracies were obtained for every neural network type that are shown in Table 3.

Thus, according to the minimum inaccuracy for neural network model of loss coefficient in the first layer logic sigmoid was chosen as an activation function. In the second layer a linear activation function and training function Levenberg-Marquardt optimization with Bayesian regularization were chosen.

Table 3. The inaccuracies of two-layers neural network with different architecture and different training functions for loss coefficient ζ

NN training functions	tansig, tansig	logsig, purelin
trainbfgs	0,092	0,044
trainbr	0,041	**0,040**
trainlm	0,070	0,063

To choose the network architecture on the chosen training function a calculation experiment was conducted. Two-layer neural networks with 10, 15, 20, 25, 30 neurons in the first (hidden) layer and 1 neuron in the output layer were created. The results are shown in Table 4.

Table 4. The inaccuracies of two-layer neural network with different neurons number in the output layer for loss coefficient ζ

Neurons number in the hidden layer	10	15	20	25	30
Average error	0,069	0,057	0,054	0,040	**0,037**

The least inaccuracy 3,7% in the calculation experiment on neural networks is registered for two-layer neural network with 30 neurons in the hidden layer.

To increase the accuracy of the neural model on the next stage tree-layer networks were investigated. The aim of the study was to calculate the optimal ratio of the neurons number in the first hidden layer 10, 15, 20, 25, 30 and in the second hidden layer 10, 15, 20, 25, 30 by means of selection of different combinations of neurons number in the first and in the second layers. In the third output layer one neuron was used. In the first and the second layers activation functions were logical sigmoid and in the second layer it was linear activation function.

The inaccuracies are defined as the root-mean-square deviations. The results are shown in Table 5.

Table 5. The inaccuracies of three-layer neural network with different neurons number in the first and second hidden layers

Neurons number	Second layer				
First layer	10	15	20	25	30
10	0,053	0,047	0,036	0,030	0,021
15	0,021	0,029	0,021	0,020	0,033
20	0,029	0,022	0,069	0,021	0,025
25	0,033	0,020	0,037	0,018	0,030
30	0,222	**0,018**	0,032	0,027	0,034

Results

As a result of training of three-layer neural network a minimum inadequacy 1,8% was achieved if the network architecture was as follows: 30 neurons in the first layer; 15 neurons in the second hidden layer; one neuron in the third layer; in the first and the second layers activation functions were logical sigmoid and in the third layer it was linear activation function with the Levenberg-Marquardt optimization method and Levenberg-Marquardt optimization with Bayesian regularization.

The results of loss coefficient modeling for different vane diffusers are shown on Figures 8-11.

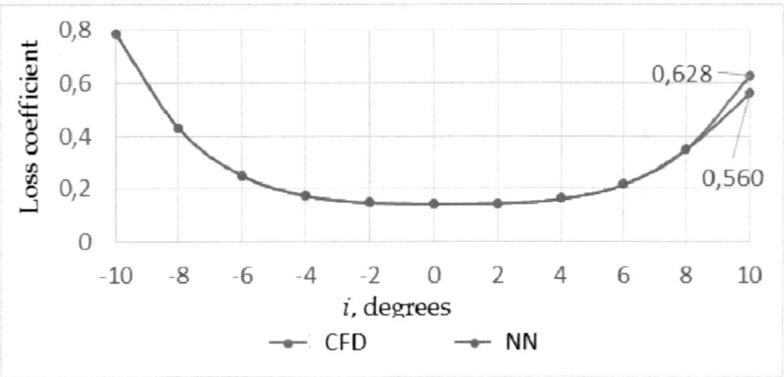

Figure 8. Loss coefficient for b=0,025; l/t=1,95883; a_{v3}=15; a_{v4}=35; z=20.

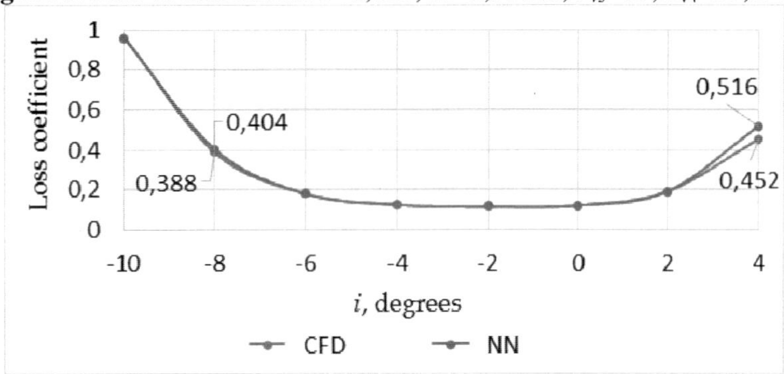

Figure 9. Loss coefficient for b=0,045; l/t=1,95883; a_{v3}=20; a_{v4}=40; z=20.

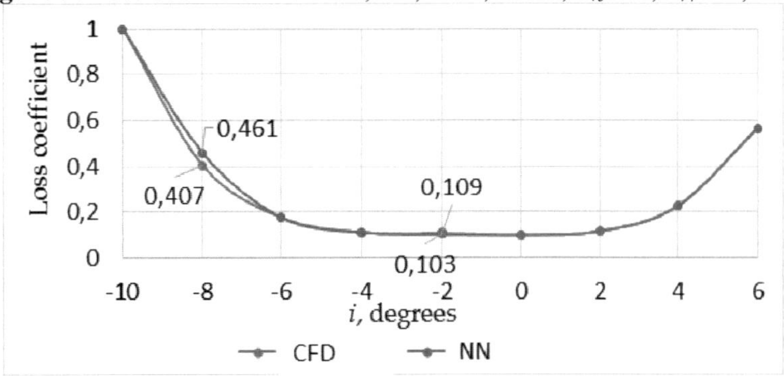

Figure 10. Loss coefficient for b=0,06; l/t=1,96986; a_{v3}=20; a_{v4}=30; z=17.

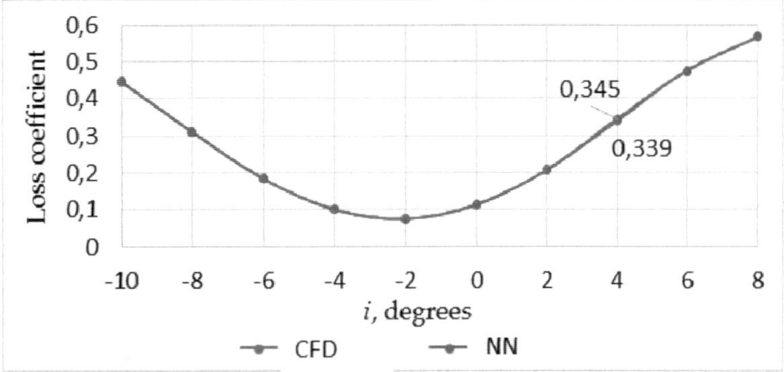

Figure 11. Loss coefficient for b=0,08; l/t=1,95883; a_{v3}=20; a_{v4}=40; z=20.

Conclusions

The article presents the analysis of the existing methods of creating mathematical models and possibilities of their use in engineering. As a result of the analysis, the authors suggest a computer model on the basis of neural networks that is one of the most efficient methods of studying of complex technical systems. This computer model allows to conduct series of calculation experiments with the aim to analyze and compare modeling results with the real object behavior.

The conducted calculation defines an optimal neural network with the acceptable error limit and adequate tracking of the gas-dynamic processes of energy conversion in a vane diffuser of the centrifugal compressor. It also defines the number of layers, activation functions and neurons number on every hidden layer.

To improve the quality of modeling the initial data were processed by the above-mentioned methods. It allowed to increase substantially the modeling quality.

In Creating and Training of Neural Network the authors analyzed different combinations of neural networks, inaccuracies of ready models and the structure of the necessary network. Based on it a neural network with the 2% inaccuracy was created.

The article presents an illustrative example of a mathematical model that is used on the basis of the calculation methods of influence of geometrical parameters on energy characteristics of VD.

The mathematical model of a vane diffuser in the existing Universal Modeling Method consists of a dozen algebraic equations with two dozen empirical coefficients. The results of physical experiments are used to identify the model. The algebraic model works satisfactorily. But in order to improve the accuracy of the calculations, new experiments are needed, which is expensive and time-consuming. As an alternative, the authors carried out a computational experiment. The characteristics of vane diffusers are well simulated by neural network equations satisfactory. The next step of the authors is the inclusion of the VD neural model into the Universal Modeling Method programs in order to find and eliminate the shortcomings.

Nomenclature

b	blade height, channel width in the direction of rotor axis;
c	absolute flow velocity;
c_p	specific heat capacity at constant pressure;
D	diameter;
h_p	polytropic pressure;
h_d	dynamic pressure;
h_l	lost pressure in stage flow part;
i	incidence angle (angle of attack);
k	isentropic coefficient;
l	blade length;
p	pressure;
r	radius;
t	distance between the vanes;
T	temperature;
z	number of blades;
α	angle between absolute velocity and circumferential direction;
α_v	angle between blade midline tangent and circumferential direction;
δ_v	blade thickness;
ζ	loss coefficient;
η	polytrophic efficiency;
ξ	recovery coefficient;
ρ	gas density;

Funding: The research was performed by a Grant of the President of the Russian Federation for young PhD MK-1893.2020.8.

Acknowledgments: *The results of the work were obtained using computational resources of Peter the Great Saint-Petersburg Polytechnic University Supercomputing Center (www.spbstu.ru).*

References

[1] Galerkin Yu, Drozdov A, Solovyeva O and Kabalyk K 2019 *E3S Web of Conferences*
[2] Galerkin Yu, Rekstin A, Drozdov A, Soldatova K, Solovyeva O, Popova E 2020 E3S Web of Conferences

[3] Galerkin Y B, Nikiforov A G, Solovyeva O A, Popova E Y, Rekovets A V 2020 *BMSTU Journal of Mechanical Engineering* **7** 29–42
[4] Solovyeva O, Drozdov A 2020 *E3S Web of Conferences*
[5] Rekstin A F, Drozdov A A, Solovyova O A and Galerkin Yu B 2018 *Proc. Int. Conf. Oil and Gas Engineering* (Omsk)
[6] Rekstin A F, Soldatova K V, Galerkin Yu B 2019 *Proc. Int. Conf. on Compressors and their Systems* (London)
[7] Rekstin A, Popova E, Ucehovscy A 2018 *Proc. Int. Conf. Oil and Gas Engineering* (Omsk)
[8] Prasad V V, Kumar L M, Reddy M B 2011 *Science Insights: An International Journal* **1** 6-10.
[9] Kabalyk K, Kryłłowicz W 2016 *Transactions of the institute of fluid-flow machinery* **131** 41–53.
[10] Marenina L N 2016 *Compressor technology and pneumatics* **3** 27-35
[11] Meduri U K, Selvam K, Nawrocki G 2015 *Proc. ASME Turbo Expo* (Montréal)
[12] Galerkin Y, Voinov I, Drozdov A 2017 *Proc. Int. Conf. Compressors and their Systems* (London)
[13] Kortikov N, Borovkov A, Voynov I, Kirillov A, Drozdov A 2018 *Proc. Int. Conf. Energy* (St.Petersburg)
[14] Borovkov A I, Voinov I B, Nikitin M A, Galerkin Yu B, Rekstin A F, Drozdov A A 2019 *Proc. Int. Conf. Oil and Gas Engineering* (Omsk)
[15] Borovkov A I, Voinov I B., Galerkin Yu B., Drozdov A A, Soldatova K V 2019 *Proc. Int. Conf. Compressors and their Systems* (London)
[16] Borovkov A, Voinov I, Galerkin Yu, Nikiforov A, Nikitin M, Solovyeva O and Kabalyk K 2019 *E3S Web of Conferences*
[17] Galerkin Yu B and Solovyova O A 2014 *Compressor technology and pneumatics* **3** 35-41
[18] Galerkin Yu B and Solovyova O A 2014 *Compressor technology and pneumatics* **4** 15-21.
[19] Galerkin Yu B, Rekstin A F, Solovyeva O A 2019 *Proc. Int. Conf Oil and Gas Engineering* (Omsk)
[20] Galerkin Y, Soldatova K, Solovieva O 2015 *Proc. Int. Conf. Compressors and their Systems* (London)
[21] Petukhov E P, Galerkin Yu B, Rekstin A F 2019 *Proc. Int. Conf. Compressors and their Systems* (London)
[22] Petukhov E P, Drozdov A A, Galerkin Yu B 2019 *Compressor technology and pneumatics* **3** 37-48
[23] Petukhov E P, Galerkin Y B, Rekstin A F 2019 *Proceedings of Higher Educational Institutions. Machine Building* **8** 51–64
[24] Nikiforov A, Popova D, Soldatova K 2017 *Proc. Int. Conf. Compressors and their Systems* (London)
[25] Nikiforov A, Kuchumov A, Terentev S, Petukhov E and Kabalyk K 2019 *E3S Web of Conferences*
[26] Nikiforov A, Avramenko D, Kuchumov A, Terentev S, Galerkin Yu, Solovyeva O 2019 *Proc. Int. Conf. Compressors and their Systems* (London)
[27] Galerkin Yu B 2010 Turbocompressors (Moscow: LTD information and publishing center KHT) p 596
[28] Seleznev K P and Galerkin Yu B 1982 Centrifugal compressors (Leningrad: Mechanical engineering)
[29] Seleznev K P and Galerkin Yu B 1986 Theory and design of turbocompressors (Leningrad: Mechanical engineering)
[30] Ris V F 1981 Centrifugal compressors machines (Leningrad: Mechanical engineering) p 351
[31] Nikiforov A G, Popova D Yu, Soldatova K V, Solovyeva O A 2015 *Compressor technology and pneumatics* **3** 18-21
[32] Nikiforov A G, Popova D Yu, Soldatova K V, Solovyeva O A 2015 *Compressor technology and pneumatics* **4** 14-16

IOP Conf. Series: Materials Science and Engineering 1180 (2021) 012025 doi:10.1088/1757-899X/1180/1/012025

[33] Nikiforov A G, Popova D Yu, Soldatova K V, Solovyeva O A 2015 *Compressor technology and pneumatics* **6** 30-33

Numerical simulation and optimization of centrifugal compressor return channels: methods and results

Y.B. Galerkin[1], L.N. Marenina[2], A.A. Drozdov[3], O.A. Solovyeva[4] and K.V. Soldatova[2]

[1] Leading research center "Digital Design and Modeling (Smart Design)" Peter the Great St.Petersburg Polytechnic University, Polytechnical st. 29, St.Petersburg, Russia
[2] R&D Laboratory "Gas dynamics of turbo machines" Peter the Great St.Petersburg Polytechnic University, Polytechnical st. 29, St.Petersburg, Russia
[3] Higher School of Power Engineering Institute of Power Engineering Peter the Great St.Petersburg Polytechnic University, Polytechnical st. 29, St.Petersburg, Russia
[4] Higher School of Hydraulic and Power Engineering Construction Engineering Institute Peter the Great St.Petersburg Polytechnic University, Polytechnical st. 29, St.Petersburg, Russia

Email: marenina_ln@mail.ru.

Abstract. The authors actively use their engineering Universal Modelling Method for gas-dynamic design of centrifugal compressors for industrial partners. Currently, there are two approaches to improving the Method: improving the preliminary design and increasing the accuracy of gas-dynamic characteristics calculation. Computational Fluid Dynamics (CFD) methods give good results for the flow path stator part. Recently, massive CFD calculations of the vaneless diffusers (VLD) characteristics were generalized by a system of algebraic equations, which replaced the previous more complex mathematical model in the Method. Then, based on CFD optimization of a large series of return channels (RCh), corrections were made to the preliminary design of this element of the flow path.
This paper presents the results of the joint CFD optimization of a vaneless diffuser and a return channel. Stator elements have many geometric parameters. For a stage with the flow rate coefficient of 0.0597 and the loading factor of 0.60, only the VLD relative radial length D_4/D_2, the number and the inlet angle of the return channel vanes were optimized. Engineering calculations and design experience predicted an optimal value of D_4/D_2 within the interval 1.9 - 2.0. CFD optimization demonstrated almost linear reduction of the total head loss towards the end of the investigated range D_4/D_2 = 2.3. After careful optimization of the U-bend, the optimization of D_4/D_2 was repeated. The influence of on the loss coefficient has decreased, but the value of D_4/D_2 = 2.3 is still far from optimum. It significantly exceeds technically acceptable radial size. The result obtained influenced the design plan of future calculation experiments with a large series of stator elements of the stages in a practically significant range of design parameters.

1. Introduction

The total capacity of industrial centrifugal compressors in developed countries reaches dozens of GW. It is important to create a compressor flow path with minimal energy consumption. The development of CFD technologies formally solves both the direct and inverse problems of gas dynamics. The problem is that CFD calculation of gas dynamic characteristics of a compressor does not guarantee their compliance with actual characteristics. The authors of [1 - 4] presented data on good agreement of CFD-calculated characteristics with test results. Other publications [5 - 11] and own experience of the authors of this paper [12 - 14] show that the characteristics of centrifugal compressors according to CFD calculations do not correspond to test results.

The authors believe that CFD technologies are currently effective as an auxiliary tool, but not as a primary design tool. The final goal of the authors of this paper is to optimize and calculate the stator elements of centrifugal stages in the range of design parameters Φ_{des} (flow rate coefficient), $\psi_{T\,des}$ (loading factor). This will make it possible to formulate design recommendations and obtain dataset from a computational experiment for creating a mathematical model. The first stage of the research presented below made it possible to work out the optimization technique and evaluate the possible results. The authors optimized the stator elements of a stage with design parameters $\Phi_{des} = 0.0597$ and head $\psi_{T\,des} = 0.60$.

2. State of the art

In recent years among researchers from different countries, CFD calculations of return channels have become widespread. Of particular interest are the papers in which the calculated results are compared with experimental data. In [15 - 19], various aspects of CFD modeling of a high-flow-rate stage with $\Phi_{des} = 0.15$ and its return channels are analyzed. The authors have a modern test rig at their disposal, which made it possible to compare the stage characteristics and structure of the flow. Satisfactory compliance of the stage characteristics was obtained, and good results of comparison of the flow structure were attained as well. CFD calculations, in particular, correctly estimated the efficiency of stage variants with different radial lengths of a vaneless diffuser. The authors of [20] experimentally confirmed the CFD calculations of the RCh with additional vanes at the outlet and improved the RCh characteristic at $\Phi > \Phi_{des}$. The authors of [21] by CFD-optimization of RCh increased the efficiency of a two-stage compressor by 0.7% (experimental confirmation). RCh with a vaneless diffuser is optimized by many parameters.

Computational gas dynamics methods are able to improve stator elements of the flow path of centrifugal compressors. The authors of [22, 23] designed and investigated return channels of stages with design parameters $\Phi_{des} = 0.015$–0.15 and $\psi_{T\,des} = 0.45$–0.70. The Direct Optimization program optimizes the main dimensions of RCh. The RCh with a constant vane height $b_5 = b_6$ and a smaller number of vanes in comparison with the preliminary design turned out to be the optimal ones. The loss coefficients of optimized RCh are less by up to 30%. Corresponding adjustments were made to the preliminary design methodology.

The main conclusion that follows from the publications of recent years is that the results of CFD calculations of return channels and vaneless diffusers are confirmed experimentally.

3. The objectives of the research

Universal Modeling Method [24, 25] makes the initial design of the VLD + RCh. Figure 1 shows the configuration.

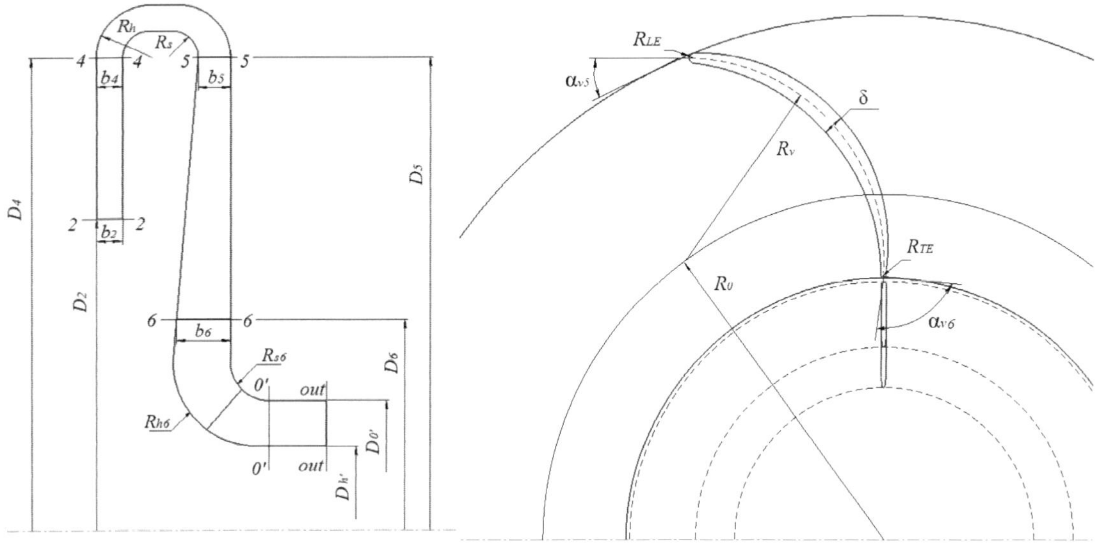

a) b)

Figure 1. The shape and dimensions of VLD and RCh in compressors designed using the Universal Modeling Method: meridional (a) and radial (b) planes

This engineering tool is in design practice since mid-1990-s. More than 400 centrifugal compressors with unit capacity of up to 25 MW designed using the Method effectively operate in different industries. The efficiency of best multistage compressors was more than 87% [20], which indicates good aerodynamic characteristics of an initial design of vaneless diffusers and return channels. To improve this flow path elements would be complicated but important.

The meridional form of VLD and RCh is circular arcs and straight lines. The choice of all dimensions is described in [25]. The new recommendation follows of the research presented in [21]. For 15 stages with design parameters Φ_{des} = 0.015–0.15, $\psi_{T\,des} = 0.45 - 0.70$, the optimal configuration is $\bar{b}_5 = \bar{b}_6$. It is accepted for return channels under investigation. In the meridional plane, the optimization parameters are:

- the relative radial length of the VLD and RCh: $\bar{D}_4 = \bar{D}_5$;

- the relative radii $\bar{R}_s$, $\bar{R}_h$ that determine the friction losses in the U-bend and the local velocity gradients.

In the radial plane, the inlet angle of the vanes α_{v5} is subject to optimization to ensure a favorable flow inlet. The number of vanes z is optimized to minimize the sum of friction and separation losses.

Stator elements of the stage with design parameters Φ_{des} = 0.0597, $\psi_{T\,des}$ = 0.60, $\bar{D}_h$ = 0.35 were chosen as a specific object of optimization. Basic dimensions and parameters of stator elements according to preliminary design: $\bar{b}_4$ = 0.039, $\bar{D}_4$ = 1.73; $\bar{b}_5 = \bar{b}_6$ = 0.085, z = 22; α_{v5} = 14.65°; α_{2-3des} = 30.8°, ζ_{SE} = 0.1988.

4. Influence of the radial length on the loss coefficient of the SE

The Workbench of the ANSYS 19 software package was used. The parameterized model of the flow path was made in DesignModeler. In the TurboGrid mesh generator, design meshes for the main and splitter vanes were built. The values y + <20 were set in accordance with the requirements of the correct modeling of the boundary layer using the SST turbulence model (Shear-Stress-Transport). The total number of elements is 893,000. The loss coefficient ζ_{SE} was the object for optimization in the

Direct Optimization program. An optimization method depends on the number of objectives, the set limits and the desired number of design points. For the research conducted in the work, the MOGA (Multi-Objective Genetic Algorithm) method was used.

The experience of RCh optimization in [20, 21] pointed out certain difficulties of multiparameter optimization. CFD calculations have a certain margin of error, so to find a truly optimal variant requires significant resources. Optimal solutions for return channels in [21] were obtained by optimizing two or three geometric parameters, and only then the others. Accordingly, the diameters $\bar{D}_4 = \bar{D}_5$, the number of vanes z, and the inlet angle of the vanes α_{v5} were optimized. The range of variation was chosen to be $\bar{D}_4 = \bar{D}_5$ = 1.40–2.14. Figure 2 illustrates the influence of optimization parameters on the SE loss coefficient:

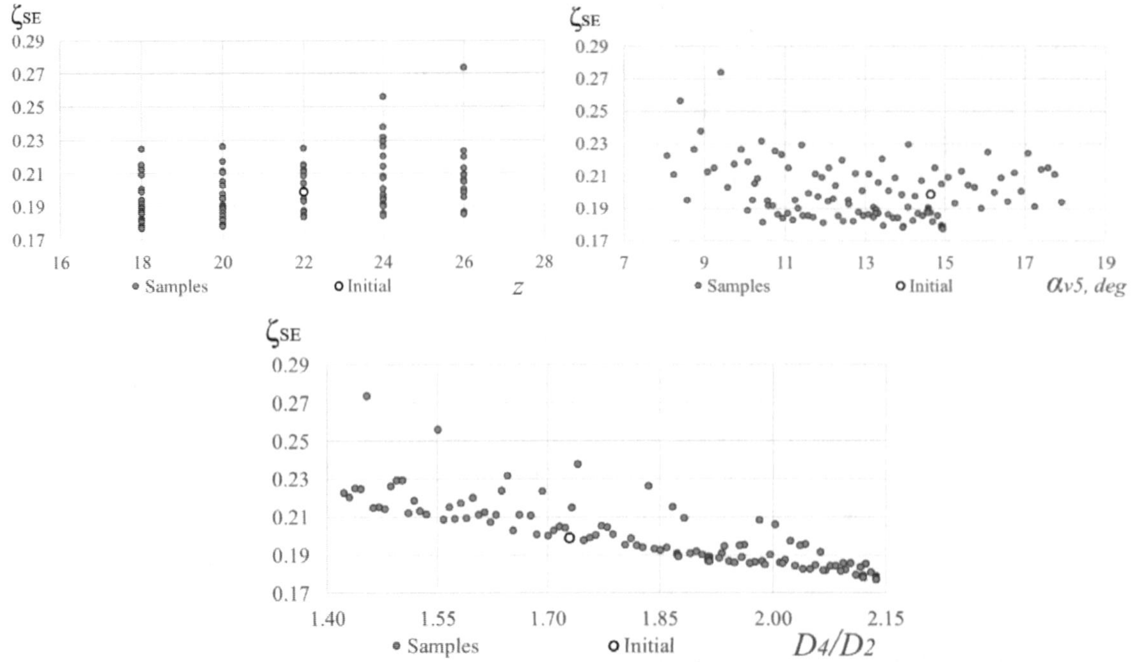

Figure 2. Influence of optimization parameters on the SE loss coefficient

Beyond expectations, the optimum of $\bar{D}_4$ was not found in the selected range. The loss coefficient decreases monotonously with increasing of $\bar{D}_4$. For investigation of the optimized stator elements operation in off-design modes, the loss coefficient and the outlet flow angle characteristics was calculated. Figure 3 shows these characteristics for the initial variant of the SE and the variant with $\bar{D}_4$ = 1.9. Hardly bigger $\bar{D}_4$ could be in real compressors.

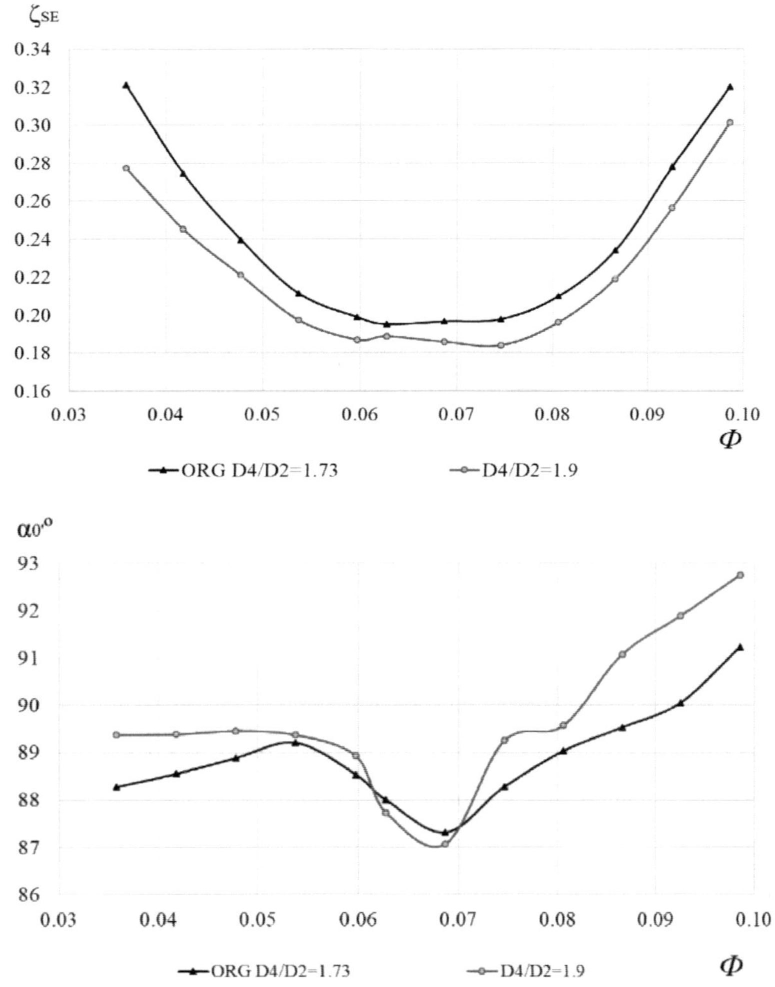

Figure 3. Loss coefficients of SE (above) and outlet flow angle (below). Stage 2DI-0.0579-0.60-0.35, $\bar{D}_4$ = 1.73 and 1.9

The range of $\bar{D}_4$ was extended up to 2.3. The loss coefficient diminished again. Figure 4 shows the dependence of the loss coefficient (design flow rate) on the radial size of the stator elements.

The linear dependence $\zeta_{SE} = f\left(\bar{D}_4\right)$ indicates the remoteness of the optimum. At the same time, the value $\bar{D}_4$ = 2.3 is far beyond the constructive limitation acceptable for real compressors.

Table 1 shows a comparison of loss coefficients and efficiency for SE with $\bar{D}_4$ = 1.73, 1.90, 2.20.

Table 1. Loss coefficients, efficiency of SE, and loss of stage efficiency for different $\bar{D}_4$

	ζ_{SE}	ζ_{RCh}	η_{SE}	$\Delta\eta_{SE}$
$\bar{D}_4$=1,73	0.1988	0.1907	0.772	0.0808
$\bar{D}_4$ =1,9	0.1869	0.2376	0.7857	0.076
$\bar{D}_4$ =2,2	0.1758	0.2911	0.7987	0.0715

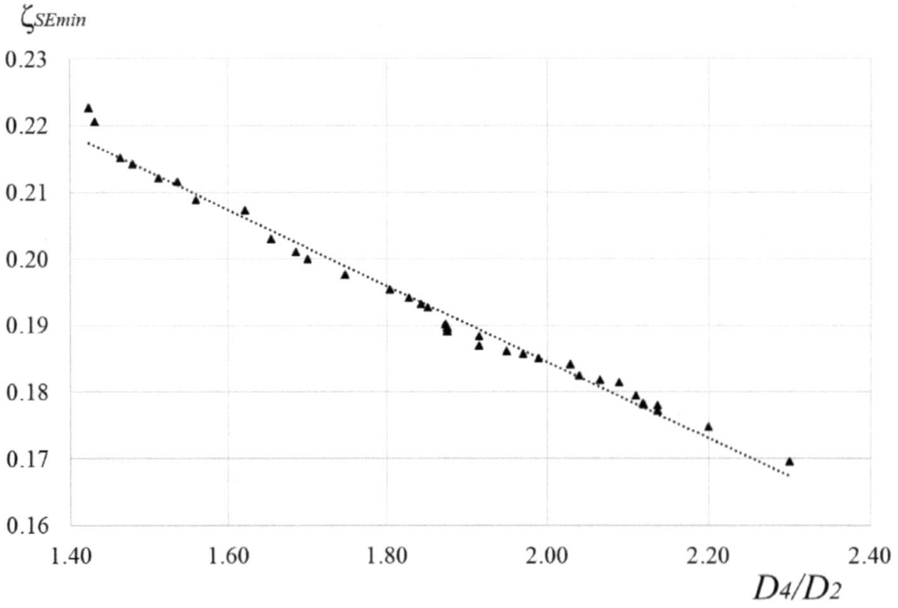

Figure 4. The minimized loss coefficient of SE, depending on $\bar{D}_4$.
The stage 2DI-0.0579-0.60-0.35 before the U-bend optimization

The stator element polytrophic efficiency is:

$$\eta_{SE} = \frac{\lg(p_{0'} / p_3)}{\frac{\kappa}{\kappa - 1}\lg(T_{0'} / T_3)} \quad (3)$$

The loss of efficiency of a stage in the SE is:

$$\Delta\eta_{SE} = \frac{\zeta_{SE}}{2\psi_T}\bar{c}_{2-3}^2 = \frac{\zeta_{SE}}{2}\frac{\psi_T}{\cos\alpha_{2-3}^2} \quad (4)$$

The radial length $\bar{D}_5$ influence obvious: the losses in the longer RCh are greater. However, VLD+RCh of greater radial length are more effective due to flow deceleration in longer VLD. Flow structure in the radial plane does not reveal principal differences between the original SE variant and the variant with a larger radial dimension $\bar{D}_5 = \bar{D}_4$ - Figure 5.

The flow pattern in Figure 6 points to a possible reason for the unexpected monotonous influence of $\bar{D}_4$ on the loss coefficient. A significant flow separation zone occurs in the U-bend. In this case, the pressure loss depends on the kinetic energy at the inlet to the U-bend. The smaller the radial length of the VLD, the greater the kinetic energy. Obviously, the size of the U-bend needs optimization.

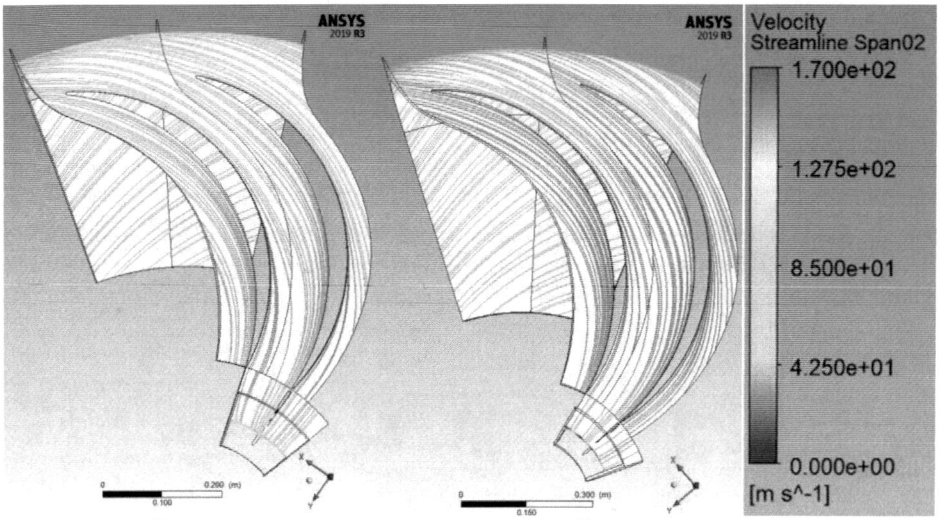

a) Initial, $\bar{D}_4$ = 1.73; z = 22; α_{v5} =14.65° b) $\bar{D}_4$ = 2.2; z = 18; α_{v5} = 13.33°

Figure 5. SE of stage 2DI-0.0579-0.60-0.35. Streamlines at the vane-to-vane mean surface.

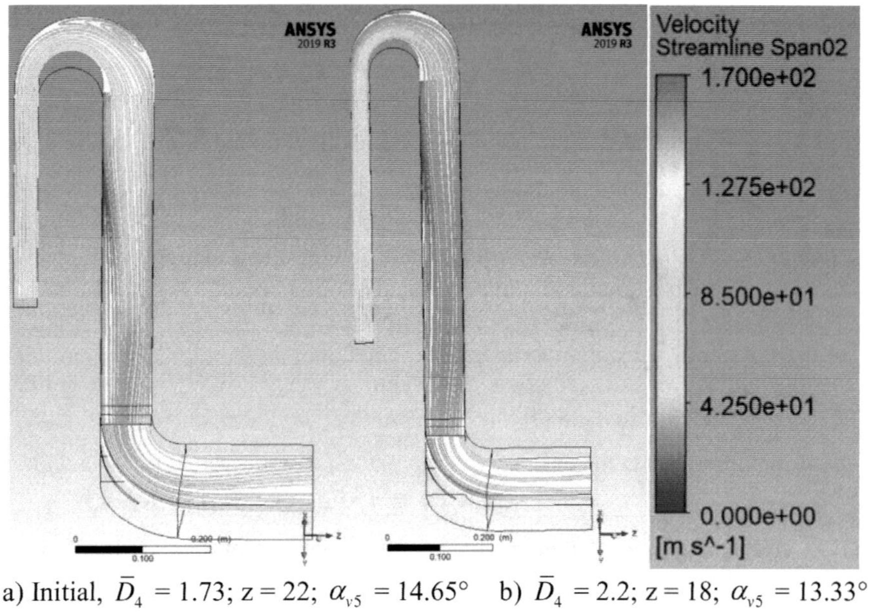

a) Initial, $\bar{D}_4$ = 1.73; z = 22; α_{v5} = 14.65° b) $\bar{D}_4$ = 2.2; z = 18; α_{v5} = 13.33°

Figure 6. SE of stage 2DI-0.0579-0.60-0.35. Streamlines in the meridional plane

5. Optimization of the U-bend of SE with $\bar{D}_4$ = 1.9

In [23] is recommended to correlate the optimal value of the radius of curvature of the channel with its width, i.e. R_s / b_4, R_h / b_4. The optimization was carried out for the inner radius of curvature R_s, auxiliary parameter a connected with R_h / b_4, the number of vanes z, the inlet vane angle α_{v5}. The outer radius of curvature $R_h = f(a, R_s)$ was calculated by the formula $R_h = a(b_4 + b_5 + 2R_s)/2$. For a = 1, the outer contour is described by one radius R_h, as in the case of the U-bend in the section 5. In accordance with the chosen MOGA method, the optimization has converged when 118 SE variants were calculated. The parameters of the initial and the optimized SE variants are presented in Table 2.

Table 2. Parameters of the initial and best optimized SE variants

№	α_{v5}, deg	z	R_s / b_4	R_h / b_4	ζ_{SE}	$\alpha_{0'}$, deg	$\Delta\tilde{c}_{av}$
118	14.00	18	1.90	2.82	0.1759	88.50	0.297
Initial	13.25	18	1.23	2.49	0.1869	88.92	0.230

Decrease of the loss coefficient is associated with better organization of the flow in the meridional plane, Figure 7. The maximum radial size of the SE with an optimized U-bend is slightly larger. The SE loss coefficient decreased by 6.4%. The loss of efficiency in the SE decreased by 0.47%: such a saving deserves attention.

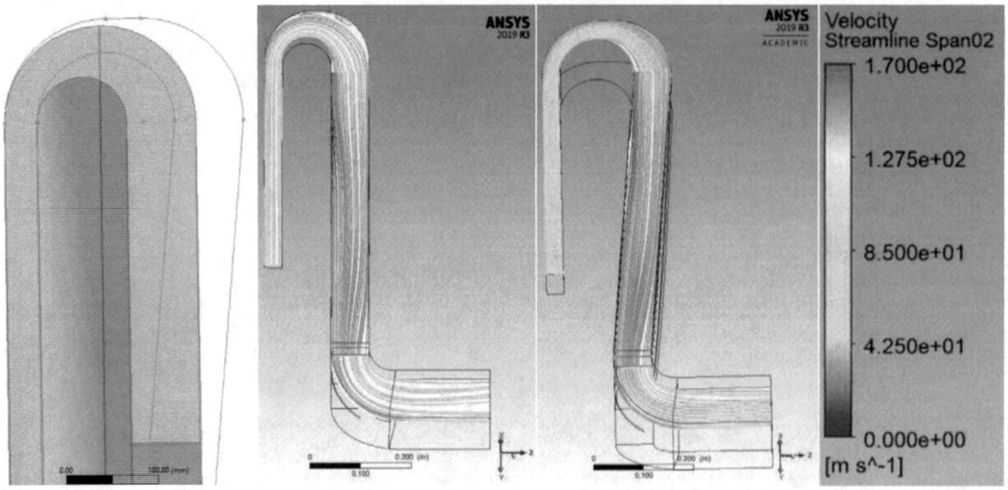

Figure 7. On the left is the U-bend shape. Preliminary design is shown in gray, the design after optimization in red. In the center are streamlines of the original version, on the right are streamlines after U-bend optimization

6. Influence of the radial length on the loss coefficient of the SE after U-bend optimization

With optimized values of R_s / b_4 and R_h / b_4, the number and inlet angle of the SE vanes with $\bar{D}_4$ = 1.73 and 2.3 were optimized. The separation zones in the meridional plane have practically disappeared. The U-bend with R_s / b_4 = 1.90 and R_h / b_4 = 2.82 can be considered optimal for all investigated $\bar{D}_4$. Table 3 shows the parameters of the SE with the optimal U-bend at $\bar{D}_4$ = 1.73, 1.9 and 2.3 and with the U-bend from after the preliminary design (ORG):

Table 3. Parameters of the SE with the optimal U-bend and with the U-bend after the preliminary design

№	1	2	3	4	5	6
$\bar{D}_4$	z	α_{v5}, deg	ζ_{SE}	$\Delta\eta_{SE}$	ζ_{SE} (ORG)	$\Delta\eta_{SE}$ (ORG)
1.73	22	14.65	0.1841	0.0749	0.1988	0.0808
1.9	18	13.25	0.1769	0.0719 (+0.3%)	0.1869	0.076 (+0.48%)

2.3	18	13.33	0.1667	0.0678 (+0.41%)	0.1696	0.069 (+0.7%)

In rows 4 and 6, the increment in the stage efficiency due to the increase in the radial size of the SE is given in brackets. With an optimal U-bend, the efficiency gain is smaller. However, the growth rate is still significant. With a lot of design practice [24, 25] the authors were unable to use vaneless diffusers with $\bar{D}_4 > 1.86$. Anyway the result in Table 3 deserves attention.

Conclusion

The authors propose in the near future to continue the search for the optimal dimensions of the stator elements of stages with vaneless diffusers in the practically significant range of design parameters $\Phi_{des} = 0.015$–0.15, $\psi_{T\,des} = 0.45$–0.70. From the study of the stage with parameters $\Phi_{des} = 0.0597$, $\psi_{T\,des} = 0.60$ it follows that the optimal radial size $\bar{D}_4$ for high-flow stages can be much larger than the constructively acceptable values. In this case, the geometric parameter $\bar{D}_4$ cannot be optimized. It is necessary to optimize stator elements with several fixed values of $\bar{D}_4$ in a constructively acceptable and practically significant range. Before doing this work, it is necessary to pay attention to the role of the return channel geometry parameters that were not studied in the present work. For example, high-flow stages may require the 3D shape of vanes. It is also necessary to study profiles and midline shape of vanes.

Acknowledgments

The research was performed by a Grant of the President of the Russian Federation for young PhD MK-1893.2020.8. The calculations were performed using computational resources of Peter the Great Saint-Petersburg Polytechnic University Supercomputing Center (www.spbstu.ru)

References

[1] Guidotti E 2014 *Towards Centrifugal Compressor Stages Virtual Testing*. Thesis for the Degree of Ph. D. p. 100. Università degli Studi di Bologna

[2] Mosdzien M, Enneking M, Hehn A, Grates D, Jeschke P 2018 Influence of blade geometry on secondary flow development in a transonic centrifugal compressor *Journal of the Global Power and Propulsion Society*, Vol 2, Iss 1

[3] Sorokes J, Hutchinson B 2000 The Practical Application of CFD in the Design of Industrial Centrifugal Compressors *Challenges and Goals in Pipeline Compressors* PID.V. 5.

[4] Sorokes J M, Nye D A, D'Orsi N, Broberg R 2000 Sidestream optimization through the use of computational fluid dynamics and model testing turbomachinery sympos Proceed. Texas: A&M.

[5] Abel M, Newton P, Martinez-Botas R F, Wohr M, Muller M, Leweux J 2018 3D computational analysis of a compressor for heavy duty truck engine turbochargers. *Proceedings of ASME Turbo Expo 2018. Turbomachinery Technical Conference and Exposition.* June 11-15, Oslo

[6] Kryłłowicz W, Świder P, Kozanecki Z, Kabalyk K, Kozanecki Jr Z 2017 Technical and Aerodynamical Aspects of a High Pressure Synthesis Gas Turbocompressor Modernization. 12th European Conference on Turbomachinery Fluid Dynamics and Thermodynamics, April 3 – 7 Stockholm, Sweden.

[7] Matas R, Syka T, Lunacek O 2017 Numerical and experimental modelling of the centrifugal compressor stage – setting the model of impellers with 2D blades. *EPJ Web of Conferences 11th International Conference on Experimental Fluid Mechanics* Volume 143, 02073 DOI:10.1051/epjconf/201714302073 Czech Republic

[8] Hazby H, Casey M, Robinson C, Spataro R 2017 The design of a family of process compressor stages. *Proceedings of 12th European Conference on Turbomachinery Fluid dynamics & Thermodynamics* ETC12, April 3-7, 2017; Stockholm, Sweden Paper ID: ETC2017-134

[9] Le Sausse P, Fabrie P, Arnou D and Clunet F 2013 CFD comparison with centrifugal compressor

measurements on a wide operating range. *EPJ Web of Conferences*, Volume 45 (2013) DOI:10.1051/epjconf/20134501059

[10] Xinqian Z, Meijie Z 2018 Criteria for the Matching of Inlet and Outlet Distortions in Centrifugal Compressors. *Applied Thermal Engineering* Volume 131, 25 February 2018, Pages 933-946

[11] Elfert M, Weber A, Wittrock D, Peters A, Voss C, Nicke E 2016 Experimental and numerical Kowalski S C, Pacheco J E, Fakhri S, Sorokes J. M. 2012 Centrifugal stage performance prediction and validation for high mach number applications. *Proceedings of the Forty-First Turbomachinery Symposium* September 24-27, Houston, Texas

[12] Kortikov N, Borovkov A, Voynov I, Kirillov A, Drozdov A Modeling the gas-dynamic characteristics of the low-flow and mid-flow model stages for an industrial centrifugal compressor MATEC Web Conf. Volume 245, *International Scientific Conference on Energy, Environmental and Construction Engineering* DOI: https://doi.org/10.1051/matecconf/201824504019

[13] Borovkov A I, Voinov I B , Galerkin Yu B, Drozdov A A, Soldatova K V 2019 Experimental characteristic simulation for two-stage pipeline centrifugal compressor. *International Conference on Compressors and their Systems* IOP Conf. Series: Materials Science and Engineering 604 (2019) 012052. doi:10.1088/1757-899X/604/1/012052

[14] Borovkov A, Voinov I, Galerkin Y, Nikiforov A, Nikitin M, Solovyeva O and Kabalyk K 2019 Issues of gas dynamic characteristics modeling: a study on a centrifugal compressor model stage *E3S Web of Conferences* 140, 06003 International Scientific Conference on Energy, Environmental and Construction Engineering https://doi.org/10.1051/e3sconf/201914006003

[15] Bisping J, Rossbach T, Grates D, Hildebrandt A, Jeschke P 2018 Influence of diffuser diameter ratio on the performance of a return channel within a centrifugal compressor stage. Proceedings of GPPS Forum 18 Global Power and Propulsion Society Montreal

[16] Franz H, Rube C, Wedeking M, Jeschke P 2015 Numerical investigation of the return channel of a high-flow centrifugal compressor stage. Proceedings of ASME Turbo Expo 2015: Turbine Technical Conference and Exposition Montréal, Canada

[17] Rube C, Rossbach T, Wedeking M, Grates D R, Jeschke P 2016 Experimental and Numerical Investigation of the Flow Inside the Return Channel of a Centrifugal Process Compressor. Journal of Turbomachinery Vol. 138 / 101006-1; DOI: 10.1115/1.4032905

[18] Yagi M, Nishioka T, Kobayashi H, Nishida H, Yamamoto S 2015 Effects of return channel with splitter vanes on performance of multistage centrifugal compressor Turbine Technical Conference and Exposition GT2015. – Montreal, Canada

[19] Nishida Y, Kobayashi H, Nishida H, Sugimura K 2013 Performance improvement of a return channel in a multistage centrifugal compressor using multiobjective optimization. Journal of Turbomachinery, Vol. 135, 031026 1-8, 201313.

[20] Marenina L, Galerkin Yu, Drozdov A 2020 Stator elements optimization of centrifugal compressor intermediate type stage by CFD methods The 2nd International Conference on High Speed Turbomachines and Electrical Drives E3S Web of Conferences 178, 01013 https://doi.org/10.1051/e3sconf/202017801020

[21] Galerkin Yu, Rekstin A, Marenina L, Drozdov A, Solovyeva O, Semenovskiy V 2020 Optimization of return channels of high flow rate centrifugal compressor stages using CFD methods Energies, 13(22), 5968; https://doi.org/10.3390/en13225968

[22] Galerkin Y, Danilov K, Popova E 1995 Universal Modelling for Centrifugal Compressors-Gas Dynamic Design and Optimization Consepts and Application. Yokohama International Gas Turbine Congress, Yokohama.

[23] Galerkin Y 2010 Turbocompressors. Workflow, calculation and design of the flow path; LLC "Information and Publishing Center" KHT", Moscow, Russia, p. 596.

[24] Galerkin Yu, Rekstin A, Drozdov A, Soldatova K, Solovyeva O, Popova E 2020 The optimal gas dynamic design system of industrial centrifugal compressors based on Universal modeling method. The 2nd International Conference on High Speed Turbomachines and Electrical Drives

E3S Web of Conferences 178, 01013 (2020) https://doi.org/10.1051/e3sconf/202017801028
[25] Soldatova K 2017 Creation of a new mathematical model of the flow path of centrifugal compressors and a database of model stages Dr. Tech. Sciences thesis. SPbPU. P. 357. Russian.

2D transient analysis of suction process for coupled vane compressor

W C Poh, B Elhadidi, K T Ooi

Nanyang Technological University, Singapore

waichang001@e.ntu.edu.sg

Abstract. Although CFD has been widely applied to the study of positive displacement compressors, the primary challenge remains in the generation of solution domain of complicated geometry and accurate handling of the deformation of the working chamber during the compressor operation. In this paper, a 2D transient CFD simulation employing the overset mesh method is performed on the Coupled Vane Compressor (CVC) with air as the working fluid. Initial results revealed the presences of flow reversal and circulation zones at the suction channel resulting in the reduction of effective suction flow area. This has significantly reduced the mass intake. Proposed modification to the suction channel design has resulted in an increase of mass flow rate of 6.97% when operating at 1500rpm. Simulation approach, results and design modification are presented and discussed in this paper.

1. Introduction

The advances in computer technologies had boosted the use of CFD in analyzing many engineering problems, including the performance and design improvement studies of compressors [1, 2, 3]. However, the primary challenges of implementing CFD studies are the difficulties facing the ability to cope with the generation of complicated geometry and accurate handling of the deformation of the working chamber during the compressor operation. Various existing modelling approaches such as the dynamic mesh with remeshing method was presented for rolling piston compressor [4], innovative solution domain generations are also available in the literature for screw compressor [5] and sliding vane compressor [6]. Nevertheless, these methods could only cater to the relatively uniform deformations of the working chamber geometry of widely known compressors. Although CFD was shown to be employed successfully in the modelling of compressors, a few numbers were focused on the flow at the discharge port [7] and the discharge valve [8] of rolling piston compressors.

In this paper, the transient 2D simulation was employed to model the working processes of the Coupled Vane Compressor (CVC) [9] using the overset mesh methodology to demonstrate its capability in obtaining the solution domain without the need to remesh or generate mesh files at every angular position. Further overset mesh guidelines are available in the literature [10] and its successful application in modelling a single screw expander [11]. Flow analyses were then conducted based on the results obtained at the suction chamber to improve its design and to enhance its performance.

IOP Conf. Series: Materials Science and Engineering 1180 (2021) 012027 doi:10.1088/1757-899X/1180/1/012027

2. Simulation setups

2.1. Solution domain

Figure 1(a) illustrates the 3D schematic of CVC showing the suction, discharge and working chambers. The midplane representation of CVC as a 2D schematic is depicted in **Figure 1**(b) with the geometry of the coupled vane shown in **Figure 1**(c). The main dimensions of CVC are listed in **Table 1** with the designed ideal volumetric displacement of 160 cm^3/rev.

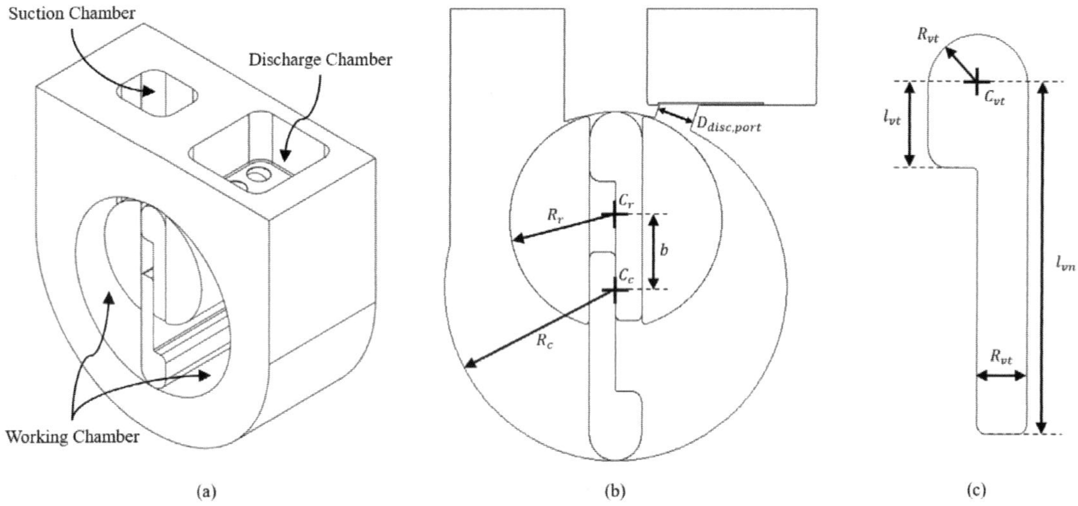

Figure 1. (a) Isometric view of CVC, (b) 2D solution domain of CVC, (c) Coupled vane.

Table 1. Main dimensions of CVC.

Nomenclature	Description	Dimension
C_r	Centre axis of rotor	–
C_c	Centre axis of cylinder	–
C_{vt}	Centre axis of vane tip	–
R_r	Radius of rotor	20.25 mm
R_c	Internal radius of cylinder	32.50 mm
$D_{disc,port}$	Diameter of discharge port	7.00 mm
b	Distance between C_r and C_c	12.50 mm
R_{vt}	Radius of vane tip	5.00 mm
l_{vn}	Length of vane from C_{vt}	34.00 mm
l_{vt}	Length of vane tip from C_{vt}	8.00 mm

In the overset mesh approach, the component meshes are generated individually, as illustrated in **Figure 2**(a) while **Figure 2**(b) shows the assembly of all the components to form the complete solution domain with the arbitrary constant clearances of 50 μm between components. Different values of clearances (100 and 50 μm) were tested and the clearances can be reduced further with the overset method as needed. At least four elements across the clearances between each component are required such that the overset interface can be applied. Mesh regeneration is not required since each component mesh is displaced relative to the center of rotation based on the user-defined functions for the motion of the components while preserving the mesh quality and the specified clearance value. Thus, design modifications can be made easily by changing either one of the component mesh.

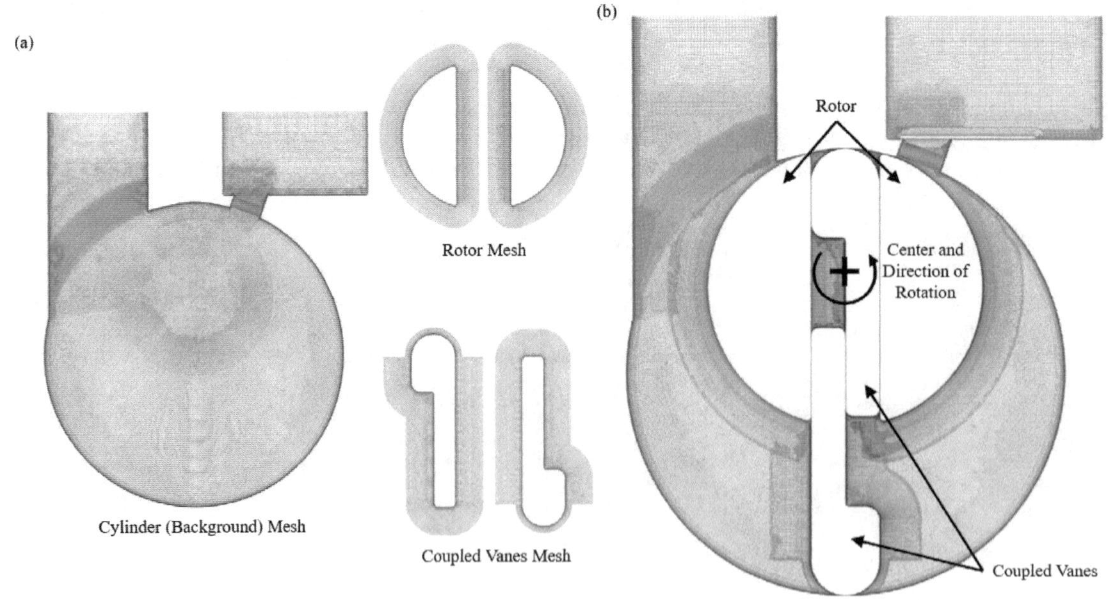

Figure 2. (a) Solution domain for individual components, (b) An assembled 2D solution domain.

2.2. Simulation parameters

ANSYS Fluent 2019 R3 computational code was used to perform analyses with air as the working fluid, modelled as an ideal gas under adiabatic consideration. The $k - \omega$ SST turbulence model was used due to the presence of adverse pressure gradient at the clearances and better prediction for the near-wall flow conditions. Three operating speeds, i.e. 1500 rpm, 3000 rpm and 4800 rpm were simulated with the boundary conditions listed in **Table 2**. Simulations were carried out at angular steps of 0.009° with up to 30 iterations per time step to ensure numerical convergence.

Table 2. Boundary conditions.

	Suction Chamber	Discharge Chamber
Type	Pressure Inlet	Pressure Outlet
Absolute Pressure	101,325 Pa	506,625 Pa
Total Temperature	300 K	475 K

3. Results and discussions

3.1. Basic simulation results

Mesh independence test shows the relative difference in mass flow rate per unit length from Mesh 3 is 1.92% as compared to Mesh 2, as shown in **Table 3**. Hence, Mesh 3 is used for further simulations.

Table 3. Mesh independence test.

Mesh	No. of Elements	Mass Flow Rate per Unit Length [kg·s^{-1}/m]	Relative Difference
1	345,329	0.038478	–
2	481,034	0.040122	4.27%
3	571,952	0.040893	1.92%

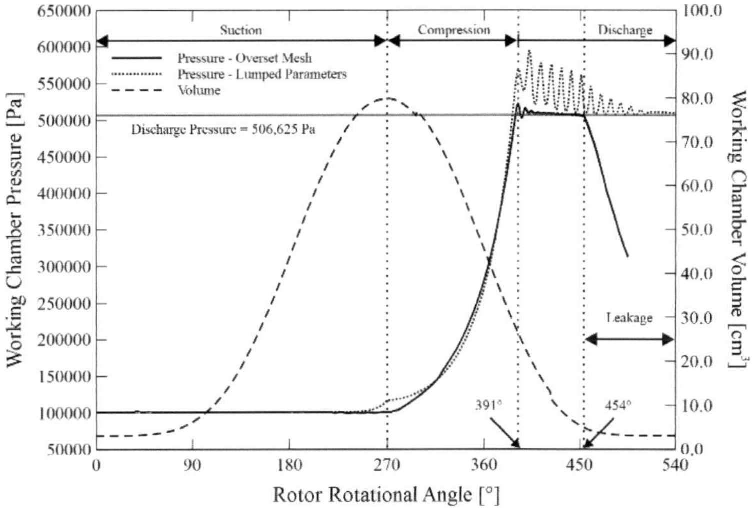

Figure 3. Variation of working chamber pressure from Shakya [9] and CFD at 1500 rpm.

Comparison of the working chamber pressure from CFD and that of the lumped parameters model from Shakya [9] is illustrated in **Figure 3** along with the variation of working chamber volume. The results from both methods are comparable in predicting the working chamber pressure. Meanwhile, the deviation at the discharge process is due to the non-uniform reed valve model used in the lumped approach [9] as compared to the 1D torsional spring model used in this study. Furthermore, the effect of internal leakage that caused the reversed flow during the discharge process is present in the CFD result whereas the lumped parameters approach assumed the working fluid is discharged completely in a single compression cycle.

Table 4 presents the mass flow rate per unit length (which only represents the 2D midplane of CVC) whereas the variations of the working chamber pressure at the three operating speeds show that the size of the suction port chosen is able to cater for these speeds. However, for the discharge process, higher overpressure was observed at higher operating speed, which may be caused by the inability of discharge port size and/or reed valve to respond to higher operating speeds, as shown in **Figure 4**. Furthermore, the internal leakage effect diminishes with the increased operating speed.

Table 4. Mass flow rate per unit length at different operating speeds.

Operating Speed [rpm]	Mass Flow Rate per Unit Length [kg·s^{-1}/m]
1500	0.040893
3000	0.103303
4800	0.217469

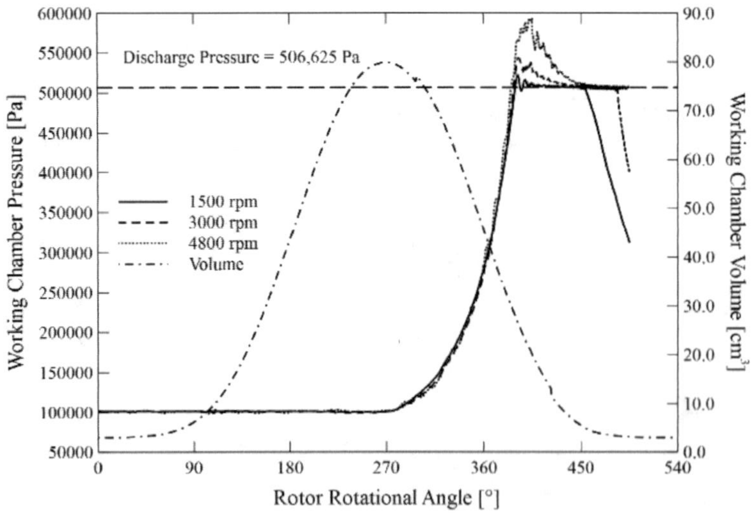

Figure 4. Variation of working chamber pressure at different operating speeds.

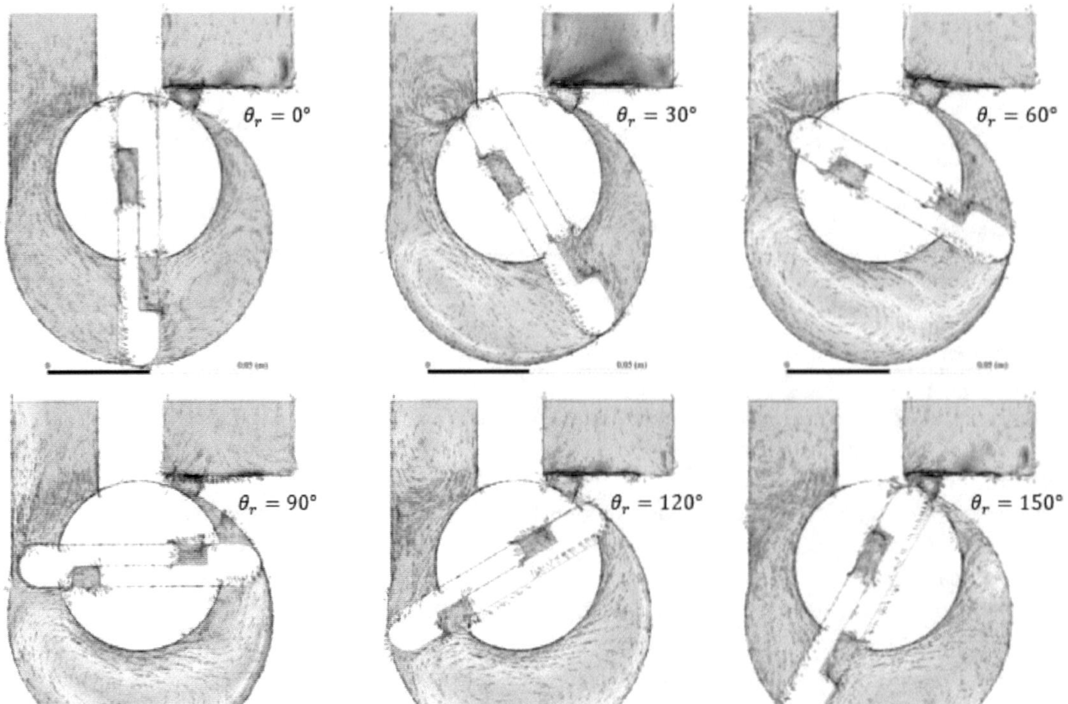

Figure 5. Typical velocity flow fields.

Figure 5 shows the typical velocity flow fields at every 30° rotor rotational angle for the operating speeds of 1500 rpm, 3000 rpm and 4800 rpm. There are circulations and backflows during the suction process up to 90° rotor rotational angle due to suction port design. This has resulted in the reduction in the effective suction flow area throughout the entire suction process.

3.2. Effect of suction valve

To improve the suction process by reducing the reversed flow, a non-return suction valve has been added and the results of the typical velocity flow fields are shown in **Figure 6**. Notable differences in the flow field can be observed from rotor rotational angle of 30° to 90° where the backflow is trapped in the suction channel instead of flowing out of it as seen previously in **Figure 5**.

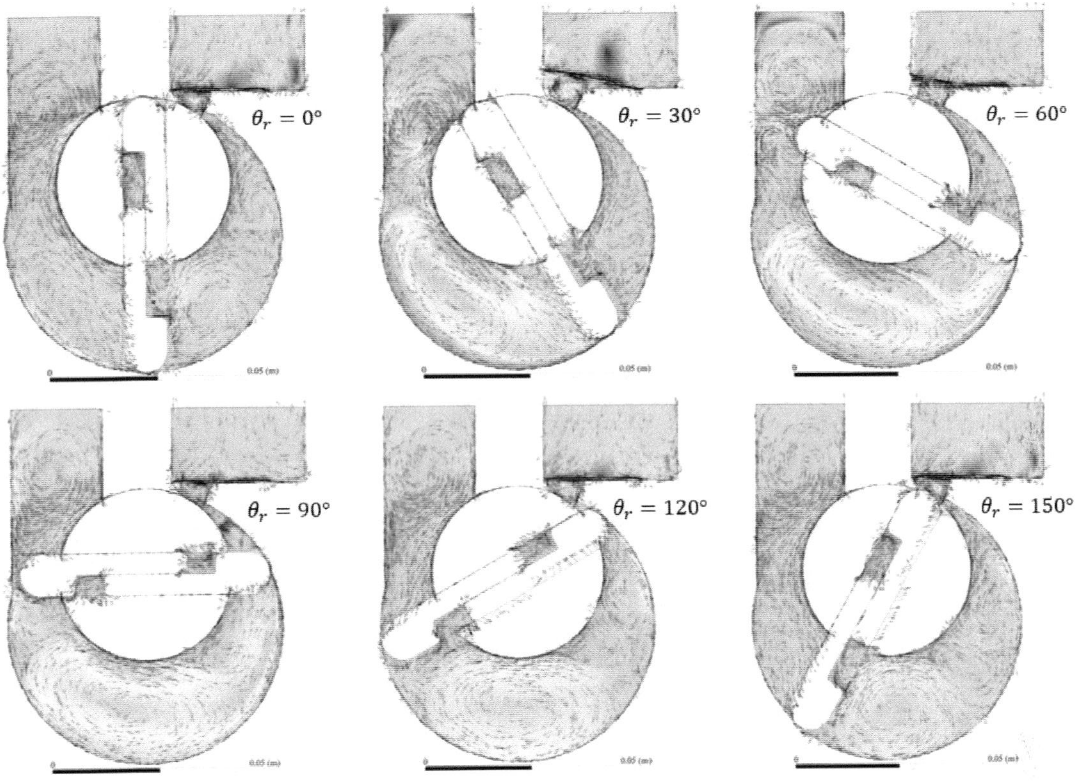

Figure 6. Typical velocity flow fields with suction valve.

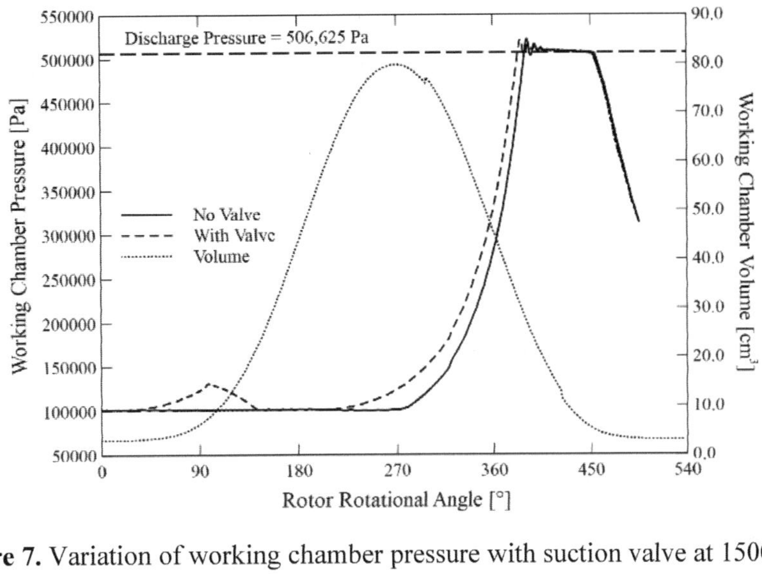

Figure 7. Variation of working chamber pressure with suction valve at 1500 rpm.

Figure 7 shows the working chamber pressure between the default and modified designs at 1500 rpm. The accumulated working fluid in the suction chamber causes the rise in working chamber pressure and peaked around 100° rotor rotational angle due to the additional mass flow caused by the inclusion of the one-way flow suction valve. The pressure then decreases due to the expansion of the working chamber volume during the suction process. Consequently, the compression process occurs earlier at 225° rotor rotational angle from the extra mass leftover from the leakages. Similar trends also occur at 3000 rpm and 4800 rpm. The mass flow rate per unit length (which only represents the 2D midplane of CVC) for the three operating speeds improved, but the increment diminishes with the rise in operating speed, as presented in **Table 5**. 3D investigations of the current improvement are being carried out to account for the 3D nature of the suction flow entering the working chamber and the discharge flows effects through the discharge ports into the discharge chamber.

Table 5. Mass flow rate improvements from suction valve.

Operating Speed [rpm]	Mass Flow Rate per Unit Length [kg·s^{-1}/m]		Improvement
	No Valve	With Valve	
1500	0.040893	0.043745	6.97%
3000	0.103303	0.108973	5.49%
4800	0.217469	0.222244	2.20%

4. Conclusions

The transient 2D simulation was performed on the Coupled Vane Compressor (CVC) utilizing the overset mesh method. Basic simulation result shows that it is comparable to that from the lumped parameters approach. Furthermore, the flow field shows the occurrences of circulation zones and undesirable reversed flow which has reduced the effective suction flow area and resulted in the reduction of mass intake. A design modification has been carried out by adding a non-return valve at the suction has effectively mitigated the problem. This has resulted in an improvement of 6.97% in the mass flow rate at 1500 rpm and 2.20% at 4800 rpm. This exercise has shown that design modifications can be made swiftly for performance improvement. A more comprehensive transient 3D modelling using the same overset approach is currently underway.

Acknowledgements

The authors would like to express their gratitude and thanks to Nanyang Technological University, Singapore for its continuous support towards the research.

References

[1] H. Ding and H. Gao, "3-D Transient CFD Model For A Rolling Piston Compressor With A Dynamic Reed Valve," in *International Compressor Engineering Conference*, 2014.

[2] S. R. Rane, A. Kovačević and M. Kethidi, "CFD Modelling in Screw Compressors With Complex Multi Rotor Configurations," in *International Compressor Engineering Conference*, 2012.

[3] G. Bianchi, S. Rane, A. Kovačević, R. Cipollone, S. Murgia and G. Contaldi, "Grid generation methodology and CFD simulations in sliding vane compressors and expanders," in *IOP Conference Series: Materials Science and Engineering*, 2017.

[4] B. Farkas, V. Szente and J. M. Suda, "Dynamic meshing strategies to model fluid flow in rolling piston compressors," in *The 16th International Conference on Fluid Flow Technologies*, Budapest, Hungary, 2015.

[5] S. Rane, A. Kovačević, N. Stošić and M. Kethidi, "Grid deformation strategies for CFD analysis of screw compressors," *International Journal of Refrigeration,* vol. 36, no. 7, pp. 1883-1893, 2013.

[6] G. Bianchi, S. Rane, A. Kovačević and R. Cipollone, "Deforming grid generation for numerical simulations of fluid dynamics in sliding vane rotary machines," *Advances in Engineering Software,* vol. 112, pp. 180-191, 2017.

[7] W. Geng, C. H. Liu and Y. Z. Wang, "The Performance Optimization of Rolling Piston Compressors based on CFD Simulation," in *International Compressor Engineering Conference*, 2004.

[8] Q. Tan, S.-l. Pan, Q.-k. Feng, X.-l. Yu and Z.-l. Wang, "Fluid-structure interaction model of dynamic behavior of the discharge valve in a rotary compressor," *Proceedings of the Institution of Mechanical Engineers, Part E: Journal of Process Mechanical Engineering,* vol. 229, no. 4, pp. 280-289, 2015.

[9] P. Shakya and K. T. Ooi, "Introduction to Coupled Vane compressor: Mathematical modelling with validation," *International Journal of Refrigeration,* vol. 117, pp. 23-32, 2020.

[10] W. M. Chan, R. J. Gomez, S. E. Rogers and P. G. Buning, "Best practices in overset grid generation," in *32nd AIAA Fluid Dynamics Conference and Exhibit*, 2002.

[11] N. Casari, E. Fadiga, M. Pinelli, S. Randi, A. Suman and D. Ziviani, "Investigation of flow characteristics in a single screw expander: A numerical approach," *Energy,* vol. 213, p. 118730, 2020.

Geometric Optimisation of a Coupled Vane Compressor

Q J Chen [1]**, K T Ooi**[2]

[1,2] School of Mechanical and Aerospace Engineering, Nanyang Technological University, Singapore, Singapore

[1]Chen1179@e.ntu.ed.sg, [2]Mktooi@ntu.edu.sg

Abstract. The coupled vane compressor (CVC) is a rotary compressor with two coupled vanes which slide diametrically through the rotor during operation. As compared to most other rotary vane compressors, this CVC works well with small rotor-to-stator diameter ratio, typically less than that for all existing rotary compressors (>0.7). Hence, CVC has probably the most compact design among all rotary compressors, thus saving materials and manufacturing cost. This compressor is intended for use in room air-conditioners and refrigerators. Currently, mathematical model and measurement have been carried out to determine the working characteristics of this compressor. In this paper, a genetic algorithm with constraints handling has been devised and linked with the verified mathematical model of the CVC to search for a set of geometrical parameters which produces highest COP when working in the refrigeration cycle. During this optimisation search process, the significant design variables such as rotor radius, cylinder radius, vane thickness and valve reed thickness are allowed to vary.

In this paper, the details on the application of the optimization process, constraints handling and descriptions, objective function definitions as well as the details of optimization implementation will be presented and discussed. Results obtained will be shown and deliberated. The results show that this optimization study reconfirms the value of optimisation in compressor design.

1. Introduction

Rotary vane compressor can be found widely in their use as part of the cooling cycles in air-conditioners and refrigerators. They have a significant amount of space in their cylinders taken up by the rotor. This reduces the working volume per cylinder size for these cylinders. The CVC[1] is able to counter this with its small rotor-to-stator diameter ratio, allowing for it to be more compact. Due to this fact, it has the potential to consume less materials for its manufacturing than other rotary vane compressors.

This research aims to build upon this by optimizing the CVC. The simulator for the CVC produces a unique set of values of efficiencies for each set of geometric parameters input. A multi-objective optimization has been written whereby the efficiency values, COP, and minimized volume are the objectives, while the geometric parameters are the variables. This program will aim to reduce the amount of raw material used in manufacturing the CVC while also maximizing the performance of the CVC. Pareto optimality can be obtained for this with the plot for the inverse of volume versus the other objectives, that is the scaled operating parameters such that each is appropriately represented.

2. Coupled Vane Compressor

The coupled vane compressor is a novel compressor which improves on current designs by reducing the amount of material required for its manufacturing while ensuring that performance does not suffer as drastically if such a reduction of geometric parameters were made on other compressors. This is done by reducing the need for the rotors of the compressor to be a minimum ratio when compared to the cylinder.

2.1. Design

The compressor is designed with a twin-vane system running through the rotors diametrically, as shown in Figure 1. They slide along each other and allows for the overall vane to vary in length. To ensure that the rotors maintain contact with the cylinder walls and that there is no leakage from the compression chambers, the centrifugal force from the rotation of the vane will balance the various pressure forces acting on the vane tip and other surfaces.

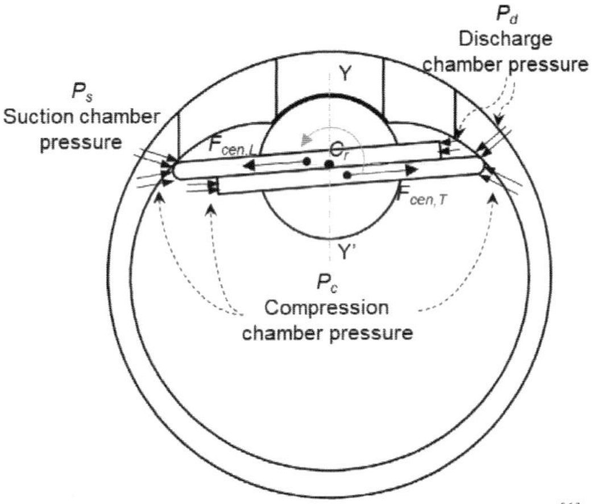

Figure 1: Coupled Vane Compressor Schematics[1]

2.2. Mathematical Model

The mathematical model is the design of the simulation program for the CVC, as created by Ooi and Shakya[1]. It is arrived at through the derivations of the geometry of the working chamber, the thermodynamic properties of the working fluid, and analysis of the primary flows through the suction and discharge ports and of the secondary leakages through internal clearances. To allow for the program to run iterations at a faster rate, the heat transfer model for this simulation has been excluded.

3. Method

To carry out the multi-objective optimization as required by this aim of this paper, genetic algorithm, a form of evolutionary algorithm, has been chosen as it is able to handle the requirements of outputting and analyzing multiple objectives, as well as its ability to handle multiple inputs.

3.1. Genetic Algorithm

Genetic algorithm is a form of optimization in computer science which brings inspiration from the process of natural selection and Charles Darwin's theory of evolution. It replicates the evolutionary processes of nature through operations such as selection, crossover, and mutation [2]. It is able to obtain, through multiple iterations and generations, better results from a certain set of variables through the above-mentioned processes. The genetic algorithm chosen here is the NSGA-II [3].

Modifications were made to the base genetic algorithm code provided from the GreenPlan-IT planning tool [4], which allowed for the program to be used for the CVC simulation program. These modifications include the adjusting of the variable constraints to fit the new program and selecting the appropriate objectives to represent the fitness for the selection of 'parents' for the next generation of iterations.

The process of the NSGA-II code begins with the random creation of an initial population within the constraints as mentioned above. The variables chosen in this optimization process are the cylinder radius, rotor radius, cylinder height, vane length, vane thickness, and discharge valve reed thickness. This initial population will then be fed into the simulation program. The cylinder volume, COP, and overall efficiency are the outputs. These values form the two objectives which we will be maximizing and are used for ranking of the solution through non-dominated and crowding distance sorting. The two objectives chosen are the inverse of cylinder volume, and the product of the COP and the overall efficiency. The inverse of cylinder volume is chosen as the main goal of this optimization program is to reduce the size of the novel compressor and the amount of raw material required for its manufacturing. The product of COP and overall efficiency ensures that the overall performance is also pushed to be improved despite a decreasing compressor size.

$$maximise\ Obj_i = f(x_{raw}, x_{rcr}, x_{slr}, x_{vtr}, x_{rr}), \quad i = 1,2 \tag{1}$$

$$where\ Obj_1 = \frac{1}{Cylinder\ Volume} \tag{2}$$

$$and\ Obj_2 = COP \times overall\_efficiency \tag{3}$$

Through ranking of the objectives from the first generation of results [5], a 'parent' population is created which 'favors' the members which have a higher value. Selection for a reduced mating pool is done based on a tournament selection of randomly selected members using their ranking.

From the reduced pool, a next generation is created through crossover and mutation of the binary-coded chromosomes. This forms the 'child' population which is the same size as the initial 'parent' population.

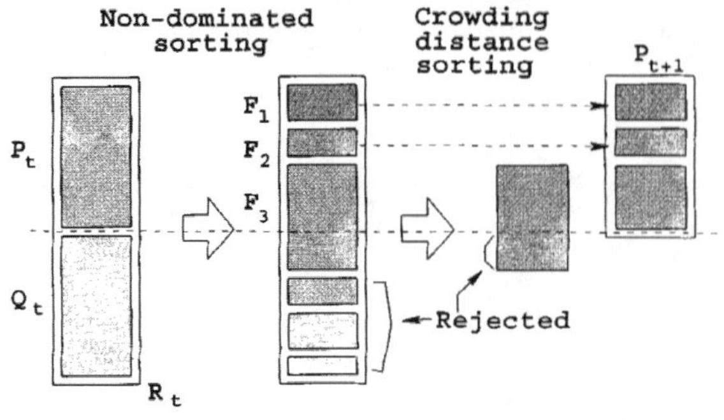

Figure 2: Halving of Global Population[3]

The 'child' population is then combined with the 'parent' population to form a global population. The global population is ranked, after which a halving process occurs, as shown in Figure 2. A new 'parent' population is created.

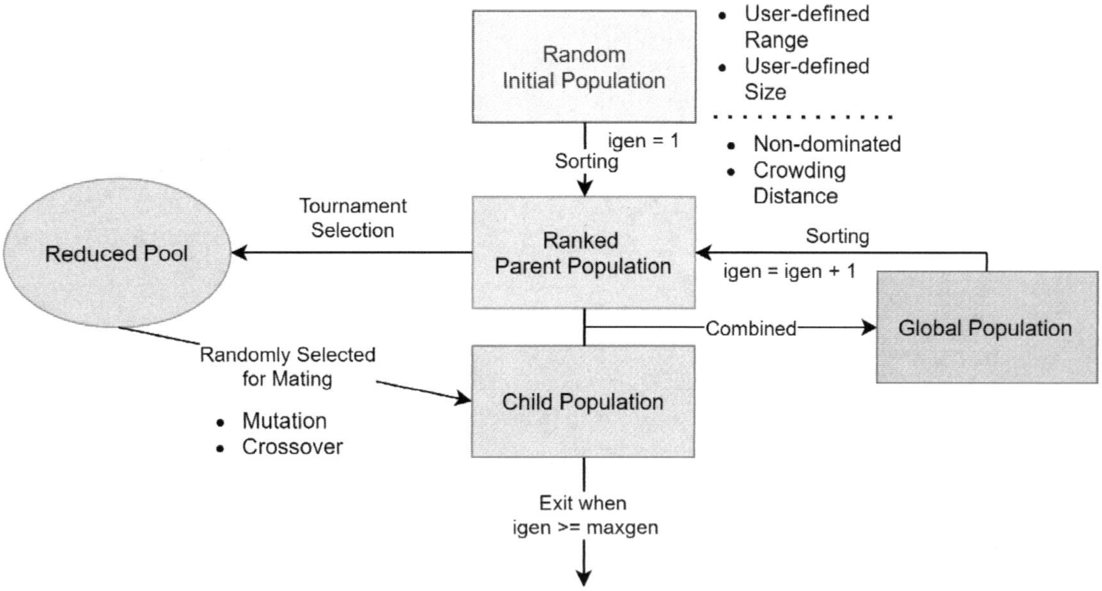

Figure 3: NSGA-II Procedure

This whole process is repeated over multiple iterations, as shown in Figure 3, until the maximum number of generations is reached or if the user chooses to end the program. Results for the variables and their objective values are recorded.

The constraints implemented for the iterations to be discussed is as follows:

$x_{ll} \le x_{var} \le x_{ul}$ represents the general constraint equation where x_{ll} and x_{ul} are the lower and upper limits of the variable respectively, and x_{var} is the value of the parameter to be evaluated.

$$0.0150mm \le x_{raw} \le 0.0350mm \quad (4)$$

$$0.55 \le x_{rcr} \le 0.80 \quad (5)$$

$$0.50 \le x_{slr} \le 2.00 \quad (6)$$

$$0.20 \le x_{vtr} \le 0.30 \quad (7)$$

$$0.0030 \le x_{rr} \le 0.0050 \quad (8)$$

Where raw is the cylinder radius, rcr is the rotor to cylinder radius ratio, slr is the slenderness ratio, vtr is the ratio of vane thickness to rotor diameter, and rr is the ratio of discharge valve reed thickness to cylinder thickness. Conversion from these decision variables to the raw geometrical parameters of the CVC are shown in Figure 4. The reason for the storage of variables in mostly ratio is to prevent a situation in which the output optimized compressor has an impossible geometry. One example of an impossible geometry created through mating would be a solution in which the rotor radius is larger than the cylinder radius. This can easily be solved by setting rcr < 1.

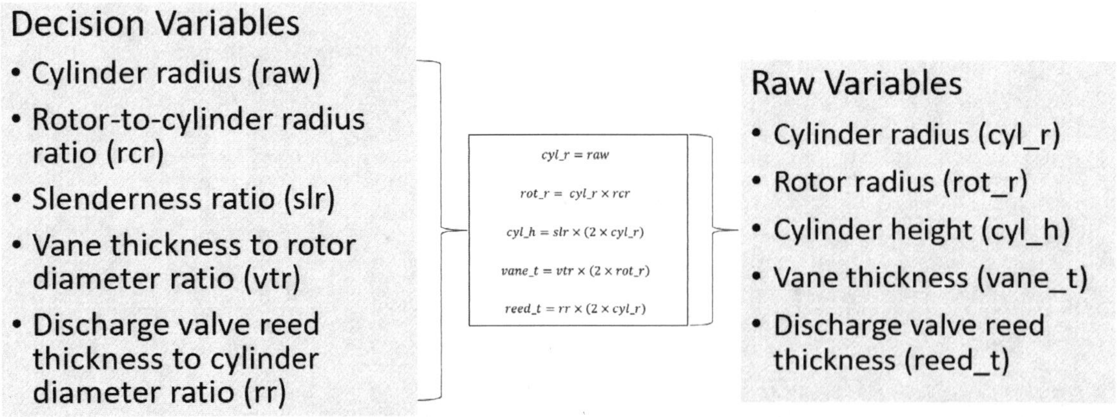

Figure 4: Conversion of Decision Variables

These values are then converted to the actual values of the rotor radius, cylinder radius, cylinder height, vane length, vane thickness, and reed thickness. Calculations had to be done to obtain the maximum and minimum allowable values of vane length based on the rotor and cylinder radius, and clearance gap length. Vane length is then taken to be the average of the two.

$$L_{min} \leq L_{vane} \leq L_{max} \tag{9}$$

$$L_{max} = \sqrt{r\_cyl^2 - \left(r\_cyl - rotr + gap + \frac{vane\_t \times \cos\beta}{2}\right)^2} - \frac{vane\_t \times \sin\beta}{2} \tag{10}$$

$$\beta = \cos^{-1}\left(\frac{r\_cyl - rotr + gap}{r\_cyl \times \frac{1 - vane\_t}{2 \times r\_cyl}}\right) \tag{11}$$

$$L_{min} = 1.01 \times (2 \times r\_cyl - 2 \times rotr + gap) \tag{12}$$

$$L_{vane} = \frac{L_{max} + L_{min}}{2} \tag{13}$$

There are 30 generations iterated, each with a population size of 20. The value for the probability of mutation is the inverse of the length of a chromosome, in this case 1/150 or 0.667%. Crossover probability is fixed at 90%. The variables for this genetic algorithm are binary coded into the chromosomes.

4. Results

Fitness of each generation are obtained and plotted as shown. Note that the opacity of the marker shows the density of data that falls on that point, while the size of the marker represents the generation number – the larger the marker, the greater the generation. Obj2 is the product of overall efficiency and COP. Inverse of Cylinder Volume represents the inverse of the size of the compressor.

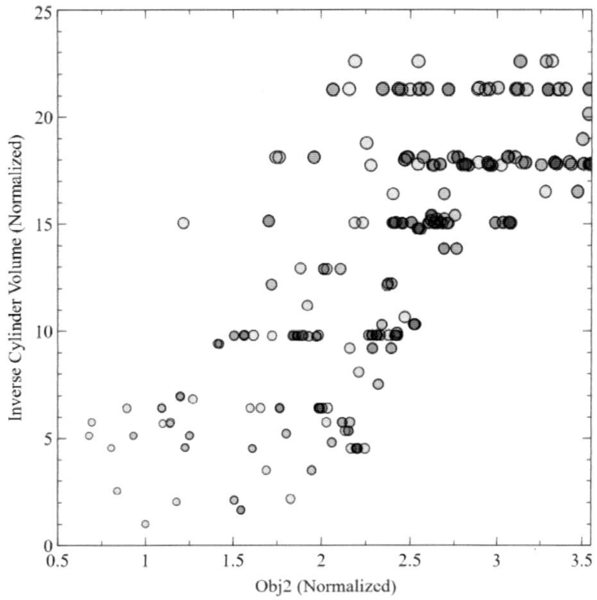

Figure 5: Scatterplot of Both Objectives

The graphs show the change in each objective by dividing its value with the one of the lowest overall fitness values obtained – in this case, a datapoint obtained in generation 1, population member 7 (Inverse Volume = 0.00337, Obj2 = 0.500).

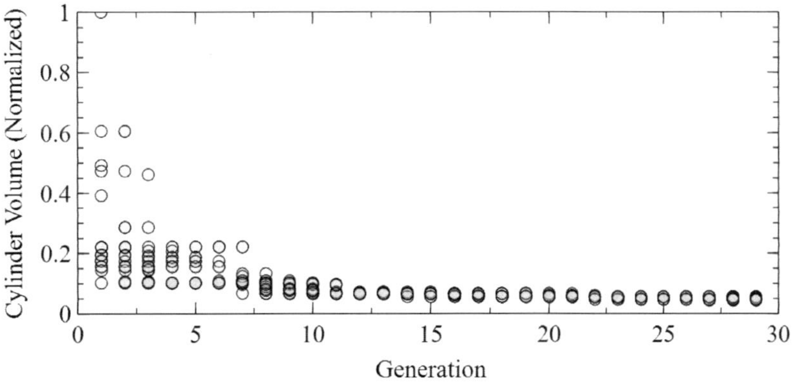

Figure 6: Cylinder Volume Change (Normalized) across Generations

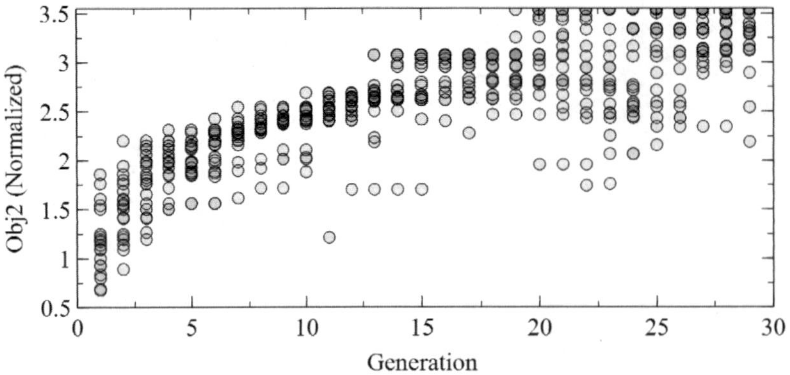

Figure 7: Obj2 Change (Normalized) across Generations

Objective values are plotted here against the generations to show their improvement through the iterations. Obj2 can be seen to be increasing while Cylinder Volume decreases, as is the purpose of the optimization.

30th Generation		Solution 4	Solution 20
Objectives	Inverse Cylinder Volume	0.071783	0.076182
	Obj2	1.769241	1.659964
Decision Variables	raw	0.01613	0.01582
	rcr	0.55156	0.66425
	slr	0.52836	0.52794
	vtr	0.23078	0.21320
	rr	0.00321	0.00321

Table 1: Final Objective Values and Geometrical Parameters

Table 1 shows the final optimized values obtained which belongs to solution 4 and 20 of the last generation. Each solution does not dominate the other for the objective values used. The final decision variables highlight that compressor design tends towards one with a generally smaller rcr. Solution 4 has a rcr value close to the minimum value of 0.55 and it has Obj2 being maximized. Solution 20 minimizes the volume of the cylinder with a lower Obj2 value. During this trade-off, the rcr value is observed to be higher at 0.664.

The general decrease in the rcr value could be due to the maximization of the maximum working chamber volume to cylinder size of the CVC. This maximization is shown to be beneficial to the defined performance of the compressor.

Another interesting parameter change to take note of during the optimization of the CVC is the tendencies for the other decision variables to approach each of their minimum set limits. These decreases brought about a better value for both the objectives identified.

It is important to note that results obtained from the optimization are from the mathematical simulation of the compressor. Physical testing should be carried out to confirm the accuracy of the calculations carried out.

5. Discussion

The results show that there is a general increase in the fitness of the algorithm, with the compressor size decreasing and Obj2 increasing. As seen from Figure 5, the front for each generation can be seen to be tending towards the upper right corner where the fitness can be maximized. The fitness for the inverse of volume can be seen to have increased in value by about 23 times when compared to the weakest member (gen 1, member 7) across all 30 generations. This means that volume decreased by around 95% as can be seen from Figure 6. Figure 7 shows that Obj2 increased by around 3.5 times when compared to the same weakest member.

Both these results are significant in showing that NSGA-II optimization, through the varying of geometrical parameters, provides great value towards the creation of better compressor designs.

A general decrease is observed for the final geometrical parameters of the optimized CVC. This decrease is less distinct in the rcr as the final value tends to fluctuate between around 0.66 to 0.55. This could be due to the trade-off between the two objectives identified where a lower value of rcr favors a greater value for Obj2 and a lower value of rcr favors a greater value for the inverse of cylinder volume.

These results show that an increase in the working chamber volume of the compressor with respect to its cylinder size generally improves the performance of the CVC. This study shows that an optimized geometry for the CVC should be one where the rcr is relatively smaller and the working chamber to cylinder size is larger. Reduction in vane thickness, which increases working chamber volume without a change in cylinder size, is also seen to contribute towards a better performance.

6. Conclusion

The decrease of the size of the CVC without a decrease in its performance, could mean the reduction in the amount of raw materials and cost required for manufacturing. This can be achieved with a design optimization software. NSGA-II has proven to be able to achieve this and reconfirms the value of such multi-objective optimization in compressor design.

In conclusion, the geometry of an optimized CVC tends towards one with a smaller rotor-to-cylinder ratio and larger working chamber to cylinder size ratio. This might be because a larger working chamber volume allows the compressor to work at a higher efficiency.

The user-defined optimization parameters such as mutation and crossover probability can be tested at other values to evaluate how they affect diversity and convergence of the results. This will allow fitter and more diverse Pareto fronts to be obtained. As optimization is carried out through computer simulations from the mathematical model of the CVC, further work can be done through the manufacturing and physical testing of a prototype to check the accuracy of the simulation program used for the optimization.

Acknowledgements

The author would like to express his gratitude to Dr. Pradeep Shakya for the creation and provision of the mathematical model and simulation program of the CVC. In addition, he would like to thank the academic staff at NTU, and his friends and family for their guidance, patience, and unwavering faith in him.

References

[1] Ooi K.T. and Shakya P. 2018 *The 24th International Compressor Engineering Conference, Purdue.* A New Compact Rotary Compressor: Coupled Vane compressor.
[2] Seyedali M. 2018 *Evolutionary Algorithms and Neural Networks* pp 43-55
[3] Deb K., Pratap A., Agarwal S., Meyarivan T. 2002 *A fast and elitist mulitobjective genetic algorithm: NSGA-II*
[4] San Francisco Estuary Institute 2020 *GreenPlan-IT Optimization Tool User Manual*
[5] Zhang X., Tian Y., Cheng R., and Jin Y. 2012 *An Efficienct Approach to Non-dominated Sorting for Evolutionary Multi-objective Optimisation*

IOP Conf. Series: Materials Science and Engineering 1180 (2021) 012029 doi:10.1088/1757-899X/1180/1/012029

Dynamic modelling and experimental validation of Coupled Vane Compressor

P Shakya[1], K T Ooi[2]

[1, 2] School of Mechanical and Aerospace Engineering, Nanyang Technological University, 50 Nanyang Avenue, Singapore 639798

[1] pradeep005@e.ntu.edu.sg
[2] mktooi@ntu.edu.sg

Abstract. The newly developed Coupled Vane Compressor (CVC) consists of a unique feature that is a couple of vanes cut through the rotor diametrically. Theoretically, this allows the rotor of any diameter which can accommodate the coupled vanes, to be able to operate CVC. Thus, unlike other rotary compressors, CVC design is mostly free of geometrical constraints imposed on the size of its rotor, making it one of the most compact compressors. Nevertheless, CVC suffers from frictional losses similar to other rotary compressors and therefore, studies of dynamic forces acting in CVC is of great interest. In this paper, dynamic forces acting on its components and frictional losses occurring due to rubbing of various components were studied and analysed. Unlike in other vane compressors, in which frictional losses are mostly due to the rubbing of vane tips against cylinder wall and vane sides against vane slot in the rotor, in CVC, the loss also occurs due to rubbing between the coupled vanes. A CVC prototype of 44 cm^3 was developed and the compressor power consumption was experimentally studied in an open-type test circuit using air as an operating fluid. The predicted and measured power consumed by CVC were compared and the predicted results were found to have maximum discrepancy of ±15% with the measured results.

1. Introduction

A new compact rotary vane compressor, namely, Coupled Vane compressor (CVC) was introduced by Ooi and Shakya [1]. As seen from figure 1, a set of two vanes cut diametrically through the rotor. The rotor has rotational motion about its centre, C_r. During the operation, vanes extend radially outward due to centrifugal force and gas pressure forces causing the vane tips to rub against the cylinder wall. The vanes partition the space inside compressor cylinder into various working chambers. The detailed mathematical models describing the working chamber, thermodynamics and mass flow have been studied by Ooi and Shakya [2, 3].

The accurate predictions of the motion and forces acting on various components of a compressor are essential to study its frictional losses and to optimize its design. In various studies by Yanagisawa et al. [4], Ooi and Wong [5] and Ooi [6], rotary vane compressors such as rolling piston compressors are reported to have high frictional losses at the vane tip due to its rubbing against the roller, vane sides with the vane slot and between the roller and eccentric. Ooi [6] reported that frictional losses due to the rubbing of roller against eccentric and against the endface accounted for more than 60% of the total frictional loss. In a study of frictional losses in sliding vane machine by Badr et al. [7] and Aradau and

Costiuc [8], it was found that the highest frictional losses were contributed by vane tip rubbing against cylinder wall.

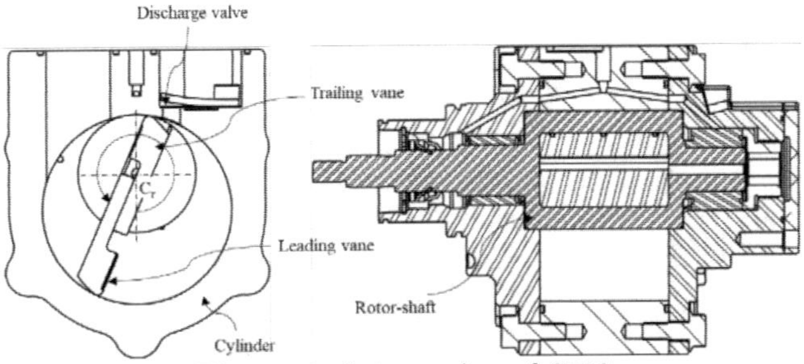

Figure 1. Schematics of CVC

To mitigate high frictional losses in rotary vane, revolving vane compressor was developed by Teh and Ooi [9]. The frictional losses were reduced by using a rotating cylinder, as opposed to a fixed cylinder. An embodiment of revolving vane compressor was predicted to have the mechanical efficiency of up to 95%. In a swing vane compressor studied by Hu et al. [10], nearly 30% reduction in frictional losses at 3000 r min^{-1} compared to the same in a sliding vane compressor was predicted. Their design consisted of vane tip connected to cylinder in a vane slot via a hinge joint. Ma et al. [11] conducted a structural optimization of swing vane compressor using complex method and reported an improvement in energy efficiency ratio by 8.5% compared to their base design. However, only three design variables including cylinder radius, rotor radius and the compressor height were selected, and their optimal values tended towards the upper limit of the scope of design variables. By adding a second swing vane, Xu et al. [12] proposed a double swing vane compressor. Their analysis showed lower frictional losses and smoother load fluctuations compared to the same of a swing vane compressor.

In a rotating spool compressor, Bradshaw et al. [13] studied the frictional losses due to spool seal, rubbing of vane tip seals against the cylinder, endface against the cover plate, shearing of fluid at top dead centre region and between vane side and rotor. In their investigation, the design of spool compressor was optimized with reportedly 15% increase in the efficiency. In a study of U-vane compressor by Lim and Ooi [14], the rubbing of vane side with cylinder wall contributed to over 35% of the total frictional loss.

This paper focuses on the study of dynamics of CVC mechanism and analysis of frictional losses due to rubbing of vane tips against cylinder wall, vane sides against vane slot in the rotor and between the vanes.

2. Mathematical models

In this section, the mathematical formulations for the kinematics and dynamics of vanes and rotor are shown in considerable details.

2.1. Kinematics model

Referring to figure 1, the position vectors of the trailing vane tip and leading vane tip with respect to C_r and their derivative with respect to the rotor angle θ_r are derived. Their magnitudes are shown in equations (1) - (4).

$$r(\theta_r) = \begin{cases} R_r & (\theta_{r,st} \geq \theta_r > 360° - \theta_{r,st}) \\ -b\cos\theta_r + \sqrt{-(b\sin\theta_r)^2 + R_c^2} \end{cases} \tag{1}$$

$$\frac{dr(\theta_r)}{d\theta_r} = \begin{cases} 0 \quad (\theta_{r,st} \geq \theta_r > 360° - \theta_{r,st}) \\ b\sin\theta_r + 0.5(-b^2\sin 2\theta_r)(-(b\sin\theta_r)^2 + R_c^2)^{-0.5} \end{cases} \tag{2}$$

$$r(180° + \theta_r) = \begin{cases} R_r & (180° - \theta_{r,st} \leq \theta_r \leq 180° + \theta_{r,st}) \\ -b\cos(180° + \theta_r) + \sqrt{-(b\sin(180° + \theta_r))^2 + R_c^2} \end{cases} \tag{3}$$

$$\frac{dr(180° + \theta_r)}{d\theta_r} = \begin{cases} 0 & (180° - \theta_{r,st} \leq \theta_r \leq 180° + \theta_{r,st}) \\ b\sin(180° + \theta_r) + 0.5(-b^2\sin 2(180° + \theta_r))(-(b\sin(180° + \theta_r))^2 + R_c^2)^{-0.5} \end{cases} \tag{4}$$

2.2. Vane Dynamics model

The forces acting on a vane can be analysed using a free body diagram. The various forces acting on the free body of the trailing and leading vane at any arbitrary rotor angular position θ_r are shown in figures 2 and 3.

From figures 2 and 3, it is noted that, for CVC to compress the gas without internal leakage through the vane tip, it is required that the vane tip be in sealing contact with the inner cylinder wall. To ensure the vane tips are in sealing contact throughout the operation of the compressor, firstly, the vane tip pressure forces $\boldsymbol{F}_{tp,p1}$, $\boldsymbol{F}_{tp,p2}$, $\boldsymbol{F}_{tp,p3}$ and $\boldsymbol{F}_{tp,p4}$ are required to be smaller than the vane neck pressure forces $\boldsymbol{F}_{H,p2}$ and $\boldsymbol{F}_{H,p4}$. At the trailing vane tip, trailing side pressure p_1 is always less than or equal to leading side pressure p_2. This implies that the curved tip area facing the leading side should be made smaller than the curved tip area facing the trailing side. Similarly, at the leading vane tip, trailing side pressure p_3 is always less than leading side pressure p_4. Again, this implies that to minimize the vane tip pressure forces, leading tip curved surface area exposed to p_4 should be smaller than the trailing side curved surface area. Finally, the vane neck pressure forces $\boldsymbol{F}_{H,p2}$ and $\boldsymbol{F}_{H,p4}$ should be allowed to be as large as appropriate to ensure that the vane tip remains in contact with the cylinder wall.

The rotation of trailing vane and leading vane about the rotor centre C_r causes two additional rotational forces to exist for each vane. The first one is centrifugal force ($\boldsymbol{F}_{cen}$). And the second one is coriolis force ($\boldsymbol{F}_{cor}$) which is due to both rotation and translation of the vanes with respect to C_r. $\boldsymbol{F}_{cen}$ assists the vane to radially extend out of the rotor slot and towards the inner cylinder wall. Therefore, it is important that vane centre of mass is designed such that the direction of centrifugal force with respect to C_r is always pointed towards the vane tip to avoid vane retracting from the vane-tip contact.

The pressure forces are computed using chamber pressures from the thermodynamics model and respective force areas. The rotational forces are determined using the vane mass, its centre of mass and operating speed. The 'unknown' forces are the contact forces between the vane and other components. They are evaluated by assuming static and dynamic equilibrium at any moment of time.

The rotor provides rotating force, $\boldsymbol{F}_{N,rot}$, which is the main driving force for the compressor. The vane rotation is assumed to be proportional to the frictional forces, $\boldsymbol{F}_{f,rot}$ and $\boldsymbol{F}_{f,vn}$. Equation (5) is derived using the force equilibrium on a free body diagram of trailing vane (as shown in figure 2) along $\mathbf{r}(\theta_r)$. Equation (5) represents the force balance assuming the forces pointing towards the rotor centre to be positive.

$$\left(F_{\parallel,1} + F_{f,vn} + F_{f,rot}\right) = F_{r,1} + F_{H,p2} + F_{H,p4} - F_{tp,p1}\cos\left(\frac{\alpha_1}{2}\right) - F_{tp,p2}\cos\left(\frac{\alpha_1}{2}\right) \quad (5)$$

Equation (6) represents the force balance which are orthogonal to $\mathbf{r}(\theta_r)$.

$$\left(F_{\perp,1} + F_{N,vn} - F_{N,rot}\right) = F_{p1} - F_{p2} + F_{tp,p1}\sin\left(\frac{\alpha_1}{2}\right) + F_{tp,p2}\sin\left(\frac{\alpha_1}{2}\right) + F_{cor,1} \quad (6)$$

Equation (7) can be derived by considering the sum of moments at the centre of mass of trailing vane assuming the anticlockwise rotation of its free body as a positive.

$$\begin{aligned}\left(-F_{\perp,1}\{r(\theta_r) - r_{CM,1}\} + F_{N,vn}\{l_{vn} - r(\theta_r) + r_{CM,1}\} + F_{N,rot}\{R_r - r_{CM,1}\}\right) \\ = \left(F_{p2} - F_{p1}\right)\left[R_r + \left\{\frac{r(\theta_r) - R_r}{2}\right\} - r_{CM,1}\right] \\ - \left\{F_{tp,P2}\sin\left(\frac{\alpha_1}{2}\right) + F_{tp,P1}\sin\left(\frac{\alpha_1}{2}\right)\right\}\left(r(\theta_r) - R_r - r_{CM,1}\right)\end{aligned} \quad (7)$$

Similarly, equations (8), (9) and (10) are the force balance equations along $\mathbf{r}(180°+\theta_r)$, orthogonal to $\mathbf{r}(180°+\theta_r)$ and sum of moments at the centre of mass of leading vane respectively.

$$\left(F_{\parallel,2} - F_{f,vn} - F_{f,rot}\right) = F_{r,2} + F_{H,P2} + F_{H,P4} - F_{tp,p3}\cos\left(\frac{\alpha_2}{2}\right) - F_{tp,p4}\cos\left(\frac{\alpha_2}{2}\right) \quad (8)$$

$$\left(F_{\perp,2} + F_{N,vn} - F_{N,rot}\right) = F_{p3} - F_{p4} - F_{tp,p3}\sin\left(\frac{\alpha_2}{2}\right) - F_{tp,p4}\sin\left(\frac{\alpha_2}{2}\right) - F_{cor,2} \quad (9)$$

$$\begin{aligned}\left(-F_{\perp,1}\{r(\theta_r + \pi) - r_{CM,2}\} + F_{N,vn}\{l_{vn} - r(\theta_r + \pi) + r_{CM,2}\} + F_{N,rot}\{R_r - r_{CM,2}\}\right) \\ = \left(F_{p4} - F_{p3}\right)\left[R_r + \left\{\frac{r(\theta_r + \pi) - R_r}{2}\right\} - r_{CM,2}\right] \\ + \left\{F_{tp,p4}\sin\left(\frac{\alpha_2}{2}\right) + F_{tp,p3}\sin\left(\frac{\alpha_2}{2}\right)\right\}\left(r(\theta_r + \pi) - R_r - r_{CM,2}\right)\end{aligned} \quad (10)$$

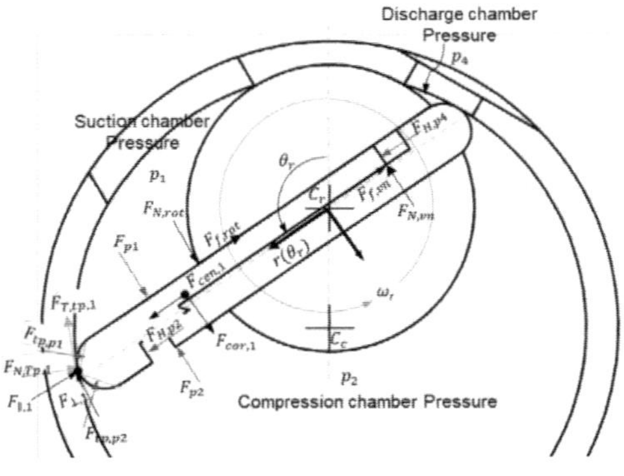

Figure 2. Free body diagram illustrating the forces acting on trailing vane

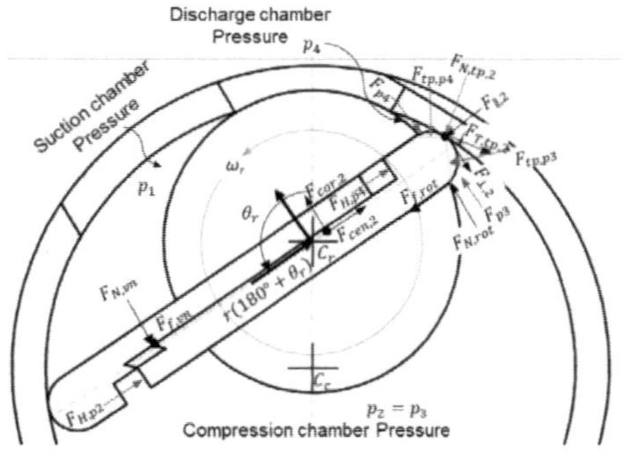

Figure 3. Free body diagram of leading vane

The six unknowns $F_{//,1}$ $F_{//,2}$, $F_{\perp,1}$, $F_{\perp,2}$, $F_{N,\,rot}$ and $F_{N,\,vn}$ are determined by solving six simultaneous equations (5 - 10). Then, the resultant force at the trailing and leading vane tips are determined using equation (11) and (12). γ_1 and γ_2 are the angles made by the $F_{tp,1}$ and $F_{tp,2}$ with respect to $\mathbf{r}(\theta_r)$ and $\mathbf{r}(180°+\theta_r)$.

$$F_{tp,i} = \sqrt{F_{\parallel,i}^2 + F_{\perp,i}^2} \tag{11}$$

$$\gamma_i = \tan^{-1}\left(\frac{F_{\perp,i}}{F_{\parallel,i}}\right) \tag{12}$$

where, $i = 1, 2$

Since the vane tips follow the circular cylinder wall, the force normal to the circular path can be determined using equation (13).

$$F_{N,tp,i} = F_{tp,i}\cos(\gamma_i - \alpha_i) \tag{13}$$

The frictional forces at the vane tips are proportional to the normal forces derived in equation (13). The frictional forces $F_{f,tp1}$ and $F_{f,tp2}$, $F_{f,rot}$, $F_{f,vn}$, can be determined using equations (14), (15) and (16).

$$F_{f,tp,i} = \mu_f F_{tp,i}\cos(\gamma_i - \alpha_i) \tag{14}$$

$$F_{f,rot} = \mu_f F_{N,rot} \tag{15}$$

$$F_{f,vn} = \mu_f F_{N,vn} \tag{16}$$

2.3. Journal bearing model

The rotor-shaft is supported at its two ends by hydrodynamically lubricated journal bearing. Assuming constant viscosity and incompressible flow of lubricating oil into the converging region of a journal bearing, Hirani et al. [15] developed a model to evaluate the torque required to overcome friction due to fluid shearing (shown in equation 17).

$$\mathcal{T}_{b,j} = \left[\frac{\pi\mu_{oil}U_j l_b R_j}{\delta_b\sqrt{1-\varepsilon^2}}\left(\frac{2+\varepsilon}{1+\varepsilon}\right) + \frac{\delta_b\varepsilon}{2R_j}W\sin\varphi\right]R_j \tag{17}$$

2.4. Rotor endface friction

Fluid shear at the endfaces of rotor can be modelled assuming Couette flow, constant oil viscosity and clearance gap. The torque required to overcome the fluid shear can be determined using equation (18).

$$\mathcal{T}_{enf,j} = \frac{\mu_{oil}}{\delta_{r,enf}}\frac{\pi\omega_r(R_r^4 - R_{sh}^4)}{2} \tag{18}$$

2.5. Power loss due to friction

To this end, the frictional forces acting on the components of CVC have been derived. The power loss due to friction at the vane tips is shown in equations (19). The power loss due to the friction between the rotor and vane and between the two vanes is shown in equations (20) and (21) respectively. The power loss at the journal bearing and due to the endface friction can be determined as shown in equations (22) and (23) respectively.

$$\mathrm{P}_{tp,i}(\theta_r) = \left|\left(F_{f,tp,i}\omega_r\right)\left[\{\cos\alpha_i \cdot r(\theta_r)\} + \left\{\sin\alpha_i \cdot \frac{dr(\theta_r)}{d\theta_r}\right\}\right]\right| \tag{19}$$

$$\mathrm{P}_{r,vn}(\theta_r) = \left|F_{f,rot}\omega_r\left\{\frac{dr(\theta_r)}{d\theta_r} + \frac{dr(\theta_r+\pi)}{d\theta_r}\right\}\right| \tag{20}$$

$$\mathrm{P}_{vn}(\theta_r) = \left(2\left|F_{f,vn}\omega_r \cdot \left\{\frac{dr(\theta_r)}{d\theta_r} + \frac{dr(\theta_r+\pi)}{d\theta_r}\right\}\right|\right) \tag{21}$$

$$\mathrm{P}_b(\theta_r) = \sum_{j=1}^{2} \omega_r \cdot \mathcal{T}_{b,j} \tag{22}$$

$$\mathrm{P}_{enf} = \sum_{j=1}^{2} \omega_r \cdot \mathcal{T}_{enf,j} \tag{23}$$

The total power loss in CVC is the sum of all the power losses obtained from equations (19) to (23). The total power loss is obtained in equation (24). The corresponding energy loss due to friction is determined using equation (25). Finally, mechanical efficiency of the compressor can be derived, as shown in equation (26), using the energy lost due to friction and the total indicated work obtained from the pressure-volume curve.

$$\mathrm{P}_{f,total}(\theta_r) = \sum_{i=1}^{2} \mathrm{P}_{tp,i}(\theta_r) + \mathrm{P}_{r,vn}(\theta_r) + \mathrm{P}_{vn}(\theta_r) + \mathrm{P}_{enf}(\theta_r) + \mathrm{P}_b(\theta_r) \tag{24}$$

$$E_f = \frac{1}{\omega_r}\int_{0^\circ}^{540^\circ} \mathrm{P}_{f,total}(\theta_r)\, d\theta_r \tag{25}$$

$$\eta_{mech} = \frac{E_{comp}}{E_{comp} + E_f} \times 100\% \tag{26}$$

3. Results and Discussions

3.1. Simulation results

A simulation program was developed in ForTran for the prediction of performance of CVC and REFPROP [16] was used to calculate thermodynamic properties of the working fluid. The simulation study of CVC is performed using the parameters shown in Table 1.

Table 1. Operating condition and the main dimensions of CVC

Operating condition		Main CVC dimensions	
Volumetric displacement, $V_{c,max}$	44 cm^3	Rotor radius, R_r	15.5 mm
Operating speed, ω_r	3000 r min^{-1}	Cylinder radius, R_c	27.5 mm
Working fluid	R1234yf	Length of compressor, l_c	30 mm
Evaporating temperature, T_{evap}	7.2°C	Distance between rotor-to-cylinder centre, b	13 mm
Condensing temperature, T_{cond}	54.4°C	Vane thickness, t_{vn}	6 mm
Compressor inlet temperature, T_{in}	33°C	Vane tip length, $l_{tip,vn}$	12.3 mm
		Discharge port diameter, d_{dis}	11 mm
		Vane length, l_{vn}	32.3 mm

The variation of position vectors $r(\theta_r)$ and $r(180° + \theta_r)$, with respect to the rotor centre C_r, are shown up to 180° in figure 4(a), since the compressor is symmetrical about its mid-plane containing the two centres C_r and C_c (see figure 1). After 180°, the trailing vane becomes leading vane and vice-versa. This implies that after 180° rotor angle, leading vane assumes the position of trailing vane at 0° and vice-versa with the trailing vane. Until about 30° rotor angle which is half of the sealing arc angle, trailing vane remains inside the rotor and its vane tip is in contact with the sealing arc, hence there is no radial motion [3]. The same applies to leading vane after 150° rotor angle.

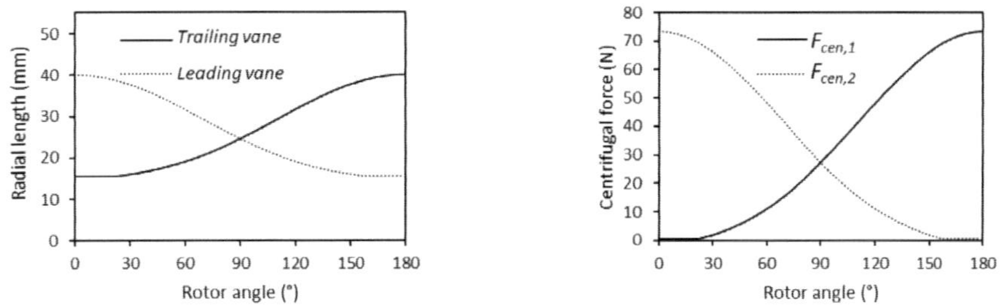

(a) (b)

Figure 4. (a) Variation of radial lengths over 180° rotor angle (b) Variation of centrifugal force for trailing and leading vane ($F_{cen,1}$ and $F_{cen,2}$)

The variations of centrifugal forces are shown for half a revolution in figure 4(b). For the operation of CVC, a necessary design criterion is that centrifugal force acting on the vane must always be in radially outward direction. It can be seen from the figure 4(b), centrifugal force acting on the vane is positive throughout the revolution. This implies that, for selected dimensions of CVC, centrifugal forces aid the vane tips to be in contact with the inner wall of cylinder.

The working chamber pressure and the respective pressure forces acting on the vane body are shown in figures 5(a) and (b) respectively. The pressure forces acting on the trailing and leading faces of the trailing vane and leading vane (F_{p1}, F_{p2}, F_{p3} and F_{p4} respectively) are shown in figure 5(c). Similarly, the pressure forces acting on the tips of the vane are shown in figure 5(d). The pressure forces acting at the neck of the vane are shown in figure 5(e).The force area at neck of the vane is designed to be the same as the force area at the rear end of the other vane. Therefore, the pressure force acting on the neck of the trailing vane ($F_{H,p2}$) is of the same magnitude as the force acting on the rear end of the leading vane (see figure 5(e)).

The net reaction forces on the trailing and leading vane tips from the cylinder wall are shown in figure 5(f). The direction of the reaction forces ($F_{//,1}$ and $F_{//,2}$) are antiparallel to the position vectors $r(\theta_r)$ and $r(180° + \theta_r)$. It was found that for the selected design dimension of CVC, $F_{//,1}$ and $F_{//,2}$ were negative when the vane tip receded into the vane slot in the rotor. The negative reaction force implies that the vane tip detaches from cylinder wall. For the trailing vane tip, this occurs at rotor angles 0° to half the sealing arc angle, $\theta_{r,st}$, at which the vane tip is inside the rotor slot. For the leading vane tip, this occurs at rotor angles close to 180° - $\theta_{r,st}$ at which the vane tip is entering into the vane slot. Upon closer inspection of forces acting on the vane, it was found that this occurred due to two reasons. First, the pressure force at the vane necks ($F_{H,p2}$ and $F_{H,p4}$) were weaker at these positions because of the expansion of the gas exposed at the vane neck at those positions. Secondly, it was assumed that the pressure at sealing arc of CVC was equal to the average of the discharge chamber pressure and the suction chamber pressure. This meant that higher pressure forces were acting on the vane tip to push the vane away from the cylinder wall. This phenomenon can be remedied using either of or combination of two approaches: by changing the length of the vane such that the centrifugal force on the vane is higher which causes larger force to push the vane against cylinder wall. The second way is by modifying the design of sealing arc such that the depth at which the rotor cuts into the cylinder wall (δ_{rc}) is lowered such that net reaction force acting on the vane during the transition from the vane slot to the working chamber is larger. In practice, vane tip detachment from cylinder wall will cause internal leakage and vane chattering.

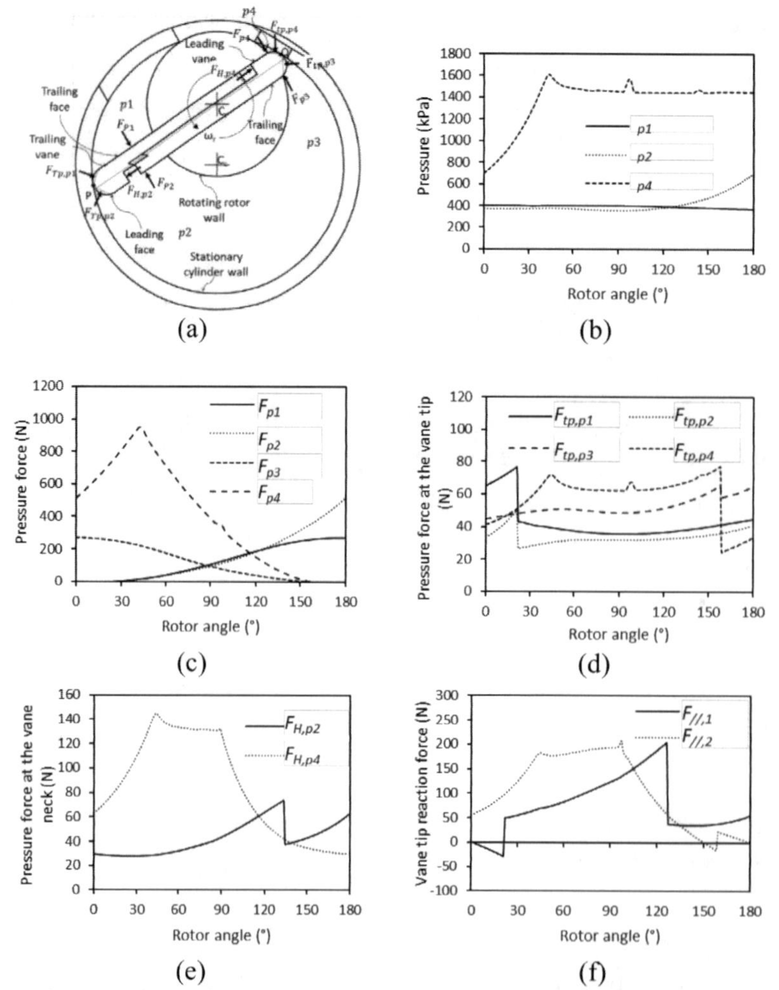

Figure 5. (a) Various working chamber pressures. (b) Variation of working chamber pressures. (c) Variation of pressure forces. (d) Variation of pressure forces at the trailing vane tip ($F_{tp,p1}$ and $F_{tp,p2}$) and at the leading vane tip ($F_{tp,p3}$ and $F_{tp,p4}$). (e) Variation of the pressure forces at the neck of the trailing ($F_{H,p2}$) and leading vane ($F_{H,p4}$). (f) Variation of resultant force pushing the trailing and leading vane ($F_{//,1}$ and $F_{//,2}$)

After reaction forces were evaluated, normal forces and frictional forces acting between the rubbing components were determined. figure 6(a) shows frictional force acting between various rubbing components of CVC.

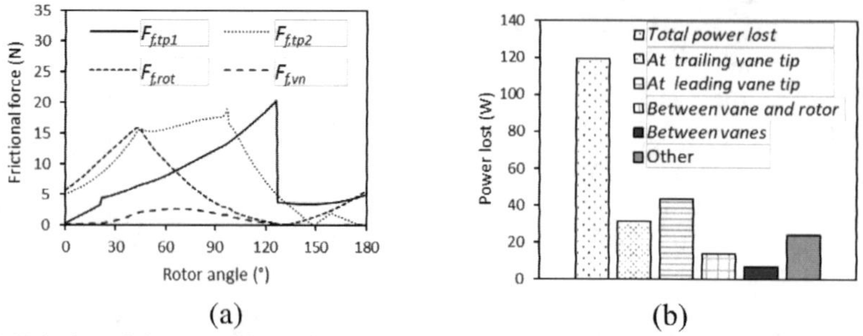

Figure 6. (a) Frictional force at the trailing vane tip ($F_{f,tp1}$) and cylinder wall, leading vane tip and cylinder wall ($F_{f,tp2}$), trailing vane and vane slot in the rotor ($F_{f,rot}$) and between the vanes ($F_{f,vn}$). (b) Power lost per revolution due to the rubbing of CVC components

As shown in figures 2 and 3, a couple of normal forces ($F_{N,rot}$) are assumed to act between the trailing vane and vane slot and leading vane and vane slot, and as such these forces have same magnitude but opposite direction. In figure 6(a), $F_{f,rot}$ is the frictional force acting between the trailing vane and vane slot and there exists frictional force between the leading vane and vane slot of same magnitude but opposite direction. Similarly, a couple of normal forces ($F_{N,vn}$), which have same magnitude but opposite direction, is also assumed to exist between the trailing and leading vanes. $F_{f,vn}$ in figure 6(a) is the corresponding frictional force due to the rubbing of rear end of trailing vane against leading vane.

The individual power lost per revolution (figure 6(b)) shows that the most significant power loss occurred at the vane tips. In figure 6(b), the other losses included the fluid frictional losses at the rotor endface, sealing arc, lower and upper bearing. By summing all the individual power losses, the total power loss due to component rubbings in a working cycle of CVC was found to be 353 W. The indicated power from the indicator diagram was 1116 W. Using equation (29), the mechanical efficiency of CVC prototype was determined to be approximately 76%. The mechanical efficiency of CVC can be improved by optimizing the design of vanes.

4. Experiment setup

The detailed description of the test circuit has been presented by Shakya and Ooi [3] and the picture of testbed is shown in figure 7. A CVC prototype with the maximum suction chamber volume of 44 cm^3 was designed and experimentally measured. For simplicity and to save cost, an open-test circuit using air as the working fluid was used. The compressor was powered by ABB 2.2 kW two-pole induction motor and a frequency controller was used to control the operating speed of the compressor. The power input into the compressor was measured using Fluke MDA-510 scope meter. The atmospheric air was directly drawn into the compressor through its suction port and discharged into the discharge tank. A needle at the outlet of the flowmeter was used for the flow control. The uncertainties of various measuring devices used in the testing of CVC are shown in Table 2.

Table 2. Uncertainties of power measuring Devices

Operating frequency, Δf	± 2.050 Hz
Futek scope meter, ΔI_l	± 0.215 A

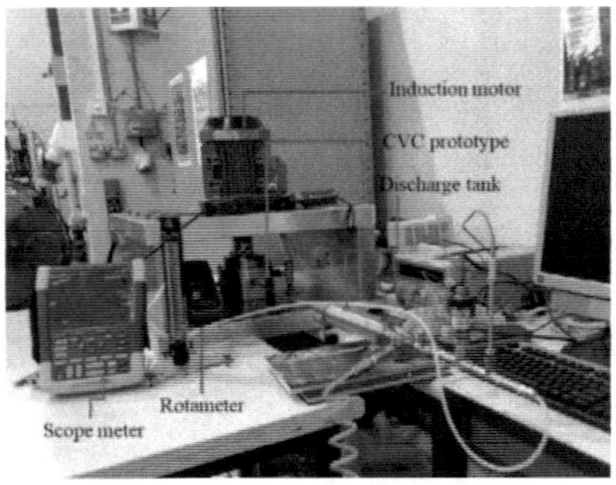

Figure 7. Comparison of the measured and predicted power input

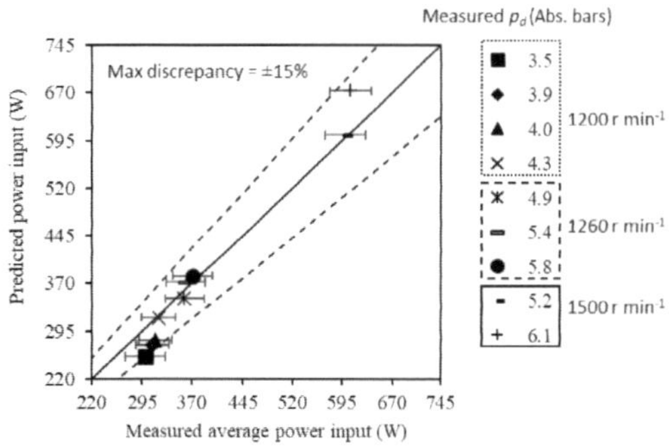

Figure 8. Comparison of the measured and predicted power input

4.1. Results and comparison

The comparison between the predicted and the measured power input to CVC is shown in figure 8. The error bars represent the uncertainties determined for measured power inputs. The frictional coefficient assumed for the prediction was 0.2. For the selected operating conditions, the predicted result was found to have the maximum discrepancy of ±15% with the measured result.

5. Conclusions

CVC's unique feature is that a couple of vanes cut through the rotor diametrically. Therefore, theoretically, any rotor size which can accommodate the vanes will be able to operate CVC. Thus, allowing it to have the smallest rotor-to-cylinder diameter ratio as compared to all existing rotary vane compressor and hence making it a very compact rotary compressor.

In this paper, the frictional losses due to rubbing components of CVC were analysed. The prediction shows the maximum discrepancy of ±15% with the measurement. The vane tip cylinder friction is the most significant frictional loss. This accounts for nearly 56% of the total frictional loss in CVC. Therefore, for good compressor efficiency of CVC, the optimization of compressor dimensions including vane dimensions are to be carried out.

For further investigation, influence of oil on the frictional losses will be studied and the heat generated due to friction between the rubbing components of CVC will be determined and its effect on the fluid flow and thermodynamics will be analysed.

Acknowledgements

The authors acknowledge their appreciation to the lab technicians at Heat transfer, Manufacturing lab and Computer Aided Engineering laboratories at Nanyang Technological University, Singapore. The authors would also like to express their sincere gratitude to SINGA for funding this PhD study and NTUitive Pte Ltd for funding this project.

Nomenclature

b	distance between the rotor centre and the cylinder centre	[m]
F	force	[N]
f	motor operating frequency	[Hz]
g	acceleration due to gravity	[m s^{-2}]
l	length	[m]
P	power	[W]
p	pressure	[Pa]
R	radius	[m]
r	radial coordinate	[m]
W	load acting on the bearing	[N]

Greek symbols

α	contact angle at the vane tip	[rad]
γ	angle of the resultant force on the vane tip	[rad]
ε	eccentricity ratio	[-]
η	efficiency	[-]
μ	frictional coefficient	[-]
T	torque	[N·m]
θ	rotation angle	[rad]
ω	angular speed ; natural frequency	[rad s^{-1} ; Hz]

Subscripts (only abbreviated subscripts are covered)

b	bearing
b,f	bearing friction
$cen,1$	centrifugal force at the vane 1 (trailing vane)
$cen,2$	centrifugal force at the vane 2 (leading vane)
$cor,1$	coriolis force at the trailing vane
$cor,2$	coriolis force at the leading vane
$comp$	compression
d	discharge
dis	discharge chamber
H	at the neck part of vane
j	journal
l,vn	leading vane
N,rot	normal force on the vane by rotor
N,vn	normal force at the point of contact between the vanes
oil	oil as the fluid
$p1$	pressure force at the suction chamber on trailing face of trailing vane side
$p2$	pressure force at the compression chamber on leading face of trailing vane side
$p3$	pressure force at the compression chamber on trailing face of leading vane

f	friction	*p4*	pressure force at the discharge chamber on leading face of leading vane
f,rot	frictional force between rotor and vane	*r*	rotor
f,vn	frictional force between the vanes	*r,enf*	endface of the rotor
enf,vn	vane endface	*r,st*	at the start of the suction with respect to the rotor centre
gap	gap	*s*	suction

References

[1]. Ooi, K. T. and Shakya, P., *A new compact rotary compressor: Coupled Vane compressor.* International compressor engineering conference. Purdue University 2018.
[2]. Ooi, K. T. and Shakya, P., *Simulation studies of a coupled vane compressor.* IOP Conference Series: Materials Science and Engineering. 2019: IOP Publishing. 1.
[3]. Shakya, P. and Ooi, K. T., *Introduction to Coupled Vane compressor: Mathematical modelling with validation.* International Journal of Refrigeration, 2020. **117**: p. 23-32.
[4]. Yanagisawa, T., Shimizu, T., Chu, I. and Ishijima, K., *Motion analysis of rolling piston in rotary compressor.* International compressor engineering conference. Purdue University 1982.
[5]. Ooi, K. T. and Wong, T. N., *A computer simulation of a rotary compressor for household refrigerators.* Applied Thermal Engineering, 1997. **17**(1): p. 65-78.
[6]. Ooi, K., *Design optimization of a rolling piston compressor for refrigerators.* Applied thermal engineering, 2005. **25**(5): p. 813-829.
[7]. Badr, O., Probert, S. and O'Callaghan, P., *Multi-vane expanders: vane dynamics and friction losses.* Applied Energy, 1985. **20**(4): p. 253-285.
[8]. Aradau, D. and Costiuc, L., *Friction power in sliding vane type rotary compressors.* International Compressor Engineering Conference. Purdue University 1996.
[9]. Teh, Y. and Ooi, K., *Theoretical study of a novel refrigeration compressor–Part I: Design of the revolving vane (RV) compressor and its frictional losses.* International Journal of Refrigeration, 2009. **32**(5): p. 1092-1102.
[10]. Hu, X., Qu, Z., Yang, X. and Sun, J., *Theoretical study on frictional losses of a novel automotive swing vane compressor.* International Journal of Refrigeration, 2013. **36**(3): p. 758-767.
[11]. Ma, J., Chen, X. and Qu, Z., *Structural optimal design of a swing vane compressor.* Frontiers in Energy, 2019. **13**(4): p. 764-769.
[12]. Yang, X., Dong, C. and Qu, Z., *Design and dynamic analysis of a novel double-swing vane compressor for electric vehicle air conditioning systems.* International Journal of Refrigeration, 2017. **76**: p. 52-62.
[13]. Bradshaw, C. R., Kemp, G., Orosz, J. and Groll, E. A., *Development of a loss pareto for a rotating spool compressor using high-speed pressure measurements and friction analysis.* Applied Thermal Engineering, 2016. **99**: p. 392-401.
[14]. Lim, Y. D. and Ooi, K. T., *Performance analysis of a U-Vane compressor.* Applied Thermal Engineering, 2020. **178**: p. 115570.
[15]. Hirani, H., Rao, T., Athre, K. and Biswas, S., *Rapid performance evaluation of journal bearings.* Tribology international, 1997. **30**(11): p. 825-834.
[16]. Lemmon, E. W., Huber, M. L. and McLinden, M. O., *NIST reference fluid thermodynamic and transport properties–REFPROP.* 2002, version.

Effects of Lubricating oil on the performance of a Four-Intersecting-Vane Rotary Expander

A A Murthy*, S Norris, A Subiantoro

Department of Mechanical Engineering, The University of Auckland, Auckland, New Zealand

*E-mail: amur812@aucklanduni.ac.nz

Abstract. Lubrication plays an important aspect in machinery. It seals internal leakage gaps, reduces friction and cools components. The present investigation is intended to test the effects of lubrication to an expander's performance. Polyester lubricating oil grades (Grade 48 and Grade 68) were tested at various operating conditions. The expander prototype is of the rotary mechanism with four vanes comprising two intersecting bars. An experimental test rig was designed and built. The pressure diagram as function of rotational angle was recorded by six pressure transducers arranged in the working chamber, based on which the features of the expansion process of the four-intersecting-vane rotary expander were analysed. To start with, the working fluid was compressed air, but the eventual intended application is in a vapour compression refrigeration system. The experimental study revealed that the rotary vane expander worked steadily at different operating conditions. The experimental results showed that the expander flow rate was generally lower with a thinner oil (POE grade 46), whereas the volumetric efficiency increased from 23.2% to 27.2%. The adiabatic efficiency decreased from 22.5% to 18.1% with a higher viscosity lubricating oil (POE grade 68).

Keywords: ***Rotary vane expander; Expansion; Oil, Friction, Efficiency***

1. Introduction

In recent years reducing energy usage has become a key aspect in all applications and the trend will continue in the coming years, and this includes refrigeration systems [1]. This problem can be effectively solved by using more energy efficient systems and cleaner energy generation methods. One such method is to improve the energy efficiency of a refrigeration system by recovering the waste energy during the expansion process. Conventional throttle valves do not have the ability to produce work, and thus the idea to use an expander as a substitute has been considered. An expander can be categorised into volumetric machines, e.g. rotary, reciprocating piston, screw and scroll expanders, and dynamic machines, e.g. radial and axial turbines [2]. There have been a number of experimental investigations of different expander mechanisms, including screw [3], scroll [4], reciprocating [5, 6], swing piston [7], rolling piston [8], rotary vane [9], cross vane

[10] and revolving vane [11] machines, all aimed at improving the expander efficiency. Among the potential concepts, the rotary mechanism is generally more compact, does not require suction/discharge valves, operates with less noise, has less vibration, but tends to suffer from high leakage and/or significant friction. Therefore, lubrication is crucial to achieve high performance and good reliability in a rotary machine [12].

Various studies are available in the literature on lubrication of compressors. Okoma and Onoda [13] discussed various means of oil supply for pressure differential system, gas carry system, vane pump system and rotary compressors. In a pressure differential system, an oil cap encloses the crankshaft suction and an oil delivery pipe from the oil cap is hungover down to an oil sump. On the other end of the oil gallery inside the crankshaft it is connected to a suction pressure, and so oil from an oil sump under the pressure of discharge is drawn into the oil gallery inside the crankshaft through the oil delivery pipe. However, this system is sensitive to load changes, so the oil supply rate becomes unstable. De Paepe et al. [14] studied the influence of oil atomization on the working process of a screw air compressor. They concluded that oil atomization considerably increased heat transfer. However, it did not significantly affect the compressor performance. It was observed that lowering the oil temperature improved the performance of the expander, while changing oil flow rate only gave small gains. Valenti et al. [15] conducted an experimental study on the thermal effect of lubricating oil in a medium size rotary vane air compressor. The results showed that oil droplets with diameters on the order of 100 μm led to a large reduction in both temperature rise and compression work. It is to note that the lubrication design of an expander has its unique challenges. This is due to the low discharge pressure and temperature working conditions For example, in a refrigeration system with R22 as the refrigerant, the expander's discharge pressure and temperature are 3.5 to 4.1 bar (abs) and -5 to 1°C, respectively [17]. This is different from a compressor, in which the discharge pressure and temperature are much higher. Therefore, the selection of the lubricant oil for the expander is important, which is based on the operating conditions such as pressure and temperature. It is to note that the oil viscosity increases with decrease in temperature. Excessive increase or decrease in the viscosity can result in permanent modifications of characteristics of oil, which can damage the lubricating system of the prototype. The survey above shows that compressor lubrication is relatively well-studied. However, to the best of the authors' knowledge, no study has been conducted on expander lubrication [16]. This study aims at filling the knowledge gap. Most expander mechanisms (scroll, screw, piston etc.) need lubrication but the study focuses on the influences of lubricating oil on a rotary expander with four vanes comprising two intersecting bars, which will be referred to as the four-intersecting-vane expander from here on. This prototype has been tested before in a refrigeration system [17]. The mechanism is different from conventional rotary vane machines because the four vanes are formed by two intersecting long bars. Furthermore, while the rotor is circular like those in conventional rotary vane machines [e.g. 7, 17, 22], the stator's inner wall is not circular, as shown in Figure 1a. This configuration results in reduced frictional loss but may amplify the internal leakages. More detailed descriptions of the mechanism can be found in [17]. The lubricating oil is supplied to the inlet pipe of the expander and its influence on the performance of expander specifically volumetric efficiency, adiabatic efficiency and output torque are studied. The study provides researchers and engineers with useful information to choose the appropriate lubricating oil to optimize the expander performance.

2. The Expander Prototype

Figures 1 (a) & (b) show the expander prototype. It mainly consists of a circular rotor with four vanes. The vanes are made of two intersecting bars. The inner wall of the stator is not circular but is designed to minimize the gaps between the vane tips and the wall. The main geometric parameters of the expander are tabulated in Table 1. As shown in Figure 1, the four vanes and the line contact between the rotor and the stator separate the space into several working chambers. A suction port is located near the line contact between the rotor and the stator. High pressure fluid enters the expander through the suction port to the working chamber Volume 1. The volume of each working chamber changes as the rotor turns. The suction process ends, and the expansion process starts when the vane is at 90°. As the chamber volume increases further, the fluid pressure decreases and the maximum chamber volume is reached when the vane is at 225°, the discharge process begins and continues up to 340°. The discharge port does not extend to 360° in order to decrease leakage flow from the suction port to the discharge. So, from 340° to 360°, the fluid might undergo a brief compression process. The developed expander prototype volume ratio is 8.3. If the volume ratio is too large for a given set of working conditions, the working fluid is trapped longer than necessary leading to a pressure drop below the outlet pressure (over-expansion). Conversely, a low volume ratio results in under-expansion. Both over- and under-expansion affect the expander's efficiency.

Table 1. Main dimensions of the expander prototype

Item	Value
Radius of the rotor	27.7 mm
Radius of the stator	40.7 mm
Length of the cylinder	38.6 mm
Thickness of vanes	8 mm
Angle range of inlet port	10° – 90°
Angle range of outlet port	225°– 340°

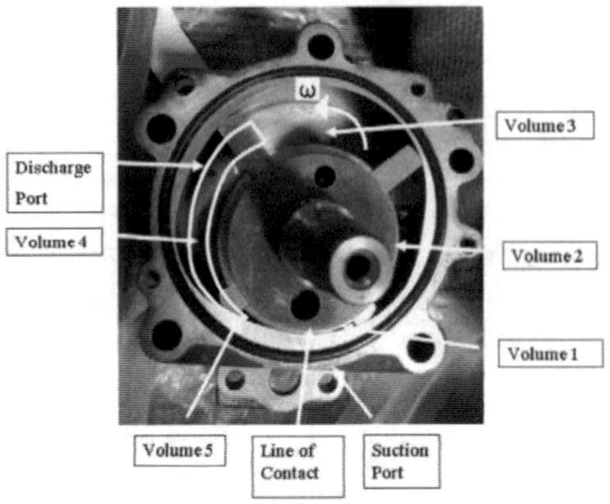

Figure 1(a). The Expander Prototype

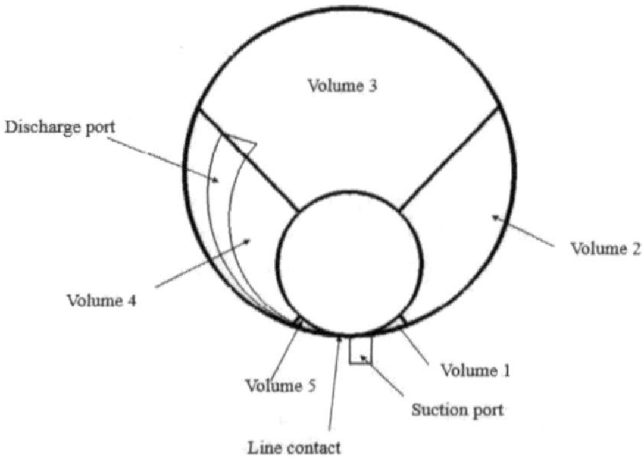

Figure 1(b). Schematics of expander prototype

3. Experimental Setup

Figure 2 shows the experimental test rig. The experiments were carried out to test the main characteristics of the expander, specifically the volumetric efficiency, adiabatic efficiency the instantaneous output torque at different pressures and rotational speeds. Compressed air was used as the working fluid. The maximum pressure available in the compressed air tank in the lab was 8 bar(g), hence, the expander was tested up to a suction pressure of 5 bar(g), to ensure a stable air flow. The expander suction pressure was controlled using a pressure regulator and the working fluid was discharged to the atmosphere. A dynamometer was used to consume power produced by the expander. The output torque was calculated by multiplying the measured force at the load cell with the length of the lever arm (i.e. 200 cm). The power output from the expander was calculated from the torque and rotational speed of the expander. The flow meter used in this experiment was the GSVT 54 float Fischer and Porter rotameter, and the speed was measured using an optical sensor. The accuracies of the measuring devices are tabulated in Table 2. The operating conditions of the experiments are listed in Table 3. All the measurement instruments were connected to a CR1000 measurement data logger designed and manufactured by Campbell Scientific. The specifications of the lubricating oils are given in Table 4.

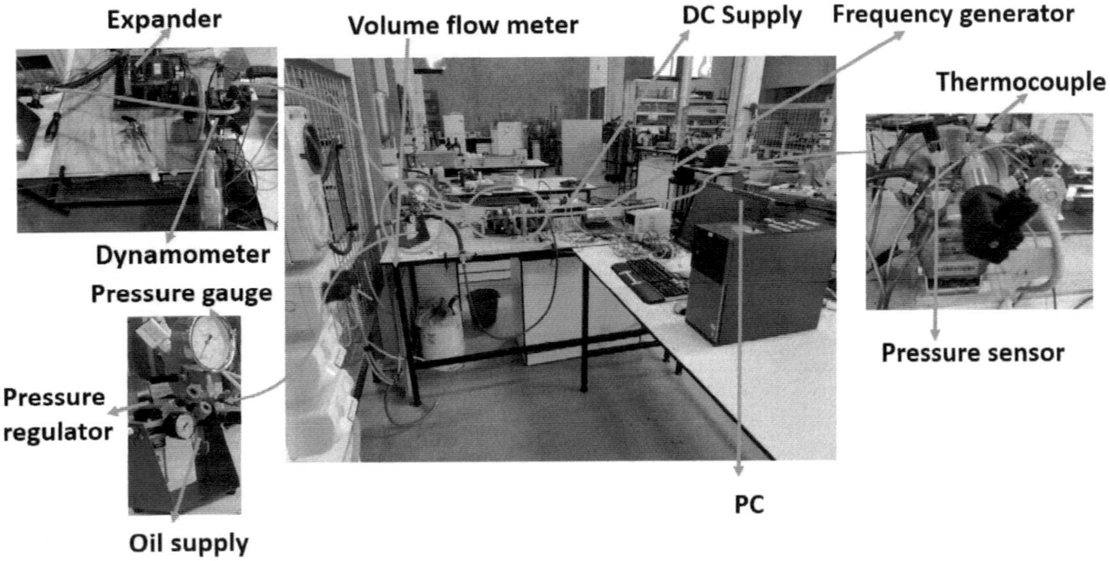

Figure 2. The experimental rig

Table 2. Accuracy of the measuring devices

Devices	Accuracies	Unit
Thermocouple (T-type)	± 0.5	°C
Load cell (HTC Sensor TAL220)	± 0.01	g
Frequency generator FS range: 1 to 10 MHz (Black star Apollo 100)	0.02% ± 0.01%	Full scale (FS)
Pressure transducer FS range: 0 to 25 bar(g) (Danfoss AKS 33)	± 0.3%	Full scale (FS)
Speed sensor (Optek/ TT electronics OPB705WZ)	± 5	Revolutions per minute

Table 3. Main testing parameters

Item	Value
Inlet pressure	4-5 bar(g)
Discharge pressure	0 bar(g)
Operating speed	295-1350 rpm

Working fluid	Compressed air/oil mixture

Table 4. Typical physical characteristics of the lubricating oil [23, 24]

ISO Fluid type	Kinematic viscosity (40°C)	Kinematic viscosity (100°C)	Density (kg/l)	Pour point (°C)
POE Grade 46	46.8	7.3	0.978$^+$	-48
POE Grade 68	72.3	9.8	0.978$^+$	-39

$^+$@20°C

3.1. Definition of parameters

The main parameters to evaluate the expander performance were power output, isentropic efficiency and volumetric efficiency.

Volumetric efficiency is the ratio of the theoretical mass flow rate to the actual mass flow rate, which is defined according to equation 1.

$$\eta_v = \frac{\dot{m}_{th}}{\dot{m}_{act}} \tag{1}$$

The theoretical mass flow rate is calculated using equation 2

$$\dot{m}_{th} = \rho_{suction} \times V_{suction} \times N \times n \tag{2}$$

where $\dot{m}_{th}$ is the theoretical flow rate if there is no leakage, and $\dot{m}_{act}$ is the net volume flow rate into the expander under the actual experimental conditions, $V_{suction}$ is the suction volume, which is the volume of the chamber at 90°, N is the rotational speed of the expander, n is the number of vanes and $\rho_{suction}$ is the suction fluid density.

The adiabatic efficiency of the expander was calculated using equation 3:

$$\eta_{ad} = \frac{h_{in} - h_{out}}{h_{in} - h_{out,is}} \tag{3}$$

where h_{in} is the inlet enthalpy, h_{out} is the outlet enthalpy and $h_{out,is}$ is the outlet enthalpy for isentropic expansion to the same outlet pressure. Thermophysical properties of the working fluid were obtained using the CoolProp database [18].

Every experimental study includes uncertainties coming from the inaccuracies in the measurement devices. To compute the uncertainty in this study the following equation 4 was used.

$$\delta R = \sqrt{\Sigma_i \left(\frac{\partial R}{\partial x_i} \delta x_i\right)^2} \tag{4}$$

where ∂R is the error propagated on generic indirect measured parameters and ∂X_i is the error of the directly measured parameter relevant to ∂R.

4. Results and discussions

To study the internal working processes of the four-intersecting-vane rotary expander, six pressure sensors were arranged in the prototype to measure the chamber pressures at different positions as tabulated in Table 5. Figure 3 shows the variation of pressure in the working chambers of the prototype corresponding to the rotational angle when the expander rotational speed was 1120 rpm. The ideal pressure-angle diagram if the pressure was constant during suction and discharge and the expansion process was isentropic is also shown and called the "ideal" case. It can be observed that the actual process of the expander differed from the ideal case. In the experiments, the pressure dropped during suction showing that the suction flow was not able to cope with the increase in volume. This might indicate that the suction port was too small. In the expansion stage, the pressure dropped less rapidly than that in the ideal/isentropic case. This was mainly because of internal leakages and heat transfer into the expanding fluid. At the end of the discharge process, the pressure rose slightly because the discharge groove only extended out to 340°. Therefore from 340° to 360°, the fluid was slightly compressed.

Table 5. Pressure and temperature sensor position

Sensor	Position (°)
P1	0°-30°
P2	30°-120°
P3	120°-225°
P4	225°-330°
P5	330°-360°
P6	Suction port (10°)
T1	Suction port
T2	Discharge port

To study the performance of the expander and how lubricant affects it, experiments were conducted using the test rig and measurement devices described in section 3. The experimental results are plotted in Figures 4 to 6. Figure 4 shows the volumetric efficiency of the expander at various suction pressures, operational speeds and oil grades. In this study, the average uncertainty of the volumetric efficiency data is ±0.16%. The larger inlet pressure increased pressure difference across the internal leakage paths. This results in the increased leakage flow rates and decrease in the volumetric efficiency [21]. As shown in Figure 4, the volumetric efficiency increases with an increase in the rotational speed of the expander. The leakage rate contribution to the total expander flow rate reduced with increasing operational speed, so a higher rotational speed led to a higher volumetric efficiency [19]. A more viscous oil decreased the leakage flow rate and improved the volumetric efficiency. The average oil flow rate for POE grades 48 and 68 lubricants at suction pressures of 4 bar(g) and 5 bar(g) were 0.28 g/s and 0.32 g/s, and 0.34 g/s and 0.37 g/s, respectively. The highest volumetric efficiency measured was 27.5% for POE 68 oil at a suction pressure of 4 bar (g). For all the cases, it is evident that a higher volumetric efficiency was achieved at higher operational speeds. This was because the leakage flow rate was mostly affected by the pressure difference and gap size, but was relatively independent of the rotational speed. However, the increase in operational speed increased the expander flow rate, hence, the leakage became less significant as the speed increased, which was also observed by other researchers [8]. The relatively low volumetric efficiency values suggest the severity of internal leakages in the expander and the need for a proper lubrication system. This could be accomplished using an active lubrication system, such as with an oil pump. The lubricating oil pump is used to circulate an amount of oil from a reservoir to the oil injector.

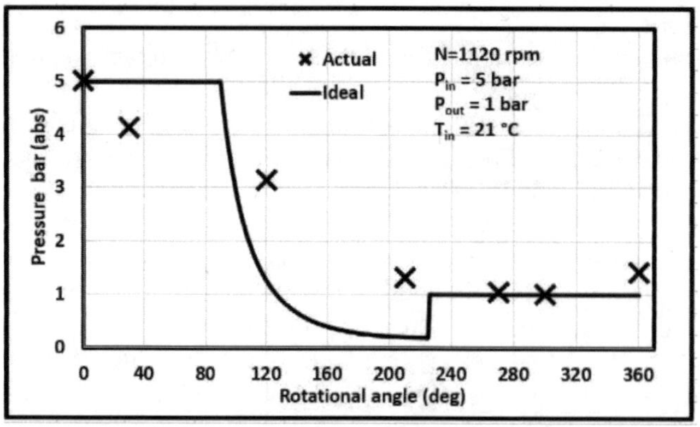

Figure 3. Pressure change versus rotational angle

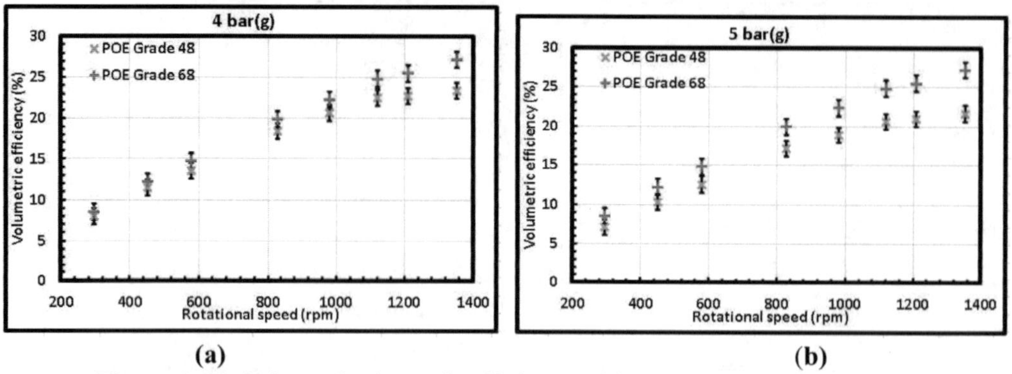

Figure 4: Variations of volumetric efficiency with respect to rotational speed

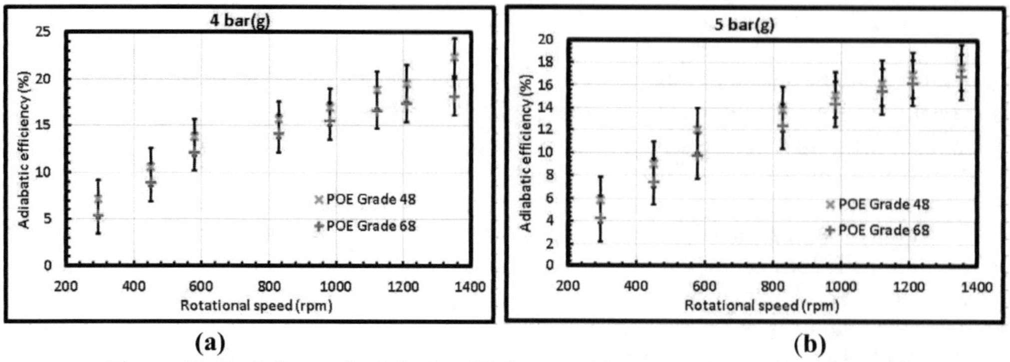

Figure 5: Variations of adiabatic efficiency with respect to rotational speed

The variation of the expander's adiabatic efficiency with suction pressure, rotational speed and different oil grades is presented in Figures 5 (a) & (b). In this study, the average uncertainty of the adiabatic efficiency data is ±1%. It can be observed that the increase in rotational speed improved the adiabatic efficiency. From Figures 5 (a) & (b), the highest adiabatic efficiency of 28% was obtained with POE 48 oil at an inlet pressure of 4 bar(g). A more viscous oil resulted in more frictional losses. It is interesting to note that expanders operate at lower temperatures as compared to compressors. Therefore, the actual operating viscosity of the oil is higher compared to that of compressors. Since the expander was tested under an open system arrangement, the exit pressure of the fluid was equalised with the atmospheric pressure. Another factor that affected the adiabatic efficiency was the fact that the expander's expansion process was fixed from 90° to 225°, which corresponds to an isentropic expansion from 19 bar to 1 bar. This meant over or under-expansion processes may happen as the inlet pressure was varied. Adiabatic efficiency was also influenced by the internal leakages in the expander, which were quite significant in this study as discussed above. It is noted that the maximum adiabatic efficiency measured in this study is relatively higher compared to all the expanders operating with compressed air as the working fluid in the open literature [16]. The results also show that oil viscosity has contradicting effects to the expander efficiency. A more viscous oil increased the volumetric efficiency but decreased the adiabatic efficiency.

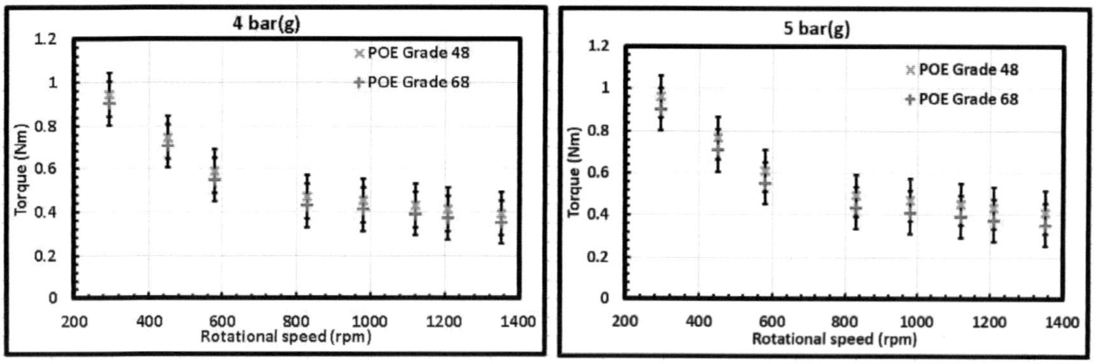

(a) (b)

Figure 6. Output Torque versus Rotational Speed

Figure 6 shows the variation of average output torque with suction pressure, operational speed and oil grades. The average output torque increased with the increase in suction pressure. This is because a higher inlet pressure led to more force applied on the vane to rotate the expander. However, the increase in rotational speed decreased the average output torque. This was due to the high frictional losses at higher rotational speeds as a result of the higher rubbing velocity and the larger frictional forces [20]. A higher oil POE grade 68 lubricant reduced the output torque because more friction losses were generated. Overall, leakage and frictional losses were found to be the main factors for the low performance of the expander and hence further improvement in the prototype is needed to improve its performance.

5. Conclusions

An experimental study has been conducted to determine the performance of a rotary vane expander with four intersecting vanes using compressed air/oil mixture as the working fluid. The study was carried out at suction pressures of 4-5 bar (g), a discharge pressure of 0 bar (g) and rotational speeds of 295-1350 rpm. The expansion process of the expander was analysed using six pressure sensors. The effects of the lubricating oil and working conditions on the expander efficiencies were analysed and the following conclusion were drawn:

1) The volumetric efficiency increases with increasing the operational speed of the expander but decreases with increase in suction pressure. Higher viscosity oil increased the volumetric efficiency. The highest volumetric efficiency was obtained for lubricating POE oil with grade 68. When the rotational speed was 295 rpm, the volumetric efficiency of the expander was around 8.4%. Further increasing the rotational speed improved the volumetric efficiency of the expander to 27.5% at the rotational speed of 1350 rpm.
2) The expander adiabatic efficiency of the expander depended on the viscosity of oil, rotational speed and inlet pressures. The highest adiabatic efficiency was 22.4% for the POE 48 at an inlet pressure of 4 bar (g). The adiabatic efficiencies varied from 7.8% to 22.4% under the tested conditions.

3) The output torque produced by the four-intersecting-vane expander prototype is sensitivity to the viscosity of the lubricant oil used. The expander torque increased with operational speed and increased with increasing suction pressure but decreased with more viscous oils. When POE oil grade 48 is used, the highest average torque readings is 1.1 Nm for 295 rpm at a suction pressure of 4 bar (g).
4) A more viscous oil reduced internal leakage and increased volumetric efficiency, but it increased frictional losses and so decreased the adiabatic efficiency and output torque.

In summary, the current study provides an insight into the performance of the prototype four intersecting vane expander. Leakage and frictional losses should be reduced to further improve the performance of the expander. This can be achieved by the proper selection of lubricating oil. Moreover, it was shown that the effect of lubricant could be significant and needs further investigations. The expander prototype will be installed into a refrigeration system and experimented with in the future.

Nomenclatures

h	Enthalpy	(kJ/kg)
$\dot{m}$	Mass flow rate	(kg/s)
p	Pressure	(bar)
P	Power	(W)
V	Volume flow rate	(m^3/s)
T	Temperature	(K)
Greek		
ω	Operation speed	(rad/s)
ρ	Density	(kg/m^3)
η	Efficiency	-
θ	Angle	(deg)
$\partial F_i/\partial x_i$	Derivative of the calculated data with respect to the independent measured variable	-
Δx_i	Uncertainty of the directly measured parameter relevant to the calculated data	-
Δy	Uncertainty of the calculated data	
Subscript		
in	Inlet	
is	Isentropic	
out	Outlet	
th	Theoretical	

6. Acknowledgments

The authors would like to acknowledge the contributions of Martin Ryder and Alan Eaton to the setting up the test rig.

References

[1] IEA, 2013. World Energy Outlook, *International Energy Agency, Paris.* https://www.iea.org/reports/world-energy-outlook-2013

[2] Qiu G, Liu H, Riffat S, 2011. *Applied Thermal Engineering*, ***31*(*16*)**, 3301-3307. https://doi.org/10.1016/j.applthermaleng.2011.06.008
[3] Li G, Lei B, Wu Y, Zhi R, Zhao Y, Guo Z, Ma C, 2018. *Energy*, ***147***, 279-285. https://doi.org/10.1016/j.energy.2018.01.034
[4] Wang M, Zhao Y, Cao F, Bu G, Wang Z, 2012. *International Journal of Refrigeration*, ***35*(*4*)**, 757-771. https://doi.org/10.1016/j.ijrefrig.2011.11.014
[5] Fukuta M, Anzai F, Motozawa M, Terawaki H, Yanagisawa T, 2014. *International Journal of Refrigeration*, ***42***, 48-56. https://doi.org/10.1016/j.ijrefrig.2014.02.005
[6] Ferrara G, Ferrari L, Fiaschi D, Galoppi G, Karellas S, Secchi R, Tempesti D, 2016. *International Journal of Refrigeration*, ***72***, 147-155. https://doi.org/10.1016/j.ijrefrig.2016.07.014
[7] Xia C, Zhang W, Bu G, Wang Z, Shu P, 2013. *Applied Thermal Engineering*, ***59*(*1-2*)**, 559-567. https://doi.org/10.1016/j.applthermaleng.2013.05.050
[8] Jiang Y, Ma Y, Fu L, Li M, 2013. *Energy*, ***55***, 916-924. https://doi.org/10.1016/j.energy.2013.03.053
[9] Haiqing G, Yitai M, Minxia L, 2006 *Applied Thermal Engineering*, ***26(2-3)***, 237-243. https://doi.org/10.1016/j.applthermaleng.2005.05.011
[10] Yap K S, Ooi K T, Chakraborty A, 2014. Introduction of the novel cross vane expander-compressor unit for vapour compression cycle. *International Compressor Engineering Conference Purdue*, ***2253***, 1-7. https://docs.lib.purdue.edu/icec/2253
[11] Subiantoro A, Tiow O K, 2009. *HVAC&R Research*, ***15(4)***, 801-816. https://doi.org/10.1080/10789669.2009.10390865
[12] Kim H J, Lancey T W, 2003. *International Journal of Refrigeration*, ***26(7)***, 800-808. https://doi.org/10.1016/S0140-7007(03)00050-1
[13] Okoma K, 1988. Study of lubrication mechanism for horizontal type rolling piston rotary compressor. *International Compressor Engineering Conference*, ***596***, 1-8. https://docs.lib.purdue.edu/icec/596
[14] De Paepe M, Bogaert W, Mertens D, 2005 *Applied Thermal Engineering*. ***25*(*17-18*)**, 2764-2779. https://doi.org/10.1016/j.applthermaleng.2005.02.003
[15] Valenti G, Colombo L, Murgia S, Lucchini A, Sampietro A, Capoferri A, Araneo L, 2013. *Applied Thermal Engineering*, ***51*(*1-2*)**, 1055-1066. https://doi.org/10.1016/j.applthermaleng.2012.10.040
[16] Murthy A A, Subiantoro A, Norris, S, Fukuta M, 2019. *International Journal of Refrigeration*, ***106***, 427-446. https://doi.org/10.1016/j.ijrefrig.2019.06.019
[17] Murthy A A, Norris, S, Subiantoro A, 2020. *Applied Thermal Engineering*, ***183(1)***, 1-12. https://doi.org/10.1016/j.applthermaleng.2020.116209
[18] Bell I H, Wronski J, Quoilin S, Lemort V, 2014. *Industrial & Engineering Chemistry Research*. ***53*(6)**, 2498-2508. https://doi.org/10.1021/ie4033999
[19] Jia X, Zhang B, Pu L, Guo B, Peng X, 2011. *International Journal of Refrigeration*, ***34*(3)**, 732-741. https://doi.org/10.1016/j.ijrefrig.2010.12.005
[20] Subiantoro A, Yap K S, Ooi K T, 2013. *Applied Thermal Engineering*, ***50*(*1*)**, 393-400. https://doi.org/10.1016/j.applthermaleng.2012.06.009
[21] Subiantoro A, Ooi K T, 2012. *International Journal of Refrigeration*, ***35(6)***, 1734-1743. https://doi.org/10.1016/j.ijrefrig.2012.04.015
[22] Thakur R, Murthy AA, Subiantoro A, 2019. Proceedings of the 25th IIR International Congress of Refrigeration, Montreal, paper ID 187, 1-8. http://dx.doi.org/10.18462/iir.icr.2019.0167
[23] Emkarate RL46H, https://www.climalife.co.uk/docs/EmkarateRL46H-Typical-Properties-data-sheet.pdf
[24] Emkarate RL68H, https://www.mandtsystems.com/documents/POE_Datasheet.pdf

Experimental Investigation on a Sliding-Vane Expander for Steam Applications

I Costanzo[1], S Murgia[1], G Valenti[2], F Fasani[1]

[1]Ing. Enea Mattei SpA, [2]Politecnico di Milano

ida.costanzo@matteigroup.com

Abstract. There are many industrial plants that use steam as an energy vector at different pressure levels, in which pressure reduction is commonly performed by a throttling valve, resulting a totally irreversible process. It is possible to recover this energy through an expander that generates mechanical or electric energy. This work is focused on the analysis of a highly innovative steam expander based on sliding-vane technology. The design principles of the sliding-vane expander and two experimental campaigns are presented. The first experimental campaign allows to assess the reliability of the system with steam as working fluid, the second campaign leads to assess the stability during long-run and the performance of the system. A maximum expander mechanical power of 28 kW is measured. The pressure-Volume diagram of the expansion process is also measured and presented.

1. Introduction

Steam is adopted widely in the industrial sector as an energy vector for direct and indirect thermal uses. Typically, steam is generated in fired boilers at relatively high pressure, transported across the industrial plant and, ultimately, throttled in valves to the desired pressure close to the point of use. Despite this approach is very effective from the control and the investment perspectives, pressure reduction by throttling is an irreversible process that, from a thermodynamic perspective, destroys exergy [1]. Rian and Ertesvåg demonstrate through the analysis of a case study that the destruction of steam exergy by throttling can be avoided largely by the substitution of throttling valves with steam expanders to generate mechanical or electric energy [2]. Additionally, Frate et al. prove that the use of expanders for the pressure reduction of steam can be technically and economically viable under many circumstance [3]. In their turn, Tian et al. analyze numerically the adoption of a twin-screw steam expander as pressure reduction means under fluctuating operating conditions, determining that volumetric and isentropic efficiencies are significant in all the analyzed conditions [4].

This work investigates experimentally the performances of a sliding-vane steam expander under diverse working conditions. Sliding-vane expander, thanks to their low speed, good off-design performance, high expansion ratio, low cost and simple manufacturing, are a viable option for the energy recovery systems. A first prototype of sliding-vane steam expander, able to work with saturated steam, has been realized and tested. The authors investigated deeply sliding-vane expander technology in ORC based energy recovery systems in previous works [5, 6], studying in great detail different working fluids [7] and validating an automated control system [8].

The following sections describe the experimental campaigns organized and executed in this work, present the experimental results discussing them, and finally draws the conclusions.

2. Experimental campaign

The experimental campaign has the purpose to perform a reliability and performance determination on the adopted sliding-vane expander, using saturated steam from an industrial boiler, as the working fluid. The test rig and the test procedure are described in the sections below.

2.1. *Test rig*

The layout of the test rig employed in this work is divided in three main areas. The first area is dedicated to the steam generation, where the low-pressure steam generator is located, with all the equipment needed to produce saturated steam at variable pressure. The second area, separated from the first one through a wall, is the expander room, where the expander with its ancillaries and all the devices aimed to control and to measure the operating conditions are installed. The third area, separated from the previous ones through walls, is the control room, where the operator can monitor the first and the second room in safe conditions, without entering in contact with steam lines. The first two areas are pictured in Figure 1.

Figure 1. Steam generator area (left) and expander area (right) of the test rig employed in this work.

2.1.1. Steam generator

The steam is generated through a diathermic oil boiler, the main technical specification of which reported in Table 1. Gas burner, fed by natural gas, provides the thermal power required to heat up the diathermic oil. The burner is embedded into the thermal oil heater which conveys and leads the hot gases through oil filled bundle tubes. The hot oil, produced by the thermal oil heater, is conducted into the steam generator and it transfers heat to feed water. The diathermic oil is driven by two centrifugal pumps. The oil line directly communicates with a closed expansion vessel pressurized with nitrogen. Nitrogen tank feeds the expansion vessel partially filled with diathermic oil. Two vertical pumps, the evaporator and the expander realize an open Rankine cycle using feed water as working fluid.

Table 1. Steam generator technical specifications

Description	Value
Manufacturer	ICI CALDAIE
Thermal Power	465 kW
Rated Pressure	12 bara
Steam flow	680 kg/h

2.1.2. Expander room

The expander room is characterized by two main lines, as illustrated by the process flow diagram in Figure 2. The steam line (in blue) and the oil line (yellow). The first one represents the working fluid flow from the boiler, as saturated steam, to the expander, that can be also vented directly to the ambient when the conditions are not compatible with the expansion process, for example during the boiler startup. The flow regulation is managed acting electrically on two pneumatic valves. Then, after the expansion process, the steam is discharged into the ambient. The oil line provides lube oil to the expander in order to reduce the friction between moving parts during the expansion process. The oil flow rate is managed by a gear pump coupled with an electric motor controlled through an inverter. The oil used for this application is a blend compatible with sliding-vane machines and the temperatures of the steam expansion process.

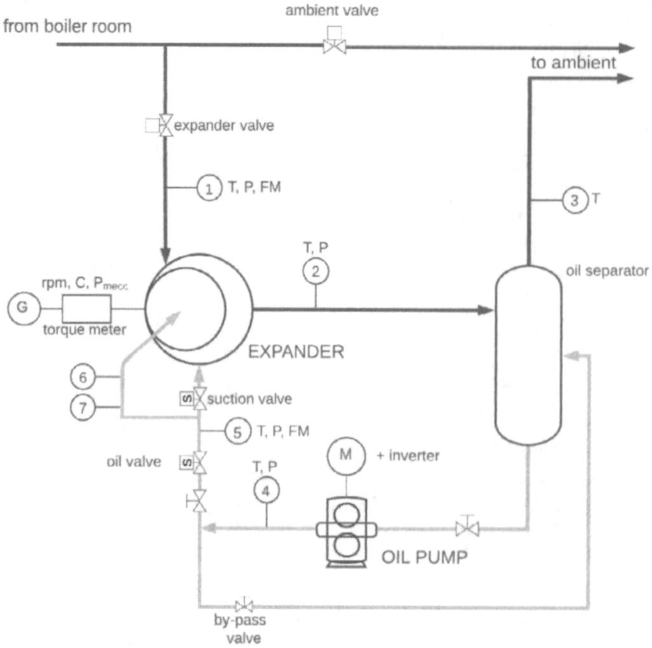

Figure 2. Process flow diagram of the expander area, with the steam (blue) and the oil (yellow) lines.

2.1.3. Sliding-vane expander

The component that realizes the steam expansion process is a sliding-vane expander. This is a positive displacement rotary machine composed by an external cylindrical housing, referred as stator, and an internal circular rotor, smaller than the stator. In the expander, the rotor is eccentrically offset to be tangent to the stator along a tangency line (or arc). The contact line between rotor and stator represents the only sealing between the suction and pressure side of the machine. Radial or tilted slots are obtained in the rotor and will host the vane, which are free to slide along the slot direction itself. During the rotation, under the effect of centrifugal force, vanes slide outwards their slots until their tip gets in contact with the internal stator surface. Hence, each couple of adjacent vanes, thanks to rotor eccentricity, generates a volume, refereed as cell, which is gradually reduced (compressors) or increased (expanders), thus determining a variation of the fluid pressure. The intake and exhaust ports are designed for a proper feeding and evacuation of the working fluid. In the expander a small amount of lubricant is injected, to guarantee lubrication and sealing during the expansion process. The lubricant is also used for the bushings lubrications, that sustain the rotor during the expander rotation. The expander schematic and main geometrical characteristics are reported in Figure 3. The expander is directly coupled with a generator, that converts the shaft mechanical power into electric power.

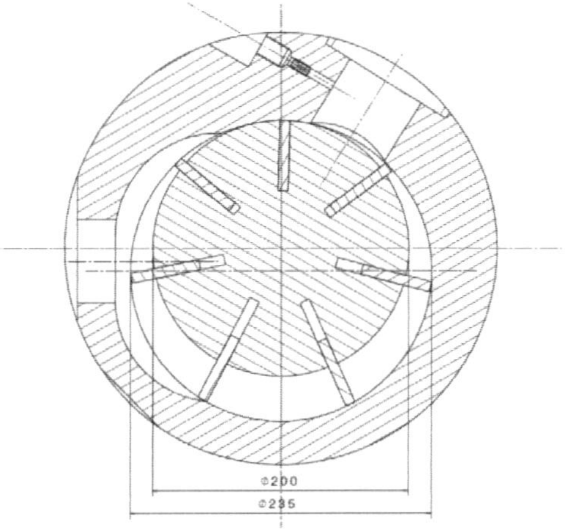

Figure 3. Schematic of the sliding-vane expander main dimensions in millimeters.

2.1.4. Instrumentation

The diagram in Figure 1 shows where the instruments reported in Table 2 are placed. The number reported to indicate the position is used also to report the results in the following section. Regarding the diagram, the letter T, P and FM means respectively the instrumentations used: thermocouple, pressure transducer and flow meter, while the progressive number indicates different position on the system, in particular:

1- steam T and P, expander inlet;
2- steam T and P, expander outlet;
3- steam T, separator outlet;
4- oil T and P, pump outlet;
5- oil T, P and FM, injection line;
6- bushing T, motor side
7- bushing T, side opposite to the motor

Four piezoelectric pressure transducers are placed along the end cover of the expander, allowing to investigate deeply the expansion process by reconstructing the indicated cycle.

Table 2. Instrumentation list

Instrument	**Model**	**Unit**	**Uncertainty**
Relative pressure transducer	Remag PR100	barg	0.04
Thermocouple	Tersid T type	°C	0.5
Piezoelectric pressure transducer	Kistler	-	-
Steam flow meter	Ital control Meters M23	kg/h	2% on read value
Oil flow meter	Omega FLR6115D	l/min	0.289
Torque meter	Kistler 4504B1K1N1	N m	0.1
RPM meter	Kistler 4504B1K1N1	rpm	12

2.2. *Test procedure*

For the experimental campaigns, a standard procedure has been identified and followed for both the data collection and the start-up as well as shut-down of the system, as described next.

2.2.1. Start-up procedure

Regarding the start-up procedure, the following is executed.

1. Instrumentation checking.
2. Setting the oil circuit. The valve that opens the oil to the expander is closed, so that the oil circulates only through the pump and the separator, increasing its pressure. The pump is turned on and the by-

pass valve is regulated to match the desired oil injection pressure, considering further pressure drops when the injection valve is open.

3. Start-up of the boiler. As the manufacturer suggests, the boiler is turned-on with the steam valve completely closed, otherwise all the flow will be discharged to the ambient.
4. Valve checking. For safety reasons and for the next steps, the valves inside the expander room are set to discharge all the steam flow to the ambient. This means that the expander valve must be close in position while the ambient valve must be in open. In this phase there is still no steam flow in the expander room.
5. Steam valve opening. Once the steam boiler has reached the design pressure, the steam valve is partially opened. For the previous step, the pressurized steam is discharged into the ambient. At this point, the system is ready to direct the steam flow to the expander.
6. Expander valve opening. Once the operator is ready to run the steam expander, the expander valve is opened and the steam flow starts to go into the expander.
7. Oil valve opening. As soon as the expander valve is opened, the oil valve is opened to lubricate the machine. In this conditions, the incoming steam flow is not enough to overcome all the mechanical resistances and produce positive power. In fact, most of the steam flow rate goes to the ambient branch due to the lower resistance to the flow.
8. Ambient valve closing. Since the flow rate is not enough for the expander, the ambient valve is turned to close position, directing the whole steam flow into the expander.
9. Connection to the grid. With the previous step, the expander starts to rotate and is able to rotate faster and faster. In this condition the expander is connected to the electric grid starting to produce electric power. After the start-up procedure, the steam expander reaches stable condition in terms of temperature and pressure. Nevertheless, during the test, it could be required to regulate the pump rotational speed to respect the constraint on the bushings temperature.

2.2.2. *Shut-down procedure*

Regarding the shut-down procedure, the following is executed.

1. Ambient valve opening. Operating this valve, a fraction of the steam flow goes into the ambient.
2. Expander valve closing. To shut-down the expander, the whole steam flow must be directed to the ambient. Hence, there is no more steam flow incoming into the expander.
3. Disconnection from the grid. Since there is no more steam flow to the expander, no more power will generated, hence, it is disconnected from the grid to avoid power consumption.
4. Oil valve closing. When the expander is shut down, lubrication is no more needed. Moreover, a heavy and unnecessary lubrication of the expander creates problems to the machine.

3. Results and discussion

This section presents the results of the tests conducted following the procedures described above. The aim of test is to measure the system performance and stability. Two experimental campaigns are performed, as described in the next two sections.

3.1. *First experimental campaign*

During the first experimental campaign, a first running of the system is performed, in order to collect the data and characterize the system zero condition. Temperatures in the main points during the start-up are presented in Figure 4 as a function of time, while all temperatures and pressures in all points at steady state in Table 3; other measured values at steady state are presented in Table 4. After the first start-up of the system, the temperature stabilizes and the conditions across the expander are almost constant. During this steady phase, a mechanical power of about 17.4 kW is measured at the expander shaft, and converted to electric power through the electric motor.

The temperature increases during the expander start-up but stabilizes after the first 200s. During the start-up phase the lubrication system must be adjusted in order to control lubrication and sealing and, consequently temperature, during the expansion process.

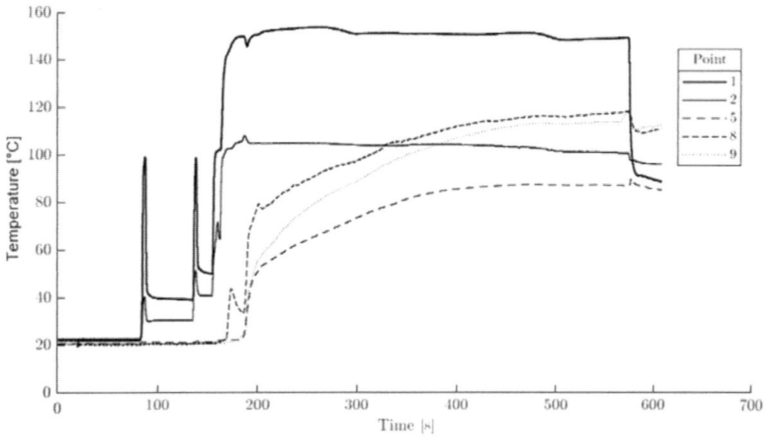

Figure 4. Temperature diagram during first experimental campaign as a function of time.

Table 3. First experimental campaign temperature and pressure at steady state.

Point	Temperature [°C]	Pressure [barg]
1	150.9	4.41
2	103.5	0.23
3	88.2	-
4	99.9	5.5
5	86.9	5.32
6	115.4	-
7	111.9	-

Table 4. First experimental campaign additional measurements at steady state.

Data	Value	Unit
Mechanical power	17.4	kW
Torque	109.9	Nm
Rotational speed	1513	RPM
Steam flow rate	547	kg/h
Oil flow rate	4.02	l/min

3.2. *Second experimental campaign*

During the second experimental campaign, some adjustments to the system have been implemented: thermal insulation was improved to avoid condensation, leading to higher temperature and pressures of the steam flow during the test. Furthermore, a fine tuning of the lubricant injection pressure has been performed, leading to a better lubrication condition of the expander. The improvements can be quantified by comparing data in Table 5 vs Table 3.

As previously, Table 5 and Table 6 show, respectively, the temperature as well as pressure of all points and the additional measurements, all at steady states.

All the mentioned improvements allowed for a better performance, reaching a higher mechanical power produced by the expander thanks to the steam expansion process, as presented in Figure 5 that show its measure during the startup as a function of time.

Table 5. Second experimental campaign temperature and pressure at steady state.

Point	Temperature [°C]	Pressure [barg]
1	156.4	6.14
2	113.6	1.11
3	98	-
4	101.5	15.46
5	88.3	15.03
6	134.4	-
7	122.4	-

Table 6. Second experimental campaign additional measurements at steady state.

Data	Value	Unit
Mechanical power	28	kW
Torque	175.8	Nm
Rotational speed	1521	RPM
Steam flow rate	1034.1	kg/h
Oil flow rate	12.87	l/min

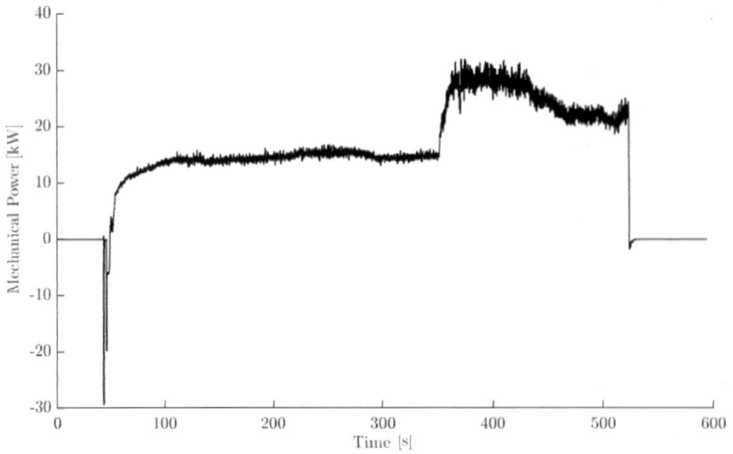

Figure 5. Mechanical power diagram during the second experimental campaign as a function of time.

Figure 6 illustrates the pressure evolution inside the expander as measured by the four piezoelectric sensors that are able to collect pressure during the entire expansion process as well as part of the suction and discharge. By means of the volume evolution inside the expanders, determined from the geometrical features of the machines, it is possible to calculate the expansion indicated power as the area enclosed by the pressure-volume curve and, consequently, the mechanical efficiency of the expander. In this testing conditions, a mechanical efficiency of 85% is reached.

Additionally, Figure 6 shows that an under-expansion happens inside the machine: in the last part of the process, the expander outlet pressure is higher than the atmospheric pressure. This pressure gap is due to the steam-lubricant separator and the steam delivery line.

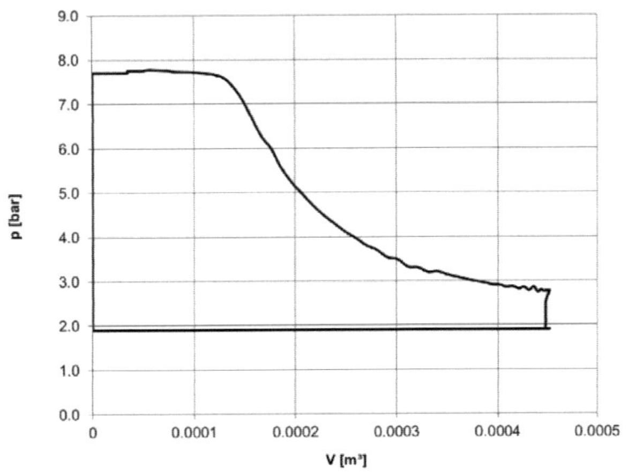

Figure 6. Measured pressure-Volume diagram of the expansion process

4. Conclusions and future works

An experimental study is carried out on sliding-vane expander using steam as working fluid. A defined test procedure has been implemented in order to operate the system in safe conditions. This work draws conclusions as follows.

- During the first experimental campaign, a first running of the system is performed, in order to collect the data and characterize the system zero condition. The system is able to correctly work and recover energy by the steam expansion through the sliding-vane expander. A mechanical power of 17.4 kW is reached.
- During the second experimental campaign, some adjustments to the system have been implemented (increased insulation to avoid condensation, fine tuning of the lubricant injection system), leading to higher temperature and pressures of the steam flow during the test. Thanks to these improvements, a mechanical power of 28 kW is reached.
- The measurement of the pressure inside the expansion chambers let the calculation of the mechanical efficiency of the expander, that states around the 85%.

Future works will focus on the system improvement, with the implementation of an automatic control system, able to adapt the expander operation to the steam flow conditions.

References

[1] TJ Kotas. The Exergt Method of Thermal Plant Analysis, Butterworths-Heinemann, Repring, 2013

[2] A.B. Rian, I.S. Ertesvåg. Exergy analysis of a steam production and distribution system including alternatives to throttling and the single pressure steam production. Energy Conversion and Management, 52, pp. 703-712, 2013. doi: 10.1016/j.enconman.2010.07.049

[3] G.F. Frate, L. Ferrari, R. Lensi, U. Desideri. Steam expander as a throttling valve replacement in industrial plants: A techno-economic feasibility analysis. Applied Energy, 238, pp. 11-21, 2019. doi:10.1016/j.apenergy.2019.01.005

[4] Y. Tian, Z. Xing, Z. He, H. Wu. Modeling and performance analysis of twin-screw steam expander under fluctuating operating conditions in steam pipeline pressure energy recovery applications. Energy, 141, pp. 692-701. doi: 10.1016/j.energy.2017.09.128

[5] Murgia, S., Colletta, D., Costanzo, I., Contaldi, G., 2016. Experimental investigation on ORC-based lowgrade energy recovery systems using Sliding-Vane Rotary Expanders, 3rd International Rotating Equipment Conference

[6] Murgia, S., Valenti, G., Colletta, D., Costanzo, I., Contaldi, G. 2017. Experimental investigation into an ORC-based low-grade energy recovery system equipped with sliding-vane expander

using hot oil from an air compressor as thermal source, Energy procedia 129C, 339-346
[7] Murgia, S., Costanzo, I., Valenti, G., Manfreda, P.P, 2019. Optimization of sliding-vane expanders for a low-enthalpy ORC energy recovery system, 11th International Conference of Compressors and their System, London City University
Murgia, S., Costanzo, I., Valenti, G., Fasani, F., Fumagalli, C., 2020. Test and Validation of a Control System for a Small-scale Energy Recovery System based on Sliding-vane Technology

IOP Conf. Series: Materials Science and Engineering 1180 (2021) 012032 doi:10.1088/1757-899X/1180/1/012032

Optimal Designed Experiments for Reliable Model Calibration of a fixed-speed Scroll Compressor with R410A and R32

Christian Vering*, Daniel Stopp, Tim Klebig, Valerius Venzik, Dirk Müller

RWTH Aachen University, Institute for Energy Efficient Buildings and Indoor Climate, Mathieustraße 10, 52074 Aachen, Germany

Email*: cvering@eonerc.rwth-aachen.de

Abstract. Mainly to provide heat, the building sector accounts for about 18 % of overall global CO2 emissions. By exchanging conventional technologies in the near future, heat pumps are a key technology to systematically decarbonise the heat supply chain. The efficiency of a heat pump strongly depends on its compressor's efficiency. Thus, the compressor is the subject of current research. The compressor performance depends on the fluid with its inlet and outlet conditions, an appropriate oil, and the mechanic compression principle. Based on that, the performance is expressed by the isentropic and volumetric efficiency. To describe a compressor in detail and fluid independent is challenging. Therefore, many theoretical studies have been conducted, while in comparison experimental studies are rare. To overcome this imbalance, this study focuses a detailed experimental investigation of a 4 kW Copeland scroll compressor with R410A and R32 on a fully automated compressor test stand. Within this paper, we present a method for in-depth uncertainty analysis, which ensures the usability and comparability of our experimental results to other experiments. We store our results in optimal calibrated simulation models based on optimal experimental design. This allows us to reduce experimental effort up to 70 % compared to full factorial experimental designs. To prove the method in a compressor and two fluids, we apply it to a fixed-speed scroll compressor with R410A and R32. The test stand is able to conduct all experiments automatically and part load behaviour of both refrigerants can be summarized in a model according to recent Literature with an overall uncertainty below 8 % for R410A. For R32, the method fails. Utilizing this method, we aim at open-source experiments in order to accelerate the solution of complex research questions.

Keywords: In-depth uncertainty analysis, compressor efficiency, model calibration, fully automated test stand

1. Introduction and State of the Art

Mainly to provide heat for space heating and domestic hot water, the building sector accounts for about 18 % of overall global CO2 emissions [1]. By exchanging conventional technologies in the near future, efficient heat pumps are a key technology to systematically decarbonize the heat supply chain. The efficiency of a heat pump strongly depends on its compressor's efficiency. Thus, the compressor is the subject of current research [2].

In general, the compressor performance depends on the refrigerant with its inlet and outlet conditions, an appropriate oil, and the mechanic compression principle and commonly the performance is expressed by the isentropic and volumetric efficiency [3]. Conventional refrigerants are hydrofluorocarbons (HFCs) like R134a or R410A, which have a high global warming potential (GWP). In order to reduce the GWP of refrigerants, different directives and legislations restrict the use of refrigerants with a high GWP. Thus, leading for example to the use of dihalogenoalkanes such as R32 with a GWP of 675. Since R410A and R32 differ in thermodynamic properties, their performance in the same compressor might differ as well [4].

To accurately qualify the compressor performance and use the insights for optimal compressor design, there is a need for combined investigations of experiments and modelling. In addition, both the complexity of experiments and the amount of data to be processed increases. Following the metaphor of the audio format MP3, we assume that in the future measurement data will also have to be stored efficiently and accurately in a defined compressed way in order to allow reasonable further processing such as comparison to other test stand results at other institutions or to serve as basis for calibration of further physical models.

In this paper, we therefore propose a method to store raw data efficiently in optimal calibrated models. To ensure the quality and reliability of the data, we do an in-depth uncertainty analysis regarding the measurement data uncertainty, the test stand bias and the calibration process error. Hence, the overall uncertainty of the final calibrated model, which stores the raw data, has three parts: (1) measurement uncertainty of the equipment, (2) bias uncertainties of the test stand and (3) storage uncertainty of the calibration process. Several compressor test stands already exist in the recent literature [5 - 7] as well as modelling approaches [3, 8, 9], which serve an overview of the recent state of the Art for uncertainty quantification and common modeling approaches.

Regarding experiments, Tello Oquendo et al. [5] measure scroll compressors with enhanced vapor injection in a climatic chamber, whereas Cueves and Lebrun [7] focus their work to investigate the impact of rotational speed on compressor efficiency in calorimeters. Yang et al. [10] investigate water-injected process-gas screw compressors to study the effects of rotational speed, injected water mass flow, and discharge pressure on the screw compressor performance. In [11], Zhang et al. study the potential of twin screw compressors for R513A as a R134a replacement. All of the aforementioned experimental studies have test stands according to the standard EN 13771-1 (2017) [5, 7] or give an overview of measurement accuracies of all sensors [10, 11]. However, a detailed in-depth analysis of measurement uncertainty according to GUM considering the test stand bias, was not reported to the best of the authors' knowledge. In addition, they do not share their measurement data stored in models open-access.

With respect to compressor modelling and storage of data, there is a common normative modelling approach according to DIN EN 12900 to model the compressor behavior regarding its input power and its refrigerant mass flow [9]. Besides DIN EN 12900, many different modelling approaches exist in the literature, which differ in black box, gray box and white box models.

In pioneering work, Corberan et al. [12] question the use of different models and polynomial robustness of 10C method according to DIN EN 12900 by analyzing polynomial approaches with less than 10 coefficients. They find that is it not necessary to employ a 10 coefficient polynomial for scroll compressors. For mass flow rate for example, three coefficients are sufficient. However, in order to fit corresponding surfaces, many experiments must be conducted. We therefore propose the optimal experimental design framework to calibrate a model based on the optimal number of experiments. To show the robustness of our method, we compare two common refrigerants (R410A and R32) within the same compressor.

The literature emphasizes, that many theoretical and some experimental studies exist, which show the need for reliable compressor models based on valid experimental data. However, there is a lack of detailed in-depth uncertainty analysis of the presented test stands. In order to make results of different institutions easier comparable and exchangeable, we introduce an extended approach to estimate the uncertainty of measurements. Based on measurement results, we store our results in models that are calibrated by the optimal number of experiments using optimal experimental design. This yields a calibrated model with uncertainty specifications, which easily can be exchange with others.

We therefore introduce the MoCaVal framework for automated model calibration and validation, which was recently presented by Vering et al. [13] and extend the framework iMoCaVal to a complete bias characterization. In addition, we conduct and evaluate experiments with R410A and R32 to investigate the fluid influence on model calibration and validation.

This publication systematically contributes to an extended characterization of compressor measurement data stored in models and therefore allowing direct comparability between different test stands. The paper is structured as follows:

- Chapter 2 shows the framework iMoCaVal of in-depth uncertainty analysis and optimal experimental design.
- Chapter 3 describes the compressor use case, investigated fluids, and first results obtained by the test stand and an extended bias specification.
- Chapter 4 presents measurement and calibration results for R410A and R32 and contains a discussion afterwards.
- Chapter 5 gives a summary and an outlook based on the obtained results.

2. Optimal Experimental Design applied to Compressors

In a previous paper, Vering et al. presented a framework for automated model calibration and validation (MoCaVal), which was applied to compare conventional calibrated simulation models from the literature to a model that was automatically set up using machine learning [13]. To ensure a slight overview, we present the highlights of MoCaVal in Chapter 2.1 leaving out the machine learning part. To assess the quality of the measurement data within MoCaVal, an uncertainty quantification was introduced using the Guide to the Expression of Uncertainty in Measurement (GUM) [14].

According to Figure 1, we extend MoCaVal by an extended in-depth uncertainty analysis to iMoCaVal in this paper, which also takes the test stand bias during measurements due to statistical analysis into account (Chapter 2.2). To make our results easily available and exchangeable, we store the data in optimal calibrated models, to the price of calibration uncertainty. We therefore use an optimal experimental design (OED) method to estimate the optimal number of experiments to sufficiently calibrate a compressor model (Chapter 2.3). This method is applied to two refrigerants in a scroll compressor which yields two data models that can be used open-access for usage or comparison.

2.1. *Introduction to MoCaVal and its extension to iMoCaVal*

According to Figure 1, iMoCaVal is a framework for automated in-depth uncertainty analysis, model calibration and validation for refrigerant compressors up to 30 kW nominal compressor power.

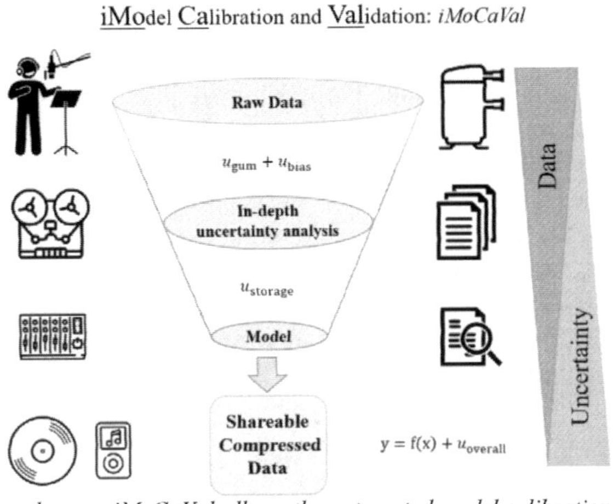

Figure 1: For a specific device under test, iMoCaVal allows the automated model calibration and validation for refrigerant compressors to efficiently and accurately store data and make it reliably shareable.

A detailed description of the test stand, which consists of a vapor compression process without evaporator, is provided in Vering et al. [13] including the accuracy of the measurement equipment.

Within the framework iMoCaVal, we differentiate between three parts: (I) raw data, (II) in-depth uncertainty analysis (III) model output.

(I): From an input point of view, the device under test needs to fulfill the requirements from Table 1 and theory on the subject of model discrimination must be available to exploit the potential of the framework.

(II): Using the test stand and different model formulations, iMoCaVal determines optimal experimental designs and performs necessary experiments. Special focus in this paper is the extension of measurement uncertainty according to GUM by in-depth statistical uncertainty analysis.

(III) As a result, MoCaVal provides calibrated models of the compressor indicating the quality of the model.

Within this paper, we prove the applicability of MoCaVal to different fluids with additional in-depth analysis. We therefore conduct experiments with two different refrigerants from thermodynamic point of view (R410A and R32) with respect to their working envelopes in the same scroll compressor to find the limitations of MoCaVal. To ensure comparability and usability of measurement results, we introduce an in-depth uncertainty analysis. Based on this, we store the measurement data in simulation models from the literature. As it is common, we use the 10C model according to DIN EN 12900 (Eq. 1) and compare it to an advanced model based on similarity theory from Miranda-Mendoza et al. (Eq. 2 and Eq. 3), which was introduced in the literature to model the performance of the halogenated refrigerants R134a, R450A, R1234yf and R1234ze [8, 9]:

$$f(C_\mathrm{i},T_\mathrm{i}) = C_1 + C_2T_\mathrm{e} + C_3T_\mathrm{c} + C_4T_\mathrm{e}^2 + C_5T_\mathrm{e}T_\mathrm{c} + C_6T_\mathrm{c}^2 + C_7T_\mathrm{e}^3 + C_8T_\mathrm{e}^2T_\mathrm{c} + C_9T_\mathrm{e}T_\mathrm{c}^2 + C_{10}T_\mathrm{c}^3 \tag{1}$$

$$\eta_\mathrm{is}(\Pi,a_\mathrm{i},n,V_\mathrm{u},T_\mathrm{i},h_\mathrm{i}) = \Pi^{a_o}\cdot\left(\frac{n^3V_\mathrm{u}}{\Delta h_\mathrm{ec,is}^{1.5}}\right)^{a_1}\cdot\left(\frac{\frac{T_\mathrm{e}+T_\mathrm{c,is}}{2}-T_\mathrm{oda}}{T_\mathrm{oda}}\right)^{a_2} \tag{2}$$

$$\lambda(\Pi,\mathrm{b_i},\rho_\mathrm{i},p_\mathrm{i},n,V_\mathrm{u}) = \Pi^{b_o}\cdot\left(\left(\frac{\rho_\mathrm{e}}{p_\mathrm{e}}\right)^{1.5}n^3V_\mathrm{u}\right)^{b_1} \tag{3}$$

Where $f(C_\mathrm{i},T_\mathrm{i})$ denotes one function for electrical power Pel and one function for refrigerant mass flow $\dot{m}_\mathrm{ref}$, η_is is the isentropic efficiency, λ the volumetric efficiency, $C_\mathrm{i}, a_\mathrm{i}, b_\mathrm{i}$ are calibration parameter, T_e ($p_\mathrm{e}, \rho_\mathrm{e}$) is the temperature (pressure, density) at compressor inlet and T_c ($T_\mathrm{c,is}$) the compressor outlet temperature (compressor outlet temperature at an isentropic compression). The compressor volume is V_u, the enthalpy difference between inlet state and isentropic outlet state is $\Delta h_\mathrm{ec,is}$, Π is the pressure ratio that Miranda-Mendoza et al. determined using similarity theory and T_oda is the outdoor air temperature. For the sake of completeness, we denote n as rotational speed, although it is set constant due to fixed speed compressor investigations.

Obviously, the refrigerants investigated by Miranda-Mendoza to exchange R134a by R450A, R1234yf or R1234ze differ from the refrigerants from this paper (R410A/R32) in terms of molar mass and saturation-pressure characteristics. Nevertheless, we pose the hypothesis that it is possible to cover the behavior of R410A and R32 within the same physical model to at least some extend. To assess the quality of measurement, ensure further usability, and comparability of our results, we conduct an in-depth uncertainty analysis initially, which bases on the following method.

2.2. In-depth uncertainty analysis and test stand bias

Most measurands Y cannot be measured directly. They are determined from other quantities X_i.

$$Y = f(X_1, X_2, \dots, X_N) \tag{4}$$

For an uncertainty analysis it is therefore necessary to consider uncertainties of all inputs. Type B evaluation of the GUM [14] provides a method to estimate the standard uncertainty $u(x_i)$ of an input quantity Xi based on available information on its possible variability. Those information in general define lower and upper bounds a_- and a_+ for X_i. If there is no specific knowledge about the contribution of Xi within the interval a_- to a_+, one can only assume that it is uniformly distributed.

In any case of a distribution more similar to a normal distribution, a rectangular distribution is an estimation to the safe side. The expected value is the midpoint of the interval,

$$\mathrm{x_i} = \frac{a_- + a_+}{2}, \tag{5}$$

and its variance is given by

$$u^2(x_i) = \frac{(a_+ - a_-)^2}{12}. \tag{6}$$

By assuming the intervall length to be $2a$ the standard uncertainty can be expressed as

$$u(x_i) = \frac{a}{\sqrt{3}}. \tag{7}$$

The combined standard uncertainty $u_c(y)$ is based on a first-order Taylor series approximation of Eq. 4. In case that all input quantities are independent, $u_{\mathrm{gum}}(y)$ is the positive square root of the combined variance $u_{\mathrm{gum}}^2(y)$ given by

$$u_{\mathrm{gum}}^2(y) = \sum_{i=1}^{N} \left(\frac{\partial f}{\partial x_i}\right)^2 u^2(x_i) \tag{8}$$

where f is the function given in Eq. 4 and $u(x_i)$ the standard uncertainty evaluated as described above. This method to assess the combined standard uncertainty is used for all target variables.

For a better understanding we show how it is applied to the isentropic efficiency. The isentropic efficiency is calculated from the enthalpies of the different states.

$$\eta_{\mathrm{is}} = \frac{h_{2\mathrm{is}} - h_1}{h_2 - h_1} \tag{9}$$

The uncertainty of the isentropic efficiency is therefore a function of those enthalpies and its uncertainties. The enthalpies are derived from equations provided by the NIST REFPROP database [15]. The uncertainties given by this database don't include specific estimates for the enthalpies' uncertainties. They were thus initially approximated by the percentage uncertainty of the heat capacity in the *MoCaVal* introduction. Here, the uncertainty estimation is applied to an approximate function to describe the isentropic efficiency which includes only heat capacities and temperatures.

$$\eta_{\mathrm{is}} = \frac{\overline{c_{\mathrm{p},1\to2\mathrm{is}}}\,(T_{2\mathrm{is}} - T_1)}{\overline{c_{\mathrm{p},1\to2}}\,(T_2 - T_1)} \tag{10}$$

In this way the specific uncertainties for heat capacities and temperatures given by the NIST REFPROP database and the manufacturers' data sheets can be used.

Furthermore, to quantify the repeatability, which means the test stand bias during a measurement of a quasi steady-state operating point, exemplarily two measurement points are selected and measured several times. Based on these detailed measurements, a test stand bias u_{bias} can be recalculated.

As introduced, storing the measurement data in a calibrated model for the easy open-access use finally leads to the price of a further uncertainty u_{storage}, which needs to be considered as well. We therefore geometrically sum up all parts of uncertainties to ensure an usability and comparability of our measurements:

$$u_{\mathrm{overall}}^2 = u_{\mathrm{gum}}^2 + u_{\mathrm{bias}}^2 + u_{\mathrm{storage}}^2 \tag{11}$$

2.3. Optimal Experimental Design for Experimental Effort Reduction

To calibrate a simulation model accurately, reliable measurement data is necessary. Additionally, to compare data to other measurements or to further exploit the data by external researchers, an in-depth uncertainty analysis allows an universal use of the data. The data is extracted from measurements, which need to be conducted. As experiments are time and resource consuming, they must be carefully planned, executed and evaluated.

In order to fully qualify a compressor in terms of energy, the entire operating envelope must be measured. According to the classical design of experiments, the analysis of the influencing variables results in a full factor plan, which can involve considerable measurement effort. With respect to the models described by Equations 1-3, the number of experiments from the full factorial plan significantly exceeds the number of calibration parameters. To minimize the time and resources required for model calibration, Optimal Experimental Design (OED) allows the number of experiments to be reduced.

In this work, we combine OED with in-depth uncertainty analysis to calibrate a model that accurately predicts compressor behavior and can be reliably used in a variety of applications, especially by external users who may not have the ability to perform experiments. In this context, this procedure might lead to "open-access" experiments, which to our opinion is the future of cooperation in research.

To reduce the experimental effort from the full factorial plan to a minimized number of experiments with OED, it is necessary to define criteria that can evaluate what information content a measurement point has. Such criteria can be chosen when model discrimination has already been completed. This means that the model to be calibrated must already be known in advance. Therefore, the choice of the initial model is crucial for the quality of the final results. In this paper, we choose existing models from the literature according to Equation 1-3 and select the criteria D-optimal OED, which is widely applied in the literature [16].

The variance-covariance matrix serves as basis for the optimization algorithm, which reduces the number of experiments. The variance-covariance matrix is also called Fisher Information Matrix (FIM) and is calculated based on all candidate combinations from the full factor plan. The D-optimal Design maximizes the determinant of the FIM, which determines the model calibration parameter in the best possible way. In contrast, I-optimal Design for example would be applied when the prediction accuracy for the target variable should be optimized. In this context, we want to emphasize that there is a wide range of optimization methods that cannot all be considered here. We use the D-optimal design, which is not necessarily the optimal choice. The choice of the correct optimal design should be investigated in following works.

3. Optimal Experimental Design applied to Compressors

Within this contribution, we calibrate models based on measurements for a scroll compressor for the two refrigerants: R410A and R32. Before the calibration can be performed, we want to introduce the two refrigerants and show how the recording of raw data works in a fully automated way.

3.1. R410A and R32 and their resulting envelopes

The device under test is the Copeland compressor ZHI18K1P-TFM with 4 kW nominal power and R410A as refrigerant. R410A is a quasi-azeotropic mixture of 50 % R32 and 50 % R125 and has a global warming potential (GWP) of 2088, which is why a phase-out of R410A is necessary in the near future. In contrast, R32 has a GWP of 675.

The thermodynamic behavior of both refrigerants and R134a is summarized in Figure 2 in their corresponding pressure-enthalpy diagrams. While the evaporation enthalpy of R32 is about 20 % higher compared to R410A, the resulting operating envelope of R32 is smaller due to higher compressor outlet temperatures at similar pressure ratios.

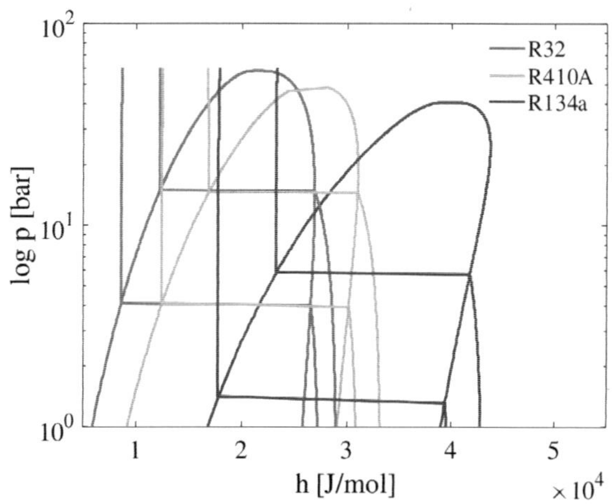

Figure 2: Comparison of R410A (light blue), R32 (dark blue) and R134a (red) in a pressure-enthalpy-diagram considering two-phase region and different isothermal curves.

With respect to the models from Equation (2) and (3), which are calibrated in [8] for R134a, we compare the fluid behavior additionally in Figure 2. Obviously, the refrigerants have different thermodynamic behavior, which might complicate the model calibration. That means for our measurements, that the envelope of R32 will be smaller as well as its resulting full factor plan for OED, because of its high compressor outlet temperature even at small pressure ratios. The operation envelopes deduced for the full factor plan are shown in Figure 3 (a) for R410A and Figure 3 (b) for R32.

As introduced in Equation 1-3, the target variables for the model calibration are the electrical power, refrigerant mass flow rate (Eq. 1) and the volumetric (Eq. 2) and isentropic efficiency (Eq. 3). The limits for the operation envelope for the inlet pressure is for R410A $p_{1,min} = 4$ bar and $p_{1,max} = 14$ bar. The limits for the compression ratio are 2.5 and 6.5, but since the outlet pressure p_2 is also limited between 20 and 44 bar or up to 130 °C compressor outlet temperature according to Table 1, not every combination of p_1 and π can be measured.

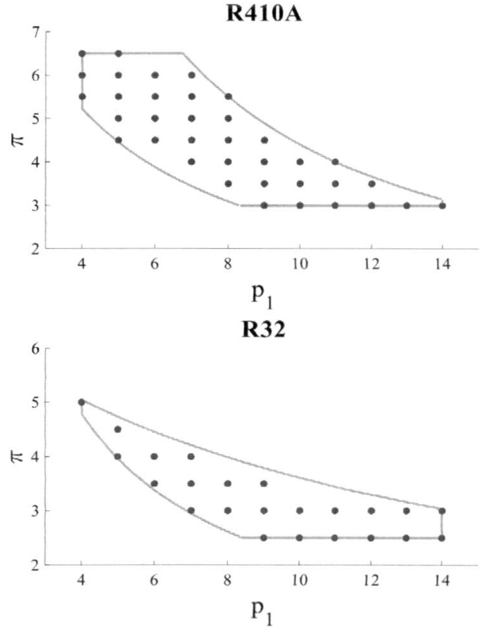

Figure 3: Operation envelope of R410A (a, top) and R32 (b, bottom) as a function of inlet pressure and pressure ratio yielding 36 operating points for R410A and 23 operating points for R32.

In Figure 3a, the R410A candidate set of the full factor plan is shown. Within the limits, 11 factor steps are planned for p_l and 8 factor steps for π, so that the step size is 1 bar for p_l and 0.5 for π, respectively. This results in 36 operating points to cover the factor space uniformly.

In Figure 3b, the R32 candidate set of the full factor plan is shown. Within the limits, again 11 factor steps are planned for p_l and due to the vapor pressure characteristics 6 factor steps for π, so that the step size is 1 bar for p_l and 0.5 for π, respectively. This results in 23 operating points to cover the factor space uniformly.

3.2. Raw data of R410A and R32 measurements

Within the operational envelope, we can conduct experiments fully automatically for both refrigerants by implementing an experimental design. During a measurement, the test stand controller ensures steady-state conditions by adjusting the inlet temperature and pressure as well as outlet pressure as function of valve positions at the heat sink (water) and in the refrigeration cycle.

To reach a steady-state the inlet temperature and the inlet and outlet pressure are controlled. A measurement point is defined as quasi steady-state if the control variables lay within a certain corridor for a predefined period. The limit for the inlet temperature is ± 1 K. The tolerance for both pressures is ± 2 % absolute.

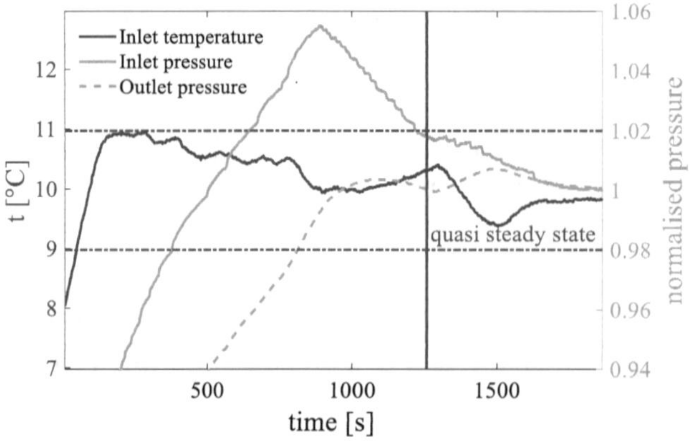

Figure 4: Trend of the three control variables inlet temperature and inlet and outlet pressure (normalised) during one entire measurement of an operating point with R410A from start-up to quasi steady-state measurement for about 2000 s.

A measurement of one certain operating point is valid if the quasi steady-state condition is held for 600 seconds. Figure 4 shows an automatic start-up and holding of an operating point for about 2000 s with the evaluated period of 600 s in quasi steady-state. The initial overshoot of the inlet pressure is caused by its dependency on the (strongly increasing) outlet pressure in the given system design. After an operating point was successfully measured, the next operating point is considered automatically.

The Figure shows of how the trends of those control parameters look like. As temperature has a valid offset ± 1 K, it is shown on an absolute scale as a function of time. The pressure scale is normalized by their setpoints to show their validity in a corridor of ± 2 % absolute. Since this paper aims at comparing both fluids in the same compressor and calibrating models based on optimal designed measurements, there is no optimization of the test stand control to further speed up measuring individual operating points. This can be part of future investigations.

Figure 4 shows that the test stand is able to conduct experiments fully automatically and hold a set point within quasi steady-state conditions using pre-defined corridors for our target values. In addition, it becomes clear that so far we are not able to set and maintain real steady-state conditions using the control parameter, which will be part of future investigations.

Nevertheless, we will use the results of 600 s quasi steady-state measurements to calculate the mean values and standard deviations for all target values obtained in this period and evaluate it as source of uncertainty (test stand bias) in the already introduced in-depth uncertainty analysis in Chapter 4.1.

Besides temperature and pressure levels during the whole measurements, the test stand records large number of further parameters. Some of them serve to monitor the performance of compressor and test stand. Others are required to evaluate the target variables afterwards. Therefore, Figure 5 shows for completeness two of those parameters – outlet temperature and refrigerant mass flow. An integral overview of the test stand is presented in Vering et al. [13].

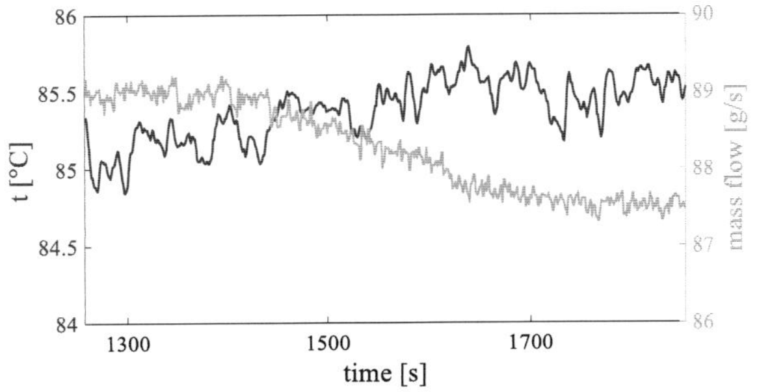

Figure 5: Trend of the outlet temperature and refrigerant mass flow during one measurement with R410A within an evaluation period in quasi steady-state conditions.

4. Results

Using the theory of Chapter 2 and the raw data of Chapter 3, we are able to calculate the measurement uncertainty by an in-depth uncertainty analysis in Chapter 4.1. After that, we use the data and the models from Eq. 1-3 to calibrate them according to D-optimal experimental design in Chapter 4.2. Lastly, we discuss our findings in Chapter 4.3

4.1. In-depth uncertainty analysis

Figure 4 According to Equation 11, the overall uncertainty of calibrated models denotes the geometric sum of measurement uncertainty, test stand bias and storage uncertainty due to calibration process. While in a simplified form the measurement uncertainties are already introduced in Vering et al. [13] in Chapter 2.2 the already existing model (Eq. 9) could be extended a detailed model (Eq. 10). It will be applied to measurement data. Therefore, the test stand bias must be qualified. This is done by repeatability tests for different operating points. Within the scope of this paper, the repeatability tests are performed for two measurement points using R410A. These points are located in different areas of the compressor envelope. In addition, one is located on the edge of the envelope whereas the other one lies more in the center. In this way the repeatability should be ensured to be valid for the whole compressor envelope. The selected measurement points are summarized in Table 1. Detailed analysis of further operating points, refrigerants and compressor types are recommended for further studies.

Table 1: Measurement points for repeatability tests. One taken from the edge of the envelope (Point 1), one taken from the centre of the compressor envelope (Point 2).

	Point 1	**Point 2**
Compression ratio π [-]	5.5	3.5
Inlet pressure p_1 [bar]	4	11

As introduced, the evaluation of quasi steady state operating conditions of each measurement lasts 600 s. With a sample rate of one sample per second this leads to 600 samples for each measurement. For Point 1 three measurements are performed. Point 2 is even measured five times to ensure repeatability over a wider range of measurements. All those measurements are evaluated separately before interpreting them all together using isentropic efficiency as a use case.

The test stand conducted all eight repeatability tests fully automatically and stored the results in result files, which are evaluated in the following. According to Table 3, the deviation during the individual measurements is comparatively small, especially for the point located near the center of the compressor envelope. For both points the fluctuation lies within less than 3 %. Evaluated for all measurements of the same point the fluctuation is around 5 % in both cases. Table 2 shows the fluctuation ranges and standard deviations for both points analyzed individually as well as combined.

Neglecting measurement uncertainty in the bias quantification, the standard deviations for the isentropic efficiency of all repeatability measurements are plotted in Figure 6. The measurement averages are indicated as three (five) separate operating points in blue in Figure 6a (Figure 6b). Calculating the average of all three (five) operating points for 600 s quasi steady-state conditions yields the overall average in red. Using error bars, we denote the standard deviation of each 600 s quasi steady-state conditions. The standard deviation (std.) is here chosen as a model for the test stand bias, which is very accurate with respect to this 8 measurements. As an estimate to the safe side, we use the higher standard deviation as test stand bias for all measurement points, which now is defined according to Eq. 11 as $u_{\text{bias}} = 1.52$.

Table 2: Fluctuation and standard deviations (std.) for individual measurements and evaluated all together for the isentropic efficiency.

	Point 1	Point 2
Max. fluctuation in measurements	2.35 %	1.85 %
Max std. in measurements	0.56 %	0.36 %
Fluctuation all measurements	5.19 %	4.91 %
Max std. all measurements	1.52 %	1.23 %

As introduced, Figure 6 shows the average quasi steady-state operating point with its resulting standard deviation as test stand bias. In addition, we plot the measurement uncertainty according to Eq. 10 indicated by the box around those measurement points. Thus, in this area we trust the measurement data according to GUM method.

It can be seen that the combined uncertainty of the isentropic efficiency is of the same order of magnitude compared to the standard deviations of a single measurement (blue) and smaller than the standard deviation of all series of measurement (red). It amounts to 0.49 % and 0.63 %, respectively. What is remarkable about the combined uncertainties is that they don't vary within the measurements of one measurement point. They are not sensitive to small variations in the location of a measurement point.

It is apparent, that the uncertainty estimation taking into account heat capacities and temperatures is significantly smaller than the one in [13] using Eq. 9. This shows that the assumption that the uncertainty of the enthalpies is equal to the heat capacities' uncertainty has a large influence on the combined uncertainty of the isentropic efficiency.

It is therefore of enormous importance to make reasonable estimates and assumptions for the input variables' uncertainties. Again as an estimate to the safe side, we use the uncertainty model according to Eq. 10 with the worse uncertainty results, which yields the uncertainty of GUM to be: $u_{\text{gum}} = 0.63$.

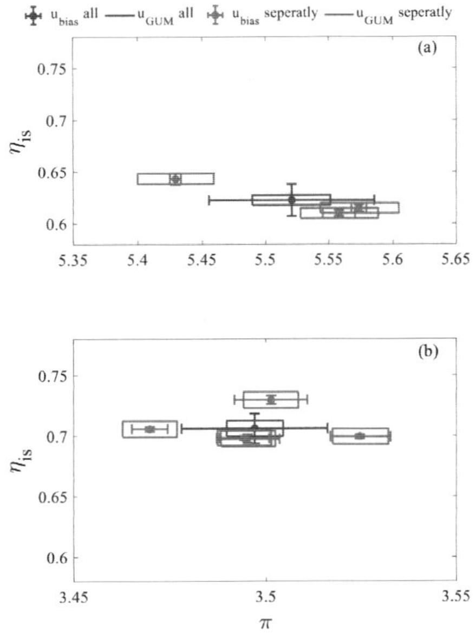

Figure 6: Standard deviations and uncertainties of the isentropic efficiency determined with Eq. 10 for (a) measurement point 1 and (b) measurement point 2 for R410A.

4.2. OED for test time reduction

Following Eq. 11, we can assess the overall uncertainty of a representative and reliable model, by adding the term u_{storage}, which is a result of model calibration process. If we want to fit a model to real measurement data in general, there is a natural deviation between model result and measurement data due to numerical issues or under- and overfitting the data using a model. We therefore measure the difference between measurement data and model result and compute u_{storage} as standard deviation. Giving this information allows external researcher to reliably compare the obtained data stored in our model. Which is a step towards open source experiments.

To calculate u_{storage} the models of Eq. 1-3 must be calibrated to measurement data. To reduce the influence of under and overfitting, we use OED with the D-optimal design to optimally design the experimental plan. Detailed analysis for R410A were already presented in Vering et al. [13], which showed a significant reduction of experimental effort, while simultaneously maintaining a certain degree of accuracy. This approach is extended by R32 in this paper.

Based on the operating envelopes in Figure 3, we show the results of D-optimal design for both refrigerants and the three modelling approaches from Eq. 1-3 in Figure 7. Using the method of full factor plans, measuring R410A needs 36 quasi steady-state operating points. In order to calibrate the models, D-Optimal design reduces number of experiments for C10 to 10 experiments, η_{is} to 3 experiments and λ to 2 experiments as absolute minimum. The most of all measurement points lie on the edge of the envelope. It is interesting that an increase in the number of experiments does not necessarily lead to an increase in the variety of measurement points. To reduce the scatter of measurements and thus to increase reliability, points already measured at the edge of the envelope are first approached several times before the D-optimal design proposes further measurement points.

The full factor plan results in 36 measurement points. Applying D-optimal design to the compressor envelope reduces the number of experiments to 10, 3 and 2, just as for R410A. This behavior fits the mathematical expectations. Compared to R410A, however, other operating points and boundary ranges are targeted, which underlines the functionality of the D-optimal design. Using this experimental setups, the test stand can automatically execute all experiments. The obtained measurement results can subsequently be used for calibration to determine u_{storage}. The calibration results are shown in the Annex Table A1.

R410A: Compared to the full factor (FFP) plan, which required 51 experiments, Vering et al. reduced experimental effort significantly, while always ensuring the prediction accuracy of the model below 11 % at worst, which was about 2 % less accurate compared to worst deviation between model and experimental data using FFP.

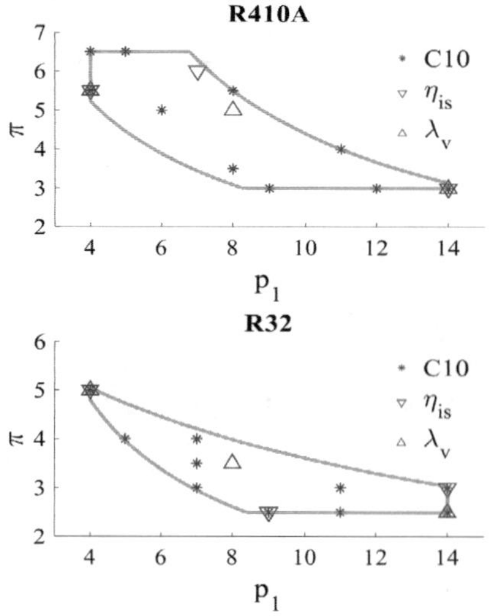

Figure 7: D-optimal resulting experimental plans to calibrate the models according to Eq. 1-3 for R410A and R32.

Finding a reasonable trade-off, they were able to reduce the experimental effort by about 75 % to only 12 experiments, which means two experiments above the absolute minimum (10) proposed by OED. Compared to FFP, the models still maintain an average prediction accuracy below 5 % for the different modelling approaches, which was at maximum only 1 % less accurate than FFP. In order to investigate the fluid dependency of OED for compressor model calibration and hence $u_{storage}$ for the final assessment, we conduct the same method for the same compressor using R32 measurements in the following.

Table 3: Evaluation of model accuracy for modelling approaches of Eq. 1-3 for R410A and R32. For both refrigerants, there is a comparison of full factor plan to a D-optimal experimental design.

	R410A (36)	**R410A (10/3)**	**R32 (23)**	**R32 (10/3)**
10C mean(P_{el}/m)	0.3/0.21	**0.6/0.33**	0.08/0.11	0.1/0.12
10C max(P_{el}/m)	1.82/0.7	2.55/1.26	0.29/0.44	0.51/0.94
η_{is} (mean)	2.36	**3.76**	30.94	34.57
η_{is} (max)	6.28	7.4	33.6	38.1
λ (mean)	3.78	**1.98**	1.26	5.16
λ (max)	5.13	3.64	2.04	6.63

R32: Compared to the calibration of R410A, which fits the uncertainties of the initial Miranda-Mendoza paper, it is not possible to calibrate the model for isentropic efficiency with sufficient uncertainty, while it is possible to fit the volumetric efficiency sufficiently to measurement data.

These results confirm the understanding that volumetric efficiency is dependent on the pressure ratio, but not so much on the refrigerant. However, modeling of isentropic efficiency is difficult due to the strong refrigerant dependence due to the molar mass, isentropic compression exponent and flow losses that are dependent on the density as well. Thus we confirm our initial hypothesis, that it is possible to model R410A within the Miranda-Mendoza formulation, while it is not possible to model the behavior of R32 sufficiency. Hence, we chose an average storing uncertainty for R410A for the discussion in the next Chapter only.

4.3. Discussion

Using an in-depth uncertainty analysis and applying it rigorously to uncertainty in measurement, test stand bias and calibration uncertainty, we are now able to calculate the final overall uncertainty of final calibrated simulation modes.

According to the GUM notation, we determined as overall uncertainties from Eq. 11 for the models from Eq. 1-3 for

$$P_{\mathrm{el}}(C_{\mathrm{i}}, T_{\mathrm{i}}, u_{\mathrm{i}}) = f(C_{\mathrm{i}}, T_{\mathrm{i}}) \pm u_{\mathrm{P,overall}} \tag{12a}$$

$$\dot{m}_{\mathrm{ref}}(C_{\mathrm{i}}, T_{\mathrm{i}}, u_{\mathrm{i}}) = f(C_{\mathrm{i}}, T_{\mathrm{i}}) \pm u_{\dot{\mathrm{m}},\mathrm{overall}} \tag{12b}$$

$$\eta_{\mathrm{is}}(\Pi, a_{\mathrm{i}}, n, V_{\mathrm{u}}, T_{\mathrm{i}}, h_{\mathrm{i}}, u_{\mathrm{i}}) = \eta_{\mathrm{is}}(\Pi, a_{\mathrm{i}}, n, V_{\mathrm{u}}, T_{\mathrm{i}}, h_{\mathrm{i}}) \pm u_{\eta,\mathrm{overall}} \tag{13}$$

$$\lambda(\Pi, \mathrm{b_i}, \rho_{\mathrm{i}}, p_{\mathrm{i}}, n, V_{\mathrm{u}}, u_{\mathrm{i}}) = \lambda(\Pi, \mathrm{b_i}, \rho_{\mathrm{i}}, p_{\mathrm{i}}, n, V_{\mathrm{u}}) \pm u_{\lambda,\mathrm{overall}} \tag{14}$$

The calculated minimal and maximal overall uncertainties for R410A are summarized in Table 4. The corresponding calibration parameter for external user are listed in the appendix.

Table 4: Overall minimal and maximal uncertainties for power consumption, refrigerant mass flow rate, isentropic and volumetric efficiency.

	Min (abs / %)	**Max (abs / %)**
P_{el} (W)	100.97 / 2.05	150.98 / 2.74
$\dot{m}$ (g/s)	0.7720 / 1.32	1.6237 / 1.72
η_{is} (-)	0.0485 / 6.56	0.0574 / 7.84
λ (-)	0.0303 / 3.66	0.0469 / 5.18

Within Chapter 4.1 we showed that our test stand is able to conduct experiments fully automatically with a low measurement uncertainty as well as low test stand bias. We also found the control parameter not to be optimized yet, since the target values are not fully formed steady-state, but quasi steady state. In addition, the duration of a quasi steady-state period is not optimized, yet. Both aspects should be considered in further investigations to further decrease the test stand bias.

According to recent findings in the literature [12], we can also state that it is easier to calibrate a model for the electrical power consumption or the refrigerant mass flow than for isentropic or volumetric efficiency, which is indicated by all mean and maximum deviations in Table 3. In addition, model discrimination itself is of high importance to get sufficient prediction accuracy due to model calibration based on experimental data. Nevertheless, Optimal Experimental Design is a promising approach to reduce experimental effort significantly. For further studies, we recommend the investigation of other, more complex compressor types and models such as inverter driven or EVI-setup to prove the robustness of calibration. Finally we should focus further optimal designs or Gaussian Processes to find the best one for reliable compressor model calibration.

Both aspects increase the quality of efficient storage of measurement data, which is absolute requirement for efficient data exchange between different institutions. Efficient storage in calculation models yields an easier and universal comparison to other measurement results and a more universal usage of measurement data by external stakeholder. By increasing complexity of research topics and increasing amounts of data, it is thus our next step towards cooperation in research to conduct experiments in the most open-access way. One promising approach here might be living labs, which even take non-experts into account to conduct experiments. We therefore highly recommend to share as much knowledge as possible to increase quality of research findings in the future and to accelerate research even at increasing complexity.

5. Conclusion

Within this paper, we present a method for efficient compressor model calibration and reliable storage of measurement data using in-depth uncertainty analysis.

The overall uncertainty to store measurement data reliable and efficient within a model, remains as geometric sum of measurement uncertainties, test stand bias, and uncertainty of calibration. For measurement uncertainty we compared to different approaches and found a more accurate model using isobaric heat capacity and temperatures for uncertainty estimation of enthalpies instead of using uncertainty for enthalpies. The test stand bias was determined by 8 repeatability tests, which indicated a robust reproducibility and the worst test stand bias to be below 2.45 %.

Lastly, D-optimal Design of Experiments reduces experimental effort up to 75 % (55 %) from 36 (23) to 10 (10) experiments for R410A (R32), respectively, while maintaining the uncertainty compared to the full factor plan for both refrigerants. Based on experimental results, we calibrated simplified models from the literature to fully describe the behavior of the power consumption, the refrigerant mass flow, isentropic, and volumetric efficiency. We showed that the average uncertainty due to storage has hardly decreased for 10C (η_{is} / λ) from 0.37 % to 0.40 % for power consumption and from 0.52 % to 0.62 % for refrigerant mass flow rate (η_{is}: from 3.88 % to 4.10 % / λ from 1.48 % to 1.79 %) for R410A, respectively. With respect to R32 we showed that it was not possible to calibrate the model of isentropic efficiency sufficiently. The calibration of 10C and volumetric efficiency leads the same magnitude of deviations as R410A. Within a detailed discussion, we deduce to further improve the quality of data storage by optimization of test stand bias quantification and performing more detailed model discrimination a-priori. In addition, we recommend to investigate the influence of compressor type, model complexity and optimal design method or even Gaussian Processes. All following investigations should aim at what we call open-access experiments, which can be conducted for example in living labs to be able to answer the ever more complex research questions more quickly. This might even include external stakeholders to take part in experiments to allow a further look at the respective subject matter.

Acknowledgements
This project was funded by the European Regional Development Fund (ERDF) (ERDF-0500029).

References

[1] IEA. Renewables 2019, www.iea.org/reports/renewables-2019, Paris: IEA; 2019. Available at: ww.iea.org/reports/renewables-2019.

[2] Roskosch, D., Venzik, V., Schilling, J., Bardow, A. and Atakan, B. (2021), Beyond Temperature Glide: The Compressor is Key to Realizing Benefits of Zeotropic Mixtures in Heat Pumps. Energy Technol. 2000955.

[3] Rasmussen, B. D., Jakobsen, A. (2000), Review of Compressor Models and Performance Characterizing Variables, International Compressor Engineering Conference, Paper 1429.

[4] Xu, X., Hwang, Y., & Radermacher, R. (2013). Performance comparison of R410A and R32 in vapor injection cycles. International Journal of Refrigeration, 36(3), 892-903.

[5] Tello O., Fernando M.; Navarro Peris, E.; Gonzálvez M., José; Corberán, J. M. (2016): Performance of a scroll compressor with vapor-injection and two-stage reciprocating compressor operating under extreme conditions. International Journal of Refrigeration 63, S. 144–156.

[6] Brendel, L.; Braun, J.; Groll, E. (2020): Test Stand to Investigate a Vapor Compression Cycle at Varying Orientation and First Experimental Results. IIR International Rankine 2020 Conference - Heating, Cooling and Power Generation.

[7] Cuevas, C. ; Lebrun, J. (2009), Testing and modelling of a variable speed scroll compressor. Applied Thermal Engineering 29, 469-478.

[8] Mendoza-Miranda, J. M.; Salazar-Hernández, C.; Carrera-Cerritos, R.; Ramírez-Minguela, J. J.; Salazar-Hernández, M.; Navarro-Esbrí, J.; Mota-Babiloni, A. (2018): Variable speed liquid chiller drop-in modeling for predicting energy performance of R1234yf as low-GWP refrigerant,

International Journal of Refrigeration 93, 144–158.
[9] DIN EN 12900:2013-10: Kältemittel-Verdichter - Nennbedingungen, Toleranzen und Darstellung von Leistungsdaten des Herstellers.
[10] Yang, Q., Liu, C., Zhang, Q., Liu, G., Zhao, Y., & Li, L. (2018). Experimental investigation of the water–injected process-gas screw compressor. Proceedings of the Institution of Mechanical Engineers, Part E: Journal of Process Mechanical Engineering, 232(1), 3-11.
[11] Zhang, Z., Wang, Y., Wu, X., Pan, X., & Xing, Z. (2020). Theoretical and experimental research on the performance of twin screw compressor using R513A as R134a replacement. Proceedings of the Institution of Mechanical Engineers, Part E: Journal of Process Mechanical Engineering, 0954408920951141.
[12] Corberan, J. M. et al. (2021), A critical analysis of the characterization of scroll compressors mass flow rate, IIR Compressors conference, 2021, Paper 0415.
[13] Vering, C. et. al. (2021), Automated Model Calibration and Experimental Validation of Compressor Simulation Models: Comparison of Optimal Experimental Design and Machine Learning, submitted paper to International Journal of Refrigeration.
[14] JCGM 100:2008, Evaluation of measurement data — Guide to the expression of uncertainty in measurement.
[15] Lemmon, E.W., Bell, I.H., Huber, M.L., McLinden, M.O. NIST Standard Reference Database 23: Reference Fluid Thermodynamic and Transport Properties-REFPROP, Version 10.0, National Institute of Standards and Technology, Standard Reference Data Program, Gaithersburg, 2018
[16] Pukelsheim, F. (2006). Optimal design of experiments. Society for Industrial and Applied Mathematics.

Table A1: Model parameter for Eq. 1-3 after calibration.

	R410A		R32	
	P_{el}	$\dot{m}_{\mathrm{ref}}$	P_{el}	$\dot{m}_{\mathrm{ref}}$
C1	24911	-53.38	-959.9	58.52
C2	1146.7	-4.0002	-200.9	0.4207
C3	-1268.2	8.247	270.9	0.3734
C4	21.29	-0.08908	-3.788	-0.02756
C5	-43.09	0.2625	9.183	0.08151
C6	24.9	-0.1557	-5.029	-0.0103
C7	0.1331	-6.32e-04	-0.0265	-5.145e-04
C8	-0.3991	0.0023	0.077	0.00127
C9	0.4022	-0.00243	-0.0995	-9.27e-04
C10	-0.1456	9.34e-04	0.0479	6.054e-05

	R410A		R32	
	η_{is}	λ_{v}	η_{is}	λ_{v}
a0	-0.3209	-	0.08395	-
a1	-0.00508	-	0.03521	-
a2	-0.0187	-	0.08711	-
b0	-	-0.08337	-	-0.08808

IOP Conf. Series: Materials Science and Engineering 1180 (2021) 012033 doi:10.1088/1757-899X/1180/1/012033

Baseline testing of a variable-speed water-cooling Chiller according AHRI standard 550/590

A J Hoess[1], N P Salts[2], D Ziviani[3], J E Braun[4] and E A Groll[5]

Purdue University, School of Mechanical Engineering, Ray W. Herrick Laboratories, West Lafayette, IN 47906-2099, USA

[1] ahoess@purdue.edu
[2] nsalts@purdue.edu
[3] dziviani@purdue.edu
[4] jbraun@purdue.edu
[5] groll@purdue.edu

Abstract. A new experimental test facility was designed to test vapor compression chillers of up to 145.9 refrigeration tons (approximately 513 kW). To implement such a large-capacity test facility in a university laboratory research environment, the system design is based on regenerative shell and tube heat exchangers. The capacity load on the chiller can be controlled via a bypass valve that allows a flow change over the internal heat exchangers. In addition to the internal sensors of the chiller, a variety of sensors in the system ensures a detailed recording of data as well as an automation of the test procedure.

The test facility is constructed and commissioned to execute the performance according the AHRI Standard 550/590. The chiller has been tested in full load and part load rating conditions to achieve a characteristic performance line of the machine. Recorded data is averaged and evaluated according to the requirements of the AHRI standard. The results of the baseline testing and the data processing are presented in this paper. A detailed analysis of the outcomes allows a comparison with the manufacturer rated efficiency values.

Finally, the potential instabilities of the test systems that caused problems during the test runs and their reasons are discussed, as well as possibilities to overcome them. The given ability of the test facility for a rapid change between different load conditions allows the arrangement of test cycles to investigate the performance degradation of the twin-screw chiller over time.

1. Introduction

Over the lifespan of a chiller, performance degradation and various faults can occur. Automated fault detection and diagnosis (FDD) approaches (either model-based or data-driven) can help monitoring the health state of the chillers and predict or detect anomalies in the chiller operation. Data-driven modelling relies on historical data of the equipment and can be quite powerful in identifying anomalies in the operation patterns of chillers. In the literature, only a limited number of studies deals with time-dependent performance analyzis of chillers and prediction of performance degradation. For instance, Ding and Fu [1] analysed the impact of the ambient temperature on the capacity of the screw chiller as well as the different ways of adjusting the control system to maintain optimal operating conditions. For this a chiller with constant water inlet temperatures on the evaporator inlet was operated at different ambient temperatures between -10 °C and 40 °C. Based on the collected data, a model-predictive control strategy was developed to run the chiller over a wide ambient temperature range. Another modeling approach was developed by Tiang et al. [2] to provide a transient coefficient of performance prediction model. More specifically, an artificial neural network model using Levenberg-Marquardt-algorithm was developed and validated with daily operating data from a 1043.7 kW screw compressor chiller. The model is used to further analyze the chiller performance under various operating conditions. Sala-Cardoso et al. [3] described an approach to predict load ranges and usual capacity trends based on previous chiller performance maps and data-driven multi-criteria orchestration strategies with the aim to operate the chiller based on load predictions in an energy effective mode by using a multi-layer pre-training strategy, a load forecasting model for the refrigeration systems. The chillers for the validation had a thermal capacity of between 130 and 430 kW. Recently, Bao et al. [4] adopted the integrated part load value (IPLV) for chillers operating in the Hong Kong area in terms of climate and temperature conditions. Data from 20 office buildings was used to develop a uniform building cooling load profile using the Monte Carlo analysis and the ASHRAE bin method. With this profile the weighting factors were optimized to the local conditions of Hong Kong and validated with the building data.

A performance mapping and related model of a screw compressor chiller is not available yet in a way that it can be used for life-cycle control and maintenance predictions. To be able to develop an accelerated life cycle testing of chillers, an accurate dataset at rated conditions is needed as a reference. As next step, a holistic system monitoring with multiple sensors is needed to map the operating conditions of the test facility and the equipment. It is therefore even more important that the measurements are carried out professionally and properly according to the specifications published in the AHRI 550/590 standard [5].

In this paper, the new experimental test facility is employed to obtain the baseline performance data of a 145.9 refrigeration tons variable-speed screw chiller, and the challenges encountered during the experimental work are discussed in detail.

2. System setup and commissioning

Before the system can be tested it needs to be set up and commissioned. The existing system was based on a 316.5 kW-chiller unit with turbo compressor. The following paragraphs include a description of the new chiller unit as well as the system set up and its control possibilities.

2.1. Chiller

The observed test unit is a water-cooled chiller with a R-134a vapor compression cycle. Core component of the chiller is a semi-hermetic screw compressor that is controlled by a variable speed drive. Due to its internal control logic, two operation modes with different maximum capacities can be chosen. The chilled water mode enables a maximum cooling capacity of 513.1 kW according to the specification at a supply temperature of 12.2 °C and a return temperature of 6.7 °C. In the second mode, the so-called ice making mode, 341 kW capacity can be achieved at a temperature level of 0 °C for the supply and -3.9 °C for the return.

The variable speed drive does not allow an external control of the compressor speed. That is why the capacity control during the baseline tests must be realized by a flow and temperature adjustment of the test system. Another factor that must be considered during the test process are the internal functions and thresholds of the chiller system that can affect the test feasibility and the behavior of the system in steady state operation. The speed control unit does not just control the compressor speed dependent on the capacity demand, but also logs 78 important system data and sensor recorded vales time stamped on a memory card. Both heat exchangers of the chiller are refrigerant to water shell-and-tube heat exchangers.

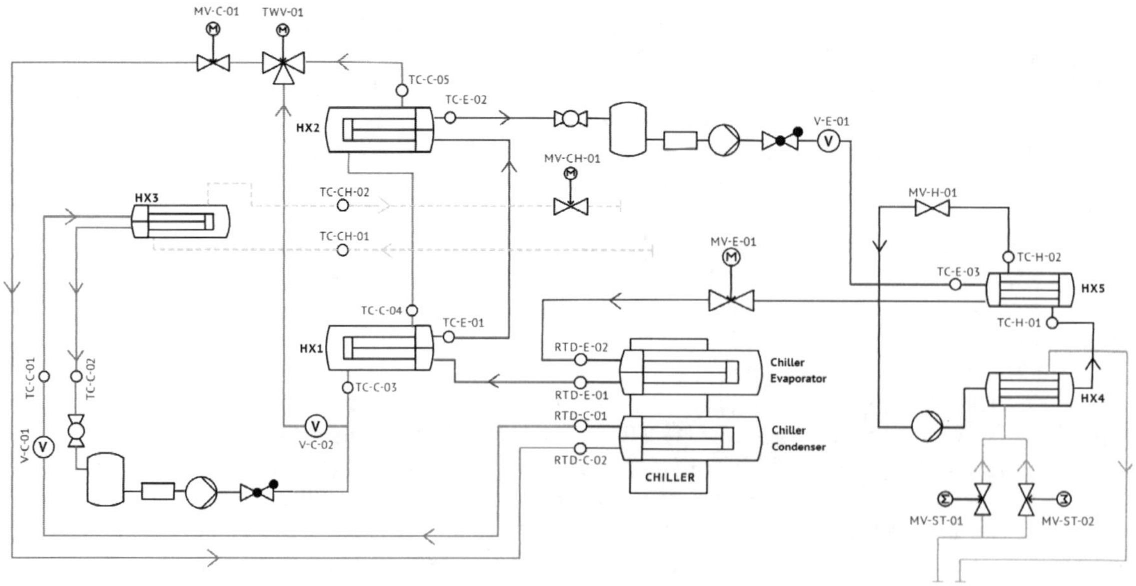

Figure 1. PI-diagram of the test stand.

2.2. System and sensors

The test rig was set up to allow a capacity control of the chiller by adjusting the coolant temperatures and the flow over the chiller heat exchangers. To solve this task within an area and with a low need of external heat dissipation and heat supply, internal heat exchangers were used to recover the heat. The following cycle description refers to the PI-diagram in Figure 1. In the flow diagram, all components and sensors of the test stand are displayed. For further simplification, the system is divided into two main cycles. On the left side, the condenser water cycle in red and on the right side the evaporator water cycle in blue.

The actual load control during the test runs is implemented by the condenser water cycle. After the cooling water is brought to the aimed mass flow by the condenser pump unit, the condenser cooling water passes the two serial heat exchangers HX1 and HX2 in this shell-and-tube heat exchangers the main heat load of the condenser water cycle is dissipated and at the same time it is recovered as load for the evaporator water cycle. A bypass line to the heat exchangers in combination with a three-way valve allows to control the flow over the internal heat exchangers, with this function, the load on the evaporator cycle which is also the load on the chiller can be controlled dependent on the aimed capacity for the test run. After the fluid is cooled down in the internal heat exchangers, it flows back to the chiller at aimed entering temperature conditions. Once the cooling medium gets heated up in the chiller condenser it passes the shell-and-tube heat exchanger HX3. This heat exchanger is connected to the building chilled water cycle displayed in green. HX3 helps to discharge the heat that is brought into the system by the compressor and that cannot be recovered in HX1 and HX2. The capacity of HX3 can be adjusted by a motor valve in the chilled water line that regulates the flow. After the surplus heat is dissipated, the cycle begins new. The condenser water cycle is equipped with five temperature sensors that give feedback of the important temperatures in the systems. A flow meter in the main flow, in combination with a flow regulation valve allows to regulate the mass flow over the heat exchangers, especially the condenser. A separate mass flow meter in the bypass line enables the detailed observation and control of the flow over the internal recovery heat exchangers.

In the evaporator water cycle, the water is cooled down in the chiller first. After leaving the chiller, the coolant flows into the two internal shell-and-tube heat exchangers HX1 and HX2 which are connected in series. In that heat exchanger set, heat is recovered from the higher temperature level in the condenser water cycle to simulate load on the evaporator water cycle. Once the heat of the condenser cycle is absorbed by the evaporator cycle the water passes the pump unit which maintains the necessary mass flow. Before entering the chiller again, the evaporator cooling water passes the shell-and-tube heat exchanger HX5. This heat exchanger is connected to an auxiliary cycle (drawn in black) includes a pump and a flow regulation valve and that again is connected to the building heating steam network displayed in gray. This indirect coupling to the steam heating cycle is not necessary for the load simulation on the evaporator but it provides an additional heat source during the start-up procedure of the system and for faster load changes during the test implementation. Furthermore, the evaporator water cycle is equipped with a flow regulation valve that allows to adjust the flow over the heat evaporator based on the measured data of the flow meter. Also, five temperature sensors provide real time data of the temperature conditions in the cycle.

3. Baseline testing

Purpose of the baseline testing is to gather reference data for future performance degradation tests. This reference includes the capacities of the chiller and the related efficiency during operation. In the baseline test, the characteristic values such as efficiency, capacity, current draw, and the overall system behavior are tested and tracked for different load conditions. With this information, a compressor mapping can be conducted. Base for this mapping is for example the values of temperature, pressure, mass flow that are logged either by the chiller system itself or by the sensors in the test stand.

3.1. Test implementation

The implementation of the baseline test targets test runs at the three part-load steps 25%, 50%, and 75% of the maximum chiller cooling capacity as well as in full load operation. If the three part-load steps cannot be achieved exactly, AHRI 550/590 allows interpolation and extrapolation as explained in Section 3.2. This is specifically important because the speed control of the chiller has an automatic shut down for cooling loads of lower than 30% of maximum that cannot be set down further which makes it impossible to collect test data at 25% load.

A valid test run must have a minimum length of 15 minutes in steady state where the mass flow and the temperatures remain constant in the system. During this time, all operational data points are logged with a related time stamp from the test software on a test protocol or from the chiller on an internal memory card of the variable speed drive. The data must be brought together in one file manually for the test evaluation after the runs. All part-load test runs are executed with a fixed setpoint of 6.7 °C for the evaporator water leaving temperature. The evaporator entering water temperature is load dependent and calculated for a pump volume flow of 15.8 kg/s. This flow rate is given by the given fix speed water pump for the evaporator cycle. Since this flow rate is not sufficient to reach 100% load at the test conditions, the full load operation is aimed at lower temperatures. This is not standard conform but the measured data should at least give an idea of the full load efficiency in comparison to the technical data in the specification for this lower temperature range**.** The condenser entering water temperature is staggered in equal steps so that a sufficient heat dissipation is ensured. Table 1 shows the aimed temperature dependent on the target loads for the part load and the full load testing. All temperatures must be controlled within an allowed deviation of ±1.1 K.

Table 1. Target conditions during the baseline testing calculated

Target load (%)	Target capacity (kW)	Setpoint temperature (°C)	Evaporator inlet temperature[a] (°C)	Condenser inlet temperatures (°C)
30	127	6.7	8.3	18.3
50	257	6.7	10.6	18.3
75	383	6.7	12.6	23.9
100	440	2.2	8.8	26.7

[a] calculated based on a flow rate of 13.8 kg/s

3.2. Data analysis

An important task related to the baseline testing is the data analysis. A basic data set is evaluated in the same way for all test runs in steady state. As a preparation for this the analysis, the time-dependent data that was recorded during the test run must be converted into a csv-file and gets reviewed. This time related data includes time-stamped values of all important temperatures, mass flows, pressures, and the power consumption during the test runs. Time periods of this raw data that indicate a steady-state behavior and fulfill the standard requirements get highlighted and if they are at least 15 minutes long, they get extracted into separate files. Then, average values are calculated over the steady state time which are used for the actual evaluation and further computation. It is the aim for a proper analysis according AHRI 550/590 to cover not just the full load behavior during the test runs but also part loads. For the part load conditions the target part loads are 25 %, 50 %, and 75% of the full load.

At first, the cooling capacity is calculated as a core value of the whole analysis. In equation (1) it is shown how the cooling capacity $\dot{Q}_E$ is a function of the mass flow over the evaporator $\dot{m}_E$ the specific heat c_p of the coolant, and the temperature difference between inlet $T_{E,i}$ and outlet $T_{E,o}$ of the evaporator. The cooling capacity allows the exact classification of the test section and the validation of the relative chiller load which is calculated in equation (2) as the ratio of the calculated cooling capacity over the maximum cooling capacity $\dot{Q}_{E,max}$ that is specified as 513.1 kW for the chiller, times the factor of 100.

$$\dot{Q}_E = \dot{m}_E \cdot c_p \cdot \left(T_{E_i} - T_{E_o}\right) \tag{1}$$

$$\text{Load} = \frac{\dot{Q}_E}{\dot{Q}_{Emax}} \cdot 100\% \tag{2}$$

Based on the capacity provided from equation (2) and the power consumption of the chiller that is measured by a power meter, the efficiency of the tested device can be calculated in two ways. The more common efficiency metric is the cooling coefficient of performance COP_R. As equation (3) displays, the COP_R is defined as the ratio of the cooling capacity in kW over the electrical power draw of the chiller $\dot{W}_{el}$ in kW.

$$COP = \frac{\dot{Q}_E}{\dot{W}_{el}} \tag{3}$$

The AHRI 550/590 standard defines an additional efficiency metric. This efficiency for is calculated as the ratio of the power consumption of the chiller in kW over the cooling capacity $\dot{Q}_{ev}$ in TR. Therefore, it is not the inverse value of the COP_R as it may be assumed from simply looking at the equation (4).

$$\text{Efficiency} = \frac{\dot{P}_{el}}{\dot{Q}_{ev}} \tag{4}$$

By using the efficiency according to the AHRI 550/590 standard [5], an integrated part load value *IPLV.IP* can be computed. The IPLV.IP is a value to express the part load efficiency in all conditions by combining the measured efficiencies as in equation (5), where A is the efficiency value for 100% load, B for 75%, C for 50% and D for 25%.

$$IPLV.IP = \frac{1}{\frac{0.01}{A} + \frac{0.42}{B} + \frac{0.45}{C} + \frac{0.12}{D}} \tag{5}$$

For the case, that the target part load steps cannot be achieved exactly in a steady state condition it is possible to interpolate linear between the lowest and the highest measured point. If the 25% load cannot be achieved, linear extrapolation is not allowed but the AHRI standard provides a correlation. The efficiency value calculated from the test data following equation (4) therefore gets corrected by the degradation correction factor C_D as in equation (6). The degradation correction factor (equation 7) again needs a load factor *LF*. The load factor calculated in equation (8) is the ratio of the target load (*Load%* times the maximum capacity $\dot{Q}_{max}$) over the load measured in the test run $\dot{Q}_{Test}$.

$$\text{Efficiency}_{CD} = \text{Efficiency}_{\text{Test}} \cdot C_D \tag{6}$$

$$CD = (-0.13 \cdot LF) + 1.13 \tag{7}$$

$$LF = \frac{Load\% \cdot \dot{Q}_{max}}{\dot{Q}_{Test}} \tag{8}$$

3.3. Baseline test results

The baseline test results are a mix of directly measured values and calculated values form the data analyses. All results give a feedback of the operating conditions and characteristics during the test and allow a comparison with specified values from the manufacturer. During the baseline tests, seven test runs were performed within a load range of 34% to 90%. The tests were conducted during steady state. The detailed test conditions and results are displayed in Table 2. Based on these values, the efficiency values for 50% and 75% can be interpolated while the value for 25% can be extrapolated with the method described in paragraph 3.2. A 100% load operation could not be achieved. Potential reasons for this are mentioned in paragraph 3.4.

Table 2. Baseline test results.

Load (%)	Evap. mass flow (kg/s)	Evap. EWT (°C)	Evap. LWT (°C)	Cond. mass flow (kg/s)	Cond. EWT (°C)	Cooling capacity (RT)	Elec. power draw (kW)	COP (-)	Efficiency (kW/RT)
34	15.40	9.73	7.04	18.71	15.95	49.70	14.15	12.36	0.2846
42	15.41	10.00	6.66	18.17	18.06	61.46	21.31	10.14	0.3467
48	15.42	10.56	6.72	17.68	17.94	70.73	26.45	9.40	0.3740
56	15.43	11.00	6.58	17.25	18.41	81.54	32.27	8.89	0.3958
74	15.45	12.48	6.62	17.31	23.83	108.25	58.74	6.48	0.5426
79	15.44	12.74	6.52	17.00	23.74	114.73	63.87	6.32	0.5567
90	15.46	14.23	7.14	18.42	26.87	130.94	86.07	5.35	0.6573

Figure 2 shows a diagram with all efficiency values of the chiller. The small green dots are the values that were analyzed from the measurements conducted in the test runs while the bigger red dots are the interpolated or extrapolated values. Since a 100% measurement could not be implemented, the value for this operation mode was linear extrapolated (gray dot). An extrapolation for full load is not intended in the standard but it was implemented to allow a calculation of the IPLV.IP value. The yellow dot is the efficiency for full load mentioned in the specification. The value for this is much higher and probably more realistic but this could just be proven by measurement. The IPLV.IP calculated from the values is 0.4191 which is 5.3% higher than in the specification. It is important to mention that this does not mean that the specification is wrong but more that the linear extrapolation for the 100% point is not suitable. A 100% measurement must be implemented to proof this. The 5.3% deviation from the IPLV.IP value in the specification shows that the efficiency measurements are overall pretty close to the specified values.

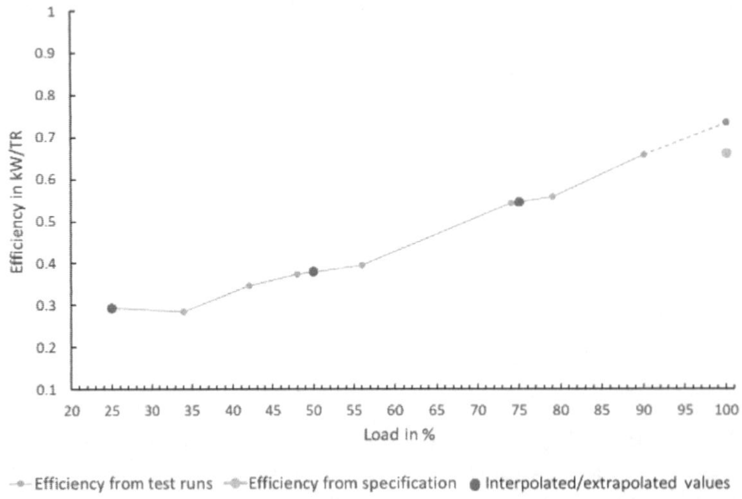

Figure 2. Summary of the testing unit efficiency

3.4. *Difficulties during test implementation*

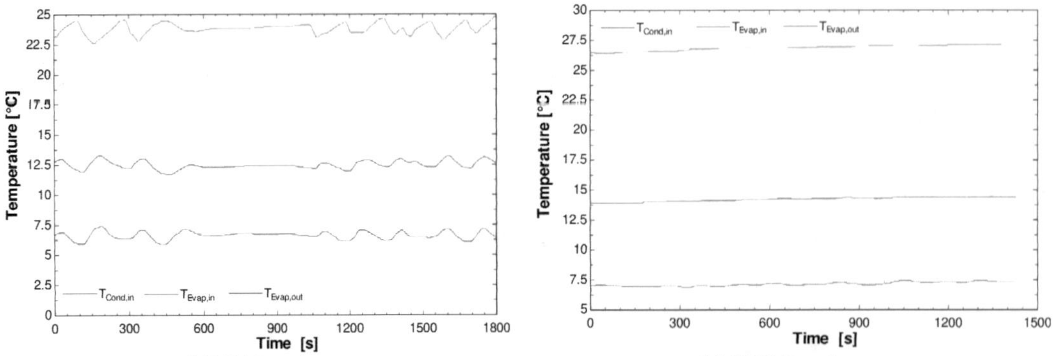

a) 74% Load b) 90% Load

Figure 3. Fluid temperatures during selected test runs

As already stated in Section 3.1, there were some difficulties during the test implementation due to the large capacity of the chiller. The test stand was initially built and used in previous tests for chillers with a maximum cooling capacity of 317 kW. For those test devices, a flow rate of nearly 14 kg/s was always sufficient. However, for the current chiller was not sufficient. According to the AHRI 550/590 standard conditions a flow rate of 22.1 kg/s is needed. The currently installed pump was not able to provide this mass flow. Therefore, higher load conditions could not be achieved under standard rated conditions. Another harmful factor that appeared was measurable at a part-load condition between about 70 and 80% load. In this load range, the variable speed drive adjusted the compressor speed whenever a steady state came up for a few seconds. Due to this control behavior, the VSD experienced a hunting situation where the temperatures developed a fluctuation of ±1 K. An example of this control instability behavior is displayed in Figure 3(a) where the condenser inlet temperature (red), the evaporator inlet temperature (black), and the evaporator outlet temperature (blue) show a consistent rise and fall of the values.

The third and last problem is the heat dissipation over the campus chilled water line. The heat exchanger HX3 (Figure 1) has enough capacity in theory but the effectiveness went down drastically in the practical use. For this, not enough heat could be conducted and therefore the temperatures in the system rose during higher load conditions what did not allow the system to come to a steady state. Figure 3b shows a test run where the temperatures were higher than aimed and could not be decreased. Also, a slight increase of the temperature can be seen in this diagram.

The test system has two internal heat exchangers that recover heat from the system. Since more heat is added to the system than can be recovered an additional heat dissipation is necessary. HX3 dissipates heat from the system via the building chilled water loop. The effectiveness of the heat exchanger HX3 is generally low but decreases sharply at higher load conditions as table 3 shows. Due to this, all temperatures in the system start to raise (Figure 3) so that the target temperatures cannot be achieved anymore. One option to solve this problem would be the installation of an additional heat exchanger in the condenser loop.

Table 3. Effectiveness of HX3 in the measured load conditions.

Load (%)	**HX3 effectiveness** (%)	$\dot{Q}_{h,max}$ **(kW)**	$\dot{Q}_{c,max}$ **(kW)**	**Q (kW)**
34	36.89	862.01	60.74	22.41
42	37.54	1062.61	69.11	25.94
48	30.62	1108.41	105.50	32.31
56	24.39	1082.34	149.50	36.46
74	19.54	1755.57	335.32	65.52
79	17.03	1779.06	414.02	70.52
90	11.80	2062.72	724.45	85.50

4. Conclusion and further steps

The seven implemented tests in which a steady state measurement was conducted show that the general quality of the measurements is sufficient and that the values are close to the stated values in the specification. Nonetheless, the problems that appeared during the test implementation must be solved to achieve a reliable data base for further tests of the chiller unit. A test stand redesign is necessary for this. Such a rework of the test stand has to include an improvement of the heat dissipation over the campus chilled water system in the first place. This can be realized over an additional shell-and-tube heat exchanger to support the existing heat exchanger HX3 (Figure 1). The test data from the implemented measurements can help to calculate the design data for the new heat exchanger and to size it properly. Furthermore, the improvement of the pump system to achieve the AHRI 550/590 flow rate of 0.043 kg/s per kW chiller capacity can be carried out during the redesign of the rig. For this both, the pump in the evaporator loop and in the condenser loop should be replaced by pumps with higher performance. A higher flow rate could allow a test implementation of all load conditions according AHRI 550/590 standard and with a constant evaporator water outlet temperature.

Once the redesign is done, the baseline test can be reimplemented in all load areas. Based on this data, following test can build up.

5. Nomenclature

A	area	(m^2)
AHRI	air-conditioning, heating, and refrigeration institute	
C_d	degradation correction factor	
C	condenser	
COP	coefficient of performance	
EWT	entering water temperature	(°C)
E	evaporator	
HX	heat exchanger	
IPLV	integrated part load value	
k	thermal conductivity	(W/m^2*K)
LF	load factor	
LWT	leaving water temperature	(°C)
MV	motor valve	
$\dot{m}$	mass flow rate	(kg/s)
P	power	(kW)
$\dot{Q}$	cooling capacity	(kW)
$\dot{Q}_E$	cooling capacity	(RT)
RTD	resistance temperature detector	
T	temperature	(°C)
TR	tons of refrigeration	
TC	thermo couple	
TWV	three-way valve	
VSD	variable speed drive	
ε	effectiveness	(%)
Subscript		
c	cool side	
h	hot side	
i	inlet	
max	maximum	
min	minimum	
o	outlet	
test	evaluated in test conditions	
E	evaporator	
el	electric	

6. References

[1] Ding G and Fu L, 2005, Performance analysis and improvement of air-to-water chiller for application in wide ambient temperature range. *Applied Thermal Engineering*, 25 (2005), Elsevier, pp. 135-145.

[2] Tian C, Xing Z, Pan X, and Tian Y, 2019, A method for COP prediction of an on-site screw chiller applied in cinema. *International Journal of Refrigeration, 98 (2019)*, Elsevier, pp. 459-467

[3] Sala-Cardoso E, Delgardo-Prieto M, Kampouropoulos K, and Romeral L, 2020, Predictive chiller operation: A data-driven loading and scheduling approach. *Energy and Buildings, 208 (2020),* Elsevier

[4] Bao Y, Lee SH, Jia J and Lee WL, 2021, Developing an Integrated Part Load Value for Chillers of Office Buildings in Hong Kong. *International Journal of Refrigeration, 2021,* Elsevier

[5] ANSI/AHRI Standard 550/590 (2018), Standard for Performance Rating of Water-chilling and Heat Pump Water-heating Packages Using the Vapor Compression Cycle, *Air-Conditioning, Heating and Refrigeration Institute*. Arlington, VA (USA).

IOP Conf. Series: Materials Science and Engineering 1180 (2021) 012034 doi:10.1088/1757-899X/1180/1/012034

Application of LDV and PIV techniques for flow measurements in the suction port of a screw compressor

J. M. Nouri[1,3], D. Guerrato[2], N. Stosic[1] and Y. Yan[1]

[1] Department of Engineering, Mechanical Engineering and Aeronautics
City, University of London, EC1V OHB, U.K.
[2] Delphi Diesel System, Courteney Rd,
Gillingham, Kent, ME8 0RU
[3] Corresponding Author: j.m.nouri@city.ac.uk

ABSTRACT

The application of LDV and PIV systems to measure the flow variation in the suction port of a new designed optical screw compressor has been found to be very successful. Time-resolved mean velocity measurements were made over a time window of 1° (Δθ) at a rotor speed of 1000 rpm, a pressure ratio of 1 and a gas temperature of 55°C. The LDV investigation on the suction port has shown a very stable and slow flow with almost no influence of the rotor motions everywhere except close to rotors at Z=-10mm where the rotor motions start to show its influence and formed a more complex flow with a wavy axial flow profile. The PIV results confirmed flow trend measured by LDV and showed no influence of rotors' movement on the mean flow structure. The PIV results within the vertical plane also showed relatively strong horizontal stream flow vectors upstream of the inlet suction port and became almost vertically downwards at the inlet port. Despite the current success, further improvements have been recommended to give better optical accessibility as close as possible to the rotors entrance.

1. INTRODUCTION

Screw compressors are well known for their simple design, low cost and high efficiency over a wide range of speeds and pressures, which make them suitable to be used in many industrial applications like air compression, refrigerant compression, food process, pharmaceutical, pneumatic transport and turbo charging [1-3]. Although the analytical methods for performance prediction of the screw compressors are well established and their design optimisations and basic operations are well known, there are limited information in the literature on the flow behaviour within compressors, from suction port to rotors' chamber to discharge port, both experimentally and CFD simulation [4-10]. Obviously CFD flow simulation approach is the way forward for fully characterising the real time fluid flow (including multiphase), which would substantially reduce the times and costs of designs and production compared with an experimental based approach [1,6,7], however, the experimental approach by means of flow visualization, and flow velocity field measurement (using LDV and PIV techniques) is essential to not only provide the real time flow dynamics but also to provide accurate data for validating CFD simulations [4,5,9,10]. Clearly, the flow process within the rotors' working chambers depends greatly on the inflow structure coming from the suction port, the shape of the suction port and the flow trajectory as it enters into the rotos' chambers [6]. Equally, The compressors performance is highly affected by flow behaviors within the suction port, rotors' chamber and discharge port, particularly flow losses in the suction port that led to a decrease in compressors efficiency [7].

A number of the previous investigations have been carried out [4,5.8-12] to show the flow behavior within the rotor's chambers, the discharge cavity and leakage flow using different experimental methods and CFD simulations. The flow within suction port of screw compressors remained fairly unexplored except those that showed the CFD can effectively be used for analysising the suction flow like that reported by [1] who also visualised the flow by high speed camera and compare them with the CFD predictions. Thus, the present report is an experimental investigation focusing on the flow measurements within the suction port of a screw compressor by designing a new optical suction port and applying LDV and PIV techniques to allow measurements of mean, rms and vector velocities as close as possible to the rotors. Time-resolved (cycled-resolved) velocity variation have been made in great details to be used for the validation of CFD codes. Detailes of the flow configuration will be given in the next section followed by the results and discussed section. Then, the report ends with a summary of the main findings.

2. FLOW CONFIGURATION AND INSTRUMENTATION

LDV and PIV measurements have been made at the inlet of an axial suction port of a standard screw compressor and their setups are shown in Figure 1. The compressor had 'N' type rotor profiles with a 5/6-lobe configuration, and was running at a rotor speed of 1000 rpm, a pressure ratio of 1 and a gas temperature of 55°C. Measurements at the suction port require a completely different setup than those used for rotors and discharge chambers by [5,9,10]. Also, the running conditions are quite different since there is no high pressure gradient, and therefore the motion of the rotors has little impact on the air flow except in its vicinity; in this study it is at Z=-10mm. To make the suction port optically accessible, the suction pipe has been replaced by a proper designed transparent cylinder that is fitted on the top of the machine inlet port with the same inner diameter, as shown in Figures 1 and 2, while the other end of the cylinder is fitted by a flat transparent window so that it would provide sufficient optical access into the suction port for both LDV and PIV measurements. The modified optical suction port with adopted coordinate system is shown in Figure 2 with the Z-axis is placed on the axis of the vertical cylinder and its zero is located at the entrance of the compressor inlet port on the top of the rotor chamber. The X-axis is aligned with the rotor's shaft with its positive direction towards the main shaft, and the Y-axis is perpendicular to X and Z plane, and its direction goes from the female to the male rotor. The dark grey spots shown in Figure 2 correspond to LDV measuring points at three axial planes of Z=-10, 5 and 75mm with Z=-10mm being the nearest plane to the rotors. X- and Y-components velocity (referred to as axial velocity, V_x, and horizontal velocity, V_y, respectively) were easily measured from the top flat window, but the vertical velocity (Z-component, V_z) through the diagonal planes of the vertical cylinder, and was only measured at Z=75mm; V_z measurements at Z=5 and Z=-10mm was not possible because the laser beams were not able to access those locations due to the presence of the metallic compressor walls; this shortfall can be improved and it is recommended to modifying the inlet casing by inserting small optical windows on diagonal planes or replacing the inlet port wall upstream of the rotors entrance by an appropriate plexiglass pipe of the same internal diameter (to preserve its geometrical integrity), to allow closer optical access to allow closer optical access to the rotors for both LDV and PIV measurements.

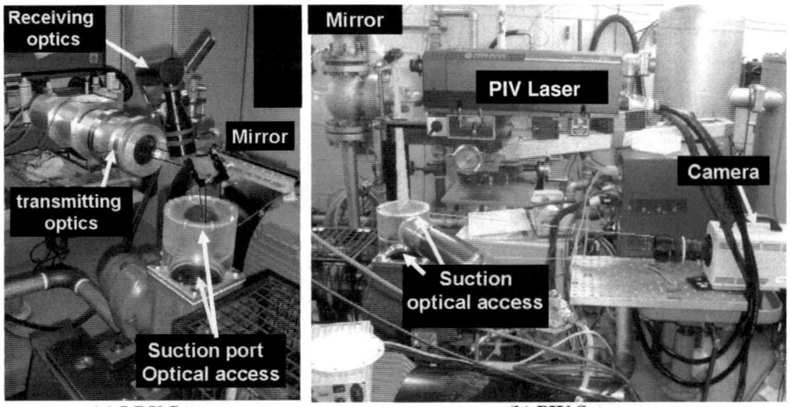

(a) LDV Setup (b) PIV Setup

Figure 1. Modified optical compressor at suction port with transparent inlet pipe for the LDV and PIV measurements.

The LDV and PIV systems were the same as those used by [4,13,14] who described the LDV and PIV systems fully and will not be repeated here. A conventional silicon oil atomizer with an average droplet size of 1-2 μm were used as the flow tracer. This is a most common air flow tracer to be used for LDV technique, and it has been shown to be suitable for PIV measurement by [4,13,14], which allowed good data processing when using the cross-correlation method. The seeding has been introduced into the suction port from the inlet feeding pipe, as shown in Figure 2. One of the disadvantages of using the silicon oil droplets is the fouling of optical windows, however, in the suction port it was found not to be a major issue mainly due to the uniform and stable flow nature of the suction port. It has been shown by [15,16] that a better result can be obtained by using CO_2 solid particles blown through the suction port as seeding.

LDV measurements include angled-resolved mean velocities of, axial V_x, V_y and V_z components along the diagonals X-plane (X-X') variations and Y-plane (Y-Y') variation, as shown in Figure 2, at different Z stations. Although LDV measurements provides very accurate information about mean and rms velocities, but it is a point measuring system and to describe the flow characteristics across the all fluid domain in full details would be a very time consuming task. Thus, PIV system is an alternative measuring technique that can capture instantaneous flow field velocity (2D or 3D) in a plane and provide useful information regarding spatial gradient and Reynolds stress everywhere within that measuring plane. Therefore, a 2-D PIV system has been employed to map the flow within the inlet suction port through horizontal and vertical planes as shown schematically in Figure 3. A comparison between LDV and PIV measurements has been made by current authors [17] and showed a very good agreement in mean velocities in a flow region of high velocity gradient and turbulent, which gives confidence in the accuracy of the measured PIV results. The transmitting PIV laser sheet was passed horizontally or vertically through the optical cylinder and top window directly or through a 45° mirror. As a consequence, all three velocity components can be measured but this requires different PIV setups. The arrangement of transmitting and receiving PIV system may differ for the above vertical and horizontal setups; some of these setups are difficult to arrange due to physical presence of the compressor unite, which may limit the range of measurements.

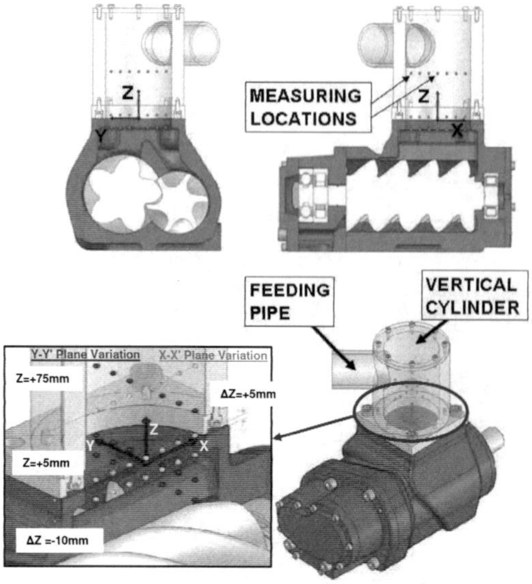

Figure 2. Optical compressor and suction port views with adopted coordinate system.

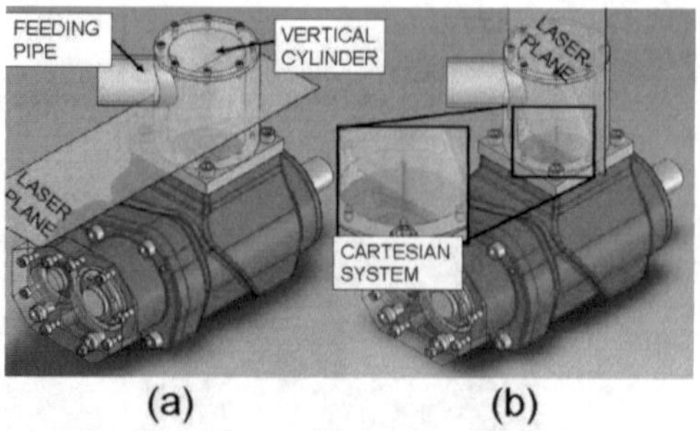

Figure 3. Schematic presentation of the PIV optical arrangement; (a) horizontal plane, (b) horizontal plane.

RESULTS AND DISCUSSION

Before going into details, a description of the inlet flow geometry is useful to help to understand the measured flow behaviour. Figure 4 shows a schematic description of the suction port from optical cylinder down to rotors. The details of the geometry of the suction port in Figure 4(a) shows that there is a bridge above the rotors that split the incoming flow towards the male and female rotors suggesting that the suction flow process takes place above the rotors around the bridge. With the present optical arrangement, unfortunately it is not possible to measure flow around the bridge; the closest location that measurement is possible is at Z=-10mm which can still show the effect of rotors rotation on the suction flow. On the other hand, the inlet flow in the optical cylinder (Figure 4(b) is expected to be mainly vertically downwards towards the suction port. Considering above flow feature, it is expected that the flow within the optical cylinder to be of a uniform and stable nature except at Z=-10mm where the flow would be exposed to the rotors' rotation and be affected by their cyclic motion. Since there are five opening/closing of the discharge port operations during a full rotational cycle, the flow velocity measurements over one opening/closing cycle are presented (i.e. 72° of the main shaft rotation) with the assumption that the flow over the other four are considered to be repeating and similar. It should be noted that the opening of the discharge port corresponds to main shaft angle of θ=0° and this is the same for all LDV and PIV measurements. The following results present a small sample of data obtained in this experiment with the main focus on the behaviour of the mean velocities as a function of the shaft angular position, θ, within the suction port; full set of results are presented in the thesis of [4], and it is planned to publish them in a separate paper.

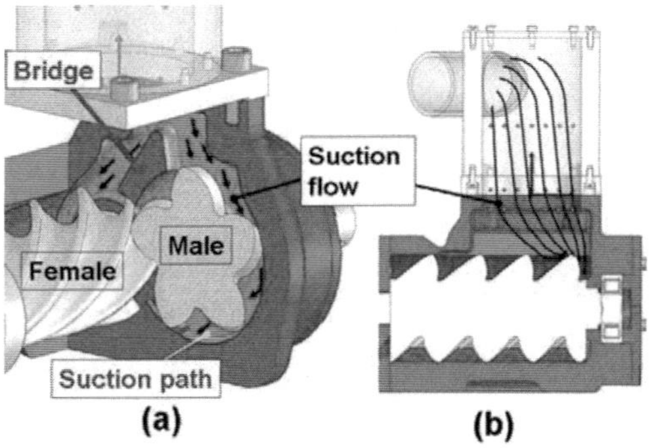

Figure 4. Schematic description of suction flow.

2.1. LDV Results

The velocity profiles presented in Figures 5 and 6 are cycle-averaged over one-degree time window (Δθ=1°) and are obtained from a continuous measurement with respect to shaft rotation over many 360° till sufficient samples are collected to provide accurate (statically) mean velocity over 1° time window. First the results of the mean vertical flow velocity, V_z, at Z=+75mm are presented and discussed. Then the mean velocity variations of axial (V_x) and horizontal (V_y) velocity components at Z=-10mm +5mm would be presented, which are the main focus of the current study as Z=+5mm represents the upstream flow as it goes into the suction port, and Z=-10mm represents flow within the suction port and as close as to the rotors.

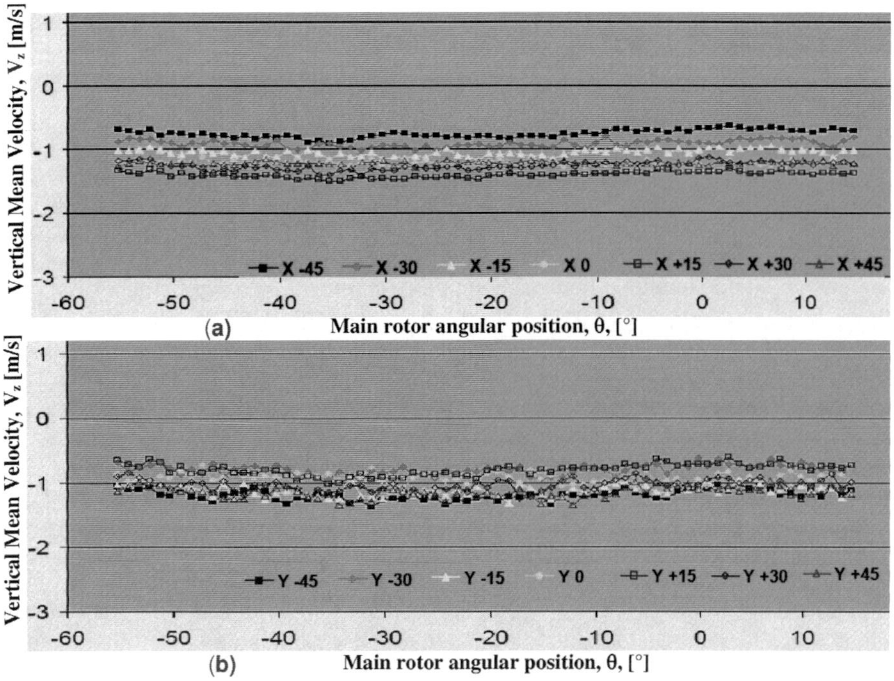

Figure 5. Variation of vertical (V_z) mean velocity as a function of rotor angle, θ, at Z=+75mm along: (a) X-X' plane at Y=0mm, (b) Y-Y' plane at X=0mm.

The variation of the mean vertical velocity component, V_z, at Z=+75mm along the X-X' and Y-Y' planes is presented in Figure 5. The results show uniform downflow (negative V_z values towards the rotors) velocity profiles along both X-X' and Y-Y' planes at different X and Y locations, respectively, with no observable impact of rotors motion as would be expected at z=+75mm. The downflow vertical velocity along X-X' plane, Figure 5(a), varies at different X locations so that it is around -0.7m/s at X=-45mm (near cylinder wall) and then increases gradually with X to a maximum value of around -1.4m/s at X=+15 and then drops to -1.2m/s at X=+45mm; note that this variation along X-X' is not due to rotors motion and it is due to the fact that the incoming flow from the top feeding pipe into the cylinder hasn't yet developed fully at this Z location . Similar downflow can be seen along Y-Y' plane at different Y locations, Figure 5(b), with smaller variation in V_z values with Y than that observed along X-X' plane, and no obvious trend with an overall average value of around -1m/s.

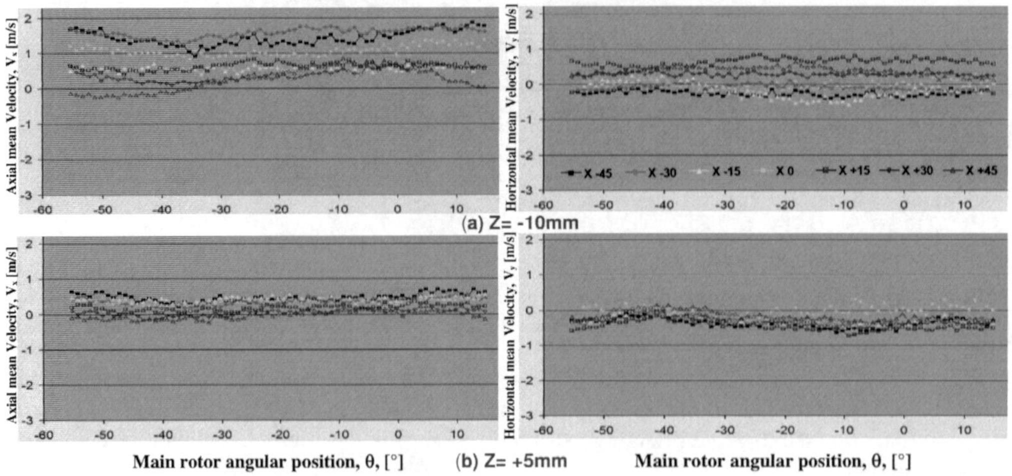

Figure 6. Variation of axial (V_x) and horizontal (V_y) mean velocities as a function of rotor angle, θ, along the X-X' plane at Y=0mm, and at: (a) Z=-10mm, (b) Z=+5mm.

Figures 6 shows the mean axial, V_x, and horizontal, V_y, velocity profiles as a function of the shaft angular position along the X-X' plane at Z=-10mm (Figure 6(a)) and Z=+5mm (Figure 6(b)); the graphs on the left column of Figure 6 represent the axial velocity profiles, while those on the right column are the horizontal velocity profiles. Starting from Z=+5mm, Figure 6(b), upstream of the suction port entrance, the results show uniform axial and horizontal velocity profiles with θ suggesting little or no influence of rotors motion at this vertical plane. Also, the profiles at different X locations seem to be similar with the axial velocity tend to be positive up to 0.7m/s especially at negative X locations, while the horizontal velocity tend to be negative up -0.7m/s suggesting a slow reverse flow motion from male to female rotor along the X-X' plane at this vertical location. Further downstream at Z=-10mm within the suction port and closer to the rotors, Figure 6(a), both axial and horizontal velocities are influenced by the rotors motion and are changed with both X positions of the control volume and also the shaft angle, θ. The latter was expected due to the close proximity of the measuring point with rotors where their movement interacts with the fluid motion and creating difference in pressure between the open rotors' working chambers and the suction port. The axial velocities are mainly positive with the highest values up to 1.8m/s at X=-30 and -40mm, and that the profiles look like a wave, which may indicate that the flow is moving towards a working chamber that is opening and absorbing the fluid, but since there is more than one working chamber connected to the suction port for each single shaft angular position, the interpretation of the flow would more difficult. Unlike the axial velocity, the horizontal velocity has positive (towards male rotor) and negative (towards female rotor) values for +X and -X locations, respectively, with almost zero velocity at centre, X=0; this is more pronounced for flow between shaft angle, θ, of -30° to 10° with V_y values up to +0.8m/s and -0.6m/s, respectively. The fact that the velocity changes from negative to positive values, suggests that there is a rotational flow movement with its centred close to the Z-axis. As mentioned before, measurements of main vertical velocity, V_z, was not possible at Z=-10mm +5mm and due to obstruction of the laser beams by the suction port wall.

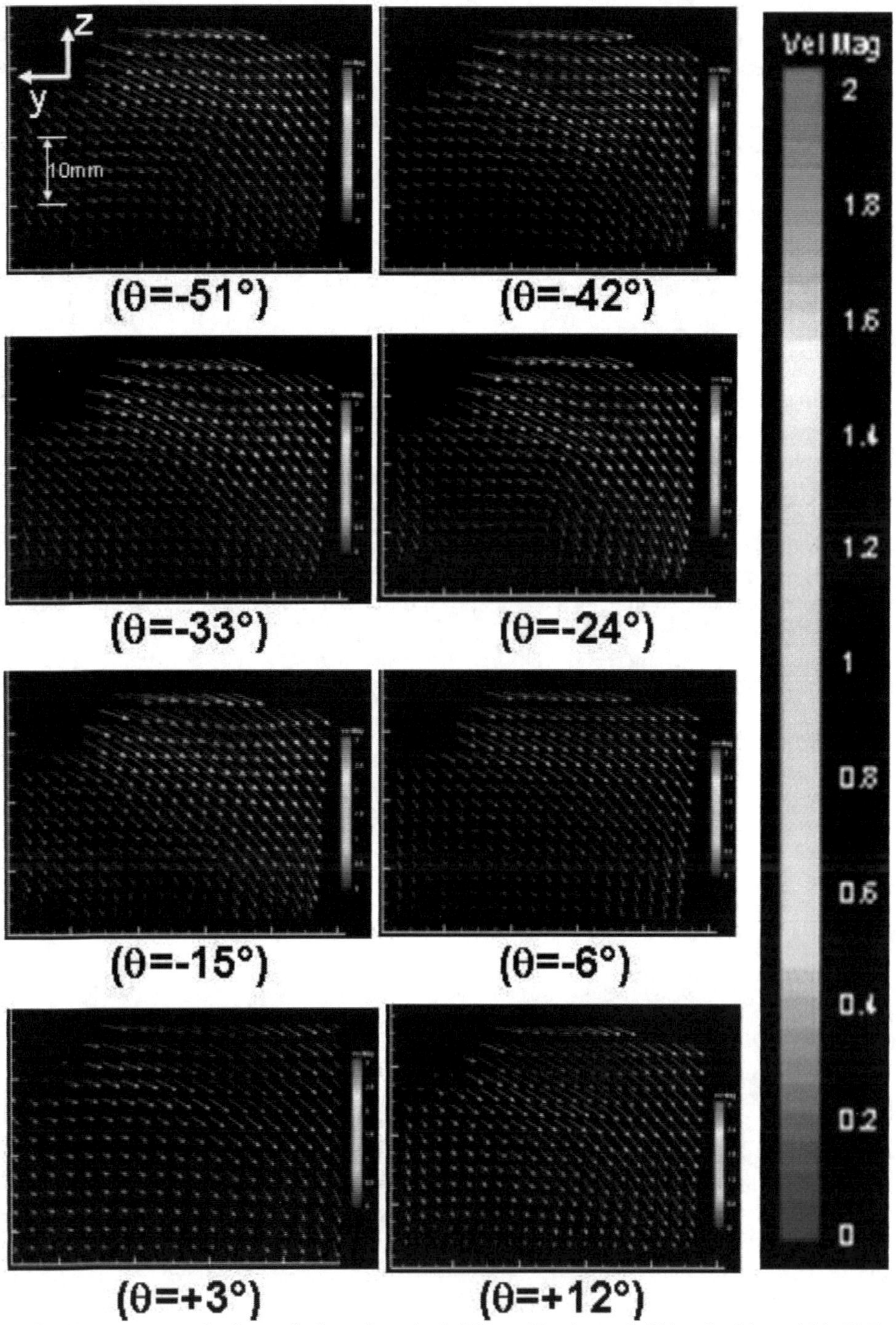

Figure 7. Mean vector velocity variation of vertical (V_z) and horizontal (V_y) velocities within Y-Z plane at different shaft angles, θ, and at X= 0mm.

The variations of the mean axial, V_x, and horizontal, V_y, velocities with shaft angle, θ, along Y-Y' plane at Z=-10mm and +5mm, not presented here, showed similar results to that of Figure 6. Although, the trends were similar, there were differences and the results showed more dependency on rotor angle. Overall, the results near the rotors show the presence of a complex flow in the suction that includes an axial component moving down towards the working chambers at the front of compressor, with the tendency of flow on the left of compressor pointing to the centre of the female rotor while the more flow on the right was pointing to the male rotor; similar to a simplified flow description presented in Figure 4. This is due to the geometry of the suction port that had been designed in such a way that the suction process takes place in front of the rotors.

2.2. PIV results

PIV setup are shown in Figures 1, 2 and 3 and measurements were recorded for different planes with respect to the shaft angle. Then, using purposely written special software, velocity results were averaged to acquire the mean vector velocity over a time window of 1° (Δθ); here 30 images were used to form the mean velocity vector field with an average standard deviation of 10% of the mean value in accordance with LDV results. As mentioned before, the suction port is more accessible with flow being stable with low mean velocities, and therefore less window fouling. Figure 3 presents the two measuring configurations used, first Figure 3(a) shows the arrangement for measuring the flow at the green laser plane (X-Y) localized at Z=75mm. This plane is the same used for LDV measurements. This position of 75mm is particularly chosen so to be far from the rotors. Figure 3(b) shows the vertical laser sheet, which is perpendicular to the rotors for measurements in the Y-Z plane. The sample results of eight selected PIV images at different main shaft angle, θ, are shown in Figure 7.

Figure 7 shows the distribution of mean vector velocity $\left(\sqrt{V_y^{\,2} + V_z^{\,2}}\right)$ of axial and vertical velocity components within Y-Z plane covering 40mm above the suction port entrance at eight different shaft angles, θ, and at X= 0mm. The expectation was to see the flow to be vertical and uniform but the results show a more complex flow structure caused by the lack of flow development in the cylinder as explained with LDV results above. Again, similar flow structure can be seen at all, θ, with no influence of shaft movement. At the top of the graphs, there are relatively strong stream with almost horizontal vectors pointing to right, which suggests the flow is dominated by V_y velocity component. Then as the flow moves downwards, the velocities become more and more vertical. At the very bottom, near the suction port, flow is almost vertical. Due to the fact that PIV system needs two directions of operation, in this case the horizontal and vertical, it was not possible to obtain results nearer to the rotors than the ones shown at the bottom of each graph on Figure 11; such measurement is possible by LDV in back scattering mode. This shows how a one-point measurement system such as LDV is far more adaptable to an environment with limited optical access. On the other hand, if access is available, PIV exhibits an unbeatable solution with high spatial resolution, thus, it is recommended to modifying the suction port to provide more optical access for PIV measurements.

3. CONCLUSION

Mean flow velocities have been measured within the suction port of a standard optical screw compressor using laser LDV and PIV techniques. Time-resolved velocity measurements were made over a time window of 1° (Δθ) at a rotor speed of 1000 rpm, a pressure ratio of 1 and a gas temperature of 55°C. A summary of the main findings is presented below:

- The application of LDV and PIV systems to characterise the flow variation in the new designed optical screw compressor was found to be very successful. To improve the optical accessibility, it is recommended to modifying the inlet port casing by inserting small optical windows on diagonal planes or replacing the inlet port wall by an appropriate plexiglass pipe of the same internal diameter, to allow closer optical access to the rotors for both LDV and PIV measurements.
- The LDV mean vertical velocity component at Z=+75mm along the X- and Y-planes showed uniform downflow velocity profiles with values up to 1.4m/s and no impact of the rotors motion. Similarly, at Z=+5mm, the measurements of the mean axial and horizontal velocity components showed no influence of the shaft movement everywhere with low velocity values with uniform profiles.
- Close to rotors, Z=-10mm, the LDV mean axial and horizontal velocity components have been influenced by both the transverse locations along X- and Y-planes and also by the rotor motions due to the close proximity of the measuring point with rotors. Tracking of the measured axial and horizontal flows showed the presence of a complex flow in the suction port that includes an axial component moving down towards the working chambers with more air to flow towards the male rotor.

- The PIV distribution of mean vector velocity of axial and vertical velocity components within Y-Z vertical plane revealed an unexpected complex flow structure due to lack of flow development in the inlet cylinder, as was also shown with LDV results. The results showed similar flow structure at all measured shaft angles with no influence of rotor motions, and that the flow consisted of relatively strong stream flows with almost horizontal vectors 40mm upstream of the inlet suction port. As the flow moves downwards, the velocities become more and more vertical so that at the inlet port the flow becomes almost vertical.

ACKNOLEGEMENT

Financial support from EPSRC is gratefully acknowledged. The authors would like to thank Tom Fleming, Michael Smith, Jim Ford and Grant Clow for their valuable technical support during the course of this work.

REFERENCE

[1] Arjeneh M, Kovacevic A, Rane S, Manolis M and Stosic N 2015 Numerical and Experimental Investigation of Pressure Losses at Suction of a Twin Screw Compressor. *IOP Conf. Ser. Mater. Sci. Eng.* ***90*** *012006.*
[2] Stosic N 2004 Review article: Screw compressors in refrigeration and air conditioning. HVAC and R Research, **10(3)**, pp 233–263.
[3] Stosic N 1998 On gearing of helical screw compressor rotors. *Proc IMechE*, **213(C)**, pp 587-594.
[4] Guerrato D 2012 Cycle-Resolved flow Characteristics Within a Screw Compressor. *Ph.D. Thesis*, City, University of London.
[5] Guerrato D, Nouri J M, Stosic N and Arcoumanis C 2007 Axial flow and pressure characteristics within a double screw compressor. *3rd Int. Conf. Optical and Laser Diagnostics, ICOLD*, London, 23-25 May 2007.
[6] Pascu M, Heiyanthuduwage M, Mounoury S, Cook G and Kovacevic A 2013 A study on the influence of the suction arrangement on the performance of twin screw compressors. *Proceedings of the ASME 2013 International Mechanical Engineering Congress and Exposition*, IMECE2013 San Diego, California, USA, November 15-21, 2013.
[7] Mujic E, Stosic N and Smith I K 2000 The CFD Analysis of a screw compressor suction flow. *15th Int. Compressor Eng. Conf., Purdue*, USA 25-28 July 2000.
[8] Mujic E, Kovacevic A, Stosic N and Smith I K 2006 The Influence of discharge ports on rotor contact in screw compressor. *18th Int. Compressor Eng. Conf., Purdue*, USA 17-20 July 2006.
[9] Nouri J M, Guerrato D and Stosic N 2020 Turbulent flow measurements near the discharge port of a screw compressor. *Int. J Flow, Turbulence and Combustion*, **104**, pp 927–946.
[10] Nouri J M, Guerrato D, Stosic N and Arcoumanis C 2008 Axial Flow Characteristics within a Screw Compressor. *ASHRAE Research Journal*, **14 (2)**, pp 259-274.
[11] Sun S, Singh G, Kovacevic A and Bruecker C 2020 Experimental and Numerical Investigation of Tip Leakage Flows in a Roots Blower. *Design,* **4(3)**, doi:10.3390/designs4010003.
[12] Vimalakanthan K., Read M., Kovacevic A.: Numerical Modelling and Experimental Validation of Twin-Screw Expanders. Energies, 13, 2020, doi:10.3390/en13184700.
[13] Marchi A, Yan Y, Nouri J M and Arcoumanis C 2013 PIV Investigation on Flows Induced by Fuel Sprays from an Outwards Opening Pintle Injector for GDI engines. *Int. Conference ILASS*, Ghania, Greece 2013.
[14] Marchi A 2009 Internal flow and spray characteristics of an outwards opening pintle-type gasoline-injector. *Ph.D. Thesis,* City, University of London.
[15] Peng X, Xing Z, Zhang X and Shu P 2000 Experimental study of oil injection and its effect on performance of twin screw compressor. *15th Int. Compressor Eng. Conf., Purdue*, USA 25-28 July 2000.
[16] Toyama T, Matsuura H and Yoshida Y 2006 Visual Techniques to quantify behaviour of oil droplets in a scroll compressor. *18th Int. Compressor Eng. Conf., Purdue*, USA 25-28 July 2006.
[17] Nouri J M, Guerrato D, Stosic N and Yan Y 2021 Flow measurements within the discharge port of a screw compressor using LDV and PIV methods. To be presented at *10th International Symposium on Turbulence, Heat and Mass Transfer THMT-21*, 6-9 July 2021, St. Petersburg, Russia.

Research and Solution of Vibration Problem of an Air-cooled heat exchanger of Reciprocating Compressor

Junchao Ye[a,b], Xiaoling Yu[a], Shiyi Fan[a], Qian Lv[a],Xiaolin Wang[a,c]
a School of Energy and Power Engineering, Xi'an Jiaotong University, Xi'an, 710049, China.
b Sinopec Petroleum Machinery Co., Ltd. Sanji Branch, Wuhan 430040, China;
c School of Engineering, University of Tasmania, Hobart, TAS 7001, Australia.
xlingyu@mail.xjtu.edu.cn

Abstract: Excessive vibration of complex structures in a multi-excitation environment is an arduous problem. Especially in a reciprocating compressor, how to quickly and accurately position the vibration sources, analyze the excitation features and take effective solutions is of great significance to the safe and stable operation of the compressor. In this paper, the problem of excessive vibration of an air-cooled heat exchanger used in a variable speed CH4 compressor was studied. The air-cooled heat exchanger is composed of tubes, tube boxes and support frame. Its vibration is evoked by complex periodic gas pulsation in the tubes. The vibration model of the heat exchanger was simplified as a multi-freedom model with periodic excitations using the lumped parameter method. The model has advantages of clear physical concept, easy solution and high accuracy of result. In the model, the air-cooled heat exchanger was simplifies as three mass blocks connected by six springs. And then, modal test of the heat exchanger was carried out on site. The vibration modes and natural frequencies of the exchanger were obtained according to the test results. After that, above mentioned multi-freedom model was further simplified as a single freedom vibration model, so that it was convenient to propose the vibration control measures. The practical application results proved that these measures were effective. This research provided a practical and fast analytical method for solutions of compressor vibration problems.

Key words: air-cooled heat exchanger, complex structural vibration, modal test, lumped parameter method

0 Introduction

Reciprocating compressor are core equipment for natural gas gathering and transportation. The inter-stage cooling and post-cooling of the compressor are air-cooled or water-cooled heat exchangers. Among them, air-cooled heat exchangers are favored by most manufacturers and users of reciprocating compressors because of simple structure, convenient design and manufacturing, and no additional water consumption. The periodic suction and exhaust of gas will cause the vibration, fatigue and even cracking of the air-cooled heat exchanger, thus greatly threatens the stability and safety of the compressor. Especially for large-scale air-cooled heat exchangers, due to their large mass and large span of tube box support, the vibration problem is more prominent and more serious. Therefore, it is very important to analyze and test the vibration of the air-cooled heat exchanger and take anti-vibration measures to ensure the safe operation of the compressor.

In order to describe the dynamic response of air-cooled heat exchanger, various mathematical models have been established,such as lumped parameter model, multi-body dynamics model, and finite element model are the most general models.The lumped parameter model is used for one-way analysis,it can extract obvious vibration characteristics from complex structural parts to establish a model, and ignore other minor or insignificant characteristics.It's widely used in practical engineering problems.

The difficulty to establish a lumped parameter model is how to effectively and accurately calculate the lumped parameter. Generally,people are more likely to use methods of the 3D model analysis [1~2] and modal test analysis [3~5]. This paper proposed a new method based on the combination of modal test [6] and finite element static analysis to slove an engineering problem, greatly accelerate the calculation speed. After field test , comparison and verification, this method was proved to be good.

1 Descriptions of the air-cooled heat exchanger vibration problem

The compressor is composed of three rows with two stage compression, and key parameters are shown in Table 1. At 300 rpm or so, severe vibration happened at the heat exchanger, while it operates normally at other operating speeds. As shown in Figure 1, the air-cooled heat exchangeris mainly composed of three parts from the bottom to the top, i.e. heat exchanger of the second stage exhaust gas, of the first stage exhaust gas, and of cooling water of the engine, respectively. Vibration mainly occurs in the direction of tube length with a maximum value of 36mm/s (RMS) (2mm (RMS)) at the tube box, which is far over the limits in standard of ISO 13706 " Petroleum, petrochemical and natural gas industries - Air-cooled heat exchangers ". After excluding the possibility of vibration of the compressor and the engine causing the vibration of the air-cooled heat exchanger, we focused on reasons of excessive gas pulsation in pipeline and structure resonance.

Table 1 Compressor parameters

Parameters	Values	Parameters	Values
Suction pressure/MPa	0.7 ~ 3.0	Stroke /mm	280
Suction temperature /℃	-5 ~ 40	Operating speeds r/min	300 ~ 400
Discharge pressure/MPa	6.3	Volume flow rate ×104 / Nm3/d	10.7-56.1
Number of stages	2	Driver	Natural gas engine
Number of first stage cylinders	2	Driver power /kW	630 @ 440RPM
Number of second stage cylinder	1		

Fig.1 Size of the air-cooled heat exchanger

2 Vibration tests of the air-cooled heat exchanger

Between 300~400 rpm, we tested the vibration of the heat exchanger[7], then analyze the frequency spectrum of vibration signals. The analysis results show that, between 300~330 rpm, there was an obvious peak around 11Hz (maximum about 60m/s^2); between 340~400 rpm, the peak value at this frequency became smaller and smaller (minimum about 1m/s^2), then gradually became equal to the harmonics at other frequencies at the Z-direction (tube bundle length direction).

At the same time, we tested the pulsation in the pipe neer the heat exchanger, analyze frequency spectrum of pulsation signals. The analysis results show that, between 300~400 rpm, the peak values of the pulsation were within the API618 standard ranges[8] (maximum was about 20 KPa). The frequency multiplication components are very obvious at each multiple speed, and decrease in turn.The amplitudes of harmonics greater than 60Hz are small(less than 1Kpa) ,which can be ignored.

According to the test results above, it was speculated that vibration resonance of the exchanger had occured,so the modal test of the exchanger was proposed.

Since the vibration mainly occurs in Z direction, the modal test was only conducted in Z direction, and sensors were set on the middle and lower part of the head, which are shown in Figure 2. The node diagram of modal analysis was established in the LMS Test.lab system.

(a) Layout of sensors on primary tube box

(b) Layout of sensors on secondary tube box

Fig. 2 layout of sensors on tube boxes

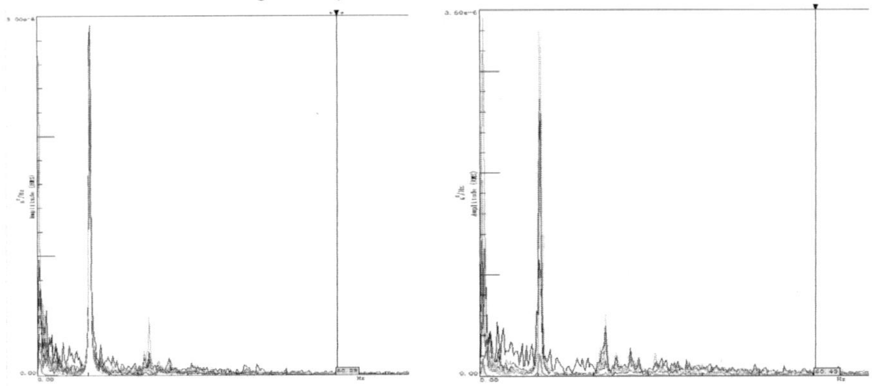

(a) Autocorrelation spectrum of primary tube box (b) Autocorrelation spectrum of secondary tube box

Fig. 3 Response spectrum of vibration in z-direction

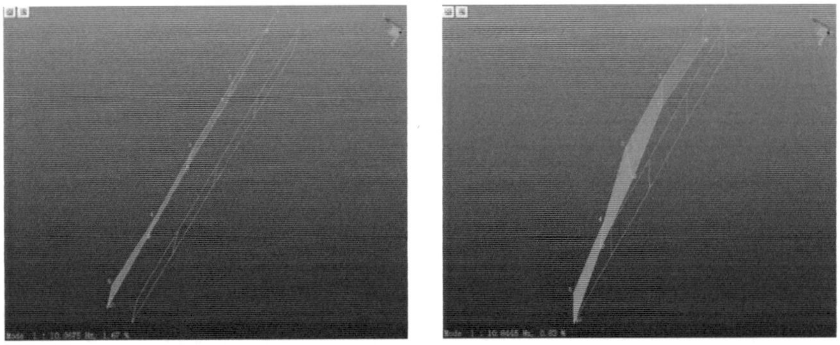

(a) Mode of primary tube box (10.9675 Hz) (b) Mode of secondary tube box (10.845Hz)

Fig. 4 Vibration mode of the tube boxes

The test results of the autocorrelation spectrum of each sensor are shown in Figure 3, the results tell that a remarkable peak occurs around 11 Hz, the amplitude of the 2nd and 3rd harmonics of the primary and secondary tube box is gradually reduced. At the same time, the higher-order modes will only appear at the frequency over 60 Hz in the spectrogram.As the harmonics greater than 60Hz of the pulsation excitation can be ignored, the higher-order modes can be ignored also. The vibration mode of the exchanger around 11 Hz shown in Fig. 4 were drawn according to the response of sensors, There is only one node in the mode of vibration around 11 Hz,it means that it's the first mode of the exchanger.

From the analyse results above, it can be determined that the reason for the excessive vibration of the heat exchanger is that a first-order resonance happed at exchanger in Z direction at a frequency around 11 Hz, and the the excitation source is the pulsation harmonic around 11Hz(2nd times of fundamental excitation frequency between 300~330 rpm). After test another exchanger in the same way, we got the same conclusion.

3 Theoretical analysis of Air-cooled heat exchanger Vibration

The 3D model of the support frame of the air-cooled heat exchanger is shown in Figure 5, and heat exchanger's tube boxes of the secondary exhaust gas, of the primary exhaust gas, and of cooling water are all fixed on the support frame by bolts.

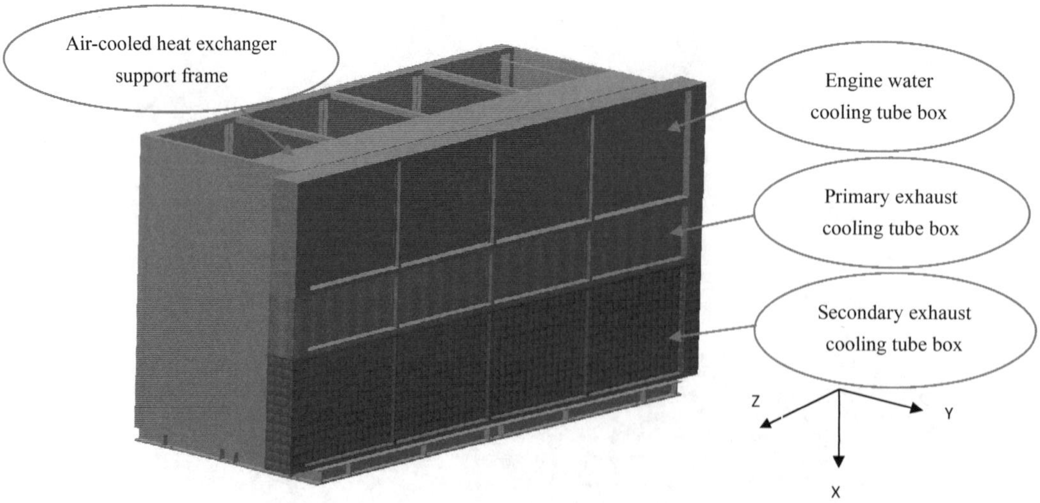

Fig.5 3D model of the air-cooled heat exchanger

Natural frequency of a single freedom vibration system can be calculated by (1) [9]:

$$f = \frac{\omega}{2\pi} = \frac{1}{2\pi}\sqrt{\frac{K}{M}} \tag{1}$$

Where f is natural frequency, ω is circular frequency, K is stiffness ,and M is mass. It's easy to see that f is positively correlated with K , but inversely correlated with M.

The Air-cooled heat exchanger is a welded assembly with complex structures. If modal analysis is conducted according to the real structure such as Fig.5, it will consume a lot of computing resources and time, the fact is it's no need to obtain high level harmonics at all. Therefore, the real air-cooled heat exchanger should be simplified in the modal analysis reasonably. So we ignoring the fan, small accessories, etc., which have negligible effects on the natural frequency of the exchanger, then stablish 3D model of support frame as shown in Figure 6:

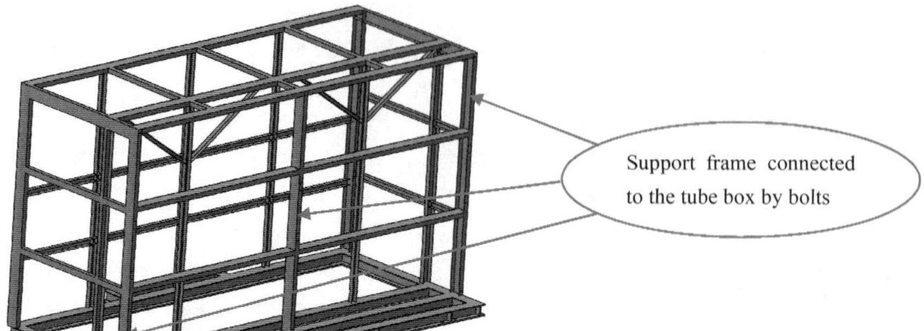

Fig.6 Air-cooled heat exchanger support frame

According to the Air-cooled heat exchanger data, the mass distribution is shown in Table 2:

Table 2 Mass distribution of the heat exchanger

Serial number	Name	Weight /kg	Media weight /kg
1	Engine water cooling tube box	2,060	300
2	Primary exhaust cooling tube box	1,526	4
3	Secondary exhaust cooling tube box	2,397	3

According to Table2, and take three tube boxes as research object, the heat exchanger can be simplified as a multiple freedom vibration system using lumped parameter method,the three tube box can be simplified as three lumped masses without stiffness,the support frame and the connection bolts can be simplified as six springs without mass,establish a multiple freedom model as shown in Figure 7:

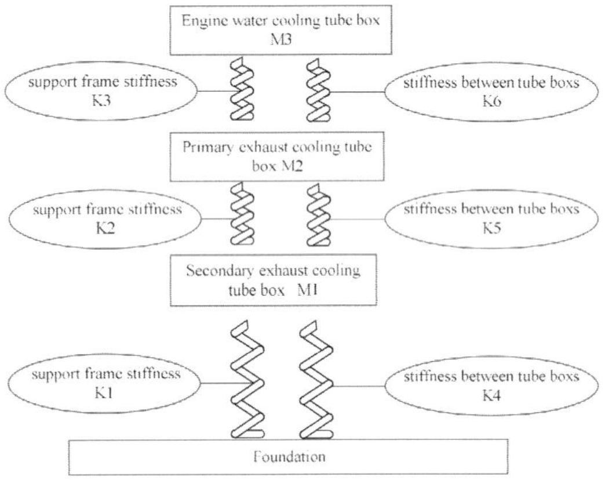

Fig.7 Lumped model of Air-cooled heat exchanger

Calculate each support frame stiffness with FEM, the results are shown in Figure 8[8]:

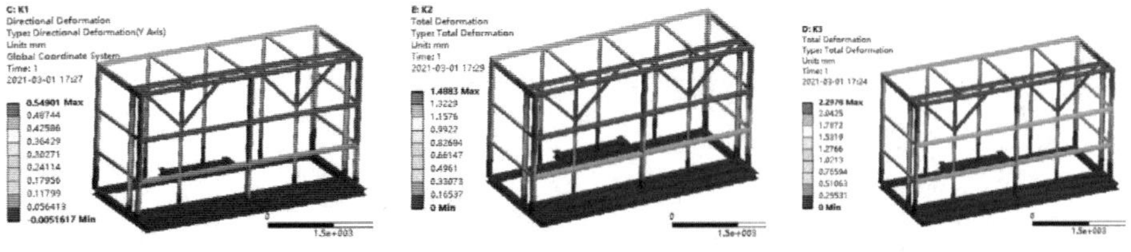

support frame stiffness K1 support frame stiffness K2 support frame stiffness K3

Fig.8 stiffness calculation of Air-cooled heat exchanger structure

The results of K1, K2, K3 are 7.02E6 N/M,3.12E6 N/M,3.51E6 N/M, respectively. But K5, K6, K7 are difficult to calculate, so they need to be further simplified and equivalent.

The equation of free vibration of a multiple freedom vibration system without damper is as following.

$$M\ddot{x}+Kx=0$$

Where, M is the structural mass matrix, K is the structural stiffness matrix, x is the displacement of each freedom.

Make $x=\phi pe^{i\omega t}$, then multiply both sides by ϕ^T, and the transformation will be:

$$\omega^2\phi^T M\phi p=\phi^T K\phi p$$

Where, p is response in modal coordinates, ϕ is modal coordinates, $e^{i\omega t}$ is complex form of vibration,and t is time.

Assume the mass of the multi-degree-of-freedom system is not coupled, and the stiffness is coupled, the degree of freedom is n, then the formula can be expressed as:

$$\omega^2\phi^T\begin{bmatrix} m_1 & 0 & \dots & 0 \\ 0 & m_2 & & \dots \\ \dots & & \dots & 0 \\ 0 & \dots & 0 & m_n \end{bmatrix}\phi p=\phi^T\begin{bmatrix} k_{11} & k_{12} & \dots & k_{1n} \\ k_{21} & k_{22} & & \\ \dots & & \dots & \dots \\ k_{n1} & k_{12} & \dots & k_{nn} \end{bmatrix}\phi p$$

According to the principle of energy conservation, the maximum kinetic energy and maximum elastic potential energy of each degree of freedom are equal and mutually converted in the free undamped vibration system:

$$E_m = E_k$$

$$E_m = \frac{1}{2}\dot{x}^T M\dot{x} = \frac{1}{2}\phi^T M\phi(\omega p)^2$$

$$E_k = \frac{1}{2}x^T Kx = \frac{1}{2}\phi^T K\phi(p)^2$$

Make the natural frequency of a certain order of the system be ω_x, vibration shape is ϕ_x ,and $\phi_x=[\varphi_1 \ \ \varphi_2 \ \ \dots \ \ \varphi_n]^T$. According to the orthogonality of the mode shape, the formula can be transformed into:

$$E_m=\frac{1}{2}\omega_x^2(m_{11}\varphi_1^2+\dots+m_{nn}\varphi_n^2)p^2$$

$$E_k=\frac{1}{2}(\varphi_1(k_{11}\varphi_1+\dots+k_{1n}\varphi_n)+\varphi_2(k_{21}\varphi_1+\dots+k_{2n}\varphi_n)+\dots\varphi_n(k_{n1}\varphi_1+\dots+k_{nn}\varphi_n))p^2$$

If the kinetic energy of the system is lumped to the first degree of freedom, the formula can be transformed into:

$$E_m=\frac{1}{2}\omega_x^2(m_{11}(\frac{\varphi_1}{\varphi_1})^2+\dots+m_{nn}(\frac{\varphi_n}{\varphi_1})^2)\varphi_1^2p^2=\frac{1}{2}\omega_x^2(m_{11}+\Delta m)\varphi_1^2p^2$$

Where $\Delta m=m_{22}(\frac{\varphi_2}{\varphi_1})^2+\dots+m_{nn}(\frac{\varphi_n}{\varphi_1})^2$

So the equivalent mass of M is $M=m_{11}+\Delta m$ (2)

Similarly available, when the potential energy is lumped to the first degree of freedom, the formula above can be transformed into:

$$E_k=\frac{1}{2}((k_{11}(\frac{\varphi_1}{\varphi_1})+\dots+k_{1n}(\frac{\varphi_n}{\varphi_1}))+\dots+\frac{\varphi_n}{\varphi_1}(k_{n1}\frac{\varphi_1}{\varphi_1}+\dots+k_{nn}\frac{\varphi_n}{\varphi_1}))\varphi_1^2p^2=\frac{1}{2}(k_{11}+\Delta k)\varphi_1^2p^2$$

Where $\Delta k = k_{12}(\frac{\varphi_2}{\varphi_1}) + \ldots + k_{1n}(\frac{\varphi_n}{\varphi_1}) + \ldots + \frac{\varphi_n}{\varphi_1}(k_{n1}\frac{\varphi_1}{\varphi_1} + \ldots + k_{nn}\frac{\varphi_n}{\varphi_1})$

So the equivalent stiffness of K is $K = k_{11} + \Delta k$ (3)

Because $E_m = E_k$,the formula 2 can be transformed into:

$$f = \frac{\omega_x}{2\pi} = \frac{1}{2\pi}\sqrt{\frac{k_{11} + \Delta k}{m_{11} + \Delta m}} \quad (4)$$

Because the structure of the engine water cooling tube box is the same as the Secondary exhaust cooling tube box, the static deformation should be the same. Define the Primary exhaust cooling tube box M2's vibration amplitude as 1, according to Fig 4,the vibration mode of the three tube boxes should be $\phi_x = [1.5, 1, 0.5]^T$.

According to Rayleigh Law and formula 2,take the Primary exhaust cooling tube box M2 as a lumped point, M1, M2, M3 are equivalent to M[9~11]; According to formula 3,K1, K2, K3 are equivalent to D1; K4, K5, and K6 are equivalent to D2; so the total equivalent stiffness should be D=D1+D2. The heat exchanger shown in Figure 7 can be simplified to the equivalent single degree of freedom model shown in Figure 9.

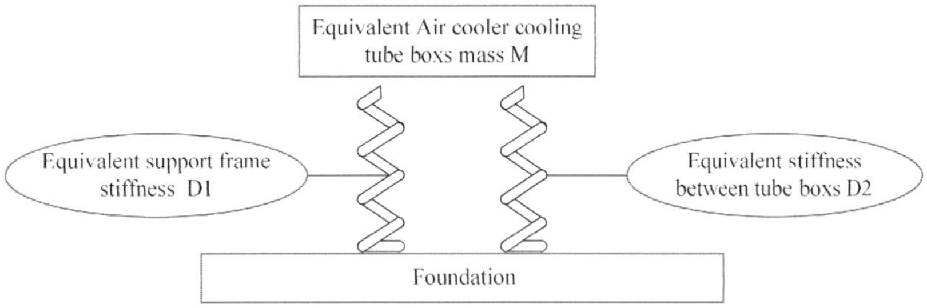

Fig.9 Equivalent lumped model of Air-cooled heat exchanger

According to formula 2, the equivalent mass is M=7440kg, then according to formula 1, the equivalent stiffness of the heat exchanger is D=D1+D2=3.53E7 N/M. According to K1, K2, K3, and formula 3, the equivalent frame stiffness is D1=3.41E6 N/M, so the equivalent stiffness between tube boxes D2=3.19E7 N/M can be obtained.

It is found that the modification of the tube box's mass is very expensive, but the structure stiffness modification will be much easier, so the modification plan is proposed as shown in Figure 10:

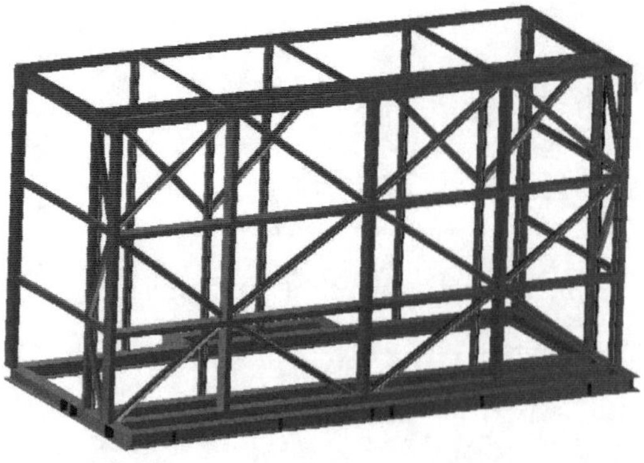

Fig.10 Support frame (modified)

The FEM is also used to calculate the stiffness of the modified air-cooled heat exchanger support frame as shown in Figure 11[12~13]:

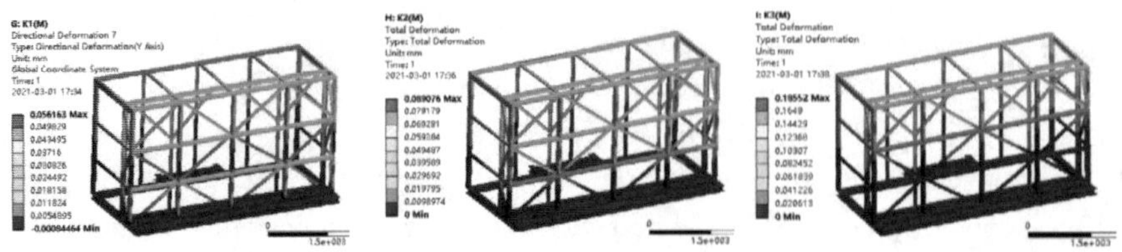

support frame stiffness K1 support frame stiffness K2 support frame stiffness K3

Fig.11 Structural stiffness calculation

After modification, K1, K2, K3 are 1.05E8 N/M, 1.64E8 N/M, and 5.92E7 N/M, respectively.

The first-order mode of the modified Air-cooled heat exchanger is still assumed as $[1.5,1,0.5]^T$, and the welding coefficient is taken as 0.95. According to formula 3, the equivalent stiffness of the modified structure is D1=7.79E7 N/M, and the equivalent mass M and D2 was not changed. So natural frequency of the air-cooled heat exchanger in Z-direction was calculated to be 19.34Hz, which is 2.4 times greater than the highest fundamental excitation frequency required by API618 standard.

4 Solution verification

Support frame of the Air-cooled heat exchanger was improved according to above modeling results, and then vibration of the improved air-cooled heat exchanger was re-tested., The test results are shown in Figure 12:

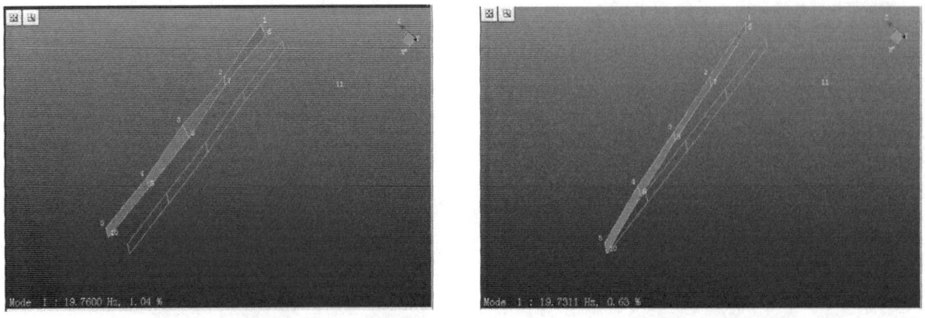

(a) Mode of primary tube box（19.76 Hz） (b) Mode of secondary tube box (19.73 Hz)

Fig. 12 Vibration mode of the tube boxes

The error of first order frequency of calculation result and the test one is: (19.76-19.34)/19.76=2.1%<10%, which meets the API618 standard requirement.

The compressor is restarted under different load. The vibration velocity value (RMS) at tube box under any speed is less than 5mm/s, which is less than the API618 standard requirement. The Air-cooled heat exchanger vibration problem is successfully solved.

5 Conclusions

In this paper, the vibration problem of the air-cooled heat exchanger of reciprocating compressor was tested, analyzed and treated. Through on-site testing, it is determined that the cause of the air-cooled heat exchanger vibration is the low-order mechanical resonance caused by the pulsation. Due to the complex structure of the Air-cooled heat exchanger, it is difficult to perform vibration modal analysis based on the real structure of the air-cooled heat exchanger. This paper proposed a lumped parameter method, by which the heat exchanger was simplified a multiple freedom model composing several mass connected by springs. The equivalent mass and equivalent spring stiffness are calculated based on the modal test results. Using this model, improvement design was made for the air-cooled heat exchanger, and vibration of the improved air-cooled heat exchanger was tested. The test results show that the first-order natural frequency of the heat exchanger measured agree well with the calculated one, with a deviation of 4% only. The Air-cooled heat exchanger vibration is reduced from 36mm/s (RMS) to 5 mm/s (RMS). Therefore, the vibration modeling method for complex structures proposed in this paper has good engineering practicability and can effectively solve engineering vibration problems .

Acknowledgements
This article was supported by the National Natural Science Foundation of China (52076166).

References:

[1] Bayles D J , Lowery R L , Boyd D E . A non-linear dynamic lumped-parameter model of a rectangular plate[J]. Journal of Sound & Vibration, 1972, 21(3):329-337.

[2] Fu C C . Computer analysis of a rotating axial-turbomachine blade in coupled bending-bending-torsion vibrations[J]. International Journal for Numerical Methods in Engineering, 2010, 8(3):569-588.

[3] Schmidt Be Rg R , Pal T . SOLVING VIBRATION PROBLEMS USING MODEL ANALYSIS. 1986.

[4] Bennett W H , Kwatny H G . Continuum modeling of flexible structures with application to vibration control[J]. Aiaa Journal, 2015, 27(9):1264-1273.

[5] HM Zhu, Chen W F , Zhu R P , et al. Dynamic analysis of a flexible rotor supported by ball bearings with damping rings based on FEM and lumped mass theory[J]. Journal of Central South University, 2020:1-18.

[6] Seto K , Ren M , F Doi. Modeling and Feedback Structural Acoustics Control of a Flexible Plate[J]. Journal of Vibration & Acoustics, 2001, 123(1):18-23.

[7] Yuan Wen Cao et al. Research on the ICP Acceleration Sensors Detecting Roadbed Compaction Degree Test[J]. Advanced Materials Research, 2011, 1325 : 275-279.

[8] API Standard 618——5th ed,Washington:American Petroleum Institute,2007.

[9] ERHAN AY et al. Calculation of the Modulus of Elasticity with Using Natural Frequency Formula and Updated Finite Element Analysis Of Automotive Rear Lamp Lens[J]. International Journal of Automotive Science and Technology, 2019, : 27-31.
[10]Minyuan Huang.Multi-Degree of Freedom Vibration Theory and Engineering Application[C]. Advanced Building Materials. Part 3.2011:2473-2476.
[11] Chang Li et al. The Vibration Model Establishment and Solution of Bending-Torsion-Axial-Swing Coupled the Helical Gear Transmission System Based on Lumped Parameter Approximation Method[J]. Advanced Materials Research, 2015, 3683 : 743-747.
[12] Kabe A M. Stiffness matrix adjustment using mode data[J]. AIAA Journal, 1985, 23(9): 1431-1436.
[13] Jaishi B, Ren W X. Structural finite element model updating using ambient vibration test results[J]. Journal of Structural Engineering, 2005, 131(4): 617-628.

Research on fault diagnosis and early warning of reciprocating compressor based on stacked convolutional autoencoder optimized by gradient differential evolution

H Li [1], **XM Liu** [2], **ZW Mao** [1], **JJ Zhang** [1]

[1] Diagnosis and Self-Recovery Engineering Research Center, Beijing University of Chemical Technology, Beijing, 100029, China

[2] State Key Laboratory of Compressor Technology (an hui laboratory of compressor technology), Hefei, Anhui, 230031, China

Corresponding author's(ZW Mao) e-mail address: maozw1990@126.com

Abstract. The reciprocating compressor is one of the key equipment in the process industrial field. Due to its complex structure and motion state, the bearing bush of the connecting rod is prone to wear failure. In the early stage of wear failure, the monitoring signal signs are very weak. As a result, it has produced bad results that identify the fault signs by using traditional data processing and spectrums analytical methods. Aiming at the early fault identification of the bearing bush, unsupervised feature mining based on auto-encoder principle and super-parameter optimization based on Gradient-Differential-Evolution are utilized, and an early-warning-model based on Gradient-Differential-Evolution and Stacked-Convolutional-Autoencoder is proposed. In order to study the sensitivity of the vibration signal and piston rod settlement signal to the early stage of wear failure, the two signals are input into the early warning model for comparison. In addition, they are fused to verify the improvement ability of multi-source signal on early warning. Moreover, to verify the early fault recognition performance of the proposed methods, the proposed method is compared with the other two early-warning-models based on Stacked-Autoencoder and Convolutional-Neural-Networks. The actual fault case analysis results show that based on the Gradient-Differential-Evolution optimization model, the difficulty of parameter setting can be effectively reduced and the proposed method has significant advantages to detect the early warning timely and effectively.

1. Introduction

Reciprocating compressor is widely used in oil exploitation, natural gas exploitation, transportation and other process industrial enterprises. it is the key power machinery for safe and stable production [1,2]. Due to the complex structure and many wear-prone parts, especially under the variable load condition, the temperature and pressure of the compression medium in the cylinder continue to change rapidly. Therefore, the stress of key moving parts is complex and failures occur frequently [3,4], especially connecting rod. In case of failure, it will bring huge losses and negative impacts on economy, environment and society [5,6]. Meantime, reciprocating compressor is developing towards high speed, high efficiency and high degree of automation, its structure complexity and value are further increased [1]. The safety risk will also be further increased. Therefore, it is of great significance to find and diagnose faults in the early stage.

Connecting rod is one of the most important moving parts of reciprocating compressor. Its bearing bush is a necessary connection pair to ensure the good operation of working parts. Unfortunately, its stress state is very complex. Due to poor lubrication, oil impurities, material deformation, processing defects and other reasons [7], the bearing bush is prone to wear. Therefore, it is very important to study the early warning diagnosis method of reciprocating compressor connecting rod bearing wear fault for timely finding fault, guiding on-site maintenance and avoiding malignant accident.

On the one hand, oil analysis is a traditional detection method in the research of wear fault detection methods for connecting rod small end bearing bush [8,9], but its off-line application mode and more complex detection and analysis process limit the wide application of this method. Although the oil on-line monitoring technology developed in recent years, the actual engineering application still needs to be further developed due to the low accuracy of wear particle detection and analysis. On the other hand, temperature monitoring is also a common wear monitoring method, because wear usually causes temperature rise. In recent years, passive wireless temperature sensor has been applied in the wear monitoring of connecting rod big end bearing bush [10]. However, it is difficult to install the sensor due to the narrow space of the small end bearing bush of the connecting rod. What is more, scholars focus on dynamic simulation computing [11-15] here are few reports on the detection method of connecting rod bearing bush wear failure based on vibration signal or piston rod settlement signal. Because understanding such signals requires a lot of artificial experience and strong feature extraction capabilities.

Fortunately, with the rapid development of machine learning, the ability of classification based on features is improving. Diagnosis is no longer limited by feature extraction capabilities, and the diagnostic methods based on vibration signal and distance signal have made breakthrough progress. Based on vibration signal, Potocnik [16] applied machine learning methods such as neural network (NN), support vector machine (SVM) to fault diagnosis of reciprocating compressor. Qi et al. [17] extracted five dimensionless features of several types of vibration fault signals of reciprocating compressor, followed by probabilistic neural network (PNN), and realized the fault diagnosis of piston rod, piston ring and suction valve of reciprocating compressor. Yang [18] et al. and Li [19] et al. based on convolution neural network to extract signal features, followed by classifier, realized fault diagnosis of reciprocating compressor. Van [20] used the deep confidence network (a method of machine learning) to realize the fault diagnosis of reciprocating air compressor in different valves, including the fault diagnosis of suction leakage, discharge leakage and spring deterioration. However, the feature extraction ability of the above papers needs to be strengthened.

In recent years, Stack-Autoencoder [21] (SAE) has become a widely used unsupervised feature extraction method, which can be effectively used for feature extraction of various signals. Based on this, Stack-Convolutional-Autoencoder [22] (SCAE) has a good feature expression ability for one-dimensional signal by introducing convolution operation. Chen [23] et al. used a One-Dimensional-Stack-Convolutional-Autoencoder (1D-SCAE) for fault detection and diagnosis of multivariable processes. By denoising high-dimensional signals and learning the deep features of multivariable signals in one-dimensional SCAE, they achieved fault detection and diagnosis in process industry. Li [24] et al. proposed a framework based on convolutional autoencoder and K-means by using a Stack-Convolutional-Autoencoder for image feature learning. Based on the computer vision method, Fu [25] et al. Grouped all variables, extracted each group of features by using the improved convolution autoencoder, and classified and diagnosed the aeroengine fault after feature fusion. Zhou [26] et al. introduced residual learning mechanism to optimize the learning ability of the model based on learning deep features of convolutional autoencoder, and achieved the fault diagnosis of gear teeth and bearings.

These study attempts to explore the performance of SCAE in the early recognition of reciprocating compressor connecting rod bearing wear fault, and make full use of the unsupervised feature mining ability of SCAE to extract early fault features. In addition, Gradient-Differential-Evolution algorithm (GDE) is proposed to solve the problem of neural network super parameter optimization combined with the operation characteristics of neural network itself, especially for the shallow plank model to adaptively obtain super parameters, so as to reduce the artificial experience required to set parameters.

In conclusion, the fault early warning model based on GDE-SCAE is established. Combined with the actual field fault case data, the sensitivity of various monitoring signals to the early fault of connecting rod bearing bush and the positive role of multi-source signal fusion are studied. In the meantime, the effectiveness and superiority of the proposed method are verified through the comparative analysis of typical methods.

2. Data acquisition

The actual fault case of a new hydrogen reciprocating compressor in a petrochemical plant was studied. The unit model: 3HHE-VL-3; rated power: 4365KW; rated speed: 300R / min; inlet design pressure (1 / 2 / 3): 2.1/4.2/9.9MPa; outlet design pressure (1 / 2 / 3): 4.3/8.8/16.1MPa. In addition, the unit is equipped with reciprocating compressor vibration online monitoring system, and the unit structure and vibration measuring point layout are shown in Figure 1. At about 6:00 on April 29 of a year, the professional inspection personnel caught the abnormal vibration of cylinder 3, and then found that the bearing bush of connecting rod small end of cylinder 3 was seriously worn.

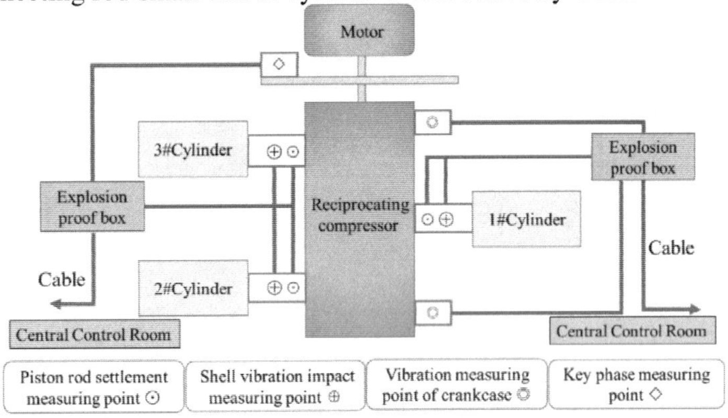

Figure 1. Sketch map of unit structure and vibration measuring points

The on-line vibration monitoring system records the data of fault occurrence and development process completely, and exports a certain amount of monitoring sample data through a special data interface. The distribution of data samples in each period is shown in Figure 2. Taking the fault discovery time as the node, the sample number before fault discovery is 657, and the sample data after fault discovery is 945. Figure 3 and Figure 4 respectively show the waveforms of vibration signal and piston rod settlement signal of 3# cylinder block in 85 hours, 60 hours, 35 hours, 20 hours, 5 hours before fault discovery and at fault discovery.

It can be seen from the waveform that the vibration and piston rod settlement waveforms have changed significantly when the fault is found. Manual inspection can find fault through waveform. In fact, if the manual inspection is carried out at any time, it is not difficult to see that the waveform is abnormal one day before the fault is found. However, two days before the fault was found and earlier, the signal waveform and amplitude were not significantly abnormal. Therefore, it is difficult to find early faults through manual inspection. The research goal is to use the powerful feature mining ability of deep learning network to identify early fault features and give early warning.

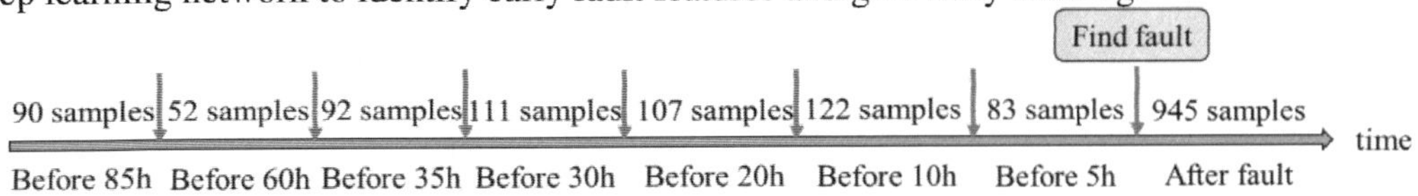

Figure 2. Distribution of the number of data samples obtained

3. The construction of model

3.1. Feature extraction principle of stack convolutional autoencoder

Convolutional autoencoder (CAE) is an unsupervised algorithm, which is often used for feature extraction. The structure diagram of CAE is shown in Figure 5, which consists of encoding and decoding. The encoding part consists of a convolution layer and a pooling layer, and the decoding part consists of an upper sampling layer and a deconvolution layer. Due to the characteristics of the autoencoder, it is required the input X after encoding and decoding should get $\hat{X}$ approximately equal to X, so a convolution layer is usually added at the top of the convolution autoencoder to make the dimension of $\hat{X}$ the same as X. The output characteristic H obtained by input X in coding is convoluted by X. Therefore, it can be considered that the information contained in it comes from X.

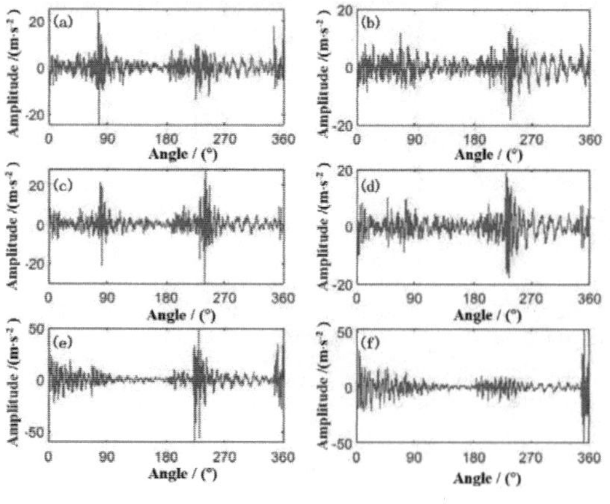

Figure 3. The waveform of vibration signal of faulty cylinder at 85 hours (a), 60 hours (b), 35 hours (c), 20 hours (d), 5 hours (e) and at the time of fault detection (f)

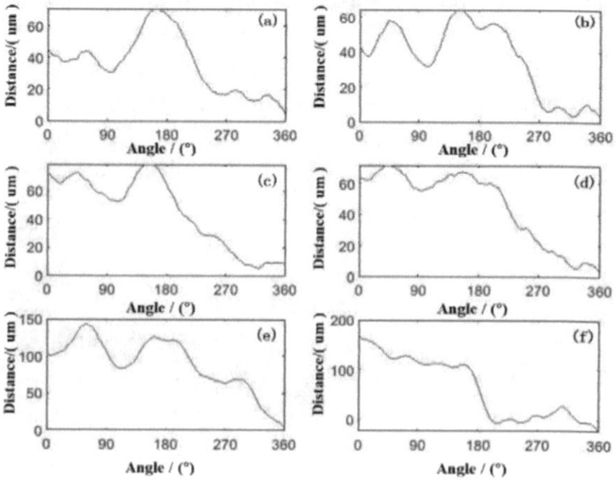

Figure 4. The waveform of piston rod settlement of the faulty cylinder at 85 hours (a), 60 hours (b), 35 hours (c), 20 hours (d), 5 hours (e) and at the time of fault detection (f)

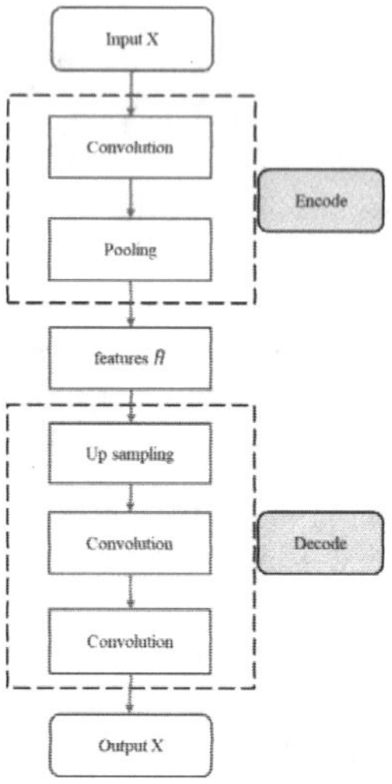

Figure 5. Structure of a single CAE

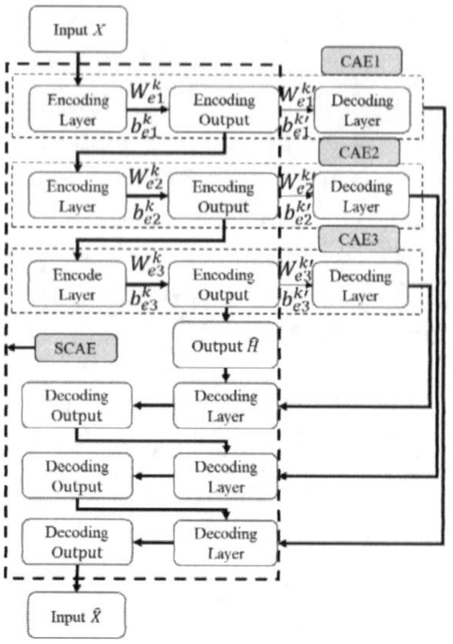

Figure 6. Structural relationship between CAE and SCAE

Stacked-convolutional-autoencoder (SCAE) is composed of multiple coding layers of CAE. A three-layer SCAE structure is shown in Figure 6, which is composed of the coding parts of three CAE. Get the weight and bias of each CAE as $W_{e1}^k, W_{e2}^k, W_{e3}^k$ and $b_{e1}^k, b_{e2}^k, b_{e3}^k$.

3.2. The structure of the model and the way of super parameter optimization.

3.2.1. The structure of the model

Establish the SCAE model as shown below, and name the super parameters. If there is no special explanation, in this study, 'Keras' is used for training. Batch Normalization (BN) layer and Dropout layer are added after pooling layer. The layer with weights is regularized by L2 with the penalty factor 0.001. In convolution layer, bias are not used; all convolution steps are set to 1; convolution kernel size is '3'; padding mode is 'same'; activation function is 'ReLU' , which only used in the encoding part. In classification layer, 'Softmax' function is used and outputs node is set to 2. The optimization method is Adam with learning rate 0.001.

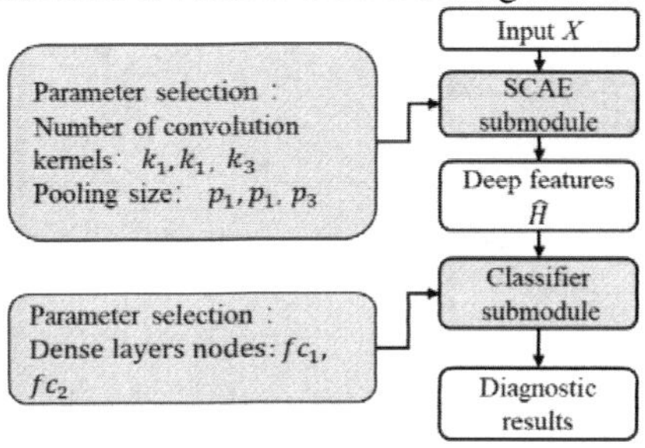

Figure 7. Structure and parameters of the model. The SCAE submodule, three layer Convolution-Max pooling and Up sampling - Deconvolution structure, have convolution kernel number k_1, k_2, k_3 and pooling size p_1, p_2, p_3. The Classifier submodule, two layer Fully-Connection layer, have nodes of fc_1, fc_2.

3.2.2. the way of super parameter optimization

It is difficult to study the optimal value of a single super parameter in isolation by using the method of control variables because of the coupling effect of many super parameters of the model. Therefore, it is necessary to optimize multiple super parameters simultaneously. Differential evolution (DE) algorithm has been widely studied and applied in combinatorial optimization. It shows better effect than simulated annealing algorithm, genetic algorithm, particle swarm optimization and other optimization algorithms [27,28]. The parameters involved in DE algorithm are: group a with size of NI, mutation parameter α, selection probability parameter CR and maximum number of iterations MAX.

However, the parameter α and CR need to be set manually. They remain unchanged in the iterative process, which makes the model difficult to converge. In order to solve this problem, some scholars have proposed an improved scheme that changes adaptively with the iteration process, and the improved scheme that carries out local optimization and fine-tuning. Unfortunately, the above improved scheme has the following shortcomings.

- To control parameters by adding a parameter, such as an adaptive adjustment scheme. It requires manual setting of adjustment parameters.
- The convergence is enhanced by increasing the number of operations, such as fine tuning and local optimization. The training time of neural network is long, and each training is full of randomness. As a result, the cost of increasing the number of operations is difficult to accept.

To solve these two problems, we propose an improved scheme based on gradient (GDE), which introduces an 'or' type constraint on the selection probability parameter CR. The details are as follows. The last evolutionary direction of genes in memory individuals consists of - 1 (decrease), 0 (unchanged) and 1 (increase). If this evolution is still in the same direction, the selection probability parameter CR is invalid, and the gene would be selected.

In order to verify the effectiveness of the improved method, two standard verification functions, Ackley and Sphere, are selected to explore. Three methods, Standard DE, Decreasing DE, Gradient DE (our method GDE), are selected to compare. The variation parameter α and selection probability parameter CR are all 0.5. The parameter quantity is 5-9, and the base number of iterations is 45. Each additional parameter increases the number of iterations by 10. Decreasing DE: the probability parameter CR decreases linearly to 5%, 10%, 15%, 20% and 25% of the initial value. The optimal solution is taken as the optimal solution decreasing with iteration.

It can be seen that the proposed method can achieve good results under the above parameters. Therefore, the flow chart of using the GDE to optimize the super parameters of SCAE model is shown in Figure 7. It should be pointed out that, the gradient matrix preserves the last evolutionary direction of all genes of all individuals and is initialized as all 0 matrix. Individual gene refers to the super parameter in SCAE module. The optimization objective is the classification accuracy of the model, that is:

$$acc = f(k_1, k_2, k_3, p_1, p_2, p_3, fc_1) \tag{1}$$

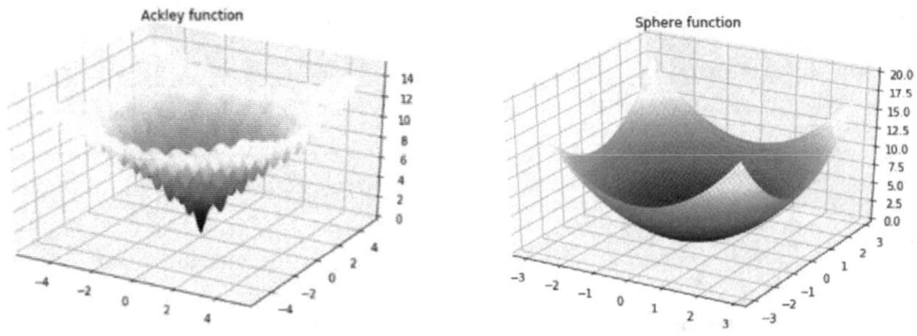

Figure 8. Ackley function and Sphere function.

Table 1. Ackley function's comparison results

Ackley/loss/number of parameters	5	6	7	8	9
Standard DE	0.048	0.047	0.044	0.041	0.041
Decreasing DE	0.049	0.047	0.044	0.044	0.042
Gradient DE	0.040	0.043	0.042	0.041	0.040

Table2. Sphere y function's comparison results

Sphere/loss/number of parameters	5	6	7	8	9
Standard DE	6e-5	1e-4	1.6e-4	2.4e-4	3.3e-4
Decreasing DE	2e-5	9e-5	2.1e-4	2.4e-4	2.7e-4
Gradient DE	0	7e-5	9e-5	1.7e-4	2.6e-4

In order to reduce the randomness of neural network, the average accuracy of each training is taken as the average accuracy of five times. According to the number of optimized super parameters, i, j, r in the flow chart are determined to be integers not greater than this. The boundary condition refers to the artificial individual gene range, namely the solution space. In the process of mutation, if the new individual gene exceeds this range, the gene will be generated randomly again within the specified range.

4. Case analysis

4.1. Fault early warning based on GDE-SCAE

The monitoring information of reciprocating compressor mainly comes from vibration and piston rod settlement signal. In order to deeply study the sensitivity of two kinds of signals to early fault, and explore the effect of two kinds of signal fusion analysis on improving the ability of early fault identification. In this paper, three groups of fault diagnosis tests based on GDE-SCAE are carried out, whose input data are vibration signal, piston rod settlement signal and their fusion signal.

According to existing experience, the solution space of SCAE model is given according to table 1. Randomly select values to initialize group A. In addition, $NI = 20$, $CR = 0.5$, $\alpha = 0.3$, and $MAX = 100$ are set. If there is no special specification, the following tests all adopt this.

The obtained data are divided into 8 groups according to the statistical results of sample data in Figure 2. The first group of data is taken as normal sample data, and the subsequent groups of data are taken as fault sample data in different tests, so as to verify the ability of early fault discrimination. Five-fold-cross validation is used to obtain the accuracy index. The average of five training accuracy is the final index.

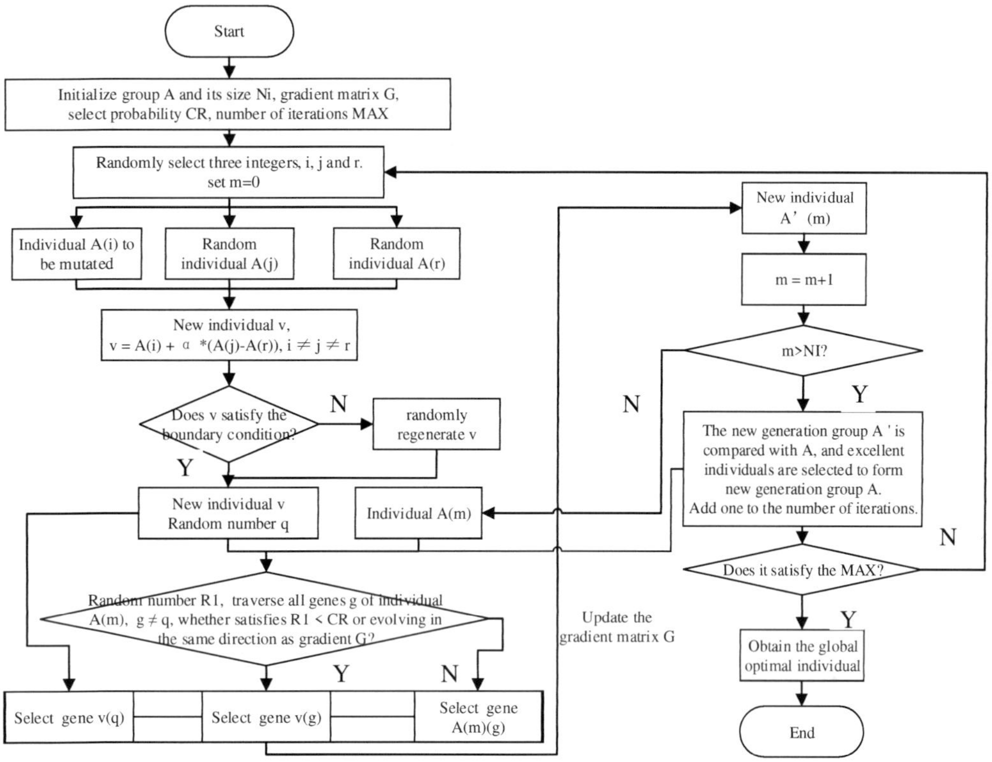

Figure 9. The flow chart of GDE optimization algorithm

In the field of mechanical equipment fault diagnosis, multi-source fusion data is verified to improve the accuracy of equipment state change judgment. Considering that the physical meaning of vibration signal and piston rod settlement signal is different, we use a simple fusion method, which inputs the two kinds of signals into the model respectively to complete feature extraction and then fusion. That is to say, the vibration signal and the piston rod settlement signal are respectively extracted to form feature vectors, and then the feature vectors of the two kinds of signals are spliced to form a fusion signal before the full connection layer.

Using the vibration signal of the fault cylinder, the settlement signal of the piston rod and the fusion signal of the two kinds of signals, the GDE-SCAE model is trained and the parameters are optimized. Under the same parameter, five-fold-cross validation was carried out, and the average accuracy was taken as the final accuracy. The results of super parameter optimization are shown in Table 4. The accuracy of fault early warning is shown in Figure 11.

The following conclusions can be seen from the results of fault early warning accuracy in Figure 11. The early warning model based on GDE-SCAE has a very good ability of mining abnormal features, which can detect faults at least 60 hours earlier than manual inspection. Vibration signal is more sensitive to fault than settlement signal. Fusion analysis is helpful to improve the ability of early fault identification.

4.2. comparative analysis

In order to verify the performance of fault early warning based on GDE-SCAE, the proposed method is compared with GDE-SAE and GDE-CNN.

In GDE-SAE method, SAE is composed of multiple AE stacked. Both SCAE and SAE have encoding and decoding structures. The difference between them is that each layer of SAE coding and decoding layer is full connection layer, while each layer of SCAE coding and decoding part is convolution and pooling layer.

The super parameters to be optimized in GDE-SAE include: the number of nodes in each middle layer (k_1, k_2, k_3). The number of nodes in the full connection layer is j_1. The default solution space is shown in Table 3. The optimal solution is shown in Table 5. The accuracy of fault early warning diagnosis is shown in Figure 12.

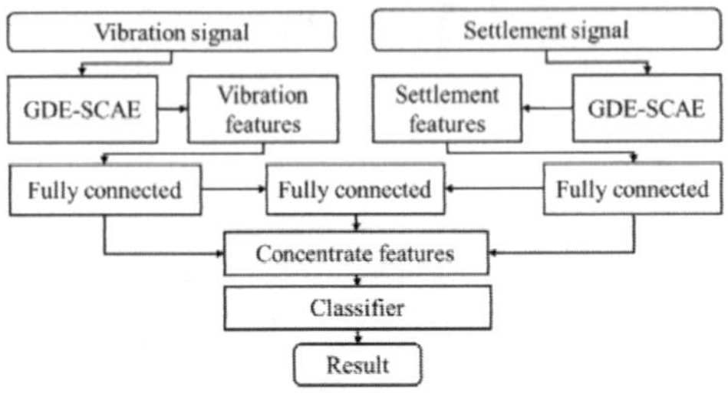

Figure 10. A method of feature fusion

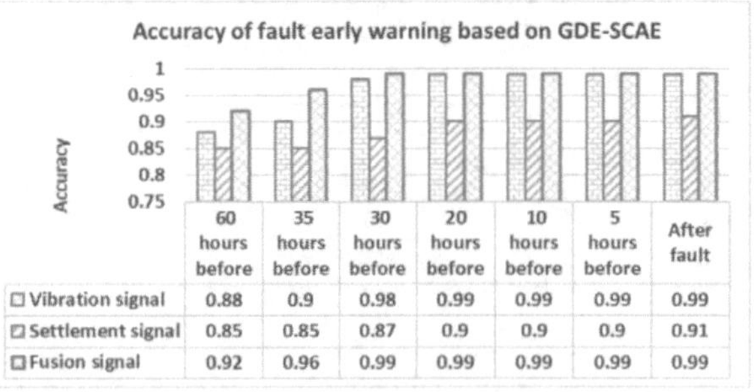

Figure 11. Accuracy of fault early warning based on GDE-SCAE

Table 3. Presupposition solution space of SCAE/CNN model (Num. of convolution kernels (Nc); Pooling size (Ps); Nodes of Fully-Connection layer (FC); Nodes of SAE layer (NS))

Type (SCAE)	Default solution space	Type (CNN)	Default solution space	Type (SAE)	Default solution space
Nc	[1~8]	Nc	[1~8]	NS	[1024,512,256,128,64,32,16,8]
Ps	[2,4,8,16,32,64]	Ps	[2,4,8,16,32,64]		
FC	[64,32,16,8]	FC	[64,32,16,8]	FC	[64,32,16,8]

Table 4. Optimal solution space of SCAE model

Parameters	Vibration signal	Settlement signal	Fusion signal
k_1, k_2, k_3	2, 1, 1	3, 3, 3	(2, 1, 1); (3, 3, 3)
p_1, p_2, p_3	16, 2, 2	16, 2, 2	(16, 2, 2); (16, 2, 2)
j_1	32	16	32; 16

Table 5. Optimal solution space of SAE model

Parameters	Vibration signal	Settlement signal	Fusion signal
k_1, k_2, k_3	1024,512,64	256,64,16	(1024,512,64); (256,64,16)
j_1	32	32	32; 32

Table 6. Optimal solution space of CNN model

Parameters	Vibration signal	Settlement signal	Fusion signal
k_1, k_2, k_3	2, 2, 2	5, 4, 2	(2, 2, 2); (5, 4, 2)
p_1, p_2, p_3	4, 2, 2	4, 2, 4	(4, 2, 2); (4, 2, 4)
j_1	32	32	32; 32

In the GDE-CNN method, CNN is composed of multiple convolution pooling layers, and then the fully-connection layer is used for feature processing. SCAE and CNN have the same point that their structures are stacked by multi-layer convolution pooling layers, and their solution spaces need to be optimized are the same. The difference is that SACE has encoding and decoding structure, while CNN only has feature extraction (encoding structure) part.

The parameters to be optimized for GDE-CNN are as follows. The number of convolution kernels in each layer (k_1, k_2, k_3), pooled size (p_1, p_2, p_3), the number of nodes in the fully- connection layer j_1. The default solution space is shown in Table 3. The optimal solution is shown in Table 6. The results of fault early warning diagnosis accuracy are shown in Figure 13.

Comparing with figure 11, figure 12 and Figure 13, we can see the following conclusion. The main advantage of de-scae is that it has the strongest ability to identify faults in the early stage. In general, the fusion of two kinds of signals is more accurate than single signal input, especially in the early stage of fault. Signal fusion plays a significant role in improving the ability of fault recognition.

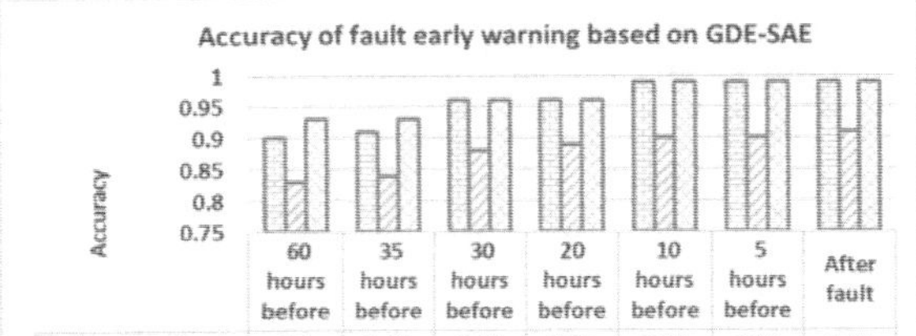

	60 hours before	35 hours before	30 hours before	20 hours before	10 hours before	5 hours before	After fault
Vibration signal	0.9	0.91	0.96	0.96	0.99	0.99	0.99
Settlement signal	0.83	0.84	0.88	0.89	0.9	0.9	0.91
Fusion signal	0.93	0.93	0.96	0.96	0.99	0.99	0.99

Figure 12. Accuracy of fault early warning based on GDE-SAE

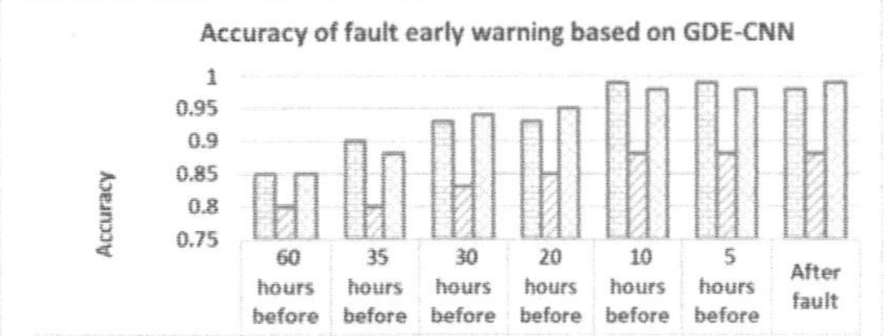

	60 hours before	35 hours before	30 hours before	20 hours before	10 hours before	5 hours before	After fault
Vibration signal	0.85	0.9	0.93	0.93	0.99	0.99	0.98
Settlement signal	0.8	0.8	0.83	0.85	0.88	0.88	0.88
Fusion signal	0.85	0.88	0.94	0.95	0.98	0.98	0.99

Figure 13. Accuracy of fault early warning based on GDE-CNN

5. Conclusion

In this paper, an early warning method of reciprocating compressor connecting rod bearing bush fault based on Stack-Convolution-Autoencoder (SCAE) is constructed. The sensitivity of vibration signal and piston rod settlement signal to the failure of connecting rod bearing bush is explored. It further verifies the better effect of multi-source signal fusion in early fault identification of complex mechanical equipment. What is more, a Gradient-Differential-Evolution (GDE) method is proposed to solve the problem of neural model super parameter selection, which reduces the difficulty of super parameter selection. This model can be directly applied to other units of the same type. When the unit type is different, this early warning model framework is still applicable, but transfer learning is needed, that is, the feature extraction part (SCAE module) needs to be retrained.

Compared and verified the fault early warning ability of GDE-SCAE, GDE-SAE and GDE-CNN, the analysis and research results are as follows. The GDE-SCAE method has a strong ability of early abnormal feature mining, and can identify early faults about 60 hours earlier than manual spectrum analysis method. Under five-fold-cross validation and multi-source signal fusion, the accuracy can reach 94%. Among the three model methods, the multi-source signal fusion is helpful to improve the fault identification ability of the model, and the proposed GDE-SCAE method has the best comprehensive recognition ability, especially in the early stage of fault.

6. Reference

[1] JJ Gao 2012 *Machine fault diagnosis and self-healing* (Higher Education Press)

[2] Pichler K, Lughofer E, Pichler M, Buchegger T, Klement E P and Huschenbett M 2016 Fault detection in reciprocating compressor valves under varying load conditions *J. Mechanical Systems & Signal Processing* **s70–71** 104-119

[3] Drewes E 2002 Condition monitoring for reciprocating compressors *J. Hydrocarbon Processing* **81(9)** 77-79

[4] Q Zhang, Y Shu, L Wang and CD Xie 2018 Research on Fault Diagnosis Method of Reciprocating Compressors *J. Fluid Machinery* **3**

[5] JX Ge 2016 Common faults and early warning in reciprocating compressor *J. Petrochemical Industry Technology* **23(11)** 101-101

[6] QG Shen and SY Zheng 2006 *Condition Monitoring and Faults Diagnosis* (Chemical Industry Press)

[7] Ebrahimi S, Salahshoor E and Moradi S 2017 Vibration performance evaluation of planar flexible multibody systems with joint clearance *J. Journal of the Brazilian Society of Mechanical Sciences & Engineering* **2** 1-15

[8] CS Zhang 2011 State monitoring and fault diagnosis technology of reciprocating compressor *J. China Plant Engineering* **12** 51-53

[9] CL Huang and SP Dong 2000 Monitoring the Condition of a Reciprocating Compressor by Various Methods *J. Lubrication Engineering*

[10] Townsend J and Badar M and Szekerces J 2016 Updating temperature monitoring on reciprocating compressor connecting rods to improve reliability *J. Engineering Science & Technology An International Journal* **19(1)** 566-573.

[11] Q Tian, C Liu, Margarida M and Paulo F 2011 A new model for dry and lubricated cylindrical joints with clearance in spatial flexible multibody systems *J. Nonlinear Dynamics* **64(1)** 25-47.

[12] Flores P 2010 A parametric study on the dynamic response of planar multibody systems with multiple clearance joints *J. Nonlinear Dynamics* **61(61)** 633-653

[13] SG Xiao, SL Liu, MM Song, N Ang and HL Zhang 2020 Coupling rub-impact dynamics of double translational joints with subsidence for time-varying load in a planar mechanical system *J. Multibody System Dynamics* **48(4)** 451-486.

[14] SG Xiao, SL Liu, HZ Wang, Y Lin, MM Song and HL Zhang 2020 Nonlinear dynamics of coupling rub-impact of double translational joints with subsidence considering the flexibility of piston rod *J. Nonlinear Dynamics* **100(2)** 1203-1229.

[15] DT Hoang and HJ Kang 2018 A survey on Deep Learning based bearing fault diagnosis *J. Neurocomputing* **335**

[16] Potocnik P and Govekar E 2017 Semi-supervised vibration-based classification and condition monitoring of compressors *J. Mechanical systems and signal processing* **93(Sep.)** 51-65.

[17] P Qi and B Ma 2016 Application of probabilistic neural network in multi fault diagnosis of reciprocating compressor *J. Machinery Design & Manufacture* **4** 238-241.

[18] HB Yang, A Nie, JA Zhang, LL Chen, HL Zhang, SL Liu and J P Amezquita-Sanchez 2018 Fault diagnosis of reciprocating compressor based on convolutional neural network based on original vibration signal *J. Machine Design and Manufacturing Engineering* **47(9)** 67-70

[19] S Li, HQ Wang, LY Song, PX Wang, LL Cui and TJ Lin 2020 An adaptive data fusion strategy for fault diagnosis based on the convolutional neural network *J. Measurement* **165**

[20] Van TT, Faisal A, Tiedo T, Andrew B and Gang N 2018 Single and combined fault diagnosis of reciprocating compressor valves using a hybrid deep belief network *J. Proceedings of the Institution of Mechanical Engineers Part C Journal of Mechanical Engineering Science*

[21] Peng Z, Kessissoglou N J and Cox M 2005 A study of the effect of contaminant particles in lubricants using wear debris and vibration condition monitoring techniques *J. Wear* **258(11)** 1651-1662

[22] LC Jiao, SY Yang, F Liu, SG Wang and ZX Feng 2016 Neural network for 70 years: revie w and prospect *J. Chinese Journal of Computers* **39(08)** 1697-2816

[23] SM Chen, JB Yu and SJ Wang 2020 One-dimensional convolutional auto-encoder-based feature learning for fault diagnosis of multivariate processes *J. Journal of Process Control* **87**

[24] FF Li, H Qiao and B Zhang 2018 Discriminatively boosted image clustering with fully convolutional auto-encoders *J. Pattern Recognition* **83**.

[25] XY Fu, H Luo, SS Zhong and L Lin 2019 Aircraft engine fault detection based on grouped convolutional denoising autoencoders *J. Chinese Journal of Aeronautics* **32(02)** 296-307.

[26] XK Zhou and JB Yu 2020 Gearbox Fault Diagnosis Based on One-dimension Residual Convolutional Auto-encoder *J. Journal of Mechanical Engineering* **56(07)** 96-108.

[27] LB Deng, CL Li and GJ Sun 2020 An adaptive dimension level adjustment framework for differential evolution *J. Knowledge-Based Systems* **206**

[28] T Huang, DT Duan, YJ Gong, L Ye, Wing W Y Ng and J Zhang 2020 Concurrent optimization of multiple base learners in neural network ensembles: An adaptive niching differential evolution approach *J. Neurocomputing* **396**.

Acknowledgments

This work was supported by the state key laboratory of compressor technology (an hui laboratory of compressor technology) (SKL-YSJ201811/ SKL-YSJ201808); the Fundamental Research Funds for the Central Universities (JD2107) ; the Double First-rate Construction Special Funds (ZD1601).

Maxwell velocity slip and Smoluchowski temperature jump boundary condition for ANSYS CFX

T. Jünemann[1], H. Pleskun[2] and A. Brümmer[3]
Chair of Fluidics, Faculty of Mechanical Engineering, TU Dortmund University, Germany
E-mail: [1]timo.juenemann@tu-dortmund.de,[2]heiko.pleskun@tu-dortmund.de, [3] andreas.bruemmer@tu-dortmund.de

Abstract. Multi-dimensional numeric flow simulation is a common but non-trivial approach to simulate the operation behavior of screw machines. Current challenges of computational fluid dynamics (CFD) simulations of screw machines are especially related to turbulence modelling and meshing. An adaption of CFD simulations to screw vacuum pumps is difficult since the Navier-Stokes equations have limited validity for low pressure regimes. However, the scope of application can be extended to higher Knudsen numbers (e.g. lower pressure regimes) by the use of velocity slip and temperature jump boundary conditions at solid surfaces. Even if the complete simulation of screw vacuum pumps is still challenging, these boundary conditions can be used to examine isolated effects like clearance flows or inhomogeneous pressure distributions in working chambers. A common commercial CFD-Solver is ANSYS CFX, which also has been successfully applied to screw machines, but it does not provide a Maxwell velocity slip and a Smoluchowski temperature jump boundary condition. In the presented work these boundary conditions are implemented in Ansys CFX using user-defined expressions in a similar way as it has been implemented in the rhoCentralFOAM solver of the open source toolbox OpenFOAM. The boundary conditions are validated by comparison with stationary OpenFOAM results, DSMC results and measurements for a hypersonic plate flow. In addition, the boundary conditions are verified by experimental and DSMC results of a gas flow in a clearance between a rotational shaft and a plane counter plate.

1. Introduction

Dry running screw vacuum pumps have become an important part of vacuum systems. In recent years, they replaced multistage roots pumps and oil-lubricated rotary vane pumps because they are compatible with soil and small amounts of liquid. Due to oil-free operation, sealing between different working chambers is realized by small clearances. Leakages through the clearances can be described as one of the major loss mechanisms of a screw vacuum pump. Screw vacuum pumps are mainly used as fore pumps, which means the suction pressure ranges from a few pascals up to the atmospheric pressure. The circumferential speed of the pistons can reach up to 80 m/s. Considering that, the flow mechanism in the clearance can be mainly described as a pressure driven Poiseuille flow and a Couette flow driven by the rotation of the pistons. Rotatory positive displacement vacuum pumps are mainly simulated with chamber simulation tools whereby classic screw compressors can also be simulated by the use of computational fluid dynamics (CFD) solving the Navier-Stokes-Fourier (N-S-F) equations [1]. In order to augment the CFD-approach to vacuum pumps transitional non-equilibrium of rarefied gas flows need to be considered. The level of gas rarefaction is characterised by the Knudsen number, which is the ratio of gas molecular mean free path and a characteristic length. The N-S-F equations produces reasonable results for Knudsen numbers up to 0.01, for Kn ranges up to 0.1 velocity slip and temperature jump boundary conditions should be applied at the walls. For larger Kn numbers the N-S-F equations aren't valid. The first-order conventional slip boundary condition was developed by Maxwell[2] in 1879. The temperature jump condition developed by Smoluchowski [3] is driven by the surface normal heat flux. Both boundary conditions has been implemented in the open source CFD software OpenFOAM, which is described in [4] and [5]. One of the most commonly used commercial CFD solvers which has been successfully used on screw machines is ANSYS CFX [6], which does not natively provide non-equilibrium boundary conditions for rarefied gases. However for subsequent research on screw vacuum pumps it is sensible to implement Maxwell velocity slip and Smoluchowski temperature jump boundary conditions into ANSYS CFX. In addition this is reasonable for detailed investigation of flow mechanisms in screw machines like it has been done in [7] and [8]. During the preparations of this paper several ways to implement the boundary conditions presented in [5] have been tried out. Therefore, the aim of the presented paper is to show the approach of best numerical stability. It's realized with user defined expressions to calculate the coefficients for the *finite-slip* and the *defined-heat-transfer-coefficient* boundary conditions. The thermal slip is represented due to a fictional moving wall and a corrected wall heat flux. The boundary conditions are validated by comparison with stationary OpenFOAM results, DSMC results and measurements for a hypersonic plate flow provided by [9]. In addition, the boundary conditions are verified by experimental and DSMC results of a gas flow in a clearance between a rotational shaft and a plane counter plate presented in [10, 11].

2. Non-equilibrium boundary condition in CFX

The non-equilibrium slip/jump boundary conditions are implemented based on the equations for Maxwell slip and Smoluchowski jump presented in [4, 5]. Before getting into details some basic quantities are introduced, which are formulated in a way that these also work for incompressible flow. Since ANSYS CFX first solves the momentum and mass equation and in a second step the energy equation, it seems to be sensible to formulate the quantities based on density ϱ and static pressure p. The most important quantity for rarefied gases is the molecular mean free path which is,

$$\lambda = \frac{\pi}{4}\frac{\bar{c}\mu}{p} \tag{1}$$

using the model of hard sphere molecules. It's calculated from the static presure p, the dynamic viscosity μ and the mean molecular velocity

$$\bar{c} = \sqrt{\frac{8}{\pi}\frac{p}{\varrho}} \tag{2}$$

of the gas.

2.1. Maxwell Viscous Slip

The Maxwell viscous slip depends on the wall shear stress and therefore can be calculated using the *finite slip* boundary condition, which is natively provided by ANSYS CFX [12]. Considering the CFX Modelling guide this boundary condition is available only for laminar flows. This option causes the fluid to slip at the wall when the wall shear stress is greater than a critical stress τ_c.

$$|\boldsymbol{\tau_w}| < \tau_c \rightarrow \boldsymbol{u}_{FiniteSlip} = \mathbf{0} \tag{3}$$

ANSYS CFX simulates the slip by using a moving wall with the wall speed computed as follows:

$$\boldsymbol{u}_{\text{FiniteSlip}} = U_S \cdot \frac{\boldsymbol{\tau_w}}{|\boldsymbol{\tau_w}|}\left(\frac{|\boldsymbol{\tau_w}| - \tau_c}{\tau_n}\right)^m e^{\frac{-Bp}{\tau_n}} \tag{4}$$

where U_S is the slip speed, τ_n is a normalising stress which is by default $\tau_n = 1\text{Pa}$, m is a positive power, B is a pressure coefficient, and p is the static pressure. As it is shown the finite-slip velocity vector $\boldsymbol{u}_{FiniteSlip}$ has the same direction as the wall shear stress vector $\boldsymbol{\tau}_w$.

The Maxwell viscous slip referring to [4] is given as follows

$$\boldsymbol{u}_{\text{MVS}} = \frac{\sigma_u - 2}{\sigma_u}\frac{\lambda}{\mu}\boldsymbol{S}\cdot(\boldsymbol{n}\cdot\boldsymbol{\tau}) = \frac{2-\sigma_u}{\sigma_u}\frac{\lambda}{\mu}\boldsymbol{\tau_w} \tag{5}$$

where σ_u is the tangential momentum accommodation coefficient, $\boldsymbol{\tau}$ the deviatoric stress tensor, $\boldsymbol{n}$ is the unit normal vector defined as positive in the direction pointing out of the flow domain and $\boldsymbol{S} = \boldsymbol{I} - \boldsymbol{nn}$ is the tensor which removes the normal components form every vector field. The tangential momentum accommodation coefficient normally is varying in-between 0 and 1. The tensor product $-\boldsymbol{S}\cdot(\boldsymbol{n}\cdot\boldsymbol{\tau})$ is resulting into the wall shear stress vector $\boldsymbol{\tau}_\text{w}$ tangential to the wall surface. Therefore Eq. 5 can be represented by Eq. 4 setting $\tau_c = 0$, $m = 1$ and $B = 0$. The slip speed is given with

$$U_S = \frac{2-\sigma_u}{\sigma_u}\frac{\lambda}{\mu}\cdot 1\,[\text{Pa}]\,. \tag{6}$$

Velocity Slip and Temperature Jump BC for ANSYS CFX

2.2. Maxwell Thermal Slip

The thermal slip is modelled by an additional wall velocity added to the real wall velocity, thereby the Maxwell thermal velocity slip is calculated as follows considering the local gradient of the static temperature T:

$$\boldsymbol{u}_{\mathrm{MTS}} = \frac{3}{4}\frac{\mu}{\varrho}\frac{\boldsymbol{S}\cdot\boldsymbol{\nabla}T}{T} \tag{7}$$

Since the thermal slip is only a model for the fluid velocity in the Knudsen layer close to the wall and the real fluid velocity at the wall is still equal to the wall velocity, this modelling approach leads to inaccuracy in the conservation of energy. The N-S-F energy equation is

$$\frac{\partial}{\partial t}(\varrho E) + \boldsymbol{\nabla}\cdot\left(\varrho\boldsymbol{u}\left(E+\frac{p}{\varrho}\right)\right) = \boldsymbol{\nabla}\cdot(\boldsymbol{\tau}\cdot\boldsymbol{u}-\boldsymbol{q}) + Q - \varrho\boldsymbol{u}\cdot\boldsymbol{g} \tag{8}$$

where $E = e + \frac{1}{2}|\boldsymbol{u}|^2 + G$, is the total energy, e is the thermal energy, ϱ the fluid density, G the specific potential energy, $\boldsymbol{q}$ the heat flux vector, Q a volumetric energy source and $\boldsymbol{g}$ the specific gravity force. All other quantities have already been shown. As the energy equation shows, the additional modelled wall velocity is resulting in an additional friction work due to wall shear stress, which should be compensated to fulfil the conservation of energy. The compensation is done by an additional heat flux at the wall which is calculated as follows from wall shear stress and thermal slip velocity.

$$q_{MTS} = \boldsymbol{u}_{\mathrm{MTS}}\cdot\boldsymbol{\tau}_{\mathrm{w}} = \boldsymbol{u}_{\mathrm{MTS}}\cdot\frac{\boldsymbol{u}_{\mathrm{FiniteSlip}}}{U_S}\cdot 1\,[\mathrm{Pa}] \tag{9}$$

If the finite-slip model is used, the wall shear stress variable *wall shear* is not calculated in CFX and can not be use in the user defined expressions and therefore a workaround is needed which recalculates the wall shear stress from the finite-slip solution. In [5] the additional viscous heat generation is modelled with a modified wall temperature such as the temperature jump. For ANSYS CFX this approach does not work since the dynamic viscosity and thermal conductivity at the wall depends on the wall temperature so that the wall temperature would depend on itself due to the Smoluchowski temperature jump and viscous heat boundary condition. For this reason the modelling approach of the Smoluchowski temperature jump does also differ from [4, 5].

2.3. Smoluchowski Temperature Jump

In order to simulate the temperature jump the *defined heat exchange parameter* option of ANSYS CFX is used. This option models a heat exchange from an environment temperature to the fluid domain with a given heat exchange coefficient. In this approach the environment temperature is set to the wall temperature and the heat transfer coefficient is calculated in order to ensure that the Smoluchowski temperature jump occurs. In reference to Smoluchowski the temperature jump depending on the wall heat flux q_w is given as follows

$$T - T_{\mathrm{w}} = \frac{2-\sigma_T}{\sigma_T}\frac{2\gamma\lambda}{(\gamma+1)Pr}\boldsymbol{n}\cdot\boldsymbol{q} = \frac{2-\sigma_T}{\sigma_T}\frac{2\gamma\lambda}{(\gamma+1)Pr}q_{\mathrm{w}} \tag{10}$$

where Pr is the Prandlt number, γ is the ratio of specific heat and σ_T is the thermal accommodation coefficient which is in-between 0 and 1. The equation can be rearranged to the following form

$$q_{\mathrm{w}} = k \cdot (T - T_{\mathrm{w}}) \tag{11}$$

where k is the heat transfer coefficient given by

$$k = \frac{\sigma_T}{2 - \sigma_T} \frac{(\gamma + 1)\, Pr}{2\gamma\lambda}. \tag{12}$$

The ratio of specific heat is calculated as follows from the specific heat capacity C_p at constant pressure, universal gas constant $\Re$ and the molar mass M

$$\gamma = \frac{C_p}{C_p - \frac{\Re}{M}} \tag{13}$$

to allow a calculation also for incompressible fluids. It may be sensible to limit k for $\lambda \to 0$ to improve numerical stability since k otherwise becomes infinity, however in this paper it was not necessary.

3. Results

The presented boundary conditions are validated by two exemplary test cases. The first one is a hypersonic flow over a flat plate and the second one is a clearance between a rotating shaft and a plane counter plate. Both simulations are performed with perfect gases of constant ratio of specific heat capacities γ, constant Prandtl number Pr and temperature depending dynamic viscosity $\mu(T)$. The dynamic viscosity is modelled with the Sutherland-Model which reads as follows

$$\mu(T) = \mu_0 \frac{T_0 + T_S}{T + T_S} \left(\frac{T}{T_0}\right)^{\frac{3}{2}} \tag{14}$$

where μ_0 is the reference viscosity at reference temperature T_0 and T_S is the Sutherland temperature. All CFD simulations with the Maxwell/Smoluchowski boundary conditions use the values $\sigma_u = \sigma_T = 1.0$.

3.1. Hypersonic flat plate

The mesh and the boundary conditions applied on the flat plate case are shown in Fig. 1. The dimensionless boundary conditions and dimensions of the fluid domain are shown in Tab. 1. The boundary conditions, such as free stream Mach number Ma_∞ wall temperature T_w and fluid parameters are set equal to those in the experiment in [9]. The applied gas is nitrogen. The results are presented for the reverse Knudsen number $Kn_x^{-1} = x/\lambda_\infty$, where x is the distance from the leading edge and λ_∞ the mean free path of the gas at free stream conditions. The angle of attack of the free stream is zero. The mesh has 80000 cells with a grading to the leading edge and to the plate resulting into a minimal cell dimension of $\Delta x/\lambda_\infty = \Delta y/\lambda_\infty = 0.14$. The results of ANSYS CFX including the boundary conditions are compared to experiments of [9]

Table 1. Hypersonic flat plate - geometry and boundaries

$Kn_x^{-1} = \frac{x}{\lambda_\infty}$	Ma_∞	$\frac{T_w}{T_\infty}$	$\frac{T_S}{T_\infty}$	γ	Pr	σ_u	σ_T	$\frac{L_0}{L}$	$\frac{H}{L}$
0...250	6.1	3.575	1.331	1.4	0.71	1	1	0.1	0.5

and to dsmcFOAM+[13] and rhoCentralFoam[14] results. For that the results shown in [4] have been reproduced including the viscous heating presented in [5]. Thereby it has been noticed that the sign of the thermal creep has been presented wrongly in [4, 5] and in OpenFOAM [14]. The thermal creep has to be calculated with a positive signed temperature gradient, since it represents a flow from the cold to the warm regions in the fluid. However for this publication the sign has been fixed in a local fork of the OpenFOAM source code in which also the viscous heating has been implemented. Changing the sign does effect the presented results only in minor way, since these are less affected by thermal creep CFX and rhoCentralFoam simulations have been done for the same mesh. For the DSMC simulation no grading has been used, the collision cell size is $\Delta x = \Delta y = \Delta z = 0.5\lambda_\infty$.

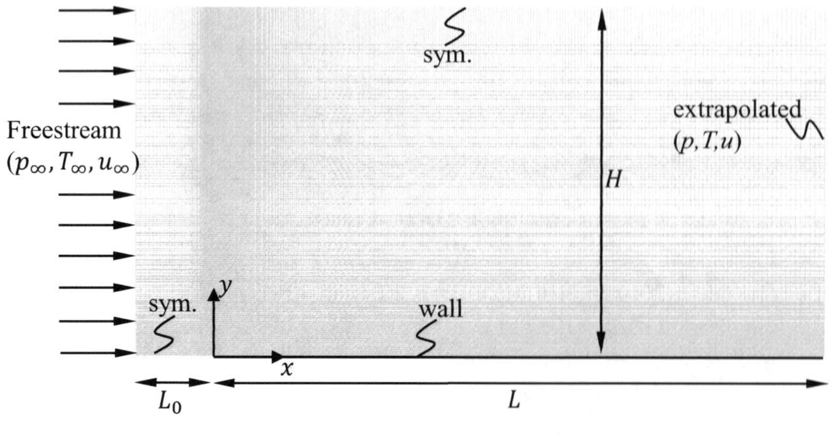

Figure 1. Metcalf et al. case [9], computational mesh and boundary conditions

The results for the static pressure $p(y = 0)$ at plate surface are shown in Fig. 2. The pressure is normalized to the pressure p_∞ of the free stream. No clear differences between CFX and rhoCentralFoam can be observed. However the deviation to DSMC results are distinctly higher, whereby the DSMC are closer to measurements, which has also been observed in [4].

In Fig. 3 the near wall temperature distribution is shown. The temperature is normalized to the free stream. Again CFX is compared to DSMC, rhoCentralFOAM and measurement results. If the results are compared to the wall temperature (Tab. 1), a temperature jump can be detected. In the opposite to the pressure distribution a deviation between CFX and rhoCentralFoam occurs, which possibly leads back to the different implementation of the viscous heat term. However the CFX results are closer to those of DSMC. In general all simulations do not give a good agreement to the measurements even for DSMC, which has also been observed in [4].

For the slip velocity, shown in Fig .4, there are only minor differences between CFX and rhoCentralFOAM, where CFX is slightly closer to DSMC. Measurements for the velocity slip are not provided in [9].

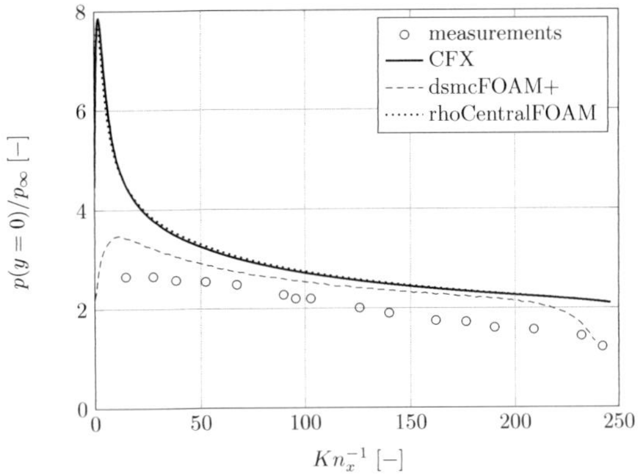

Figure 2. Metcalf et al. case [9], surface pressure

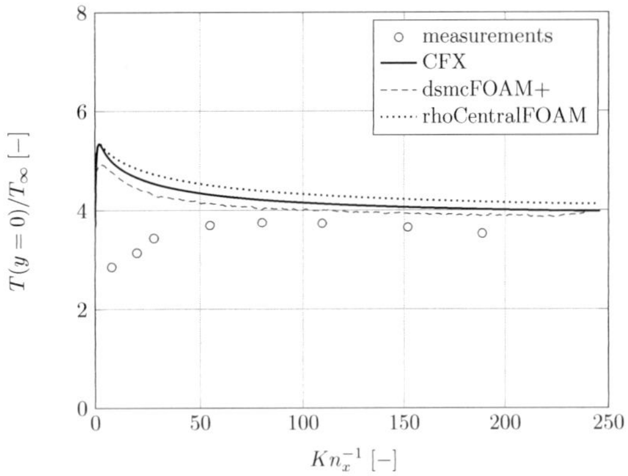

Figure 3. Metcalf et al. case [9], near wall temperature

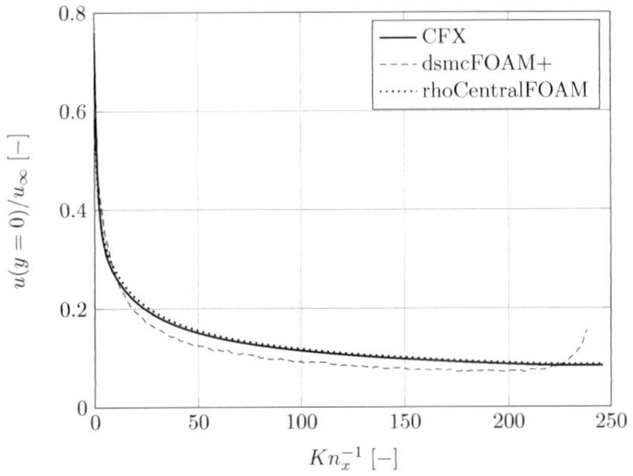

Figure 4. Metcalf et al. case [9], near wall velocity

3.2. Rotating Shaft

In order to verify the boundary conditions for clearance flows in screw vacuum pumps, simulations are compared to measurements of Huck et al. [10]. Huck set up an experiment with a clearance formed by a rotating shaft with a radius of $R = 75\,\text{mm}$ on the one side and a plane contour on the other side. The clearance height between the plane contour and the shaft was set to $h = 0.3\,\text{mm}$. The inlet and outlet pressure of the clearance as well as the circumferential speed of the shaft could be varied while the mass flow rate $\dot{m}$ through the clearance was measured. For details of the experimental setup and methods of measurement, see Ref. [10]. The comparison of simulation and experimental results is done for a variation of a dimensionless wall velocity $U_0 = U_{shaft}/\bar{c}_{in}$ at a constant pressure ratio $\Pi = p_{in}/p_{out}$. In addition the DSMC results from [11] are shown. The mesh and the applied boundary conditions for CFX are shown in Fig. 5. A grading is applied normal to the walls and the minimal cross sectional flow area leading to a minimal cell dimension of $\Delta x/h = 0.02, \Delta y/h = 0.00277$ and 32830 cells. In front and behind of the shaft a channel of height H is modelled. The overall length of the fluid domain is given with L. The width of the quasi-2D fluid domain is given width $\Delta z = 1mm$. On the surface of the shaft a wall velocity U_{shaft} is applied. In order to allow mass flow in both directions at the inlet as well as on the outlet an opening boundary condition is applied. The dimensionless boundary conditions as well as the geometry parameters are given in Tab 2. The measurements and simulations have been done with dry air.

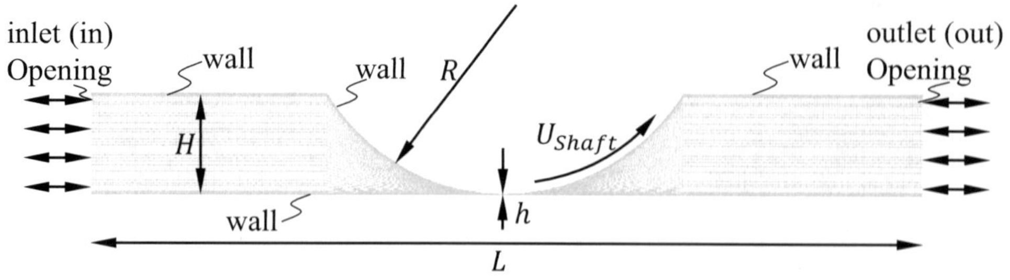

Figure 5. Huck et al. case [10], computational mesh and boundary conditions

The results are presented as dimensionless reduced mass flow rate,

$$C_0 = \frac{\pi \bar{c}_{in}}{2 p_{in} h \Delta z} \dot{m} \tag{15}$$

as a function of the Knudsen number $Kn_h = \lambda_{in}/h$ and for different dimensionless wall velocities U_0. The wall temperature is set to the inlet temperature since all measurements have been done at environment temperature. In the case of a back flow, the outlet temperature is also set to the inlet value. All results are shown in Fig. 6 with the classifications of the different flow regimes of rarefied gas flows. The diagram is split into a logarithmic and a linear part since negative values have to be shown also. The N-S-F results using Maxwell/Smoluchowski boundary conditions are giving a good agreement with the measurements in continuum and slip flow regime for Knudsen numbers upto $3 \cdot 10^{-2}$. For higher Knudsen numbers deviations occur though it is the slip regime considering the nominal Knudsen number Kn_h. This is explained due to higher local Knudsen number downstream of the clearance where the slip boundary condition is not valid any more. However this effect is reduced for higher shaft velocities leading to a good agreement of the CFX results over the full shown range of Knudsen number for the maximum rotational speed. Considering the DSMC results it can be observed

Velocity Slip and Temperature Jump BC for ANSYS CFX

Table 2. Rotating shaft - geometry and boundaries

$\frac{R}{h}$	$\frac{L}{h}$	$\frac{H}{h}$	Kn_h	Π	U_0	$\frac{T_{out}}{T_{in}}, \frac{T_w}{T_{in}}$	$\frac{T_S}{T_{in}}$	γ	Pr	σ_u	σ_T
250	$1e^3$	120	$1e^{-3} - 1$	8	$-8.64e^{-2}...8.64e^{-2}$	1	0.379	1.4	0.71	1	1

that these perfectly merge with CFX results in the slip regime, which also confirms the quality of the boundary condition in CFX.

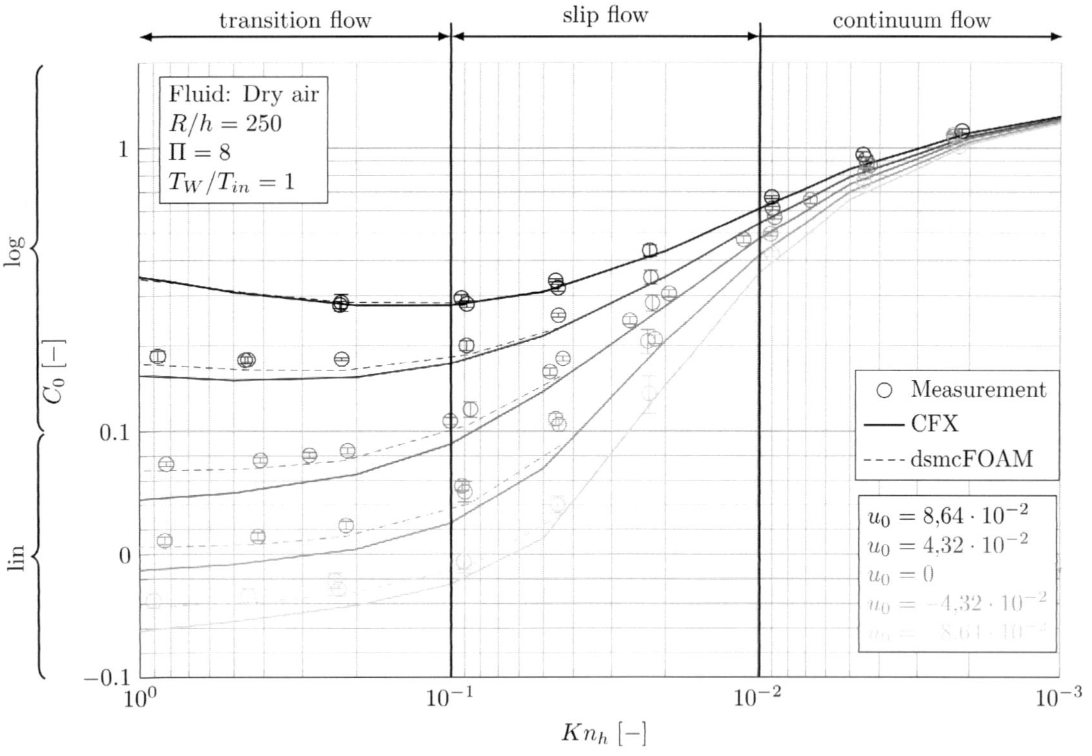

Figure 6. Huck et al. case [10], dimensionless mass flow for varying Knudsen numbers and circumferential speeds

4. Conclusion

This paper does provide an approach to implement a Maxwell velocity slip and Smoluchowski temperature jump boundary condition into ANSYS CFX. The approach has been validated with simulations of the rhoCentralFoam solver of the open source CFD Software OpenFOAM and verified with DSMC simulations and measurements of two test cases. The first test case verifies the boundary condition for hypersonic flow, where thermal creep and the temperature jump can not be neglected. In this case both N-S-F solvers are giving similar results with minor deviations in temperature jump due to different modelling of viscous heat generation. The second case verifies the boundary condition for the usage in screw vacuum pumps especially for moderate Knudsen numbers and high positive wall velocities. Since the circumferential speed of screw vacuum pumps is typically high a usage of ANSYS CFX for such machines could be possible, if the shown boundary conditions are used. Especially inside the working chamber minor uncertainties are expected compared to the clearance flows since the Knudsen number in working chambers is smaller. However the wall velocity in the clearances is higher and so the boundary conditions should work for a wide range of machine suction pressure. Detached from the simulation of a complete screw machine the boundary condition in CFX can be used to improve models for clearances and working chamber in a chamber model simulation. Exemplary applications are mentioned in [15] and [8]. Finally one restriction of boundary condition should be mentioned, which has not been proved in this work. The used finite slip boundary condition is restricted to laminar flow, however in vacuum pumps also turbulences can occur so a turbulence modelling is needed. The presented approach may work if a shear stress turbulence model is used and the near wall laminar sub layer is resolved, but this has not been tested yet. This is not a major restriction since the laminar sub layer should always be resolved for accurate clearance modelling [7]. At least it could be sensible to limit the heat transfer coefficient for low mean free paths for improved numerical stability of the temperature jump.

References

[1] Kovacevic A, Stosic N and Smith I K 2006 *Screw Compressors: Three Dimensional Computational Fluid Dynamics and Solid Fluid Interaction Effects* (Springer-Verlag)

[2] Maxwell J C 1878 *Proc. R. Soc. Lond.* **27**(185-189) 304–308

[3] Smoluchowski von Smolan M 1898 *Annalen der Physik* **300** 101–130

[4] Le N T, White C, Reese J M and Myong R S 2012 *International Journal of Heat and Mass Transfer* **55** 5032–5043

[5] Le N T P, Vu N A and Loc L T 2017 *AIAA Journal* **55** 474–483

[6] Kovacevic A, Rane S, Stosic N, Jiang Y, Lowry S, and Furmanczyk M Influence of approaches in cfd solvers on performance prediction in screw compressors

[7] Utri M and Brümmer A Fluid flow through front clearances of dry running screw machines using dimensionless numbers *ICEC* (Purdue e-Pubs) p 2583

[8] Pleskun H, Jüenemann T, Bode T and Brümmer A 2021 Modelling of inhomogeneous chamber states in rotary positive displacement vacuum pumps *Proceedings of the 12th International Conference on Compressors and their Systems* (IOP Publishing)

[9] Metcalf S, Lillicrap D and Berry C J 1969 A study of the effect of surface temperature on the shock-layer development over sharp-edge shapes in low- reynolds-number high-speed flow *Proceedings of the 7th International Symposium on Rarefied Gas Dynamics* pp 619–634

[10] Huck C, Pleskun H and Brümmer A 2018 *Journal of Vacuum Science & Technology A* **36** 031606

[11] Huck C 2020 *Verdünnte Strömungen in Spalten von Vakuumpumpen mit bewegter Berandung* Ph.D. thesis TU Dortmund University

[12] CFX 2021 Cfx modelling guide

[13] White C, Borg M K, Scanlon T J, Longshaw S M, John B, Emerson D R and Reese J M 2018 *Computer Physics Communications* **224** 22–43

[14] OpenFOAMFoundation 2021 Openfoamv8 URL `https://openfoam.org/`

[15] Jünemann T and Brümmer A 2020 *Journal of Vacuum Science & Technology B* **38** 044201

IOP Conf. Series: Materials Science and Engineering 1180 (2021) 012038 doi:10.1088/1757-899X/1180/1/012038

A numerical CFD simulation method using static grid based on momentum compensation for screw pumps

Di Yan[1], **Qian Tang**[2]

[1] School of Machinery and Automation, Wuhan University of Science and Technology, Wuhan, Hubei Province, 430081, P. R. China;
[2] College of Mechanical and Vehicle Engineering, Chongqing University, Chongqing, 400030, P. R. China

Email: dyan@wust.edu.cn

Abstract. Three-dimensional CFD numerical simulation is the most powerful tool for performance analysis of screw machines. This paper compares and analyzes two CFD numerical calculation methods based on moving and static grid, and proposes a static grid calculation method based on momentum compensation, which takes into account the wall velocity and fluid momentum on the rotor surface when the screw rotor rotates. That is, while the static grid is used, the rotation characteristics of the fluid domain between the rotors are considered. Compared with the commonly used static grid method, this method greatly improves the accuracy of the calculation, the calculation error of mass flow rate is reduced from 2.32% to 1.14%. At the same time, it is simple and easy to implement. It is also suitable for modelling screw machines with special-shaped rotors, which are still challenge for moving grids. The accuracy and reliability of the method are verified by comparison with the calculation results of the moving grid and the experiments.

1. Introduction

Due to the complexity of geometry and working process of screw machines, there are higher requirements for the grid when performing 3D CFD numerical simulation. The generation and processing of the grid for the fluid domain has always been a thorny problem. With the development of numerical methods and grid technology, the CFD numerical simulation of screw pumps has evolved from a static grid generated directly from the pump chamber to a structured moving grid generated by dividing the geometric domain by a special algorithm. The moving grid can be updated and changed in real time with the rotation of the screw rotors, which can reflect the real working process of the screw pump. It is also the hotspot and frontier of the numerical simulation of screw machines [1]. Since the emergence of structured moving grid technology that can reflect the dynamic meshing process of screw rotors, it has been widely used in the analysis of screw machines, especially for screw compressors, and can accurately simulate the fluid dynamics and thermodynamics characteristics during the internal compression process.

Now the most powerful tool for analysing screw pumps is the three-dimensional CFD modelling based on the structured moving grid. However, although the moving grid method is accurate and reliable, it also has some shortcomings. First of all, special algorithms and programs are needed to realize the meshing and dynamic update of the geometry, which is difficult to realize on the current general commercial meshing software. In the calculation process, due to the need to update and iterate a large amount of dynamic grid data, a lot of computing resources are required, and it takes a long time to complete a simulation calculation case, the calculation efficiency is relatively low. Moreover, for some complex screw machines, such as variable pitch and variable cross-section screw pumps and claw-type screw rotors with pointed hooks, the generation of accurate and reliable moving grids is still a huge challenge [2]. In addition, in the three-dimensional CFD numerical simulation of the screw vacuum pump, the calculation based on the moving grid is often difficult to converge due to the low inlet pressure and the rarefied gas [3]. In these cases, the moving grid method cannot show its advantages.

In these cases, the static grid method is a practical option. It is easy to operate the calculation and the convergence is easy to achieved. Especially when modelling screw pumps, such as multiphase pumps and vacuum pumps, the static grid method are simple and easy to use. It is very practical for evaluating the performance of screw pumps. Since the static grid is generated by directly dividing the geometric model of the flow channel extracted from the pump chamber, the CFD numerical simulation based on the static grid can obtain the leakage, pressure distribution, rotor torque, rotor axial force and radial force, power consumption, etc. Before the structured dynamic grid technology was developed, a large number of simulation calculation cases were based on static grids [4-8].

Compared with the thermodynamic chamber mathematical models which neglect kinetic energy and simplify the analysis of the main and leakage flow [9], the CFD numerical simulation based on static grid can be used to investigate the influence of the geometric characteristics of the pump and rotors on the pump performance, and the pressure analysis and leakage situation of different rotation positions can be obtain. The CFD method is more intuitively. These static grid-based calculation examples have played a good role in the introduction of CFD numerical simulation methods into the analysis and research of positive displacement screw machines, and have also enabled researchers to further improve the understanding of performance characteristics and internal flow of the screw machines.

However, CFD models based on static grids have certain limitations when simulating screw pumps. By using static mesh, approximate pressure gradients and the leakage velocity field can be obtained; but such results do not take into account the velocity field of the main flow and neglect the transient nature of the working process in a screw pump. Some important parameters cannot be obtained, such as the real-time mass flow rate, rotor torque and pressure fluctuation. More so, in the case of a multiphase pump the pressure field will differ significantly from the result obtained on a static mesh and therefore power calculation will be inaccurate. Compared with the moving grid calculation method, the static grid method has larger calculation errors.

Although there are many calculation cases based on static grids and moving grids, there is no comparative analysis of the calculation results and errors of these two methods. This paper performs

IOP Conf. Series: Materials Science and Engineering 1180 (2021) 012038 doi:10.1088/1757-899X/1180/1/012038

CFD simulation on the same screw pump with static grids and moving grids of the same cell number and cell size. By comparing and analyzing the calculation results of the dynamic and static grids, the difference of these two methods is obtained. At the same time, a static grid simulation method based on momentum compensation is proposed to improve the accuracy and feasibility of static grid CFD simulation in the analysis of screw pumps. The proposed method is aimed to improve the accuracy and feasibility of static grid CFD simulation in the analysis of screw pumps, and provide a reliable method for the analysis of special-shaped screw rotors and the screw pumps under special working conditions.

2. Numerical Methods

Positive displacement screw pumps operate on the basis of changing the size and position of a working domain which consequently causes change in the pressure of the domain which causes transports of the fluid. To calculate performance of such a pump, quantities such as mass, momentum, energy etc. need to be modelled. Conservation of these quantities can be represented by a general transport equation for a control volume.

In order to account for the deformation of the working domain, the conservation equation needs to account for the velocity of the domain boundary. This could be done by replacing the velocity in the convective term with the relative velocity $(\boldsymbol{v}-\boldsymbol{v_b})$, where $\boldsymbol{v_b}$ is the velocity vector at the cell face [10]. The grid velocity $\boldsymbol{v_b}$ and the grid motion are independent of the fluid motion. However, when the grid velocities are calculated explicitly and used to calculate the convective fluxes, the conservation of mass and other conserved quantities may not necessarily be preserved. To ensure full conservations of these equations, the space conservation law needs to be satisfied. Space conservation can be regarded as mass conservation with zero fluid velocity. The unsteady terms in the governing equations involving integration over a control volume Ω, which is now changing with time, need to be treated in a way consistent with the space conservation equation with a deforming and/or moving grid [10, 11].

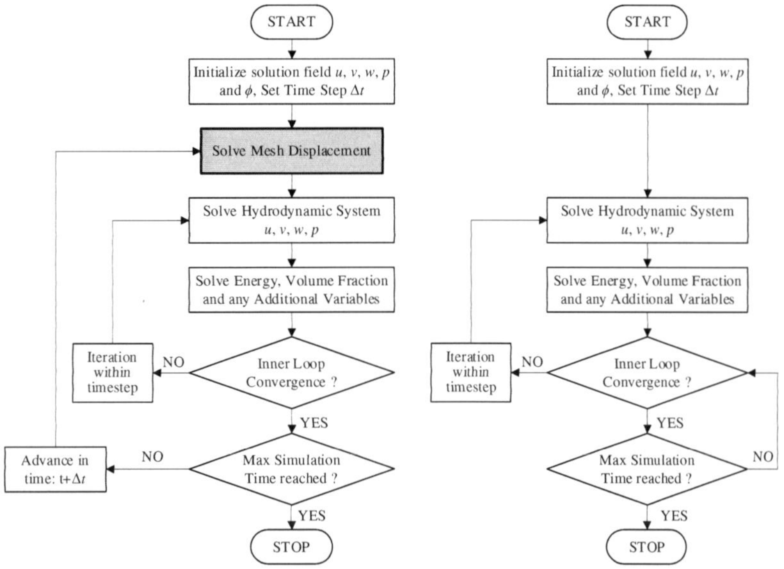

Figure 1 CFD calculation flow chart (left) based on moving grid; (right) based on static grid

Figure 1 shows the flow charts of CFD modelling based on moving mesh and static mesh. It can be observed that compared with moving mesh method, one key step is missed——solve the grid displacement. The CFD modelling flow chart of based on moving mesh includes the momentum calculation of the grid cell and the data transfer. The static mesh method neglects the flow of fluid in the pump chamber driven by the rotation of the rotors [12, 13].

When using static mesh method, a rotation speed was added to the rotor surfaces according the real rotation of rotors. However, when the screw rotors rotate, the rotational speed of the entire fluid domain is ignored, that is, the momentum of the fluid is not considered.

The velocity decomposition of the fluid in the screw pump is shown in Figure 2. Its momentum term is

$$M_S = \rho v_z \tag{1}$$

The velocity of the fluid in the axial direction v_z is related to the rotation speed $\boldsymbol{n}$ of the rotors and the lead $\boldsymbol{S}$,

$$v_z = \frac{n}{60} S \tag{2}$$

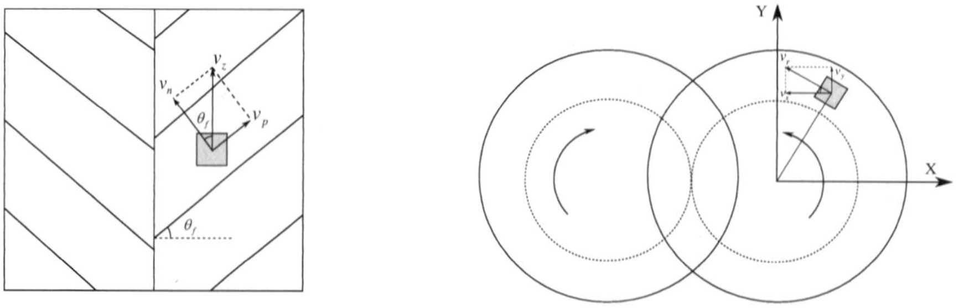

(a) **(b)**

Figure 2 Velocity of infinitesimal fluid element with the rotating of rotors: (a) axial direction; (b) circumferential direction

When the screw rotor rotates, the infinitesimal fluid element rotates around the axis relative to the screw rotor. Its speed is

$$v_r = r\omega \tag{3}$$

Among them, r is the radial distance between the position of the fluid element and the centre of the rotor, ω is the rotation speed of the screw rotor, and the rotation speeds of the male and female screw rotor are respectively

$$\omega_1 = \frac{2\pi n}{60}, \qquad \omega_2 = -\omega_1 \frac{z_1}{z_2} \tag{4}$$

3. Geometry and Grid Generation

3.1. Geometry of Screw Pump

The rotor profile of the screw pump is involute-cycloid profile. Table 1 shows the geometry parameters of rotors.

IOP Conf. Series: Materials Science and Engineering 1180 (2021) 012038 doi:10.1088/1757-899X/1180/1/012038

Table 1 Geometry parameters of screw rotors used in study

	Number of Lobes	Pitch Radius (mm)	Root Radius (mm)	Tip Radius (mm)
Male Rotor	2	42	35	70
Female Rotor	3	63	35	70
Centre distance	105*mm*			
Thread pitch	61*mm*			
Radial clearance	0.24*mm*			
Inter-lobe clearance	0.12*mm*			

3.2. Grid Generation

In order to compare the numerical calculation results based on moving and static mesh, six different rotation positions are taken to generate the static mesh, while keep the same size and cell number as the moving mesh, as shown in Figure 3. At the same time, this grid generation method can be used to investigate how the rotor position influences the calculation results when using static mesh.

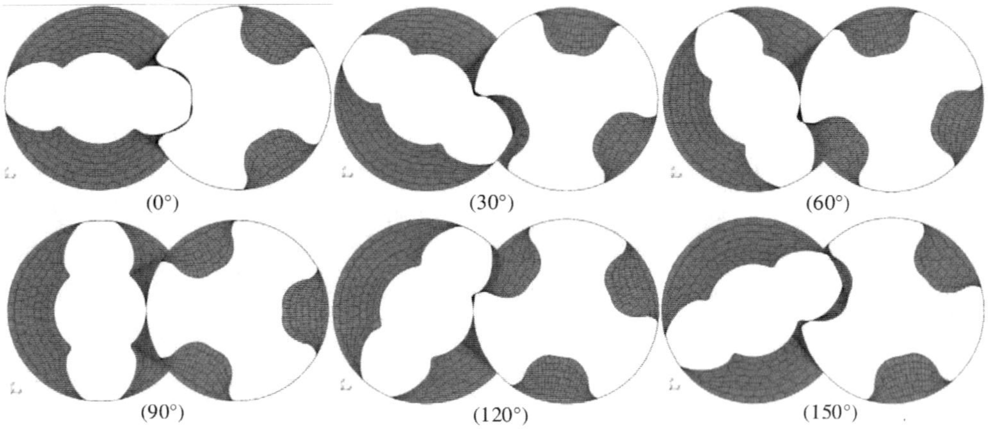

Figure 3 Static mesh of rotor domain in different rotation position

4. Simulation results and discussion

By using the same numerical scheme and parameter settings, it is found that the pressure distribution of the flow field calculated by using the static mesh is very close to the calculation result of the moving mesh, and the pressure gradient of the pump chamber calculated by these two methods are completely consistent. Therefore, it is an accurate and reliable method to use static mesh to calculate the pressure distribution inside the screw pump.

The main difference in the calculation results of these two methods is reflected in the velocity field distribution. Figure 4 shows the velocity distribution on cross section of rotors based on static and moving grid. It can be observed that the rotor section velocity calculated by static grid is lower than the result of moving grid calculation, among them, the maximum velocity calculated by the static grid is 15.390m/s, and the maximum velocity calculated by the moving grid is 23.147m/s. The high-velocity area corresponding to the static grid is mainly distributed on the wall of the rotor, which is caused by the rotation speed of the screw rotor set by the boundary conditions, and the velocity in the area far away from the rotor wall in the pump chamber is extremely low. However, the high-speed area calculated by moving grid is distributed throughout the pump chamber.

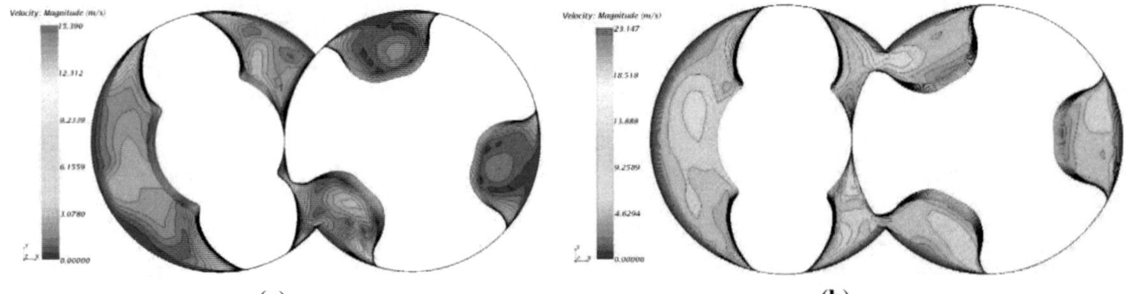

(a) (b)

Figure 4 Velocity distribution on cross section of rotors: (a) based on the static grid and rotating wall; (b) based on moving grid

It can also be observed from Figure 4 that the velocity distribution close to the wall of the casing calculated by two methods are consistent, and due to the main effect of viscosity, the fluid velocity near the wall is close to zero. At a far distance from the wall of the pump casing, the inertial force is much larger than the viscous force and plays a major role. Therefore, there is an obvious velocity gradient along the normal direction of the inner wall of the casing.

Because the rotors in the pump chamber are static when the static grid method is used for calculation, the fluid passage in the pump is fixed, mainly composed of the leakage passage of the screw rotors, including the interlobe clearance, the radial clearance and the blow-hole area. At this time, the fluid velocity direction is mainly along the axial direction of the screw rotors. When using the moving grid method for calculation, in addition to pumping fluid from the inlet to the outlet in the axial direction, there is also a rotation speed relative to the casing wall as the screw rotor rotates. Among them, the flow situation close to the inner wall of the casing can be analyzed by the boundary layer theory, which is not detailed here.

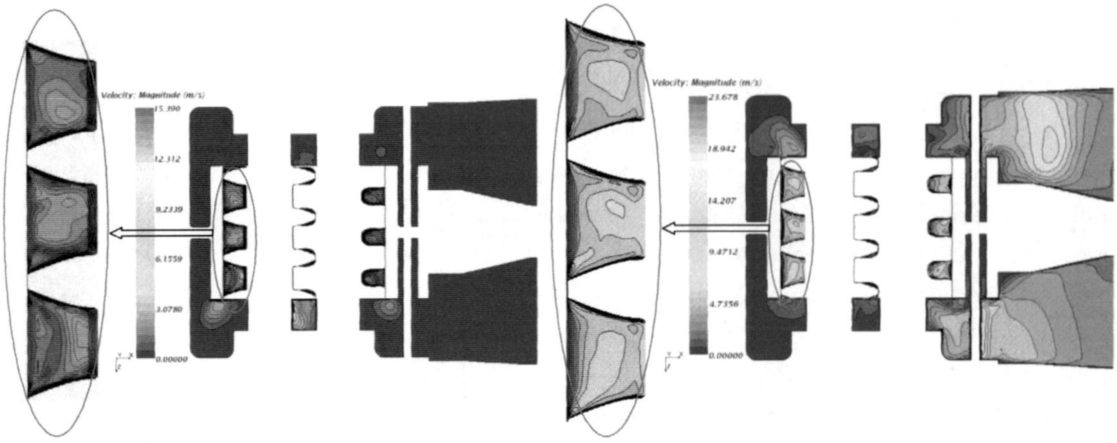

(a) (b)

Figure 5 Velocity distribution on axial cross section: (a) based on the static grid and rotating wall; (b) based on moving grid

IOP Conf. Series: Materials Science and Engineering 1180 (2021) 012038 doi:10.1088/1757-899X/1180/1/012038

It can be seen from Figure 5 that the velocity in axial section of the rotor calculated by the static grid is smaller than the calculation result of the moving grid. Among them, the maximum velocity on the rotor axial section calculated by the static grid is 15.390m/s, while it is 23.678m/s when calculated by the moving grid. It can also be observed from the figure that the velocity gradient in the inlet and outlet ports of the screw pump calculated by the static grid is almost zero, while there are obvious velocity gradients and vortex flow in the inlet and outlet area by the moving grid.

Refer to the JB/T 8091-1998 screw pump test method [14], the experimental setup is shown in Figure 6. A pressure gauge is installed at the outlet of the pump. A flow meter and a throttle valve are installed in the pipeline. The real-time flow rate and outlet pressure can be measured in this hydraulic system [15].

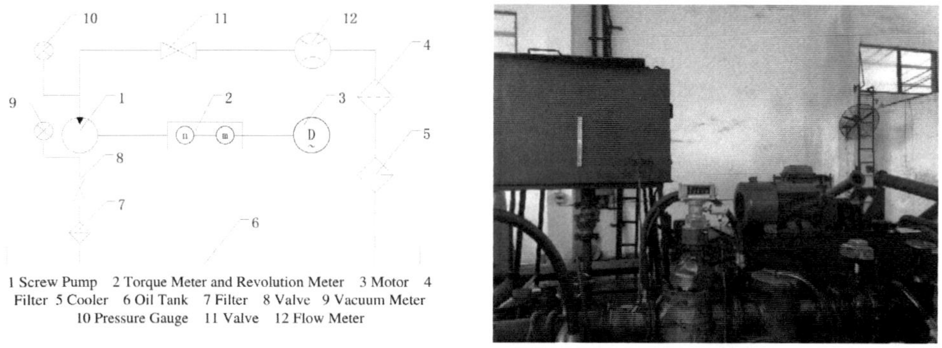

(a) (b)

Figure 6. System performance test bench (a) schematic diagram; (b) test rig

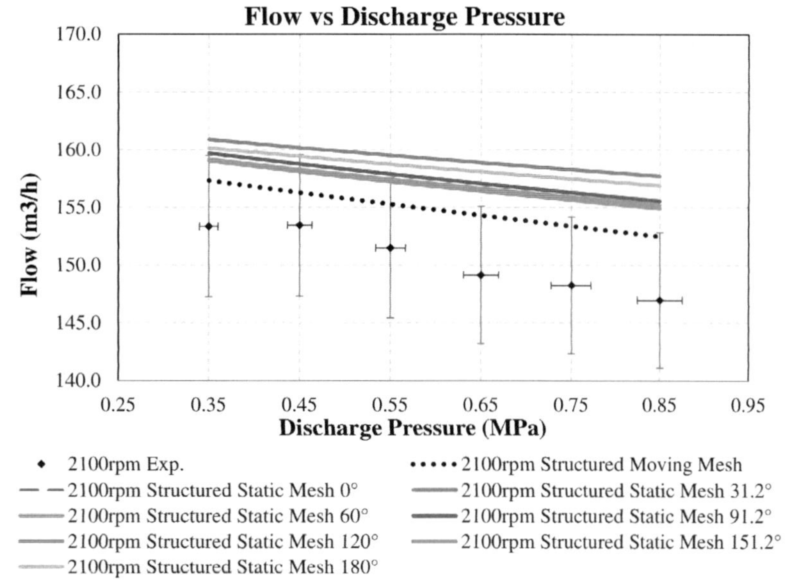

Figure 7 Mass flow rate calculated by the static mesh under different discharge pressure

Since the flow channel in static grid calculation is the leakage channel of the screw pump, the flow rate calculated by using the static grid is the leakage volume, and the flow rate of the screw pump can be obtained by subtracting the leakage amount from the theoretical flow rate. The flow rate of the screw

IOP Conf. Series: Materials Science and Engineering 1180 (2021) 012038 doi:10.1088/1757-899X/1180/1/012038

pump calculated by static grid at different rotation angle positions is shown in Figure 7. From the calculation results, it can be seen that the instantaneous leakage of the screw pump changes with the change of the screw rotation angle, and it also shows periodicity. It can also be observed from Figure 7 that the flow rate of the screw pump calculated by the static grid at different rotation angle is larger than that by the moving grid. According to the comparative analysis of these two velocity fields, this is due to the fact that the momentum of fluid transmission is ignored when using static grid in calculations. The calculated flow rate in pump chamber is low, and the corresponding leakage rate is also low, so the calculated flow rate is high.

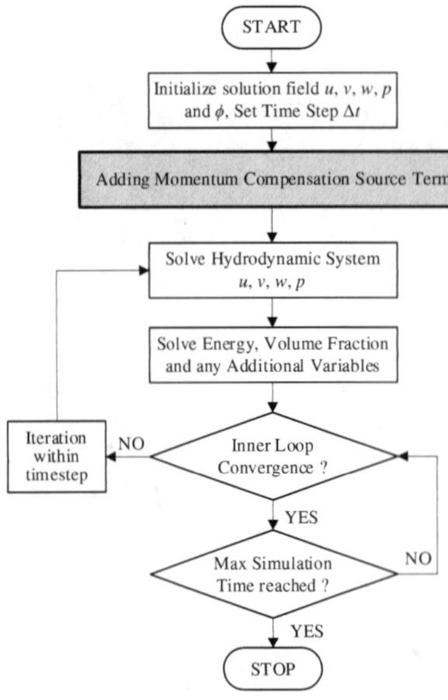

Figure 8 flow chart of static grid CFD method based on momentum source compensation

In order to improve the accuracy of calculation, a compensation momentum source term is added in the static grid CFD calculation, shown as Figure 8. The calculation results show that the velocity of the flow field obtained by this method is higher than the static grid calculation result, and the value is close to the result by the moving grid.

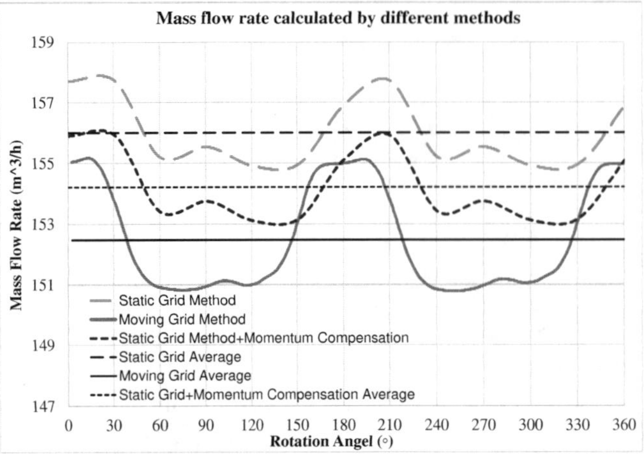

Figure 9 Mass flow rate under different rotation angle calculated by three calculation methods

IOP Conf. Series: Materials Science and Engineering 1180 (2021) 012038 doi:10.1088/1757-899X/1180/1/012038

Figure 9 shows the flow rates of the screw pump at different rotation angles calculated by three different methods. It can be seen from the figure that the period and amplitude of the flow fluctuation calculated by the static grid are basically consistent with the calculation results of the moving grid, but the average value is higher than the calculation result of the moving grid. When the momentum compensation source term is added, the calculated leakage is increased and the flow rate is reduced, which is closer to the result of moving grid calculation, and it is also closer to the real flow situation. If assuming the mass flow rate calculated by moving grid as standard, the calculation error is reduced from 2.32% to 1.14% when adding the momentum source in static grid. This result shows that adding momentum compensation source term can improve the accuracy and feasibility of the static grid calculation method.

5. Conclusions

This paper conducted comparative study of the moving and static grid CFD numerical simulation methods. The calculation process and errors of these methods were analyzed. A momentum compensation method based on the static grid calculation was proposed. It can be concluded that,

- The pressure distribution of the flow field in the screw pump calculated by the static grid and the moving grid are consistent, so the method of using the static grid method to solve the pressure distribution of the screw pump is reliable.
- The velocity of the flow field calculated by static grid is smaller than that by using the moving grid. Among them, the velocity gradient of the inlet and outlet ports in the screw pump calculated by static grid is zero, and the velocity of the rotor flow field region far away from the rotor wall is also extremely small. However, there are obvious velocity gradients and vortices in the inlet and outlet ports of the pump when calculated by moving grid.
- The period, frequency and fluctuation trend of the pump flow rate calculated by static grid at different rotation angles are the same as those of the moving grid, but the average value is higher than that of the moving grid. This is due to the ignoring of the fluid momentum during rotation when using static grid calculations. The calculated velocity in the pump chamber is low, and the corresponding calculated leakage is also low, so the calculated flow rate is high.
- By adding the momentum compensation source term on the basis of the static grid method, the calculation error of mass flow rate is reduced from 2.32% to 1.14%, the calculation accuracy and feasibility of the static grid method are greatly improved, and it provides a method reference for the analysis of the difficult-to-handle special-shaped screw rotors.

Acknowledgments: The authors wish to thank Prof. Kovacevic in City University of London for providing the license of SCORG, which is used in this study.

Funding: The authors wish to acknowledge the support of grant No. 2020CFB114 and grant No.2018A07.

References

[1] Rane Sham, Kovacevic Ahmed. Algebraic generation of single domain computational grid for twin screw machines. Part I. Implementation[J]. Advances in Engineering Software, 2017, 107:38-50.

[2] Stosic Nikola, Smith K. Ian, Kovacevic Ahmed. Vacuum and Multiphase Screw Pump Rotor Profiles and Their Calculation Models, Conference on Screw Type Machines[C]. VDI-Schraubenmachinen,

Dortmund, Germany, October 2006

[3] Christopher Huck, Andreas Brümmer. Limits of one dimensional modeling of rarefied Couette-Poiseuille clearance flow in vacuum pumps[C]. International Conference on Screw Machines 2018, IOP Conf. Series: Materials Science and Engineering 425 (2018)

[4] Hsieh Chiu-Fan, Zhou Qin-Jie. Fluid analysis of cylindrical and screw type Roots vacuum pumps[J]. Vacuum, 2015, 121: 274-282.

[5] Sun Shu-Kai, Zhao Bin, Jia Xiao-Han, Peng Xue-Yuan. Three-dimensional numerical simulation and experimental validation of flows in working chambers and inlet/outlet pockets of Roots pump[J]. Vacuum, 2017, 137:195-204.

[6] Tang Qian, Zhang Yuanxun. Screw Optimization for Performance Enhancement of a Twin-Screw Pump[J]. Proceedings of the IMechE, Part E: Journal of Process Mechanical Engineering, 2014, 228(1):73-84.

[7] Wei Jing, Zhou Dongming, Zhang Aiqiang, Yang Yuliang; Chen Dabing. Profiles evolutionary design and evaluation of mixing performance of screw elements for intermeshing counter-rotating twin-screw kneader[J]. Proceedings of the IMechE, Part C: Journal of Mechanical Engineering Science. 2016, 230(17):3076-3091.

[8] Prashanth S R , Arumugam S K , Gangradey R , et al. CFD modelling and performance analysis of a twin screw hydrogen extruder[J]. Fusion Engineering and Design, 2019, 138(JAN.):151-158.

[9] Cao Feng, Peng Yueyuan, Xing Ziwen, Shu Pengcheng. Thermodynamic Performance Simulation of a Twin-Screw Multiphase Pump[J]. Proceedings of the IMechE, Part E: Journal of Process Mechanical Engineering, 2001, 215(2):157-163.

[10] Demirdžić I, Muzaferija S. Numerical method for coupled fluid flow, heat transfer and stress analysis using unstructured moving meshes with cells of arbitrary topology[J]. Computer Methods in Applied Mechanics and Engineering, 1995, 125(1-4):235-255.

[11] Joe F. Thompson, Bharat K. Soni, Nigel P. Weatherill. Handbook of Grid Generation[M]. CRC Press, 1999.

[12] Rane Sham, Kovacevic Ahmed, Stosic Nikola, Kethidi Madhulika. Grid deformation strategies for CFD analysis of screw compressors[J]. International Journal of Refrigeration. 2013, 36(7):1883-1893.

[13] Rane Sham, Kovacevic Ahmed, Stosic Nikola, Kethidi Madhulika. Deforming grid generation and CFD analysis of variable geometry screw compressors[J]. Computers & Fluids, 2014, 99(5):124-141.

[14] JB/T 8091-1998 Testing methods for screw pumps (*in Chinese*)

[15] Yan Di, Kovacevic Ahmed, Tang Qian, Rane Sham, Zhang Wenhua. Numerical modelling of twin-screw pumps based on computational fluid dynamics[J]. Proceedings of the IMechE, Part C: Journal of Mechanical Engineering Science. 2017, 231(24):4617-4634.

Modeling and comparison of different capacity modulation strategies with focus on seasonal performance

S Inampudi[1], **F Botticella**[2] **and S Elbel**[1,2,3]

[1] Air Conditioning and Refrigeration Center, University of Illinois at Urbana-Champaign, 1206 W. Green St., Urbana, IL 61801, USA
[2] Creative Thermal Solutions, Inc. (CTS), 2209 N. Willow Rd., Urbana, IL 61802, USA
[3] Author to whom correspondence should be addressed (elbel@illinois.edu)

Abstract. Over the past few years, demand for energy efficient HVAC systems has been increasing rapidly, and this trend is likely to continue in the future. There are various modulation technologies that can respond to load variations by switching between capacity levels (tandem, two-stage compressors), PWM control (digital scroll), speed modulation (variable-speed). These technologies can be compared using a single number figure of merit, IPLV.SI which expresses part-load efficiency based on weighted operation at various partial capacities. A model is proposed for a water chiller application having a nominal cooling capacity of 8kW. System performance has been calculated using a 10-coefficient compressor model provided by the manufacturer and the most well-known correlations for heat transfer and pressure drop in a brazed plate heat exchanger. Model validation is performed with experimental data obtained in a parallel experimental investigation. Single-speed and two-stage compressor have a comparable IPLV.SI while a variable-speed compressor has 20% higher IPLV.SI.

1. Introduction

HVAC&R systems are required to operate under different loads. The load can vary due to different ambient weather conditions, increased demand during certain time of the day or the systems are oversized to account for the quick pull down. Modulating compressor technologies are a good solution for systems operating under these varying loads. The modulation also provides precise temperature and humidity control.

The major limitation of any comparison study on the modulation strategies to the best of our knowledge is that these comparisons are either limited to only a few of the existing modulation strategies or a comparison between different data sets obtained with different systems using different refrigerants and operating conditions. This makes driving any insights on the relative effectiveness of these modulation strategies difficult.

As a first step, this paper proposes a model for the water chiller application having a nominal cooling capacity of 8 kW. This paper presents the results of modeling investigations into the part-load performance of single-speed and two-stage compressor. These results are validated with experimental data obtained in the parallel experimental investigation. The validated system model is used to predict the performance of a variable-speed compressor.

The model calculates IPLV.SI (Integrated Part Load Value), a single figure of merit used to compare modulation strategies in water chillers according to AHRI Standard 551/591 (2018) [1].

2. Modulation strategies being considered

The following are the different compressor modulation strategies being considered as part of this study. As a first step, this paper compares a single-speed, two-stage, and variable-speed compressor. All the compressors are scroll and have similar capacity and belong to the same generation.

2.1. Single-speed compressor
A single-speed compressor modulates capacity by cycling i.e., turning on and off. The on/off capacity control is the simplest method of adjusting the predetermined temperature (setpoint) using a thermostat. After the set temperature is reached, the thermostat stops the compressor and circulating refrigerant in the cycle. When the thermostat detects a rise in air temperature, it turns the compressor on [2].

2.2. Two-stage compressor
This is a commercially available two-stage compressor. It is different from conventional two-stage compressors that use a low-pressure stage and a high-pressure stage. Instead, it has two internal bypass ports which enable the system to run at 67%-part load capacity. This compressor can be operated either at a high stage (100% capacity) or a low stage (67% capacity). [3].

2.3. Variable-speed compressor
This compressor modulates its capacity by varying the speed of the motor [4]. The focus will be on the inverter driver variable-speed compressor with a brushless DC motor. The selected compressor can be operated between 20 Hz and 100 Hz.

3. AHRI 551/591 (2018) standard
AHRI 551/591 is used for the determination of the part-load performance of water chillers. The standard defines a single number part-load efficiency figure of merit called Integrated Part Load Value (IPLV.SI) calculated at part load rating conditions. These part load rating conditions are shown in Table 1. IPLV.SI is the weighted average of the COP_R measured at these standard rating conditions as shown in Equation (1). These factors in Equation (1) are based on the weighted average of the most common building types and operations using average weather in 29 U.S cities.

$$IPLV.SI = 0.01 \cdot A + 0.42 \cdot B + 0.45 \cdot C + 0.12 \cdot D \tag{1}$$

$$A = COP_R \; at \; 100\%$$
$$B = COP_R \; at \; 75\%$$
$$C = COP_R \; at \; 50\%$$
$$D = COP_R \; at \; 25\%$$

If a compressor cannot be unloaded to 25%, 50%, or 75% load point, then the compressor is run at the minimum step of unloading at the condenser entering water shown in Table 3 for 25%, 50%, or 75% capacity points as required. Once the COP_{test} is calculated at these conditions using Equation (2), it is degraded to COP_R using the Equation (3), (4), (5), and (6).

$$COP_{test} = \dot{Q}_{ev,avg} / \dot{W}_{cp} \tag{2}$$

$$COP_R = COP_{test} / C_D \tag{3}$$

$$C_D = (-0.13 \cdot LF) + 1.13 \tag{4}$$

$$LF = \frac{(\%Load)(Q_{ev\;100\%})}{(Q_{ev\;min\%Load})} \tag{5}$$

$$\%Load = {(Part\ load\ net\ capacity)}\Big/{(Full\ load\ rated\ net\ capacity)} \quad (6)$$

Table 1: AHRI 551/591 part load conditions for IPLV.SI

Condition	Part load ratio (%)	Condenser Inlet/Outlet (°C)	Evaporator Inlet/Outlet (°C)
A	100	30/35	12/7
B	75	24.5/*	*/7
C	50	19/*	*/7
D	25	19/*	*/7

Table 1 shows the part load conditions. As seen in the table, condenser inlet, outlet, and evaporator inlet, outlet are mentioned for A condition while for the B, C, and D conditions, only the condenser inlet and evaporator outlet temperature are mentioned. 20% ethylene glycol by volume in water (Water Ethylene Glycol mixture, WEG) is used as the secondary fluid. Standard required that the condenser and evaporator WEG flow rate used for the A condition be used for the B, C, and D conditions.

4. System model

R410A is the refrigerant used in this study. The properties of the refrigerant and Water Ethylene Glycol (WEG) mixture are calculated using REFPROP [5] database. The effects of oil circulation are not considered in the model.

4.1. Compressor

The compressor model is developed using the ARI 10-coefficients provided by the manufacturer. Mass flow rate and compressor energy consumption are expressed as a function of the evaporation temperature T_{ev} and the condensation temperature T_{co} for a rated superheat. The variable-speed compressor is a 20-coefficient model expressed as a function of compressor speed RPM, T_{ev} and T_{co}

$$\begin{aligned}\dot{m}_{rated} = {} & M_0 + M_1 \cdot T_{ev} + M_2 \cdot T_{co} + M_3 \cdot RPM + M_4 \cdot T_{ev} \cdot T_{co} + M_5 \cdot T_{ev} \cdot RPM + M_6 \cdot T_{co} \\ & \cdot RPM + M_7 \cdot T_{ev}{}^2 + M_8 \cdot T_{co}{}^2 + M_9 \cdot RPM^2 + M_{10} \cdot T_{ev} \cdot T_{co} \cdot RPM + M_{11} \\ & \cdot T_{ev}{}^2 \cdot T_{co} + M_{12} \cdot T_{ev}{}^2 \cdot RPM + M_{13} \cdot T_{ev}{}^3 + M_{14} \cdot T_{co}{}^2 \cdot T_{ev} + M_{15} \cdot T_{co}{}^2 \\ & \cdot RPM + M_{16} \cdot T_{co}{}^3 + M_{17} \cdot RPM^2 \cdot T_{ev} + M_{18} \cdot RPM^2 \cdot T_{co} + M_{19} \cdot RPM^3\end{aligned} \quad (7)$$

$$\begin{aligned}\dot{W}_{cp} = {} & P_0 + P_1 \cdot T_{ev} + P_2 \cdot T_{co} + P_3 \cdot RPM + P_4 \cdot T_{ev} \cdot T_{co} + P_5 \cdot T_{ev} \cdot RPM + P_6 \cdot T_{co} \cdot RPM \\ & + P_7 \cdot T_{ev}{}^2 + P_8 \cdot T_{co}{}^2 + P_9 \cdot RPM^2 + P_{10} \cdot T_{ev} \cdot T_{co} \cdot RPM + P_{11} \cdot T_{ev}{}^2 \cdot T_{co} \\ & + P_{12} \cdot T_{ev}{}^2 \cdot RPM + P_{13} \cdot T_{ev}{}^3 + P_{14} \cdot T_{co}{}^2 \cdot T_{ev} + P_{15} \cdot T_{co}{}^2 \cdot RPM + P_{16} \\ & \cdot T_{co}{}^3 + P_{17} \cdot RPM^2 \cdot T_{ev} + P_{18} \cdot RPM^2 \cdot T_{co} + P_{19} \cdot RPM^3\end{aligned} \quad (8)$$

where M_0-M_{19} and P_0-P_{19} are constant coefficients

AHRI 540 (2015) recommends a mass flow correction while calculating mass flow rate for a different superheat from rated superheat. The change of superheat has been reported to have negligible effect on compressor power consumption [6].

$$\dot{m}_{corrected} = \left\{1 + F_v\left[\left({v_{rated}}/{v_{corrected}}\right) - 1\right]\right\} \cdot \dot{m}_{rated} \quad (9)$$

where:

F_v $= Volumetric\ efficiency\ correction\ factor - a\ value\ of\ one\ (1)\ is\ used\ as\ an\ approximation$

$\dot{m}_{corrected} = Refrigerant\ mass\ flow\ rate\ at\ suction\ condition, kg/s$

$\dot{m}_{rated}$ $= Refrigerant\ mass\ flow\ rate\ at\ rated\ superheat, kg/s$
$v_{corrected}$ $= Specific\ volume\ at\ suction\ condition, m^3/kg$
v_{rated} $= Specific\ volume\ at\ rated\ condition, m^3/kg$

Compressor efficiencies are used to validate the compressor model with the experimental setup. Isentropic compressor efficiency is a ratio between the isentropic work $\dot{W}_{isen}$ and real compressor power consumption $\dot{W}_{cp}$

$$\eta_{isen} = \dot{W}_{isen} / \dot{W}_{cp} \quad (10)$$

where $\dot{W}_{isen} = \dot{m}_{corrected} \cdot (h_{cp,o,isen} - h_{cp,i})$

Compressor volumetric efficiency is a ratio of the actual delivered gas volume to the theoretic swept volume of the compressor

$$\eta_{vol} = \frac{v_{corrected} \cdot \dot{m}_{corrected}}{V_{dis} \cdot f} \quad (11)$$

where V_{dis} is the displacement volume of the compressor and f is the operating frequency of the compressor expressed in $[m^3/rev]$ and $[rev/s]$ respectively.

Compressor isentropic and global efficiencies can be expressed as a function of the pressure ratio (τ) between condensation and evaporation pressure

$$\eta_{isen} = A + B \cdot \tau + C \cdot \tau^2 \quad (12)$$

$$\eta_{vol} = D + E \cdot \tau \quad (13)$$

where *A*- *E* are regression coefficients.

4.2. Brazed plate heat exchanger

The heat transfer between WEG and refrigerant occurs in Brazed Plate Heat Exchangers (BPHE). Each BPHE is divided into finite elements along the flow of the fluid, pressure drop, and heat transfer correlations are applied to this element. End plate effects are neglected in BPHE model.

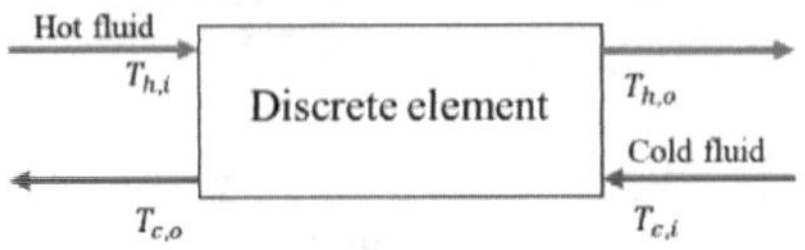

Figure 1: Counter flow in a discrete element

For each discrete element shown in Fig. 1, heat transfer is calculated using LMTD approach shown in Eq. (14) and (15) and energy balance between the two fluids shown in Eq. (16).

$$\dot{Q} = UA \cdot LMTD \quad (14)$$

$$LMTD = \frac{(T_{h,i} - T_{c,o}) - (T_{h,o} - T_{c,i})}{\ln((T_{h,i} - T_{c,o})/(T_{h,o} - T_{c,i}))} \quad (15)$$

where UA is the overall heat transfer coefficient, $LMTD$ is the log mean temperature difference. Eq. (15) represents $LMTD$ calculation for a counter flow heat exchanger.

$$\dot{Q} = \dot{m}_c \cdot c_c \cdot (T_{c,o} - T_{c,i}) = \dot{m}_h \cdot c_h \cdot (T_{h,i} - T_{h,o}) = \dot{m_{ref}} \cdot (\Delta h) \quad (16)$$

where c is the specific heat $[kJ/(kgK)]$, $\dot{m}$ is the mass flow rate $[kg/s]$ and Δh is the refrigerant enthalpy.

The procedure to calculate the heat transfer in a discrete element involves taking an initial guess of the one of the outlet/inlet temperatures. Then the other outlet/inlet temperatures and $\dot{Q}$ are found using Equation (16). Next, a new value of $\dot{Q}$ is determined using Equation (14). If the newly calculated heat flux matches the previous one, the element computation is complete. Otherwise, $\dot{Q}$ from Equation (14) becomes the new estimate for Equation (16) and another computation sequence is performed [7].

$$\frac{1}{UA} = \left(\frac{1}{htc_{weg}A_{weg}} + \frac{1}{htc_{ref}A_{ref}}\right) \quad (17)$$

where htc_{weg} and htc_{ref} are the heat transfer coefficients of WEG and refrigerant respectively.

Correlations are used to predict htc_{weg} and htc_{ref}. These correlations change according to the phase of the refrigerant and whether it is a single phase or two phase heat transfer.

4.2.1. Single phase heat transfer

Eq. (18) is used for the WEG and refrigerant heat transfer coefficients [8]. This correlation is valid for $50 \leq Re \leq 8000; 2 \leq Pr \leq 290$.

$$Nu = (-1.342 \times 10^{-4} \cdot \beta^2 + 1.808 \times 10^{-4} \cdot \beta - 0.0075) \cdot Re^{(-7.956\times10^{-5}\cdot\beta^2+9.687\times10^{-3}\cdot\beta+0.3155)} \cdot Re^{\phi/\beta} \cdot Re^{\gamma/\beta} \cdot Pr^{1/3} \cdot \left(\frac{\mu}{\mu_w}\right)^{0.14} \quad (18)$$

where Nu is the Nusselt number, Re is Reynolds number, Pr is Prandtl number and μ is the dynamic viscosity. The other factors in the Equation (18) are dependent on BPHE geometry. β is the herringbone angle measured with vertical, ϕ is the enlargement factor and γ is corrugation profile aspect ratio of the BPHE.

$$Nu = htc \cdot D_h/k_w \quad (19)$$

where D_h is the hydraulic diameter and k_w is the thermal conductivity. Heat transfer coefficient is calculated using Equations (18) and (19).

$$Re = \frac{\rho V D_h}{\mu} \quad (20)$$

$$V = \frac{\dot{m}}{\rho n W b} \quad (21)$$

where ρ is density, V is the velocity inside the BPHE, n is the channel number for corresponding side, W is the width of the BPHE and b is the corrugation depth.

4.2.2. Boiling

Han *et al.* developed the following correlation to model the boiling in R410A [9]. Nusselt number is given by

$$Nu = Ge_1 \cdot Re_{eq}^{Ge_2} \cdot Bo_{eq}^{0.3} \cdot Pr^{0.4} \quad (22)$$

where Re_{eq} is equivalent Reynolds number, Bo_{eq} is the equivalent Boiling number, G_{eq} is the equivalent mass flux and the coefficients Ge_1 and Ge_2 are functions of the BPHE geometry and are given by the below Equations (23), (24), (25) and (26).

$$Re_{eq} = \frac{G_{eq} D_h}{\mu} \quad (23)$$

$$Bo_{eq} = \frac{\dot{Q}}{A G_{eq} \Delta h_{fg}} \quad (24)$$

$$G_{eq} = \left(\frac{\dot{m}}{Wbn}\right) \cdot \left[1 - x + x\left(\frac{\rho_f}{\rho_g}\right)^{0.5}\right] \quad (25)$$

$$Ge_1 = 2.81\left(\frac{p_{corr}}{D_h}\right)^{-0.041} \cdot \beta^{-2.83} \quad (26)$$

$$Ge_2 = 0.746\left(\frac{p_{corr}}{D_h}\right)^{-0.082} \cdot \beta^{0.61} \quad (27)$$

where p_{corr} is BPHE pitch, x is vapor quality and Δh_{fg} is enthalpy of vaporization

4.2.3. Condensation

Han *et al.* developed a similar correlation for the condensation heat transfer. The proposed Nusselt number has the following form [10].

$$Nu = Ge_1 \cdot Re_{eq}^{Ge_2} \cdot Pr^{1/3} \quad (28)$$

where the coefficients Ge_1 and Ge_2 are functions of the BPHE geometry. These are different from the ones given by Equations (26) and (27).

$$Ge_1 = 11.22\left(\frac{p_{corr}}{D_h}\right)^{-2.83} \cdot \beta^{-4.5} \tag{29}$$

$$Ge_2 = 0.35\left(\frac{p_{corr}}{D_h}\right)^{0.23} \cdot \beta^{1.48} \tag{30}$$

4.3. Solving algorithm

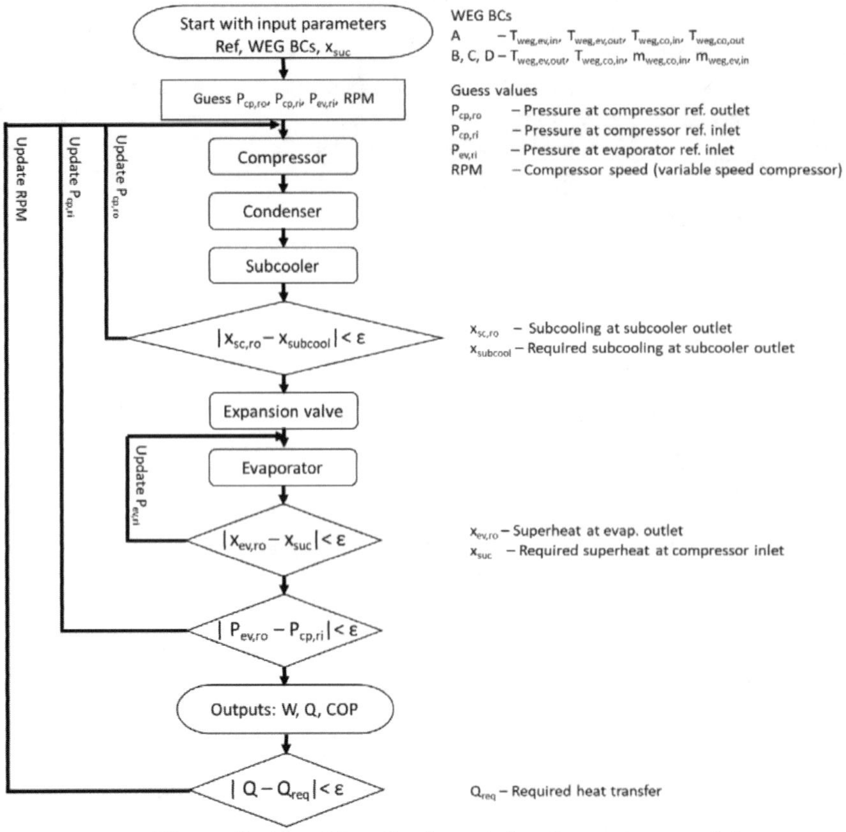

Figure 2: Algorithm for the solving the system model

A sequential iteration algorithm was designed and implemented as shown in the flow chart in Figure 2 [11]. Newton-Raphson method was used to calculate the iteration step. First, the condenser and evaporator WEG boundary conditions, as well as the required superheat and subcooling are provided as inputs. The WEG boundary conditions depend on the condition being modelled. If it is A condition, the inlet and outlet WEG temperatures are specified for the condenser and evaporator. For the B, C and D conditions, WEG inlet temperature and mass flow rate are provided as inputs.

The system parameters such as compressor suction $(P_{cp,ri})$, discharge pressure $(P_{cp,ro})$, evaporator inlet pressure $(P_{ev,ri})$, compressor speed (RPM) (for a variable-speed compressor) are initialized. Starting with the compressor modules, discharge state and refrigerant mass flow rate are determined. These are input to the condenser and subcooler modules. Once, these two BPHE modules converge, they output subcooling temperature at the subcooler outlet $(x_{sc,ro})$. The discharge pressure of the compressor is updated to match the required subcooling $(x_{subcool})$.

Then, expansion valve (modelled as an isenthalpic process) outputs the evaporator inlet temperature and pressure. The evaporator model outputs the superheat at the compressor inlet $(x_{ev,ro})$. Evaporator inlet pressure is updated to match the required superheat (x_{suc}). Then the suction pressure i.e. evaporator outlet pressure (calculated with the pressure drop calculation) is compared with the initially guess compressor suction pressure. The iteration is stopped if these two pressures match, if not the suction pressure is changed and model reiterates. In case of a variable-speed compressor, $(\dot{Q}_{ev})$ is compared to the $(\dot{Q}_{req})$ and the compressor speed is changed accordingly.

5. Results

The results from the system model are validated using the experimental data from a parallel study. The experimental results from a single-speed and two-stage are used for the system validation. Then the system model is used to estimate the performance of a variable-speed compressor.

5.1. Model validation

In a parallel experimental study, R410A Water Ethylene Glycol (WEG) chiller was built to compare the different modulation strategies.

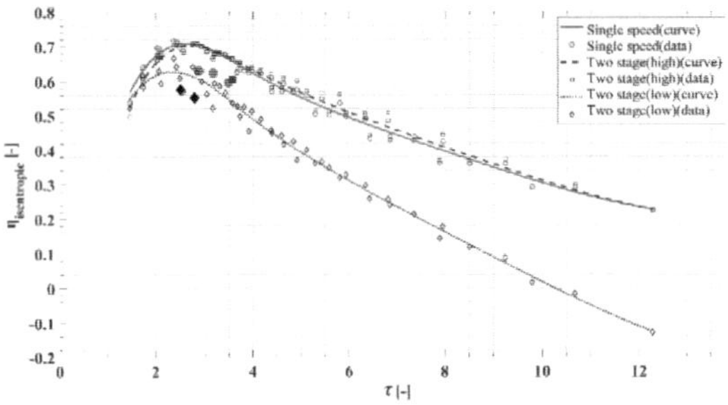

Figure 3: Compressor isentropic efficiency

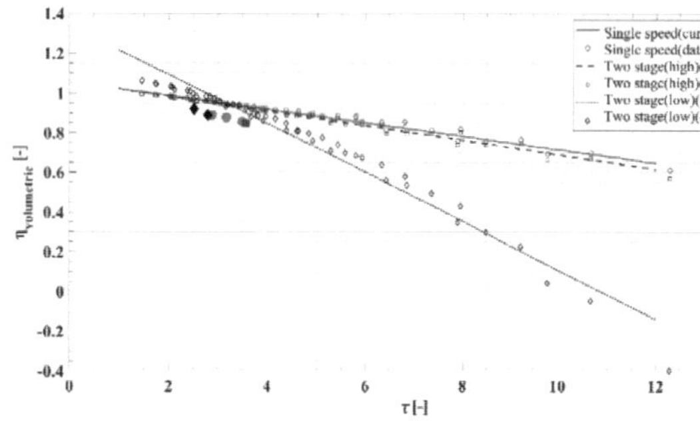

Figure 4: Compressor volumetric efficiency

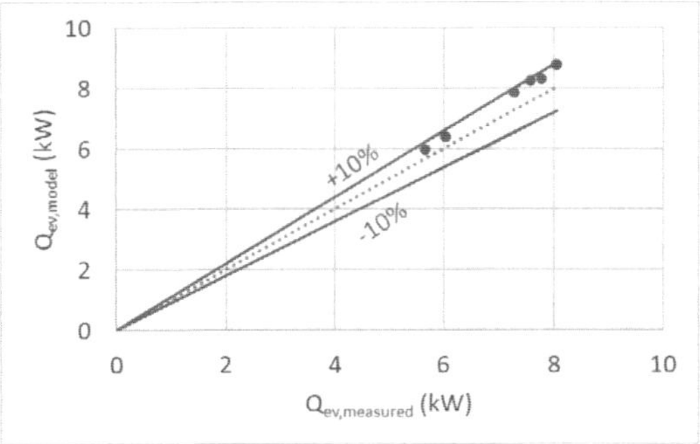

Figure 5. Evaporator cooling capacity prediction.

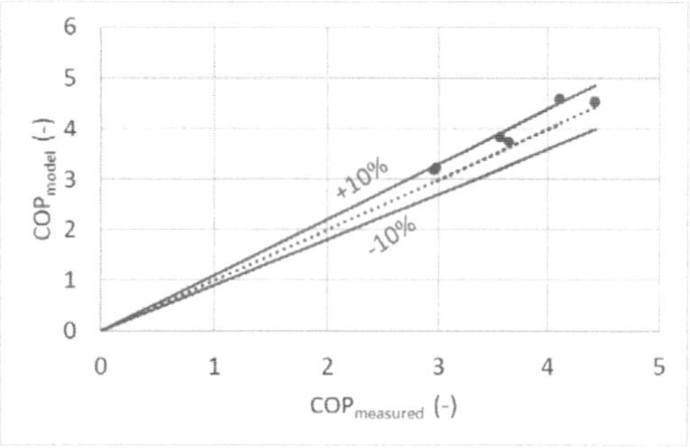

Figure 6. System COP prediction

Fig. 3 and Fig. 4 show the results of the regression fit compressor efficiencies as a function of pressure ratio (τ) with the manufacturer data. The experimental data is also shown on the same curve with solid points. Experimentally determined compressor efficiencies have a maximum difference of 11% compared to the regression fit curves.

The modelled cooling capacity and COP are compared with the experimentally determined cooling and COP in Fig. 5 and Fig. 6. Most of the predictions are within ±10%.

5.2. Single-speed compressor, two-stage and variable-speed compressor part load performance

The part load performance of single-speed and two-stage compressor are shown in Fig. 7 and Fig. 8. A single-speed compressor can only operate at one stage. As the percent load changes from 100% to 25%, the standard requires a reduced condenser WEG inlet temperature, this reduces the condensing temperature and pressure which in turn causes capacity to increase and compressor power to reduce in a single-speed compressor. A two-stage compressor can be operated at either high stage or low stage. For the A condition, the two-stage compressor is operated at a high stage. For the B, C, and D conditions, as the required load reduces, the compressor is operated at the low stage. The low stage cannot match capacity at C, D condition, thus incurring cycling losses. The shaded region is a representation of the cycling losses, which are higher in a single-speed compressor.

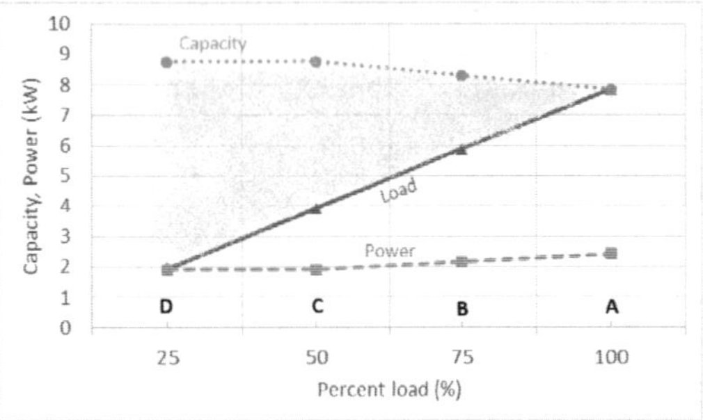

Figure 7. Variation of capacity, power with percent load for single-speed compressor

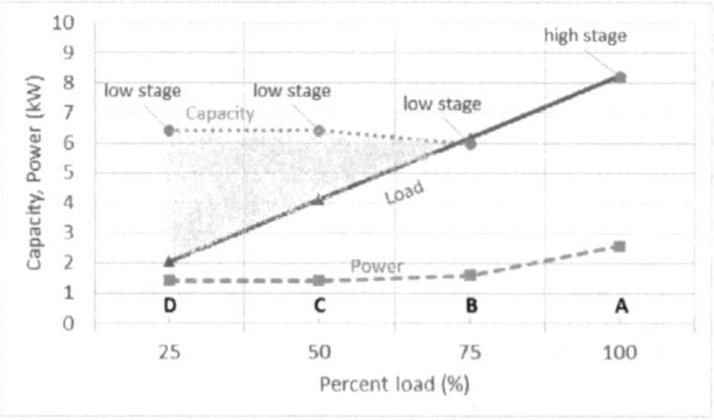

Figure 8. Variation of capacity, power with percent load for two-stage compressor

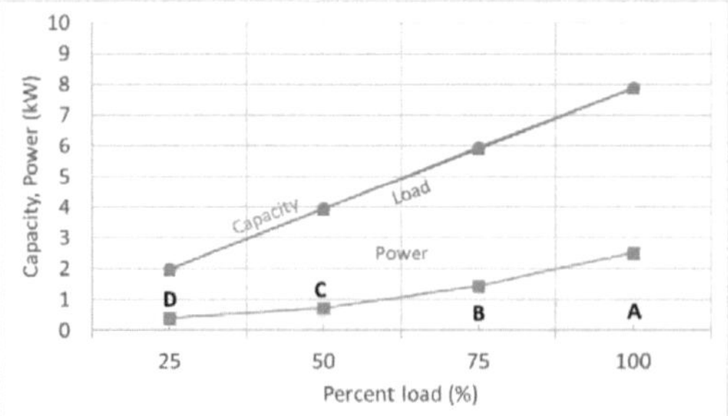

Figure 9. Variation of capacity, power with percent load for variable-speed compressor

The part load performance of variable-speed compressor is shown in Fig. 9. The compressor can vary its speed between 100 Hz and 20 Hz, thus able to match the capacity with the load. The power consumption reduces from A to D because of the lower operation speed.

5.3. Comparison between single-speed, two-stage and variable-speed compressor

Fig. 10 shows the COP_{test} given by Equation (2). All three compressors have similar capacity and power consumption at the 100% load, thus the COP_{test} are around the same. It is interesting to note that at the 75% load, though the two-stage compressor is operating at the low stage, the variable-speed compressor is running at lower frequency, the single-speed compressor has higher COP_{test} than two-stage. However, the COP_{test} for the variable-speed compressor is significantly higher at the lower percent loads because of reduced capacity and reduced compressor power. The COP_{test} are 2% higher for a single-speed compressor at the B, C and D conditions compared to a two-stage compressor.

Fig. 11 shows the COP_R given by Equation (3). Except for the 100% load when all the compressors have similar performance, the variable-speed compressor has a higher COP_R than other two compressor. This trend can also be graphically explained by the cycling losses represented in Fig. 7, 8 and 9. COP_{test} is higher for a single-speed compressor compared to a two-stage compressor at B, C and D conditions. However, only single-speed compressor undergoes cycling losses at B condition. Though both the two-stage and single-speed compressor undergo cycling losses at C and D conditions, two-stage compressor has lower degradation factor because of its operation in low stage which leads to similar COP_R in both these compressors.

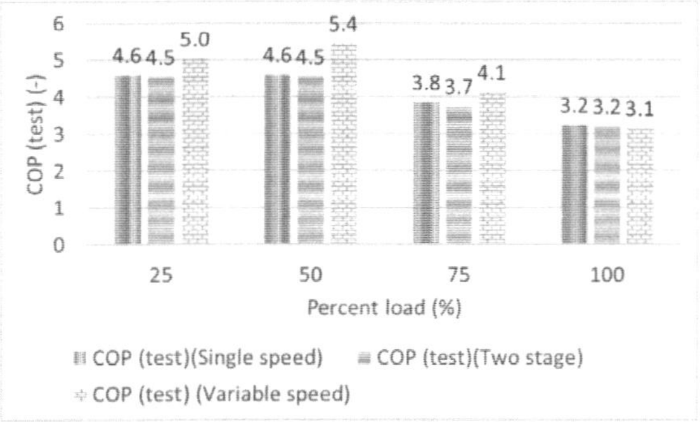

Figure 10. Comparison of COP_{test} between single-speed, two-stage and variable-speed compressor

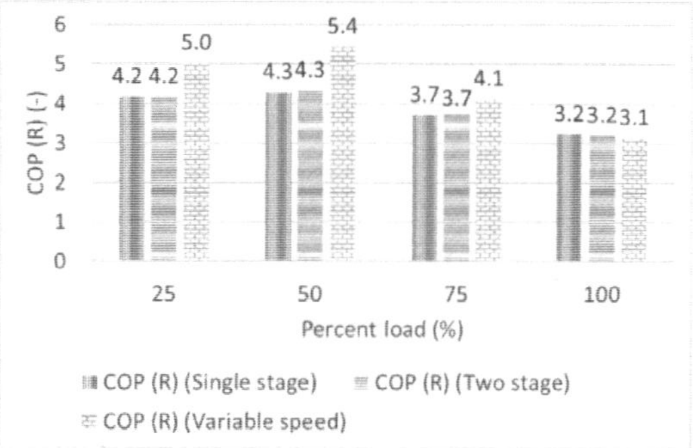

Figure 11. Comparison of COP_R between single-speed, two-stage and variable-speed compressor

IPLV.SI is calculated using the COP_R from Fig. 11 and Equation (1). Single-speed and two-stage compressors have an IPLV.SI of 4.0 while the variable-speed compressor has an IPLV.SI of 4.8. This 20% difference can be explained by the higher COP_R of the variable-speed compressor at 25%, 50% and 75% load which have the higher weighting factors than 100% load in Equation (1). Though COP_{test} are higher for a single-speed compressor compared to a two-stage compressor, IPLV.SI are the same because of higher cycling losses incurred by a single-speed compressor.

6. Conclusions

This study compared a single-speed, two-stage and a variable-speed compressor using a system model. All the compressors have similar capacity and belong to the same generation. The system model was validated with results from a parallel experimental study and then the model is used to estimate the performance of a variable-speed compressor. AHRI 551/591 was used to determine the IPLV.SI for these compressors. A variable-speed compressor has the highest IPLV.SI of 4.8 because of its modulation. Single stage compressor has a higher COP_{test} than a two-stage compressor but it is more penalized for the cycling losses which causes both the compressors to have similar IPLV.SI of 4.0.

7. Nomenclature

Symbols

A	heat transfer area [m^2]
b	corrugation depth [m]
B_o	boiling number
$BPHE$	brazed plate heat exchanger
c	specific heat [kJ/kg-K]
C_d	degradation coefficient [-]
COP	coefficient of performance [-]
D_h	hydraulic diameter [m]
f	compressor frequency [s-1]
F_v	volumetric efficiency factor [-]
G	mass flux [kg/s m^2]
Ge_1, Ge_2	Han – Lee – Kim correlation coefficients
h	enthalpy [kJ/kg]
htc	heat transfer coefficient [W/m^{2}K]
IPLV.SI	integrated part load value [-]
k	thermal conductivity [W/mK]
LF	load factor [-]
$\dot{m}$	mass flow rate [kg/s]

Greek

β	herringbone angle [°]
γ	Corrugation profile aspect ratio [-]
ε	error [-]
η	efficiency [-]
μ	dynamic viscosity [Pa-s]
ρ	density [kg/m^3]
τ	compression ratio [-]
ϕ	enlargement factor [-]
Δ	difference [-]

Subscripts

avg	average
c	cold-side
co	condenser
cp	compressor
dis	displacement
eq	equivalent

$M_0 - M_{19}$	ARI compressor coefficients	ev	evaporator
Nu	Nusselt number [-]	f	saturated liquid
$P_0 - P_{19}$	ARI compressor coefficients	g	saturated vapor
p_{corr}	heat exchanger pitch	h	hot-side
Pr	Prandtl number [-]	i	refrigerant
$\dot{Q}$	heat transfer rate [kW]	isen	isentropic
Re	Reynolds number [-]	o	outlet
T	temperature [°C]	ref	refrigerant
U	overall heat transfer coefficient [W/m^{2}K]	req	required
v	specific volume [m^3/kg]	ri	refrigerant inlet
V	fluid velocity in BPHE [m/s]	ro	refrigerant outlet
W	BPHE width [m]	sc	subcooler
$\dot{W}$	power [kW]	suc	suction
WEG	water ethylene glycol	vol	volumetric
x	vapor quality [-]	w	wall
		weg	water ethylene glycol

8. References

[1] A. H. R. I. 551/591 (SI), 2018: 2018 Standard for performance rating of water-chilling and heat pump water-heating packages using the vapor compression cycle. *Air-Conditioning, Heating, Refrigeration Institute, Arlington, VA, USA*.

[2] Ekren, O 2017 Refrigeration System: Capacity Modulation Methods. *IntechOpen, London, UK*, 119-143.

[3] Wang, B, Han, L, Shi, W, and Li, X 2012 Modulation method of scroll compressor based on suction gas bypass. *Applied Thermal Engineering*, **37**, 183-189.

[4] Qureshi, T Q, and Tassou, S A 1996 Variable-speed capacity control in refrigeration systems. *Applied Thermal Engineering*, **16(2)**, 103-113.

[5] Refprop11 (REFerence fluid PROPerties) Program, Developed by the National Institute of Standards and Technology (NIST)

[6] Standard, A. H. R. I. 540, 2015: 2015 Standard for performance rating of positive displacement refrigerant compressors and compressor units. *Air-Conditioning, Heating, Refrigeration Institute, Arlington, VA, USA*.

[7] Kinab, E, Marchio, D, Rivière, P, and Zoughaib, A 2010. Reversible heat pump model for seasonal performance optimization. *Energy and Buildings*, **42(12)**, 2269-2280.

[8] Yang, J, Jacobi, A, & Liu, W 2017. Heat transfer correlations for single-phase flow in plate heat exchangers based on experimental data. *Applied Thermal Engineering*, *113*, 1547-1557.

[9] Han, D H, Lee, K J, and Kim, Y H 2003. Experiments on the characteristics of evaporation of R410A in brazed plate heat exchangers with different geometric configurations. *Applied thermal engineering*, **23(10)**, 1209-1225.

[10] Han, D H, Lee, K J, & Kim, Y H 2003. The characteristics of condensation in brazed plate heat exchangers with different chevron angles. *Journal-Korean Physical Society*, **43(1)**, 66-73.

[11] Feng, L, & Hrnjak, P 2016. Modeling of a reversible air conditioning-heat pump system for electric vehicles. *SAE International Journal of Passenger Cars-Mechanical Systems*, **9(2016-01-0261)**, 68-74.

Acknowledgements

The authors would like to thank the member companies of the Air Conditioning and Refrigeration Center at the University of Illinois at Urbana-Champaign for their funding, Emerson / Copeland for compressor samples, Danfoss for brazed plate heat exchangers, and Creative Thermal Solutions for the technical support.

IOP Conf. Series: Materials Science and Engineering 1180 (2021) 012040 doi:10.1088/1757-899X/1180/1/012040

Theoretical Study on the Performance of Air Cooler in Cold Storage

Lin Dong*, Zhilong He

School of Energy and Power Engineering, Xi'an Jiaotong University, 710049 Shaanxi, China.

* Corresponding author, Email: 506251197@qq.com

Abstract. Air cooler is the most common air cooling and conditioning equipment that used in frozen and cold storage. However, neither application-oriented or production-oriented enterprises have a sufficient understanding about the performance of air cooler, which will restrict its performance enhancement and technological innovation to some extent. Therefore, this paper conducts a theoretical study on the performance of air cooler to guide the optimal design of air cooler. The distributed parameter model of the air cooler is first established along with some sub-models calculating the structural parameters, the physical properties of the wet air, the heat transfer coefficient and the flow resistance on the refrigerant side and the air side. Then, the proposed models are integrated to calculate the performance of air cooler under the effect of different structural parameters, such as fin spacing, fin thickness, tube spacing, outer tube diameter and finned tube materials. Results show that choosing the relatively smaller fin spacing and outer tube diameter, the larger tube spacing and finned tube materials with high thermal conductivity would contribute to the performance improvement of the air cooler, while the effect of the fin thickness is very small. However, the occupied volume, cost and processing difficulty should also be considered for enterprises in the design stage.

Main symbol table

A_i	Surface area inside the tube/m^2
A_o	Surface area outside the tube/m^2
D_0	Tube outer diameter/m
Fr_l	Liquid Froude number
Nu	Nusselt number
Pr	Prandtl number
Q	Cooling capacity/J
R	Gas constant/J·mol^{-1}·K^{-1}
Re	Reynolds number
T	Temperature/K
c	Specific heat capacity/J·kg^{-1}·K^{-1}
d	Moisture content/kg·kg^{-1}

d_0	Tube inner diameter/m
l	Length/m
h	Specific enthalpy/J·kg^{-1}
m	Mass Flow/kg·s^{-1}
p	Pressure/Pa
u	Speed/m·s^{-1}
t	Temperature/°C
v	Specific volume/m^3·kg^{-1}
x	Dryness
α	Heat transfer coefficient/W·(m^2·K)$^{-1}$
η_f	Fin efficiency
λ	Thermal Conductivity/W·(m·K)$^{-1}$
ξ	Moisture analysis coefficient
ρ	Density/kg·m^{-3}
φ	Relative humidity
Subscript	
a	Air side
g	Liquid phase
i	Inflow
l	Gas phase
o	Outflow
r	Refrigerant side
s	Saturation
w	Wall surface

1. Introduction

In recent years, the refrigeration industry has been more and more widely used in various sectors of the domestic economy and people's lives. Therefore, the frozen and refrigerated food industry also gains the rapid development, for the frozen and refrigerated food has the advantages of nutrition, hygiene and convenience.

Air cooler is the most common air cooling and conditioning equipment that used in frozen and cold storage. It belongs to the forced convection heat transfer equipment, so the heat transfer coefficient on the air side of the finned tubes can be changed by adjusting the frequency and thus wind speed of the fan, so as to reduce the power consumption and improve the cooling effect. Therefore, in order to maintain the heat transfer performance and high-efficiency running of the air cooler, the influence of the frost layer on the heat exchanger must be fully considered.

Many domestic and foreign scholars have carried on the detailed research to the air cooler performance. Stocker W. F [1] researched the law of the performance of the cooler with the thickness of the frost layer. Kondepudi S.N and D.L O'Nea [2] placed three louvered tube-fin chillers with different pitches in the wind tunnel test section. Chen et al. [3] introduced fins in the form of staggered longitudinal vortices into the elliptical finned tube cooler. Huee-Youl Ye [3,4] et al. Established the mathematical model for predicting the heat and mass transfer of finned tubes under frost conditions for the purpose of defrosting under critical operating conditions. The model is verified through experiments, and it is inferred that when the interface temperature between the air and the cold wall surface temperature is higher than the dew point temperature of the air or the freezing point of water, the frost problem can be avoided, and this conclusion can be in good agreement with the test results.

Domestic and foreign scholars mainly focus on theoretical analysis, numerical simulation, but always cannot get rid of the dilemma of accurately establishing the heat transfer model of the air cooler. This work mainly includes the following aspects:

1) A mathematical model of the performance of air cooler is established

Select the appropriate empirical correlation to compile the physical properties of the air, divide the air cooler into a three-zone model of super-cooling, two-phase, and overheating, and simulate the stable working state of the air cooler through MATLAB to obtain the temperature distribution of the air cooler, The heat transfer coefficient on the air side and the overall performance of the chiller.

2) Analysis of the calculated results

The theoretical calculation results are analyzed and used as the theoretical basis for performance optimization of air cooler.

2. Mathematical Model

In this paper, the tube-fin heat exchanger is chosen as the evaporator in the air cooler for the theoretical study. As shown in Fig.1, the counter-current heat transfer between air and refrigerant is adopted in the heat exchanger.

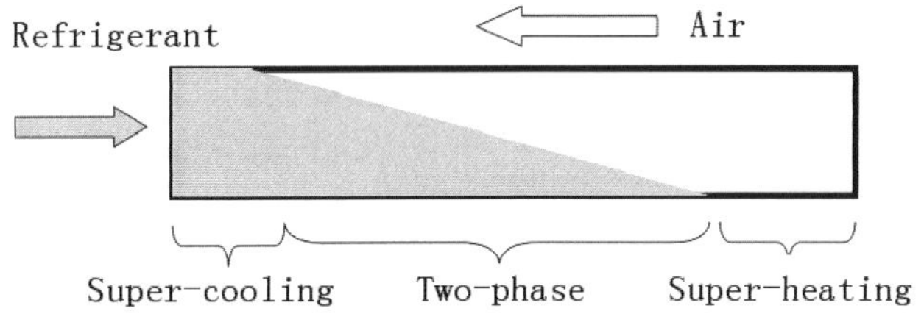

Figure 1. Schematic diagram of heat exchanger in air cooler

In order to investigate the performance of the air cooler, the distributed parameter method is used to establish the mathematical model of the air cooler. The refrigerant in the tube is divided into super-cooling section, two-phase section and super-heating section along the flow direction of refrigerant.

2.1 Assumptions

In order to simplify the calculation process, the following assumptions are made to establish the mathematical model of the heat exchanger:

1) The thermodynamic process of both refrigerant and air is treated as one-dimensional incompressible steady-state flow, and the refrigerant is equally distributed in all tubes;

2) The axial heat conduction, thermal resistance of tube wall and heat capacity caused by the condensed water film are not considered;

3) The influence of gravity on refrigerant flow can be neglected;

4) In the calculation process, the reduced air flow part due to frost deposits and the increased flow resistance and heat transfer resistance were ignored.

2.2 Governing equations

Based on the above assumptions, each micro element is chosen as the control volume, as shown in Fig.2. According to the mass and energy conservation, the differential equations describing the mass and heat transfer in the micro element can be written as following equations.

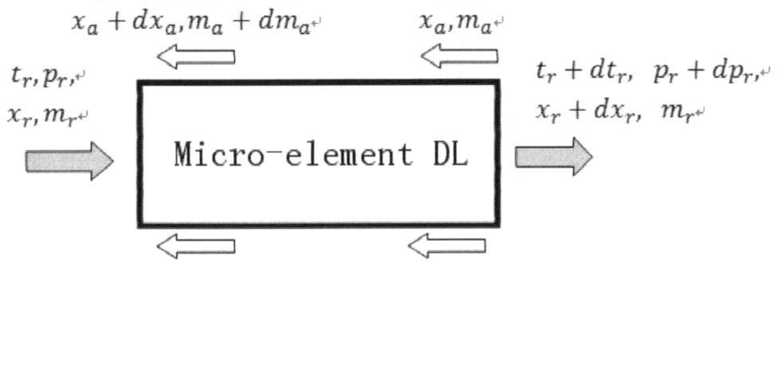

Figure 2. Schematic diagram of each micro element

The heat transfer for air:

$$(c_{pa}\cdot dt_a+\gamma\cdot dx_a)m_a=\alpha_a(t_a-t_w)A_o+\gamma\sigma(d_a-d_w)A_o \tag{1}$$

During the actual operation of the heat exchanger, the refrigerant also absorbs heat from other equipment except the tube wall, and the oil in the refrigerant also needs to be taken into account. Therefore, a modified coefficient, k, less than 1 is adopted to adjust the equation, and is assumed as 0.98 [5]. Hence, the heat balance between internal and external micro tube can be modified as follows:

$$\alpha_a(t_a-t_w)A_o+\gamma\sigma(d_a-d_w)A_o=k\alpha_r(t_r-t_w)A_i \tag{2}$$

Where γ is the latent heat of water and equals 2835 kJ/(kg·K); σ is the mass transfer coefficient of wet air and fin surface.

The moisture content of wet air can be expressed as:

$$d=622\frac{\varphi P_s}{P_a-\varphi P_s} \tag{3}$$

Where P_s is the saturated vapor pressure of wet air.[6] :

The enthalpy of wet air should be calculated according to the sate of air as follows [7]:

$$\begin{aligned}&Dry: h=1.005t+0.001\times d\times(2501+1.86t)\\&Condensation: h=1.005t+0.001\times d\times(2501+1.86t)+(d-d_s)\times h_w\\&Frost: h=1.005t+0.001\times d\times(2501+1.86t)+(d-d_s)\times h_f\end{aligned} \tag{4}$$

where h_w and h_f are the enthalpy of saturated water and frost.

For the water condensation conditions:

$$\xi=1+\frac{r_0+t_m c_p-t_w c_w}{c_{pm}}\times\frac{d_m-d_w}{1000(t_m-t_w)} \tag{5}$$

For the frosting conditions:

$$\xi=1+\frac{r_0'+t_m c_p-t_w c_f}{c_{pm}}\times\frac{d_m-d_w}{1000(t_m-t_w)} \tag{6}$$

The heat transfer coefficient of the equivalent fin surface on the air side can be calculated according to the following formula[8]:

$$\alpha_j=\xi\alpha_o\times\left(\frac{\eta_f a_f+a_b}{a_f+a_b}\right) \tag{7}$$

The heat transfer coefficient of refrigerant in the single-phase zone is calculated by the Petukhov-Popov formula[9]:

$$Nu=\frac{(f/8)\cdot \mathrm{Re}\cdot \mathrm{Pr}}{1.07+12.7(f/8)^{0.5}\cdot(\mathrm{Pr}^{2/3}-1)} \tag{8}$$

The flow resistance on the air side is related to the arrangement of the tube bundle, fin type and air flow. For the staggered tube bundle with straight fins, the flow resistance on the air side can be calculated by the following formula[11]:

$$\begin{cases} \Delta p = 0.2 \times 0.1107 \left(\dfrac{l}{d_e} \right) \left(u_{\max} \rho_a \right)^{1.7} \\ d_e = \dfrac{2\left(s_1 s_f - D_0 b - 2h_f \delta_f \right)}{2h_f + s_f - \delta_f} \end{cases} \tag{9}$$

The flow resistance on the refrigerant side consists of two parts: the frictional resistance loss Δp_l and the partial resistance loss Δp_m [10,12]:

$$\Delta p = \Delta p_m + \Delta p_l \tag{10}$$

For two-phase flows, the frictional resistance loss can be expressed as:

$$\begin{cases} \Delta p_l = \psi_r \Delta p_l^{"} \\ \Delta p_l^{"} = \dfrac{1}{2} \zeta N \left(\dfrac{l}{d_0} \right) \left(u_s \right)^2 \rho_s \end{cases} \tag{11}$$

In this paper, appropriate values will be obtained through iterative procedures and convergence conditions. The reference values of these parameters used in this program are list in Table 1.

Table 1. Structural parameters of chiller

Heat exchange tube length/mm	2680
Number of heat exchange tubes	128
Number of face tubes	16
Process number	32
Heat exchange tube diameter/mm	20
Wall thickness of heat exchange tube/mm	1.5
Heat Exchange Tube Material	Aluminum

3. Analysis and comparison of theoretical calculations and experimental results

This chapter compares the theoretical results with the experimental data, and analyzes the deviation between them qualitatively and quantitatively by means of charts to verify the applicability and accuracy of the mathematical model in this article.

Table 2. Cooling fan test conditions

Variable parameters	Temperature/°C	Drive temperature difference/°C	Headwind/ms^{-1}	Frost layer thickness	Cycle rate
Parameter value	5~-30	4~16	2.3/2.63/2.96/ 3.29/3.62	1/2/3	1.5/2/4/5

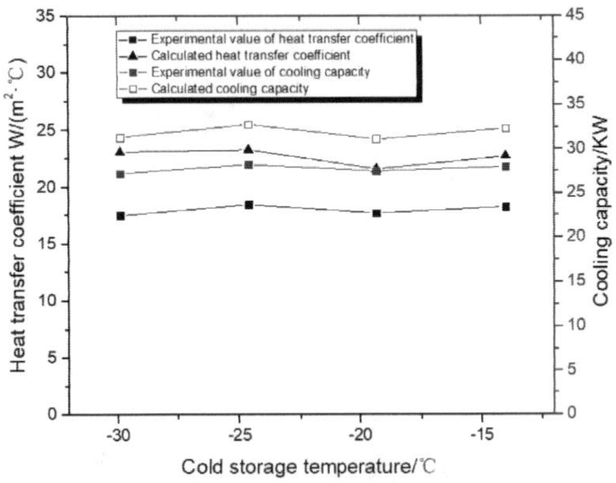

Figure 3.

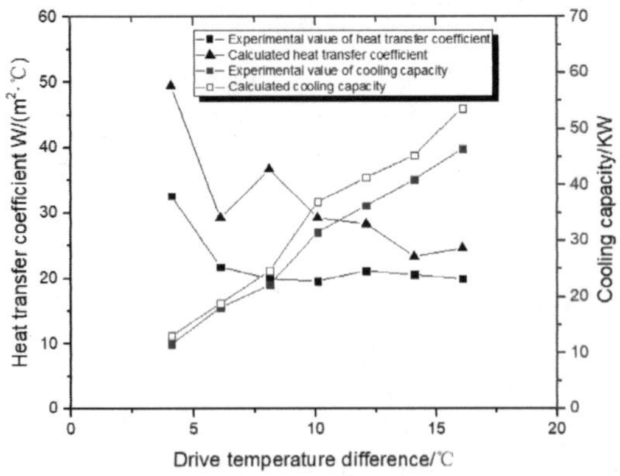

Figure 4.

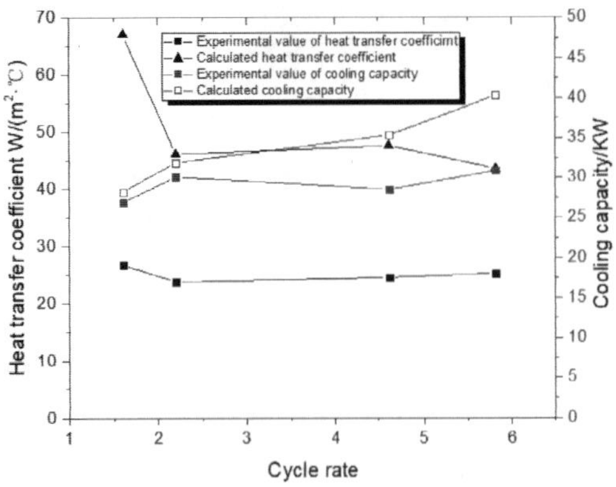

Figure 5.

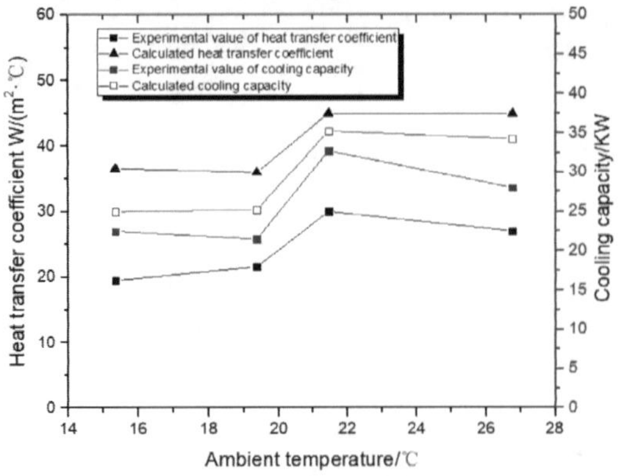

Figure 6.

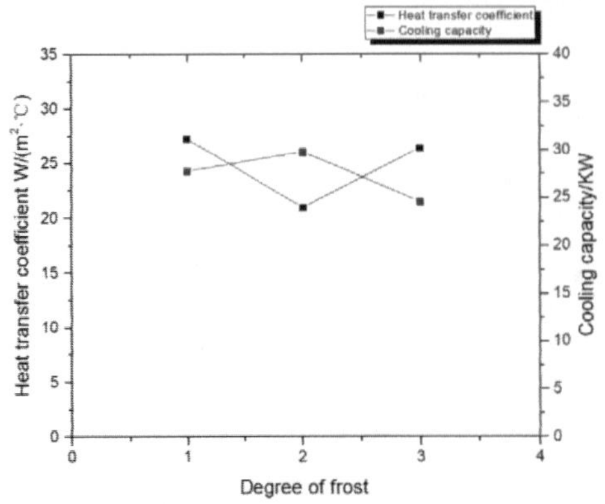

Figure 7.

As shown in Figure 3, since the density of the refrigerant becomes smaller and the viscosity of the liquid becomes larger under low temperature conditions, the resistance in the tube increases at the same mass flow rate, and the change of the evaporation temperature with the pressure is more obvious. Therefore, at low storage temperature, the average evaporation temperature increases, and the actual heat exchange temperature difference decreases, which will lead to a reduction in heat transfer coefficient.

It can be seen from the Figure 4 that as the driving temperature difference increases, the heat transfer coefficient of the cooling fan has a downward trend. When the driving temperature difference increases from 4 ℃ to 10 ℃, its heat transfer coefficient has a very obvious decrease, about 33%; And it tends to be stable when the driving temperature difference is greater than 10°C. For the cooling capacity of the chiller, its value gradually increases with the increase of the driving temperature difference.

It can be seen from the above Figure 5 that with the increase of the cycle rate, the heat transfer coefficient of the chiller slightly decreases, then increases slightly, and gradually stabilizes, while the cooling capacity tends to stabilize after the cycle rate gradually increases. The level of circulation rate mainly affects the mass flow rate and outlet dryness in the pipe, and also affects the resistance drop in the pipe.

In the range of 18 ℃ ~ 25 ℃, the heat transfer coefficient of the chiller increases first and then decreases, but the overall fluctuation is within the range of 10%, indicating that the effect of heat leakage on the chiller is not great in Figure 6. Theoretical calculation results show that no matter which ambient temperature the theoretical value of the heat transfer coefficient deviates greatly from the actual one, it is only feasible as a qualitative one. The accuracy of the cooling capacity prediction is still very high when the ambient temperature is low. When the temperature reaches 25 ℃, the error also reaches 30%.

Figure 7 shows the effect of the frost layer on the cooler. It can be seen that the cooling capacity of the air cooler first increases and then decreases, and the heat transfer coefficient is opposite, first decreasing and then increasing. This has a great relationship with the increase of the surface roughness of the frost layer on the initial frost layer, the thinning of the boundary layer, and the excessively thick frost layer on the late stage, which increases the thermal resistance and impedes the flow on the air side.

Some important factors are ignored in the model, such as the thermal resistance of the pipe wall, the pressure loss at the elbow, the frost on the air side, and the wind speed. The effect of uniformity, and the ideal treatment means such as one-dimensional treatment of heat exchange tubes are also one of the factors that cause the model to not accurately express the performance of the cooler.

4. Conclusions

1) Through theoretical calculations, it can be seen that the cooling capacity and the pressure drop on the air side continue to increase as the fin spacing decreases. The cooling capacity should be reduced as much as possible without affecting the air flow to increase the cooling capacity; when the fins become thicker, the heat exchange capacity increases slightly, and the cooling capacity increases in the range of -3%~3% Variety. An appropriate increase in thickness can help increase the strength of the heat exchange part of the cooler; the spacing of the fin tubes becomes larger, and the cooling capacity of the cooler will gradually increase, but the increase of the tube spacing will increase the volume of the cooler, and the optimal tube spacing needs to be determined according to the actual situation; the cooling capacity and the loss of flow resistance decrease simultaneously with the increase of the tube diameter. Within the range of 21mm ~ 25mm, the cooling capacity drops significantly. Therefore, the size of the tube diameter near 21mm is more suitable for the overall performance of the air cooler.

2) The accuracy of the model was verified through experiments, and some factors affecting the cooler were analysed: when the humidity in the cold storage did not change much and there was no additional humidification, there was the same heat transfer coefficient at different storage temperatures. However, in the actual cold storage, the heat transfer coefficient value should be appropriately enlarged by 5% to 25%; when the driving temperature difference is increased from 4 ℃ to 10 ℃, its heat transfer coefficient has a very significant decrease, about 33%. For the cooling capacity, its value gradually increases with the increase of the driving temperature difference; with the increase of the cycle rate, the heat transfer coefficient of the chiller decreases slightly and then increases slightly and finally stabilizes, while the cooling capacity increases. The cycle rate gradually increases and then stabilizes. The level of the cycle rate mainly affects the mass flow rate and outlet dryness in the tube, and also affects the drop in the resistance of the tube; when the ambient temperature is 18 ℃ ~ 25 ℃, the heat transfer coefficient first increases and then decreases, but

overall the heat transfer coefficient fluctuates. It is not very large in the range of 10%; as the thickness of the frost layer increases, the cooling capacity first increases and then decreases, and the heat transfer coefficient is opposite. This has a lot to do with increasing the surface roughness of the frost layer on the initial frost layer, thinning the boundary layer, and the excessively thick frost layer increasing the thermal resistance and obstructing the air-side flow.

References

[1] Stoecker W F.Industrial refrigeration handbook[M]. New York: McGraw-Hill, 1998.
[2] Kondepudi S N,O'Neal D L.Effect of frost growth on the performance of louvered finned tube heat exchangers [J]. International Journal of Refrigeration,1989,12(3): 151-158.
[3] Ye H Y, Lee K S. Refrigerant circuitry design of fin-and-tube condenser based on entropy generation minimization[J]. international journal of refrigeration, 2012, 35(5): 1430-1438.
[4] Ye H Y, Lee K S. Performance prediction of a fin-and-tube heat exchanger considering air-flow reduction due to the frost accumulation[J]. International Journal of Heat and Mass Transfer, 2013, 67: 225-233.
[5] Chen Jian, Wan Jikang, Cao Guangrong, et al. Experimental study on heat transfer coefficient of air cooler under frosting condition [J]. Journal of Refrigeration ,2001,20(2):16-19.
[6] Zang Runqing, Liu Qi, Li Xing. Comparative study on the performance of direct expansion fluid supply and gravity recirculation liquid supply refrigeration system [C].2011 Annual Conference of China Refrigeration Society. 2011.
[7] Wang C C,Lee W S,Sheu W J,et al.A comparison of the airside performance of the fin-and-tube heat exchangers in wet conditions;with and without hydrophilic coating[J]. Applied Thermal Engineering,2002,22(3):267-278.
[8] Zhang Chun Road. Simulation Principle and Technology of Refrigeration and Air Conditioning System [M].Chemical Industry Press ,2012.
[9] Yu Bing, Feng Yinshan, que Xiongcai, et al. Theoretical Analysis and Improvement of Fin Efficiency Formula in Frosting Condition of Heat Exchanger [J]. Journal of Engineering Thermophysics ,1999,20(1):74-77.
[10] Hsu Sung-wah. Calculation of Thermal Physical Properties of Wet Air [J]. Journal of Suzhou University: Natural Science Edition ,1999(3):54-59.
[11] Yuan Zengzhi. Thermodynamic parameters of wet air and their numerical calculation during cooling [J]. Fluid machinery ,1991(1):58-60.
[12] Wu YZ, Zhu RQ, Cao Xiaolin, et al. Refrigeration Principles and Equipment [M]. Xi'an Jiaotong University Press ,2010.

Development of accurate and widely applicable compressor performance maps

Cheng-Yi Lee[1], Tao Cao[1], Yunho Hwang[1]*, Reinhard Radermacher and Scott Shaffer[2]

[1]Center for Environmental Energy Engineering, Department of Mechanical Engineering, University of Maryland, College Park, MD 20742, United States
[2]GE Appliances, Appliance Park, Louisville, KY 40225-0001, United States

* Corresponding Author. Tel: (301) 405-5247, email: yhhwang@umd.edu

Abstract. The compressor performance map provides critical information, including mass flow rate, power consumption, capacities, and operating current. The map can be applied to an alternative efficiency determination method (AEDM) for HVAC&R products. However, most of the existing maps have a low prediction accuracy under low ambient temperature heating modes, as such conditions were not thoroughly considered when maps were generated. This paper provides a comprehensive study of a fixed speed compressor performance under a wide range of operating conditions, including the AHRI 210/240 standard conditions. The test matrix was developed based on realistic heat pump system operating conditions. We then quantified the existing performance map's prediction inaccuracy, particularly under the low ambient temperature heating modes. Driven by the significant deviation at lower ambient temperature conditions, a comprehensive performance map was generated based on the test results. We also found that one polynomial equation could simultaneously fulfill the needs for predicting the compressor performance under both the cooling and heating modes. Furthermore, the accuracy has improved significantly from 63% to 2% regarding the relative maximum prediction errors over existing performance maps. It suggests that to increase prediction accuracy when using the AEDM, one must provide a suitable superheat model in the capacity simulation of the heat pump systems with the fixed speed compressor.

Keywords: alternative efficiency determination method, compressor performance map, fixed speed compressor

1. Introduction

The Code of Federal Regulations part 429.70 (DOE, 2014) illustrates how we can apply alternative methods to determine energy efficiency and energy use instead of sending the product to a qualified psychrometric test room to conduct experimental tests. VapCyc (Huang et al., 2021) provides an excellent vapor compression cycle's steady-state performance simulation capability. It can be a handy tool along with a heat exchanger's steady-state performance simulation tool, CoilDesigner (Jiang et al., 2006). Utilizing these two tools after validating their accuracy with the experimental laboratory test results can serve as an alternative method to evaluate the product's energy efficiency and energy use. Industry can save lots of time and cost for experimental works by adopting the alternative efficiency determination method (AEDM). However, some inconsistencies are found when the experimental results of the heat pump systems are compared with the simulation results. When it comes to low ambient temperature heating conditions, the deviation starts to increase. Therefore, the AEDM accuracy becomes questionable. In the previous studies, Kegel (2017) and Le Lostec (2014) pointed out that the heat pump performance only relies on the manufacturer's performance curve and sometimes not reliable, typically under low ambient temperature. Hence, this study investigated the effect of ambient temperature on the condensing temperature, evaporating temperature, and suction superheat. A new comprehensive performance map is proposed to improve the model accuracy quantitatively in terms of mass flow rate, power consumption, and volumetric efficiency. Then the improved compressor performance map is applied to the AEDM to see the accuracy impoprvement after the model enhancement.

2. Test facility

2.1 Test facility design

The compressor test stand was developed based on a hot gas bypass test stand similar to Vetsch et al.'s work (2016). The main specifications are as follows. The capacity of compressor testing ranges from 0.58 kW to 7.3 kW. The condensing temperature ranges from 25 to 80 °C, and the evaporating temperature ranges from -20 to 15 °C. The isentropic and volumetric efficiencies are measured with an uncertainty below 3%. Figure 1 shows the schematic of the compressor performance test stand. The mass flow meter measuring the total refrigerant flow was placed right after the discharge port. After the mass flow rate meter, the electronic expansion valve (EEV), together with parallel needle valves, control the high side condensing pressure. To achieve more precise control of high side pressure, 0.37 cv needle valves and 0.09 cv needle valves were used parallel to the target pressure value. The electronic expansion valve (EEV) was used for fine-tuning by the PID control. After the first expansion section, the refrigerant flow splits into two paths. One directly bypasses the mass flow meter and enters the second expansion section, and the other passes through the heat exchanger and a receiver, which collects the excess refrigerant in the system, and a sub-cooler to ensure it is in the liquid phase. The heat exchanger provides the cooling capacity by condensing the vapor refrigerant and stabilizes the intermittent pressure. Two RTDs and pressure transducers were placed before the heat exchanger and after the sub-cooler to monitor the pressure difference and temperature change. To better understand the mass flow split percentage between two paths, the mass flow meter before the EEV2 section plays a role in providing the system's refrigerant flow information. EEV2 section and EEV3 section serve jointly for a low side pressure control in the system. The PID control of EEV2 aims at controlling the suction pressure, and the EEV3 controls the suction superheat. Some auxiliary components help to improve the smooth operating of the system. The sight glass helps to check if moisture was in the system. Two shredder valves help vacuum the system and charge or discharge the refrigerant. Figure 2 shows the picture of the entire setup after the construction. The entire test facility was installed inside the environmental chamber, providing temperatures from 10 °C ~ 40 °C. To expand the testing temperature range, we installed a local cooling device to supply a stable air temperature. The target airflow rate was the same as the measured one in the actual air conditioning system. The compressor was installed in a 0.3 m x 0.3 m x 0.3 m casing and connected with the 0.3 m diameter insulated flexible duct to minimize the heat loss. An R-134a vapor compression system controlled the air inlet temperature with a 1,000 W cooling capacity. A 600W heater assisted in stabilizing the air temperature through the heater's PID control. The air inlet temperature was maintained within 0.3 °C variations. Two thermocouple sensors monitored the compressor's temperature. It should be noted that the air velocity was lower than 0.2 m/s in this study.

IOP Conf. Series: Materials Science and Engineering 1180 (2021) 012041 doi:10.1088/1757-899X/1180/1/012041

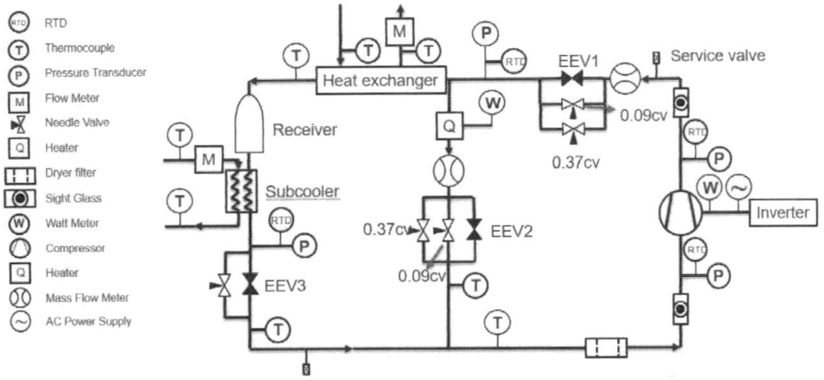

Figure 1: Test facility schematic diagram

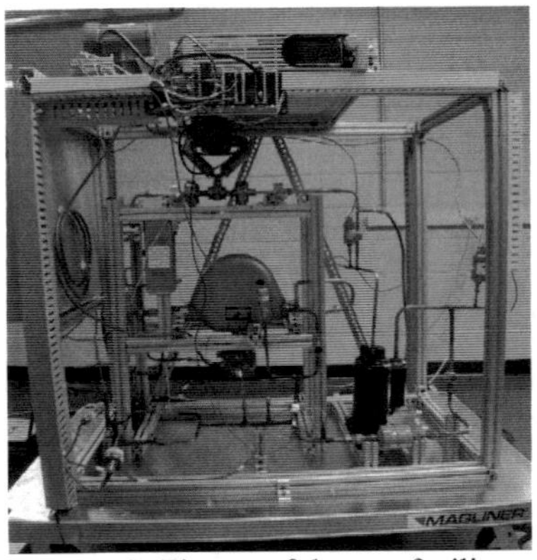

Figure 2: Picture of the test facility

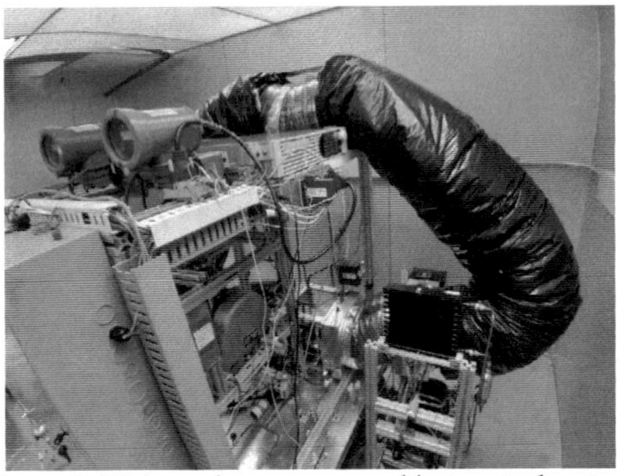

Figure 3: Picture of compressor ambient control system

2.2 Instruments and uncertainties

Table 1 lists the instrument used in our compressor test facility. The compressor discharge and suction temperatures were measured by using the RTD sensor. In addition, T-type thermocouple sensors were placed over the entire system to monitor the system's overall system behavior, including the intermittent system temperature, compressor surface temperature, and local environment temperature. The pressure transducers were in 0 ~ 3,447 kPa measurement range, except the discharge one was 0 ~ 6,894 kPa for the sake of safety. There were two mass flow meters used. One was for the total mass flow rate, located right after the compressor discharge site, and the other was for measuring the hot gas bypass mass percentage. Both of the mass flow meters had 0.11% reading accuracy. For compressor's power consumption, the wattmeter with 0.2% reading accuracy was adopted. All the test instruments were calibrated before installation to the test facility. The uncertainty analysis includes the systematic error as shown in Table 1 and the random error. Before acquiring the test data, the system was operated for one or two hours to reach the steady state and then data was recorded every second for at least half an hour. The random error was then evaluated to calculate the standard deviation of the sample mean. The absolute uncertainty was then calculated by the root sum of squares (RSS) of the systematic error and random error (Mandel, 2012).

Table 1: Instruments and Uncertainties

Instrument	Manufacturer / Model	Range	Systematic Error
RTD	Omega 1/10 DIN	-100 °C ~ 400 °C	± 1/10 * (0.3 + 0.005 * t) °C
Thermocouple	Omega T Type	-200 °C ~ 200 °C	± 0.5 °C
Pressure transducer	Setra 280E	0 ~ 3,447 kPa	± 0.11% FS
Pressure transducer	Setra 280E	0 ~ 6,895 kPa	± 0.11% FS
Mass flow meter	Micro Motion 2700+CMFS025M	0 ~ 310 g/s	± 0.35% of reading
Watt meter	Ohio Semitronics GH-020D	0 ~ 4,000 W	± 0.2% of reading
Power supply	Preen AFV-P-2500	Voltage 0~310 V Frequency 50/60	± 0.07% FS

3. Test matrix

3.1 Heating mode

3.1.1 Heating mode based on experimental data

The condensing temperature, evaporating temperature, ambient dry-bulb, wet-bulb temperature, and suction superheat were used as inputs to the VapCyc to calculate the high-side and low-side pressures of the heat pump system operating under realistic conditions. From the AHRI 210/240 standard (2017), we know the humidity ratio of the ambient air for four test conditions. Then we took a second-order polynomial curve fitting to predict each condition's wet-bulb temperature, as shown in Figure 4. A correction factor 1.1 was necessary for the VapCyc results to match the experimental laboratory data (Alabdulkarem, 2013). Based on the VapCyc simulation results, the relation between ambient temperature and the evaporating temperature was depicted as a yellow line in Figure 4. Then the red points are chosen as the test matrix for the experiment. The entire test matrix is shown in Table 2 with varied superheat for each test condition.

3.1.2 Heating mode based on fixed superheat

In the previous section, the heat pump unit's superheat values were assumed to be the same as the previous experimental work by Alabdulkarem et al. (2013). However, we do not know the exact superheat value of the current system. Therefore, the 11K superheat was used in the following test matrix to compare with the original compressor performance map, which complies with the AHRI standard 540. The entire test matrix for the fixed superheat is shown in Table 3.

ID: 156, Page 5

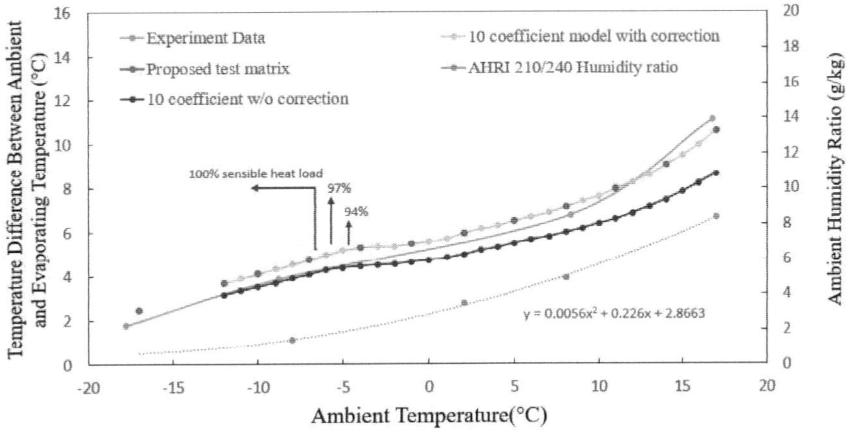

Figure 4: Temperature difference between ambient and evaporating temperature and ambient humidity ratio

Table 2: Heating mode test conditions with experimental superheat

Test#	Evaporating Temperature (°C)	Condensing Temperature (°C)	Superheat (K)	Outdoor Air Temperature(°C)
1	6.4	36.2	10.2	16.7
2	4.9	35.9	8.2	14.0
3	3.0	35.0	6.3	11.0
4	0.8	34.5	4.7	8.3
5	-1.5	33.5	3.3	5.0
6	-4.0	33.5	2.1	1.7
7	-6.5	32.0	1.0	-1.0
8	-9.2	31.1	0.5	-4.0
9	-11.8	30.2	0.5	-7.0
10	-12.6	30.0	0.5	-8.3
11	-14.1	29.4	0.5	-10.0
12	-15.7	28.8	0.5	-12.0

Table 3: Heating mode test conditions at fixed superheat

Test#	Evaporating Temperature (°C)	Condensing Temperature (°C)	Superheat (K)	Outdoor Air Temperature(°C)
1	5.6	36.8	11.0	16.7
2	2.8	35.5	11.0	14.0
3	-0.1	34.6	11.0	11.0
4	-3.1	33.6	11.0	8.3
5	-6.0	32.6	11.0	5.0
6	-9.0	31.6	11.0	1.7
7	-12.0	30.5	11.0	-1.0
8	-15.0	29.4	11.0	-4.0
9	-18.0	28.1	11.0	-7.0
10	-18.5	28.1	11.0	-8.3

3.2 Cooling mode

3.2.1 Cooling mode based on experimental data

For the cooling mode conditions, the superheat from experiment data by Alabdulkarem et al. (2013) was considered again here, and the correction factor 1.05 was applied to fit the experimental data. The yellow line in Figure 5 represents the VapCyc simulation results. Also, the AHRI standard 210/240 provides us information on the ambient air humidity ratio. Therefore, the red circle points in Figure 6 were chosen as test points. Figure 5 shows that the simulation results deviate from the experiment data by Alabdulkarem et al. (2013) because the experimental data was based on a split unitary system while the current study was for the packaged terminal heat pump unit. Otherwise, the trends of both systems are similar. Furthermore, the superheat values were consistent at 2 K while the ambient temperature varied.

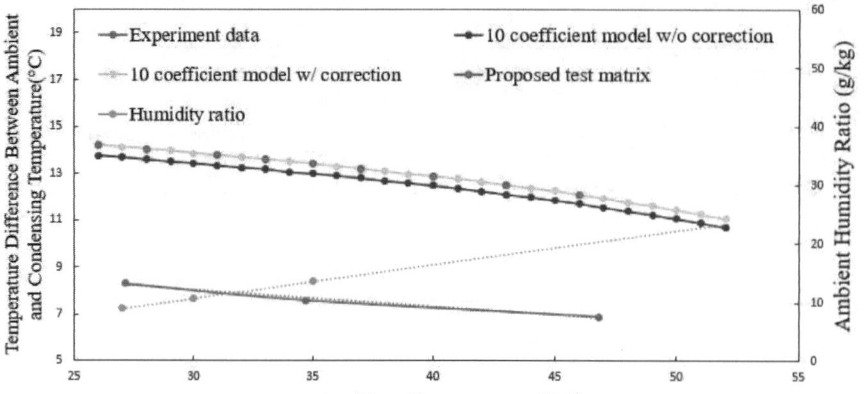

Figure 5: Temperature difference between the condensing and ambient temperature, and ambient humidity ratio

3.2.2 Cooling mode with fixed superheat

In the cooling mode, we considered the superheat as an additional test variable, while the original performance map defined the 11 K of superheat by default. Table 4 denotes the cooling mode test conditions with two superheat values (2 K and 11 K). The test matrix for experimental data superheat was combined with the fixed superheat conditions in Table 4.

Table 4: Cooling mode test conditions with experimental and fixed superheat

Test #	Evaporating Temperature (°C)	Condensing Temperature (°C)	Superheat (Experimental / Fixed) (K)	Outdoor Air Temperature(°C)
1	13.3	58.1	2 / 11	46.0
2	13.1	55.5	2 / 11	43.0
3	12.8	52.9	2 / 11	40.0
4	12.6	50.2	2 / 11	37.0
5	12.3	48.4	2 / 11	35.0
6	12.2	46.6	2 / 11	33.0
7	12.0	44.8	2 / 11	30.6
8	11.8	42.0	2 / 11	27.8
9	11.6	40.2	2 / 11	26.0

4. Test results

4.1 Heating mode results

4.1.1 Heating mode results with experimental superheat data

In Table 2, we selected 12 data from the previous VapCyc simulation results. Figure 6a shows the comparison of experiment mass flow rate results and compressor performance map output. As the ambient temperature decreases, the relative deviation starts to increase from 8.5% to 40%, suggesting that the performance prediction against lower ambient temperature underestimates the actual heating capacity. From Figure 6b, the power consumption's relative deviation increases from 5% to 9.8%, and it overestimated the actual power consumption under low ambient temperature conditions. Figure 7 shows that both isentropic and volumetric efficiencies decrease as the ambient temperature decreases, the isentropic efficiency dropped from 65.7% to 58.2%, and the volumetric efficiency from 90.3% to 85.1%. The larger temperature difference between indoor and ambient temperature causes a larger pressure ratio between the evaporator and the condenser. Therefore, the isentropic efficiency decreases when it needs more power to overcome the larger pressure ratio, and the volumetric efficiency decreases due to increased internal leakage and a lower suction density.

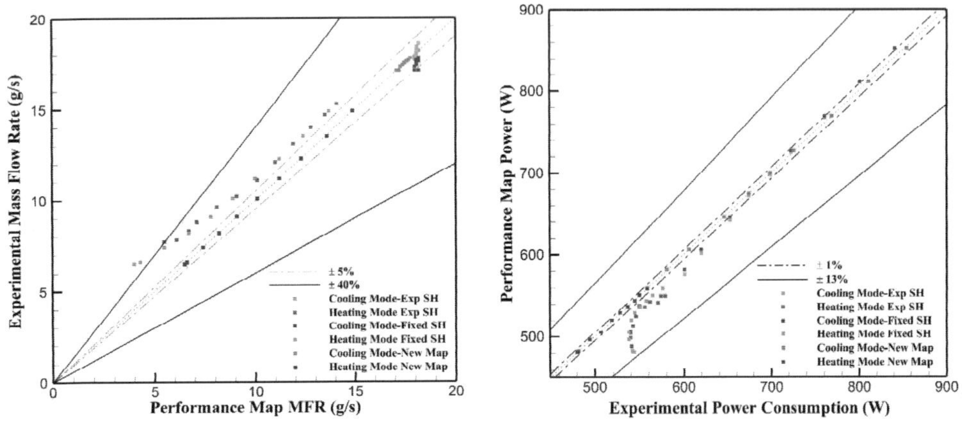

Figure 6: Parity plot of mass flow rate and power consumption for cooling mode and heating mode

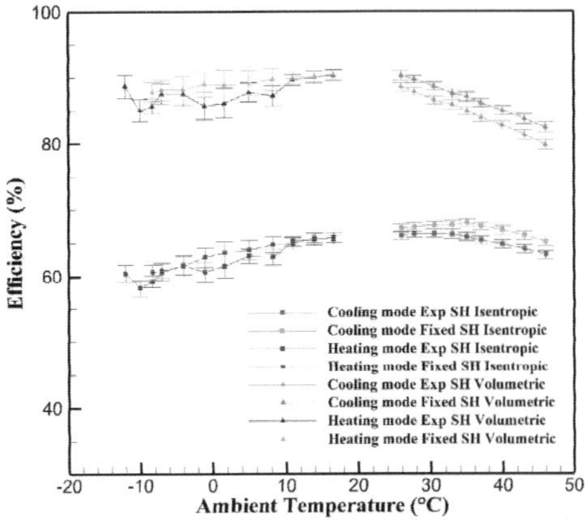

Figure 7: Isentropic efficiency and volumetric efficiency for cooling mode and heating mode

4.1.2 Heating mode results with fixed superheat

The relative deviation in fixed superheat testing results shows an even more significant value than experimental superheat data, 62.5% in mass flow rate and 11% in power consumption. The results reconfirm that the current compressor performance map did not cover the low ambient temperature conditions adequately. Moreover, most of the existing maps have a low prediction accuracy under low ambient temperature heating mode because the training data points did not include the low ambient temperature conditions. This study shows that if extrapolating the original map, the deviation could be tremendously significant and not appropriate to predict the performance in such region.

4.2 Cooling mode results

The mass flow rate with the experimental superheat is more consistent toward the performance map, which is within 2.2%. On the other hand, the fixed superheat case is within 5%, and both the isentropic and volumetric efficiencies are higher in the fixed superheat case. The power consumption has a similar accuracy, in the experimental superheat case was within 4.2%, and in the fixed superheat case was within 3.2%. Either the mass flow rate or power consumption is following the performance map values closely, indicating that the current performance map is suitable for the cooling mode conditions but not for heating mode at the low ambient temperature. The isentropic and volumetric efficiencies have a similar trend as the heat pumping mode. The temperature difference between the indoor and ambient temperature decreases means a reduced pressure ratio. As the pressure ratio decreases, the isentropic and volumetric efficiency increases by around 3% and 8%, respectively.

4.3 Repeatability tests results

So far, the characteristics of the compressor have been tested and described in the previous section. Repeatability tests were conducted with two more compressors if there is any difference between the same model products. Three test conditions were selected from Table 2. Results show that the mass flow rate deviation was within 2%, and the power consumption deviation was within 1%. Therefore, the first compressor results were taken to develop the new map in the next section.

5. Generated 10-coefficient model and validation

5.1 Combined performance map

With the fixed superheat in cooling mode and heating mode, the 19 points data can generate a high-quality, comprehensive performance map applicable to the VapCyc software. These 19 data points, including mass flow rate, power consumption, current, and capacity, were used to generate the third-order polynomial equation by Matlab regression tool. The experiment-based performance map was then applied to the VapCyc to see the deviation between simulation results and experiment results. The comparisons are shown in Figure 6a and Figure 6b. As shown, the relative deviations of the new model were improved from 62.5% to 0.7% for the mass flow rate and 13.1% to 0.4% for the power consumption in heating mode. The mass flow rate in cooling mode was improved from 1.2% to 1% and 3.2% to 0.2% in power consumption. The newly generated map can accurately predict the compressor performance under wide ambient temperature, -20°C to 50°C. The third-order polynomial equation, which comes from standard AHRI 540, is described below, and the coefficient of the manufacturer's map and newly generated map are list in Table 5.

$$X = C_1 + C_2 \cdot t_S + C_3 \cdot t_D + C_4 \cdot t_S^2 + C_5 \cdot (t_S \cdot t_D) + C_6 \cdot t_D^2 + C_7 \cdot t_S^3 + C_8 \cdot (t_S^2 \cdot t_D) + C_9 \cdot (t_S \cdot t_D^2) + C_{10} \cdot t_D^3$$

Where:

X = Power Input, or Refrigerant Mass Flow Rate

t_D = Discharge Dew Point Temperature, °C

t_S = Suction Dew Point Temperature, °C

Table 5: Coefficients of Manufacturer's Map and Newly Generated Map

Coefficients	Manufacturer's Power Map (W)	Manufacturer's Mass Flow Rate Map (g/s)	New Generated Power Map (W)	New Generated Mass Flow Rate Map (g/s)
C_1	5.2021E+02	1.8187E+01	5.2020E+02	1.4850E+02
C_2	5.3703E-01	2.5048E-01	-1.1830E+01	3.5180E+00
C_3	-3.9258E+00	-3.2905E-01	-3.5960E+00	-5.6360E+00
C_4	1.2128E-01	-1.8153E-03	2.1350E-02	1.6610E-01
C_5	-1.2666E-01	3.9346E-04	2.0930E-01	-1.2530E-01
C_6	7.2308E-02	2.4886E-03	2.7810E-02	8.6030E-02
C_7	-3.5722E-04	3.1050E-05	7.1760E-05	6.4650E-04
C_8	-9.4307E-04	2.9808E-06	-7.4150E-04	-2.1050E-03
C_9	1.1574E-03	-4.7693E-06	-6.7700E-04	1.1140E-03
C_{10}	-2.6292E-04	-6.3590E-06	1.2330E-04	-3.8550E-04

5.2 System performance comparison using VapCyc

The new performance map has been generated and compared with the old one, as shown in Table 6. Under the same superheat, the new performance map could approach the experimental data closer without using any correction factor. The correction factors of 10% and 5% were needed when using the old map for heating and cooling capacity. However, the superheat value was found by a try-and-error approach to match the capacity near the experimental value. Therefore, knowing the right superheat value for the system would be critical for the AEDM. In this study, the compressor performance map generated encompassing the low ambient temperature conditions would deliver much better prediction accuracy in application.

Table 6: Results for performance prediction

Mode	Experimental Capacity [W]	Super heat [K]	Capacity, MFR, and Power by Old Map [W]/ [g/s] / [W]	Capacity, MFR, and Power by Old Map with Correction Factor [W]/ [g/s] / [W]	Capacity, MFR, and Power by New Map [W]/ [g/s] / [W]
Heating	2,468	12.3	2,324 / 9.7 / 548	2,484 (1.1) / 10.5 / 554	2,476 / 10.5 / 551
Cooling	2,852	0.1	2,761 / 18.5 / 701	2,855 (1.05) / 19.2 / 711	2,819 / 18.7 / 700

6. Conclusions

The heat pump system has become the most critical heating method for the world to achieve the carbon-neutrality. However, the critical component, the compressor, has an unsatisfied prediction capability under low ambient temperature when the conventional compressor performance map approach is used. To better understand the compressor's realistic performance, a compressor test stand has been built, validated, and assembled with the local cooling device. The test capacity ranges from 580 W to 7,300 W. The compressor ambient temperature ranges from -20 °C to 60 °C. From the test results, we confirmed that the original compressor performance map has inaccurate results as compared to the experimental data when the ambient compressor temperature is low. In other words, the prediction of the heating performance using the system simulation tool with conventional compressor performance map would only generate a biased result as compared to the system test data. This study has constructed a new comprehensive performance model covering a wide range of operating conditions. The newly generated performance map improves the relative error of mass flow rate prediction compared to the experimental data, from 62.5% to 0.7%. On the other hand, the superheat in the system would be critical to accurately predicting the capacity in the system model simulation. Moreover, the difference in mass flow rate between experimental superheat and fixed superheat can vary from 3.3% to 6%.

Refernces

[1] Alabdulkarem, A., Y. Hwang, R. Radermacher. (2013). System Drop-In Tests of Refrigerants R-32, D2Y-60, and L-41a in Air Source Heat Pump. Air-Conditioning, Heating, and Refrigeration Institute (AHRI) Low-GWP Alternative Evaluation Program.
[2] DOE. (2014). 10 CFR Parts 429 and 431.
[3] Huang, R., V. Aute, and J. Ling. (2021). A tripartite graph based methodology for steady-state solution of generalized multi-mode vapor compression systems. Applied Thermal Engineering.
[4] Jiang, H., V. Aute, V., and R. Radermacher. (2006). CoilDesigner: a general-purpose simulation and design tool for air-to-refrigerant heat exchangers. International Journal of Refrigeration.
[5] Kegel, M. e. (2017, May 15). Performance Testing of Cold Climate Air Source Heat Pumps. Proceedings of the 12th IEA Heat Pump Conference, 18.
[6] Le Lostec, B. N. (2014). On-site Performance of Air Source Heat Pumps. 11th International Energy Agency Heat Pump Conference, (p. paper #O2.4.3.). Montreal, Canada.
[7] Mandel, J. (2012). The statistical analysis of experimental data. Courier Corporation.
[8] AHRI. (2015). AHRI Standard 540 for Performance Rating of Positive Displacement Refrigerant Compressors and Compressor Units. Virginia., Arlington: Air-Conditioning, Heating, and Refrigeration Institute.
[9] AHRI. (2017). AHRI Standard 210/240 for Performance Rating of Unitary Air-conditioning & Airsource Heat Pump Equipment. Virginia., Arlington: Air-Conditioning, Heating & Refrigeration Institute.
[10] Vetsch, B., Feichter, G., Bachmann, A., and Bertsch, S. S. (2016). Design And Realization Of An Automated Test-Stand For Variable Capacity Household Compressors. International Compressor Engineering Conference, (p. 2407).

Acknowledgments

We gratefully acknowledge the support of the Center for Environmental Energy Engineering (CEEE) at the University of Maryland and GE Appliances.

Assessing the performance limits of a variable-speed residential heat pump system

John K. Brehm[1], David Augustine[2], Davide Ziviani[1], Eckhard A. Groll[1]

[1]Ray W. Herrick Laboratories, School of Mechanical Engineering, Purdue University, West Lafayette, IN, USA

[2]QM Power, Inc., Kansas City, MO, USA

brehm3@purdue.edu; daugustine@qmpower.com; dziviani@purdue.edu; groll@purdue.edu

Abstract. Most of the ducted split-type heat pump systems in the world feature fixed-speed compressors and fans. To meet forthcoming minimum energy rating requirements, reduce operational costs of the heat pump, and increase environmental sustainability, the efficiency of heat pump systems must be improved. Variable-speed equipment offers significant advantages for load modulation and has the ability to increase the seasonal performance significantly. Additionally, novel electrical motor technologies, such as permanent magnet (PM) motors can reduce the power consumption of the drives by up to 25-55% compared to the widely used permanent split capacitor (PSC) motor.

In this study, a ducted fixed-speed heat pump system with a cooling capacity of 17.6 kW and a seasonal coefficient of performance (SCOP) rating of 4.11 is analyzed to quantify the theoretical performance limits that could be obtained by increasing the motor efficiency of the condenser fan, the evaporator fan and the compressor. Furthermore, the performance increase by replacing the fixed-speed components with variable-speed equipment is evaluated. A detailed charge-sensitive system simulation model that was previously developed by the authors and validated using experimental data obtained with the same unit was employed to carry out parametric studies to evaluate the impact of variable-speed compressor and indoor/outdoor fans on the SCOP rating.

The study showed a 0.9% improvement of SCOP with every percentage increase of compressor motor efficiency. The dependence of the evaporator fan and condenser fan motor efficiency on SCOP was lower, but considering the generally low motor efficiencies of these components, a noticeable efficiency gain can be achieved. In variable-speed equipment, the motor efficiency of the component at minimum operational point was the significant factor on SCOP. The motor efficiency at intermediate and maximum speed had only marginal effects on the SCOP. This showed the importance of equipment with high motor efficiencies at low speeds for ducted split-system heat pumps. The change of the fixed-speed system to a fully variable-speed heat pump resulted in a high gain of SCOP, proving the superiority of a variable-speed system. Overall, the model predicted an SCOP gain of 35% when changing the system to a fully variable system with state-of-the-art high efficient motors.

Keywords: Optimization, Variable-speed Compressor, Fan Motor Efficiency, Permanent Magnet Motor

1. Introduction

During the last two decades, the need for more efficient heat pump systems increased significantly due to concerns on energy consumption and awareness of environmental issues including global warming. Moreover, equipment manufacturers are facing challenges with the forthcoming more stringent minimum energy rating requirements for residential heat pumps that will be imposed by the U.S. Department of Energy by 2023. Due to the low cost, most ducted split-type heat pump systems in the U.S. feature fixed-speed components with electric motors in the lower efficiency range [1]. Variable-speed and multi-speed systems can significantly improve the systems seasonal performance due to their inherent ability to adjust the capacity to the needed loads. Fixed-speed systems generally run in an overdesigned condition because the heat pump is chosen for the highest use case, which is only reached a few days a year. Most of the time, a much lower heating or cooling capacity is needed and the unit cycles on-off, which results in inefficiencies. Variable-speed systems can reduce their capacity to the required needs, which leads to an increased efficiency due to less cycling losses.

The use of higher efficiency motors can directly reduce the consumed power consumption due to less motor losses. Especially the compressor, which is the most power consuming component in a heat pump, needs to operate with a high efficiency motor to reduce losses and increase the SCOP. The evaporator fan motor and condenser fan motor have a less significance on the systems efficiency and therefore, manufacturers often choose a less efficient motor. A change of these motors to state-of-the-art motors can significantly reduce the components power consumption. The Permanent magnet (PM) motor is a novel electrical motor technology which is able to reach very high motor efficiencies. PM motors are currently available with efficiencies of over 90% for HVAC equipment [1]. This is a big efficiency gain over current systems, which often use permanent split capacitor (PSC) or electronically commutated (EC) motors, typically with motor efficiencies in the range of 35% to 80% [1].

This study investigates the potential performance improvements of a heat pump system when changing fixed-speed components to variable-speed equipment and increasing the motor efficiency of the compressor, evaporator fan and condenser fan motor.

2. System Description

2.1. Baseline Unit

The unit considered in this study is a commercially available ducted fixed-speed heat pump system. The unit has a cooling capacity of 17.6 kW and is rated at an SCOP of 4.11 using the AHRI Standard 210/240. The heat pump is running with the refrigerant R-410A and uses a fixed-speed scroll compressor, which uses a PSC-motor. Both the evaporator fan and the condenser fan run in single speed. The evaporator fan motor has a power draw of 407 W and runs with an electronically commutated motor (ECM). The ECM is estimated to have a motor efficiency of 0.75. The condenser fan uses a PSC motor and has a power draw of 286 W. The motor efficiency of the condenser fan is assumed to be 0.65.

2.2. Modified Unit

To assess the effect of a variable-speed system, components of the baseline unit are replaced with variable-speed components. As a compressor, the latest commercially available variable-speed R-410A scroll compressor is chosen. This compressor allows to assess the performance limits of current-day high-efficiency compressors. The cooling capacity of the compressor is in a range of 6 kW to 20.9 kW. To match with the baseline unit, the compressor is modeled for a rated capacity of 17.6 kW, which translates to a rotational speed of 4500 rpm. The motor efficiency is estimated to be at 0.88 at the rated speed of 4500 rpm and 0.85 at lower speeds.

Depending on the configuration of the simulation run, the evaporator fan and condenser fan are also changed to variable-speed. In variable-speed mode, the motors are assumed to have the same motor efficiency as the fixed-speed motors. This translates to an evaporator fan motor efficiency of 0.75 at all speeds and condenser fan motor fan motor efficiency of 0.65 at all speeds. While analyzing the unit with higher motor efficiencies, the motor efficiencies are adjusted accordingly.

3. Methods

3.1. Efficiency Evaluation
In this work, the heat pump is considered to operate in cooling mode. To evaluate the efficiency of the unit, the seasonal energy efficiency ratio (SEER) defined according to the AHRI Standard 210/240 [2] is used. To match with SI-units, the SEER is converted to the seasonal coefficient of performance (SCOP). The standard tries to replicate the use of the heat pump over a cooling season and develop the average COP of the unit within the season. This is a reasonable indicator of the system's performance as it not only measures the efficiency in one operational point but in many points.

The standard differentiates between the heat pump running with a fixed-speed compressor or a variable-speed compressor. Depending on the mode, the unit is tested at a set of different temperature points which are reported in Table 1 and Table 2. In the test for fixed-speed systems, the compressor and fans run with maximum speed. In variable-speed systems, the compressor and fans run in either maximum speed, intermediate speed or minimum speed, depending on the test condition.

With temperature bins which display the expected temperature within a year of operation and the information of cooling capacity and power consumption at the different operation points, the SCOP is calculated.

Table 1: Single-speed system test points [2]

Test Name	Air Entering Outdoor Unit in °C (Dry Bulb / Wet Bulb)	Air Entering Indoor Unit in °C (Dry Bulb / Wet Bulb)	Compressor Speed	Indoor Airflow
A_{full}	35.0 / 23.9	26.7 / 19.4	Max.	Max.
B_{full}	27.8 / 18.3	26.7 / 19.4	Max.	Max.
C_{full}	27.8 / 14.4	26.7 / 13.9	Max.	Max.
D_{full}	27.8 / 14.4	26.7 / 13.9	Max.	Max.

Table 2: Variable-speed system test points [2]

Test Name	Air Entering Outdoor Unit in °C (Dry Bulb / Wet Bulb)	Air Entering Indoor Unit in °C (Dry Bulb / Wet Bulb)	Compressor Speed	Indoor Airflow
A_{full}	35.0 / 23.9	26.7 / 19.4	Max.	Max.
B_{full}	27.8 / 18.3	26.7 / 19.4	Max.	Max.
B_{low}	27.8 / 18.3	26.7 / 19.4	Min.	Min.
E_{int}	30.6 / 20.6	26.7 / 19.4	Int.	Int.
F_{low}	19.4 / 11.9	26.7 / 19.4	Min.	Min.
G_{low}	19.4 / 14.4	26.7 / 13.9	Min.	Min.
I_{low}	19.4 / 14.4	26.7 / 13.9	Min.	Min.

3.2. Simulation Model

3.2.1. ACHP+ Description
The numerical analyses are conducted by employing the simulation framework "ACHP+" which is based on the Python programming language. Such modeling framework is the result of significant efforts by the authors' research team simulating various vapor compression systems during the last two decades. In a nutshell, ACHP+ is a detailed charged sensitive model, which assesses the influence of heat exchanger sizing, compressor efficiencies, expansion work recovery devices, working fluids, etc. on the

system's performance. The model has been experimentally validated by various studies, *e.g.*, by Bahman et al. [3], proving a high accuracy of the model.

3.2.2. Compressor Implementation

The fixed-speed commercially available scroll compressor is integrated in the ACHP+ model with a 10-coefficient polynomial map developed by AHRI Standard 540 [4]. The coefficients for power input and mass flow rate are provided by the compressor manufacturer [5]. With the input of suction dew point temperature and discharge dew point temperature, the desired value for power input and mass flow rate at the modelled operation point is calculated.

For modelling the variable-speed commercially available scroll compressor, a modified 10-coefficient polynomial map is used. The compressor manufacturer provides a 20-coefficient polynomial map to additionally account for the rotational speed of the compressor [6]. With the input of suction dew point temperature, discharge dew point temperature and rotational speed, the power input and mass flow rate of the compressor can be calculated.

To model the effect of a changed compressor motor efficiency, the calculated power is multiplied with the ratio of initial motor efficiency to new motor efficiency:

$$P_{comp,new} = \frac{\eta_{init}}{\eta_{new}} * P_{comp,init} \tag{1}$$

3.2.3. Evaporator Fan Implementation

In the configuration with the evaporator fan in single-speed, the evaporator fan power at a given volume flow rate is taken from experimental data. The data is used from an analysis performed by Salts et al. [7] on the same unit.

For configurations with the evaporator in variable-speed mode, the fan power for different external static pressure dp and volume flow rate $\dot{V}$ needs to be calculated. The following equation is implemented in the model:

$$P_{evap} = \frac{dp * \dot{V}}{\eta_f * \eta_{m,evap}} \tag{2}$$

The external static pressure dp is defined by the AHRI Standard 210/240 for the full load airflow condition [2]. At lower flow rates the minimum static pressure that needs to be reached is calculated according to the standard with

$$dp = dp_{full} * \left(\frac{\dot{V}}{\dot{V}_{full}}\right)^2 \tag{3}$$

and depends on the Volume flow rate.

The evaporator fan motor efficiency η_m is set depending on the simulation run. For runs in the baseline condition the motor efficiency is set at 0.75 at all speeds. For simulations assessing the impact of a changed evaporator fan motor efficiency, the efficiency is changed, depending on the speed mode.

The fan efficiency η_f is calculated with data from the experimental analysis by Salts et al. [7] and Equation 4.2. In particular, η_f is assumed to be constant at different speeds and is found to be 0.1759.

3.2.4. Condenser Fan Implementation

The condenser fan power consumption at rated speed is calculated by dividing the rated fan power HP_{rated} by the evaporator fan motor efficiency $\eta_{m,cond}$:

$$P_{cond} = \frac{HP_{rated}}{\eta_{m,cond}} \tag{4}$$

In variable-speed mode, the fan motor efficiency at different rotational speed is calculated using the fan laws:

$$\dot{V}_2 = \dot{V}_1 * \left(\frac{RPM_2}{RPM_1}\right) \tag{5}$$

$$HP_2 = HP_1 * \left(\frac{RPM_2}{RPM_1}\right)^3 \tag{6}$$

Depending on the condenser motor efficiency at certain speed, the power input is calculated with Equation (*4*).

4. Validation

4.1. System Rating

The ACHP+ model has been validated in several experimental data. Bahman et al. [3] modelled the same heat pump unit with a different variable-speed compressor and validated the results experimentally. The model reached a good agreement with the experimental data for a wide set of operational points. The mean absolute error for cooling capacity and power consumption was under 6%, proofing the high accuracy of the model.

The ACHP+ model for the unit in baseline configuration with the fixed-speed compressor calculated an SCOP of 4.04. This value is just 1.5 % percent lower than the SCOP of 4.11 stated by the manufacturer and hence agrees accurately with the experimental data.

4.2. Evaporator Fan Power in variable-speed mode

The methodology for calculating the evaporator fan power consumption as described in Section 3.2.3 is validated with experimental data of the same evaporator fan collected in a previous experiment by Salts et al. [7]. Table 3 shows the evaporator fan power for different test conditions of the AHRI Standard 210/240 at different volume flow rates. The deviation between experimentally obtained results and calculated results is less than 6% for volume flow rates in the range between 2124 m^3/h and 2973 m^3/h, which is the range for the analysis in this study. Overall, the agreement is very high, especially considering that the experimental results become more uncertain at lower volume flow rates.

Table. 3: Deviation of evaporator fan power between experimental data and calculations in the ACHP+ model

Test	Evaporator volume flow rate in m^3/h	Evaporator fan power consumption in W		Deviation in %
		Experimentally obtained by Salts et al. [7]	Calculated	
B2	2973	406.9	406.7	0.0
B2	2549	247.5	248.3	-0.3
B2	2124	145.8	139.6	4.2
B2	1699	95.87	79.5	17.1
Ev	2973	403.9	409.5	-1.4
Ev	2549	238.3	246.9	-3.6
Ev	2124	149.3	141.7	5.1
Ev	1699	93.85	85.2	9.2
B1	2973	390.9	404.6	-3.5
B1	2549	246.1	250.8	-1.9
B1	2124	152.9	143.7	6.0
B1	1699	95.73	86.2	9.9
F1	2973	399.9	408.8	-2.2
F1	2549	243.1	249.8	-2.8
F1	2124	147	141.3	3.9
F1	1699	92.24	85.2	7.6

5. Parametric Studies

5.1. Compressor Speed Studies

The rotational speed of the variable-speed compressor varies to obtain a high SCOP. The maximum speed is set for the unit to run at the same cooling capacity as the baseline unit. 17.6 kW is reached at a rotational speed of 4500 rpm at test condition A_2 of the AHRI Standard 210/240. This is the maximum speed as per standard. The minimum speed can be optimized to increase the SCOP of the system. The intermediate speed is dictated by the minimum speed and maximum speed and is set according to the AHRI Standard 210/240 with the Equation (7).

$$Speed_{int} = Speed_{min} + \frac{Speed_{max} - Speed_{min}}{3} \tag{7}$$

Fig. 1 shows the SCOP at different minimum compressor speeds. An optimum is found at a minimum rotational speed of 2475 rpm, which is 55% of the chosen max speed. For the given minimum and maximum speeds, the intermediate speed is 3150 rpm.

5.2. Evaporator Fan Speed Optimization

The evaporator fan runs in rating condition with a volume flow rate of 2973 m³/h. In variable-speed mode the maximum evaporator fan speed is kept at this rate and the rate for test conditions with lower flow rate are adjusted. Experimental data from Salts et al. [7] showed a high SCOP at an intermediate evaporator fan flow rate of 2549 m³/h and a minimum flow rate of 2124 m³/h. These volume flow rates were transferred to the model.

5.3. Condenser Fan Speed Optimization

At rating conditions in fixed-speed mode, the condenser fan runs at a rotational speed of 1000 rpm, which results in an airflow of 6874 m³/h. To find the optimum speed in variable-speed mode, a parametric study with the condenser running at different fan speeds is conducted. Fig. 2 shows the SCOP with varying fan speeds at intermediate and minimum airflows. The condenser fan speed at maximum airflow is kept constant and the minimum and intermediate fan speeds are varied individually. The curve for the intermediate speed shows only a minor impact on the adjusted airflow. The optimum is reached at a speed of 875 rpm. At minimum airflow, the speed has a bigger impact on the systems SCOP. An optimum is found at a minimum fan speed of 775 rpm.

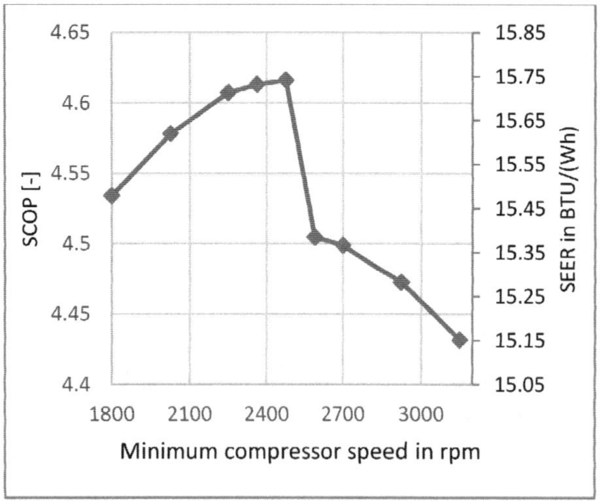

***Fig. 1:** System SCOP for variable-speed mode at different minimum compressor speeds*

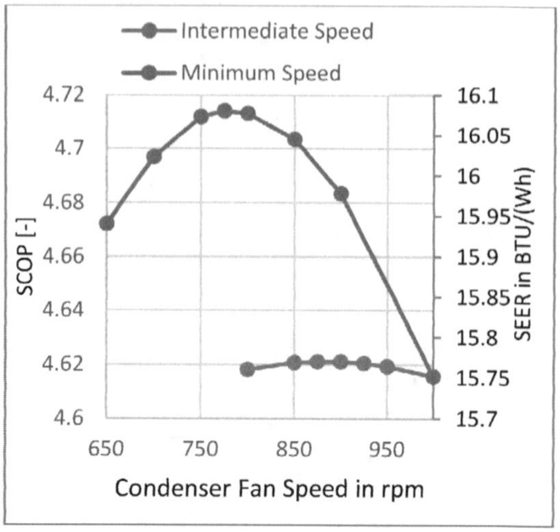

***Fig. 2:** System SCOP in variable-speed mode at different fan speeds.*

6. Results

6.1. Efficiency Evaluation on Increased Motor Efficiency in a Fixed-Speed Heat Pump

In this section, the components of the unit run all in a fixed speed condition. The dependence of the compressor motor efficiency, evaporator motor efficiency and condenser fan motor efficiency on the SCOP are evaluated individually.

6.1.1. Compressor Motor

The compressor is generally the most power consuming component in a heat pump system. Therefore, the compressor motor efficiency has a big impact on the SCOP as it directly impacts the compressor power consumption. Fig. 3 shows the impact on increasing the compressor motor efficiency of the baseline single speed heat pump. Due to a lack of detailed information on the initial motor efficiency of the compressor, a study with different motor assumptions is performed. The study shows that the increase of SCOP is fairly independent of the initial motor assumptions in the range between 0.8 and 0.9. A 1% increase of motor efficiency results in an SCOP increase of approximately 0.04 or 0.89 %. Increasing the compressor motor efficiency can increase the heat pumps SCOP greatly which speaks for using high efficiency motors in the compressor.

6.1.2. Evaporator and Condenser Fan Motor

The evaporator fan motor and condenser fan motor usually have lower efficiencies than the compressor motor. High efficiency fan motors can increase the systems SCOP. Fig. 4 shows the SCOP improvement potential of fixed-speed fans in a fixed-speed heat pump system. In the study, the motor efficiencies are increase individually while the motor efficiency of the other fan is kept constant.

An increase of the evaporator fan motor efficiency from the baseline motor efficiency of 0.75 to a state-of-the-art motor with an efficiency of 0.9 would result in an increased SCOP of 0.082 or 2%.

The condenser fan motor efficiency has a lower impact of SCOP. For the same SCOP increase of 2%, the motor efficiency needs to be increased from 0.65 to 0.9 or by 0.25.

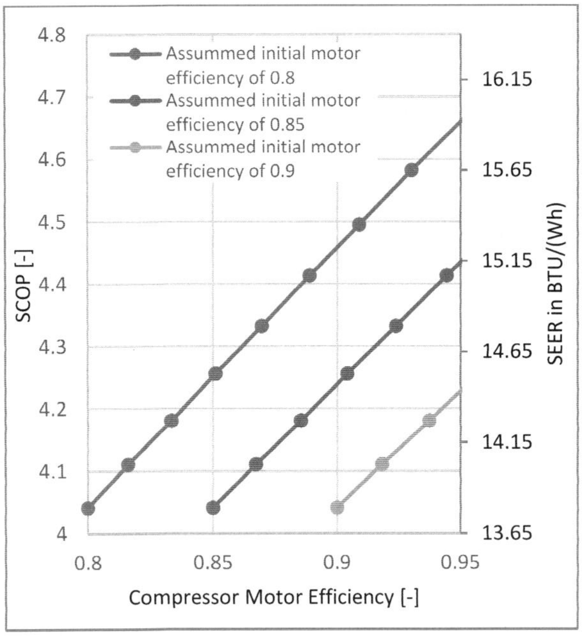

Fig. 3: SCOP with change of compressor motor efficiency in a fixed-speed system.

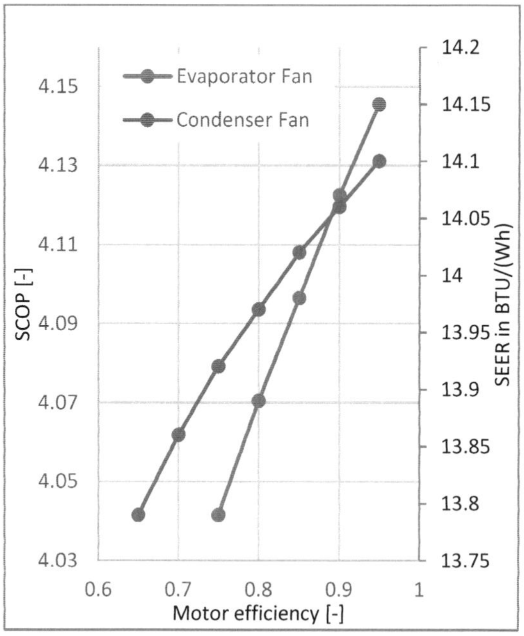

Fig. 4: SCOP with change of evaporator fan or condenser fan motor efficiency in a fixed-speed system.

6.2. Efficiency Evaluation on Increased Motor Efficiency in Variable-Speed System

In variable-speed systems, the unit runs at different compressor speeds and fan speeds throughout the AHRI Standard 210/240 tests. An electric motor usually has different motor efficiencies at different rotational speeds. In this paper, the effect of changing the motor efficiency at all speeds, as well as for certain speeds is looked at, to see the dependence of motor efficiency to SCOP. In the following section, the variable-speed scroll compressor is used and the fans are in different speed mode configurations.

6.2.1. Compressor Motor

Similarly to the fixed-speed system, the compressor motor efficiency is changed and the impact on SCOP is looked at. In this parametric study, the compressor runs at variable-speed and the evaporator fan and condenser fan in single speeds.

Fig. 5 shows the SCOP with change of motor efficiency for all speeds and for speeds individually. For simplification, an initial compressor motor efficiency of 0.88 at all speeds is assumed. The motor efficiency assumption is not that significant when looked at the change of SCOP per percentage of motor efficiency increase. A different motor efficiency assumption is expected to result in a shifted curve with the same gradient. In the simulations where the motor efficiency is changed individually, the motor efficiency at all other speeds is kept at 0.88. This allows to see the effect of motor efficiency for certain speeds on the SCOP.

An increase of compressor motor efficiency at all speeds results in an SCOP increase of approximately 0.038 or 0.85%. This result is very similar to the effect of compressor motor efficiency on SCOP in fixed-speed systems. When looked at the compressor motor efficiency at individual speeds, it shows that the motor efficiency change at maximum rotational speed and intermediate rotational speed only have a marginal effect on the systems SCOP. The motor efficiency change at minimum speed has the largest effect on the systems SCOP which shows that the manufacturers of electromotors need to put their focus on increasing the motor efficiency at low speeds.

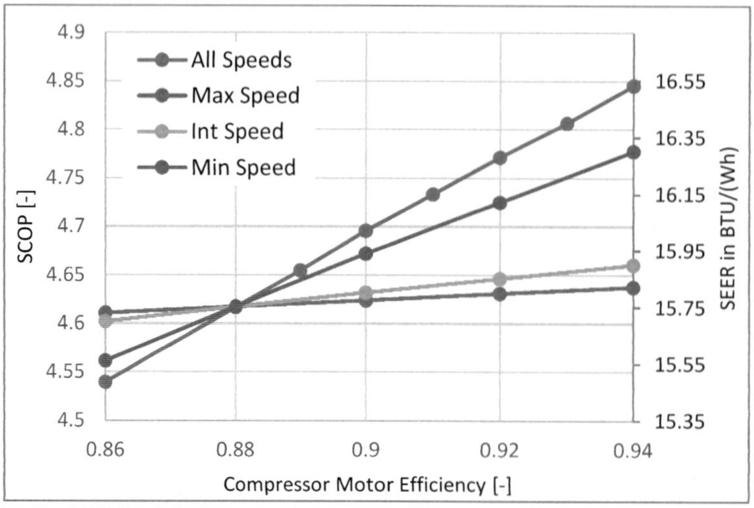

***Fig. 5:** SCOP with change of compressor motor efficiency at different speeds in a variable-speed system.*

6.2.2. Evaporator Fan Motor

The configuration in this study is a variable-speed compressor and a variable-speed evaporator fan. The condenser is running in a fixed-speed condition.

The evaporator fan motor efficiency is changed from the initial motor efficiency of 0.75 to higher values individually for certain speeds and for all speeds. Fig. 6 shows the results of the parametric study. Similarly to the compressor motor efficiency study, the evaporator fan motor efficiency maximum and intermediate speed only have a low impact on SCOP while at minimum speed the effect on SCOP is dominant. A change of evaporator fan motor efficiency at all speeds from 0.75 to 0.9 results in an SCOP increase of 0.075 or 1.5%.

6.2.3. Condenser Fan Motor

The configuration of the heat pump in this analysis is a unit with a variable-speed compressor and variable-speed condenser fan. The evaporator fan is running in fixed-speed mode.

The condenser fan motor efficiency is increased from the initial fan motor efficiency of 0.65 to higher values individually and for all speeds. The results are similar to the evaporator fan motor efficiency, showing a high dependency of SCOP to the motor efficiency at minimum speed and only marginal impact of the motor efficiency at maximum and intermediate speed. An increase of the motor efficiency from 0.65 to 0.9 at all speeds results in an SCOP increase of 0.08 or 1.7%. This shows a slightly reduced effect the condenser fan motor efficiency on SCOP compared to the evaporator fan motor efficiency.

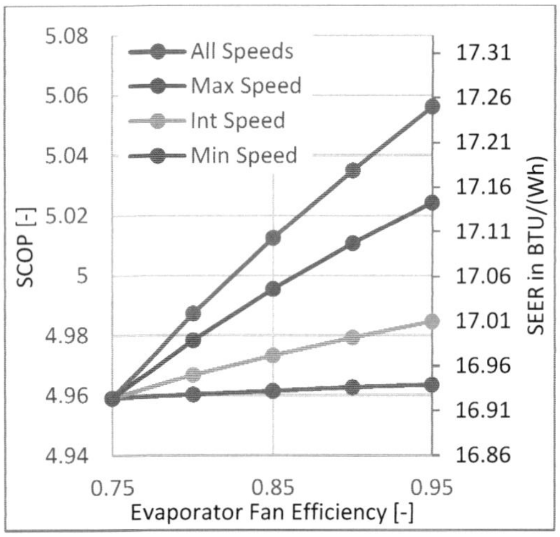

Fig. 6: *SCOP with change of evaporator fan motor efficiency at different speeds in a variable-speed system.*

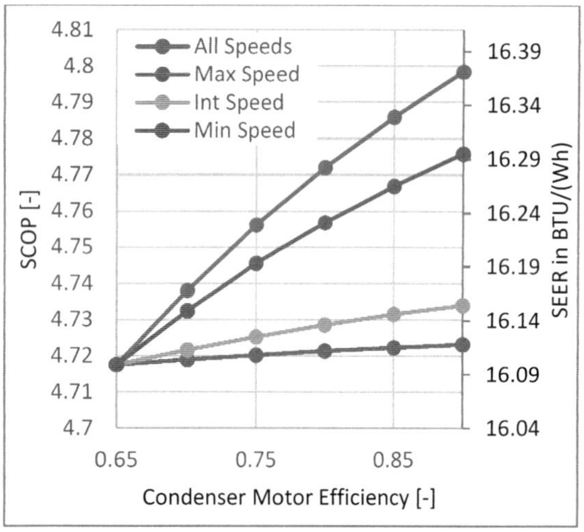

Fig. 7: *SCOP with change of condenser fan motor efficiency at different speeds in a variable-speed system.*

6.3. Comparison of System in Different Speed and Motor Efficiency Configurations

The effect on SCOP of different speed mode configurations is evaluated. In addition, the effect of changing the baseline motors to high efficiency state-of-the-art motors is analyzed. Table 4 shows the initial motor efficiencies of the compressor, evaporator fan and condenser fan motors and the motor efficiencies that can be reached with high efficiency permanent magnet motors.

Fig. 8 shows the SCOP of the unit in different configurations. In all configurations the variable-speed scroll compressor is used. In configuration 1, the compressor runs at a single speed and is tested with the single speed test of the AHRI 210/240 standard. In single-speed mode with baseline motor efficiencies, the heat pump with the advanced scroll compressor is less efficient with an SCOP of 3.89 than the baseline unit with a dedicated single speed compressor which has an SCOP of 4.11. With a change to variable-speed of the compressor, configuration 2, the SCOP rises significantly to 4.62 with an increase of 0.73 or 18.7%. This shows the potential of changing the unit to a variable-speed heat pump. The SCOP is increased because unlike with a fixed-speed system, the unit does not run with a much too high cooling capacity most of the time. This leads to less cycling which has losses and a reduced pressure ratio because of an increased evaporating temperature and reduced condensing temperature.

Changing the evaporator fan to variable-speed mode, configuration 3, further increases the SCOP significantly to 4.96, which is a rise of 0.34 or 7.4% compared to configuration 2.

Lastly, making the system to a full variable heat pump by changing the condenser to variable-speed, configuration 4, the SCOP rises by 0.13 or 2.7% to 5.09. Compared to the fully fixed speed system in configuration 1, this is an increase of 1.2 or 31%.

The heat pumps efficiency can further be improved with the use of high efficiency motors. The systems SCOP with high efficiency motors is represented with blue bars in Fig. 8. In the fully fixed-speed configuration, a significant increase of 0.37 or 9.6% can be achieved by increasing the compressor, evaporator, and condenser motor efficiency. In configuration 2, the SCOP increase with high efficiency motors is similar, at 0.37 or 8.0%. The SCOP improvement in configuration 3 is 0.26 or 5.2% and in configuration 4 0.14 or 2.8%. The results show that with the use of high efficiency motors, the heat pumps SCOP can increase significantly. For full variable systems, the advantage of high efficiency motors reduces slightly, and the highest increase can be seen when increasing the motor efficiency in a heat pump with fixed-speed fans.

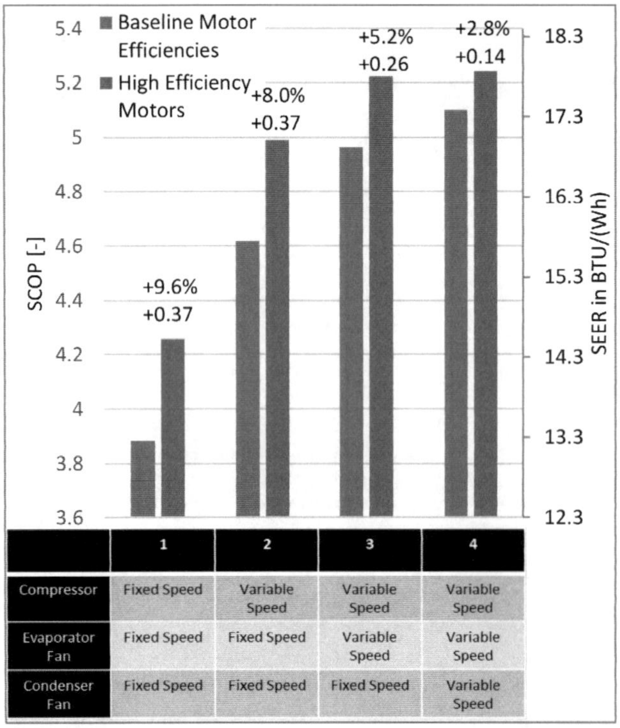

Fig. 8: *Performance evaluation of system in different speed and motor efficiency configurations*

Table 4: Initial and new high motor efficiencies of compressor, evaporator fan and condenser fan used in Section 6.3

	Motor Efficiency [-]					
	Max. Speed		Int. Speed		Min. Speed	
	Init.	**New**	**Init.**	**New**	**Init.**	**New**
Compressor	0.88	0.93	0.85	0.85	0.85	0.85
Evaporator Fan	0.75	0.93	0.75	0.83	0.75	0.83
Condenser Fan	0.65	0.92	0.65	0.82	0.65	0.82

7. Conclusion

This study quantified the theoretical performance limits of a heat-pump system by changing fixed-speed components to variable-speed and increasing the motor efficiency of the compressor, evaporator fan and condenser fan. For an evaluation of efficiency, the AHRI Standard 210/240 was used, which assesses the systems performance of a full season. The analysis was performed with the simulation model ACHP+, which showed good agreement with experimental data.

In fixed-speed mode and variable-speed mode, the SCOP improved by approximately 0.9 for every percentage of improved compressor motor efficiency. It was found that in variable-speed mode, the compressor motor efficiency at minimum speed has the highest impact on SCOP, while the motor efficiency at intermediate and maximum speed only has low relevance. The same is valid for the motor efficiency of the evaporator fan and condenser fan. This shows that manufacturers of electrical motors need to put their focus on designing motors with high efficiencies at lower rotational speeds. Current day motors mostly have their highest efficiency at maximum speed, with decreasing efficiency at lower speeds. Compared to the compressor, the improvement of SCOP of the evaporator fan and condenser fan with an increase of motor efficiency is only marginal. Nevertheless, due to generally low motor efficiencies of the fan motors, a noticeable improvement can be made with high efficiency motors.

Furthermore, it was shown that the change to variable-speed systems can improve the system significantly. Especially, the change of the compressor to a variable-speed compressor showed an improvement gain of 18%. The change to a variable-speed evaporator fan had a reduced but still significant SCOP increase and the condenser fan resulted in the lowest but still not negligible efficiency gain.

An additional change to high efficiency motors, as can be reached with state-of-the-art permanent magnet motors showed high improvement gains, especially in systems with fixed-speed fans. In heat pumps with variable-speed fans, the seasonal performance gain by changing to high efficiency motors was reduced.

Overall, the study showed that by changing a fixed-speed heat pump system to a variable-speed system with high efficiency motors, an overall SCOP improvement of 35% can be achieved.

Although fully variable-speed systems with high efficiency motors require higher initial investment, the big efficiency gain shows that the upgrade is worth the effort. High efficiency heat pumps not only have lower operational costs, which saves costs long term, but also are more environmentally friendly because of a reduced power consumption.

References

[1] **Goetzler, William; Sutherland, Timothy; Reis, Callie.** *Energy savings potential and opportunities for high-efficiency electric motors in residential and commercial equipment.* EERE Publication and Product Library, 2013.

[2] **AHRI Standard 210/240.** Performance Rating of Unitary Air-conditioning & Air-source Heat Pump Equipment. 2017.

[3] **Ammar M. Bahman, Davide Ziviani, Eckhard A. Groll.** *Development and Validation of a Mechanistic Vapor-Compression Cycle Model.* West Lafayette, IN, USA : 17th International Refrigeration and Air Conditioning Conference at Purdue, July 9-12 2018.

[4] **ANSI/AHRI Standard 540.** Performance Rating of Positive Displacement Refrigerant. 2020.

[5] **Copeland, Emerson.** ZP51K6E-PFV Rating Coefficients. [Online]

[6] **Emerson Copeland.** ZPV038CE-2E9, Coefficient Performance Calculator. [Online] [Zitat vom: 15. November 2020.] https://climate.emerson.com/online-product-information/OPIServlet.

[7] **Ludovic Chretien, Roger Becerra, Nicholas P. Salts, and Eckhard A. Groll.** *Seasonal Energy Efficiency Rating Improvement of Residential HVAC Systems Using a Low Power Inverter with a PSC Compressor.* West Lafayette, IN, USA : 17th International Refrigeration and Air Conditioning Conference at Purdue, 2018.

Acknowledgement

The authors would like to thank QM Power for sponsoring this research project.

CFD Simulation and Experimental studying of a Dry Screw Vacuum Pump

K Ma, B Guo, Y H Zhou, Y B Wang
School of Energy and Power Engineering, Xi'an Jiaotong University, Xi'an 710049, China;
Email: monkey8023@stu.xjtu.edu.cn

Abstract. Dry screw vacuum pumps are widely used in semiconductor, microelectronics and food industries due to their oil-free characteristics in the vacuum chamber during operation. The performance of screw vacuum pumps is usually predicted by one-dimensional numerical simulation or empirical methods. However, the above method cannot obtain the distribution of the fluid flow field inside the vacuum pump. In this paper, a three-dimensional model is established based on the experimental dry screw vacuum pump. The internal flow field of the dry screw vacuum pump was simulated by computational fluid dynamics (CFD) software using structured dynamic grid. The flow characteristics of the fluid inside the vacuum pump were analyzed. The distribution of the pressure field and velocity field inside the compressor is obtained. Comparing the P-θ indicator diagram and the performance parameters of the vacuum pump with the experimental results. It was found that the results of the three-dimensional numerical simulation were more consistent with the trends of the experimental data. On this basis, the flow field characteristics and performance of the vacuum pump are analyzed and the flow field of the vacuum pump can be better understood.

1. Introduction

Vacuum pumps has been widely used in different industrial processes and has become an indispensable part of production and life. With the progress of information technology, the electronic and semiconductor industries have extensive demands for vacuum pumps and higher requirements for the performance of vacuum pumps. Dry screw vacuum pump has the advantages of simple and compact structure, no oil in the working chamber, long life, environmental protection, and lower pressure pulsation. Therefore, we need to understand the working principle of dry screw vacuum pumps to guide the design and manufacture of screw vacuum pumps.

The mathematical model of the screw vacuum pump can predict the working characteristics of the vacuum pump when the structural parameters is changed. So the model is of great significance to the performance research and structural development of the vacuum pump. There are two ways to establish the mathematical model of the screw vacuum pump.

One is the chamber model. By studying the changes of volume, temperature and pressure with the rotor angle, a one-dimensional mathematical model of the working process is established. Kauder et al. [1] established a complete screw mechanical performance simulation software KaSim. The software takes the chamber volume as the research object, systematically considers the thermodynamic behavior of the fluid in the cavity and the leakage in different gaps. Janicki [2] used the chamber model to perform thermodynamic simulation on the screw machinery, and described the effect of working conditions and the gap height on the performance of the machine with the influence matrix. Nadler [3] took the chamber model as the research object, used the thermodynamic simulation software KaSim for theoretical research, and introduced adsorption and desorption to calculate the backflow model. Zhao et al. [4] studied the working performance of the variable pitch screw vacuum pump. The results show that the multi-stage gradual screw vacuum pump is more energy-saving, noise-reducing, and the variable pitch screw vacuum pump has better heat dissipation and gas transport performance, etc. But the article did not conduct experimental verification.

The chamber model is simple and the calculation is convenient, but it cannot reflect the change and distribution of the flow field in the chamber. The other method is to use CFD and other computational fluid dynamics software to perform three-dimensional numerical simulation of the screw vacuum pump. The model can calculate the distribution of the flow field and is very useful for calculation and design.

Few scholars have performed numerical simulation on dry screw vacuum, so the numerical simulation of screw vacuum mainly based on the CFD numerical simulation of screw pump and screw compressor. The numerical simulation of screw machinery is mainly the difference between the meshing software and the solving software. The meshing software includes SCORG, TwinMesh, Ansys Meshing, PumpLinx, etc. In addition, many scholars divide the mesh by themselves. The solution software mainly includes Fluent, CFX, Star CCM+, etc.

Kovacevic [5] used an advanced mesh generation method and applied it to the CFD program to simulate the three-dimensional flow of the screw compressor A program was developed to make the rotor get a good mesh boundary adaptation at any angle. After that, Rane and Kovacevic did a more in-depth study on the meshing and solution method of screw machinery. They [6] proposed a new algebraic mesh generation method suitable for twin-screw machinery. They [7] used algebraic transfinite interpolation to generate the deformed mesh of the twin-screw machine to generate the initial mesh. They [8] established a grid distribution program to refine the grid at the leakage triangle of the screw compressor.

Yan et al. [9] divided the grid of the screw rotor basin and applied it to the screw pump. They compared mass flow, shaft power and efficiency at various speeds and exhaust pressures, and verified the accuracy of the simulation through experiments. Lu et al. [10] proposed a grid generation algorithm. The new grid generation method can better align the calculation grid with the main leakage flow, significantly improve the quality, and reduce the numerical deviation caused by the numerical analysis of spiral machinery with larger helix angles. Ding et al. [11] used the VOF method to simulate the transient flow of an oil-injected twin-screw compressor, and evaluated the cooling and sealing effects of oil-injection by comparing the simulation results with and without oil.

Yue et al. [12] used a dynamic grid method to simulate the flow field of a two-stage dry scroll vacuum pump. When the pressure is 10000Pa, the simulation result is consistent with the experimental pumping speed. But when the pressure was reduced, the error is large. Kennedy et al. [13] compared the one-dimensional thermodynamic model with the three-dimensional computational fluid dynamics model of an oil-free twin-screw compressor, and compared the two models with experimental data. Mierka et al. [14] proposed a new calculation strategy applied to twin-screw extruders. Wu et al. [15] established a three-dimensional model of a refrigeration screw compressor, divided the internal flow field with TwinMesh software, and simulated it with CFX software. The simulation results were compared with experimental results, and the maximum error was 2.578%.

This paper establish a three-dimensional numerical model of the flow field of the dry screw vacuum pump, and simulate the working process of the dry screw vacuum pump under different loads by using CFD technology. The flow characteristics of the fluid inside the vacuum pump are analyzed, and the pressure, velocity and temperature distributions are obtained. This paper establish a performance test bench for dry screw vacuum pumps to obtain the performance of the vacuum pump under different working conditions. The simulation results were similar to the experimental data, which proves that the model is correct. So the model can be used for reference to the establishment of three-dimensional numerical model of dry vacuum pump.

2. Geometric Model and Meshing Generation

At present, the profile of dry screw vacuum pumps is mainly composed of curves such as points, cycloids, arcs and involutes. The profile of the driving rotor and the driven rotor are the same, but the direction of rotation is opposite. The rotor profile studied in this paper is shown in Figure 1. The composition tooth curve of the profile is smoothly connected, and the first derivative is continuous; the theoretical profile has no closed volume and leakage triangle.

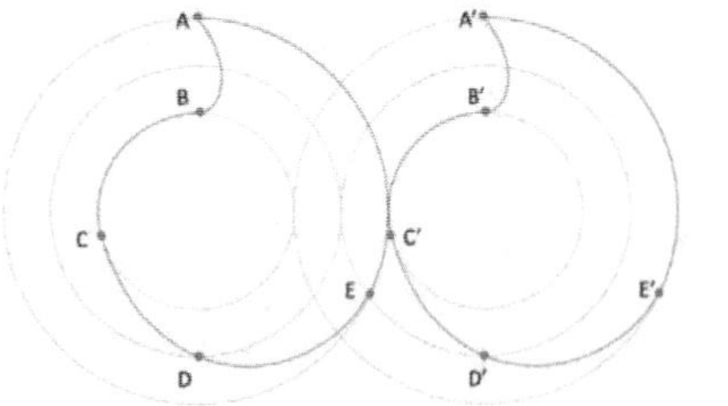

Figure 1. The rotor profile of the dry screw vacuum pump

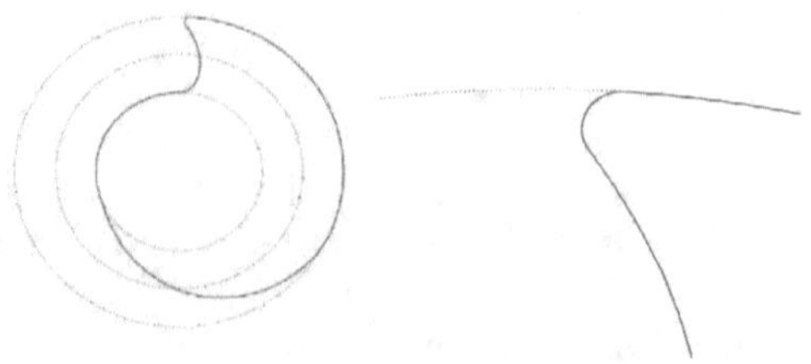

Figure 2. The modified profile and partial enlarged view

The profile studied in this paper has a sharp corner due to the existence of point A. When dividing the mesh, the curvature is too high to obtain a mesh that meets the requirements of the calculation conditions. Therefore, the profile needs to be corrected. In order to meet the grid quality required by the calculation, the sharp point is smoothly transitioned with a small arc. And the small arc is smoothly connected with the cycloid and the tooth tip arc, and the first derivative is continuous. The modified profile and partial enlarged view are shown in Figure 2.

This paper study a two-stage variable pitch vacuum pump. Two-stage variable pitch rotors generally have two equal pitch combinations or a combination of gradual pitch and equal pitch. This article selects the combination of two equal pitches. If the exhaust end surface is used as a reference, the pitch of the first section is smaller than that of the second section. The first section of the screw rotor in this paper has a pitch of 88mm and a length of 200mm; the second section of the rotor has an equal pitch of 60mm and a length of 140mm. The axial view of the two-stage screw rotor is shown in Figure 3.

The working process of dry screw vacuum pump is very complicated. In order to facilitate the simulation and reflect the working process of the compressor, this research divides the flow domain model into four parts: suction port, chamber volume, discharge port and discharge end gap. In the calculation process, these four parts connected, and exchanged data information through the interface. The fluid domain of the control volume of the vacuum pump is shown in the Figure 4.

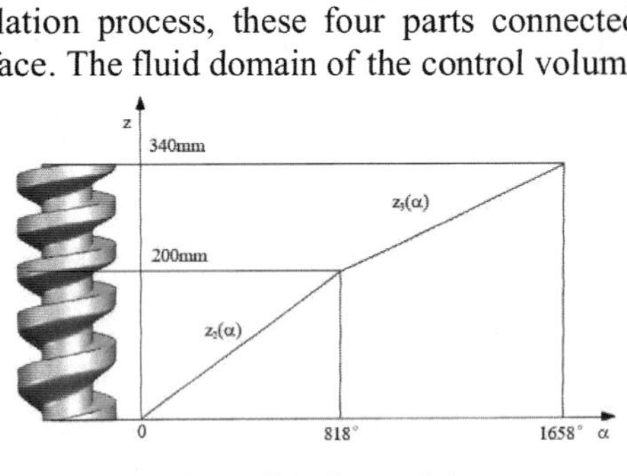

Figure 3. The axial view of the two-stage screw rotor

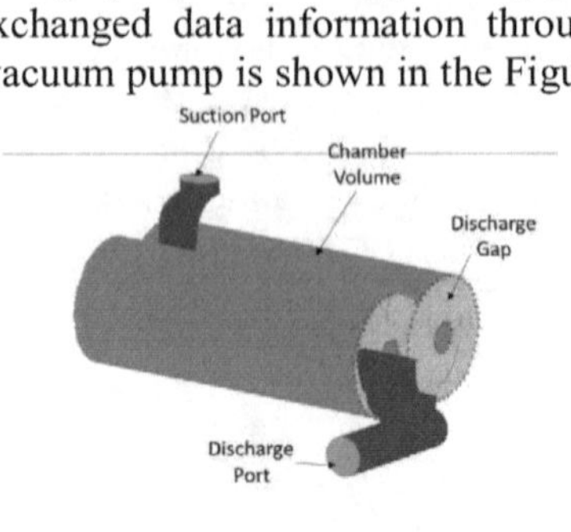

Figure 4. The fluid domain of the control volume of the vacuum pump

IOP Conf. Series: Materials Science and Engineering 1180 (2021) 012043 doi:10.1088/1757-899X/1180/1/012043

Import the inlet and outlet flow domain model and the exhaust end gap flow domain model of the dry screw vacuum pump into ANSYS-Mesh, and mesh the inlet and outlet flow domain model and the exhaust end gap flow domain respectively. For the inlet and outlet basins, due to its complex shape, the mesh size function of Proximity and Curvature is used. For the meshing of the exhaust end face gap, the sweep method is used to generate it. In order to ensure the accuracy of the calculation, there are five layers of grids along the sweep direction, and the minimum size of the grid is 0.0001mm, which meets the calculation requirements.

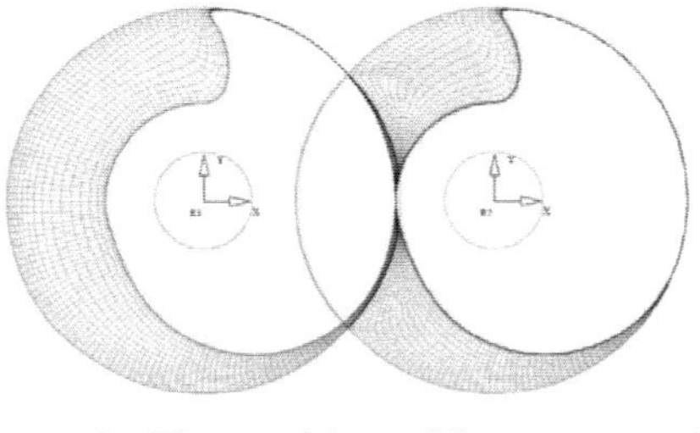

Figure 5. The meshing of the cross-section

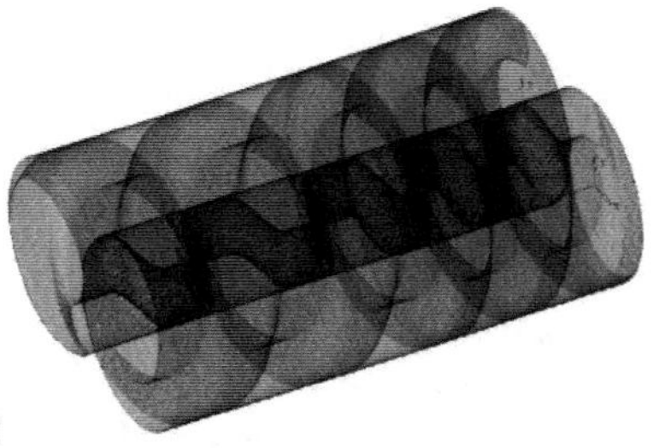

Figure 6. The flow field grid of the chamber volume

TwinMesh is a meshing tool specially designed for the simulation of internal fluid flow in twin-screw machinery. According to the profile of the compressor rotor, it can generate a high-quality hexahedral structural mesh. The grid of the rotating watershed adopts a structured division method, with 15 radial nodes, 61 Interface nodes, and 136 circumferential nodes. The meshing of the cross-seciton is shown in Figure 5.

The axial distribution assigns nodes according to the value of the spiral angle. When assigning axial nodes by the value of the helix angle, the number of axial nodes should be proportional to the value of the torsion angle. After calculation, the height in the MZ direction is 0.3333333mm, the mesh expansion rate in the MZ direction is 1.001; the height in the PZ direction is 0.48899mm, and the mesh expansion rate in the PZ direction is 1.001. The flow field grid of the chamber volume show in Figure 6.

3. Simulation Setup

3.1 Governing Equation

Fluid flow and heat transfer need to follow the laws of conservation of physics, including the law of conservation of mass, the law of conservation of momentum and the law of conservation of energy. Because the flow is in a turbulent state, the system also needs to comply with additional turbulent transport equations.

$$\frac{\partial \rho}{\partial t} + \frac{\partial}{\partial x_i}(\rho u_i) = 0 \tag{1}$$

$$\frac{\partial}{\partial t}(\rho u_i) + \frac{\partial}{\partial x_i}(\rho u_i u_j) = -\frac{\partial p}{\partial x_i} + \frac{\partial \tau_{ij}}{\partial x_j} + \rho g_i + F_i \tag{2}$$

$$\frac{\partial(\rho T)}{\partial t} + div(\rho u T) = div\left(\frac{k}{c_p} gradT\right) + S_T \tag{3}$$

3.2 Turbulence Model

The fluid movement of the vacuum pump is complicated. In the numerical simulation calculation, the selection of a reasonable turbulence model has a great influence on the accuracy of the results. For the numerical simulation of rotating machinery, the SST (Shear Stress Transport) model can obtain more accurate results.

The SST model is a low Reynolds number model, which is suitable for industrial applications. The shear pressure transmission (SST) k-ω model combines the far-field k-ε model and the near-wall k-ω model. The model also considers the transmission of turbulent shear stress, which can accurately predict the beginning of the flow and the amount of fluid separation under negative pressure gradient conditions, with higher precision and accuracy.

$$\frac{\partial}{\partial t}(\rho k)+\frac{\partial}{\partial x_i}(\rho k u_i)=\frac{\partial}{\partial x_j}\left(\Gamma_k\frac{\partial \omega}{\partial x_j}\right)+G_\omega-Y_\omega+D_\omega+S_\omega \tag{4}$$

$$\frac{\partial}{\partial t}(\rho \omega)+\frac{\partial}{\partial x_i}(\rho \omega u_i)=\frac{\partial}{\partial x_j}\left(\Gamma_\omega\frac{\partial k}{\partial x_j}\right)+G_k-Y_k+S_k \tag{5}$$

3.3 Immersed Solid Method

In order to reduce the influence of profile correction on the calculation results, this article uses the Immersed Solid method in ANSYS CFX for reference. The immersed solid method can simulate the movement of rigid solids in the fluid domain. The momentum source acts on the fluid domain, forcing the fluid and the solid to move together. The movement of fluid outside the submerged entity is like flowing around a rigid entity.

$$S_i=-\alpha\beta C\left(U_i-U_i^F\right) \tag{6}$$

The scale factor α of the momentum source term is used to control the size of the source term, and its value is used to balance the accuracy and robustness of the simulation. By default, its value is set to 10, and its value can be modified according to the convergence. The special forcing function β is used to characterize whether there is an entering solid in the flow field.

3.4 Boundary Condition

The boundary conditions of the intake and exhaust ports of the vacuum pump simulated in this paper are derived from experimental measurements. The air inlet of the vacuum pump is set to open condition, the pressure is set to 0.04bar, and the temperature is 293.15K; the condition of the exhaust port is open condition, the pressure is set to 1.0bar, and the return temperature is 338.15K.

3.5 Independence Verification

This paper takes the inlet pressure of 0.04bar and the rotational speed of 3000 $r\cdot min^{-1}$ as an example to verify the grid independence. The different node settings and the number of grids are shown in Table 1. The verification result of grid independence of the working condition with an inlet pressure of 0.04 bar and a speed of 3000 $r\cdot min^{-1}$ is shown in Figure 7.

Table 1. Grid independence verification

Number	Radial Nodes	Circumferential Nodes	Axial Nodes	Number of Meshes/ten thousand
1	15	300	680	424
2	15	333	680	497
3	16	333	680	513
4	16	333	829	665
5	16	333	1020	860

Table 1 and Figure 7 show that with the subdivision of the grid and the increasing number of grids, the solution result will be more accurate. For the last three grids, the volume flow rate and input power vary little with the grid. Considering the time cost of calculation and the accuracy of calculation, this paper finally adopts the fourth set of grids as the calculation grid.

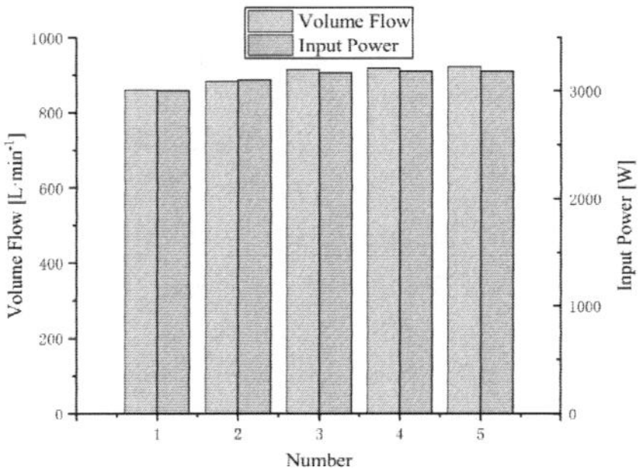

Figure 7. The results of grid independence verification

For rotating machinery, the time step is often calculated based on the speed. Since the dynamic mesh of the rotation domain in this paper is generated by the commercial rotating machinery meshing software TwinMesh, the number of mesh generation is related to the angular step of rotation. Therefore, in the simulation of this article, the time step is not only related to the rotation speed, but also related to the angle step.

$$\Delta t = \frac{n/360}{N/60} \tag{7}$$

At the same speed, the smaller the angle step, the smaller the corresponding time step. The time step requires trial calculations to obtain a suitable value. This paper tries to calculate the influence of different angle steps on the results under the same speed and the same working conditions. It is found that a time step of 2° can meet the calculation requirements and greatly improve the calculation efficiency. In the end, the time step selected in this paper is about $1.1111 \cdot 10^{-4}$ seconds.

4. Experiment Studying

This paper use a two-stage dry screw vacuum pump as a prototype to build a vacuum pump performance test bench. The system diagram is shown in Figure 8. The whole system is mainly composed of dry screw vacuum pump, muffler, motor, frequency converter and test acquisition system. The test acquisition system is composed of pressure sensor, temperature sensor, torque sensor and NI acquisition system.

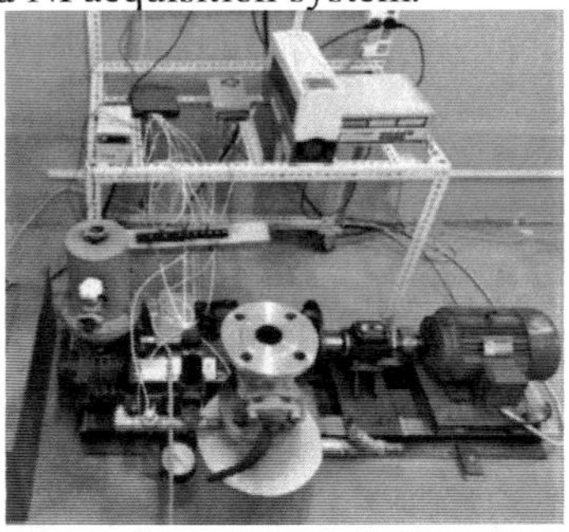

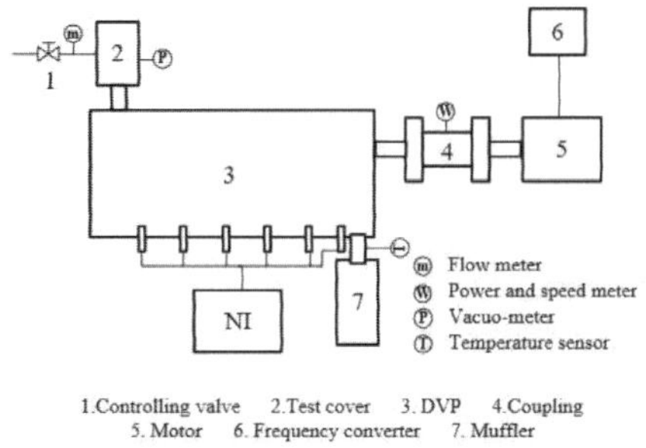

Figure 8. Vacuum pump performance test system

This experiment use a ASC510 frequency converter to change the speed of dry screw vacuum pump. The screw vacuum pump connect to the variable frequency motor through a coupling. By adjusting the frequency of the inverter to change the motor frequency, the speed of the vacuum pump can be adjusted.

Under normal circumstances, each pressure sensor can measure the pressure change of one revolution of the rotor. But considering that the sensor will sweep across the top of the screw, the measured pressure will be deviated. Therefore, there will be overlaps in the data between two adjacent pressure sensors to eliminate the above influence.

5. Results

5.1 Model validation

In order to verify the accuracy of the three-dimensional flow field model of the dry screw vacuum pump, this paper compares the experimental and simulated values of the p-θ diagram of the internal working process. As shown in Figure 9, the figure shows a comparison between the simulated value and the experimental value when the rotation speed is 3000r·min^{-1} and the suction pressure is 4kPa. There is an error between the experimental value and the simulated value, but the pressure change trend is consistent.

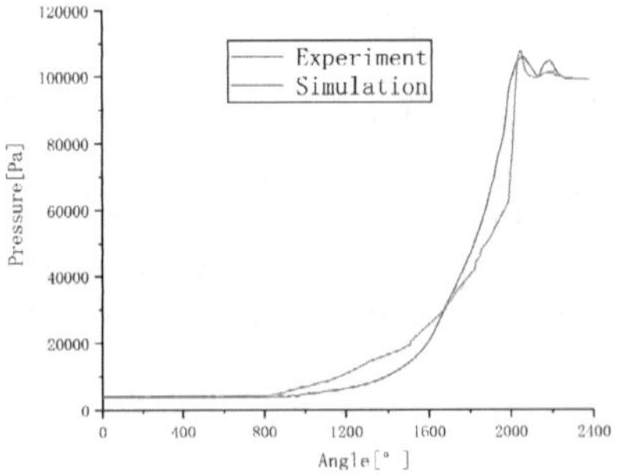

Figure 9. Comparison of simulated data and experimental data

The first part is the intake process, and the initial angle is 0° at the beginning of the inhalation process. For screw vacuum pumps, the suction orifice is generally designed to have a larger shape to ensure sufficient suction, and the suction resistance loss is relatively small. And the experimental data of the intake process cannot be obtained in this paper, so the pressure in the intake process of the experiment in this paper is always equal to the intake pressure.

The second part is the isometric process. At this time, the suction orifice has been closed, and due to the impact of the high-pressure chamber leakage, the gas pressure will slowly rise during this process. The third part is the internal compression process. In this process, the volume of the chamber will decrease with the decrease of the pitch. At the same time, considering the impact of the leakage of the high-pressure chamber, the pressure rises faster. The fourth part is the second stage of isometric process. The gas in the chamber will still rise due to the leakage of the high pressure chamber. The fifth part is the second stage of the compression process. The working chamber reaches the exhaust end surface but is not connected to the exhaust orifice. The reduction in the volume of the chamber and the end surface leakage will cause the gas pressure to rise. Similarly, here is the difference between simulation and experiment. There are three main reasons for the error. First, this article modified the original profile, which resulted in a larger fluid area at the sharp point and increased leakage from the high-pressure chamber to the low-pressure chamber. Although this article draws on the solid immersion method, it does not completely eliminate the effects of leakage. Secondly, the gap value selected in this paper is deviated from the gap value during actual operation. Third, due to the large helix angle of the dry vacuum pump rotor, a relatively orthogonal grid cannot be obtained, which leads to errors in numerical calculations. Fourth, there is a water-cooling cavity between the vacuum pump and the front cover. This article did not consider the effect of water cooling in the simulation.

The last part is the exhaust process. At this time, the chamber connect to the exhaust orifice. The over-compression phenomenon can be observed from the figure, and both simulations and experiments have such phenomena.

5.2 Simulation Result

The pressure distribution in the vacuum pump is calculated by CFD software, and the pressure change in the flow field of the vacuum pump is analyzed in detail.Choose the axial section of the vacuum pump to reflect the pressure distribution in the working chamber of the vacuum pump. Because the vacuum pump is used in this simulation, the rotor profile of the driving rotor and the driven rotor are the same, and they are both single-tooth rotors. Therefore, the pressure change cycle is periodically once every 360°. Since the angle step used in this paper is 2° when generating the rotating watershed, every 180 result files is a cycle. At 0° and 360° rotation angles, the pressure distribution in the vacuum pump chamber is the same. With 60° intervals, Figure 10 shows the changes in the internal pressure of the vacuum pump at different angles.

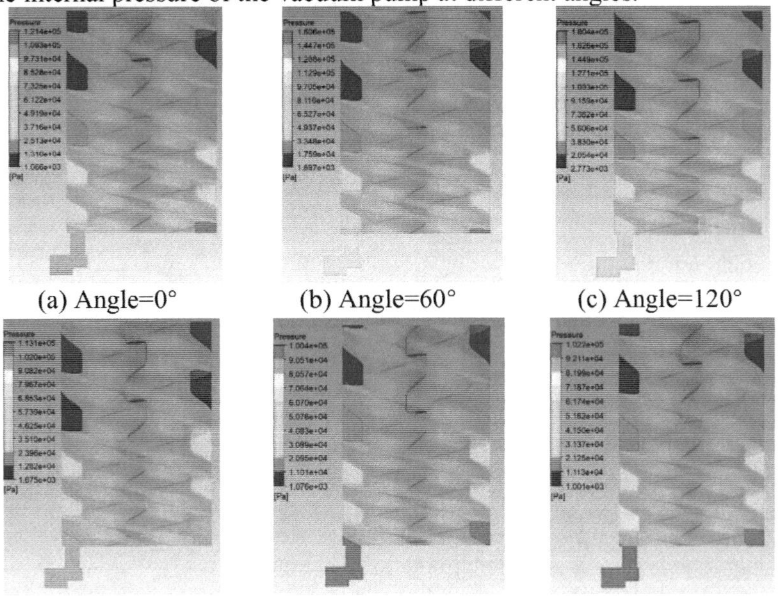

(a) Angle=0° (b) Angle=60° (c) Angle=120°

(d) Angle=180° (e) Angle=240° (f) Angle=300°

Figure 10. The pressure distribution in the vacuum pump

The pressure gradually increases along the spiral surface of the rotor from the suction port to the discharge port, and the pressure at the discharge port reaches the maximum. The exhaust port of the dry vacuum pump is only opened on one side, and only on the side of the driven rotor. When the rotor chamber is not connected to the exhaust, the rotor chamber will be over-compressed, and the pressure will be higher than the atmospheric pressure. The highest pressure is on the exhaust end side of the active rotor. When the rotor chamber is connected to the exhaust gas, the pressure in the rotor chamber is discharged and the pressure is reduced.

Figure 11 shows the velocity distribution in the rotor basin under different angles. The flow velocity in the rotor chamber is less than $10\mathrm{m}\cdot\mathrm{s}^{-1}$, and the velocity in the tip leakage channel is higher than the velocity in the chamber. The highest speed in the rotor chamber is located in the clearence where the two rotors mesh. The speed in the meshing clearence is generally higher than $100\mathrm{m}\cdot\mathrm{s}^{-1}$, in some places as high as $240\mathrm{m}\cdot\mathrm{s}^{-1}$. The speed is unreasonable in some places in the velocity counter is mainly due to the quality of the grids. The reason for the excessively high flow velocity at the meshing gap is that the gap at this position is too small, and the velocity is affected by the leakage from the high-pressure chamber to the low-pressure chamber and the rotation speed, resulting in a higher velocity than other places.

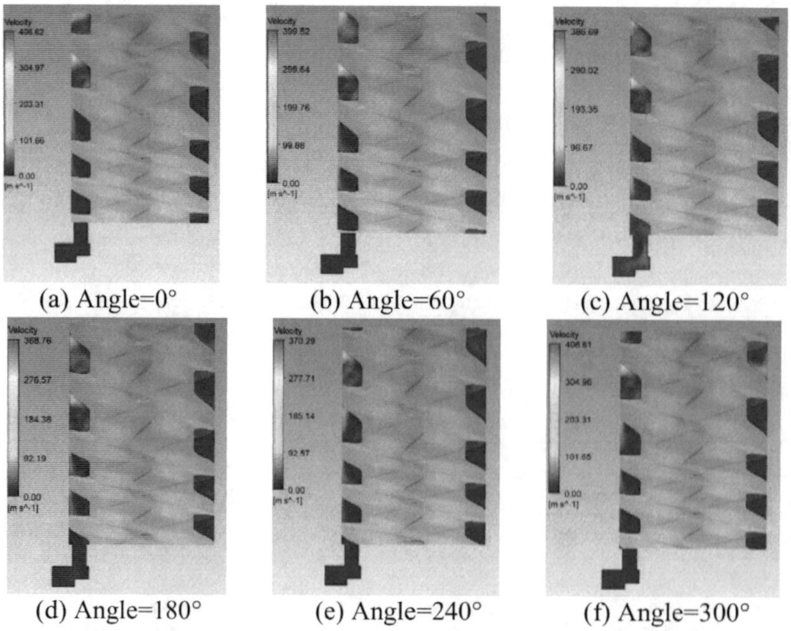

(a) Angle=0° (b) Angle=60° (c) Angle=120°

(d) Angle=180° (e) Angle=240° (f) Angle=300°

Figure 11. The velocity distribution in the vacuum pump

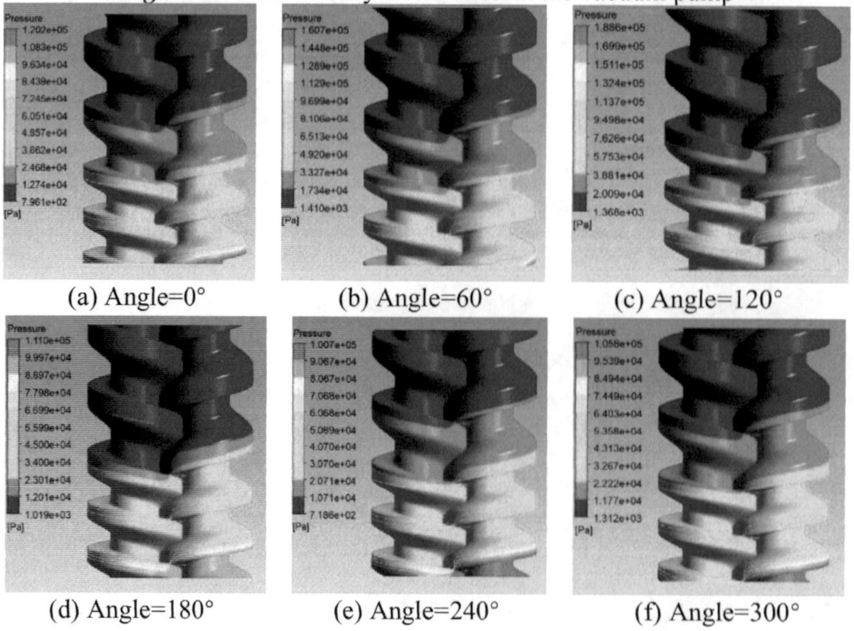

(a) Angle=0° (b) Angle=60° (c) Angle=120°

(d) Angle=180° (e) Angle=240° (f) Angle=300°

Figure 12. The pressure distribution on the rotor surface

Figure 12 shows the pressure distribution on the rotor surface. There is a very distinct color boundary on the pressure counter, and this boundary is the dividing line of different chamber. The suction chamber of the rotor occupies the first two chambers. After going through multiple processes, the gas is finally discharged. The pressure change range of the front part of the rotor is small, and the pressure at the rear end of the rotor changes drastically. This is also the place where the leakage is the most serious.

IOP Conf. Series: Materials Science and Engineering 1180 (2021) 012043 doi:10.1088/1757-899X/1180/1/012043

6. Conclusions

In this paper establish a three-dimensional numerical model of the dry vacuum pump and numerically simulate the internal flow field of the vacuum pump by the fluid analysis software CFX. Numerical simulation has obtained the pressure and velocity distribution of the flow field inside the vacuum pump. The rotor meshing line divides the rotor flow field into different chambers. Because the meshing clearence between the rotors along the contact line is too small, and with the increase of the pressure difference between the high-pressure chamber and the low-pressure chamber and the influence of the speed, the leakage effect becomes more and more obvious. The leakage velocity at the gap of the vacuum pump is greater than the leakage velocity between the teeth. The clearence is the most important factor affecting the performance of the vacuum pump, and the maximum leakage velocity of the clearence exceeds 240 m/s. Therefore, various leakage clearences of the vacuum pump are reasonably reduced during the design.

References

[1] Kauder K and Janicki M 2003 Adiabatic modelling and thermodynamic simulation af rotary displacement machines with the program KaSim

[2] Janicki M. 2007 Modellierung und Simulation von Rotationsverdrängermaschinen Dissertation, Universität Dortmund

[3] Nadler K. 2017 Modellierung und Analyse von Schraubenvakuumpumpen im Blower-Betrieb. Dissertation, Universität Dortmund

[4] Zhao F, Zhang S and Sun Kun 2018 Thermodynamic performance of multi-stage gradational lead screw vacuum pump. Applied Surface Science 432 97-109

[5] Kovacevic A 2010 Boundary adaptation in grid generation for CFD analysis of screw compressors International Journal for Numerical Methods in Engineering 64 401-426

[6] Rane S and Kovacevic A 2017 Algebraic generation of single domain computational grid for twin screw machines Advances in Engineering Software 107 38-50.

[7] Rane S, Kovacevic A and Stosic N 2015 Analytical Grid Generation for accurate representation of clearances in CFD for Screw Machines British Food Journal 90

[8] Rane S, Kovacevic A and Stosic N 2016 Computational Fluid Dynamics in Rotary Positive Displacement Screw Machines International Symposium on Transport Phenomena & Dynamics of Rotating Machinery

[9] Di Y, Kovacevic A and Qian T 2016 Numerical modelling of twin-screw pumps based on computational fluid dynamics Journal of Mechanical Engineering Science 1989-1996 (vols 203-210) 231.24(2016).

[10] Lu Y, Kovacevic A and Read M 2019 Numerical Study of Customised Mesh for Twin Screw Vacuum Pumps

[11] Ding H and Jiang Y 2017 CFD simulation of a screw compressor with oil injection Iop Conference 232

[12] Yue X, Zhang Y and Su Z 2017 CFD-based analysis of gas flow in dry scroll vacuum pump Vacuum 139 127-135

[13] Kennedy S, Wilson M and Rane S 2017 Combined Numerical and Analytical Analysis of an Oil-free Twin Screw Compressor IOP Conference Series: Materials Science and Engineering 232

[14] Mierka O,Theis T and Herken T 2014 Mesh Deformation Based Finite Element - Fictitious Boundary Method (FEM-FBM) for the Simulation of Twin-screw Extruders

[15] Wu H, Huang H and Zhang B 2019 CFD Simulation and Experimental Study of Working Process of Screw Refrigeration Compressor with R134a Energies

IOP Conf. Series: Materials Science and Engineering 1180 (2021) 012044 doi:10.1088/1757-899X/1180/1/012044

Theoretical study on the leakage characteristics of Disc-Seal Single Screw Pump used in oily sludge waste heat source absorption refrigeration system

Z L Wang*, H W Shi, S Y Wang, Z M Wang, M M Hao and J Wang

College of New Energy, China University of Petroleum (East China), Qingdao 266580, China
Email address: wangzengli25@ upc.edu.cn
* Corresponding Author

Abstract. The disc-seal single screw pump (DSSP) used in the field of oily sludge transport has a huge advantage. However, there is no research on the leakage characteristics of the DSSP at present, which makes its application limited. Therefore, the mathematical calculation model of leakage amount of each leakage channel was established in this paper according to the geometric model of the leakage channel. By using this mathematical model, the influence rule of working parameters on the leakage characteristics was analyzed. The results show that the total leakage amount per unit time of the DSSP varies periodically with the rotation of the screw, and the total leakage per unit time of the DSSP will increase with the increase of outlet pressure and screw rotation velocity. When the dynamic viscosity coefficient of the medium increases, the corresponding leakage will be reduced, in which the transmission medium dynamic viscosity is the main factor affecting the total leakage of the DSSP. The study of the leakage characteristics of the DSSP can provide a theoretical basis for the optimal design of its structure in the future.

1. Introduction

With the rapid development of economy and society, the demand for oil in all countries has increased greatly. All kinds of petrochemical enterprises to produce a large amount of sewage sludge and the oily sludge composition is complicated that mainly contains crude oil and solid impurities [1-3]. The large solid particles of impurities may be blocked and accelerated wear of pumping unit in the process of pumping. In addition, as the growth of the world's energy needs, the amount of oil transported by sea has increased rapidly, and the risk of oil spill pollution from oil tankers, especially VLCCS, has increased [4-6]. Once offshore oil spill occurs, the huge offshore oil film will seriously block the oxygen exchange between seawater and atmosphere, break the original stable Marine ecological balance, cause ecological pollution and serious economic losses that are difficult to recover, and eventually endanger human beings ourselves. Therefore, efficient cleaning and rapid recovery of offshore oil spill is particularly important for ecological environment and human beings. In addition, in the current situation of energy shortage, power supply shortage and environmental problems becoming more and more serious, absorption refrigeration technology with its unique advantages has been widely concerned. So the absorption refrigeration system uses waste heat recovered from oily sludge as heat source will have great application prospects. During the oily sludge waste energy recovery process in the absorption refrigeration system, the oil sludge need to be pumped. Due to the high viscosity of oily sludge, and many large particle impurities such as gravel, garbage, animal carcasses contained in the oily sludge, the traditional fluid pump does not meet its pumping needs. The traditional pumping device is difficult to be applied to the high-viscosity and multi-impurity sewage oil transport on the sea due to the disadvantages such as narrow working chamber and easy wear [7-9], so it is in urgent need of a fluid pumping device with strong medium adaptability and large discharge capacity. For this reason, Johansson put forward the DSSP for the first time in 1991 [10]. This new type of screw pump is widely used in high-viscosity, multi-impurity sewage sludge transportation field and other industrial fields due to its advantages such as large displacement, compact structure and strong adaptability to solid particles.

The DSSP is mainly composed of sealing disc, screw rotor, pump shell and driving gears (as shown in figure 1), and its core working parts are meshing pair formed by sealing disc and pressurized section groove [11]. The pressurization process is realized by the change of the volume of the pressurization cavity formed by the engagement of the sealing disc and the pressurization groove and the DSSP has been widely studied for its unique advantages in the field of high viscosity and multi-impurity sludge transport. The surface equations of meshing pair was obtained by Wang et al. through space meshing principle and the coordinate transformation method [12]. The influence of working parameters on meshing characteristics of the DSSP was studied and the mathematical calculation model of suction capacity of the DSSP was established through the mesh characteristic analysis model [13]. The formula of the volume of the groove in the pressurized section was obtained and the numerical calculation model of the theoretical displacement of the DSSP was established by Wang et al. by using the calculus method [14]. However, the leakage characteristic is a key factor that affecting the conveying capacity of the DSSP, and there are no published papers on the leakage characteristic of the pump. Therefore, it is necessary to establish a mathematical model for calculating the leakage rate of the DSSP and conduct a thorough study on its leakage characteristics, so as to lay a theoretical foundation for improving the pump delivery efficiency and optimizing the structure of the pump.

Based on this, the leakage characteristics of the DSSP were studied in this paper. The mathematical calculation model of the leakage rate of each leakage channel was established according to the geometric model of the leakage channel. And the variation law of the leakage rate of each leakage channel and the total leakage rate of the screw pump during the rotation process was obtained. Finally, the influence of the working parameters of the DSSP on its leakage characteristics was analyzed.

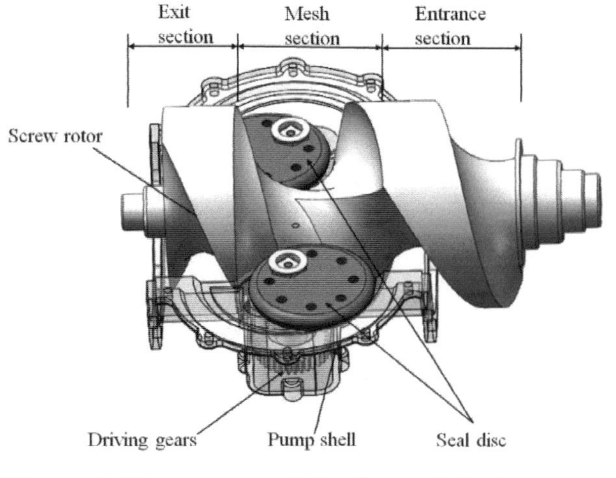

Figure 1. The structure of the disc-seal single screw pump

2. The leakage channel of DSSP

2.1 Backflow pressurized leakage channel

The geometric model of the backflow pressurized leakage channel is shown in figure 2. The leakage channel is mainly the leakage hole formed between the screw slot, the outer edge line of the sealing disc and the inner wall surface of the casing. The flow pattern of fluid leakage through orifice plate depends on the type of orifice plate. According to the geometric structure, the fluid is almost in line contact with the hole wall when the fluid flows through the suction and backflow pressurized leakage channel, so the leakage channel can be regarded as a model of thin-walled orifice plate.

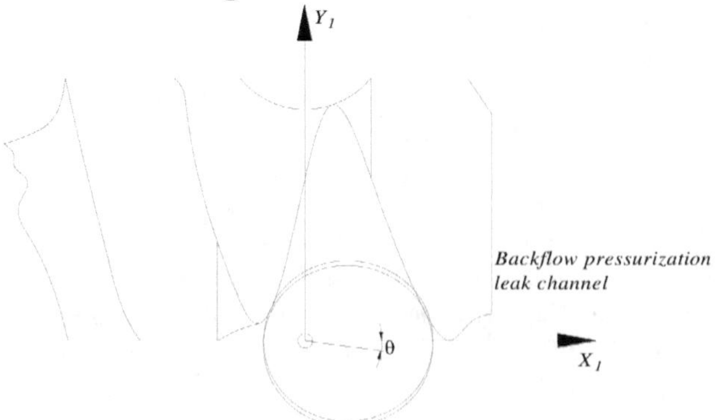

Figure 2. Backflow pressurized leakage channel

2.2 Sealing line gap leakage channel

Sealing line gap leakage channel is the gap between sealing disc and screw rotor in meshing movement. During the meshing movement of the screw rotor and the sealing disc, the contact parts of both take on the shape of curves. Therefore, the clearance leakage model of the sealing line gap can be simplified to the flow model of the zoom nozzle(as shown in figure 3).

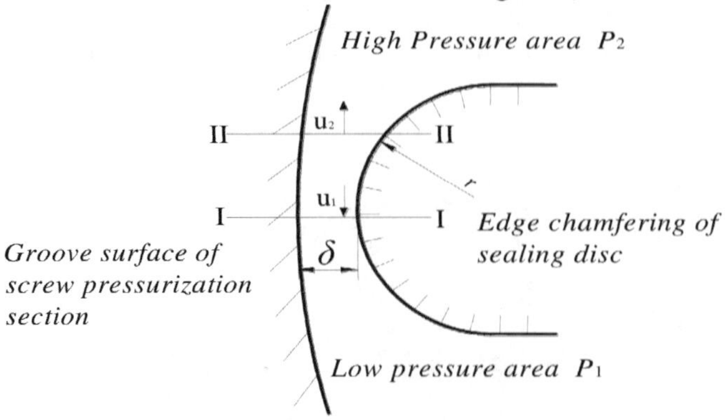

Figure 3. Sealing line gap flow model

During the working process of the screw pump, the relative velocity between the screw groove and the sealing disk is u (the velocity direction of the sealing disc is perpendicular to the paper surface, and the screw velocity direction is along the tangent direction of the groove surface). The upper part of the sealing disc is the high-pressure area and the lower part is the low-pressure area. The average pressure and velocity of the fluid medium in the high-pressure area are respectively shown by p_1 and u_1, and the average force and velocity of the fluid medium in the low-pressure area are respectively represented by p_2 and u_2. The relative motion velocity of the screw rotor is equal to the velocity in the low-pressure region. The width of engagement gap leakage between the sealing disc and the screw rotor is expressed by δ , and the length of the leakage channel is expressed by L. By ignoring the influence of the height of the two sections, the equation can be obtained as follows:

$$p_1 + \rho g h_1 + \frac{1}{2}\rho\alpha_1 u_1^2 = p_2 + \rho g h_2 + \frac{1}{2}\rho\alpha_2 u_2^2 + p_{\substack{Local\\loss}} \qquad (1)$$

Where α_1 , α_2 respectively represent the correction coefficients of liquid kinetic energy on two different sections. The main function of α_1, α_2 is to correct the pressure loss caused by pipe diameter shrinkage in the shrinkable tube flow. In the flow model of the zoom nozzle, the pressure energy loss is mainly caused by the sudden reduction of the liquid cross section, and the liquid viscosity could be ignored in this case, the value of α_1, α_2 is 1 here. $p_{\substack{Local\\loss}}$ represents the local pressure loss in the flow process, and its mathematical expression can be obtained as:

$$p_{\substack{Local\\loss}} = K \cdot \frac{1}{2}\rho u_2^2 \qquad (2)$$

Where K represents the local loss factor of the rounded inlet, which is determined by the ratio of the radius of the sealing disc to the clearance distance. After looking up the table1, K is selected here as 0.04.

Table 1. The local loss factor of the rounded inlet

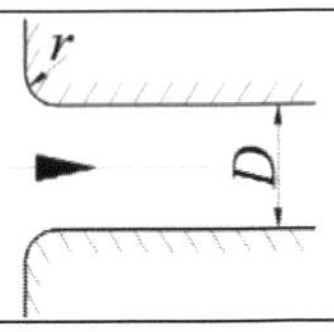	r/D	0.02	0.06	≥0.15
	K	0.28	0.15	0.04

Combined with the geometric model of the sealing line gap leakage channel, then the mathematical expression of sealing line gap leakage can be obtained.

2.3 The leakage channel between the inner wall of the pump shell and the screw rotor/the seal disc

All of the leakage channel between the inner wall of the pump shell and the screw rotor/the seal disc can be equivalent to a parallel plate model. In this paper, the leakage channel between the inner wall of the pump shell and the screw rotor is taken as an example to illustrate. As shown in figure 4, in combination with the working principle of the DSSP, there are two flow forms in the leakage zone formed by the gap between the inner wall of the pump shell and the screw rotor. One is a differential pressure flow caused by the pressure difference between a high pressure area and a low pressure area, the other part is the shear flow leakage caused by the relative movement of the screw rotor and the sealing disc.

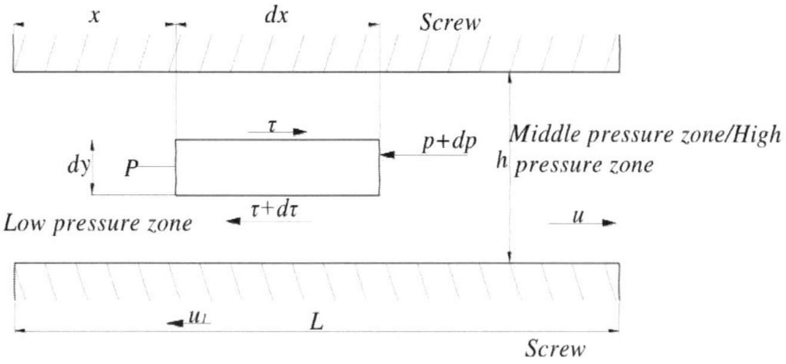

Figure 4. Schematic diagram of flow in parallel plate model

A fluid element is selected as the research object between the leakage channels as shown in figure.4. The length, width and height of the fluid micro element are represented by dx, dy and dz, where the direction of Z is perpendicular to the paper. During the working process of the DSSP, the inner wall of the pump shell is in a static state, the distance between the two plates is expressed as h, the width of the parallel plate is *b*, and the length of the parallel plate is *L*. Taking the fluid microelement as the research object, the compressive stress acting on the *yz* plane can be expressed as *p* and *p* + *dp*, and the shear stress acting on the *xz* plane can be expressed as τ and $\tau + d\tau$. The force analysis of fluid microelement is carried out and we can get the following expression:

$$pdy + d\tau dx = \left(p + dp\right)dy + \left(\tau + d\tau\right)dx \tag{3}$$

According to Newton's law of internal friction, the expression of τ can be obtained as follows:

$$\tau = \mu \frac{du}{dy} \tag{4}$$

Substitute the expression for τ into equation 2 and the following formula can be obtained:

$$\frac{d^2u}{dy^2} = -\frac{1}{\mu} \cdot \frac{dp}{dx} \tag{5}$$

By integrating the above equation, the velocity of the leakage fluid passing through the parallel plate can be obtained as follows:

$$u = -\frac{1}{2\mu} \cdot y^2 \cdot \frac{dp}{dx} + C_1 y + C_2 \tag{6}$$

Where C_1 and C_2 are the integral constant.

Combined with the geometric model between the inner wall of the pump shell and the screw rotor, then the mathematical expression of the leakage quantity between the inner wall of the pump shell and the screw rotor can be obtained.

3. Result and discussions

In order to calculate the leakage of the DSSP, the structural parameters of the DSSP used for theoretical research are shown in table 2, and the working parameters are shown in table 3. Based on the mathematical model of the leakage channel, the influences of working parameters such as outlet pressure, screw speed and liquid dynamic viscosity on the leakage characteristics are analyzed.

Table 2. The structural parameters of DSSP

Parameters	value
Screw radius /mm	100
Center distance /mm	100
Radius of sealing disc /mm	60
Eccentricity /mm	30

Table 3. The working parameters of DSSP

Parameters	value
Outlet pressure /kpa	800
Inlet pressure /kpa	101.315

Screw speed / n / r.s^{-1}	10
Dynamic viscosity coefficient	1

3.1. Leakage characteristic analysis

There are many leakage channels in the working process of DSSP. In combination with the above analysis of the mathematical models of each leakage channel, 0.3s-0.4s cycle is selected to analyze and study the leakage amount, and the changes of leakage amount in different leakage channels during the rotation of screw pump are shown in figure 5.

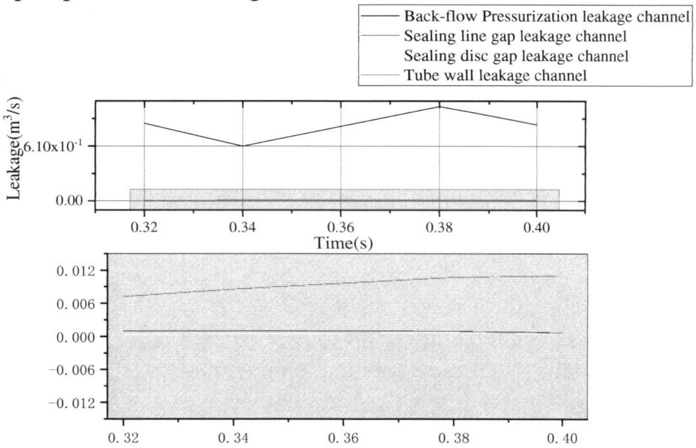

Figure 5. The amount of leakage in a cycle of different leakage channels

It can be seen from figure.5 that the leakage amount of the backflow pressurization leakage channel is much higher than that of other leakage channels in a rotation cycle, so it can be explained that the backflow pressurization leakage channel is the leakage channel with the largest proportion in the working process of the DSSP, and is also the key factor to improve the pressure of the medium in the working cavity.

In order to further analyze the leakage characteristics of the DSSP in the working process,the total leakage amount of DSSP is obtained by adding the leakage amount of each leakage channel in figure 5 . As shown in figure 6, according to the relationship between the leakage amount and the time step of the screw rotor in the figure, it can be seen that the total leakage amount changes periodically with the rotation of the screw rotor. The total leakage of screw pump in a rotation period decreases first, then increases and then decreases.

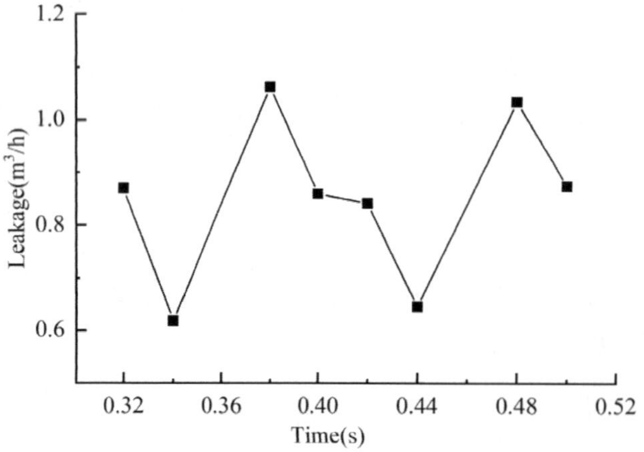

Figure 6. Variation law of total leakage

3.2. The influence of each working parameter on the leakage quantity

The change of working parameters during the operation of the DSSP will also have an influence on the change rule of the leakage quantity . In order to further study the leakage characteristics of the DSSP, this paper analyzes the influence law of outlet pressure, screw speed and medium viscosity of screw pump on the leakage quantity. In order to increase the contrast, a single variable was selected for each parameter based on the basic value. The results are shown in figure 7 to figure 9.

It can be seen from figure 7 that in the two rotation cycles selected, the variation trend of the leakage amount under different outlet pressures is similar. At the same time , the greater the outlet pressure, the greater the leakage, this is because that the DSSP is mainly dependent on the backflow leakage of the medium to achieve pressurization. When the inlet pressure is constant, the higher the outlet pressure is, the greater the pressure gradient along the axis direction of the screw rotor is, and the leakage amount caused by the pressure difference also increases accordingly.

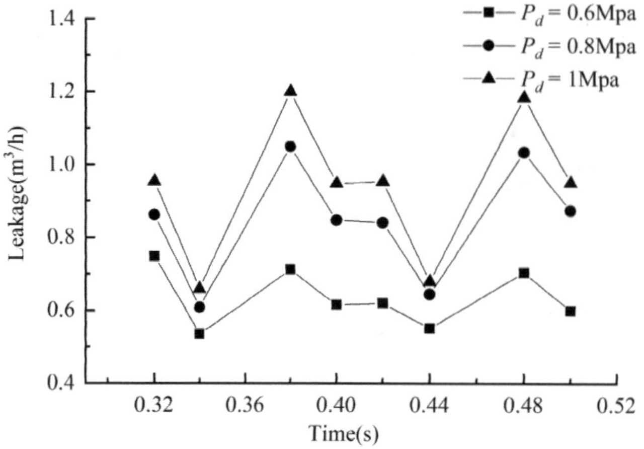

Figure 7. Variation law of total leakage under different outlet pressure

The influence of screw speed on leakage of DSSP is shown in figure 8. It can be seen from figure 8 that the change trend of leakage quantity of DSSP is similar under different screw speeds. And the leakage quantity will increase with the increase of screw speed at the same time. This is because the shear flow caused by the relative velocity between the moving parts on both sides of the clearance will also affect the leakage amount. Therefore, the increase of screw speed will increase the speed of the relative moving parts of each clearance, and the leakage amount caused by the shear flow will also increase accordingly.

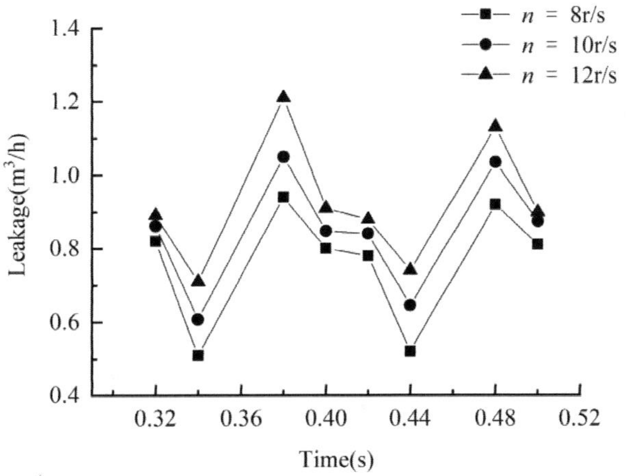

Figure 8. Variation law of total leakage at different screw speed

The influence rule of the dynamic viscosity of the conveying medium on the leakage of the DSSP is shown in figure 9. According to the changing trend of the curve in the figure, the variation law of leakage amount is similar under different media conditions. And the larger the dynamic viscosity coefficient of the medium, the smaller the total leakage of the DSSP. This is because the leakage channels of disc seal single screw pump are narrow clearance, fluid through each leakage channel resistance loss will increase with the increase of medium viscosity. As a result, the pressure drop is reduced and the leakage of DSSP is eventually reduced.

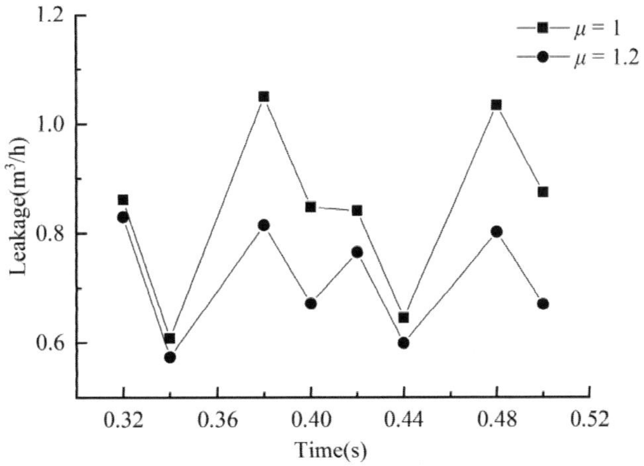

Figure 9. The change law of total leakage under different dynamic viscosity coefficient of medium

4. Conclusion

In this paper, the theoretical calculation mathematical model of the leakage amount of each leakage channel is established according to the working principle and the leakage geometry model of the DSSP. Based on the above mathematical model, the leakage characteristics of DSSP were studied and the following conclusions can be drawn from the research:

1) In the working process of the DSSP, the total leakage of the screw pump varies periodically. In a rotation period, the total leakage of DSSP will show a trend of decreasing first, then increasing and then decreasing.
2) The leakage through the backflow pressurization leakage channel is much larger than that of other leakage channels, which has the greatest impact on the transmission efficiency of the DSSP.
3) The variation trend of leakage is similar under different working parameters. The total leakage amount per unit time will increase with the increase of outlet pressure and screw speed. When the transmission medium's dynamic viscosity coefficient increases, the corresponding leakage will be reduced, and the transmission medium's dynamic viscosity is the main factor that affecting the leakage of the DSSP.

Acknowledgements

This work is supported by National Natural Science Foundation of China [NO. 51706247] and the Fundamental Research Funds for the Central Universities and the Opening Fund of National Engineering Laboratory of Offshore Geophysical and Exploration Equipment [NO. 20CX02312A].

References

[1] Evans M N C, Chidinyane T M, Oluwademilade M F, et al. Biosurfactant assisted recovery of the C5-C11 hydrocarbon fraction from oily sludge using biosurfactant producing consortium culture of bacteria[J]. J Environ Manag, 2017, 196:261-269.
[2] Lin B C, Wang J, Huang Q X, et al. Effects of potassium hydroxide on the catalytic pyrolysis of oily sludge for high-quality oil product[J]. Fuel, 2017, 200: 124-133.
[3] Wang J, Han X, Huang Q X, et al. Characterization and migration of oil and solids in oily sludge during centrifugation[J]. J Environ Technol, 2018, 39(10):1350-1358.
[4] Zhang C, Han L, Shi X. Modified Assessment Methodology for Mechanical Recovery Capacity for Oil Spill Response at Sea [J]. Aquatic Procedia, 2015, 3:29-34.
[5] Yu F, Li J, Cui S, et al. A hindcast method to simulate oil spill trajectories for the Bohai Sea, Northeast China[J]. Ocean Engineering, 2016, 124:363-370.
[6] Singh A, Asmath H, Chee C.L, et al. Potential oil spill risk from shipping and the implications for management in the Caribbean Sea[J]. Marine Pollution Bulletin, 2015, 93(1-2):217-227.
[7] Du X H. Research on mechanics properties and the wear failure of the screw-bushing pairs of oil extraction progressing cavity pump[D]. Daqing: Daqing Petroleum Institute, 2010.
[8] Lin S D. Research on Mechanical analysis and wear of single screw pump[D]. Xian: Xi'an Shiyou University, 2014.
[9] Shi J Q, Fang J K, Guo Y, et a1. Test research on oil screw pumps wear mechanism[J]. J Sci Eng, 2012, 12(05):1151-1154.
[10] Johansson A. Screw pump. WO 91/14869, 1991.
[11] Shen Y F. Study on the meshing characteristics and working mechanism of disc sealed single screw pump[D]. Qingdao:China University of Petroleum(East China), 2018.
[12] Wang Z L, Shen Y F, Wang Z M, et a1. Research on the meshing characteristics of disc-seal single screw pump[J]. J Xian Jiaotong Univ, 2018, 52(05):69–74+101.
[13] Wang Z L, Wang H, Wang Z M, et a1. Study on the suction capacity of disc-seal single screw pump used in energy recovery systems[J]. Int J Refrig, 2020, 112:333-340.
[14] Wang Z L, Shi H W, Wang S Y, et a1. Study on the operating characteristic of disc seal single screw pump used in energy recovery systems[J]. Int J Refrig, 2020, 118:336-344.

Design optimization and test of the novel FeTu 'compander', utilising organic fluids within a closed cycle for HVACR applications

J Subert[1], J P Fenton[1,3], K Hinchliffe[1] and I M Arbon[2]

[1] FeTu, The Wharf, Gas Works Lane, Elland. HX5 9HH, UK
[2] Engineered Solutions, Merkland, Pinmore, Girvan, KA26 0TE, UK
[3] Author to whom any correspondence should be addressed: info@fetu.co.uk

Abstract: HVACR performance is influenced by the efficiency of the compression and expansion process. The ability to operate oil-free; with non-toxic, inert, natural working fluids such as air or CO2 holds high environmental value.

This paper presents data of the FeTu™ compander operating in a closed cycle, for means of heating and cooling. A novel system, where compression and expansion is a continuous and simultaneous process; inherently regenerating energy to assist the compression process by expanding within a positive displacement chamber. This study documents the oil-free performance with organic working fluids pertaining to HVACR applications.

Commercial 1D modelling software; GTSuite (GTS) is used to optimise cooling capacity (Q) and Coefficient of Performance (CoP). Different volume ratios (VR), pre-charge pressures, temperature inputs (ΔT) and running speeds are analysed. A multi-faceted test rig is built to compare and validate all performance data against the GTS fluid model which, once calibrated, can then be used with a high degree of certainty in predicting the performance of other applications and ΔTs. Increasing the system pre-charge pressure has a proportional impact on capacity, so this phenomenon is tested to consider the conflicting impacts of higher pressure against increased viscosity, with respect to volumetric efficiency.

The CoP of the system ranges from 1 to 5 across typical HVACR ΔTs. Low VR offers highest CoP but lowest Q, so the optimal trade-off is explored. The strengths & weaknesses are then comparatively documented against incumbent vapour cycle systems.

1. Introduction

1.1. Literature Review

The use of fluorinated refrigerants (F-gases) specifically Chlorofluorocarbons (CFC's), hydrofluorocarbons (HFC's) and hydrochlorofluorocarbons (HCFC's) have a known and significant detrimental effect on the environment. Under the 1987 Montreal Protocol, both CFC's and HCFC's have been phased out in developed countries - January 1996 and 2020 respectively [1].

HFC's have been widely used as a replacement refrigerant for HCFC's, as they do not contain chlorine atoms, it has no effect on the O-Zone layer. However, whilst HFC's have zero O-Zone Depletion Potential (ODP), their Global Warming Potential (GWP) is very high. Furthermore, in comparison to chlorine containing refrigerants, HFC's last longer in the atmosphere thus their usage is regulated via the Kyoto Protocol [2]. R134A, a widely used HFC, has a GWP 1430 times that of CO_2

[3]. Chen et al [4] states the 'leakage of R134A in automotive air conditioners can reach up to 6%-9% of the total charge'. Refrigerant leakage associated with use in supermarkets is estimated between 9% and 25% although these rates have been reducing [5]. HFC's directly contribute to global warming, their phase out coupled with The Paris Agreement and the United Kingdom's own legislation to achieve net zero by 2050 are driving the demand for natural refrigerant alternatives.

Emani and Mandal [6] state the basic properties necessary to be considered an effective refrigerant. Characteristics include being harmless, non-flammable, non-explosive, non-corrosive and non-toxic. The refrigerant should demonstrate a high Coefficient of Performance (COP) within the working temperature range. The pressures used should enable practical sizing of both the pipes and the compressor. Critically, any ideal working fluid should be zero GWP and very low ODP.

Air is one such fluid, free and abundant, it is non-toxic, non-flammable, zero ODP and zero GWP. Yang and Yang [7] describe the widespread use of air cycle machines as part of an aircraft's environmental control system, citing its high reliability and low maintenance workload. Shengjun, Zhenying and Lili [8] agree that their ease of maintenance, high reliability and no leakage concerns are distinct advantages. However, low energy efficiency has prevented wide uptake in the refrigeration and air conditioning sector up to date.

The FeTu open cycle 'compander' discussed by Fenton et al [9] sees high COP, overcoming the performance limitations historically associated with conventional air cycle HVACR systems. However, it is predicted that closed cycle holds a distinct advantage, giving the ability to pre-charge the system with pressure, scaling the capacity for the same geometric sized machine.

This paper investigates the closed cycle performance using air, for typical HVACR applications, testing different sized machines, Built In Volume Ratios (BIVR), external temperature differentials (ΔT) and pre-charge pressures. The results are simulated using a mathematical approach and a GT Suite model which validate each other. A closed cycle, proof of concept test is also completed in order to verify the real-world capability of the machine.

1.2. FeTu Concept

The FeTu™ Roticulating™ unit is a positive displacement 'compander' in which compression and expansion occur simultaneously with two pairs of opposed double-acting chambers. Driveshaft rotation operates a spherical rotor which defines the four chambers and allows continuous flow of fluid via the suction and discharge ports. Each of the four chambers reduce in volume by 250:1 [9]. The device is designed with micron level internal clearance, and although lubricant free, precision manufacturing allows high volumetric efficiency, recorded at up to 97% in compressor tests of Zhang [10].

By designing a BIVR between Chamber ***A*** and ***B*** of Figure 1 and 2, rotation of the driveshaft allows the 'compander' principle to operate by creating a positive or negative pressure ratio in the flow path which joins the chambers, creating a change in temperature. A Heat eXchanger (HX) is utilised in the flow path for heat addition or extraction, forming a simple HVACR system with only two moving parts. The 'compander' system can run open or closed cycle, is scalable, modular and at variable speed to invoke analogue heat transfer rates (Q), dependant on demand. [9].

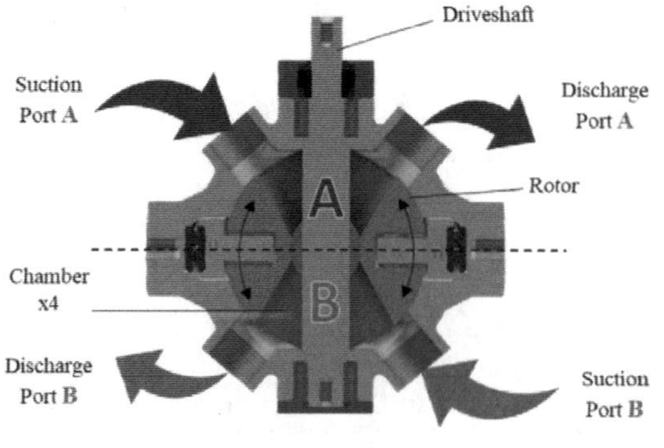

Figure 1: Internal Operation [9].

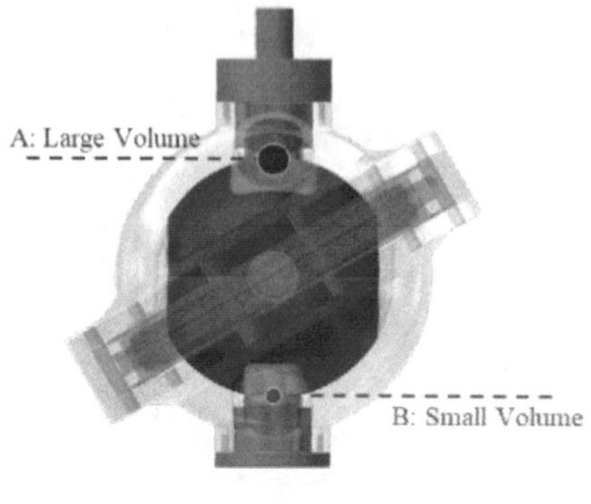

Figure 2: Showing Volume Ratio of A – B.

2. Test Methodology

2.1. Proof of Concept Test

The first 'compander' prototype that was manufactured was an 84mm, BIVR:1.464 unit. To verify that the closed cycle HVACR system is operational in a real-world application, a small, in house test rig was designed with two HX, four Omega pressure transducers (0-10bara, ±0.05%) and four type K thermocouples (0 to 200°C, ±0.4%), in Figure 3 and 4. Data was recorded using an Omega data capture unit.

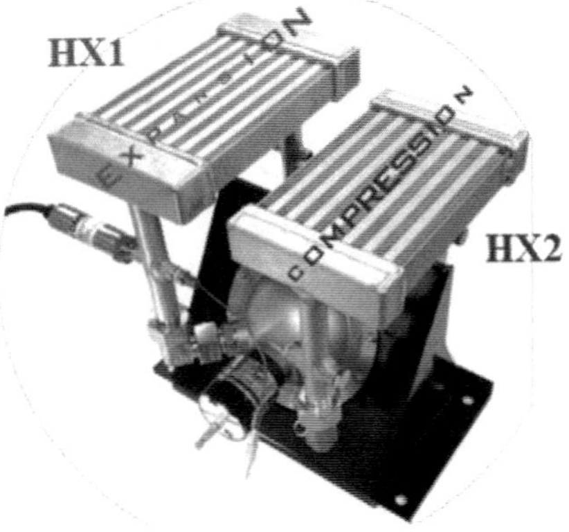

Figure 3: Closed cycle test rig.

Figure 4: Closed cycle schematic.

2.2. Mathematical Approach

The process followed in this analysis is broken into four discrete instances in time within a 360deg shaft rotation, schematised by Figure 5 below. There are two compressions (HX2) and two expansions (HX1) each revolution, as well as two instances of four locked volumes (top dead centre -TDC). The Volume, Pressure, Temperature and Mass are calculated at each moment based on Equations 1-37. Q, W and COP are then calculated. The knowns and unknowns of the method are summarised in Table 1:

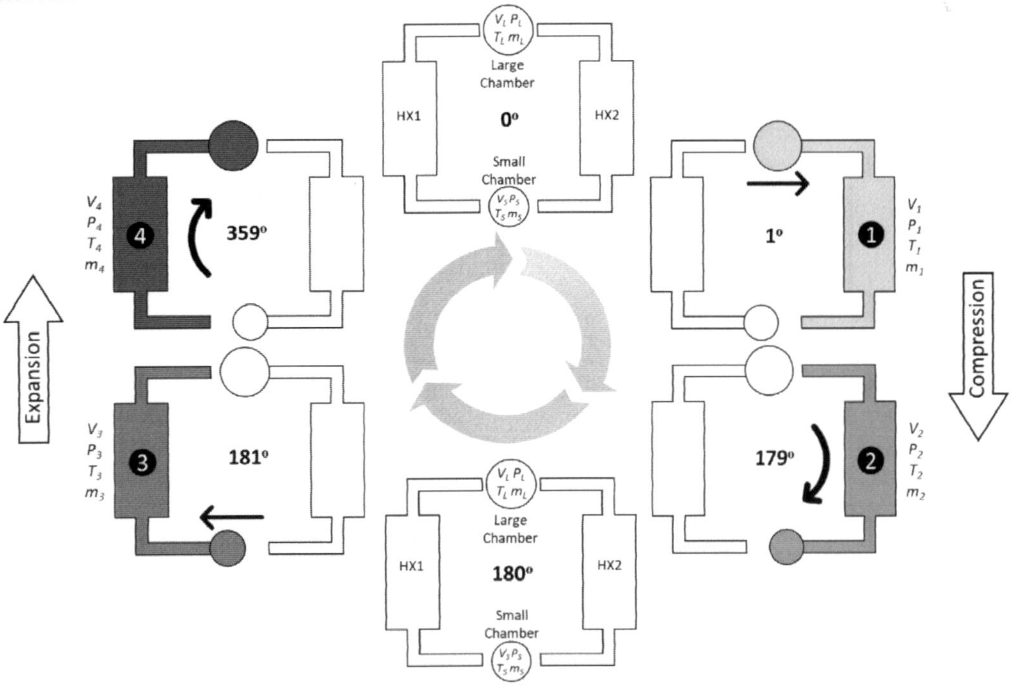

Figure 5: Mathematical Approach - Cycle Analysis.

Table 1: Knowns and Unknowns pre-process.

Inputs (Knowns)	Outputs (Unknowns)
Gas Properties: R and Cv	Polytopic index of compression and expansion: n
Motor Speed ω (r/min) & Frequency: $F = \frac{2\,x\,\omega}{60}$ (s^{-1})	System Pressures (Pa) and Pressure Ratio: PR
Volumes of chambers and HXs: V_L, V_s, V_{HX} (m^3)	System Temperatures and Temperature Ratio: TR
Pre-Pressure in the system: P_{pc} (Pa)	Stabilised Mass: m_{T2} (kg) and Mass flow $\dot{m}$ (kg/s)
Temperature at Pre-charge: T_{pc} (K)	Work in/out: W (J)
External Temperatures: $T_{HX\ External}$ (K)	Cooling/Heating Power in/out: Q (J)
Initial Total Mass after Pre-charge: m_{T1} (kg)	Coefficient of Performance: COP

2.2.1. Mass flow in HX1 and HX2 are equal at stabilisation.

$$\dot{m}_{HX1} = \dot{m}_{HX2} \tag{1}$$

$$\frac{2\,x\,\omega}{60}\ x\ m_L = \frac{2\,x\,\omega}{60}\ x\ m_s \tag{2}$$

$$m_L = m_s \tag{3}$$

$$PV = mRT\ \therefore\ m = \frac{PV}{RT}\ (gas\ law) \tag{4}$$

$$\frac{P_L V_L}{RT_L} = \frac{P_S V_S}{RT_S} \tag{5}$$

$$\frac{P_S}{P_L} = \frac{T_S}{T_L}\ x\ \frac{V_L}{V_S} \tag{6}$$

$$P_L = P_4\ , T_L = T_4, P_s = P_2\ , T_s = T_2 \tag{7}$$

$$\frac{P_2}{P_4} = \frac{T_2}{T_4}\ x\ \frac{V_L}{V_S} \tag{8}$$

$$P_2 = P_4\ x\ \frac{T_2}{T_4}\ x\ BIVR \tag{10}$$

Assume that the Heat Exchanger transfers all heat available to system.

$$T_{HX1\ External} = T_4 \tag{11}$$

$$T_{HX2\ External} = T_2 \tag{12}$$

$$P_2 = P_4\ x\ \frac{T_{HX2\ External}}{T_{HX1\ External}}\ x\ BIVR = P_4\ x\ TR\ \ x\ BIVR \tag{13}$$

2.2.2. Conservation of Mass - the total mass at rest is equal to the total mass at stabilization. Total mass before operation:

$$m_{T1} = \frac{P_{PC}\ x\ V_{Total}}{R\ x\ T_{PC}} \tag{14}$$

Total mass after stabilisation:

$$m_{T2} = m_{HX1} + m_{HX2} + m_L + m_S \tag{15}$$

$$m_{T2} = \frac{P_4 V_{HX1}}{RT_4} + \frac{P_2 V_{HX2}}{RT_2} + \frac{P_4 V_L}{RT_4} + \frac{P_2 V_s}{RT_2} \tag{16}$$

$$m_{T2} = m_{T1} \tag{17}$$

$$R\ x\ m_{T1} = P_2\left(\frac{V_{HX2}+V_S}{T_2}\right) + P_4\left(\frac{V_{HX1}+V_L}{T_4}\right) \tag{18}$$

Sub in Equation 13 into 18:

13. $$P_2 = P_4\ x\ TR\ \ x\ BIVR \tag{13}$$

18. $$R\ x\ m_T = P_2\left(\frac{V_{HX2}+V_S}{T_2}\right) + P_4\left(\frac{V_{HX1}+V_L}{T_4}\right) \tag{18}$$

$$R\ x\ m_T = \left(P_4\ x\ TR\ x\ BIVR\ x\left(\frac{V_{HX2}+V_S}{T_2}\right)\right) + P_4\left(\frac{V_{HX1}+V_L}{T_3}\right) \tag{19}$$

$$P_4 = \frac{R\ x\ m_T}{\left(TRx\ BIVR\ x\left(\frac{V_{HX2}+V_S}{T_2}\right)\right)+\left(\frac{V_{HX1}+V_L}{T_4}\right)} \tag{20}$$

Finally, calculate P_1 with Equation 13.

2.2.3. Internal Energy Balance. There is a mixing of two locked volumes at stage 1-2 and 3-4. This occurs at a constant internal energy (U) since it is instantaneous with zero compression or expansion.

$$U_3 = U_{HX1} + U_s \quad (22)$$

$$U = m \; x \; Cv \; x \; T \quad (23)$$

$$m_3 \; x \; Cv \; x \; T_3 = (m_{HX1} \; x \; Cv \; x \; T_4) + (m_s \; x \; Cv \; x \; T_2) \quad (24)$$

$$T_3 = \frac{(m_{HX1} \; x \; T_4)+(m_s \; x \; T_2)}{(m_3)} = \frac{(m_{HX1} \; x \; T_4)+(m_s \; x \; T_2)}{(m_{HX1} + m_s)} \quad (25)$$

$$\therefore T_1 = \frac{(m_{HX2} \; x \; T_2)+(m_L \; x \; T_4)}{(m_{HX2} + m_L)} \quad (26)$$

Table 2: Summary of Equations.

	1	2	3	4
V	$V_{HX2} + V_L$	$V_{HX2} + V_s$	$V_{HX1} + V_s$	$V_{HX1} + V_L$
P	$\frac{m_1 R T_1}{V_1}$	$P_4 \; x \; TR \; x \; BIVR$	$\frac{m_3 R T_3}{V_3}$	$\frac{R \; x \; m_T}{\left(TR \; x \; BIVR \; x\left(\frac{V_{HX2}+V_S}{T_2}\right)\right)+\left(\frac{V_{HX1}+V_L}{T_4}\right)}$
T	$\frac{m_{HX2}T_2+m_L T_4}{m_1}$	$T_{HX2\ External}$	$\frac{m_{HX1}T_4+m_s T_2}{m_3}$	$T_{HX1\ External}$
m	m_2	$\frac{P_2 V_2}{R V_2}$	m_4	$\frac{P_4 V_4}{R V_4}$

2.2.4. Polytropic Index Calculation. To find *Q* capacity, the polytropic index each side is calculated. This will be influenced by the amount of heat transfer during the compression and expansion.

$$\frac{P_3}{P_4} = \left(\frac{V_4}{V_3}\right)^{n_{3-4}} \quad (27)$$

$$ln \frac{P_3}{P_4} = n_{3-4} \; ln \; \frac{V_4}{V_3} \quad (28)$$

$$n_{3-4} = \frac{\left(ln \frac{P_3}{P_4}\right)}{\left(ln \frac{V_4}{V_3}\right)} \quad (29)$$

$$\therefore n_{1-2} = \frac{\left(ln \frac{P_1}{P_2}\right)}{\left(ln \frac{V_2}{V_1}\right)} \quad (30)$$

2.2.5. Work. Note: for rate of Work (*W*) multiply by *F* (s^{-1})

$$\Delta W_{3-4} = \int_{V3}^{V4} PdV = \int_{V3}^{V4} \frac{P_3 V_3^{n_{3-4}}}{V^{n_{3-4}}} dV = P_3 V_3^{n_{3-4}} \int_{V3}^{V4} \frac{1}{V^{n_{3-4}}} = \frac{P_3 V_3^{n_{3-4}}}{1-n_{3-4}} [V^{1-n_{3-4}}]_{V3}^{V4}$$

$$\Delta W_{3-4} = \left(\frac{P_3 V_3^{n_{3-4}}}{1-n_{3-4}}\right)\left(V_4^{1-n_{3-4}} - V_3^{1-n_{3-4}}\right) \quad (31)$$

$$\therefore W_{1-2} = \left(\frac{P_1 V_1^{n_{1-2}}}{1-n_{1-2}}\right)\left(V_2^{1-n_{1-2}} - V_1^{1-n_{1-2}}\right) \quad (32)$$

$$W_{Total} = W_{3-4} + W_{1-2} \quad (33)$$

2.2.6. Cooling/Heating Capacity. Note: for rate of *Q* (W) multiply by *F* (s^{-1})

$$\textbf{1st law}: \Delta Q_{3-4} = \Delta U_{3-4} + \Delta W_{3-4} = m_{HX1} Cv(T_4 - T_3) + \frac{P_3 V_3 - P_4 V_4}{n_{3-4} - 1}$$

$$= m_{HX1} Cv(T_4 - T_3) + m_{HX1} R \frac{T_3 - T_4}{n_{3-4} - 1} \quad (34)$$

$$Q_{3-4} = m_{HX1} \; x \left(Cv + \frac{R}{1-n_{3-4}}\right) x \; (T_4 - T_3)$$

$$\therefore Q_{1-2} = m_{HX2} \; x \left(Cv + \frac{R}{1-n_{1-2}}\right) x \; (T_2 - T_1) \quad (35)$$

2.2.7. COP.

$$COP_{HX1} = \frac{Q_{3-4}}{W_{Total}} \quad (36)$$

$$COP_{HX2} = \frac{Q_{1-2}}{W_{Total}} \quad (37)$$

2.3. *GTSuite Model*

FeTu have currently manufactured three unit sizes: 84mm, 120mm, 215mm. A GTSuite model has been designed to replicate each sizes performance, utilised within a closed 'compander' cycle.

The model has four chambers, bound by a common shaft, which oscillate with set volume curves seen on Figure 6 (215mm), each 180°. All chambers expand and fully evacuate fluid twice every 360° rotation. The top and bottom chamber pairs operate 180° antiphase. The bottom two chambers are lesser in volume to the top two by a factor of BIVR. The volume flow dictated by the curves are emitted though suction and discharge ports of given opening areas, again plotted though 180° as seen in Figure 6.

The HXs have a given volume and set so they can accept/reject max Q to a given input temperature, replicating idealised performance and validating the method in Section 2.2. Other inputs include: speed (r/min), leakage, dead zones volumes and internal heat migration - set to zero for idealised comparison.

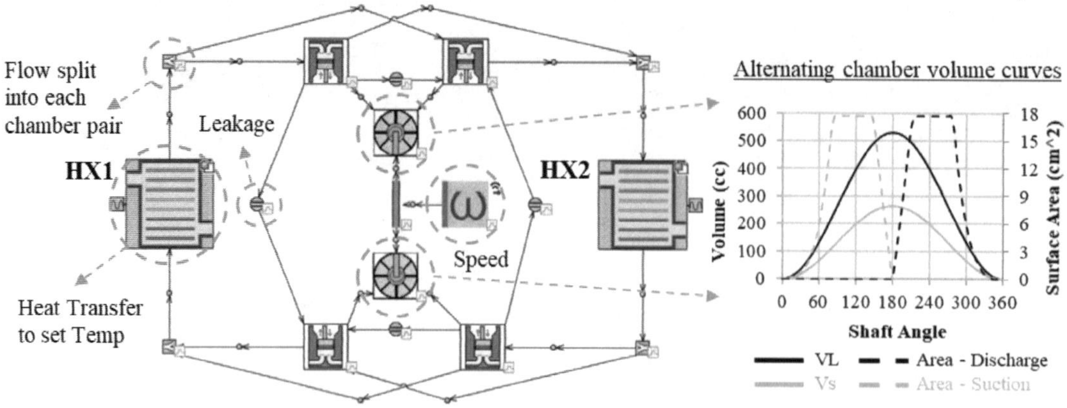

Figure 6: GTSuite Model with Volume and Port Area Curves.

3. Results

3.1. *Proof of Concept Test Results*

Figures 7 and 8 show real test data, with proof of concept, using the 1.46 BIVR, 84mm test rig (Figure 10). When run at 800r/min, HX1 becomes low and HX2 becomes high pressure and temperature.

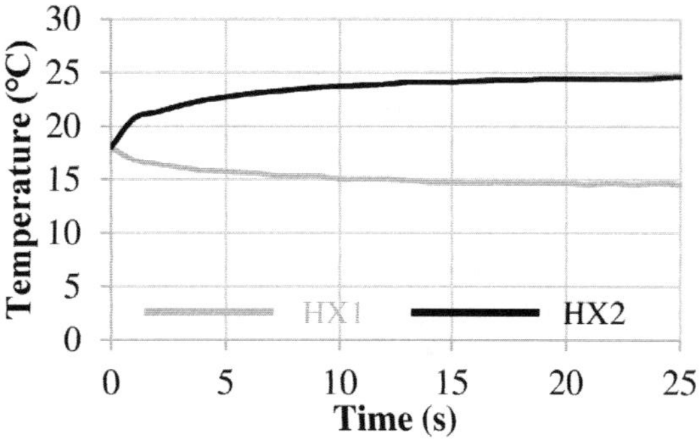

Figure 7: Temperature vs Time – real test.

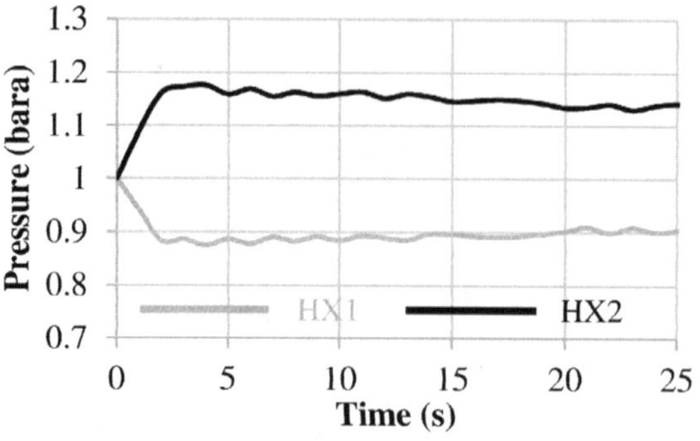

Figure 8: Pressure vs Time – real test.

3.2. *Validation of Mathematical and GT Model*

The results of the GTSuite simulations of Section 2.3 and the mathematical model of Section 2.2 agree entirely as seen in Figures 9 – 12 which plot stabilised P, $\dot{m}$, Q and COP over different BIVRs (1-3).

- Rotor D: 120mm
- Speed: 3000r/min
- T: HX1 & HX2 = 20°C
- Pre-charge: 10bara
- V: HX1= 4L, HX2 = 4L

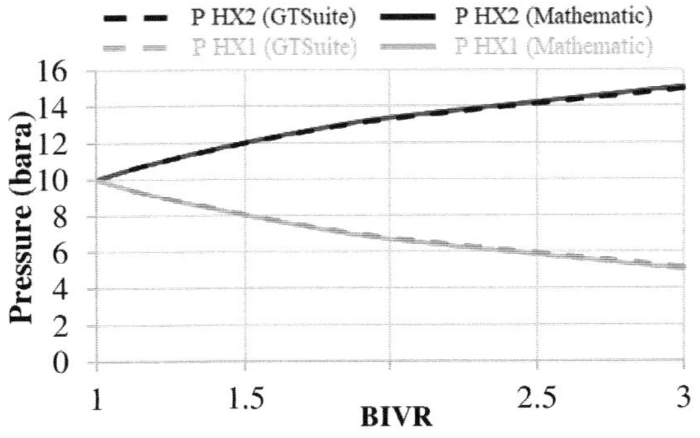

Figure 9: Pressure vs BIVR – Comparison.

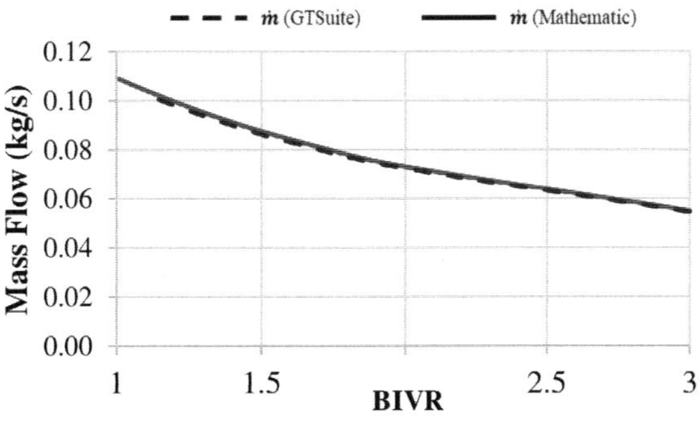

Figure 10: Mass Flow vs BIVR – Comparison.

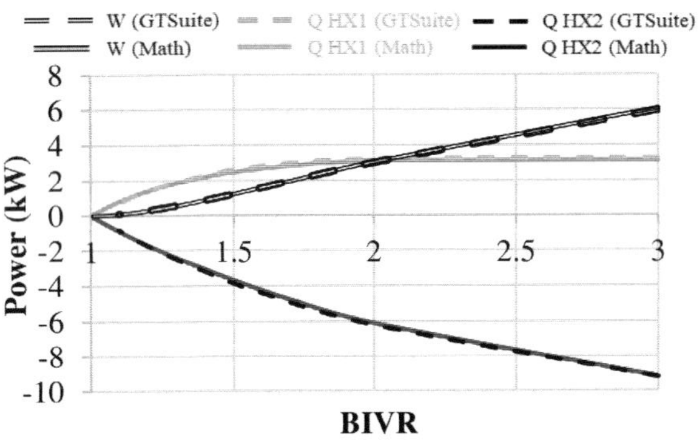

Figure 11: *Q* and *W* vs BIVR – Comparison.

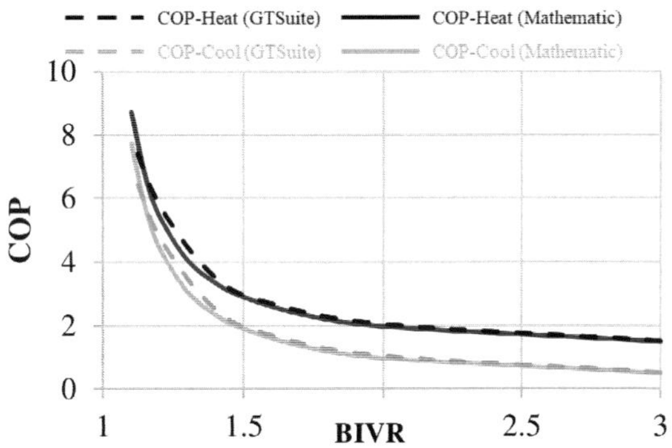

Figure 12: *COP* vs BIVR – Comparison.

3.3. Pre-charge Pressure Investigation

An investigation was completed to observe the effect of pre-charging the system with different pressures. *PR*, *Q*, *W* and *COP* were simulated across a range of BIVRs in Figures 13-16 respectively:

- Rotor D: 120mm
- Speed: 3000r/min
- T: HX1 & HX2 = 20°C
- Pre-charge: 1, 5, 10 bara
- V: HX1= 4L, HX2 = 4L

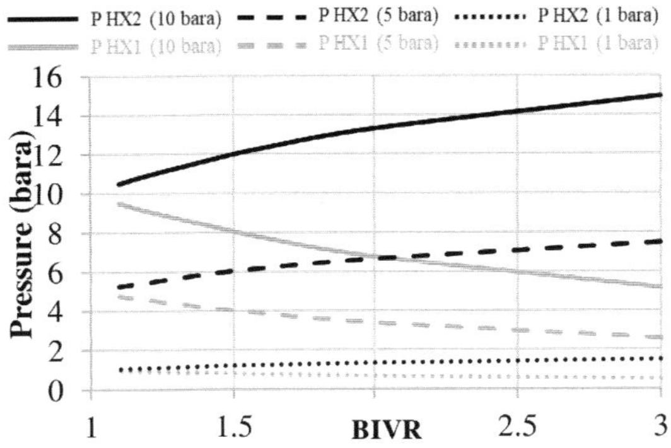

Figure 13: Pressure vs BIVR – Different Pre-charge.

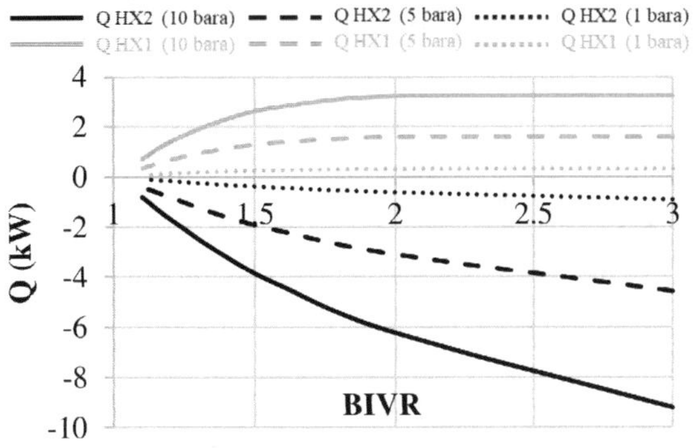

Figure 14: Temp vs BIVR – Different Pre-charge.

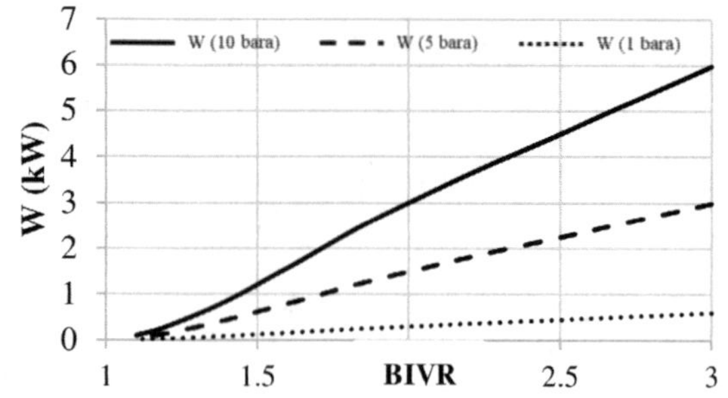

Figure 15: Work vs BIVR – Different Pre-charge.

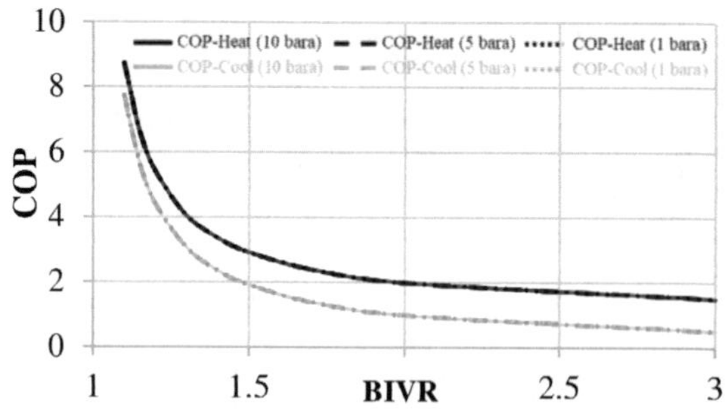

Figure 16: COP vs BIVR – Different Pre-charge.

3.4. Δ Temperature Investigation

An investigation was completed to observe the effect of different ΔT externally in Figures 17 and 18. Note: $\Delta T = T_{L\ External} - T_{R\ External}$. Test Conditions:

- *T*: HX1= 15, 20, 25°C
- *T*: HX2T = 25, 20, 15°C
- Rotor D: 120mm
- Speed: 3000r/min
- Precharge:1,5,10 bar
- *V*: HX1= 4L, HX2 = 4L

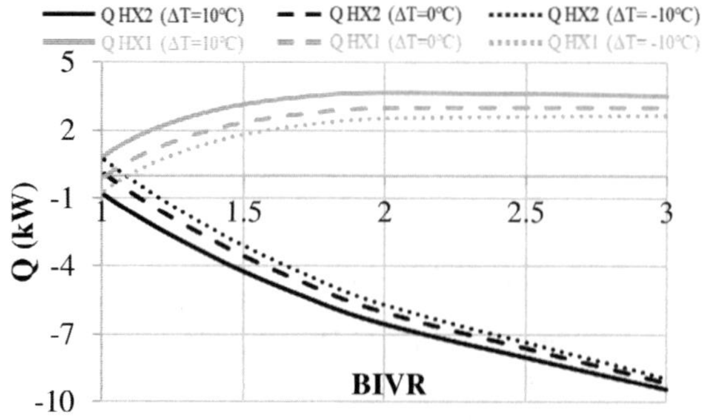

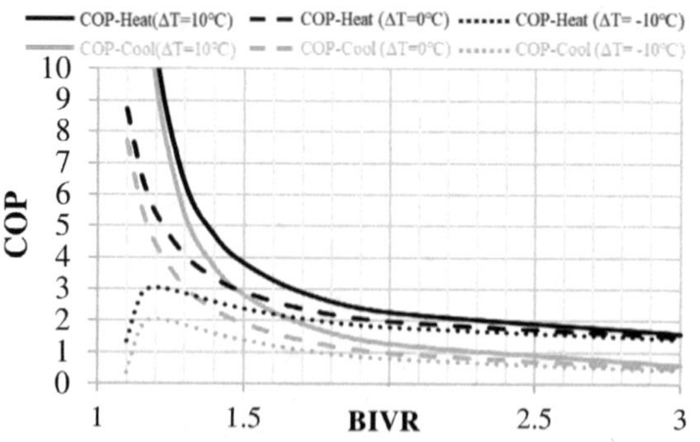

Figure 17: *Q* vs BIVR – Different *ΔT*.

Figure 18: *COP* vs BIVR – Different *ΔT*.

3.5. Rotor Size Investigation

The three currently manufactured units (84, 120, 215mm) of FeTu have been simulated across the BIVR range of 1-3 in Figures 19 and 20, plotting mass flow and Q. Case:

- Rotor D: 84, 120,215mm
- Speed: 3000r/min
- *T*: HX1&HX2 = 20°C
- Pre-charge: 10bara
- *V*: HX1 & HX2 2L, 4L, 10L

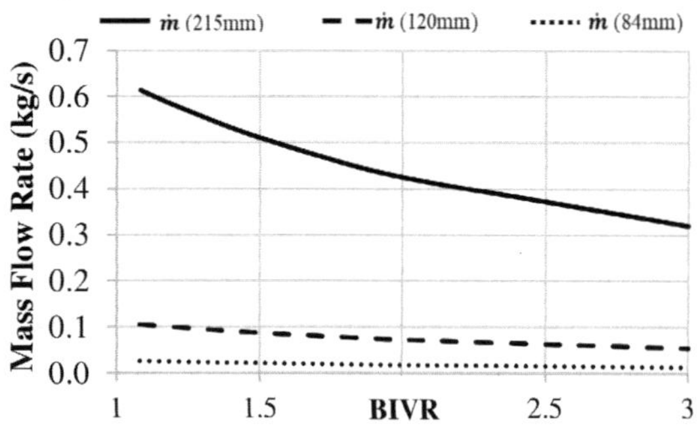

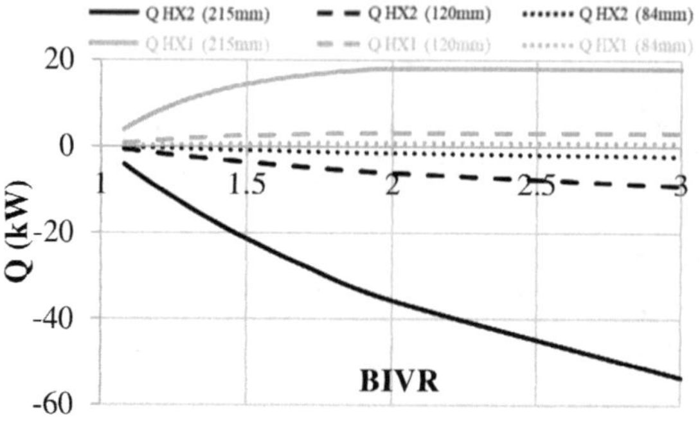

Figure 19: Mass Flow vs BIVR – Different Rotor D.

Figure 20: *Q* vs BIVR – Different Rotor D.

The temperature pre- and post-exchanger is a good indication of the min/max temperature attainable in the external cooling and heating. Figures 22 shows the temperature at probe points: A, B, C, D for each rotor size for the case above based on Figure 21 test arrangement.

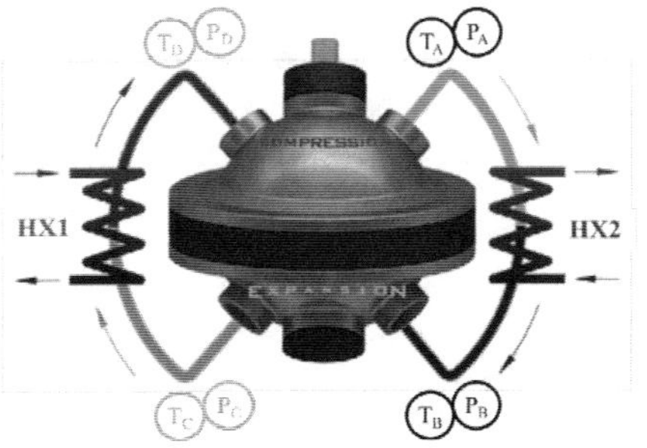

Figure 21: Closed cycle schematic T locations.

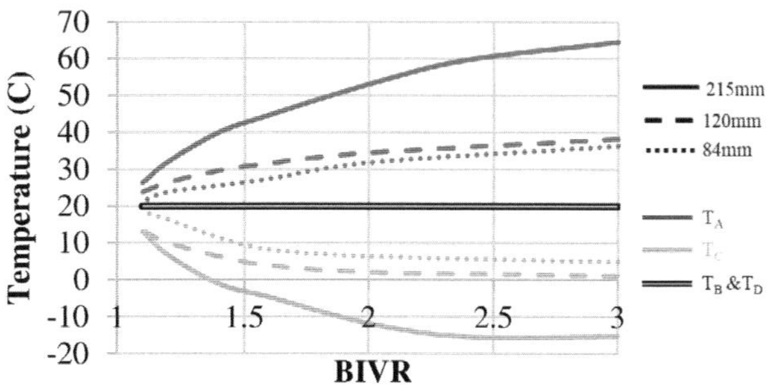

Figure 22: T_A, T_B, T_C and T_D vs *BIVR*.

4. Discussion

4.1. Validation (Section 3.1, 3.2)

The FeTu closed-cycle 'compander' is a new derivation of the TRL6 proven FeTu oil-free compressor. Although basic, the proof-of-concept test in Section 2.1 successfully proves that the theorised cycle behaves as expected; from start-up HX1 drops in pressure and temperature, and in contrast HX2 rises.

Figures 9-12 clearly show the agreeing results of both calculation methods - Section 2.2 and 2.3, giving complete validation. Although GTSuite provides a more extensive output than the mathematic approach, the more complex input settings and long simulation time means that it is less time efficient.

Running a *BIVR* of 1.0 creates a fluid pump system with zero pressure change, this is exhibited in Figure 9 where the HX1 and HX2 maintain the pre-charge pressure state (10bara). As *BIVR* trends away from 1.0 this promotes the 'compander' principle, lowering HX1 and raising HX2 in pressure. Greater *BIVR* leads to a larger PR. When there is temperature addition of ambient 20°C (as the case denotes), the rise and drop in pressure is of equal magnitude about the initial pre-charge pressure condition.

Increasing the *BIVR* means keeping V_L the same but reducing V_s to a smaller volume, because of this, the total stabilised mass flow of the system reduces, as seen in Figure 10. It is important to know that in the GTSuite model at stabilised conditions, the mass flow through HX1 and HX2 are equal.

Figure 11 shows that the Q for both heating and cooling tends to 0 at *BIVR*=1, this is because there is zero compression and expansion. From *BIVR*=1-1.5, there is a sharp increase in absorbed heat (Q_{HX1}) and rejected heat (Q_{HX2}). Q_{HX1} then starts to plateau after 1.5 meaning that for a cooling application, there is no longer an advantage of increasing the Volume/Pressure Ratio past *BIVR*=1.5. Q_{HX2} however does not plateau, therefore there appears to be no peak *BIVR* for heating. Using the 1st law of thermodynamics in a lossless system; $Q_{HX2} - Q_{HX1} = W$, dividing through by W and using Equation 36 and 37 give: $COP_{HX2} - COP_{HX1} = 1$. This law is upheld in both modelling techniques as seen on Figure 12 that shows Heating COP_{HX2} exactly 1 higher than Cooling COP_{HX1}. Figure 11 shows that the W tends to 0kW at low *BIVR* which leads to inordinately high *COP* (Figure 12). Higher *BIVR* sees the lowest COP values.

4.2. Investigations (Section 3.3-3.5)

The investigation of Section 3.3 shows that the pre-charge pressure increase proportionally scales the internal density, mass-flow, and heat capacity (Figure 14). The cost of this being extra input work as seen in Figure 15. Figure 16 shows that the thermodynamic *COP* is unaffected by internal pre-charge pressures.

The ΔT in HVACR systems is critical to the system performance. A large ΔT in favour of the operational HX typically results in higher levels of performance. This is seen in Figure 17 which shows larger cooling and heating potential at elevated ΔT. Changing external temperature subtlety, as investigated, has minimal impact on TR and as such PR (Equation 8). Because of this, W doesn't change as much as Q and therefore larger ΔT results in a greater *COP* (Equation 36 and 37) - seen in Figure 18.

The rotor Diameter increases the volumetric capacity of the machine, due to the spherical design: by a cubic factor. Larger machines running at the same speed will therefore exhibit a greater internal flow rate (Figure 19), resulting in an increase in Q (Figure 20). The W to compress and expand the fluid increases proportionally keeping *COP* constant with rotor size. Figure 22 shows that the larger rotor sizes produce a larger cooling/heating potential in terms of temperature change.

5. Summary

The paper explores two comparative modelling approaches, which agree, furthermore the application is verified by physical test. All objectives of the study were met. The success of this foundation work shows encouraging potential to function well within a HVACR scheme. Whilst open-cycle systems using air as the working fluid are commonplace to aviation; the ability to pre-charge the FeTu closed-cycle system, significantly increases the cooling and heating capacity for a given size. Due to its unique operation, high volumetric efficiency and small number of flow stages, the 'compander' seems to overcome *COP* limitations of air cycle HVACR, able to achieve a *COP* in excess of 5.

A complementary, mechanical GTSuite Model, which considers physical losses such as heat migration, port geometry, leakage and bearing friction is in development. Whilst not yet validated, simulations show interesting trends: As pre-charge pressure increases, the bearing loads remain constant but Q increases vastly, meaning that mechanical losses become a smaller factor, elevating the *COP*. However, larger pressures become port sensitive which limits *COP*, meaning each rotor diameter has an optimal pre-charge value for a given set of operational conditions. Furthermore, the extra W due to real losses, removes the unrealistically high *COP* at low *BIVR* (Figure 20), typically peaking around *BIVR*:1.2-1.6 and diminishing toward *BIVR*:1.0, depending on conditions.

Although successful, the data supporting Section 2.1 is scarce, so a large amount of performance data for the closed-cycle operation will be collected throughout 2021. Using this to calibrate the combined GTSuite thermo/mechanical model will allow FeTu to predict and optimise a wide range of applications.

Other natural working fluids will be tested, Supercritical CO_2 is of specific interest (Heat-to-Power).

6. References:

[1] Tsvetkov O B, Laptev Y A, Sharkov A V, Mitropov V V and Fedorov A V 2020 Alternative refrigerants with low global warming potential for refrigeration and air conditioning industries IOP. Conf. Ser.: Mater. Sci. Eng. 905 012070

[2] Dreepaul R K 2017 A study of alternative refrigerants for the refrigeration and air conditioning sector in Mauritius IOP Conf. Ser.: Earth Environ. Sci. 93 012054

[3] Bolaji B O 2020 Theoretical assessment of new global warming potential refrigerant mixtures as eco-friendly alternatives in domestic refrigeration systems J. Sci. Af. 10 e00632

[4] Chen X, Liang K, Li Z, Zhao Y, Xu J and Jiang H 2020 Experimental assessment of alternative low global warming potential refrigerants for automotive air conditioners application Case Stud. Therm. Eng. 22 100800

[5] Mylona Z, Kolokotroni M, Tsamos K M and Tassou S A 2017 Comparative analysis on the energy use and environmental impact of different refrigeration systems for frozen food supermarket application Energy Procedia 123 pp. 121 – 130

[6] Emani M S & Mandal B J 2018 The use of natural refrigerants in refrigeration and air conditioning systems: a review IOP Conf. Ser.: Mater. Sci. Eng. 377 012064

[7] Yang H and Yang C 2020 Thermodynamic characteristics and order degree of air cycle system Int. J. Refrig 118 pp 61 – 71

[8] Shengjun L, Zhenying Z and Lili T 2011 Thermodynamic analysis of the actual air cycle refrigeration system Syst. Eng. Proc. 1 pp 112- 116

[9] Fenton J, Subert J, Hinchliffe K, Bianchi G and Tassou S 2020 Air cycle feasibility using a novel single rotor compander for refrigeration and heating Proc. Int. Institute of Refrigeration Conf. Rankine 2020 Advances in cooling, heating and power generation 2020 July 27-31; Glasgow, United Kingdom, Paper ID 1167

[10] Zhang Y, Madamedon M, Copeland C, Fenton J P, Subert J, Hinchliffe K, Arbon I M and Leefe S 2019 The roticulating concept air compressor: experimental and numerical investigation IOP Conf. Ser.: Mater. Sci. Eng. 604 012070

Development of a Reconfigurable Prototype Peristaltic Compressor

M. Islam[1], C. Tubbs[2], C. Bengs[3], C. R. Bradshaw[4]

Center for Integrated Building Systems, Oklahoma State University, Stillwater, OK 74078, USA

[1] mazharul.islam10@okstate.edu, [2] colton.tubbs@gmail.com, [3] cbengs@okstate.edu, [4] craig.bradshaw@okstate.edu.

Abstract. Traditional positive displacement compressors require valves, internal volume ratios or both to operate. These two features generate the majority of the losses in compressors. The peristaltic compressor is a novel compressor architecture that feature valve-less compression with variable volume ratios, which provides the potential for highly efficient compression. This paper will present a design of a reconfigurable prototype peristaltic compressor and illustrate how the structure of the compression chamber and actuation pattern affect the compressor efficiency. The prototype utilizes a flexible diaphragm constrained and sealed against a female die to make a compression chamber. Linear actuators with male dies press the diaphragm sequentially against the female die to move and compress the fluid. Preliminary experiments were performed using air as the working fluid for volume ratios ranging from 1.143 to 8 and operational frequencies ranging from 1.28 to 2.31 Hz. These results suggest that the actuation pattern has an impact on the major parameters such as mass flow, displacement volume, and pressure ratio which influence the overall efficiency of the compressor.

1. Introduction

Refrigeration and air-conditioning systems are applications of the vapor compression cycle, which are widely used in various fields such as residential buildings, industry, aviation and transportation. Increasing demand for a more efficient vapor compression cycle has created the necessity for more innovative equipment, particularly for compressors. The compressor experiences significantly more losses than other components in a vapor compression cycle [1]. The main factors to these losses in positive displacement compressors are valves which cause leakage, port losses in the case of rotary and reciprocating compressors, and fixed volume ratios which cause over/under compression in the case of scroll and screw compressor [2].

A peristaltic compressor has the potential to operate at variable volume ratios without valves and is therefore an interesting candidate for efficient compression, particularly for the HVAC&R industry. The compressor consists of a linear cylindrical chamber. The chamber is formed by using a flexible material that serves as a diaphragm. The chamber can be compressed via electromechanical actuation methods to form a series of segmented compression pockets. Once adjacent electromechanical devices are actuated, a pocket is formed and is separated from the rest of the chamber. This compression is accomplished without the use of fixed volume ratios or valves. The ultimate goal is to evaluate this technology for HVAC&R purposes and understand its full potential in various applications.

A preliminary thermodynamic model of the peristaltic compressor has been developed, presented in a companion work by Islam & Bradshaw [3] using various diaphragm shapes, materials, actuation speeds and volume ratios which was leveraged to generate the design of a reconfigurable prototype by allowing the estimation of the required physical dimensions of the compressor. The prototype allows for a feasibility evaluation of this compression technology, providing an easy way to vary the volume ratio and diaphragm shape/materials to achieve the optimum capacity and volumetric efficiency. While the ultimate goal is refrigerant compression, air is currently being used as a working fluid, for experimental simplicity.

2. Literature Review

In recent years, diaphragms have been examined as a mechanism to be used as a compressor in vapor compression cycles. Previous work showed that the diaphragm offered the potential of a less leaky mechanism and ability to generate higher pressures [4]. The peristaltic actuation mechanism first introduced in a miniature compressor where a cylindrical elastic tube used to compress synchronously and discharge gas by using electrostrictive ceramic blocks. The efficiency of the system is very low due to energy consumption by the circuit and bigger size of the capacitors [5]. Saif et al. [6] developed an analytical model to predict the deformation and internal stress of the diaphragm. The model was studied under a circular cavity and a parabolic cavity condition. When the actuation occurred, in the first condition, a two-sided membrane coming towards to contact on a floor and for the second case, the upper membrane went down to the bottom layer of the cavity to compress the area. The result of the experiment indicated that the parabolic condition gives higher pressure. That implies, mass flow rate will be better when the compression chamber has a fixed gap area and the diaphragm actuation fills the area.

A computational analysis of a mesoscopic peristaltic compressor introduced a new actuation design where an electric field actuated a diaphragm to make a sinusoidal wave which moved the fluid through a channel. The compression ratio can be controlled by changing the depth of the cavity. The authors derived a geometric model, energy balance and diaphragm dynamics of the system. The theoretical analysis of the paper showed that the peristaltic compressor can perform the required work for a compressor. The analysis of physical testing is missing to evaluate the result and the author did not examine the heat transfer of the system [7].

Sathe et al. [8] developed a similar concept using progressive actuation of a single circular diaphragm in a compressor. An electrostatic actuation force was applied on the diaphragm in order to pull it down by using a DC voltage. In this segment, the inlet valve opens and refrigerant enters the chamber. After the suction stroke, a reversed applied voltage pulls off the diaphragm and opens the discharge valve to complete the compression stroke. A prototype was used to test the actuation of the diaphragm. The highest pressure rise of the prototype is 1.5 kPa, but the predicted pressure from the simulation was 30 kPa for R134a. The main limitation of this method is the sustainable pressure of the diaphragm [9].

Pressure and discharge volume rely on the actuation pattern of the diaphragm. Pressure rise to its maximum capacity when compression and discharge happen simultaneously. Performance of the compressor also increases when the forward stroke of the diaphragm occurs because the flow rate of the fluid is maximized at that point [10]. The main drawback of using electrostatic actuation is that, it may not be suitable for liquid because the electrostatic process may ionize the liquid. The displacement of the electrostatic actuation is also comparatively smaller than other actuation techniques. Chen et al. [11] proposed a rhombic micro displacement amplifier for a piezoelectric actuator which can improve the displacement by no more than 100 μm.

All the research for electrostatically actuated peristaltic compressors that use a diaphragm have some limitations and they are the theoretical model of microscale compressors. The efficiency of a compressor can be increased by reducing the irreversibility. The peristaltic compressor shows promise, as it can compress without an internal volume ratios and valves which leads the system to reduce the irreversibility [1]. Previous work in this area developed only rudimentary prototypes for feasibility purposes and did not highlight these features. This work will present a more comprehensive prototype design that allows the ability to study the influence of variable volume ratios, valve-less operation, and easily reconfigurable geometry.

3. Reconfigurable Peristaltic Compressor Mechanism

The actuation of the diaphragm occurs sequentially along a linear axis by using 10 linear actuators. The flexibility of the diaphragm allows the process to perform the compression without any discharge and suction valves. The internal volume ratio of the compression chamber can also be modified by changing the actuation sequence. Preliminary testing is performed using air as the working fluid in the compression chamber. Figure 1 presents a schematic of the proposed re-configurable peristaltic compressor prototype. The diaphragm is constrained and sealed against a female die which creates a cylindrical compression chamber. A male die (piston) is attached to a linear actuator and presses the diaphragm against the female die to provide compression by sealing the chamber synchronously.

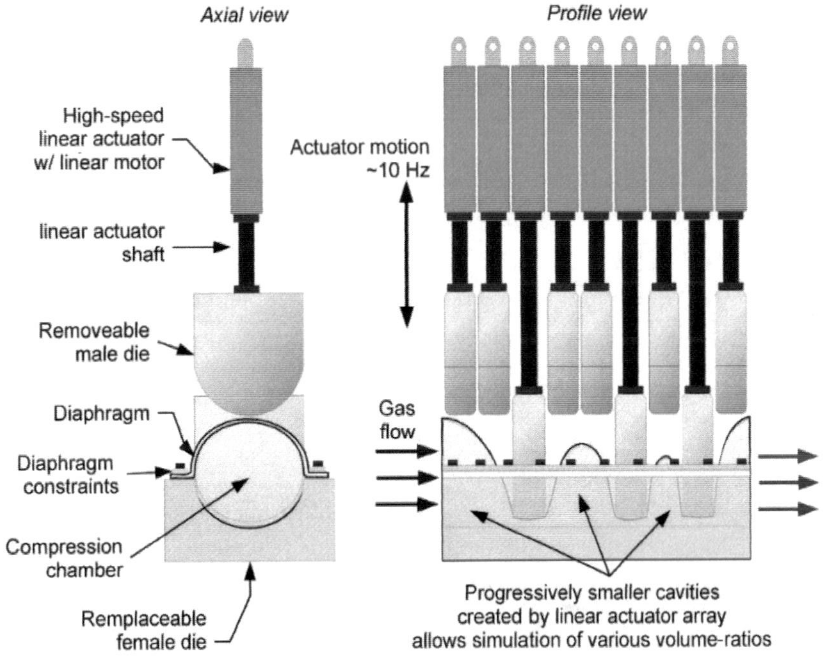

Figure 1: Reconfigurable prototype platform of a peristaltic compressor that has been designed for this study [12].

The actuators, which play a key role in the operation of the Peristaltic Compressor, are controlled by a programmable logic controller (PLC). These actuators are programmed using Rockwell Automation's Studio 5000 programming language. The program was designed to provide a straightforward and intuitive way to operate the compressor. As such, there are only three operations the user needs to perform: turning on the servos, selecting the appropriate motion pattern, and starting and stopping the compression cycle.

The user can control the compression ratio of the current test by selecting an actuator as a pivotal piston which indicates when the fluid from the compression chamber will begin to discharge or by selecting an actuator as valve which will create several compression pockets. In order to achieve different volume ratios, we present two different actuation modes: Mode1 and Mode 2. Each mode has several patterns that can be used to operate the compression process.

3.1. Mode 1 Actuation

For the first mode, the 1st and 10th piston will operate as the suction and discharge valves of the compressor. For example, consider a selection of the 7th piston as the pivotal piston, which is illustrated in Figure 2. Initially, all the pistons are up and the working fluid fills the diaphragm. Then, at step 2, the 1st and 10th pistons isolate the fluid within the compression chamber. At the start of compression in step 3, the 2nd to 6th piston will press the diaphragm sequentially and compress the fluid to a significantly smaller volume. Then in steps 4 and 5, when the system hits the 7th piston, simultaneously 7th to 9th piston will press the diaphragm in a sequence and the 10th piston will rise to allow fluid to discharge from the system.

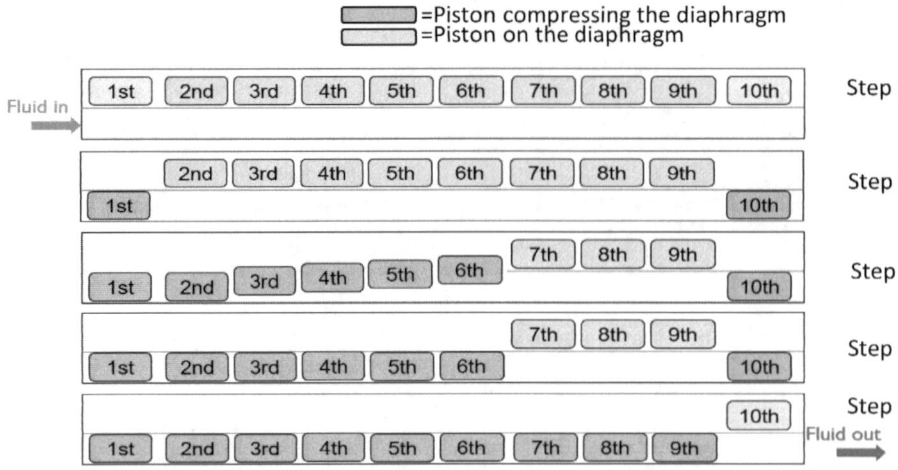

Figure 2: Schematic diagram of the operation with pivotal piston 7 selected for Mode 1.

3.2. Mode 2 Actuation

For the second mode, a new compression pocket will be formed by using an actuator to separate two pockets of adjacent compression, this actuator is referred to as the valve. For example, suppose piston 4 is the valve and piston 7 is the pivotal piston, as illustrated in Figure 3.

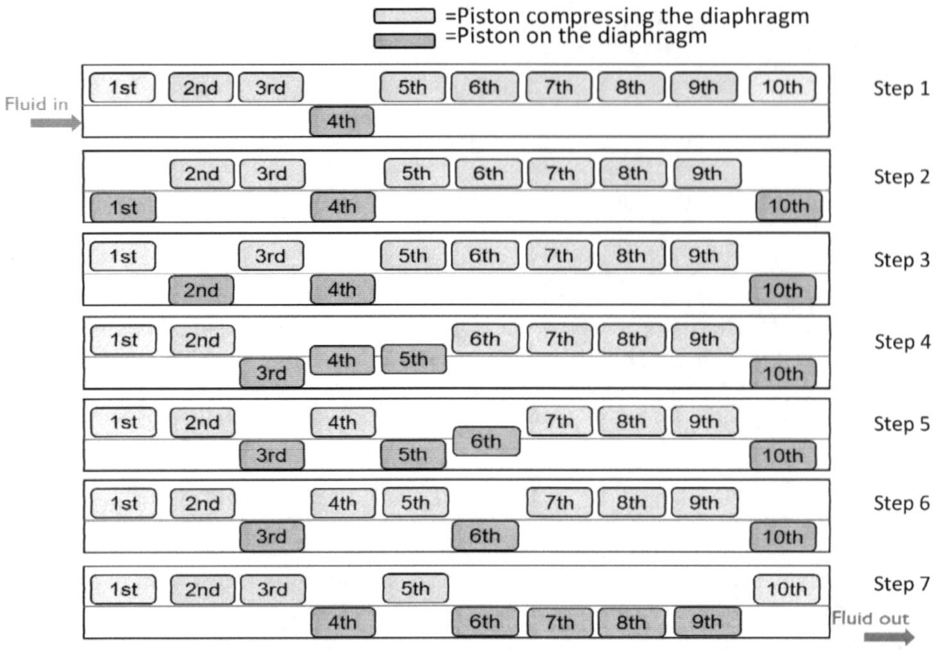

Figure 3: Schematic diagram of the operation with pivotal piston 7 and valve 4 selected for Mode 2.

At the start of the process, the 4th piston presses the diaphragm. Then at step 2, the 1st and 10th pistons descend and creates two different compression pockets. The First pocket between the 1st and 4th pistons isolate the fluid, and compression starts at step 3 with the descension of the next piston. In step 4, the 3rd piston presses the diaphragm, and the valve piston starts to rise to move the compressed fluid into the next compression pocket. The actuation of the second pocket will act as in Mode 1. In the second pocket, 5th and 10th piston isolate the fluid and sequentially start to compress the fluid. In step 6-7, when the system hits the pivotal piston 7, simultaneously 7th to 9th pistons completely descend while the 10th piston ascends to discharge the fluid. In the meantime, 1st and 4th piston create another pocket to start the next cycle. This pattern will continue until the compressor is stopped.

These two modes can be controlled to achieve a large variety of volume ratios by choosing different pivotal pistons and valves. The volume ratios of the compression chamber that can be achieved vary from 1.143 to 8. The results from this large variety of volume ratios will allow for an efficient study of the influence of the volume ratio on relevant compressor metrics such as the volumetric efficiency.

4. Reconfigurable Prototype Design

The detailed design and manufacture of the prototype is presented in this section. The working platform for the compressor prototype is a movable cart. Atop this cart, the compressor frame was mounted. The frame contains the list of components shown in the SOLIDWORKS render of the final compressor design.

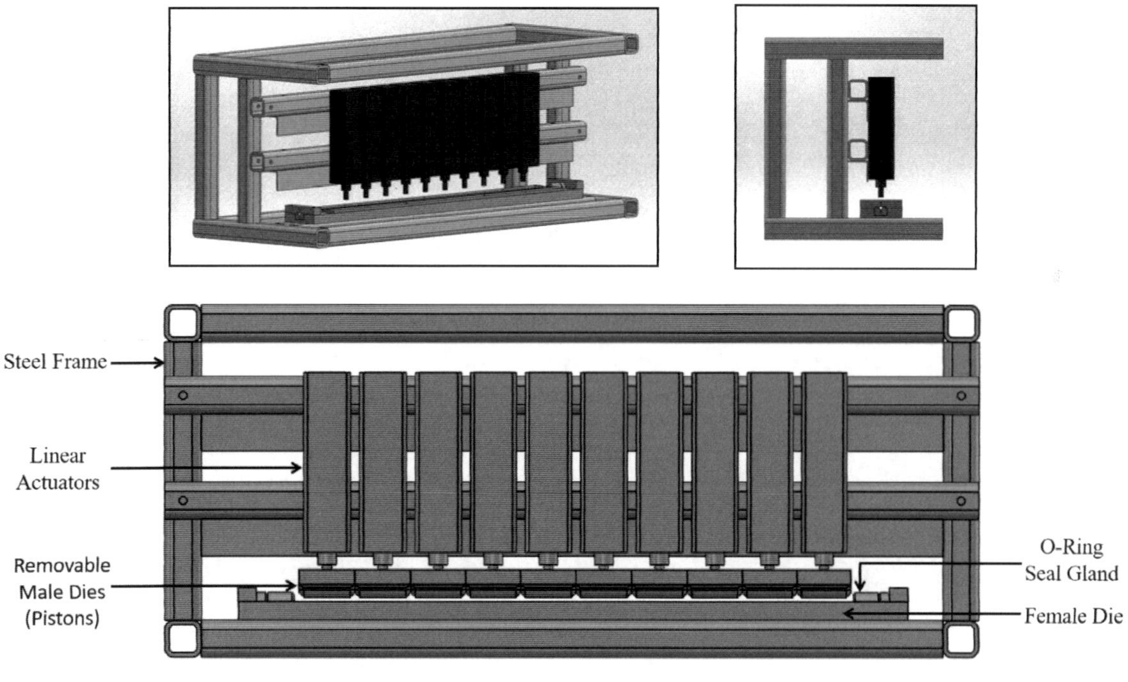

Figure 4: Prototype Compressor Design.

This design contains the following: A steel frame, a machined aluminium channel and sealing components, and linear actuators. Male dies (Pistons) are connected to the actuators, and the diaphragm is attached and sealed with the female die. Understanding the seal paths of a component that would eventually be used with a flexible diaphragm while being "modular" in terms of component redesign or replacement was challenging. The advantage to making the compression chamber modular will be clearer in the future, as the research group will have the ability to change chamber profiles/dimensions with the replacement of one component rather than completely redesigning and manufacturing all parts. The chamber dimensions for the parts were taken from the thermodynamic model, and the remaining work focused on diaphragm sealing. This was accomplished by the use of O-ring seals at each end of the compression chamber, with the seal bars running the length of the chamber to seal the diaphragm along the edges. A cross-sectional diagram of these components is shown below in Figure 5. The O-ring gland was designed with Marco gland design documentation for an outside diameter seal type based on a fixed bore diameter of 0.0254 [m] using a -020-size ring.

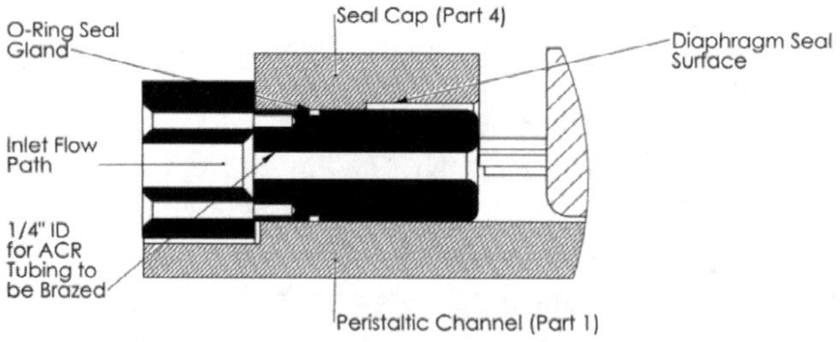

Figure 5: Part design highlighting seal paths.

All the parts installed on the steel frame as designed. At the inlet and outlet of the peristaltic compression chamber, both a pressure transducer and a resistance temperature detector (RTD) are mounted in a pipe network and used to measure pressure and temperature on either side of the compressor. An Omega mass flowmeter with analog output is used at the inlet of the compressor to measure the mass flowrate. Each of these sensors are integrated into the analog I/O module, which reads the data to be processed and displays it in the Studio 5000 software. A Recovery tank is used to collect the compressed fluid allow a continuous discharge pressure from the system through a metering valve. The metering valve is mounted at the outlet of the compressor to manage fluid flow and backpressure. By regulating the metering valve, the pressure ratio of the compressor can be controlled. Following Figure 6 shows the schematic of the prototype with the measuring sensors.

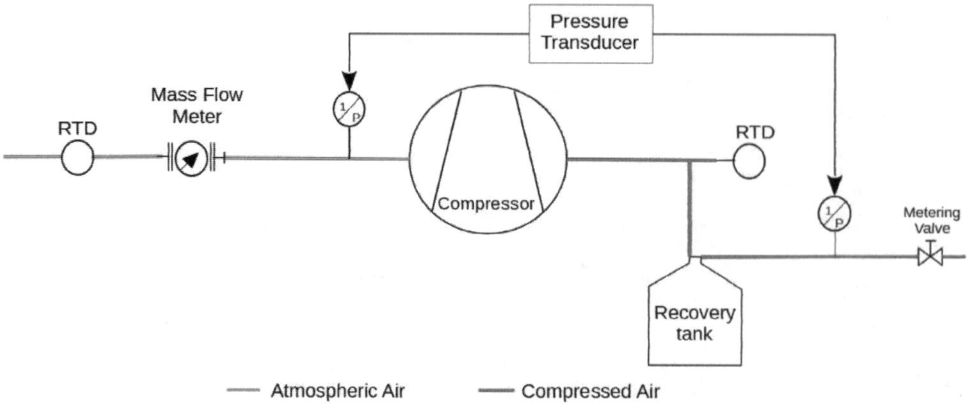

Figure 6: Schematic of the compressor setup with the sensors.

A 40-Durometer neoprene rubber sheet of 0.125" thickness is used as the diaphragm, and it is constrained against the female die with the seal bars and seal caps. Figure 7 (a) shows the construction of the diaphragm in the experimental setup. When the compressor is operational, the linear actuators sequentially press the diaphragm against female die and create a vacuum pocket. The process pulls air into the compression chamber from the inlet, which is open to the atmosphere. Figure 7 (b) demonstrates how the sensors are placed at the inlet to measure the flow rate, pressure and temperature of the suction fluid. During operation, the actuators compress the air and discharge it through the recovery tank. The pressure and temperature of the discharge fluid is measured by the outlet sensors, as illustrated in Figure 7 (c).

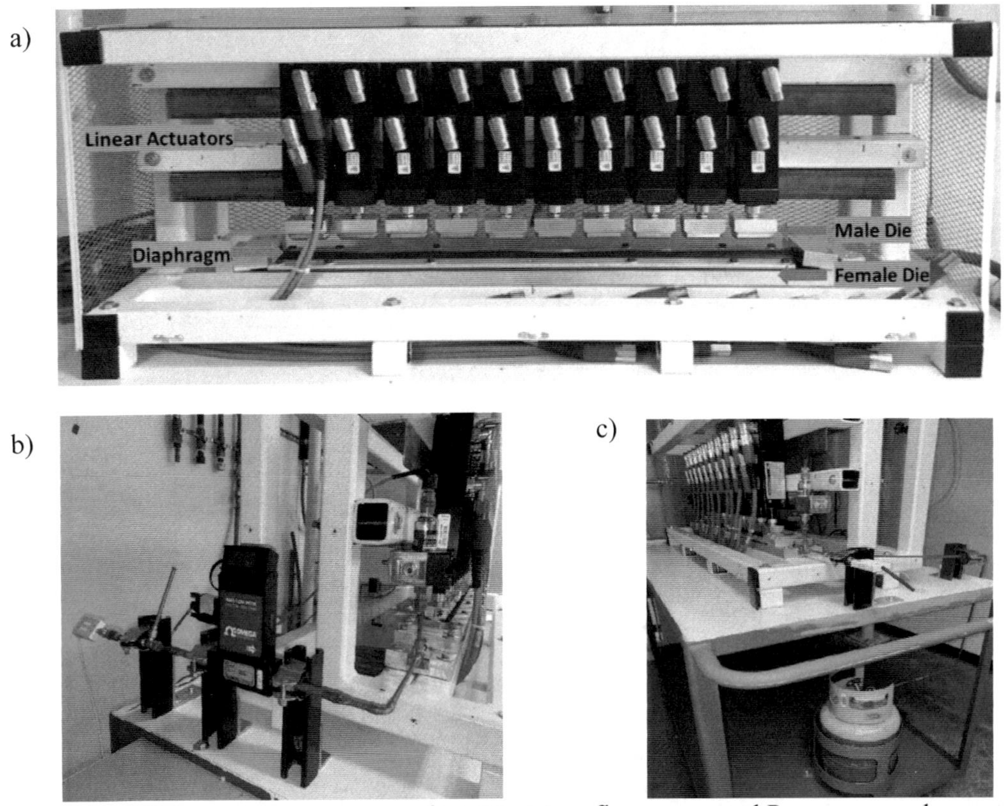

Figure 7: (a) Experimental prototype, (b) RTD, Mass-flow meter and Pressure transducer at inlet, (c) Pressure transducer, RTD and Recovery tank at outlet.

5. Preliminary Test Results

In the reconfigurable prototype compressor, various diaphragm shapes have been used which are also considered in the computational model. After an analysis, it was determined that the circular shape of the compression chamber has limitation to rebound the diaphragm properly during the linear actuation. When the diaphragm stayed flat on the female die, the mass flowrate and pressure ratio of the system increased, and the sealing pattern created a semi-circular compression chamber. An investigation also shows that the diaphragm gets stretch because of its elasticity due to linear actuation which keeps changing the displacement volume of the compression chamber. It takes around one hour of the actuation process on the diaphragm to become stable and after that, the displacement volume is measured to do the performance test of the prototype. Two different test matrices for Mode1 and Mode2 have been introduced where actuation speed, pivotal piston, valve and metering valve position (which controls the pressure ratio of the compressor) are considered.

The discharge pressure of the system is adjusted using the metering valve. The compressor is tested in a variety of metering valve positions including fully closed to test the pumping limit of the compressor prototype. The 1st and 10th pistons are acting as an inlet and discharge valve. 2nd to 9th pistons area is considered to calculate the displacement volume. The volumetric efficiency of the compressor is calculated by using the displacement volume, mass flow rate ($\dot{m}$) and compressor speed ($f[Hz]$) by using the following equation:

$$\eta_{vol} = \frac{\dot{m}}{\rho_{in}\, f\, V_{disp}}$$

Here, ρ_{in} is the density at inlet and displacement volume of the compressor is V_{disp}. Figure 8 shows how the mass flow and volumetric efficiency of the system change for different pressure ratios. Both mass flowrate and volumetric efficiency decrease by increasing the pressure ratio. When operating the system with a higher pivotal piston the fluid is compressed over a larger area and when using a smaller pivotal piston, fluid is discharged earlier than higher pivotal pistons. For that reason, the mass flowrate improves when a higher pivotal piston is used for the actuation.

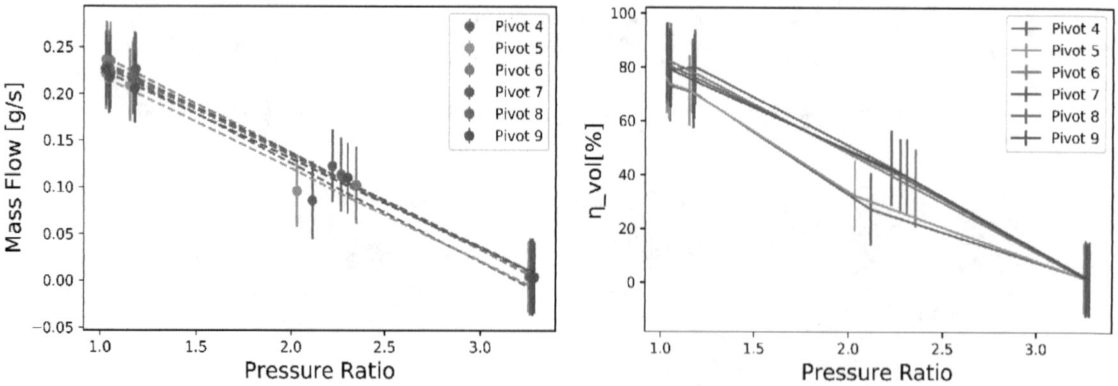

Figure 8: Variation of the Mass flow rate and Volumetric efficiency with pressure ratio for Mode1.

For Mode 2, a different actuator (3-6) is used as the valve to change the actuation pattern and displacement volume of the compression chamber. The analysis shows that the area of the first compression pocket for valves 3 and 4 is relatively small. As a result, the system could not bring in enough fluid to be compressed in the second compression pocket, which impacts the overall efficiency and mass flow rate. Valve 5 for pivotal piston 6,7,8 & 9 and valve 6 for pivotal 7,8 & 9 were chosen to analyse Mode 2, which is illustrated in the following figures.

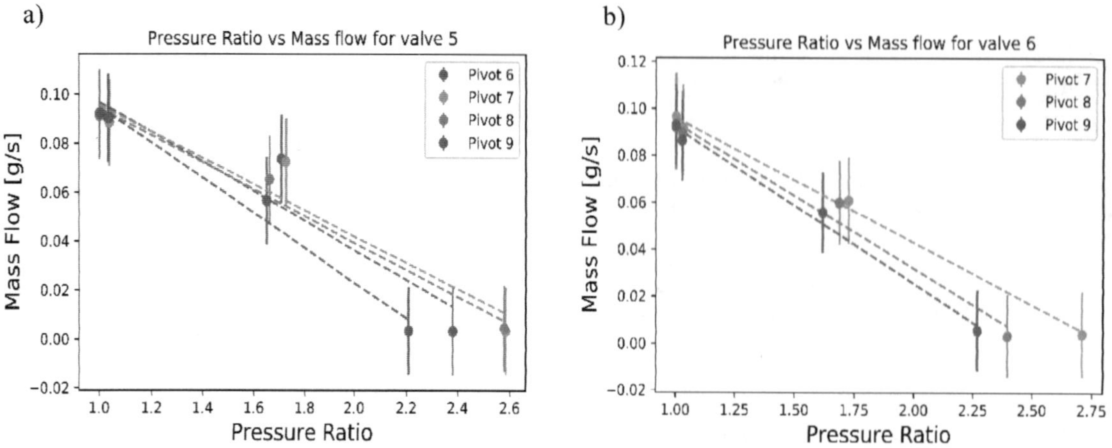

Figure 9: Variation of the Mass flow rate for Mode 2: a) When actuator 5 is used as a valve,
b) When actuator 6 is used as a valve.

a)

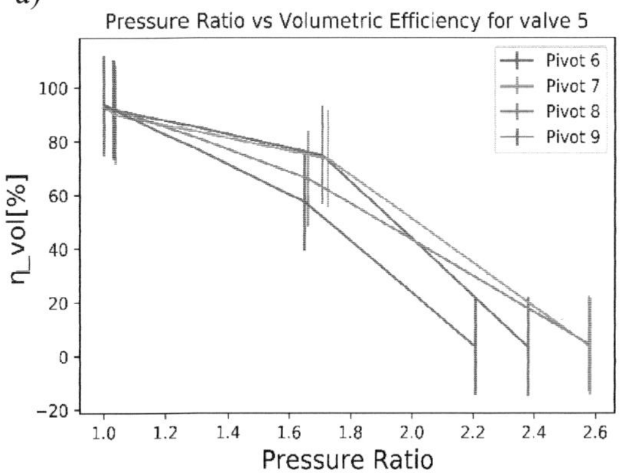

b)

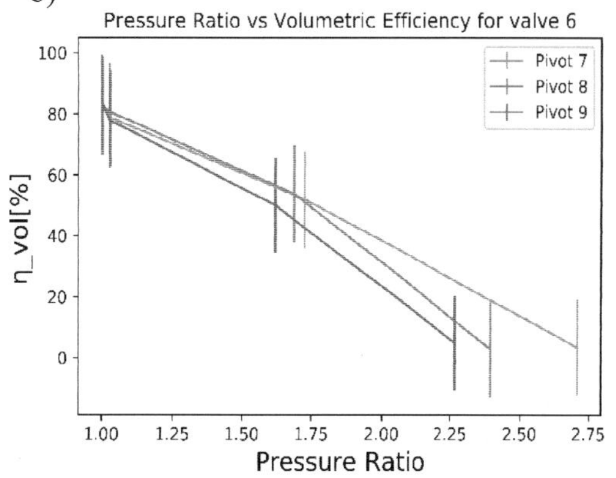

Figure 10: Variation of the Volumetric Efficiency for Mode 2: a) When actuator 5 is used as a valve, b) When actuator 6 is used as a valve.

As the analysis of Mode 2 indicates, the actuation mechanism pulls less mass, and during the compression on the second pocket, a certain amount of mass leaks backwards into the first compression pocket. In Mode 1, the fluid does not get enough space to leak backwards because of its pattern, which is described in the third section. Therefore, in the Mode 2 results, typically the pressure ratio of the system decreases by increasing the pivotal piston.

6. Conclusions

This paper presents the design and development of a reconfigurable prototype peristaltic compressor in order to help the reader understand its operation. The protype design highlights the flexibility to modify the dies or compression chamber's dimensions to find the most appropriate piston actuation pattern with results presented for two modes of operation. Preliminary test results from both modes suggest the peristaltic compressor can achieve higher volumetric efficiency by regulating the volume and pressure ratios. The design is also able to produce significant mass flow rate at relatively high volumetric efficiencies. The main drawback of the system is the material and shape of the diaphragm, which limits the pressure at the pumping limit. The material of the diaphragm needs to be soft as the elasticity of the material influences the ability to move fluid. Consequently, the diaphragm becomes worn over time and requires replacement. The flexible nature of the diaphragm results in each diaphragm having slightly different displaced volume, which may affect the displacement volume of the compression chamber. The analysis also showed that the more the fluid is compressed, the better the resulting volumetric efficiency.

The prototype compressor, and work captured in the above sections, serve as the starting point for future opportunities to advance the understanding of the novel design. The protype will be used to validate a comprehensive mechanistic thermodynamic model. When validation has been completed by using air as a working fluid, the compressor model and prototype will be modified to process refrigerants. The compressor may then be connected to a small compressor hot-gas-bypass load stand which will allow for a more detailed understanding of the novel compressor's performance in the context of a complete vapor compression cycle.

Acknowledgements

The authors gratefully acknowledge support for this work from the ASHRAE Innovative Research Grant (019-1819) and Mechanical and Aerospace Engineering department of Oklahoma State University.

References

[1] Ahamed JU, Saidur R, Masjuki HH. A review on exergy analysis of vapor compression refrigeration system. Renew Sustain Energy Rev [Internet]. 2011;15(3):1593–600. Available from: http://dx.doi.org/10.1016/j.rser.2010.11.039

[2] Pearson A. What does the future hold for compressor manufacturers? Proc Inst Mech Eng Part E J Process Mech Eng. 2015;229(2):88–95.

[3] Islam M, Bradshaw C. Development of a Mechanistic Chamber Model of a Novel Peristaltic Compressor for Air-conditioning and Refrigeration Applications. In Purdue Conference, 2021; 2021.

[4] Bradshaw C, Islam M, Tubbs C. Peristaltic Mechanism for Compression of Compressible Fluids. US Provisional Patent Application 63/019,707, 2020.

[5] Lawless WN, Arenz RW, Cross LE, Lawless WN, Arenz RW. Miniature , solid-state gas compressor Miniature , solid-state gas compressor. 1987;1487.

[6] Saif MTA, Alaca BE, Sehitoglu H. Analytical modeling of electrostatic membrane actuator for micro pumps. J Microelectromechanical Syst. 1999;8(3):335–45.

[7] Sulfridge MA, Miller NR. The Mesoscopic Peristaltic Compressor. 2001;61801(January).

[8] Sathe AA, Groll EA, Garimella S V. Analytical model for an electrostatically actuated miniature diaphragm compressor. J Micromechanics Microengineering. 2008;18(3).

[9] Sathe AA, Groll EA, Garimella S V. Optimization of electrostatically actuated miniature compressors for electronics cooling. Int J Refrig. 2009;32(7):1517–25.

[10] Mathew B, Hegab H. Analytical modeling of microscale diaphragm compressors. Appl Therm Eng.2013;51(1–2):130–6. http://dx.doi.org/10.1016/j.applthermaleng.2012.08.052

[11] Chen J, Zhang C, Xu M, Zi Y, Zhang X. Rhombic micro-displacement amplifier for piezoelectric actuator and its linear and hybrid model. Mech Syst Signal Process [Internet]. 2015;50–51:580–93. Available from: http://dx.doi.org/10.1016/j.ymssp.2014.05.047

[12] Bradshaw C. Development of a Novel Peristalitc Compressor for Air Conditioning Applications. ASHRAE 019-1819. Funding Source: ASHRAE Innovative Research Grant. Project Dates: S. 2019-present. 2017.

Research on motion and friction of rolling piston in rotary compressor

J Xu[1, 2, 3], B Yu[3]*, O X Yang[1, 3], S P Ding[1, 2, 3] and H H Zhao[3]
[1] State Key Laboratory of Air-conditioning Equipment and System Energy Conservation, ZhuHai, Guangdong 519070, China
Phone: +86-756- 8587795, Fax: +86-756-8668386
*E-mail: jvcanjian@163.com
[2] Guangdong State Key Laboratory of Refrigeration Equipment and Energy Conservation Technology, ZhuHai, Guangdong 519070, China
[3] Gree Electric Appliances, Inc. of Zhuhai,
ZhuHai, Guangdong 519070, China

Abstract. Based on the Reynolds equation of mixed lubrication, the dynamic model of rolling piston was established, and the axis locus, rotation velocity, sliding velocity at vane tip, and friction loss of the piston were numerically solved. The analysis results indicated that the numeric results of sliding velocity at vane tip are in good agreement with the experimental results. In addition, the influences of crankshaft rotation speed, working condition, piston density and compression structure parameters on motion and friction were analyzed. Taking the ratio of piston rotation velocity and revolution velocity (λ) as evaluation index, whether λ conforms to cosine curve can judge the motion state between piston and vane. The results show that there is not only sliding state but also rolling state between piston and vane tip, and keeping rolling state is helpful to reduce friction loss. The design schemes of lighter piston, thicker vane and larger height-diameter ratio cylinder are proposed to keep rolling state in a wider rotation speed range and smaller pressure difference condition, thereby improving the annual operating efficiency of R32 rotary compressor.

1. Introduction

At present, rotary compressor is widely used in refrigeration and air conditioning fields. Therefore, improving rotary compressor efficiency is very important for the energy-saving operation of refrigeration equipment [1]. The rolling piston is a key component of rotary compressor, so it is always the key research object of compressor developers. Yanagisawa and Shimizu analyzed the motion of rolling pistons by approximating the bearing pressure, and rolling piston sliding positive and negative on the vane tip is proved by experiment [2]. Further studies on friction losses indicated that the short length / diameter configuration of the cylinder is favorable to decrease total loss [3]. Wu developed an approach and a set of equations to estimate the type of motion at vane tip and roller interface, and discussed the piston rotating velocity and the relative motion at vane tip in R410A inverter rotary compressor [4]. Cho studied the partial elastohydrodynamic lubrication characteristics between vane and rolling piston， and the results showed that the shaft rotational speed and the discharge pressure significantly influence the friction force and the energy loss [5]. Further research showed that the friction between rolling piston and vane has the largest influence on the piston motion, and the sliding velocity at vane tip can be reduced by reducing the weight of rolling piston [6].

As we all know, APF has been the mainstream evaluation method of air conditioning efficiency level, and it is very significant to improve compressor efficiency under wide operating frequency range and multiple variable conditions, especially at low operating frequency ($f \leq 40$Hz). However, the operating conditions and frequency range are not wide enough in previous studies. In this paper, based on the mixed lubrication model, the motion and friction of rolling piston were solved synchronously by finite difference method, and several conclusions were obtained to improve the operating efficiency R32 compressor. This paper will provide some new ideas for compressor product developers.

2. Mathematical model and Calculation program

2.1 Mathematical model

Figure 1 shows the forces and motions of rolling piston. The rolling piston revolves around axis **o** with angular velocity ω_s, and rotates around axis $\mathbf{o_r}$ with angular velocity ω_r. The forces and moments acting on the rolling piston include: (a) the force F_g caused by pressure difference; (b) the hydrodynamic force F_{cn} and viscous force F_{cf} between rolling piston and cylinder; (c) the pressing force F_{vn} and friction force F_{vf} between rolling piston and vane tip; (d) the centrifugal force F_m of rolling piston; (e) the friction force F_r of the both ends of rolling piston; (f) the friction moment M_b between rolling piston and crankshaft; (g) the friction moment M_r of the both ends of rolling piston.

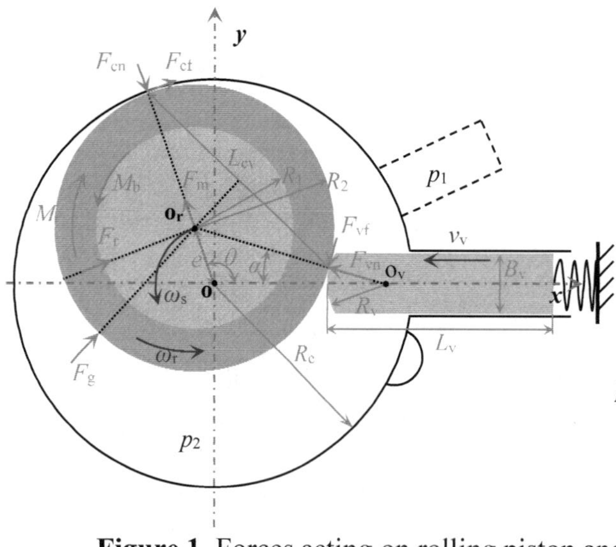

R_c - Radius of cylinder
R_1 - Inner radius of rolling piston
R_2 - Outer radius of rolling piston
e - Eccentricity of crankshaft
R_v - Radius of vane tip
B_v - breadth of vane
L_v - Length of vane
o - Axis of cylinder
o_r - Axis of rolling piston
o_v - Axis of vane tip
Not shown in the left picture:
H_c - Height of cylinder
H_e - Height of rolling piston bearing
C_e - Radial clearance at rolling piston bearing

Figure 1. Forces acting on rolling piston and main dimensions of compressor

The motion equation of rolling piston is as follows:

$$I_{\mathrm{r}}\frac{\mathrm{d}\omega_{\mathrm{r}}}{\mathrm{d}t}=M_{\mathrm{b}}-M_{\mathrm{r}}-R_2(F_{\mathrm{vf}}+F_{\mathrm{cf}}) \tag{1}$$

Where, $M_b = R_1F_{bf}$, F_{bf} is the friction force between rolling piston and crankshaft, I_r is the rotational inertia of rolling piston, and the formula of M_r can be obtained by previous paper [6].

The total load F_{xy} acting on rolling piston is as follows:

$$\overrightarrow{F_{\mathrm{xy}}}=\overrightarrow{F_{\mathrm{g}}}+\overrightarrow{F_{\mathrm{r}}}+\overrightarrow{F_{\mathrm{m}}}+\overrightarrow{F_{\mathrm{vn}}}+\overrightarrow{F_{\mathrm{vf}}}+\overrightarrow{F_{\mathrm{cn}}}+\overrightarrow{F_{\mathrm{cf}}} \tag{2}$$

Where, the first six items on the right side of "=" can be obtained through the previous papers [2-4].

Both the inner and outer wall frictions of rolling piston can be regarded as bearing systems, and figure 2 shows schematic diagram of journal bearing.

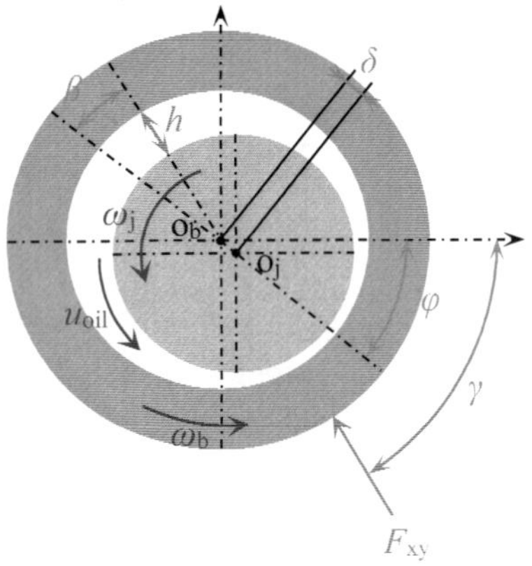

Figure 2. Journal bearing

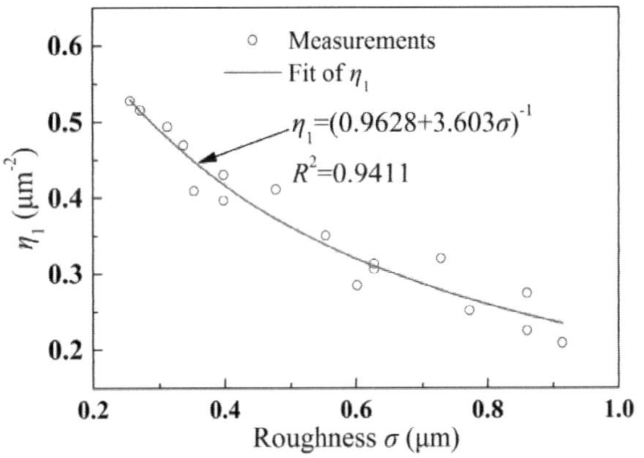

Figure 3. Actual measurement result of η_1

Based on the Reynold equation of average flow model and the contact factor [7-8], the basic control equation of mixed lubrication in journal bearing is expressed as:

$$\frac{\partial}{r^2\partial\beta}\left(\varphi_\beta h^3\frac{\partial p}{\partial\beta}\right)+\frac{\partial}{\partial z}\left(\varphi_z h^3\frac{\partial p}{\partial z}\right)=\frac{6\eta(U_{\mathrm{j}}+U_{\mathrm{b}})\varphi_{\mathrm{c}}}{r}\frac{\partial h}{\partial\beta}+\frac{6\eta(U_{\mathrm{j}}-U_{\mathrm{b}})\sigma}{r}\frac{\partial\varphi_{\mathrm{s}}}{\partial\beta}+12\eta\varphi_{\mathrm{c}}\frac{\partial h}{\partial t} \tag{3}$$

Where, $h = C_e(1+\varepsilon\cos\beta)$, $\varepsilon = \delta/C_e$, r is the radius of the bearing, p is the oil film pressure, $U_j \approx R_1\omega_j$, $U_b = R_1\omega_b$, φ_β and φ_z are the pressure flow factors in tangential and axial directions respectively, φ_s is the shear flow factor, φ_c is the contact factor, and σ is the standard deviation of combined roughness.

According to the elastic contact model proposed by Greenwood and Tripp [9], and further combined with the research conclusion of Song [10], the formula for calculating the contact pressure is as follows:

$$p_{\mathrm{t}}=\frac{\frac{32}{15}(0.075)^2\sqrt{\frac{\pi\eta_1\sigma^2}{0.075}}\int_H^\infty(s_1-H)^{2.5}e^{-\frac{s_1^2}{2}}ds_1}{\frac{1-\nu_1^2}{E_1}+\frac{1-\nu_2^2}{E_2}} \tag{4}$$

Where, $H = h/\sigma$, η_1 is the areal density of asperities (the number of rough acoustic peaks in unit area), $E_{1,2}$ are the Young's modulus of the bearing and the journal respectively, and $\nu_{1,2}$ are the Poisson's ratios. The actual measurement results of η_1 are shown in Figure 3. As can be seen from

Figure 3, there is a strong correlation between η_l and σ in the cast iron crankshaft surface, so a fit formula is given for accurately predicting contact stress at different roughness.

In the mixed lubrication state, the bearing carrying force W and the friction force F_f are composed of two parts: hydrodynamic pressure and metal contact. The calculation formulas are:

$$\vec{W} = \int_0^{2\pi}\int_0^{B} \vec{p} r \mathrm{d}\theta \mathrm{d}z + \int_0^{2\pi}\int_0^{B} \vec{p_t} r \mathrm{d}\theta \mathrm{d}z \tag{5}$$

$$F_f = \int_0^{2\pi}\int_0^{B}\left[\frac{\eta(U_j - U_b)}{h}(\varphi_f + \varphi_{fs}) + \varphi_{fp}\frac{h}{2}\frac{\partial p}{r\partial\theta} + f_0 p_t\right] r \mathrm{d}\theta \mathrm{d}z \tag{6}$$

Where, B is the width of journal bearing, f_0 is the contact friction coefficient between the bearing and the journal, and the factors φ_f, φ_{fs} and φ_{fp} can be obtained by previous paper [11].

Based on the above model, the total load should be equal to the bearing capacity, that is, formula (2) is equal to formula (5), so that the variation of oil film thickness can be obtained. The bearing friction moment can be obtained by formula (6), then the piston motion can be obtained by formula (1). Based on the variation of oil film thickness and piston motion, formulas (3) and (4) can be solved. Therefore, formulas (1) - (6) constitute a closed system of equations for solving piston motion and friction.

2.2 Calculation program

The mixed Lubrication equation and the axis locus are solved by finite difference method (hahm method), which requires the assumption of the initial value of the axis locus of the rolling piston bearing [12]. It can be known from the above model that the rolling piston rotation angular velocity ω_r is a key parameter that links various forces and moments, so the dynamic model should be solved by assuming the initial value of ω_r.

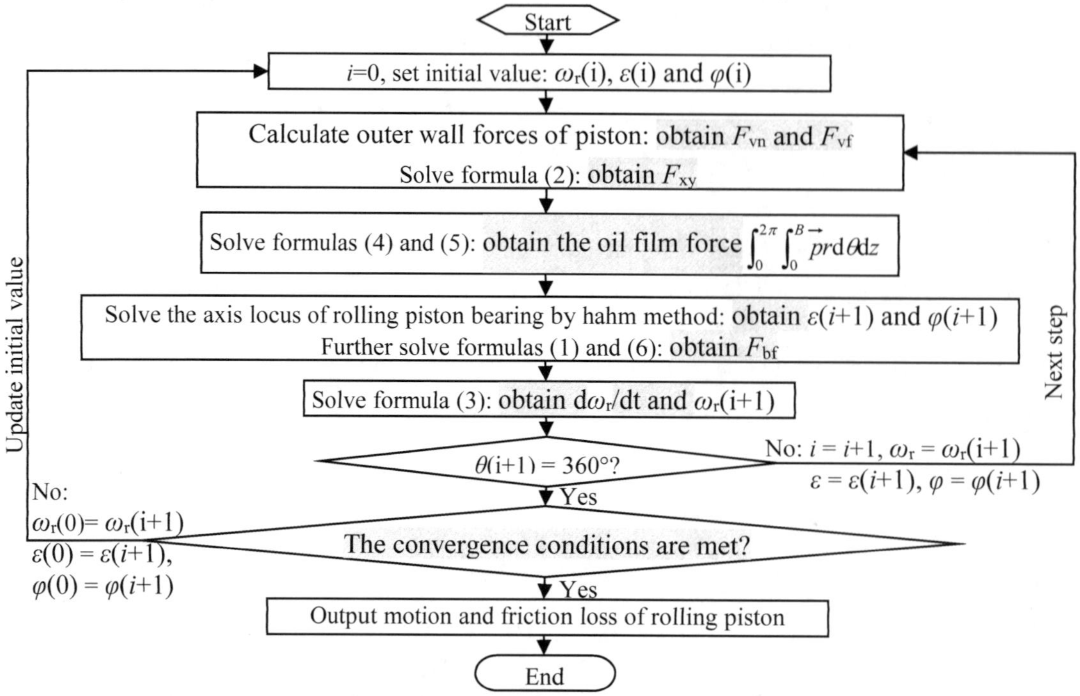

Figure 4. Flow chart of calculation program

Figure 4 shows the main flow of the calculation program: (a) set initial values; (b) solve for the angular acceleration ($d\omega_r/dt$) and the axis movement speed of rolling piston bearing ($d\varepsilon/dt$, $d\varphi/dt$); (c) obtain the new ω_r, δ, and φ; (d) continue the calculation of the next step until ω_r, ε and φ all meet the above convergence conditions (7). According to this program, the motion and friction of rolling piston were solved synchronously at every step, so the characteristics can be analyzed in a wide range of frequencies and working conditions.

The convergence condition of the calculation program is that the piston rotation velocity and axis locus meet the relative accuracy requirements:

$$\left|\frac{\omega_r(i+1)-\omega_r(0)}{\omega_r(0)}\right| < 0.01 \& \left|\frac{\varepsilon(i+1)-\varepsilon(0)}{\varepsilon(0)}\right| < 0.01 \& \frac{\varphi(i+1)-\varphi(0)}{\varphi(0)} < 0.01 \tag{7}$$

3. Conditions and Model validation

Table 3 and 4 respectively show the main design parameters and working conditions of the compressor, in which the design parameters of compressor 1, condition 1 and condition 2 are the same as those in the literature [2], so as to compare the analysis results.

Table 3. Main dimensions of compressors (in millimeters)

Compressor	R_c	H_c	R_2	R_1	R_v	H_e	Lv	Bv	C_e
1	27	23.8	24.3	14.4	6	14	24	4.7	0.013
2	18	25	14.2	10.8	3.2	16	20	3.2	0.010
3	22	18.5	16.6	11.5	3.2	11	22.5	3.2	0.010

Table 4. Working conditions

Condition	refrigerant	Oil type	Suction temperature(℃)	Suction pressure(MPa)	Discharge pressure(MPa)
1	R22	4GS	10	0.58	2.03
2	R22	4GS	18	0.74	2.83
3	R32	POE68	18.3	1.02	2.86
4	R32	POE68	18.3	1.02	3.47
5	R32	POE68	25	1.47	2.48

The comparisons between the simulation results of compressor 1 in this paper and the simulation results and experimental results in the literature [2] are shown in Figures 5 and 6. It can be seen that the simulated values of the sliding velocity v_{vs} at vane tip in this paper is more consistent with the experimental values. Under the working condition 1, the coefficient of determination R^2 between the simulated values of the new model and experimental values is 0.94, while the original is 0.9. Under the working condition 2, the coefficient of determination R^2 between the simulated values of the new model and experimental values is 0.84, while the original is 0.68. Therefore, the new calculation program in this paper can more accurately simulate the motion and friction characteristics of rolling piston.

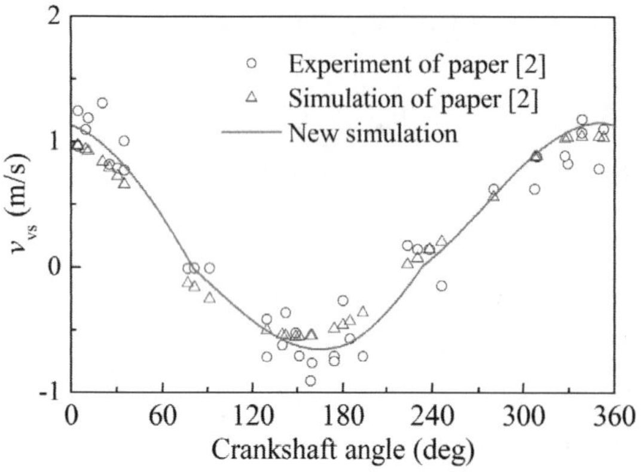

Figure 5. v_{vs} under working condition 1

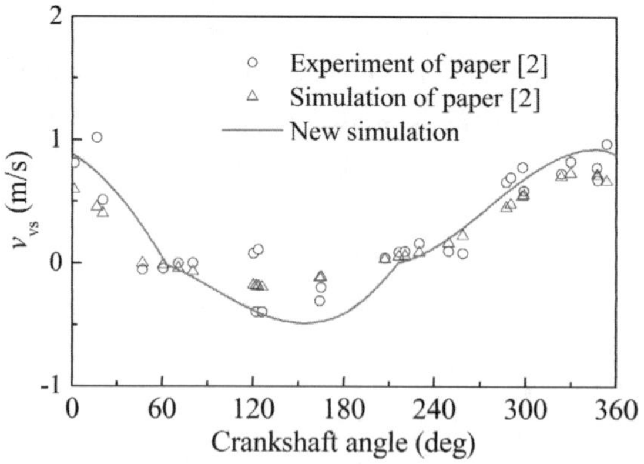

Figure 6. v_{vs} under working condition 2

4. Simulation results and Discussions

4.1 Effect of crankshaft rotation speed on movement and friction of rolling piston

Taking compressor 2 and condition3 as the analysis object and working condition, respectively, the variations of the axis locus, rotation velocity, sliding velocity and friction loss of the rolling piston with the crankshaft rotation speed are simulated. The figure 7 shows the variations of the loaded force and axis locus with the crankshaft rotation speed (operating frequency f). When F_{xy}=1500N, define the relative force as 100%.

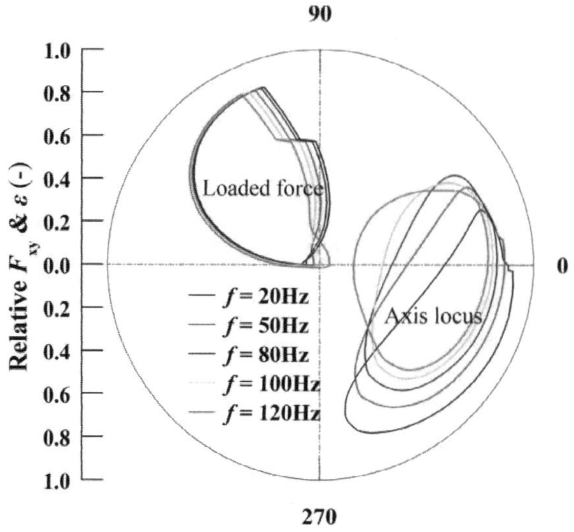

Figure 7. The relative loaded force and axis locus of rolling piston

It can be seen from Figure 7 that the eccentricity ε decreases as f increases. The reason is that the bearing capacity of the oil film increases with the increase of f, but the loaded force of the bearing changes little with f under the same working condition. On the other hand, the axis locus of the rolling piston bearing is in the angular range of -80°~40°, while the loaded force is in the range of 80°~180°. The reason is that the loaded force acts on the piston, and the piston should also be in the second and third quadrants under the force of the oil film, but there is a 180° phase difference between the journal axis locus and the bearing axis locus, so the axis locus of crankshaft is in the fourth and first quadrants.

The piston rotation velocity ω_r, the ratio of rotation to revolution angular velocity λ, the sliding velocity at vane tip v_{vs}, and the friction loss of each surface are respectively shown in Figures 8~11. It can be seen from Figure 8 that ω_r increases with the increase of f but the regularity of the change is not significant. Taking the ratio of piston rotation angular velocity and revolution angular velocity (λ) as evaluation index, as shown in Figure 9, 40Hz is the inflection point of ω_r at which the λ can maintain the "cosine" shape.

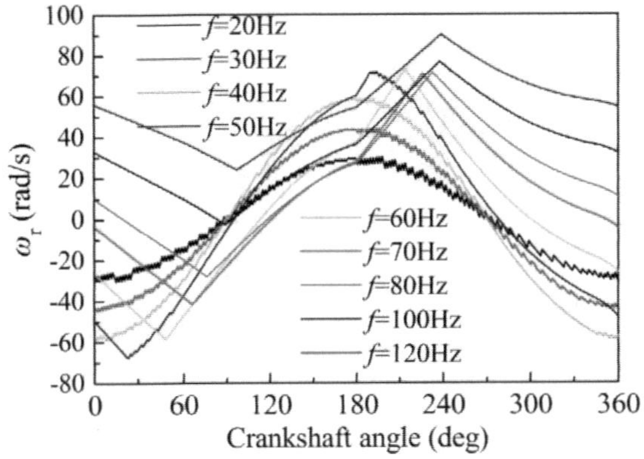

Figure 8. The variety of ω_r at different f

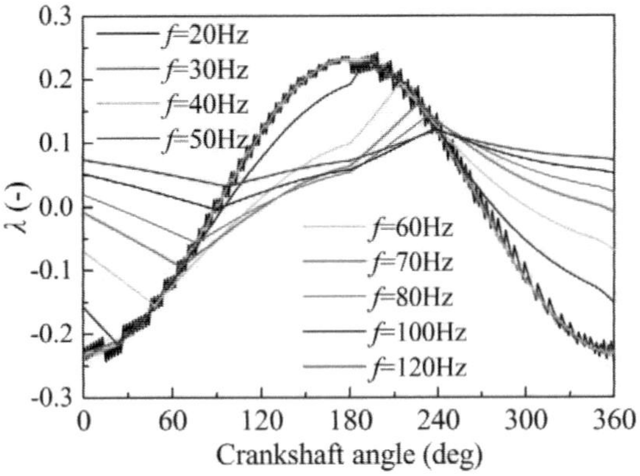

Figure 9. The variety of λ at different f

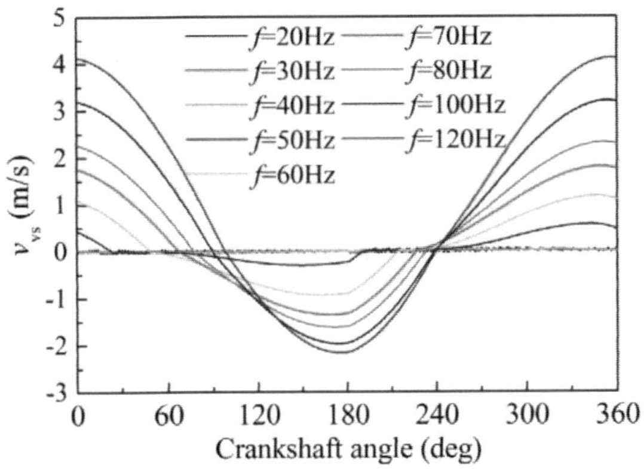

Figure 10. The variety of v_{vs}

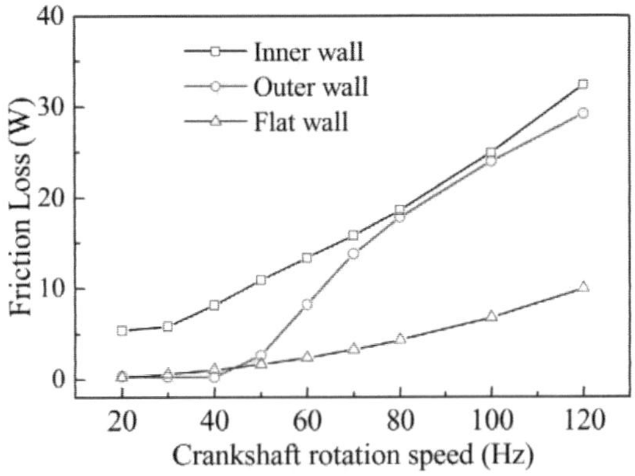

Figure 11. The variety of friction loss

Compared with Figures 9 and 10, the sliding velocity is always zero when the operating frequency is lower than 40Hz of inflection point, which indicates that the motion state between rolling piston and vane is relative rolling. When the operating frequency is higher than 40Hz, λ presents a "broken line" shape, and the sliding speed increases with the increase of operating frequency f. The inflection points of the "broken line" all fall on the "cosine" line. The reason is that when the rotation velocity changes to the rolling state between piston and vane tip, the direction of relative sliding changes reversely. So the angular acceleration of piston rotation also changes positively and negatively.

It can be seen from Figure 11, the loss of each surface of the rolling piston increases with the increase of the operating frequency f. In particular, the friction loss of outer wall also takes 40Hz as the inflection point. The friction loss is very small when $f \leq 40$Hz, and the friction loss increases rapidly when f is in 40Hz~80Hz. Therefore, the motion state of rolling piston and vane tip is the key factor affecting the friction loss of rolling piston. It is helpful to reduce the friction loss to keep the relative rolling state.

4.2 Effect of working condition on movement and friction of rolling piston

In order to optimize the performance of compressor under different working conditions, the effect of pressure difference ($\Delta p = p_2\text{-}p_1$) is analyzed. Figure 12 shows the variation of λ under working condition 5. Figure 13 shows the variation of friction loss of outer wall at different Δp.

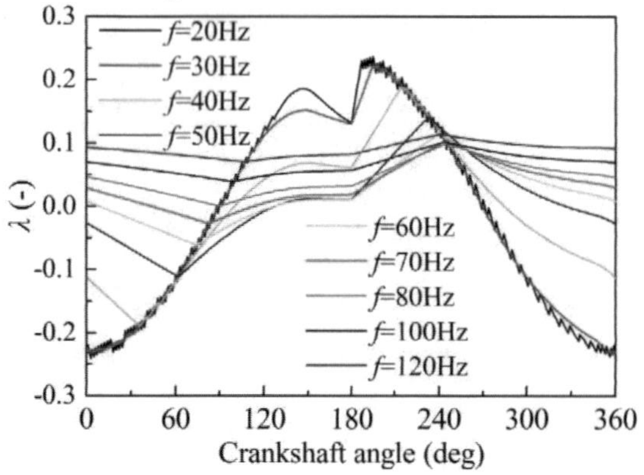

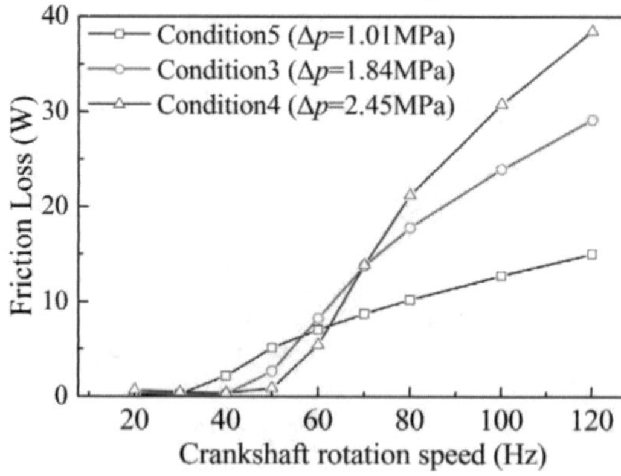

Figure 12. The variety of λ under condition 5 **Figure 13.** Friction loss of outer wall

It can be seen from Figure 12 that even if the operating frequency f is as low as 20Hz, λ can't keep "cosine" shape all the time, and the rolling piston and the vane tip always slide relative to each other in the range of 150°~200° of crankshaft angle. Because the pressing force F_{vn} and friction force F_{vf} decrease greatly with the decrease of Δp, the relative rolling state between rolling piston and vane tip cannot be maintained. A short time of relative sliding may lead to the separation of rolling piston and vane, which may cause the impact noise.

As can be seen from Figure 13, a larger Δp does not mean that the friction loss of outer wall is always greater, and it has different friction loss characteristics at different crankshaft rotation speeds. When f = 50Hz, the friction loss under condition 5 with the minimum Δp is the largest, but when f = 80Hz, the friction loss of condition 4 with the maximum Δp is the largest. This means that if the friction loss under the condition with smaller Δp is to be reduced, the sliding velocity should be reduced, such as increasing the thickness of vane B_v to increase the F_{vn}.

4.3 Effect of the material on movement and friction of rolling piston

Several new materials (cast iron, alumina ceramic and PEEK) were used in rolling piston, and the effects of different material density on motion and friction characteristics were analyzed.

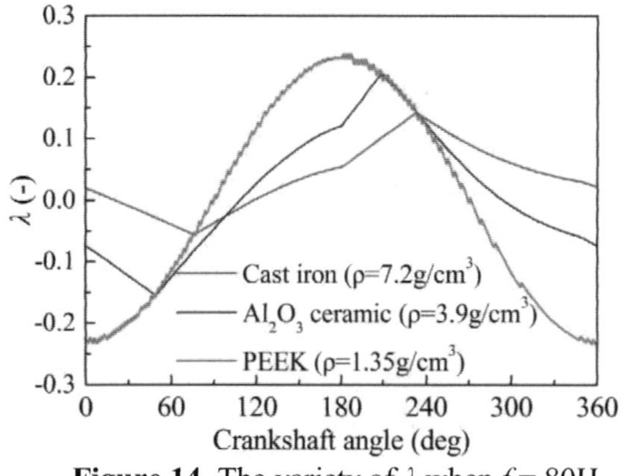

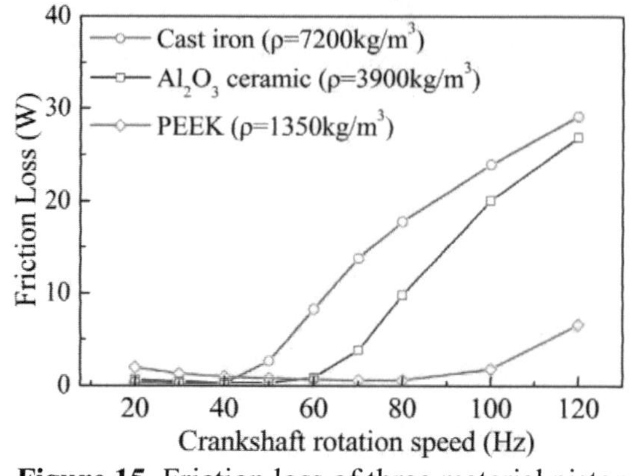

Figure 14. The variety of λ when f = 80H **Figure 15.** Friction loss of three material pistons

Figure 14 shows the comparison of λ of three kinds of material pistons in compressor 2 under condition 3. Figure 15 shows the variation of friction loss of three material pistons' outer wall. It can be seen from Figure 14 that the PEEK piston can still keep relative rolling state when f = 80Hz, and the Al_2O_3 ceramic piston is in relative sliding state, but its following performance is better than that of the cast iron piston. With the decrease of material density, the relative rolling state between rolling piston and vane is more favorable. The rotational inertia of the light piston is smaller, which makes the fluctuation of the rotation velocity better, so it better matches the movement of vane tip. As shown in Figure 15, the friction loss of outer wall decreases with the decrease of roller density, which will help to improve the mechanical efficiency of rotary compressor.

4.4 Effect of cylinder structure on movement and friction of rolling piston

It is a common engineering method to improve the mechanical efficiency by optimizing the structural parameters of cylinder. The compressor 3 is proposed, which has a larger height-diameter ratio than compressor 2. Figure 16 shows the variation of λ of compressor 3 with f under condition 2. Figure 17 shows that the friction loss of outer wall in compressor 2 and compressor 3 varies with f.

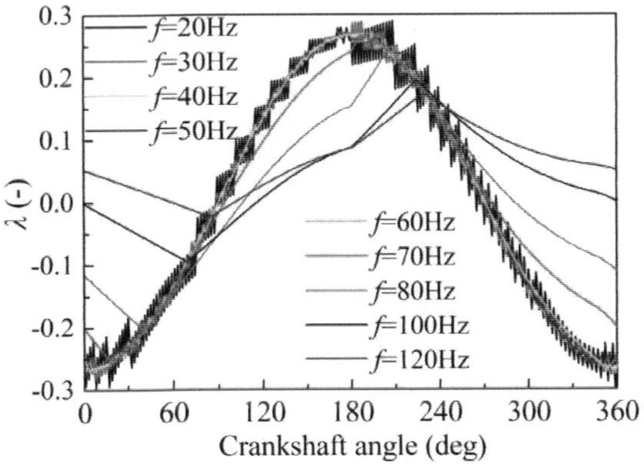

Figure 16. The variety of λ of compressor 3

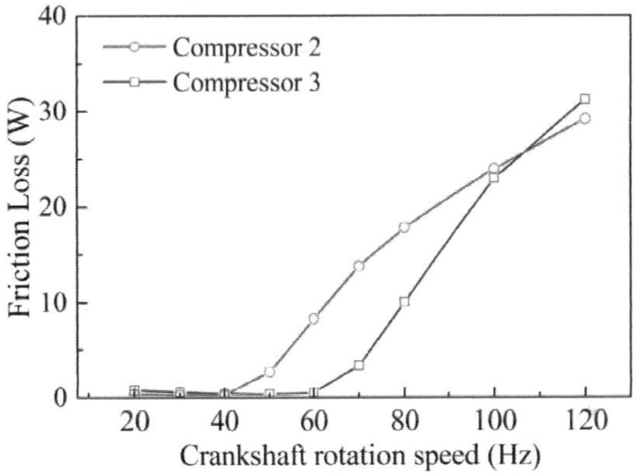

Figure 17. Friction loss in two compressors

It can be seen from Figure 16 that the inflection point of compressor 3 is 60Hz, which is significantly higher than that of compressor 2. It can be seen from figure 17 that the friction loss of outer wall in compressor 3 is significantly lower than that in compressor 2 in the operating frequency range of 40Hz - 100Hz. The reason is that the compressor with larger height-diameter ratio has thinner piston and larger pressing force F_{vn}, which makes rolling piston and vane have better follow-up and smaller sliding velocity. When the operating frequency f reaches 120Hz or even higher, the sliding velocity difference of two compressors is small, because the friction moment M_b of rolling piston bearing plays a major role. Therefore, the friction loss in compressor 3 is higher.

5. Conclusions

Based on the mixed lubrication model, the motion and friction loss of rolling piston are numerically solved. Furthermore, the effects of crankshaft rotation velocity, working condition, piston density, and compression structure parameters were analyzed.

The simulation results of rolling piston motion are in good agreement with the experimental results. It is proposed that the ratio of piston rotation velocity and revolution velocity (λ) can be used as the evaluation index, and the motion state between rolling piston and vane can be judged according to whether λ conforms to "cosine" shape.

The conclusions helpful to design high efficiency R32 rotary compressor are as follows: (1) With the increase of crankshaft rotation speed, the motion state between rolling piston and vane changes from rolling to sliding, which leads to the sharp increase of friction loss at vane tip; (2) With the decrease of pressure difference, it is more difficult to keep the rolling state between piston and vane, and the sliding velocity should be reduced by increasing the thickness of vane; (3) The smaller the density of piston material is, the better it is to keep the rolling state between piston and vane, so light material or hollow piston should be used to reduce friction loss; (4) A larger height-diameter ratio cylinder can keep rolling state in a wider range of crankshaft rotation speed, that is to say, the crankshaft rotation speed range with lower friction loss is wider.

References

[1] Wang L 2018 Compressor Market Analysis *Chinese Journal of Refrigeration Technology* **38(S1)** 6-23 (in Chinese)

[2] Yanagisawa T, Shimizu T, Chu I, and Ishijima K 1982 Motion Analysis of Rolling Piston in Rotary Compressor *International Compressor Engineering Conference* Paper 392

[3] Yanagisawa T, and Shimizu T 1985 Friction losses in rolling piston type rotary compressors III *Int. J. Refrigeration* **8(3)** 159-65

[4] Wu J 2000 Dynamic Analysis of Roller and Vane of Inverter Controlled Rotary Compressors *International Compressor Engineering Conference* Paper 1394

[5] Ihn-Sung Cho, Seok-Hyung Oh, and Jae-Youn Jung 2001 Lubrication Characteristics Between the Vane and the Rolling Piston in a Rotary Compressor Used for Refrigeration and Air-Conditioning Systems *KSME International Journal* **15(5)** 562-8

[6] Yasutaka I, Hitoshi H, and Kazuhiko M 2010 Numerical Analysis for Rotating Motion of a Rolling Piston in Rotary Compressors - Effective Factors for Characteristics of Rotating Motion of a Rolling Piston *International Compressor Engineering Conference* Paper 1963

[7] Patir N, and Cheng H S 1978 Application of Average Flow Model for Determining Effects of Three-Dimensional Roughness on Partial Hydrodynamic Lubrication *Trans. ASME Journal of Lubrication Technology* **100(1)** 12-7

[8] Wu C W, and Zheng L Q 1989 An Average Reynolds Equation for Partial Film Lubrication With a Contact Factor *Trans. ASME Journal of Tribology* **111(1)** 188-91

[9] Greenwood J A, and Tripp J H (1970). The contact of two nominally flat rough surfaces *Proceedings of the Institution of Mechanical Engineers* **185(46)** 625-33

[10] Song G F 2015 Theoretical Study on Properties of Mixed Lubrication of Inclined-Micro-Dimpled Surfaces *Zhejiang University of Technology* 10-1

[11] Patir N, and Cheng H S 1979 Application of Average Flow Model to Lubrication between Rough Sliding Surfaces *Trans. ASME Journal of Lubrication Technology* **101(4)** 220-30

[12] Huang P 2012 *Lubrication Numerical Calculation Methods* (Beijing: Higher Education Press)

IOP Conf. Series: Materials Science and Engineering 1180 (2021) 012048 doi:10.1088/1757-899X/1180/1/012048

Comparison of Two Reverse-cycle Defrosting Methods for a R290 Split Room Air Conditioner

Y J Du, J H Wu, C Wang, B W Lei
Xi'an Jiaotong University, Xi'an 710049, China
Corresponding author's e-mail address: duyanjun@stu.xjtu.edu.cn

Abstract. R290 has obvious advantages in aspects of environmental protection and thermal properties, and researches on R290 room air conditioners (RACs) are attracting more and more attention. As for R290 RAC, the suction temperature is continuously low and a large number of refrigerants are dissolved in the compressor oil sump during defrosting process, resulting in a longer time length of the start-up process after defrosting. Two reverse-cycle defrosting methods, the discharge-throttling method (DTM) and suction-discharge bypass method (SDBM) are proposed in order to solve these problems. This paper compares the two defrosting methods on an R290 RAC by experiment and the results show that the defrosting time of DTM is relatively short, but the temperature at the inlet and outlet of the indoor heat exchanger is relatively low after defrosting. The SDBM can alleviate the situation of low suction temperature in the defrosting process, but the viscosity of the oil sump is large in the early stage of defrosting and the time length of the start-up process after defrosting is longer. Both methods have their own advantages and disadvantages, and the paper will provide an experimental basis for the reliability study of R290 RACs.

1. Foreword

As a natural working fluid, R290 is gradually becoming the focus of attention because of its negligible lower global warming potential (GWP), zero ozone depletion potential (ODP) and excellent thermal performances. Meanwhile, it has a low price and is easy to obtain because it has a wide range of sources. Nevertheless, it should be noted that R290 is highly flammable, which limits its application to a certain extent. Many literatures have studied the safety of R290 RAC, and many scholars have reached a consensus: through technological innovation and charge control, the safety hazard of R290 RAC can be effectively controlled [1-3]. Tang et al. [4] studied the performance and risk assessment of R290 in a 13 kW RAC and offered some suggestions on improving the safety design of the R290 RAC.

In fact, the refrigerant charge of R290 RAC is far less than that of HFCS RAC under the same rated capacity, due to the small density of R290 and the latent heat per unit mass of refrigerant is larger. Less refrigerant charge is conducive to safety, but it also makes the dynamic migration characteristics of refrigerant in R290 RAC system different from other types of air conditioners [5]. Too much refrigerant is stored in the compressor or accumulator, which is not conducive to the normal operation of the air conditioner, because this part of refrigerant can not heat exchange effectively. Especially for R290 air conditioner with small refrigerant charge, although the mass of refrigerant in the compressor and accumulator is small, it accounts for a large proportion of the total mass. The lack of refrigerant in the system pipeline will cause many adverse consequences: the low refrigerant mass flow makes the heating capacity insufficient, the high suction superheat leads to the weak oil return capacity of the system, and the suction pressure is lower than the atmospheric pressure, which makes the system pipeline have the risk of leakage [6]. These adverse consequences will affect the reliability of air conditioning system.

The phenomenon of lacking refrigerant in the system pipeline often occurs in the start-up process, especially after the reverse circulation defrosting. When using RAC in winter, the defrosting process is inevitable. Because the system pipeline does not need to be modified and can be effectively controlled through a four-way reversing valve, reverse circulation defrosting is the most common defrosting method at present [7, 8]. In order to eliminate the possible adverse consequences due to the lack of refrigerant in the system and improve the reliability of R290 RAC under low temperature conditions, the main content of this paper is to propose and compare two new cycle structures based on reverse cycle defrosting, and analyze their advantages and disadvantages under low-temperature heating conditions.

2. Experimental apparatus and procedure

2.1 Test system and control method

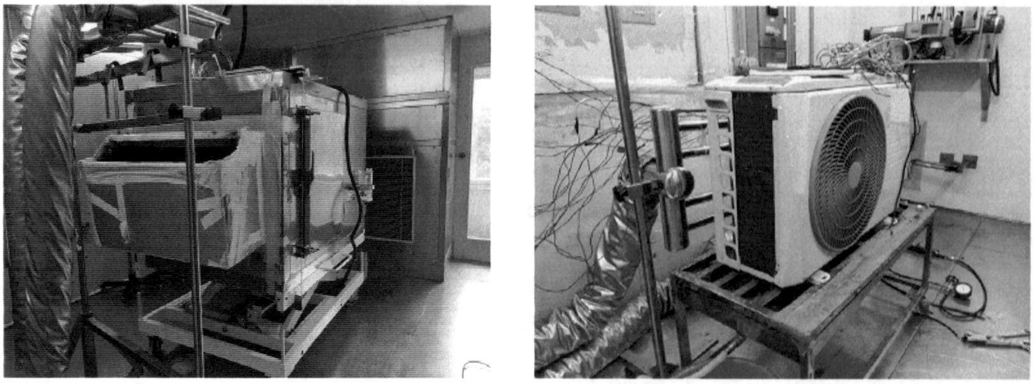

Figure 1 The multifunctional enthalpy difference laboratory

The experiment was carried out in the multifunctional enthalpy difference laboratory as shown in Fig.1, and different working conditions can be simulated by controlling the indoor and outdoor dry bulb air temperature and wet bulb air temperature (DBT/WBT). Meanwhile, the cooling or heating capacity and power consumption of the air conditioner can be accurately measured. In addition, the measured air dry bulb temperature and wet bulb temperature precision of the enthalpy difference room were with ± 0.2 ℃, the accuracy of the power consumption was 0.5%, the uncertainty of air conditioner capacity and COP were 1.53% and 3.27% respectively.

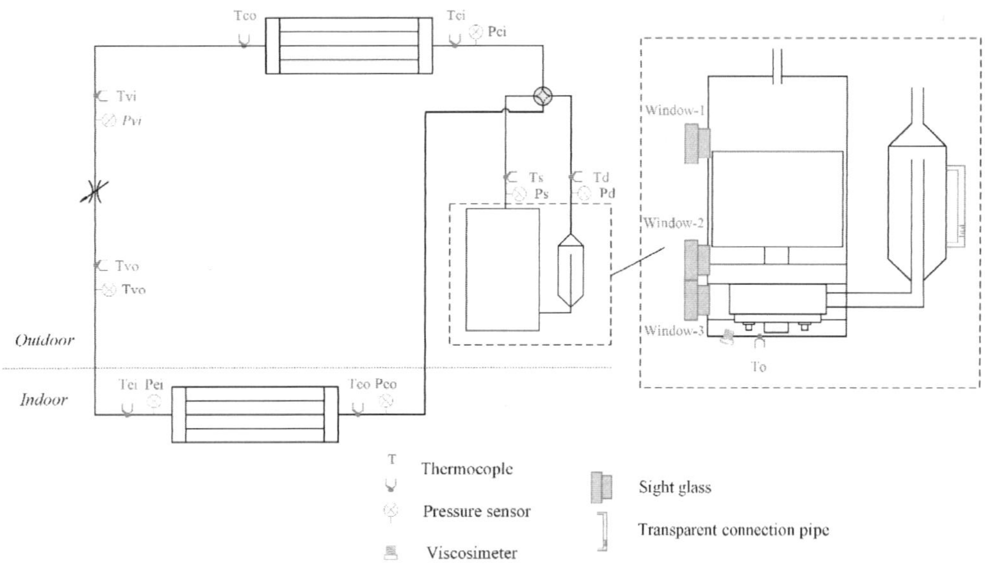

Figure 2 Schematic diagram of the experimental setup

The pressure sensor was HM-23 and the two measurement ranges were 0-1.0MPa and 0-2.5MPa respectively, meanwhile, the measurement accuracy is 0.25%. Temperature collection used the T-type thermocouples in pipeline and the K-type thermocouples in compressor, and the measurement precision was in ±0.5℃. The temperature and pressure collector was Agilent 34970A data collector with a collection accuracy of 0.02%. In total, there were 7 pressure measuring points and 9 temperature measuring points, and their positions are shown in Fig. 2. A viscometer was performed to measure the viscosity of oil sump and it was installed in bottom of compressor shell. The installation angle was inclined by 45 degrees due to measurement requirements of the viscometer. A transparent pipe was installed on the side of the accumulator to monitor the liquid level. In addition, three visual windows were set on the side of the compressor shell.

The main components of the experimental system included: a rotary compressor, an outdoor heat exchanger (ODHE), an indoor heat exchanger (IDHE), an electronic expansion valve (EEV). Among them, the stroke volume of the compressor was 18cm^3, the content of the outdoor heat exchanger was about 697cm^3, and the content of the indoor heat exchanger was about 622cm^3. In addition, the rated cooling capacity of the R290 RAC was 3500W. The oil was MO VG100 in the compressor, refrigerant charge of the R290 RAC was 435g, and the oil charge was 270g in the R290 compressor. During the experiment, the heat capacity, power consumption and COP under each working condition were collected separately. The defrosting method of air conditioner was reverse cycle defrosting, controlled by four-way valve, and the indoor and outdoor fans during the defrosting process are closed.

Table 1 Characteristic parameters of experimental compressor

parameters	Cylinder diameter /mm	Cylinder height /mm	Crank shaft eccentricity/mm	Stroke volume/cm^3	Motor efficiency/%
values	50	24	5.3	17.9	89

The experiment was carried out under the conditions of which the indoor dry bulb temperature was 20°C, the indoor wet bulb temperature was 15°C, the outdoor dry bulb temperature was - 7°C and the outdoor wet bulb temperature was - 8°C. In order to avoid the impact of cold air from the indoor heat exchanger on the comfort during the reheating process, the heat pump manufacturer will set up a "cold air prevention strategy". Before the reheating process, if the TC7 is less than 0 °C, the fan of the indoor heat exchanger will be closed temporarily after the compressor is started, until the TC7 is larger than 29 °C or the length of startup time is longer than 3.5 min, the fan of the indoor heat exchanger gradually reaches the set value. Before the reheating process, if the TC7 is larger than 0 °C, the air volume of the indoor heat exchanger fan after the compressor starts is 400 kg/h, and it gradually reaches the set value 8 min later.

2.2 New cycle structures

2.2.1 DTM

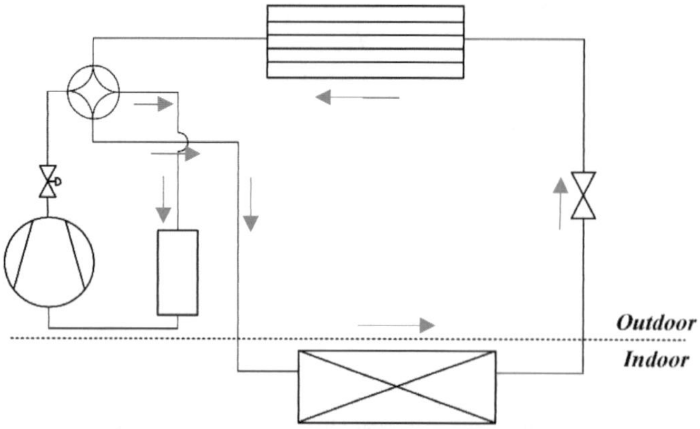

Figure 3 System diagram of DTM

As shown in Fig. 3, a solenoid valve is set at the discharge pipe to control the discharge mass flow, so that the compressor pressure and condensation pressure have a certain degree of pressure difference. On the one hand, the discharge pressure of the compressor can be increased and the surface temperature of the oil sump can be significantly increased by using discharge throttling in the defrosting process, so as to avoid the rise of liquid level in the compressor shell after defrosting. On the other hand, the existence of discharge pipe solenoid valve can reduce the quality of lubricating oil discharged from compressor during defrosting process to a certain extent.

2.2.2 SDBM

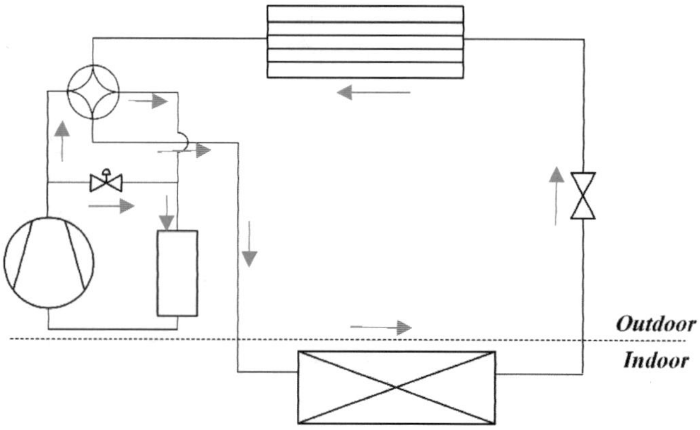

Figure 4 System diagram of SDBM

As shown in Fig. 4, a solenoid valve is arranged between the suction pipe and the discharge pipe to control the mass flow of the refrigerant by-pass. In the defrosting process, the suction and discharge bypass can effectively improve the suction temperature and suction pressure, prevent the refrigerant in the suction pipe being liquid phase, thus avoiding too much refrigerant left in the accumulator. In addition, the suction and discharge bypass will reduce the temperature difference between the ODHE and the IDHE, so as to avoid excessive refrigerant migrating to the IDHE after the defrosting shutdown, thus preventing the low suction pressure and temperature caused by insufficient refrigerant of ODHE during reheating.

3. Experimental results and analysis

3.1 Capacity and consumption

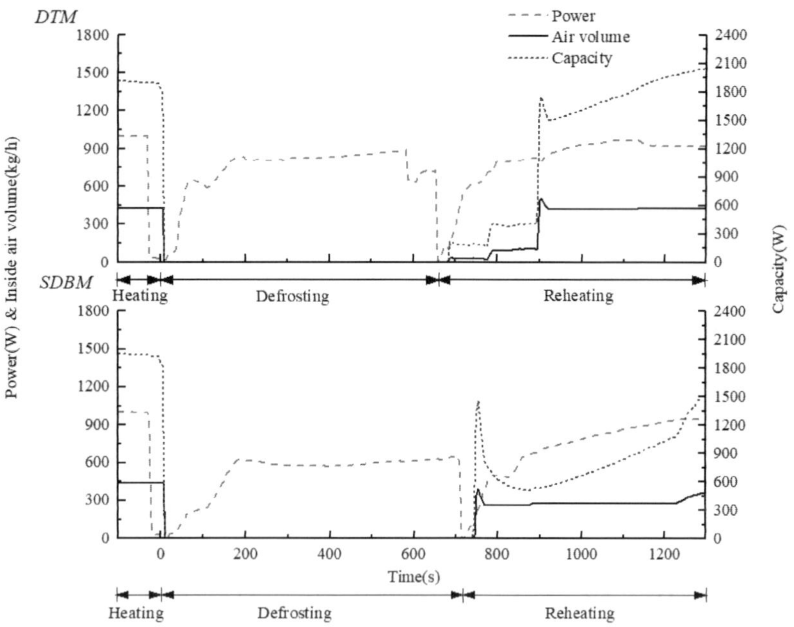

Figure 5 Variations of system power consumption, indoor air volume and capacity

The capacity and power changes of defrosting process under the two control strategies are shown in Fig. 5. It can be seen that the defrosting time of DTM was 652s and that of MST was 717s. It can be seen from Fig. 5, the power consumption of DTM was higher than that of SDBM, so the defrosting time was shorter. In addition, the anti-cold-air strategy was activated during the start-up process of reheating with DTM, resulting in the fan of the IDHE being turned on at 900s. However, the heating capacity of DTM increased very fast when the fan of IDHE was turned on.

3.2 Pressure and temperature

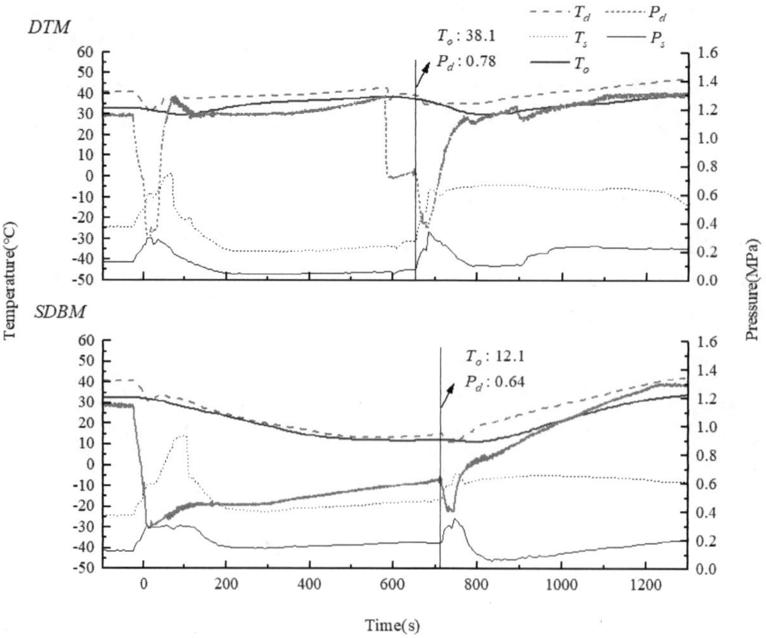

Figure 6 Variation of discharge temperature and oil sump temperature during defrosting

It can be seen from Fig. 6 that the oil sump temperature of DTM had a slight upward trend during the defrosting process, but the oil sump temperature of SDBM was in a continuous downward trend. In addition, it can be predicted that the pressure difference between suction and discharge of DTM was larger and the temperature difference between suction and discharge was also larger. After defrosting, the oil sump temperature of DTM was 38.1 ℃, and that of SDBM was 12.1 ℃. In addition, the higher suction temperature of SDBM was the main reason why it did not activate the anti-cold-air strategy. The suction pressure of DTM was very low during defrosting, but the lowest suction pressure was higher than that of SDBM during start-up process of reheating.

3.3 Viscosity and refrigerant distribution

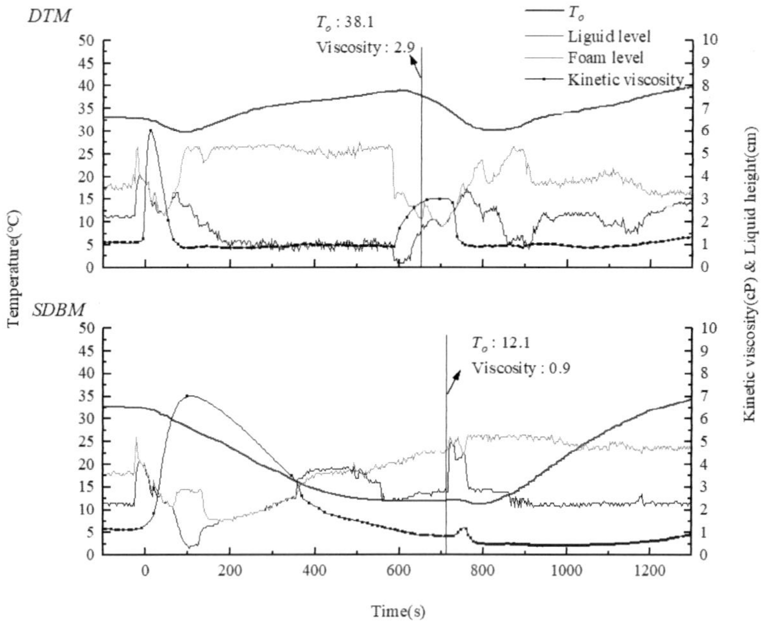

Figure 7 Change of liquid and foam level of the oil sump during defrosting process

After the defrosting started, the oil sump temperature of DTM continued to rise, the thicker foam layer was kept in the compressor oil sump and the viscosity of the oil pool was basically stable at 1cP. The oil sump temperature of SDBM continued to decline, there was almost no foam layer in the oil sump, and the viscosity was in a state of continuous decline. After defrosting, R290 distributed in the compressor was mainly dissolved in the lubricating oil in the oil sump, and R290 distributed in the accumulator was mainly in the form of liquid refrigerant.

As shown in Fig. 7, the liquid level at the end of defrosting with DTM was 1.5cm, and that with SDBM was 3cm. The total mass of oil sump was calculated by combining viscosity, temperature, pressure and liquid level height as shown in Eq. 1 – 5 [9]. Where the absolute pressure, P, is in Pa; the temperature, T, is in K; S is the solubility of the refrigerant; the kinematic viscosity, ν, is in mm2/s; a1 through a9 and b1 through b9 are empirical correlation constants and the value reference the cited document. The function K0 is zero order modified Bessel function of the second kind, and the constant φ is defined as 1.244067769[10].

$$Log_{10}\left(\frac{P}{10^5}\right) = a_1 + \frac{a_2}{T} + \frac{a_3}{T^2} + Log_{10}(S)\left(a_4 + \frac{a_5}{T} + \frac{a_6}{T^2}\right) + Log_{10}^2(S)\left(a_7 + \frac{a_8}{T} + \frac{a_9}{T^2}\right) \quad (1)$$

$$\begin{aligned} \mathrm{Log}_e(\mathrm{Log}_e(\nu + 0.7 + e^{-\nu} K_0(\nu + \varphi))) = b1 + b2\,\mathrm{Log}_e(T) + b3\,\mathrm{Log}_e{}^2(T) \\ + s(b4 + b5\,\mathrm{Log}_e(T) + b6\,\mathrm{Log}_e{}^2(T)) \\ + s^2(b7 + b8\,\mathrm{Log}_e(T) + b9\,\mathrm{Log}_e{}^2(T)) \end{aligned} \quad (2)$$

$$\rho mix = \frac{s \cdot \rho ref + (1-s)\rho oil + 1/(\frac{1}{s \cdot \rho ref} + \frac{1}{(1-s)\rho oil})}{2} \quad (3)$$

$$\mu = v \cdot \rho mix \quad (4)$$

$$m_t = \rho mix \cdot V \quad (5)$$

As shown in Fig. 8, the calculation results show that the total mass of oil sump, lubricating oil mass and refrigerant mass of DTM at the end of defrosting were 275g, 203g and 72g respectively. The total mass of oil sump, lubricating oil mass and refrigerant mass of SDBM at the end of defrosting were 274g, 147g and 127g respectively. Although the total mass of the oil sump was basically the same of these two methods at the end of defrosting, the solubility of the refrigerant in the oil sump was quite different, which leads to the difference of the refrigerant mass in the compressor. In the late defrosting period of SDBM, as the oil sump temperature continues to decreased, more and more refrigerant gradually dissolved in the oil sump, making the oil sump level rise. At the same time, the lubricating oil mass gradually decreased, and gradually discharged the compressor into the system pipeline.

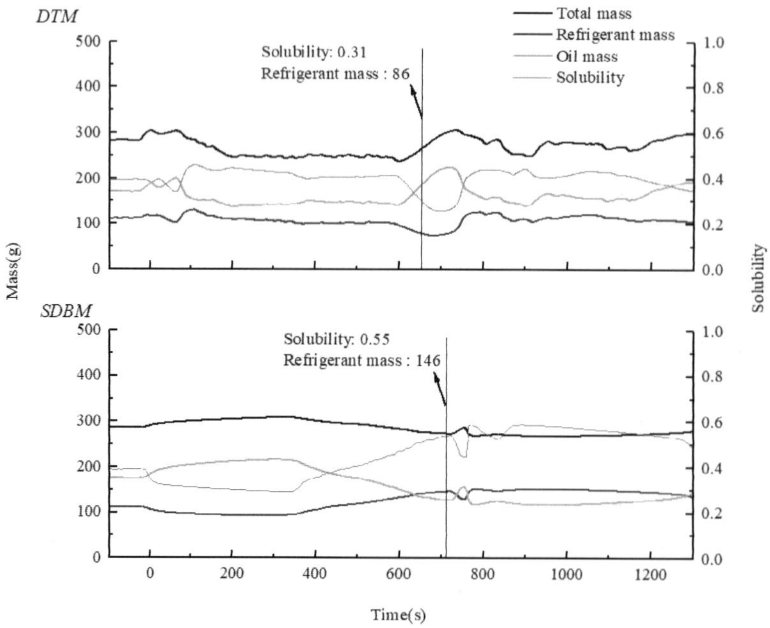

Figure 8 Change of mass and solubility of oil sump

After defrosting, the liquid level of the accumulator using DTM was 1cm and that of SDBM was 0cm. The refrigerant mass in the accumulator was calculated to be 90g with DTM and 60g with SDBM respectively considering the temperature, pressure and liquid level at the end of the defrosting process, and ignoring the mass of lubricating oil in the accumulator. At the end of defrosting with DTM, the distribution of refrigerant in the whole circulation system was about 16.5% in the oil sump, 20.6% in the accumulator, and 62.8% in the system pipeline and two heat exchangers. At the end of defrosting with SDBM, the distribution of refrigerant in the whole circulation system was about 29.2% in the oil sump, 13.8% in the accumulator, and

57.0% in the system pipeline and two heat exchangers.

Compared with SDBM, there were more refrigerants distributed in the heat exchangers with DTM at the end of defrosting, which was the main reason for its faster recovery of capacity and higher minimum of suction pressure during the start-up process of reheating. In addition, at the end of defrosting, the lubricating oil discharged from compressor with DTM was 56g less than that with SDBM.

4. Conclusion

This paper compares the two defrosting methods on a R290 RAC by experiment and the main conclusions are as follows：

1. The defrosting time of DTM is relatively short, but the temperature at the inlet of indoor heat exchanger is relatively low after defrosting. Stimulating the anti-cold-air strategy will affect the user comfort to a certain extent.
2. The SDBM can alleviate the situation of low suction temperature in the defrosting process, but the viscosity of oil sump is large in the early stage of defrosting and the time length of start-up process after defrosting is longer.
3. Compared with SDBM, there are more refrigerants distributed in the heat exchangers with DTM at the end of defrosting and the lubricating oil discharged from compressor with DTM is 56g less than that with SDBM.

Both methods have their own advantages and disadvantages, and the paper will provide an experimental basis for the reliability study of R290 RACs.

Acknowledgements

This study is supported by Guangdong Midea refrigeration equipment Co., Ltd. The author would like to thank the Midea Group for the sponsorship.

References

1. Liu, C., et al., *Effects of charge on the performance of R290 air conditioner with different expansion devices.* Applied Thermal Engineering, 2018. **140**: p. 498-504.
2. Zhou, W. and Z. Gan, *A potential approach for reducing the R290 charge in air conditioners and heat pumps.* International Journal of Refrigeration, 2019. **101**: p. 47-55.
3. Ghoubali, R., P. Byrne, and F. Bazantay, *Refrigerant charge optimisation for propane heat pump water heaters.* International Journal of Refrigeration, 2017. **76**: p. 230-244.
4. Tang, W., et al., *The performance and risk assessment of R290 in a 13 kW air source heat pump.* Applied Thermal Engineering, 2018. **144**: p. 392-402.
5. Li, T., et al., *Measurement of refrigerant mass distribution within a R290 split air conditioner.* International Journal of Refrigeration, 2015. **57**: p. 163-172.
6. Wang, C., et al., *Experimental study on startup characteristics of a R290 room air conditioner under various ambient temperatures.* International Journal of Refrigeration, 2019.
7. Amer, M. and C.-C. Wang, *Review of defrosting methods.* Renewable and Sustainable Energy Reviews, 2017. **73**: p. 53-74.
8. Song, M., et al., *Review on improvement for air source heat pump units during frosting and defrosting.* Applied Energy, 2018. **211**: p. 1150-1170.
9. Wu, J., et al., *Experimental analysis on R290 solubility and R290/oil mixture viscosity in oil sump of the*

rotary compressor. International Journal of Refrigeration, 2018. **94**: p. 24-32.

10. Seeton, C.J. and P. Hrnjak, *Thermophysical Properties of CO2Lubricant Mixtures and Their Affect on 2Phase Flow in Small Channels (Less than 1mm).* 2006.

IOP Conf. Series: Materials Science and Engineering 1180 (2021) 012049 doi:10.1088/1757-899X/1180/1/012049

Theoretical and Experimental Study on Rotary Compressor with Double Vapor Injection and Its System

H J Wei[1, 2, 3], **J Xu**[1, 2, 3], **B Yu***[3], **O X Yang**[1, 2, 3], **S G He**[3], **H H Zhao**[3] and **Y P Hu**[1, 2, 3]

[1] State Key Laboratory of Air-conditioning Equipment and System Energy Conservation, ZhuHai, Guangdong 519070, China

Phone: +86-756- 8587795, Fax: +86-756-8668386

*E-mail: jvcanjian@163.com

[2] Guangdong State Key Laboratory of Refrigeration Equipment and Energy Conservation Technology, ZhuHai, Guangdong 519070, China

[3] Gree Electric Appliances, Inc. of Zhuhai, ZhuHai, Guangdong 519070, China

Abstract. This paper takes three kinds of double vapor injection compression cycle as the research object, that is, the two-stage compression cycle with double vapor injection on low-pressure cylinder and intermediate of two cylinders (Cycle A), the two-stage compression cycle with double vapor injection on high-pressure cylinder and intermediate of two cylinders (Cycle B), and the three-stage compression cycle with double vapor injection (Cycle C). According to the law of conservation of mass and energy, the thermodynamic analysis model of each cycle was established, and the influence of saturation temperature of double vapor injection on the performance of each cycle was analyzed. The theoretical analysis results indicated that the refrigeration capacity of cycle A, cycle B and cycle C increase by 12.5%, 3.5% and 4.6%, respectively, and COP of that increase by 3.09%, 2.23% and 3.22%, respectively, compared with that of single stage compression cycle. Based on cycle A, a single-machine two-stage rotary compressor with double vapor injection and its heat pump air conditioning are made. The experimental results showed that, the heating capacity of cycle A can increase by 14.6% when the heating COP is equivalent, and the heating COP can increase by 4.4% when the heating capacity is equivalent, compared with the two-stage compression cycle with single vapor injection, under the –15°C ambient temperature heating condition.

1. Introduction

In recent years, with the advancement of "coal to clean energy" project in North China, the vapor injection compression technologies have developed rapidly [1]. The quasi two-stage compression technology with vapor injection on compression chamber was first studied and applied in scroll compressor [2]. In order to solve the problems of insufficiency and back-flow of vapor injection, some technical solutions have been developed for rotary compressor, such as "vapor injection structure on blade [3]", "vapor injection ports on the mid plate between two cylinders [4]" and "end-plate injection structure with check valve [5]". Therefore, the heating capacity and COP of heat pump air conditioning in low temperature environment are improved to a certain extent. The two-stage compression technology has the characteristics of small compression ratio and small power loss during. The two-stage rotary compressor based on the optimal volume ratio [6] and vapor injection valve technology [7] has excellent heating performance and is widely used in air source heat pump.

All the above technologies belong to single vapor injection technology, while the double vapor injection technology can further improve the performance of refrigeration cycle in theory. Jung D et al. simulated and evaluated the performance of multi-stage heat pumps using mixtures [8]. Mathison M et al. established a mathematical model to evaluate the effect of the number of injection ports on the performance of vapor injection compression cycle [9]. Theoretical and experimental study of the double vapor injection scroll compressor shows that the capability and COP have been improved compared with the single vapor injection [10-11]. In this paper, according to the working characteristics of the rotary compressor, the thermodynamic models of three kinds of double vapor injection compression cycle were established to simulate their performance. Based on the optimum cycle mode, a single-machine two-stage rotary compressor with double vapor injection and its heat pump air conditioning are made, and the performance was tested at low temperature conditions.

2. Analysis models

2.1 Rotary Compressor with Double Vapor Injection and Its System

In order to realize double vapor injection function on rotary compressor, either two-cylinder structure or three-cylinder structure can be used. When the compressor is of two-cylinder structure, one vapor injection port is set between two cylinders, and the other can be on the compression chamber of any cylinder. When the compressor is of three-cylinder structure, the two vapor injection ports are set between the two adjacent cylinders. Considering the high cost performance requirements of small air source heat pump, the main form of vapor injection cycle is the system with flash tank [12]. In order to avoid back-flow and insufficiency of vapor injection, it is better to start vapor injection immediately after the compression process [4, 6]. Based on the above characteristics of vapor injection on cylinder of rotary compressor, Figure 1 shows the schematic diagram and pressure-enthalpy diagram of three kinds of double vapor injection compression cycle.

It can be seen from Figure 1-(a), Figure 1-(c) and Figure 1-(e) that saturated vapors 7 and 10 with different pressures are produced by three expansion valves and two flash tanks in each refrigeration cycle, and the two differences are compression structure and location of vapor injection ports. In cycle A, a two-stage rotary compressor is used, and the double vapor injection ports are located on low-pressure cylinder and intermediate of two cylinders respectively. In cycle B, the double vapor injection ports are located on high-pressure cylinder and intermediate of two cylinders respectively. In cycle C, a three-stage rotary compressor with double vapor injection ports is used.

IOP Conf. Series: Materials Science and Engineering 1180 (2021) 012049 doi:10.1088/1757-899X/1180/1/012049

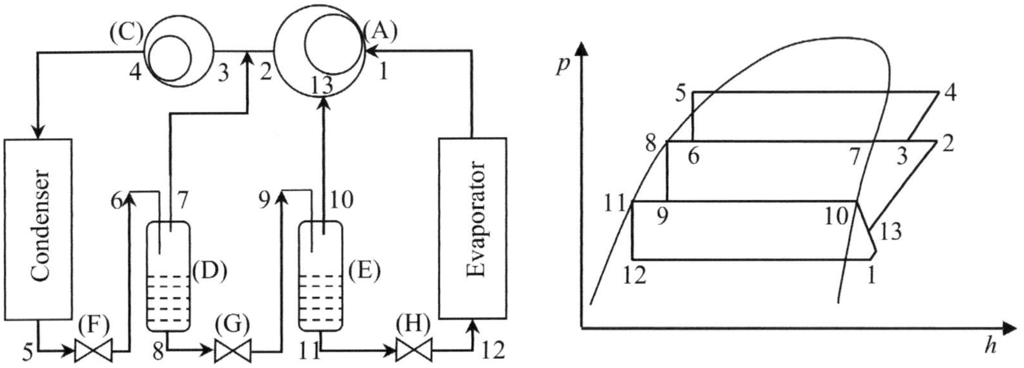

(a) Schematic diagram of cycle A (b) *p-h* diagram of cycle A

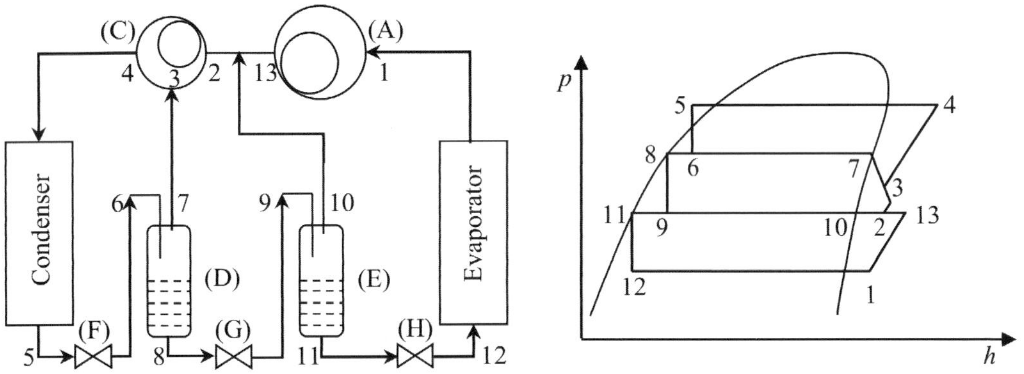

(c) Schematic diagram of cycle B (d) *p-h* diagram of cycle B

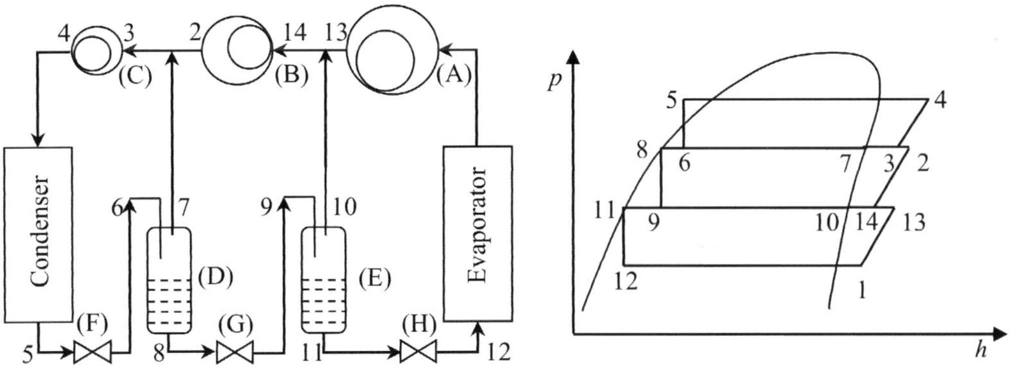

(e) Schematic diagram of cycle C (f) *p-h* diagram of cycle C

(A) – Low-pressure cylinder (B) – Middle-pressure cylinder (C) – High-pressure cylinder

(D) – 1st stage flash tank (E) – 2nd stage flash tank (F) - 1st stage expansion valve

(G) - 2nd stage expansion valve (H) - 3rd stage expansion valve

Figure 1. Schematic diagram and *p-h* diagram of double vapor injection compression cycle

Take Figure 1-(b) as an example to illustrate the change process of refrigerant state. After the suction gas 1 starts to be compressed, the vapor 10 from the 2nd stage flash tank is injected into the compression chamber of low-pressure cylinder, and then the gas is continuously compressed from the mixed state 13 to the state 2 and discharged. The vapor 7 from the 1st stage flash tank is injected between the two cylinders and mixed with the exhaust of the low-pressure cylinder to state 3. Then it is sucked into the high-pressure cylinder, compressed to state 4 and discharged into the condenser.

2.2 Model of vapor injection on cylinder

Taking the process vapor injection in cycle A as an example, the model of vapor injection on cylinder is analyzed. The vapor injection flow is a large pressure difference injection process, which can be regarded as a constant-volume adiabatic process [12]. The mass conservation equation, energy conservation equation and state equation can be expressed as:

$$\begin{cases} m_{13} = m_1 + m_{10} \\ m_{13}u_{13} = m_1 u_1 + m_{10}h_{10} \\ (T_{13}, p_{13}, s_{13}) = f(v_{13}, u_{13}) \end{cases} \tag{1}$$

Where, m_1, m_{10} and m_{13} are the suction mass, the vapor injection mass and the refrigerant mass in the compression chamber after vapor injection in each rotation cycle, respectively, h is the specific enthalpy, u is the internal energy, T is the temperature, p is the pressure, s is the specific entropy, and v is the specific volume. The relative vapor injection mass α_{10} should meet the following requirements:

$$\begin{cases} \alpha_{10} = \dfrac{m_{10}}{m_1 + m_{10}} \le x_9 \\ p_{13} \le p_{10} \end{cases} \tag{2}$$

Where, x_9 is the dryness of two-phase refrigerant behind the 2nd stage expansion valve. When $\alpha_{10} < x_9$, the refrigerant 11 in front of the 3rd stage expansion valve is the two-phase refrigerant, so its specific enthalpy can be calculated as follows:

$$h_{11} = \frac{h_9 - \alpha_{10}h_{10}}{1 - \alpha_{10}} \tag{3}$$

Process 1-13 is a pressure mutation process, and the technical work is as follows:

$$w_{1-13} = \int v\mathrm{d}p \approx (p_{13} - p_1)\frac{v_1 + v_{13}}{2} \tag{4}$$

2.3 Thermodynamic models of three cycles

The process 2/7-3 of cycle A, the process 10/13-2 of cycle B and the double vapor injection of cycle C all belong to vapor injection on the intermediate of two cylinders, and their models can be referred to the relevant literature [12]. Assuming that the compression process is isentropic and the throttling process is adiabatic, the thermodynamic models of three cycles are established.

The specific refrigeration capacity and COP of the three cycles can be expressed as follows:

$$q_c = h_1 - h_{12} \tag{5}$$

$$COP = q_c / w \tag{6}$$

The specific technical work of cycle A, cycle B and cycle C are shown in equation (7), equation (8) and equation (9), respectively.

$$w = (p_{13} - p_1)\frac{v_1 + v_{13}}{2} + \frac{h_2 - h_{13}}{1 - \alpha_{10}} + \frac{h_4 - h_3}{(1 - \alpha_{10})(1 - x_6)} \tag{7}$$

$$w = h_{13} - h_1 + (p_3 - p_2)\frac{v_2 + v_3}{2} + \frac{h_4 - h_3}{(1 - x_9)(1 - \alpha_7)} \tag{8}$$

$$w = h_{13} - h_1 + \frac{h_2 - h_{14}}{1 - x_9} + \frac{h_4 - h_3}{(1 - x_9)(1 - x_6)} \tag{9}$$

Where, α_7 is the relative vapor injection mass on high-pressure cylinder of cycle B.

3. Discussions on simulation results

In this paper, R32 is used in the double vapor injection compression cycles. According to the above analysis model, the theoretical performance of each cycle under typical low temperature heating condition was calculated. The parameters of the working condition are as follows: evaporation temperature is -20°C, condensation temperature is 45°C, superheat of suction gas is 10K, and subcooling of liquid is 7K. Figures 2, 3 and 4 show the map of COP varying with the saturation temperature of double vapor injection in three cycles respectively.

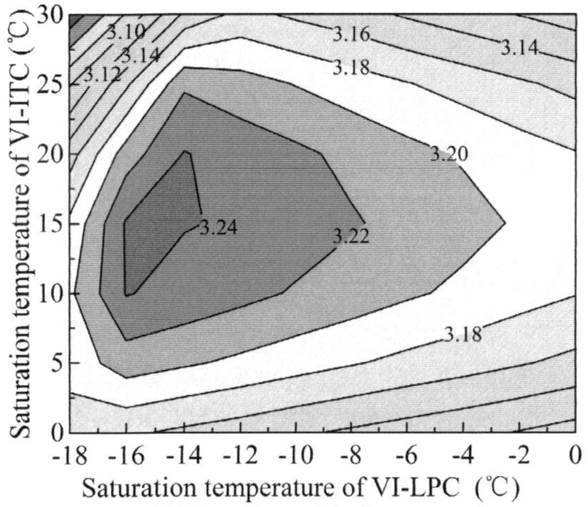

Figure 2. COP varies with the saturation temperature at cycle A

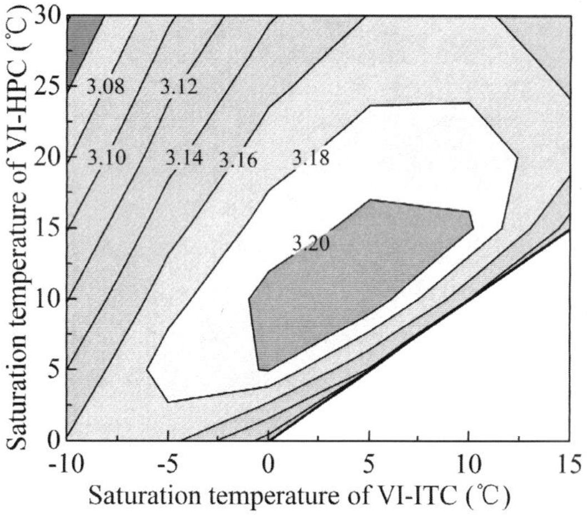

Figure 3. COP varies with the saturated temperature at Cycle

It can be seen from Figure 2 that the COP of cycle A first increases and then decreases with the saturation temperature of VI-LPC (vapor injection on low-pressure cylinder) and VI-ITC (vapor injection on the intermediate of two cylinders). The reason is that the relative vapor injection mass decreases with the increase of saturation temperature, and the refrigeration capacity and technical work decrease with the decrease of relative vapor injection mass. Therefore, COP reaches the maximum at a certain temperature. When the saturation temperature of VI-LPC is between -16°C and -13°C and the saturation temperature of VI-ITC is between 10°C and 20°C, COP is in the best range, and it's better in the region near evaporation temperature. The main reason is that there is extra technical work loss of vapor injection on cylinder compared with that on the intermediate of two cylinders. According to equation (4), the smaller the difference between p_{13} and p_1 is, the lower the technical work loss will be. However, when the saturation temperature of VI-LPC is too low, the gas separated by the 2nd stage flash tank cannot be completely injected into the cylinder. According to equation (3), the refrigerant in front of the 3rd stage expansion valve does not reach the saturated liquid state, which makes the refrigeration capacity and COP decrease rapidly.

As can be seen from Figure 3, COP is higher in the area where the saturation temperature of VI-HPC (vapor injection on high-pressure cylinder) is close to the saturation temperature of VI-ITC. The reason is consistent with the above analysis. In this area, the smaller the difference between p_3 and p_2

is, the lower the technical work loss will be. When COP is greater than 3.2, the saturation temperature area of cycle A is larger than that of cycle B.

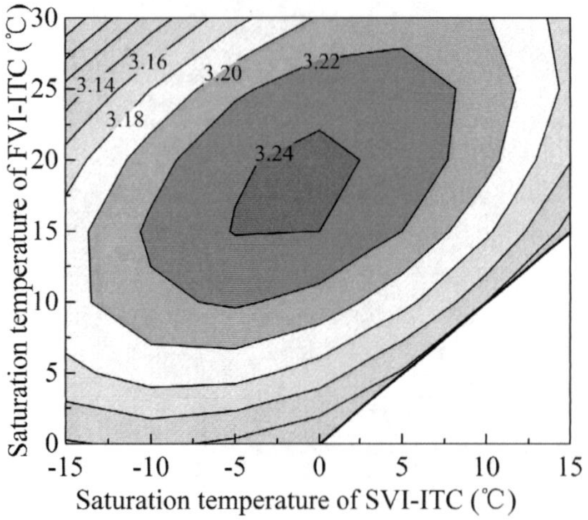

Figure 4. COP varies with the saturation temperature at cycle C

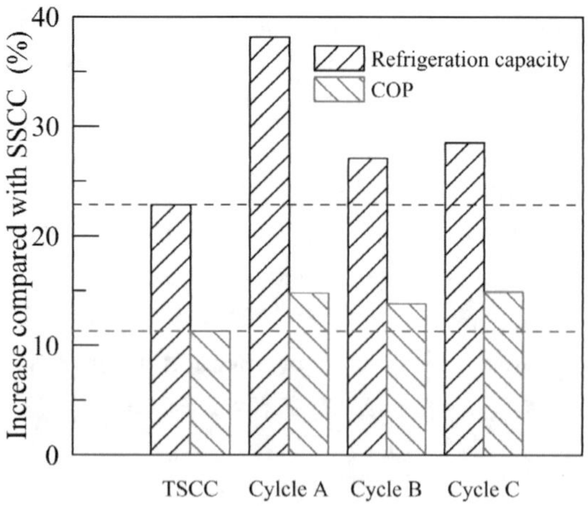

Figure 5. Increase of refrigeration capacity and COP compared with SSCC

It can be seen from Figure 4 that the effects of FVI-ITC (the first vapor injection on the intermediate of two cylinders) and SVI-ITC (the second vapor injection on the intermediate of two cylinders) on cycle C are balanced, and the COP contour presents a regular ellipse. Compared with Figures 2, 3 and 4, the highest COP of cycle A and cycle C is approximate. For the problem of how to choose the double vapor injection cycle, Figure 5 gives the comparison of the optimal refrigeration capacity and COP at the optimum COP point among these cycles.

It can be seen from Figure 5 that the refrigeration capacity and COP improvement of cycle A, cycle B and cycle C are all greater than that of TSCC (the two-stage compression cycle with single vapor injection). The refrigeration capacity of cycle A, cycle B and cycle C increase by 12.5%, 3.5% and 4.6% respectively, and COP of that increase by 3.09%, 2.23% and 3.22% respectively, compared with that of SSCC (single stage compression cycle). Compared with cycle C, the COP increase of cycle A is slightly lower, but the refrigeration capacity increase is more significant. Because the second vapor injection of cycle A is VI-LPC, when p_{10} is slightly higher than the suction pressure, vapor injection can be realized. However, cycle B and C are both VI-ITC, so the vapor pressure is affected by cylinder volume ratio. When the COP is optimal, the vapor pressure is greater than that of cycle A, so the enthalpy in front of evaporator of cycle A is even lower. At the same time, considering the complexity of compressor structure, cycle A is a better choice.

4. Prototypes of compressor and air conditioning system

Based on cycle A, a two-stage rotary compressor with double vapor injection has been developed. On the basis of the original two-stage rotary compressor QXFT-B096, the vapor injection structure is set on low-pressure cylinder. The volume of the low-pressure cylinder is 9.6cm^3, and the volume ratio of the high-pressure cylinder to the low-pressure cylinder is 0.68. On the basis of the original two-stage compression air conditioning KFR-35GW, a flash tank and an expansion valve are added. Figure 6 shows the three-dimensional model of the two-stage rotary compressor with double vapor injection, and Figure 7 the photo of the heat pump air conditioning.

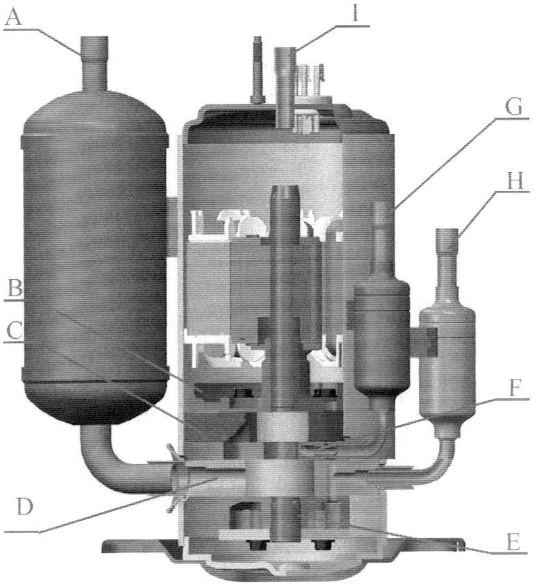

A – Suction tube
B – Main bearing
C – High-pressure cylinder
D – Low-pressure cylinder
E – Sub bearing
F – Separation plate
G – Vapor injection inlet on low-pressure cylinder
H – Vapor injection inlet between two cylinders
I – Discharge tube

Figure 6. 3-D model of two-stage rotary compressor with double vapor injection

G – Vapor injection inlet of low-pressure cylinder
H – Vapor injection inlet between two cylinders
J – 1st stage expansion valve
K – 1st stage flash tank
L – 1st on-off valve
M – 2nd stage expansion valve
N – 2nd stage flash tank
O – 2nd on-off valve
P – 3rd stage expansion valve
Q – Two-stage rotary compressor with double vapor injection
R – Outdoor heat exchanger

Figure 7. Two-stage compression system with double vapor injection (Outdoor unit)

As can be seen from Figure 6, the two vapor injection structures are as follows: (1) the vapor injection inlet G is connected with the working chamber of the low-pressure cylinder D through the vapor channel of the separation plate F to form the low-pressure vapor injection structure; (2) the inner chamber of the sub bearing is set between the low-pressure cylinder D and the high-pressure cylinder C, and the vapor injection inlet H is connected with the inner chamber to form the high-pressure vapor injection structure. For the low-pressure vapor injection structure, the position of the rolling piston changes with the eccentric movement of the crankshaft, and the vapor channel is blocked by the end face of the piston to close the vapor injection structure. For the high-pressure vapor injection structure, a valve is set between the inner chamber and the vapor injection inlet H, so as to avoid the back-flow of the gas from the inner chamber.

As can be seen from Figure 7, the air conditioning system uses three electronic expansion valves to adjust the saturation temperature of the double vapor injection, and uses two flash tanks to separate the double vapor with different temperature and pressure characteristics. An on-off valve is set between the gas outlet of the flash tank and the vapor inlet to open or close the vapor channel as required. In addition, a thin tube is connected to the vapor channel, which is used to access pressure sensor to monitor the vapor pressure.

5. Discussions on experimental results

The air enthalpy difference method is used to test the performance of the two-stage compression air conditioning system. The test condition is -15°C ambient temperature heating condition, the indoor dry-bulb and wet-bulb temperature are 20°C and 15°C respectively, and the outdoor dry-bulb temperature is -15°C. The heating performance of the air conditioning was tested at the compressor operating frequency of 110Hz and 95Hz, and the performance was compared with the single vapor injection two-stage compression system and single-stage compression system under the same condenser and evaporator. Figure 8 and Figure 9 show the variation of relative heating capacity and relative COP (compared with single-stage compression system) with the saturation temperature of double vapor injection two-stage compression air conditioning system.

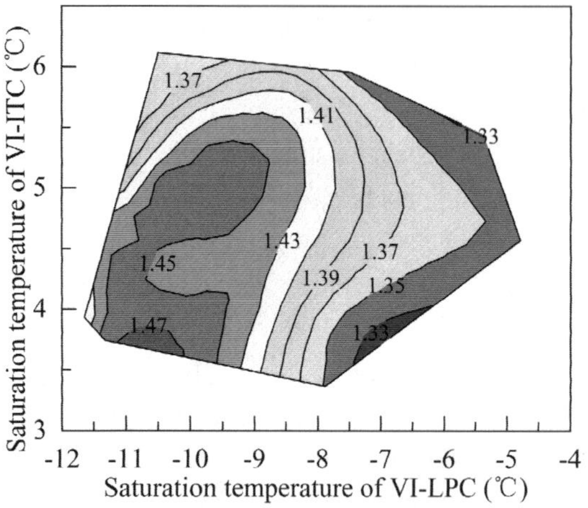

Figure 8. Experimental results of relative heating capacity

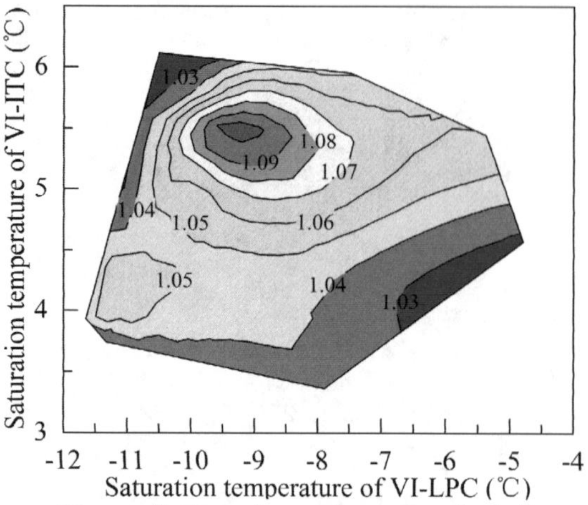

Figure 9. Experimental results of relative heating COP

As can be seen from Figure 8 and Figure 9, the maximum heating capacity of the two-stage compression system with double vapor injection is increased by more than 47%, and the maximum COP is increased by more than 10%. The reason why the experimental capacity is better than the simulated value is that the compression ratio and pressure difference of the low-pressure cylinder are greatly reduced, and the volume efficiency has been significantly improved. There is an optimal value of heating capacity with the variation of saturation temperature of VI-LPC. In the temperature range of

-12°C to -5°C, the refrigerant enthalpy at the inlet of the evaporator decreases with the decrease of the saturation temperature of VI-LPC, so the heating capacity increases. When the saturation temperature of VI-LPC is lower than -12°C, the heating capacity decreases because the vapor is not completely injected into the low-pressure cylinder. On the other hand, the heating capacity decreases with the increase of the saturation temperature of VI-ITC, because the exhaust temperature and the heating capacity decrease rapidly with the two-phase refrigerant injection.

The volume ratio of high-pressure cylinder to low-pressure cylinder is a constant value, so the state of the refrigerant entered by vapor injection inlet is affected by injection pressure [13]. When the injection pressure is too low, the separated gas of flash tank can not be completely injected. When the injection pressure is too high, some liquid can be injected. This is the reason why COP has an optimal value with the change of saturation temperature.

Table 1. Experimental results of different cycles at low ambient temperature heating condition

Compressor type	Cycle mode	Operating frequency	Heating capacity	COP	Discharge temperature
QXF-B096	`	110Hz	2625W	1.84	82.6°C
QXFT-B096	TSCC	110Hz	3308W	2.05	74.8°C
QXFTD-B096	Cycle A	110Hz	3792W	2.03	68.8°C
QXFTD-B096	Cycle A	95Hz	3355W	2.14	66.1°C

Table 1 shows the experimental results of the optimal COP, heating capacity and discharge temperature under the set operating frequency. Compared with SSCC, the maximum heating capacity of cycle A is increased by 44.5%, and the COP is increased by 10.3%. Compared with TSCC, the heating capacity of cycle A can increase by 14.6% when the heating COP is equivalent, and the heating COP can increase by 4.4% when the heating capacity is equivalent. Since the discharge temperature of the cycle A can be reduced by 6°C to 8.7°C than TSCC, theoretically, the double vapor injection system can operate at lower ambient temperature (e.g. -30°C). However, due to the limitation of experimental conditions, there is no reliable experimental result yet, which will be the next research work.

The reasons why the experimental results are better than the simulated results in Figure 5 are as follows: (1) The volume ratio of the compressor prototype is larger than simulated value in Figure 5 (0.53), which is beneficial to increase the vapor injection mass flow, so as to improve the heating capacity; (2) The efficiency of the compressor is not considered in the simulation, but the two-stage compression and vapor injection process are beneficial to reduce the compression ratio and improve the isentropic efficiency, which is beneficial to improve COP.

6. Conclusions

In this paper, the thermodynamic performance of three kinds of double vapor injection compression cycles is simulated. A two-stage rotary compressor with double vapor injection and its system are developed. The influence of saturation temperature of double vapor injection on heating performance at low ambient temperature was studied.

Under typical low temperature heating conditions, the theoretical refrigeration capacity of cycle A, cycle B and cycle C are increased by 12.5%, 3.5% and 4.6% respectively, and the theoretical COP is increased by 3.09%, 2.23% and 3.22% respectively, compared with single vapor injection two-stage compression cycle.

With the increase of saturation temperature of vapor injection, COP increased first and then decreased. The results show that there are differences in the optimal saturation temperature regions of COP among the three kinds of cycles. The region of cycle A is close to the evaporation temperature, the region of cycle B is close to the saturation temperature of VI-ITC, and the region of cycle C is affected by the equilibrium of double vapor temperature.

The experimental results showed that, the heating capacity of cycle A can increase by 14.6% when the heating COP is equivalent, and the heating COP can increase by 4.4% when the heating capacity is equivalent, compared with TSCC, under the – 15°C ambient temperature heating condition.

References

[1] Ma G, Wang L, and Liu Y 2019 Technological progress of rolling piston compressor *Refrigeration and Air-conditioning* **19(2)** 55-64 (in Chinese)

[2] Ma G Y, and Zhao H X 2008 Experimental study of a heat pump system with flash tank coupled with scroll compressor *Energy and Buildings* **40(5)** 697-701.

[3] Liu X, Wang B, Li X, and Shi W 2016 Performance of Rotary Compressor with Vapor Injection Mechanism on Blade *J. Refrigeration* **37(2)** 1-8 (in Chinese)

[4] Yan G, Jia Q, and Bai T 2015 Experimental investigation on vapor injection heat pump with a newly designed twin rotary variable speed compressor for cold regions *Int. J. of Refrigeration* http://dx.doi.org/doi:10.1016/j.ijrefrig.2015.10.024

[5] Wang B, Ding Y, and Shi W 2018 Experimental research on vapor-injected rotary compressor through end-plate injection structure with check valve *Int. J. Refrigeration* https://doi.org/10.1016/j.ijrefrig.2018.08.018

[6] Hu Y, Luo H, Wu J, Wei H and Yang O 2018 Research on two-stage rotary compressor with refrigerant injection for cold climate heat pump I*nternational Compressor Engineering Conference. Purdue* Paper 2604

[7] Hu Y S, Wei H J, Yu B, et al 2019 Research on the vapor injection of two-stage rotary compressor *IOP Conf. Series: Materials Science and Engineering* **604** 012072

[8] Jung D, Kim H K, and Kim O 1999 A study on the performance of multistage heat pumps using mixtures *Int. J. Refrigeration* **22** 402-13

[9] Mathison M M, Braun J E, and Groll E A Performance limit for economized cycles with continuous refrigerant injection *Int. J. Refrigeration* **34** 234-242

[10] Xu S, and Ma G Y 2015 Characteristic of Scroll Compressor Refrigeration System with Twice Vapor Injection *J. Refrigeration* **36(1)** 40-4 (in Chinese)

[11] Kang D, Jeong J H, and Ryu B 2018 Heating performance of a VRF heat pump system incorporating double vapor injection in scroll compressor *Int. J. Refrigeration* https://doi.org/10.1016/j.ijrefrig.2018.09.027

[12] Huang H 2018 *Technology and application of two stage compression variable volume ratio air source heat pump* (Beijing: Machine Industry Press)

[13] Yu B, Wei H, and Yang O 2019 Effect of Intermediate Pressure on Performance of Single-machine Two-stage Rotary Compressor *J. Refrigeration Technology* **42(1)** 43-7 (in Chinese)

IOP Conf. Series: Materials Science and Engineering 1180 (2021) 012050 doi:10.1088/1757-899X/1180/1/012050

Part load performance of single and two stage compressors – a comparative experimental study in a R410A chiller unit

S Inampudi[1], F Botticella[2] and S Elbel[1,2,3]

[1] Air Conditioning and Refrigeration Center, University of Illinois at Urbana-Champaign, 1206 W. Green St., Urbana, IL 61801, USA
[2] Creative Thermal Solutions, Inc. (CTS), 2209 N. Willow Rd., Urbana, IL 61802, USA
[3] Author to whom correspondence should be addressed (elbel@illinois.edu)

Abstract. The rising cooling needs for the residential and commercial air conditioning sectors and the requirement of higher SEER values increased the focus on energy efficient part load performance. Although several studies exist on the topic of capacity modulation, there is no comprehensive study that compares different modulation strategies in the same experimental facility. In a first step, this paper aims to experimentally compare two different capacity controls for scroll compressors (single speed vs two stage compressor) using the same R410A water chiller having a nominal cooling capacity of 8kW. The tests were conducted according to AHRI Standard 551/591 (2018). The Integrated Part Load Value (IPLV.SI) was used to fairly compare all the different capacity modulation strategies. Based on the experimental results, differences on the order of 8% in IPLV.SI are found.

1. Introduction

Almost all HVAC&R systems need to be designed to handle cooling loads below the target or design condition. These load variations are often caused by ambient weather variations, different levels of product loading, or the fact that systems are overdesigned to achieve quick pull-down. The range of load variations can be very substantial, in some cases even down to 30% or less of the design capacity. Different capacity control strategies employ different methods so that they can be implemented into the compressor or outside the compressor, in which case the system needs to provide adequate provisions. The modulation target is the same across the different techniques employed: the (average) refrigerant flow rate across the evaporator is reduced to adjust for the different levels of cooling capacity needed. There are many different methods of how to achieve this goal.

The main shortcoming of any literature-based comparison to the best of our knowledge is the fact that different data sets were obtained with different systems using different refrigerants, evaporation and condensation temperatures, secondary flow rates, and different levels of system controls. It is therefore nearly impossible to derive meaningful comparisons between the different studies when it comes to efficiency.

As a first step, this paper aims to experimentally compare two capacity modulation strategies, a single speed scroll compressor and a two stage scroll compressor using the same R410A WEG chiller system. These compressors when used in the experimental setup provide comparable nominal cooling capacity. The cooling capacity of a compressor is proportional to the mass flow rate provided by the compressor. The tests are conducted according to AHRI Standard 551/591 (2018) and the comparison is done using a single figure of merit called IPLV.SI (Integrated Part Load Value) [1].

2. Modulation strategies being considered

The following are the different compressor modulation strategies being considered as part of this study. As a first step, this paper compares a single speed and two stage compressor. These compressors are scroll and can provide similar cooling capacity and belong to the same generation.

2.1. Single speed compressor
A single speed compressor modulates capacity by cycling i.e. turning on and off. The on/off capacity control is the simplest method of adjusting the predetermined temperature (setpoint) using a thermostat. After the temperature reaches the setpoint, the thermostat stops the compressor and circulating refrigerant in the cycle. Since the secondary fluid continues to circulate, the temperature of the water or air gradually raises. When the thermostat detects, this rise turns the compressor on [2].

2.2. Two stage compressor
This is a commercially available two stage compressor, marketed as two-stage Copeland ultratech compressor. It is different from conventional two stage compressors that use a low-pressure stage and a high-pressure stage. In this compressor, two stage refers to its ability to operate at two stages (100% and 67%). Instead, it has two internal bypass ports which enable the system to run at 67%-part load capacity. If the compressor is operated at the same condenser and evaporator water temperature, then the capacity at the 67% stage is approximately 67% of the capacity at the 100% stage. This compressor can be operated either at a high stage (provides 100% capacity) or a low stage (provides 67% capacity). The capacity is reduced by bypassing a portion of the gas in the scroll back to suction. These bypass ports are opened or closed using a solenoid valve present in the scroll chamber [3].

3. Details of the experimental setup
Figure 1 shows the schematic of R410A Water Ethylene Glycol (WEG) chiller. 20% ethylene glycol in water (by volume) is used as the secondary fluid. Two closed WEG loops are connected to the evaporator and condenser. Variable speed pumps and electric heaters are used to control the WEG flow rate and the inlet temperature of the condenser and the outlet temperature of the evaporator. An additional heat exchanger with chilled water flowing through is included in the condenser WEG loop to reject the heat from the condenser WEG loop. All the heat exchangers used in the facility are brazed plate heat exchangers. A 0.9 L receiver is connected between the condenser and the subcooler. Superheat is controlled by an Electronic Expansion Valve (EEV) while the subcooling is a function of the charge. The geometric dimensions of the evaporator, condenser, and subcooler can be found in Table 1.

Type-T thermocouples, absolute and differential pressure transducers, and Coriolis-type mass flow meters are used to obtain the refrigerant side measurements while type-T thermocouples, differential pressure transducers, and Coriolis-type mass flow meters are used to obtain WEG measurements. Data is collected at steady-state conditions at 5s intervals for 20 consecutive minutes, and the data is averaged over the collection period.

The uncertainties of the calculated properties associated with the instrument uncertainties of measured properties are calculated by uncertainty propagation as shown in Equation (1):

$$U_Y = \sqrt{\sum_i (\partial Y/\partial X_i)^2 U_{X_i}^2} \tag{1}$$

where U_Y is the uncertainty of calculated property and U_{X_i} is the uncertainty of measured property. The uncertainty of the other sensors used in the experimental facility is presented in Table 2.

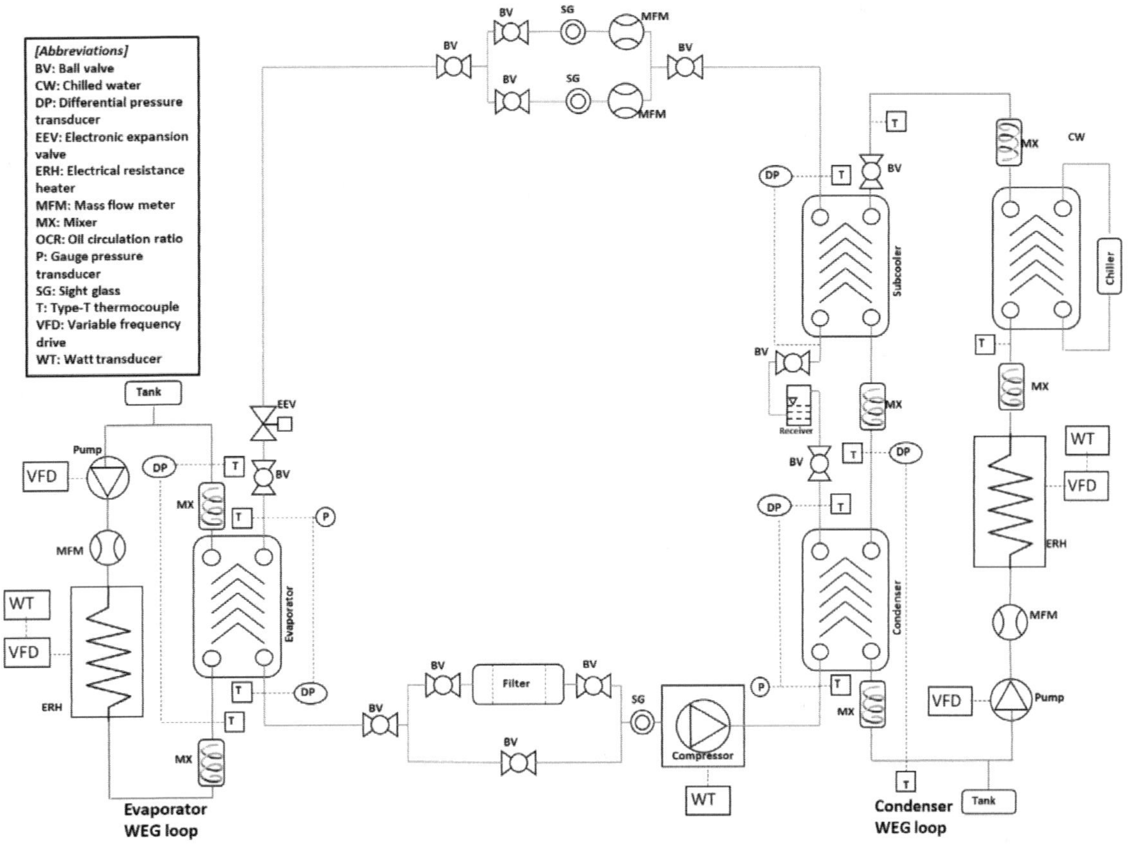

Figure 1: Schematic of the R410A experimental facility

Table 1: Dimensions of the brazed plate heat exchangers

Heat exchanger	Length (mm)	Width (mm)	Number of plates
Evaporator	311	111	28
Condenser	311	111	14
Subcooler	207	77	14

Table 2: Summary of measured and calculated property uncertainties

Instrument	Thermocouple (°C)	Mass flow meter (g/s)	Wattmeter (kW)	Capacity (kW)	COP (-)
Uncertainty	±0.1°C	±0.2%	±0.5%	±2.8%	±2.9%

Capacity is calculated on the refrigerant side and WEG side. For the WEG side, mass flow rate, temperature, and specific heat are used to calculate capacity as shown in Equation (2). For the refrigerant side, temperature and pressure are used to calculate the enthalpy which is then used with the mass flow rate to calculate the capacity as shown in Equation (3). The capacity reported is the average of the refrigerant side and WEG side capacity given by Equation (4). The difference between the two capacities is indicated by the error given by Equation (5). AHRI standard requires only the capacity measured from

WEG side, however in the current calculations, capacity is average of the refrigerant and WEG side. Equation (5) is a way to indicate the difference between these two energy balances. This error is always less than 3% for the part load rating tests. Power consumed by the compressor is measured using a Wattmeter. The ratio of the average capacity and power consumed by the compressor is used to calculate the COP_{test} as shown in Equation (6).

$$\dot{Q}_{ev,WEG} = \dot{m}_{WEG} C_p \Delta T \tag{2}$$

$$\dot{Q}_{ev,ref} = \dot{m}_{ref} \Delta h \tag{3}$$

$$\dot{Q}_{ev,avg} = \frac{(\dot{Q}_{ev,WEG} + \dot{Q}_{ev,ref})}{2} \tag{4}$$

$$\varepsilon_{Qev} = \frac{(\dot{Q}_{ev,avg} - \dot{Q}_{ev,WEG}) \cdot 100}{\dot{Q}_{ev,avg}} \tag{5}$$

$$COP_{test} = \dot{Q}_{ev,avg} / \dot{W}_{cp} \tag{6}$$

4. AHRI 551/591 (2018) standard

AHRI 551/591 is used for the determination of the part-load performance of water chillers. The standard defines a single number part-load efficiency figure of merit called Integrated Part Load Value (IPLV.SI) calculated at part load rating conditions. These part load rating conditions are shown below in Table 3. IPLV.SI is the weighted average of the COP_R measured at these standard rating conditions as shown in Equation (7). These factors in Equation (7) are based on the weighted average of the most common building types and operations using average weather in 29 U.S. cities.

$$IPLV.SI = 0.01 \cdot A + 0.42 \cdot B + 0.45 \cdot C + 0.12 \cdot D \tag{7}$$

$$\begin{aligned} A &= COP_R \; at \; 100\% \\ B &= COP_R \; at \; 75\% \\ C &= COP_R \; at \; 50\% \\ D &= COP_R \; at \; 25\% \end{aligned}$$

If a system with a compressor cannot be unloaded to 25%, 50%, or 75% load point, then the compressor is run at the minimum step of unloading at the condenser entering water shown in Table 3 for 25%, 50%, or 75% capacity points as required. Once the COP_{test} is calculated at these conditions using Equation (6), it is degraded to COP_R using the Equation (8), (9), (10), and (11). The standard does not explicitly state the effects included in the cycling losses (indicated by C_D). Refrigerant migration during OFF cycle, pressure equalization and thermal mass of the components are major causes of cycling losses [4].

$$COP_R = COP_{test} / C_D \tag{8}$$

$$C_D = (-0.13 \cdot LF) + 1.13 \tag{9}$$

$$LF = \frac{(\%Load)(Q_{ev\ 100\%})}{(Q_{ev\ min\%Load})} \tag{10}$$

$$\%Load = {(Part\ load\ net\ capacity)}\Big/{(Full\ load\ rated\ net\ capacity)} \quad (11)$$

Table 3: AHRI 551/591 part load conditions for IPLV.SI

Condition	Part load ratio (%)	Condenser Inlet/Outlet (°C)	Evaporator Inlet/Outlet (°C)
A	100	30/35	12/7
B	75	24.5/*	*/7
C	50	19/*	*/7
D	25	19/*	*/7

Table 3 shows the part load conditions. As seen in the table, condenser inlet, outlet, and evaporator inlet, outlet are mentioned for A condition while for the B, C, and D conditions, only the condenser inlet and evaporator outlet temperature are mentioned. Standard required that the condenser and evaporator WEG flow rate used for the A condition be used for the B, C, and D conditions.

5. Results

Before testing the compressor at the part load rating conditions, charge optimization was done at the A condition i.e. condenser WEG inlet temperature of 30°C for both the compressors to determine the optimum charge for the compressor. After that, the compressors are tested at the part load rating conditions at their respective optimum charge. The next section shows the charge optimization of a single speed compressor.

5.1. Charge optimization of a single speed compressor

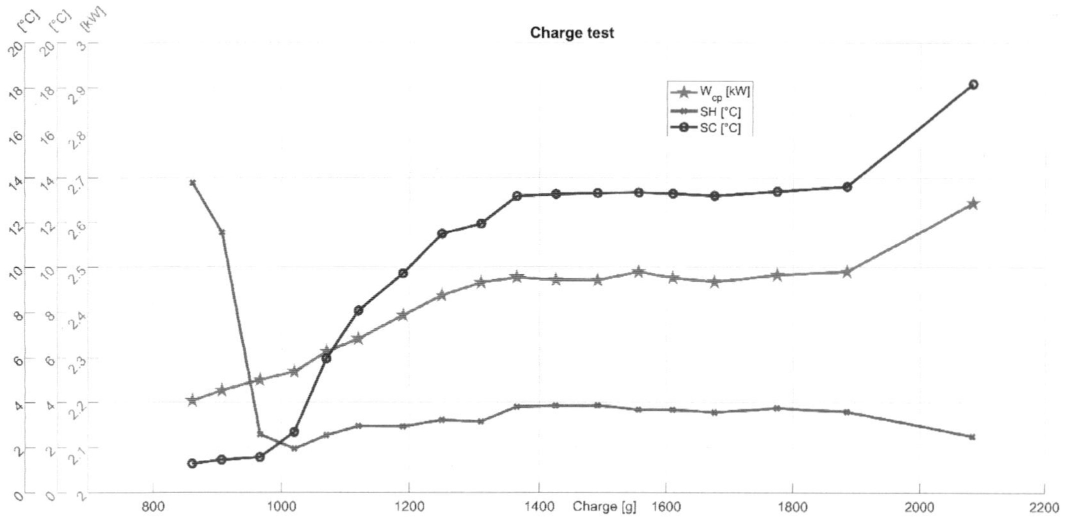

Figure 2: Variation of W_{cp}, superheat and subcooling with charge

Figure 2 shows the variation of the compressor work, superheat, and subcooling with the charge. As the charge increases, subcooling increases until it reaches a plateau. Superheat remains the same after the first few points because the EEV maintains the same superheat. The high superheat at the initial charge is because of the undercharged system. Figure 3 shows the variation of COP, average capacity, and the compressor discharge pressure with the charge. There is a range of charge from 1300 g to 1800 g where the curves remain almost constant. This range exists because of the receiver present after the condenser. As the charge increases in this range, the receiver gets filled and the amount of charge present in the remaining system remains almost constant. If the charge is increased beyond this charge, the receiver can no longer hold the charge and liquid starts backing up in the condenser. This is seen by an increase in the compressor discharge pressure and increase in the subcooling. A charge of 1500 g is selected as the optimum charge.

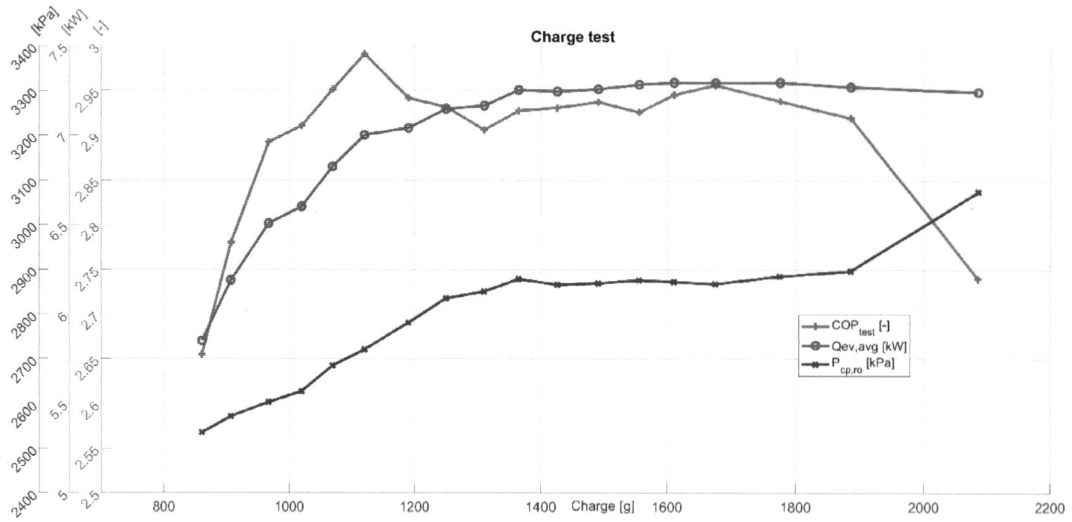

Figure 3: Variation of COP_{test}, $Q_{ev,avg}$, and $P_{cp,ro}$ with charge

5.2. Single speed compressor part load performance

As explained before, as single speed compressor does not have any modulation, it can operate only at 60 Hz. The variation of the capacity with part load rating conditions is shown in Figure 4. The highest load (corresponding to A condition) is the maximum cooling capacity provided by the system with this compressor operating at A condition (which happens to be 7.3 kW). For the other conditions, this load is multiplied by the corresponding part load ratio from Table 3. Thus load is a linear function of the part load rating conditions.

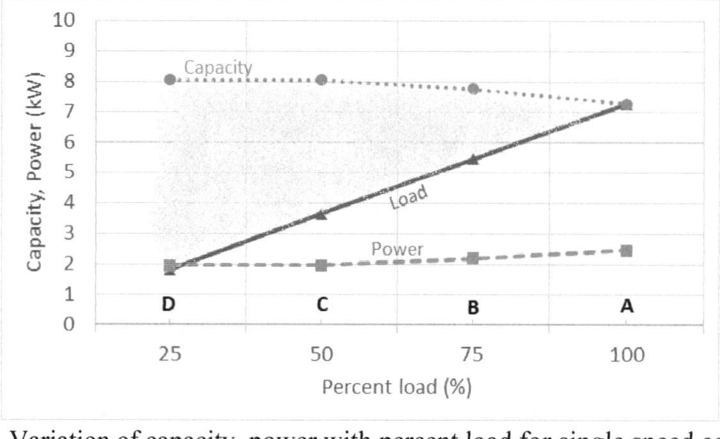

Figure 4: Variation of capacity, power with percent load for single speed compressor

As seen in Figure 4, as the percent load changes from 100% to 25%, the standard requires a reduced condenser WEG inlet temperature, this reduces the condensing temperature which in turn causes the capacity to increase. The shaded region between the capacity curve and the load curve is an indication of how much the obtained capacity differs from the required load and is a representation of the cycling losses. The capacity remains constant for the C and D conditions because the condenser WEG entering temperature is maintained the same as required by Table 3, only the percent load changes. Since a single speed compressor only operates at one stage, and thus the performance of the compressor at these percent loads is the same. As expected, the compressor power reduces as the percent load reduces because of reduced condensing temperature and compressor discharge pressure.

5.3. Two stage compressor part load performance

A two stage compressor can be operated at either high stage (100%) or low stage (67%). The performance of this compressor at different part load conditions is shown in Figure 5.

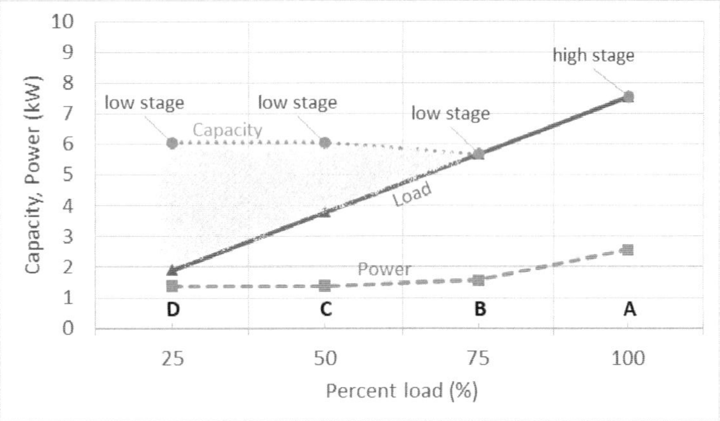

Figure 5: Variation of capacity, power with percent load for two stage compressor

For the A condition, the compressor is operated at a high stage. For the B, C, and D conditions, as the required load reduces, the compressor is operated at the low stage. The capacity is not a continuous line between A and B conditions because this compressor does not provide a continuous modulation, it provides a step modulation. At the B condition when the compressor is operated at a low stage, it provides a capacity which matches the required load. A low stage provides 67% cooling capacity compared to high stage when operated at the same conditions. For the B condition, the low stage happens to provide a capacity which is equal to the required load (75% of the load at A condition). However, at the C and D conditions, the low stage cannot match the capacity and it behaves similarly to a single speed compressor and hence there will be cycling losses. The shaded region is a representation of the cycling losses. By comparing Figures 4 and 5, you can see that the shaded region i.e. cycling losses are higher in a single speed compressor compared to a two stage compressor.

5.4. Comparison between single speed and two stage compressor

Figure 6 compares the capacities between a single speed and two stage compressor. At the 100% load, both the compressors provide a similar capacity which indicates that both the compressors are comparable. However, for the remaining conditions, the single speed compressor has higher capacities than a two stage compressor because the two stage compressor can operate at a low stage.

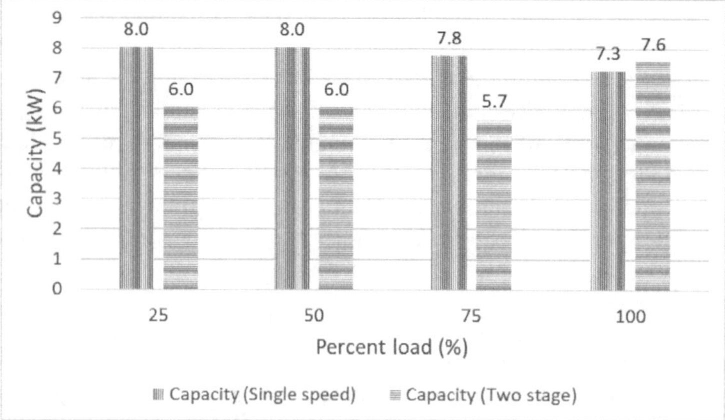

Figure 6: Comparison of capacity between single speed and two stage compressor

Figure 7 shows the COP_{test} given by Equation (6). Both the compressors have the same COP_{test} at 100% load. It is interesting to note that at the 75% load, though the two stage compressor is operating at the low stage and the single speed compressor is operating at its highest (only) stage, the COP_{test} are still comparable. However, the COP_{test} for the two stage is significantly higher at the lower percent loads because of reduced capacity and reduced compressor power. The COP_{test} for each compressor remains the same for the 50% and 25% load because the part load operating conditions are the same for these points.

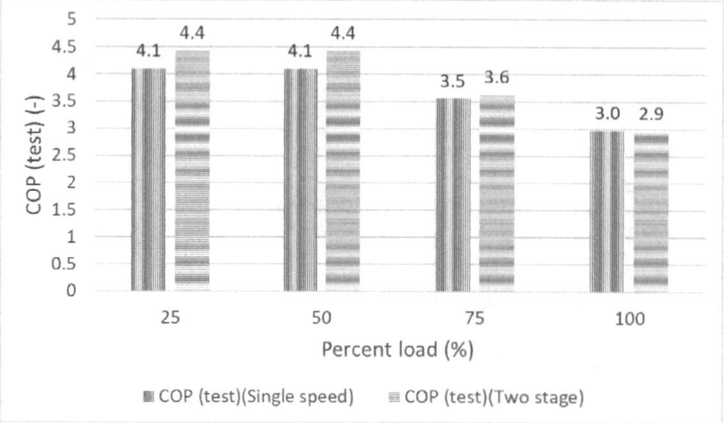

Figure 7: Comparison of COP_{test} between single speed and two stage compressor

Figure 8 shows the COP_R given by Equation (8). Except for the 100% load when both the compressors have similar performance, the two stage compressor has a higher COP_R than a single speed compressor. This is because of its ability to operate at lower stage. This trend can also be graphically explained by the cycling losses represented in Figures 4 and 5. Though the COP_{test} is the same for each compressor at the 25% and 50% load, COP_R at the 25% load is lower than 50% load because higher cycling losses reduce the COP_{test} by greater value as shown in Equation (9).

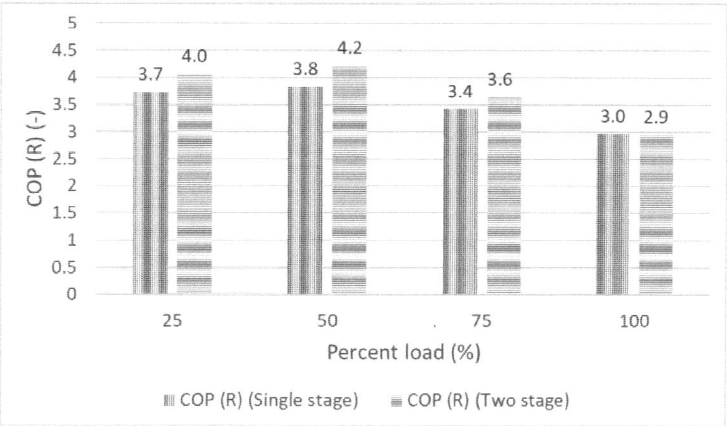

Figure 8: Comparison of COP_R between single speed and two stage compressor

IPLV.SI is calculated using the COP_R from Figure 8 and Equation (7). A single speed compressor has an IPLV.SI of 3.6 while the two stage compressor has an IPLV.SI of 3.9. This 7% difference can be explained by the higher COP_R of the two stage compressor at 50% and 75% load which have the highest weighting factors in Equation (7). A two stage compressor does not have any degradation coefficient at the 75% load and though it does incur some cycling losses at the 50% load, it is still operating at the low stage and these losses are lower than that incurred by the single speed compressor.

6. Conclusions

This study experimentally compared a single speed and a two stage compressor in the same R410A WEG chiller. Both the compressors have similar capacity and they belong to the same generation. Charge optimization was done to determine the optimum charge for both the compressor and AHRI 551/591 was used to determine the IPLV.SI for both the compressors at their respective optimum charge. Both compressors showed comparable capacities and COP_{test} at the higher modulation conditions. However, two stage compressor has a higher COP_R at the lower condition because of its ability to operate at a lower stage and better meet the required load, this reduces the C_D. Single speed compressor has an IPLV.SI of 3.6 while the two stage compressor has an IPLV.SI of 3.9.

7. Future work

As explained this paper is a first step in a series of comparison studies. The following compressor modulation strategies will be investigated as part of the future work.

7.1. Variable speed compressor

A system with this compressor modulates its capacity by varying the speed of the compressor motor [5]. The focus will be on the inverter driver variable speed compressor with a brushless DC motor. The selected compressor can be operated between 20 Hz and 100 Hz. Variable speed compressor This compressor modulates its capacity by varying the speed of the motor [4]. The focus will be on the inverter driver variable speed compressor with a brushless DC motor. The selected compressor can be operated between 20 Hz and 100 Hz.

7.2. Digital scroll compressor

In this commercially available compressor, the modulation is achieved through rapid engagement/disengagement of the upper scroll (using a piston and solenoid valve). The scrolls are separated in a periodic cycle to obtain a time-averaged capacity based on the ratio of loading and unloading times. This provides a modulation range from 10% to 100% [6].

7.3. Tandem compressors

In this configuration multiple compressors operate in parallel. The modulation is achieved by turning a different combination of compressors on and off [7]. This study will cover a tandem combination of two compressors. The study will include combinations of a single speed compressor with a variable speed, digital scroll, two stage, and another single speed compressor.

Acknowledgements
The authors would like to thank the member companies of the Air Conditioning and Refrigeration Center at the University of Illinois at Urbana-Champaign for their funding to support this project, Emerson / Copeland for compressor samples, Danfoss for brazed plate heat exchangers, and Creative Thermal Solutions for the technical support.

Nomenclature

Symbols

C_d	degradation coefficient [-]
COP	coefficient of performance [-]
C_p	specific heat [kJ/kg-K]
IPLV.SI	integrated part load value [-]
h	enthalpy [kJ/kg]
LF	load factor [-]
m	mass flow rate [kg/s]
Q	heat transfer rate [kW]
SC	subcooling [°C]
SH	superheat [°C]
T	temperature [°C]
W	power [kW]
WEG	water ethylene glycol

Greek

ε	error [-]

Subscripts

avg	average
cp	compressor
ev	evaporator
ref	refrigerant
ro	refrigerant, outlet

References

[1] A. H. R. I. 551/591 (SI), 2018: 2018 Standard for performance rating of water-chilling and heat pump water-heating packages using the vapor compression cycle. *Air-Conditioning, Heating, Refrigeration Institute, Arlington, VA, USA.*

[2] Ekren, O 2017 Refrigeration System: Capacity Modulation Methods. *IntechOpen, London, UK,* 119-143.

[3] Wang, B, Han, L, Shi, W, and Li, X 2012 Modulation method of scroll compressor based on suction gas bypass. *Applied Thermal Engineering*, **37**, 183-189.

[4] Rubas, P J, & Bullard, C W 1995 Factors contributing to refrigerator cycling losses. *International Journal of Refrigeration*, **18(3)**, 168-176.

[5] Qureshi, T Q, and Tassou, S A 1996 Variable-speed capacity control in refrigeration systems. *Applied Thermal Engineering*, **16(2)**, 103-113.

[6] Wu, G H, & Wu, K Q 2004 Design and Application Research on Digital Scroll Compressor in Air Conditioning System, *Proc. of the International Compressor Engineering Conf,* Paper 1672, West Lafayette, IN, USA

[7] Cecchinato, L 2010 Part load efficiency of packaged air-cooled water chillers with inverter driven scroll compressors. *Energy Conversion and Management*, **51(7)**, 1500-1509.

Thermodynamic analysis of volume flow rate ratio on the performance of a NH_3/CO_2 cascade refrigeration system

Q C Yang, L Liu G B Liu, Y Y Zhao & L S Li
College of Electromechanical Engineering, Qingdao University of Science and Technology, Qingdao, 266061, P.R. China

E-mail:qichaoyang@163.com

Abstract. NH_3/CO_2 cascade refrigeration system is the most potential and promising selection in low temperature applications. To meet the cooling demands of CRS at various working conditions, the capacity regulation of the screw refrigeration compressor is needed. This paper specifically studied the impacts volume flow rate ratio of high temperature cycle (HTC) and low temperature cycle (LTC) on the system performance. The simulation model of CRS including energy and exergy analysis was established. Based on the obtained results with the models, it can be found that there exists an optimum volume flow ratio that makes the maximum COP and exergy efficiency and the optimum volume flow ratio value is different under various working conditions. When the evaporation temperature is -55°C, the optimum volume rate ratio is about 2.4 and corresponding maximum COP and exergy efficiency of CRS is 1.031 and 0.346, respectively. The optimum condensing temperature of LTC is -9.75°C. When the volume flow ratio is increasing, which means the motor frequency or rotational speed of HTC compressor increases while the rotational speed of LTC compressor is fixed, the optimum intermediate temperature decreases and the discharge temperature of HTC compressor increases proportionally. The cooling capacity could be enhanced by increasing the volume flow ratio. The obtained results are helpful for the screw refrigeration compressor regulation in CRS under various working conditions and cooling demands.

1. Introduction

The cascade refrigeration system (CRS) which consists of two different refrigeration cycles was introduced to meet the demand for low temperature refrigeration which is required in the temperature range from -100°C to -30°C. A NH_3/CO_2 cascade refrigeration system uses ammonia and carbon dioxide as refrigerants in high and low temperature circuits, respectively. Both ammonia and carbon dioxide are natural refrigerant and many investigations of the NH_3/CO_2 cascade refrigeration system are attracting attention.

In the process of introducing and designing a CRS, one of the most important issues is to find an optimal intermediate temperature, which is the condensation temperature of low temperature cycle (LTC) or the evaporation temperature of a high temperature cycle (HTC), to obtain the best efficiency of the system [1]. Many researches have been conducted on the optimum condensation temperature of LTC for various cascade systems. Lee et al. theoretically studied the NH_3/CO_2 cascade refrigeration system to determine the optimal condensing temperature under given various design parameters. The useful correlations that yield the optimal condensing temperature of the CRS are proposed[2]. Dopazo et al. also investigated a NH_3/CO_2 cascade refrigeration system with numerical analysis to find optimal

conditions[3]. Hansaem Park et al. researched the optimal intermediate temperature of a cascade refrigeration system with R134a and R410A cascade refrigeration system by thermodynamic analysis. Theoretical analysis was targeted for establishing a mathematical model to enable the prediction of optimal intermediate temperatures that make the system operate with its maximum efficiency.
For actual operation of cascade refrigeration system, working conditions including the condensing temperature and evaporation temperature. An evaporative condenser or air-cooled condenser is often used in CRS. The outdoor air temperature will have an effect on the condensation pressure of cascade refrigeration system. Therefore, it is necessary to regulate the working parameters to ensure a good performance of CRS. Reciprocating or screw compressors are the main type utilized in NH_3/CO_2 CRS. For screw refrigeration compressor, sling valve regulation and variable frequency regulation are two capacity control methods. Although many theoretical and experimental studies have been conducted to investigate the optimum intermediate temperature for CRS, these studies do not provide much attention to the volume flow rates of HTC and LTC compressor and its influence on the system performance. Especially, when the screw refrigeration compressor used in CRS is regulated by variable frequency method, which caused a changing of volume flow rate ratio of HTC compressor and LTC compressor. For rolling piston refrigeration compressor used in air conditioner and two-stage air source heat pump, the selection of the cylinder volume ratio of the two compressor is one of the most important tasks, which will not only influence the initial cost but also have a significant impact on the operation cost. Ko et al. investigated the performance of a two-stage vapour heat equipped with a twin rotary compressor under ARI conditions. Three different cylinder volume ratios, 0.59,0.68 and 0.82, were tested. The results show that the performance of the system with a cylinder volume ratio of 0.59 is superior to those with other ratios[4]. Baek et al. experimentally measured the heating performances of the vapor injection CO_2 systems including the flash tank vapor injection and sub-cooler vapor injection cycles, by adjusting the cylinder volume ratio of the twin rotary compressor at the ambient temperature of −15 °C. It was pointed that the optimum cylinder volume ratio is 0.7 in order to achieve the maximum COP with the designed heating capacity[5].
To sum up, it is essential to investigate the volume flow ratio in CRS and its impacts on the system performance. This work employs thermodynamic energy and exergy analysis to investigate the effect of volume flow ratio of HTC and LTC compressor on the performance characteristics. A theoretical simulation model is established based on energy and exergy balance equations and the effects of volume flow ratio on the optimum intermediate temperature, COP, exergy efficiency are investigated.

2. Theoretical model of CRS

2.1. System description

A schematic diagram of the NH_3/CO_2 cascade refrigeration system is illustrated in figure 1. The CRS is constituted by two single stage refrigeration circuits, HTC and LTC. These two circuits are connected by a cascade heat exchanger, which serves as the condenser of LTC and evaporator of HTC. The LTC with CO_2 as the refrigerant is used for cooling and the HTC with NH_3 as the refrigerant is used to condensate the CO_2 of the LTC. The evaporator absorbs a refrigerated load Q_e from the cold space at T_{cl} to the evaporating temperature T_e. The heat absorbed by the evaporator of the LTC plus the work input to the LTC compressor equals the heat absorbed by the evaporator of the HTC. The condenser of HTC rejects a heat of Q_c from the condenser at condensation temperature T_c to ambient. Temperature difference in cascade heat exchanger represents the difference between the condensation temperature of LTC and the evaporation temperature of HTC. The corresponding temperature-entropy diagram is shown in figure 2.

2.2. Mathematical model

In order to simplify the thermodynamic analysis, the following assumptions have been made in the analysis:
(a) Refrigerants at evaporator outlet and condenser outlet is saturated state.
(b) Pressure drops and heat losses in linking pipes and system components are negligible.
(c) All components are assumed to be a steady-state and steady-flow process.

(d) The temperature difference in cascade heat exchanger and the temperature difference between cold space and evaporation temperature are both 5°C.

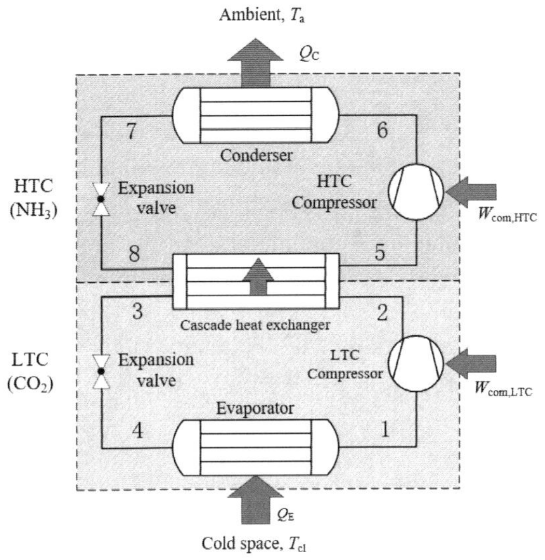

Figure 1. Schematic diagram of the NH_3/ CO_2 cascade refrigeration system.

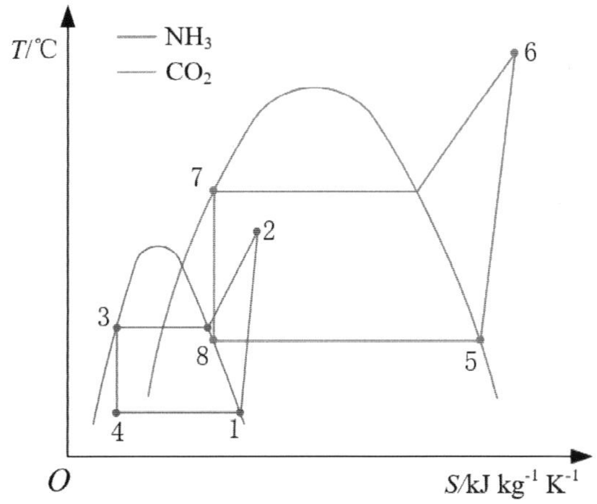

Figure 2. Temperature-entropy diagram of the NH_3/CO_2 cascade refrigeration system.

Taking into account the assumptions previously made, the theoretical model is established based on the principle of CRS and the mass, energy and exergy balances are given by equations (1)-(3), respectively.

$$\sum_{in} m = \sum_{out} m \tag{1}$$

$$Q - W = \sum_{out} m \cdot h - \sum_{in} m \cdot h \tag{2}$$

$$ED = \sum_{out} \left(1 - \frac{T_a}{T_j}\right) \cdot Q_j - W + \sum_{in} m \cdot e - \sum_{out} m \cdot e \tag{3}$$

The volume flow ratio is defined as the ratio of refrigerant volume flow rate of HTC compressor and that of LTC compressor.

$$n_V = V_{HTC} / V_{LTC} \tag{4}$$

The mass flow rate can be obtained and is the product of volume flow rate and suction state density.
The exergy analysis method is necessary and more accurate to evaluate the intensity of the irreversibility in individual components. Generally, the physical exergy of every state point in a cycle can be expressed as:

$$e = m[(h - h_0) - T_a (s - s_0)] \tag{5}$$

Where h_0 and s_0 is the enthalpy and entropy of the refrigerant at ambient temperature T_a, respectively.
The compression is adiabatic and the isentropic efficiency of compressors of HTC and LTC is related to compression ratio. Different empirical correlations to calculate the compressor's isentropic efficiency can be found from open literatures. The isentropic efficiencies of both compressors are determined by a simple empirical form from reference[3].

$$\eta_s = 0.874\text{-}0.0135 \cdot \varepsilon \tag{6}$$

The enthalpy of state 2 and state 6 were calculated based on isentropic efficiency as follows.

$$h_2 = \frac{h_{2s} - h_1}{\eta_{\text{s,LTC,comp}}} + h_1 \quad (7)$$

$$h_6 = \frac{h_{6s} - h_5}{\eta_{\text{s,HTC,comp}}} + h_1 \quad (8)$$

The mechanical and electrical efficiency are also considered in the analysis. For LTC compressor, the value of mechanical and electrical efficiency is 0.9 and 0.95 and 0.9 for both efficiencies for the NH_3 compressor[3]. Therefore, the overall compressor's efficiency was determined by the equation

$$\eta_{comp} = \eta_s \cdot \eta_{mec} \cdot \eta_{elec} \quad (9)$$

Specific equations for each component of CRS are summarized in table 1.

Table 1. Balance equations for each component of the CRS

Component	Energy balance	Exergy balance
LTC compressor	$W_{\text{comp,LTC}} = \frac{m_{\text{LTC}}(h_{2s} - h_1)}{\eta_{\text{comp,LTC}}}$	$ED_{\text{comp,LTC}} = T_{\text{a}} m_{\text{LTC}}(s_2 - s_1)$
LTC expansion device	$h_4 = h_3$	$ED_{\text{exp,LTC}} = T_{\text{a}} m_{\text{LTC}}(s_4 - s_3)$
LTC evaporator	$Q_{\text{evap,LTC}} = m_{\text{LTC}}(h_1 - h_4)$	$ED_{\text{evap,LTC}} = T_{\text{a}}\left[m_{\text{LTC}}(s_1 - s_4) - \frac{Q_{\text{evap,LTC}}}{T_{\text{cl}}}\right]$
HTC compressor	$W_{\text{comp,HTC}} = \frac{m_{\text{HTC}}(h_{6s} - h_5)}{\eta_{\text{comp,HTC}}}$	$ED_{\text{comp,HTC}} = T_{\text{a}} m_{\text{HTC}}(s_6 - s_5)$
Condenser	$Q_{\text{cond,HTC}} = m_{\text{HTC}}(h_7 - h_6)$	$ED_{\text{cond,HTC}} = T_{\text{a}} m_{\text{HTC}}(s_7 - s_6) + Q_{\text{cond,HTC}}$
HTC expansion device	$h_8 = h_7$	$ED_{\text{exp,HTC}} = T_{\text{a}} m_{\text{HTC}}(s_8 - s_7)$
Cascade heat exchanger	$Q_{\text{che}} = m_{\text{LTC}}(h_2 - h_3) = m_{\text{HTC}}(h_5 - h_8)$	$ED_{che} = T_a[m_{\text{LTC}}(s_2 - s_3) + m_{\text{HTC}}(s_5 - s_8)]$

The coefficient of performance (COP) was used to evaluate the CRS based on the first law of thermodynamics and can be defined as:

$$\text{COP} = \frac{Q_{\text{evap,LTC}}}{W_{\text{comp,LTC}} + W_{\text{comp,HTC}}} \quad (10)$$

According to the second law of thermodynamics, exergy analysis is necessary to evaluate the maximum work capacity of the CRS. In addition, from the analysis of irreversible loss, one can know the degree of a real cycle deviating from the ideal reversible cycle, the effect of each part of the irreversible loss on the system and the component in which the maximum loss occurs. Exergy efficiency of CRS is defined as the ratio of the minimum work requirement to the actual input power. The exergy efficiency of the system is given by the equation

$$\eta_{\text{II}} = \frac{W_{\text{comp,LTC}} + W_{\text{comp,HTC}} - ED_{\text{total}}}{W_{\text{comp,LTC}} + W_{\text{comp,HTC}}} \quad (11)$$

where $W_{comp,LTC}$ and $W_{comp,HTC}$ are the input power of LTC compressor and HTC compressor, respectively, ED_{total} are the total exergy destruction rate of the CRS system.

A parametric study was performed to evaluate the performance of CRS. The ambient temperature is 25°C and the temperature difference in cascade heat exchanger is regarded as 5°C. The condensing temperature of CRS is assumed to 40°C. The temperature difference between cold space and evaporation temperature is 5°C. The refrigerant volume flow rate of LTC compressor is 180 m^3/h according to an actual type screw compressor. When the volume flow ratio is varied, which means the volume flow rate of HTC compressor is changing which caused by frequency regulation or slide vale regulation method. The calculation flowchart of the simulation process is shown in figure 3. The needed refrigerant thermodynamic parameters were obtained by using REFPROP 9.0 and the model is simulated by using Matlab.

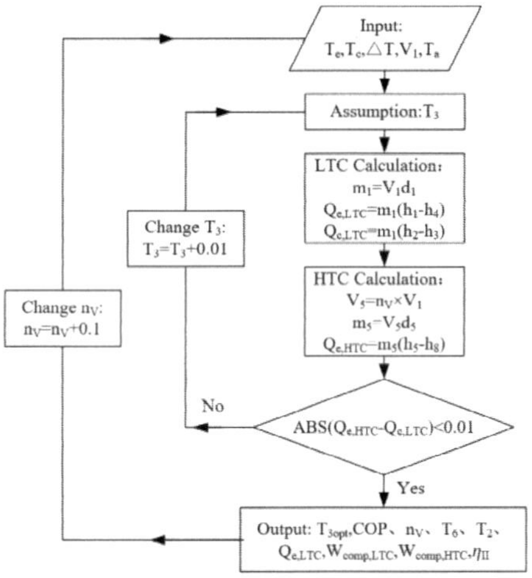

Figure 3. Calculation flowchart for optimum volume flow rate ratio and system performance.

2.3. Model validation

In order to verify the theoretical model and program, the simulation results in this research were compared with the reported data in the literature as shown in figure 3. It can be seen that there is a very good agreement between simulation results and the reported data from the literature[3, 6]. From the calculated results, the maximum difference appears to be in the condenser of HTC when the condensing temperature of LTC is -25°C and the maximum difference is 0.646%.

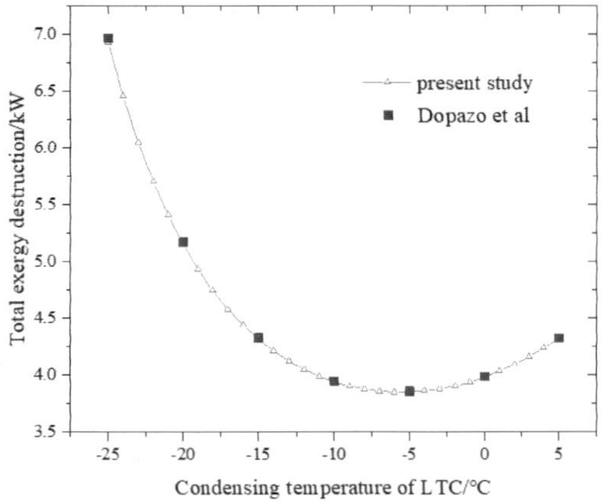

Figure 4. Comparison of the total exergy

destruction between present study and reference.

3. Results and discussion

Figure 5 shows the effect of the volume flow rate ratio on optimum condensation temperature of LTC. It can be seen that the optimum condensation temperature of LTC is decreasing with the volume flow rate ratio increases at given working conditions. The reason can be explained from the energy balance of cascade heat exchanger. When the suction volume rate of LTC compressor is kept constant, the suction volume rate of HTC is increasing with the increases of volume flow rate ratio n_V. If the intermediate temperature remains constant, the condensation heat of LTC in cascade heat exchanger is also unchanged. However, under this circumstance, the evaporation capacity of HTC in cascade heat exchanger is increasing proportionally with the increasing of suction volume flow rate of HTC compressor since the density at the suction state of HTC compressor is constant and caused an increase of mass flow rate. Therefore, in order to ensure the energy balance in cascade heat exchanger, the intermediate temperature is decreased. In addition, the optimum condensation temperature of LTC is different for different evaporation temperature when the condensing temperature of HTC is kept constant.

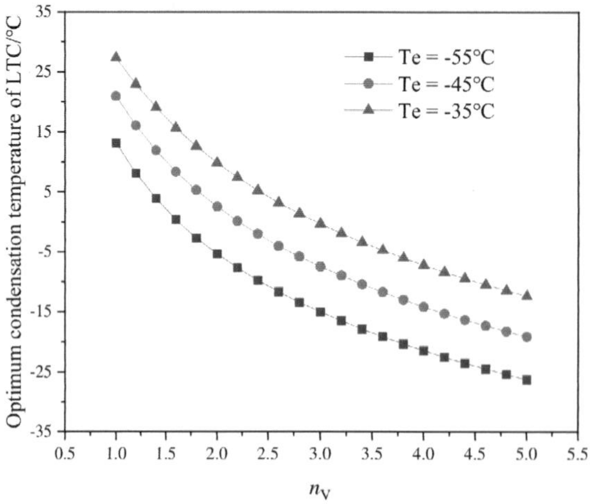

Figure 5. Effect of n_V on optimum condensation temperature of LTC.

Figure 6 depicts the variation of volume flow rate ratio on input compressor power. As figure showed, the CO_2 compressor power consumption decreases with the increasing of n_V. The pressure-enthalpy diagram illustrates that as the condensation temperature of LTC decreases, the compression isentropic of LTC compressor shifts to down, which causes a lower value of (h_2-h_1), leading to a lower input power of the compressor. The variation of input power of HTC NH_3 compressor has an opposite trend. It increases as n_V increases. The effect of n_V on total input compressor power is illustrated in figure 7. It can be learned that the total compressor power decreases slightly at first and then increases with the increasing of n_V.

The variation of discharge temperature of HTC compressor and LTC compressor with the volume flow rate ratio are shown in figure 8 and figure 9. It is observed that the discharge temperature of HTC is increasing with the volume flow rate ratio increases. The results indicate that as the volume flow rate ratio increases, the discharge temperature of HTC increase rapidly firstly and then slowly increase. At a given working conditions that the evaporation temperature of -45°C and condensing temperature of 40°C, the discharge temperature of HTC increased from 114.50°C to 170.81°C when the n_V increased from 2 to 4. When the volume flow ratio is higher, the discharge temperature of NH_3 compressor in HTC is too higher that should be controlled carefully. However, the variation of LTC compressor discharge temperature has an opposite trend. This is due to the change of optimum condensation temperature as shown in figure 5. With the increasing of n_V, the optimum intermediate temperature is decreasing and results in an increasing of HTC compressor pressure ratio and a decreasing of LTC compressor pressure ratio.

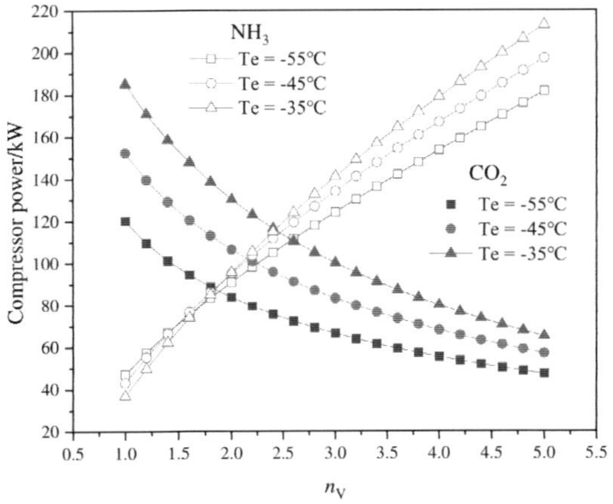

Figure 6. Effect of n_V on HTC and LTC input compressor power.

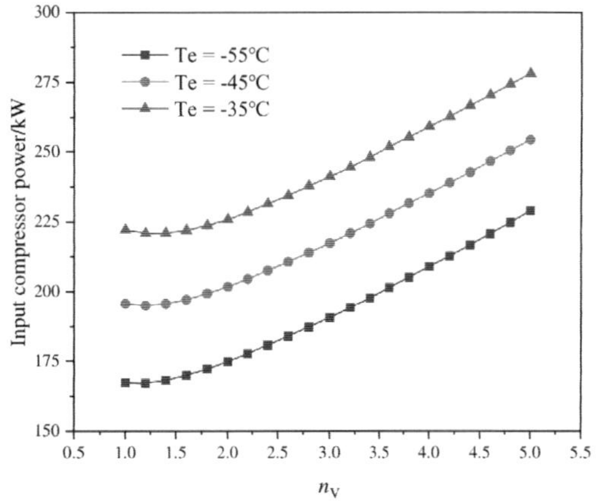

Figure 7. Effect of n_V on total input compressor power.

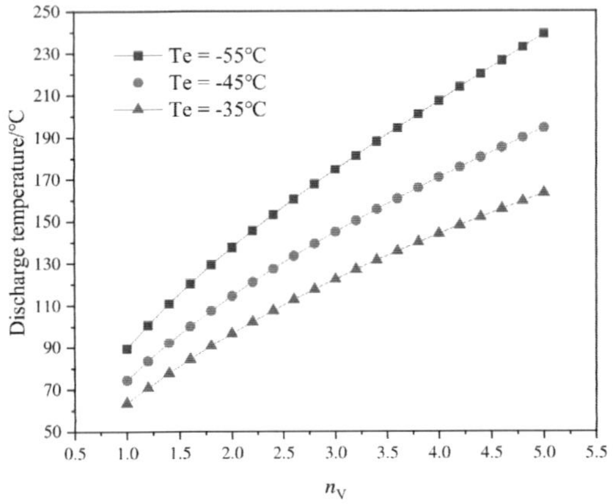

Figure 8. Effect of n_V on discharge temperature of NH_3 compressor.

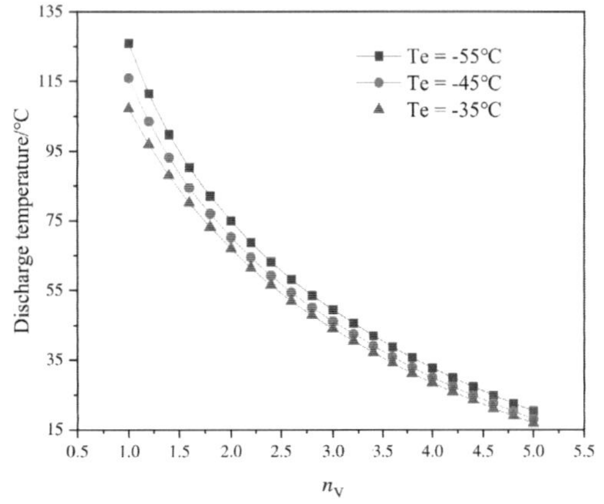

Figure 9. Effect of n_V on discharge temperature of CO_2 compressor.

Figure 10 indicates the effect of volume flow ratio on the COP of the cascade refrigeration system. It can be seen that there exists an optimum volume flow rate ratio that makes the maximum COP of CRS. In addition, the optimum condensation temperature of LTC is different for different evaporation temperature if the condensation temperature of HTC is kept constant. When the evaporation temperature is -55°C, the optimum volume flow rate ratio is 2.4 and corresponding maximum COP of CRS is 1.031 and optimum condensation temperature of LTC is -9.75°C. The COP of CRS reaches a maximum value of 1.259 and 1.537 and the corresponding volume flow rate ratio is 2.8 and 3.2 at the evaporation temperature value of -45°C and -35°C.

Figure 11 shows the variation of cooling capacity with volume flow rate ratio. It can be observed that the cooling capacity of the cascade refrigeration system is increasing rapidly firstly and then increases gradually as n_V increases. Since the volume flow rate of LTC compressor is assumed to be constant value, the optimum intermediate temperature decreases with the increasing of n_V as shown in figure 5. The pressure ratio of LTC is also reduced while the evaporation temperature is kept constant. As a result, the volumetric cooling capacity is increased. When the evaporation temperature is -55°C, the cooling capacity is increased from 178.75kW to 205.09kW and increased by about 14.73% as n_V increase from 2 to 4. The cooling capacity increased more at higher evaporation temperature. It increased by about 20.09% as n_V increase from 2 to 4 when evaporation temperature is -35°C. This means that when the LTC compressor operates at a constant evaporation temperature and speed, the cooling capacity of the CRS could be increased through increasing the rotational speed of HTC compressor. As showed in figure 10 and figure 11, although the cooling capacity is continuously to increase, a maximum COP value exists during volume flow rate ratio increasing. The discharge temperature of NH_3 compressor should be carefully controlled as the n_V increases.

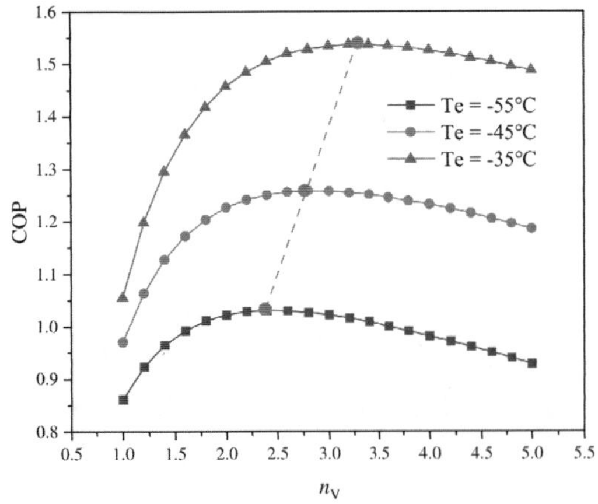

Figure 10. Effect of n_V on COP.

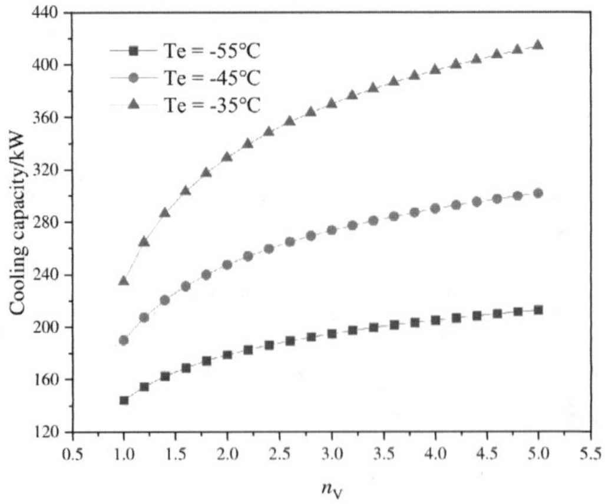

Figure 11. Effect of n_V on cooling capacity.

Figure 12 depicts the effect of n_V on total exergy destruction rate. It can be seen that the total exergy destruction rate has a minimum value as the volume flow ratio increases. Figure 13 illustrates that the impact of volume flow rate ratio on exergy efficiency of cascade refrigeration system. The exergy efficiency first increases and then decreases as volume flow rate ratio increases. There exists an optimum volume flow rate ratio that makes exergy efficiency maximum. When evaporation temperature is -55°C, the maximum exergy efficiency value of 0.346 is obtained and the corresponding volume flow rate ratio is about 2.4, which is consistent with the optimum value obtained previously as shown in figure 10. This means that the optimum volume flow ratio could make COP and exergy efficiency maximum instantaneously. It can be further noticed that the maximum exergy efficiency of the CRS at evaporation temperature of -45°C is higher than that of evaporation temperature of -55°C and -35°C.

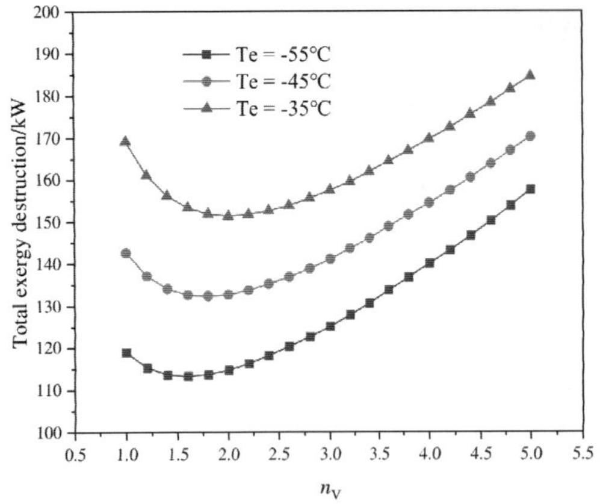

Figure 12. Effect of n_V on total exergy destruction

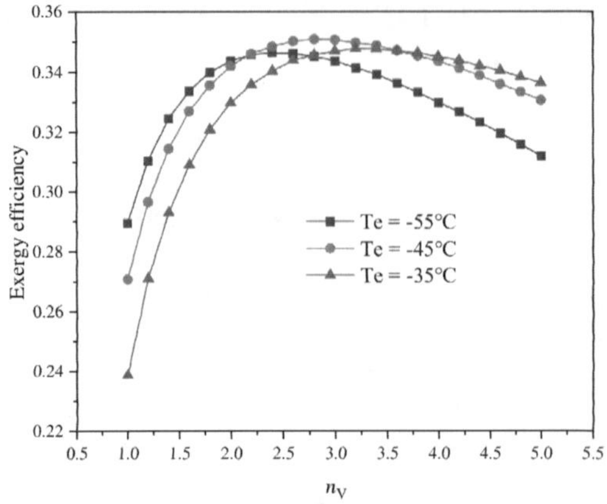

Figure 13. Effect of n_V on exergy efficiency.

4. Conclusions

The simulation model of cascade refrigeration system including energy and exergy analysis was developed to study the effect of volume flow ratio of HTC compressor and LTC compressor on the system performance. Based on the obtained results with the models, it can be found that there exists an optimum volume flow ratio that makes the maximum COP and exergy efficiency and the optimum volume flow ratio is different under various working conditions. When the evaporation temperature is -55°C, the optimum volume flow rate ratio is about 2.4 and corresponding maximum COP and exergy efficiency of CRS is 1.031 and 0.346, respectively. The optimum condensing temperature of LTC is -9.75°C. When the volume flow ratio is increasing, which means the motor frequency or rotational speed of HTC compressor increases while the rotational speed of LTC compressor is fixed, the optimum intermediate temperature decreases and the discharge temperature of HTC compressor increases rapidly and should be controlled. The cooling capacity could be enhanced by increasing the volume flow ratio. The obtained results are helpful for the screw refrigeration compressor regulation in CRS under various working conditions and cooling demands.

Acknowledgments
The authors are grateful for the support of the No.51976095 project supplied by the National Natural Science Foundation of China and the Taishan Scholar Program of Shandong (No. tsqn201812073).

Nomenclature

T	temperature [°C]	*HTC*	high-temperature cycle
ΔT	temperature difference [°C]	Subscripts	
h	specific enthalpy [kJ/kg]	comp	compressor, compression
ε	compressor pressure ratio	che	cascade heat exchanger
m	refrigerant mass flow rate [kg/s]	c,cond	condenser, condensation
n_V	volume flow rate ratio	exp	expansion device
ED	exergy destruction rate [kW]	e,evap	evaporation, evaporator
η_{II}	exergy efficiency	s	isentropic
W	input power of compressor [kW]	cl	cold space
V	refrigerant volume flow rate[m^3/s]	opt	optimum
LTC	low-temperature cycle	a	environmental temperature

References

[1] Park, H., H.K. Dong, S.K. Min, Thermodynamic analysis of optimal intermediate temperatures in R134a–R410A cascade refrigeration systems and its experimental verification. *AppliedThermal Engineering*, 2013. 54(1): 319-327

[2] Lee T, Liu C, Chen T, Thermodynamic analysis of optimal condensing temperature of cascade-condenser in CO_2/NH_3 cascade refrigeration systems. *International Journal of Refrigeration*, 2006.29(7): 1100-1108

[3] Dopazo J A, Jose F S, Sieres J, et al., Theoretical analysis of a CO_2 –NH_3 cascade refrigeration system for cooling applications at low temperatures. *Applied Thermal Engineering*, 2009. 29(8): 1577-1583

[4] Jiang, S., Wang S., Jin X., et al., Optimum compressor cylinder volume ratio for two-stage compression air source heat pump systems. *International Journal of Refrigeration*, 2016. 67: 77-89

[5] Baek, C., Heo J., Jung J., et al., Effects of the cylinder volume ratio of a twin rotary compressor on the heating performance of a vapor injection CO_2 cycle. *Applied Thermal Engineering*, 2014. 67(1): 89-96

[6] Gholamian E, Hanafizadeh P, Ahmadi P, Advanced exergy analysis of a carbon dioxide ammonia cascade refrigeration system. *Applied Thermal Engineering*, 2018. 137: 689- 699

IOP Conf. Series: Materials Science and Engineering 1180 (2021) 012052 doi:10.1088/1757-899X/1180/1/012052

A study into the impact of chloride ions on the make-up hydrogen compressors

M Fouladivanda and M A Heidary

Engineering Department, Iran Chemical Industries Investment Company, Esfahan, Iran

mojtaba.f.vanda@gmail.com, m.amin.heidary@gmail.com

Abstract. The presence of unwanted chemicals in process lines and downstream equipment causes defects and failures, which sometimes have significant impacts on systems and imposes extra costs for the production process. In this study, one pair of four-stage make-up hydrogen compressors is investigated. These reciprocating compressors which serve as a part of the hydrogen treating unit have failed approximately simultaneously. The feed of these compressors is net hydrogen-rich gas from the Pacol process. Both dry compressors have shown similar problems in the first and second stages. The simultaneous rise in the cylinder temperature and decrease in the flow has forced shut-down of the compressors for the next actions in the process. Initial inspections have revealed some kind of deposits covered the cylinder and its components. The valves, liquid tested and results have shown significant leakage of all the suction and discharge valves of the cylinders. Moreover, the thickness of all piston rings have decreased up to 50 percent which is more than allowable values. The compositions of deposits, tested by X-ray fluorescence (XRF) and X-ray diffraction (XRD), result from the analysis shown that it contains 35.2 percent (mass fractions) of Chlorine and 27.5 percent of Iron, and noticeable amount of Sulfur or Phosphor. XRD analysis has reported that the Iron Chloride Hydrate is the main part of the fouling sample. In this paper, the impact of the aforementioned chemicals on the compressor components are investigated, and some approaches are proposed to absorb or inhibit these chemicals.

1. Introduction

The most widely used type of compressors found in industrial applications are reciprocating compressors(1, 2), which are crucial elements of the process due to their high-pressure ratio. Maintenance costs of reciprocating compressors are approximately three and half times greater than those for centrifugal compressors (3). An increase in the compressor life for one week could mean a cost saving of a quarter to three-quarters of a million dollars for a refinery (4). So, reducing maintenance costs and downtime is of paramount importance for these machines.

A number of factors have a direct influence on the operating conditions and efficiency of the reciprocating compressors, for which dozens of experimental and theoretical studies have been done in the literatures. In some of these studies, compressor failure mechanism have been categorized. Leonard et al. (4) states 8.8 % of unscheduled compressor shut-downs is because of process problems. Process problems affect the mechanical parts of the reciprocating compressors, such as valves, packings, and piston rings, which cause the rest of unscheduled shut-downs.

Compressor valves can also be damaged by unwanted chemicals. It is a common problem for the reciprocating compressor's unscheduled shut-downs. According to some studies, compressor valve failure is divided into environmental and mechanical categories. Environmental causes are those elements in the valve environment that can lead to valve failures, such as corrosive contaminants, foreign material, liquid slugs, or improper lubrication.

Unwanted chemicals and contaminated process gas can harm the internal components of the compressors and would impose the process-stop beside the higher repair costs. The presence of corrosive agents, including trace quantities in the process fluids which can cause corrosion cracking, should be specified, and some precautions must be taken. (5)

For instance, the Continuous Catalyst Regeneration (CCR) reformer process inherently creates viscous liquid called green oil, which causes problems for the compressors of this unit. CCR reformer is one of the essential units in refinery plants. The polymerized long-chain hydrocarbons, generally called green oil, are the end-products of undesirable polymerization reactions taking place over the catalyst surface area. The reaction of Hyrogen chloride with hydrocarbons leads to chlorinated hydrocarbons molecules which are green or red and contain mainly C6-C18 hydrocarbons, with a potential tail above C40. These hydrocarbons are believed to be oligomers of light olefinic hydrocarbons, with some aromatic nuclei included in the structures (6).

Green oil forms quickly at high pressure and temperature. In some case studies which have been done on the CCR unit of a refinery plant, observations have shown consistent problems with the presence of downstream equipment's plugging or fouling with the existence of this green sludge. After removing Chloride from the process gas, the aforementioned problems have not been observed anymore.

Another example of compressor failure due to the process issues has been studied experimentally by Lee et al. (7). In this study, three compressors have faced the problems such as damaged valves and cracked pistons after revamping. The process and mechanical design to decrease the compressor's failure have been also reviewed the process and mechanical design to decrease the compressor's failure. The operating conditions was adjusted to reduce the light olefin. As a result, the green oil and chloride formation in process lines have been lowered. Moreover, this study has proposed the valve type alternation and the piston design improvement to withstand the large stresses.

In complement of previous studies on observed problems of the compressors due to the process issues, this research has been done on the dehydrogenation and hydrotreating process. For better understanding, a brief description of the process has been explained in the next section.

1.1. Process description

The dehydrogenation process is a fixed bed catalytic process designed to selectively dehydrogenate a high purity, normal paraffin feed to the corresponding mono-olefin product. The feed to the dehydrogenation unit is supplied from a normal paraffin separation process (such as Molex Process). The feed must be appropriately hydrotreated to be free of potential dehydrogenation catalyst poisons, such as sulfur and nitrogen compounds. (8)

The desired product from the dehydrogenation unit, founded based on Pacol license, is generally a mixture of normal mono-olefins in C10-C13 or C11-C14 carbon range. These olefins are then used with a detergent alkylation process unit to produce linear alkylbenzene (LAB) by reacting the olefins with benzene. The linear alkylbenzene is usually sulfonated to produce a biodegradable detergent in a downstream unit.

As shown in Figure 1, the dehydrogenation reactor is filled with fixed bed platinum-containing catalyst. Selective dehydrogenation of normal paraffin feed to the corresponding mono-olefin product takes place in this reactor. The flow scheme and process equipment of dehydrogenation unit are very similar to many other hydrogen recycle reaction systems; such as platforming, isomer, tatoray, etc. The net rich hydrogen gas produced is sent to the hydrotreating unit.

Potential acidic catalyst poisons such as chloride and fluoride-containing compounds must not enter the dehydrogenation reactor. These compounds could decompose on the catalyst and deposit the chloride or fluoride ions on the catalyst surface. This would produce acidic catalyst sites that would significantly increase cracking reactions, light end production, and n-paraffin isomerization. Therefore, the fluoride, chloride, and other unwanted chemicals such as Sulfur and heavy metal levels should be less than one ppmwt (Parts Per Million by weight) in the dehydrogenation feed. Separation of these undesirable compounds is done in the hydrotreating process using the hydrogen produced in the dehydrogenation unit.

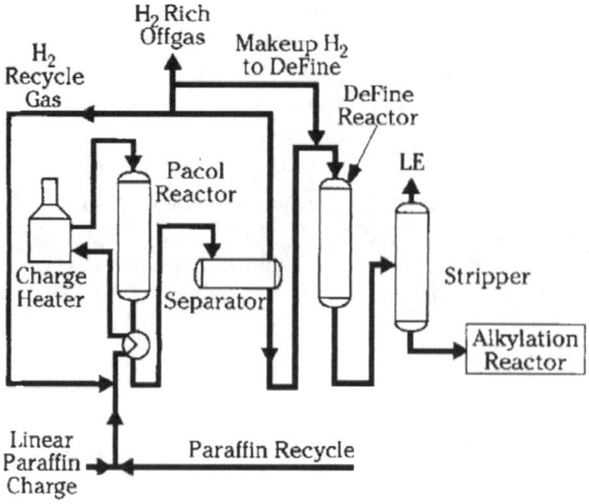

Figure 1. Paraffin Dehydrogenation Unit (8).

The hydrotreating unit is founded based on the "Unionfining license". Figure 2 provides a generic view of the flow scheme of a "Kerosene hydrotreating unit". The feed is exchanged with reactor effluent, mixed with recycle hydrogen, and then heated to reaction temperature in a fired heater. Afterwards, the combined feed flows through the reactor, which contains the catalyst that accelerates the reaction. The reactor effluent is cooled in exchange with the feed and then within a series of coolers before being separated in a vapor-liquid separator. The vapor portion is recompressed, combined with fresh hydrogen, and returned to the reactor feed. The liquid portion is fed to a fractionator, stripped of light ends, H_2S, and NH_3 (8).

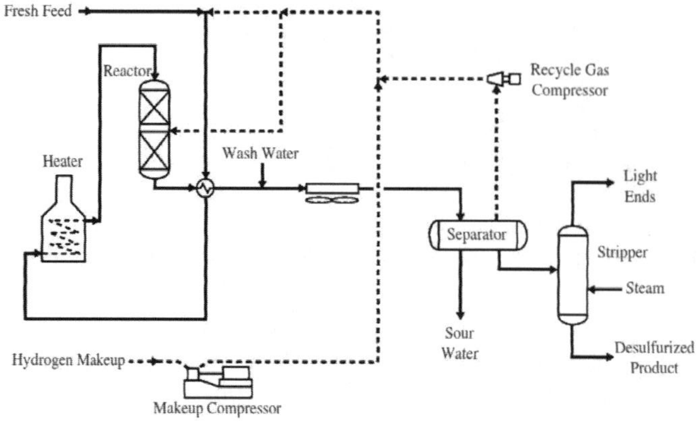

Figure 2. Kerosene Hydrotreating Unit (KHT)(8).

The hydrotreating process reduces the sulfur and nitrogen to less than 0.5 wt ppm and the metals to undetectable levels (6). Makeup hydrogen is added to the reactor effluent stream before it enters the reactor product's condenser. The makeup hydrogen is obtained from the hydrogen plant and the dehydrogenation unit. Since, kerosene hydrotreater operates at a pressure of 59 kg/cm^2, the make-up hydrogen must be boosted to unit pressure by makeup compressor. The makeup compressor for the hydrogen from the dehydrogenation unit is a four-stage reciprocating compressor. Hydrogen from the dehydrogenation unit recycle gas stream is available at a pressure of about 2 kg/cm^2 and enters the suction at the first stage of the four-stage compressor, and stripped gas from the low-pressure flash drum enters the 3rd stage suction.

The four-stages reciprocating compressor increases the make-up gas pressure from about 2kg/cm^2 , to the reactor product separator pressure of about 59 Kg/cm^2.

As previously mentioned, even the presence of a small amount of hydrogen chloride in the net hydrogen gas can cause problems with process and equipment, such as corrosion in equipment such as pipes, valves, and compressors. Classically, a chlorination agent is injected during catalyst production and regeneration in the oxychlorination process to restore the optimal metallic phase dispersion of the platinum-based catalyst and restore a typical chlorine content of 0.9-1.1 weight percent on the catalyst. This typically leads to hydrotreating make-up gas chlorine content of about one ppmwt (Parts Per Million by weight). Since the hydrogen gas from dehydrogenation unit contains chlorine, entering the make-up compressor suction, the main chlorine contributor is hydrotreating make-up gas, which leads to exacerbated issues of green oil formation, corrosion, and fouling.

1.2. Previous reviews

Similar problems have been observed in some industries. The same kind of liquid was found in a Korean refinery complex. The problem was solved by installing a chloride absorber at the outlet line of the reduction chamber. Special absorbent in a fixed-bed vessel can remove small quantities of chloride as a trapping material in liquid or gas streams (6).

A non-regenerative promoted alumina is the right choice for the removal of hyrogen chloride in the gas phase. A promoted activated alumina is designed to trap both inorganic and organic chloride impurities in a process stream. Some companies have also recommended their adsorbents which achieves up to 99% chlorine removal efficiency and reduces the operating cost of a CCR Platforming unit.

Karimi and Mohamadmoradi (9) have also done a feasibility study to investigate ways to reduce green oil production using a hierarchical analysis with the help of Expert Choice software. They have selected three activated alumina adsorbents, enhanced alumina, and the combination of carbon fiber and activated carbon. They have also indicated that the hybrid of carbon fiber and activated carbon alternative was selected as the best adsorbent with 35.3% priority over other adsorbents.

2. Observations and Discussion

According to the process mentioned earlier, hydrogen gas from the dehydrogenation unit enters the make-up compressors to enter the hydrogen cycle of the hydrotreating unit. The hydrogen-gas pressure must be increased to 59 kg/cm^2 and injected into the hydrotreating reaction loop. The make-up compressors are four-stage reciprocating types, where the pressure ratio for each stage is approximately 2.5. The first stage suction pressure is 2.58 kg/cm^2. Typically, due to the low working pressure for the first and second stages of these compressors, observations at the maintenance program did not show serious problems.

The most common problems which lead to unscheduled maintenance of these compressors, according to the maintenance history, are the increase in discharge gas temperature and a decrease in the overall flow of the compressors. In some cases, where the compressors are forced to stop operation, investigations of the components have revealed some kind of chemicals covering the valves and inside of the cylinders of first and second stages. Fouling formed in these stages is shown in Figures 3 to 5.

(a) (b) (c)

Figure 3. Fouling formation in make-up compressor section first stage, a) cylinder, b) valves, c) piston touched the cylinder liner.

The second stage's rider ring band measurement by using feeler gauge has shown that the wear rate was high, and piston rings must have been changed. The chemicals which covered the first stage were powder likes, and the second stage components were covered with some kind of sludge. In this investigation, liquid oil was found in the third stage cylinder too, and the valve plate was covered with coked hydrocarbon. The valves have been liquid tested at the workshop, and all of the valves have been rejected. The valve components observations have shown that part of the valve ring was damaged. Also, the springs were broken or were not found in the valve assembly. Moreover, all the packing rings were covered by previously-mentioned deposits.

In this case, analysis of fouling sample, taken from compressor's components, was an excellent method to diagnose and analyse these problems. This problem was repeated several times, but often, the amount of sample was not enough for laboratory tests. In one case in which enough sample was available for analysis purposes, the deposit samples were sent to an accredited laboratory for X-ray fluorescence (XRF) based on ASTM E 1621-13 standard and X-ray diffraction (XRD) based on BS EN 13925-1 standard analysis. The fouling sample of the compressor contained 35.2 weight percent of chlorine and 27.5 weight percent of iron. Also, sulfur, calcium and phosphor is detected. The test results are shown in Table 1 and Table 2.

(a) (b) (c)

Figure 4. Fouling formation in second stage of make-up compressor, a) on the valve b) in the cylinder c) on the valve cover.

(a) (b)

Figure 5. a) Fouling on the gas packing rings b) defected valve.

Table 1. XRF Semi-Quantitative fouling analysis results based on ASTM E 1621-13(10)

Component	L.O.I[a]	Fe	Cl	S	Ca	P	Si	Al
Mass Fraction(%)	30	27.5	35.2	5.3	0.26	0.26	0.08	1.4

[a] Loss Of Ignition measured by heating the sample for 1.5 hours in 950 °Celsius

Due to the hydrotreating process characteristics, identifying small amounts of sulfur, phosphor , and calcium compounds in fouling samples are predictable. But results of the analysis declared that significant amounts of chlorinated compounds have been identified in the fouling samples. Also, the presence of iron chloride compounds indicated that this chemical has caused corrosion in the compressor components resulting in the creation of aforementioed corrosion product.

Table 2. XRD fouling analysis results based on BS EN 13925-1(11)

Component	Iron Chloride Hydrate	Iron Chloride Hydrate	Sulfur
Chemical Formula	$FeCl_2.2H_2O$	$FeCl_2.4H_2O$	S

The dominant metal in the fouling composition is iron at 27.5 weight percent content. This suggests that the residue of the particulate analysis is mainly iron compounds. XRD analysis reported that the Iron Chloride Hydrate is the main part of the fouling sample. Therefore, the existing of chlorine trace in hydrogen, which was generated in the dehydrogenation unit, can form some deposits with iron compounds. It can also form green oil with hydrocarbon content of recycle gas, and settle in the compressor components.

To accurately detect the amount of chlorine in compressor components, samples of hydrogen gas, and condensates separated in the compressor stages, were analyzed. However, the chlorine content in these samples was less than the detectable level of measurement methods which have been used previously.

Among the analysis methods that have been used in this study, the following can be applicable:

- Make-up hydrogen gas analysis using Dräger method (12) based on ASTM D4490 (No chlorine compounds were detected in this analysis.)
- Analysis of hydrotreating unit feed condensate samples separated in compressor drain paths. (Test results are shown in Table 3)
- Analysis of sour water separated from hydrocarbons in hydrotreating unit separator. (Test results are shown in Table 4)

Table 3. Results of Analysis of hydrotreating unit feed and condensates samples from compressors drain paths based on ASTM D4929(13)

Test	Test Method	Sampling Points: Hydrotreating Unit Compressor Drain	Sampling Points: Dehydrogenation Unit Compressor Drain	Sampling Points: Kerosene Feed of Hydrotreating unit
Organic Chloride Content (mg/l)	ASTM D4929	<1	<1	2

Table 4. The test result of sour water separated in hydrotreating unit separator based on Standard Method 4500(14)

Test	Test Method	Sour Water Chloride Content
Ionic Chloride Content, mg/l	Standard Method 4500	1

The above analyses results showed that the chlorine detected in compressor fouling was not detectable in any of the other liquid and gas stream samples as mentioned above, so in this case, the only reliable method for chlorine detection is the analysis of the fouling in compressor components during repair procedures.

For more reliable definition of fouling source, , a more accurate analysis of the fouling has been done after 9 month of the compressor service. In this case, the sample X-ray fluorescence (XRF) has been done in two stages: before testing the loss of ignition (L.O.I) and after testing the loss of ignition (L.O.I), for which results are presented in Table 5.

Table 5. XRF Semi-Quantitative fouling analysis results based on ASTM E 1621-13(10)

Component	L.O.I[a]	Fe_2O_3	Cl	SO_3	CaO	l_2O_3	MnO	SiO_2
Before L.O.I Test (%)	54.42	34.4	8.25	2.8	0.04	01	0.08	N.D.
After L.O.I Test (%)	0	98.1	0.16	0.14	0.35	24	0.03	0.98

[a] Loss Of Ignition measured by heating the sample for 1.5 hours in 950 °Celsius

As it shown in the table, the chloride content of the samples changes before and after testing the loss of ignition (L.O.I). From this analysis, it is inferred that there are two types of chlorine compounds in the sample:

- Organic chlorinated compounds that have loss during Loss of Ignition (L.O.I) test.
- Inorganic chlorinated compounds ,which remains after the Loss of Ignition (L.O.I) test.

It is important to note that most of the chlorinated compounds in the samples are organic. These compounds can be made by the polymerisation reaction in which green oil is generated. It seems the best strategy to tackle this issue was to remove chloride from this hydrogen stream utilizing a specific adsorbent. For this reason, an action plan has been defined to implement a chloride guard bed on the hydrogen gas stream from the dehydrogenation reactor.

Some procedures to inhibit chloride in the process gas have been suggested by licensors, which are discussed in the Introduction. However, to get the best result, it is necessary to survey and encounter this problem from different aspects and utilize suitable solutions. For this reason, a closer look at the solutions and their pros and cons is of high importance. To get best approaches, the cost calculation of each proposed solution and the assessment of solution's reliability is crucial. It is necessary to mention that the fixed capital investment and maintenance costs of these chloride guards must be calculated and compared to the other solutions.

Therefore, other solutions should be considered. The first and the most vulnerable part of the reciprocating compressors due to the history of downtimes are the suction and discharge valves, so alternating the valve type for a more reliable, efficient, and robust type is an option. One of the most comfortable options for the more reliable valve is to upgrade the valve internals using robust material for rings and springs, so that compressors' internal design does not change with an upgrade in the valve's internal components. However, for severe unwanted chemical entrance into the compressors, which make more issues, it is suggested to replace ring-type valve by poppet type ones. By this change, valves utilize a high lift area and aerodynamic flow path. The valve lifetime can increase two times due to its design. In this procedure, the design of the compressor valves will be manipulated.

Wearing non-metallic parts like pressure rings, rider rings and sealing rings that can withstand harsh environments, is a solution that can improve components life and reliability. Employing a more suitable material composition such as polytetrafluoroethylene (PTFE) alloy with a high-performance filler or proprietary-blend low-creep PTFE are the preferred materials.

Because of the higher temperature and pressure in the third and fourth stages, as well as the difference in process conditions of these stages due to the addition of stripped off-gas from hydrotreating flash drum to the third stage, the above-mentioned enhanced material has been selected for these stages of the compressor. In this way, the upgrade in materials increases the lifespan of the pressure and the rider rings from 1000 hours to 10000 hours on average. Therefore, it is reasonable to use these materials for non-metal components of the first and second stages.

The aforementioned issues would happen in case it is not possible to prevent unwanted chemicals entrance in the process lines. For reducing the maintenance costs, compressor's monitoring is a priority. By diagnosing the malfunctions of compressors at the right time, unscheduled shut-downs can be prevented. For instance, by using a pressure vs. volume graph plus temperature monitoring, it is possible to identify the exact defected valves or pressure rings. By utilizing rod drop, the decrease in the rider band can be easily detected.

The initial set-up cost for compressors' condition monitoring is high. However, compared to the other unwanted and imposed maintenance costs, it will be more logical to utilize the monitoring systems. For a long-term solution, it seems that a combination of the solutions mentioned above, is a perfect choice for more reliable process conditions and continuous production. Furthermore, more analytical studies, investigations, and cost calculations are necessary to choose the best procedure to achieve the best process stability and reliability.

3. Conclusion

Chlorides existing in recycle and make-up hydrogen gas produced in dehydrogenation units causes problems such as corrosion and fouling inside downstream pipes, compressors and reactors, disrupts the process operation, poisons the catalysts, and decreases the reliability of the process.
The analysis revealed that the only reliable chlorine detection method is analyzing the foulings in compressor components during repair procedures. The composition of these sediment samples tested by X-ray fluorescence (XRF) and X-ray diffraction (XRD) indicates that they contain significant amounts of chlorine and iron compounds. XRD analysis reported that the iron chloride hydrate is the main part of the fouling samples. More accurate analysis of the fouling highlights the fact that most of the chlorinated compounds in the samples are organic.

For more reliable compressors and process conditions, according to investigations, experiences, and researches, a combination of these solutions can be applied:

- Chloride removal from process streams. It seems that the best way to prevent these problems is to eliminate the source of the problem. Similar experiences show that after removal of chloride from the process, the aforementioned problem was not observed anymore. In this way, an action plan is defined to implement a chloride guard bed on the hydrogen gas. Particular types of adsorbents loaded in a fixed-bed vessel enable removing small quantities of chloride in liquid or gas streams.
- Using a more suitable material for non-metal wearing parts like piston rings, sealing rings, and metal components that can withstand harsh environments, or contaminated process gas, can be a solution to improve components life and reliability.
- Implementing a condition monitoring system to measure rod drop, P-V diagram, and temperature monitoring to diagnose compressors problem before any catastrophic failure happens due to compressors malfunction.
- Valve type alternation from ring-type into poppet type; poppet type valves utilize a high lift area and aerodynamic flow path. The valve lifetime can be increased drastically due to its design.

One of the important points about the solutions is that with the observation of problems in both recycle and make-up hydrogen gas equipment, selecting the optimal location for the chloride guard has a significant impact on the effectiveness of chloride removal system. Because of the high volume of recycled gas in the Pacol and CCR hydrogen recycle loop, installation of chloride guard bed on hydrogen circulating gas can deteripate the advantages of this solution. Therefore, to achieve the best solution, more economic investigations are required on this subject.

4. Acknowledgments

We would like to express our special thanks to Iran Chemical Industries Investment Company (ICIIC) executives who provide us with the opportunity to accomplish this project.

5. References

[1] Bloch HP and Godse A 2006 *Compressors and modern process applications* (John Wiley & Sons) pp 3-20
[2] Bloch HP and Hoefner JJ 1996 *Reciprocating compressors:operation and maintenance* (Elsevier)
[3] Griffith WA and Flanagan EB 2001 *Int. Conf. on 30th Turbomachinery Symposium* Online, Continuous Monitoring Of Mechanical Condition And Performance For Critical Reciprocating Compressors (Texas: Texas A&M University Turbomachinery Laboratories.
[4] Leonard SM 1996 *Hydrocarbon processing Increase reliability of reciprocating hydrogen compressors* pp 67–74
[5] Api Standard 618 Fifth Edition 2007 Reciprocating Compressors for Petroleum, Chemical and Gas Industry Services (washington DC, USA: American Petroleum Institute) p 69.
[6] Karan OK Ay Ma Kahraman K and Selmen A 2013 Petroleum technology quarterly Combating green oil formation in a CCR reformer p 18.
[7] Kim B Uk Lee S and Lee S 2017 *Int. Conf. on 46th Turbomachinery Symposium* Reliability Improvement for Booster Reciprocating Compressor In CCR Reformer (Texas: Texas A&M University Engineering Experiment Station Turbomachinery Laboratory).
[8] Meyers RA 2016 *Handbook of petroleum refining processes* (McGraw-Hill Education) pp 5.13 8.37.
[9] Karimi A and Mohamadmoradi HR 2020 *Iranian Chemical Engineering* Journal Study of Reducing the Amount of Green Oil in the CCR Plant of Oil Refinery Unit 18(107) pp 6-14.
[10] ASTM E1621-13 2013 Standard guide for elemental analysis by wavelength dispersive x-ray fluorescence spectrometry (West Conshohocken, PA).
[11] BSI. BS EN 13925-1 Standard 2003 Non-destructive testing X-ray diffraction from polycrystalline and amorphous materials General principles. (British Standard Institute)
[12] ASTM D4490-96 2016 Standard Practice for Measuring the Concentration of Toxic Gases or Vapors Using Detector Tubes (West Conshohocken, PA).
[13] ASTM D4929-19 Standard Test Method for Determination of Organic Chloride Content in Crude Oil (West Conshohocken, PA).
[14] 4500 Cl-Chloride standard methods for the examination of water and wastewater 1999 *Standard Methods for the Examination of Water and Wastewater* (American Water Works Association).

IOP Conf. Series: Materials Science and Engineering 1180 (2021) 012053 doi:10.1088/1757-899X/1180/1/012053

Thermodynamic analysis and optimization of a vapor injection organic Rankine cycle system for low-grade heat recovery

D T Li*, Z L He, L T Ji and Z W Xing

School of Energy and Power Engineering, Xi'an Jiaotong University, 710049 Shaanxi, China.

* Corresponding author, Email: dantongli_xjtu@163.com

Abstract. The organic Rankine cycle (ORC) system has been widely used in waste heat recovery and geothermal utilization, but the system efficiency is limited due to the large exergy loss in the evaporation process. To improve the performance of ORC systems, in this paper, a vapor injection organic Rankine cycle (VIORC) system using positive-displacement expanders was proposed. The thermodynamic model of the proposed system was developed in views of energy and exergy. The effects of system operation parameters on system performance were investigated, and the system was optimized under different waste heat temperatures. Furthermore, the optimized systems were compared with basic ORC (BORC) systems. Results showed that compared with the BORC system, the proposed system increased the net output power by 3-8% and the exergy efficiency by 0.8-3.2% under the heat source temperature of 90-150 °C. The exergy loss analysis further showed that the reduction of the exergy loss in the evaporation process was the main contribution to the efficiency improvement of the proposed system.

Introduction

The organic Rankine cycle (ORC) system employing refrigerants as working fluids is widely used as a method of waste heat recovery due to its flexibility, low investment, and high safety [1]. However, its energy utilization efficiency is far lower than that of the Carrot's cycle.

To improve the efficiency of the basic ORC (BORC) system, the optimization for the operating parameters and working fluids were first conducted [2-3]. Although system performance was improved, the exergy efficiency of the BORC system was still low [2]. A significant exergy loss in the evaporation process limited the efficiency improvement for the BORC system [4]. To reduce the exergy loss, one effective means was to improve the cycle architecture of the ORC system [5-6]. Therefore, many novel cycle architectures have been proposed such as improved single-pressure ORC, multiple-loop ORC (ML-ORC), and multiple-pressure ORC (MP-ORC).

ML-ORC system integrated some separate ORC systems to achieve the step-utilization of waste heat for the efficiency enhancement [7-9]. The related analysis showed that a dual-loop ORC system increased the system exergy efficiency by 11.2% compared with the single-loop ORC system [7]. MP-ORC system used multiple evaporators working at different pressures to achieve the step-utilization of waste heat [10-12]. A significant improvement for the system efficiency could be achieved by reasonably configuring evaporators. However, the present ML-ORC and MP-ORC systems suffered from the high cost caused by the complex system lay-out and multiple components.

The improved single-pressure ORC system used an equivalent number of components to improve the system efficiency and thus had the advantages of low cost and high efficiency. The typical system included the organic flashing cycle (OFC) system [13,14] and the regenerative ORC (RORC) system [15-17]. The OFC system achieved an ideal heat absorption process from waste heat. However, a large throttle loss was inevitable, and thus the overall system efficiency was even lower than the BORC system [13]. The RORC system used a recuperator to reheat working fluid to increase the cycle average temperature. However, additional heat exchangers or turbines were needed in the RORC system resulting in the system's worse economic performance [17,18]. Therefore, a high-efficiency ORC system with a low cost is still expected to further promote the development of the ORC technology. As a common technology in the fields of refrigeration and heat pumps, the vapor injection technology is considered as an effective method to enhance the system performance without increasing too much additional investment cost [19]. Until now, there is no available open literature to discuss the potential of the vapor injection technology in the ORC system.

To fill this gap, in this paper, a vapor injection organic Rankine cycle (VIORC) system using the positive-displacement expanders was proposed to investigate the improvement potential when the vapor injection technology is employed in an ORC system. The proposed system had the same number of main components (heat exchanger and expander) as the BORC system. To investigate the system performance, the thermodynamic models and performance evaluation models were established. The effect of the key operating parameters was investigated. The system performance was compared with the BORC system in views of energy and exergy. The exergy loss analysis revealed the advantages of the proposed system. The proposed VIORC system was promising to enhance the performance of traditional ORC systems with a limited capital investment difference.

System description

Fig. 1 shows the schematic diagram of the proposed VIORC system and its corresponding T-s diagram. The system consists of two working fluid pumps, two gas-liquid separators, an expander, an evaporator, a condenser, and a throttle valve. Different from the BORC system, the working fluid partially evaporates in the evaporator, and thus the heat transfer amount in the equal-temperature evaporation process reduces. As a result, the average heat transfer temperature difference of the evaporator in the proposed system is smaller than that in the BORC system. Therefore, the system is expected to have higher efficiency. At the same time, the difficulty on the system design and control is also increased due to the additional flow path for the proposed system. On the aspect of system design, the evaporator area should be precisely designed to keep the working fluid in Separator I at a suitable vapor quality. On the aspect of system control, the opening degrees of the throttle valve should be controlled reasonably to keep the suitable vapor injection pressure for the expander.

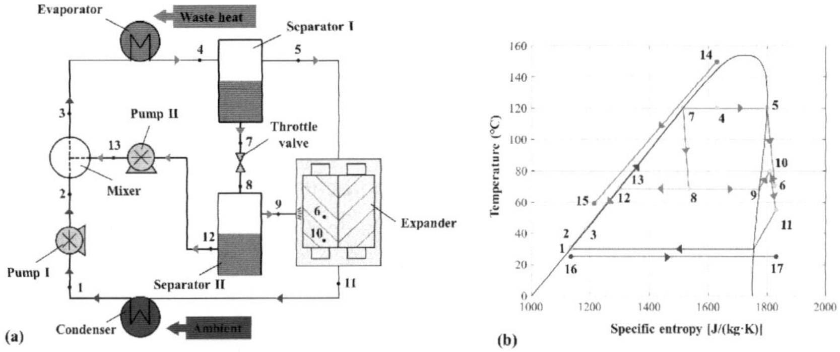

Figure 1. A schematic diagram of the proposed VIORC system and its *T-s* diagram, (a) VIORC system layout, and (b) *T-s* diagram.

The working process of the proposed VIORC system is as follows: the liquid working fluid is first pumped into a mixer to mix with the hot liquid in Separator II. Then the mixed working fluid flows into Evaporator to absorb thermal energy from waste heat and partially evaporates into the liquid-vapor mixture. Then the mixture flows into Separator I where it is separated into liquid and vapor streams. The vapor portion enters the expander to generate power. The liquid portion of the working fluid flows through a throttle valve to continue to generate vapor. The vapor is separated by Separator II and is injected into the expander through a supplemental inlet. The injected gas mixes with the original gas in the expander to raise the pressure in the working chamber. The mixed gas continues to expand to generate power. Finally, the exhausted working fluid from the expander rejects heat into the ambient through a condenser.

Mathematical model

1.1. Thermodynamic process

To establish the thermodynamic model of the proposed VIORC system, the following assumptions are made:

1) The system is running at a steady state.
2) The pressure drop in the system is ignored.
3) There is no subcooling is at the condenser outlet.
4) Waste heat generates a fixed mass flow rate of hot water.
5) All heat transfer exchangers studied have the same minimum heat transfer temperature difference.
6) The vapor injection process is assumed as a constant volume compression process in a very short time.

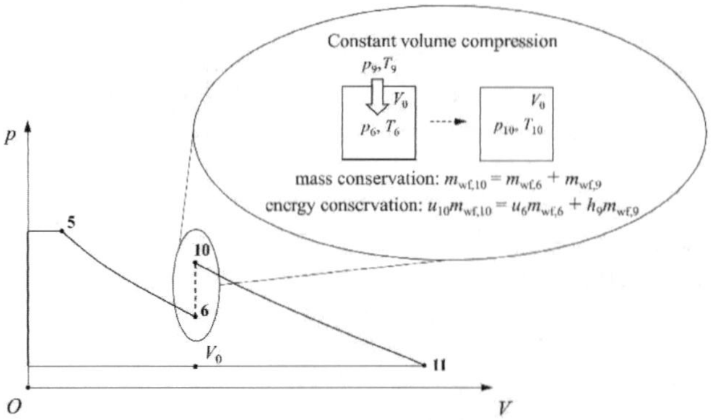

Figure 2. The vapor injection process.

Under the above assumptions, the working process of the expander is shown by a *p-V* diagram in Fig. 2. The points in the working process correspond to the state points in the *T-s* diagram in Fig. 1(b). When the vapor is injected in the expander, the state point is converted from point 6 to point 10 instantly, and the pressure at point 10 equals the vapor injection pressure of point 9. This instantaneous process is considered as the constant volume compression process due to the unchanging working chamber volume in the expander. The energy and mass conservation equations could be established for this process as shown in Fig. 2. For other components, the energy and mass balance equations are listed in Table 1. The state points of the working fluid in the cycle are shown in Fig. 1. Table 2 lists the exergy equations for each system component.

Table 1. Mass and energy balance equations for each component.

Components	Working process	Energy equations	Equation number

Pump I	Compression process 1-2	$h_2 - h_1 = (h_{2,is} - h_1)/\eta_p = W_p/((x_4 + x_8(1-x_4))\dot{m}_{wf})$	Eq. (1)
Mixing process	Liquid mixing 13-3, 2-3	$h_3 = (x_4 + x_8(1-x_4))h_2 + ((1-x_8)(1-x_4))h_{13}$	Eq. (2)
Evaporator	Constant pressure heat transfer process 3-4, 14-15	$h_4 - h_3 = Q_{eva}/\dot{m}_{wf} = \dot{m}_{hwt}(h_{hwt,14} - h_{hwt,15})/\dot{m}_{wf}$	Eq. (3)
Separator I	Gas separation 4-7, 4-5	$\dot{m}_{wf} = \dot{m}_{wf,5} + \dot{m}_{wf,7}$	Eq. (4)
		$x_4 = \dot{m}_{wf,5}/\dot{m}_{wf}$	Eq. (5)
Expander	Expansion process 5-6	$h_5 - h_6 = (h_5 - h_{6,is})\eta_{exp}$	Eq. (6)
Throttle valve	Throttling process 7-8	$h_8 - h_7 = 0$	Eq. (7)
Separator II	Gas separation 8-9, 8-12	$\dot{m}_{wf,7} = \dot{m}_{wf,12} + \dot{m}_{wf,9}$	Eq. (8)
		$x_8 = \dot{m}_{wf,9}/\dot{m}_{wf,7}$	Eq. (9)
Mixing process	Gas mixing 6-10, 9-10	$(x_8(1-x_4) + x_4)u_{10} = x_4 u_6 + (x_8(1-x_4))h_9$	Eq. (10)
		$\rho_{10}(x_8(1-x_4) + x_4) = \rho_6 x_4$	Eq. (11)
Expander	Expansion process 10-11	$h_{10} - h_{11} = (h_{10} - h_{11,is})\eta_{exp}$	Eq. (12)
Pump II	Compression process 12-13	$h_{13} - h_{12} = (h_{13,is} - h_{12})/\eta_p = W_{p,II}/((1-x_8)(1-x_4))\dot{m}_{wf})$	Eq. (13)
Condenser	Constant pressure heat transfer 11-1,16-17	$h_{11} - h_1 = Q_{con}/\dot{m}_{wf,11} = \dot{m}_{cwt}(h_{17} - h_{16})/\dot{m}_{wf,11}$	Eq. (14)

Table 2. Exergy equations for the system component.

Components	Working process	Exergy equations	Equation number
Pump I	1-2	$\dot{I}_{p,I} = T_a((x_4 + x_8(1-x_4))\dot{m}_{wf})(s_2 - s_1)$	Eq. (15)
Mixing process	13-3, 2-3	$\dot{I}_{Mixing,I} = T_a[((x_4 + x_8(1-x_4))\dot{m}_{wf})(s_3 - s_2) + ((1-x_8)(1-x_4)\dot{m}_{wf})(s_3 - s_{13})]$	Eq. (16)
Evaporator	3-4, 14-15	$\dot{I}_{eva} = T_a[\dot{m}_{wf}(s_4 - s_3) + \dot{m}_{hwt}(s_{15} - s_{14})]$	Eq. (17)
Separator I	4-7, 4-5	$\dot{I}_{sep} = 0$	Eq. (18)
Expander	5-6	$\dot{I}_{exp,I} = T_a \dot{m}_{wf,5}(s_6 - s_5)$	Eq. (19)
Throttle valve	7-8	$\dot{I}_{thr} = T_a[\dot{m}_{wf,7}(s_8 - s_7)]$	Eq. (20)
Separator II	8-9, 8-12	$\dot{I}_{sep} = 0$	Eq. (21)
Mixing process	6-10, 9-10	$\dot{I}_{Mixing,II} = T_a[x_8(1-x_4)\dot{m}_{wf}(s_{10} - s_9) + x_4 \dot{m}_{wf}(s_{10} - s_6)]$	Eq. (22)
Expander	10-11	$\dot{I}_{exp,II} = T_a((x_4 + x_8(1-x_4))\dot{m}_{wf})(s_{11} - s_{10})$	Eq. (23)
Pump II	12-13	$\dot{I}_{p,II} = T_a((1-x_8)(1-x_4))\dot{m}_{wf})(s_{13} - s_{12})$	Eq. (24)
Condenser	11-1,16-17	$\dot{I}_{con} = Q_{con} - ((x_4 + x_8(1-x_4))\dot{m}_{wf})(s_{11} - s_1)$	Eq. (25)

1.2. Thermodynamic performance
System net power output, thermal and exergy efficiencies are used to evaluate the system performance and expressed below:

The net power output is calculated by:

$$W_{net} = W_{exp} - W_{p,I} - W_{p,II} \tag{26}$$

The system thermal efficiency is calculated by:

$$\eta_{th} = \frac{W_{net}}{Q_{eva}} \tag{27}$$

The system exergy efficiency can be calculated by:

$$\eta_{exg} = \frac{W_{net}}{W_{net} + \dot{I}_{tot}} \tag{28}$$

where $\dot{I}_{tot}$ is the total irreversible loss in the system, which can be expressed by:

$$\dot{I}_{tot} = \sum \dot{I}_{comp} + \dot{I}_{Mixing} + \dot{I}_{Waste} \tag{29}$$

where the term $\dot{I}_{comp}$ represents the exergy loss rate of the components in the system, the term $\dot{I}_{Waste}$ represents the waste exergy in the heating water at the evaporator outlet.

Results and discussion

1.3. Thermodynamic performance
Table 3 shows the operating parameters of the proposed system according to the previous studies [20-22]. These parameters are determined by the user's settings. Table 4 shows the independent parameters of the proposed system. These parameters could be optimized under different optimization objectives. In this section, the effects of independent parameters on the system performance indices are first investigated.

Table 3. Operating parameters of the proposed system.

Parameter (Unit)	Symbol	Value
Heating water input temperature (°C)	$T_{hwt,in}$	90~150
Heating water flow rate (kg/s)	$\dot{m}_{hwt}$	5
Heating water pressure (bar)	p_{hwt}	10
Ambient temperature (°C)	T_a	25
Expander isentropic efficiency (%)	η_{exp}	70
Pump isentropic efficiency (%)	η_p	80
Pinch temperature difference (°C)	T_{pinch}	5
Working fluid (-)	-	R245fa

Table 4. System independent parameters.

Parameter (Unit)	Symbol
Vapor quality at Separator I(-)	x

Evaporation temperature in Evaporator (°C)	T_4
The statured temperature at the pressure of point 6	T_6

1.3.1. Effect of vapor quality. Fig. 3 shows the effect of vapor quality at point 4 on system performance indices, where $T_{hwt,out}$ represents the temperature of the heating water at the evaporation outlet. It is found that the system thermal efficiency increases with the increase in vapor quality. This is because the higher vapor quality causes a higher average evaporation temperature. However, maximum net output power and exergy efficiency exist for the system. It is because the increasing vapor quality also causes a reduction in the mass flow rate of the working fluid. With the increase in the vapor quality at point 4, the thermal efficiency is improved but the mass flow rate of working fluid reduces so that there is an optimized vapor quality to maximize the system's net output power. It should be noted that since the total exergy in the heating water is constant, the system exergy efficiency has a linear relationship with the net output power.

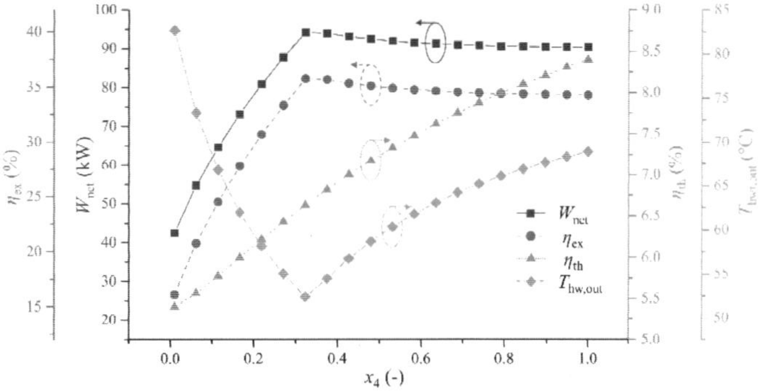

Figure 3. Effect of vapor quality at point 4 on system performance.

In Fig. 3, it can be seen that the optimal system has the minimum heating water temperature at the evaporator outlet. It illustrates that the optimal system uses as much thermal energy from the heating water as possible. Fig. 4 shows the effect of vapor quality on the heat transfer process in the evaporator. It is found that for the system with a low vapor quality (x=0.15), the minimum temperature difference ($T_{pitch,0.15}$) in the evaporator occurs at point 3 ($\mathbf{3}_{0.15}$), while for the system with a high vapor quality (x=0.5) the minimum temperature difference ($T_{pitch,0.5}$) occurs at the point 7 ($\mathbf{7}_{0.5}$). A suitable vapor quality (x=0.32) makes the heat transfer temperature difference more uniform during the liquid heat transfer process ($\mathbf{3}_{0.32}$ to $\mathbf{7}_{0.32}$). In this case, the heating water temperature at the evaporator outlet is lower so that the system has a higher net output power.

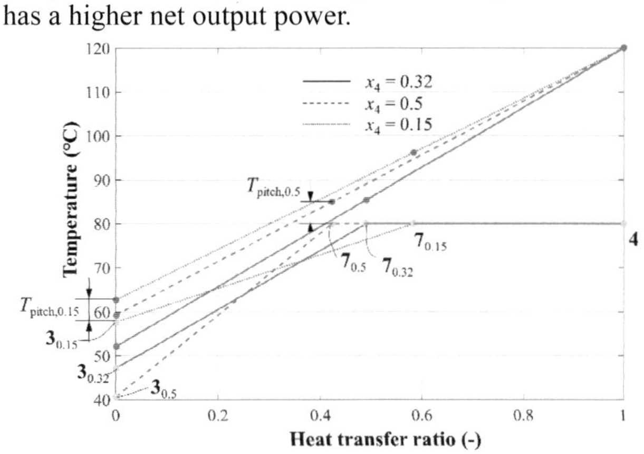

Figure 4. Effect of vapor quality on the heat transfer process in evaporator.

1.3.2. Effect of evaporation temperature. Fig. 5 shows the effect of evaporation temperature on system performance. It can be seen that as the evaporation temperature increases the system thermal efficiency and the heating water temperature at the evaporator outlet increase at the same time. It illustrates that with the increase in evaporation temperature the energy utilization efficiency increases while the energy inputting amount decreases. When the evaporation temperature is at around 85°C, the heating water temperature at the evaporator output begins to quickly increase. This increase causes a serious reduction in the net output power and system exergy efficiency so that an optimal evaporation temperature exists for the proposed system.

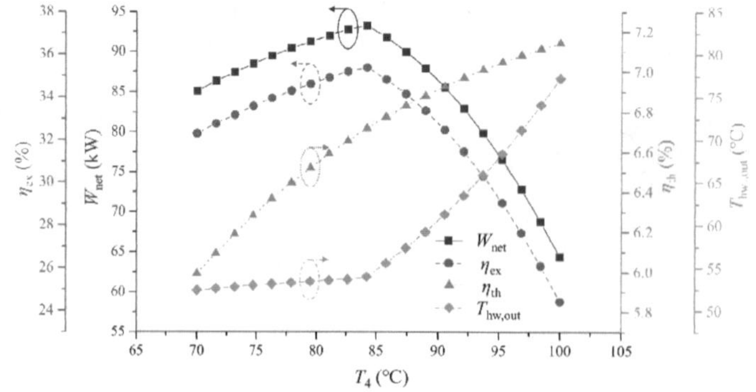

Figure 5. Effect of the evaporation temperature on system performance.

1.3.3. Effect of statured temperature at the pressure of point 6. Fig. 6 shows the effect of the statured temperature at the pressure of point 6 (T_6). It should be noted that since the vapor injection process has a constant working chamber volume and the vapor injection mass is determined by the vapor quality, the parameter T_6 directly affects the vapor injection pressure for the system. It is found that with the increase in temperature T_6 both the system thermal efficiency and the heating water temperature at the evaporator outlet increase. As a result, an optimal temperature of T_6 exists to maximize the system's net output power.

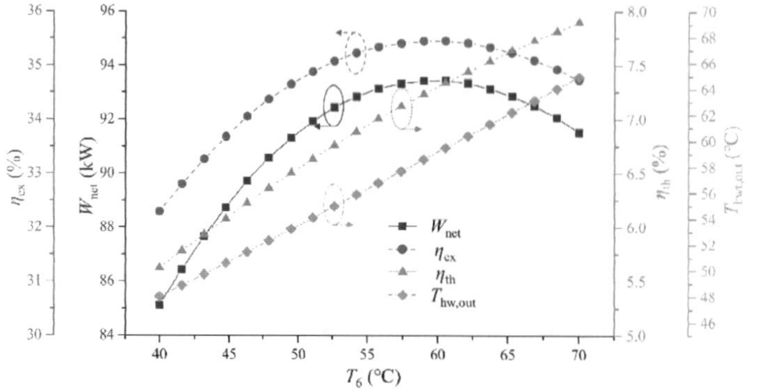

Figure 6. Effect of the statured temperature at the pressure of point 6 on system performance.

1.4. System optimization and performance comparison

The above analysis reveals that all independent operating parameters in the system have the optimal values to maximize the system's net output power. Therefore, in this section, a particle swarm optimization (PSO) algorithm is used to optimize the VIORC system with the maximum net output power as the optimization objective. Fig. 7 shows the optimized operating parameters under different waste heat temperatures. With the increase in waste heat temperature, the optimal vapor quality, evaporation temperature, and temperature T6 increase synchronously. To compare the performance of the proposed VIORC system with the basic ORC (BORC) system, the BORC systems are also optimized under the same setting parameters in Table 3. Fig. 8 shows the schematic diagram of the BORC system.

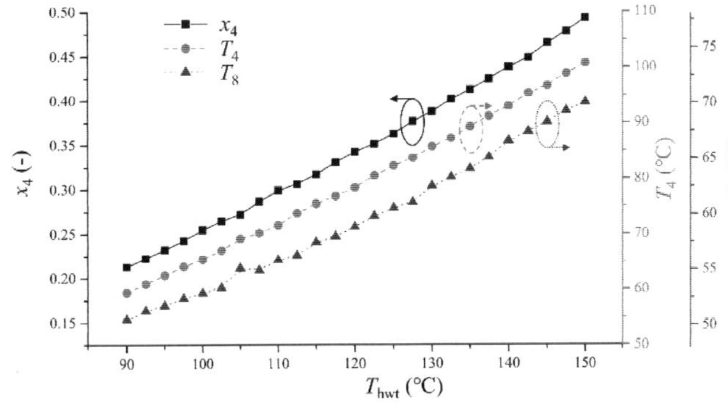

Figure 7. Optimized system-independent parameters under different waste heat temperatures.

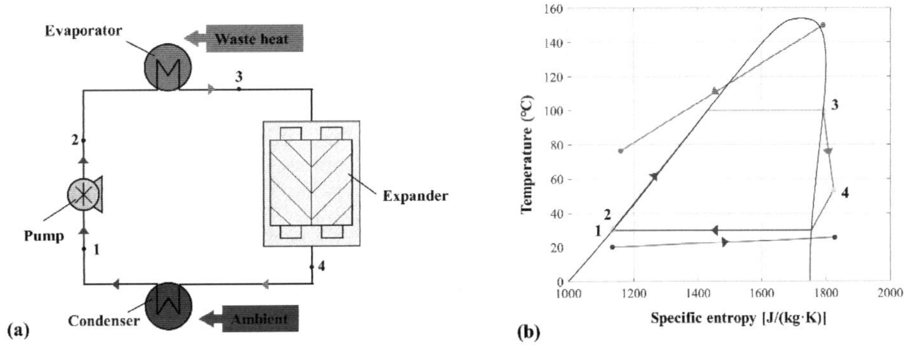

Figure 8. Schematic diagram of the basic ORC (BORC) system, (a) BORC system layout (b) *T-s* diagram.

The optimal performance of the VIORC and BORC systems is shown in Fig. 9. It is found that the net output power of the VIORC system is always higher than that of the BORC system. As the waste heat temperature increases, the increment rate becomes more significant, which changes from 3% to 8%. The heating water temperature at the evaporator outlet keeps a limited change for the VIORC system, while the outlet temperature in the BORC system continually increases. It shows that the VIORC systems have a higher utilization degree of the thermal energy in heating water and thus achieve a larger net output power.

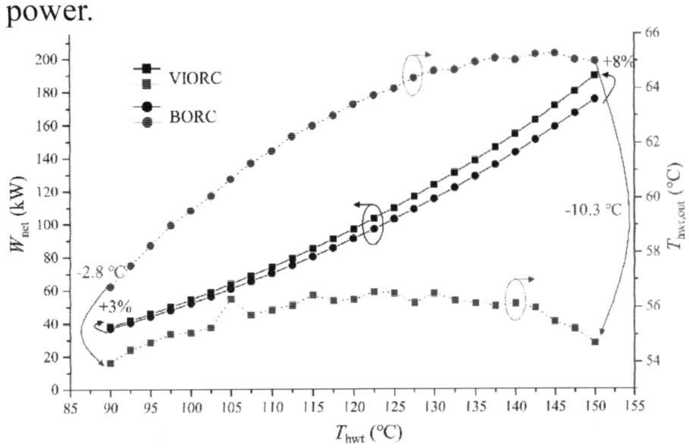

Figure 9. Optimal system performance under different waste heat temperatures.

Fig. 10 shows the efficiencies of the optimal systems under different waste heat temperatures. It is found that the thermal efficiency of the VIORC system is always lower than that of the BORC system. This is because that the VIORC systems could absorb more energies from the heating water so that the average heat absorption temperature from the heating water is lower. However, the system exergy efficiency increases by 0.8 to 3.2% for the proposed system compared to the BORC system. This is attributed to the higher energy utilization degree of waste heat.

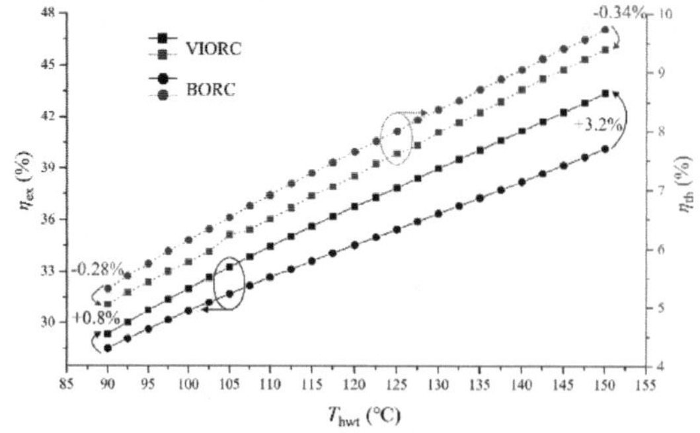

Figure 10. Efficiencies of the optimal system under different waste heat temperatures.

1.5. Exergy loss analysis

To further investigate the reason for the efficiency improvement of VIORC systems, the exergy loss distribution diagram is drawn for the VIORC and BORC systems under the waste heat temperature of 120°C as shown in Fig. 11, where the waste exergy represents the remained exergy in the heating water at the evaporator outlet. It is found that the evaporator exergy loss is reduced and exergy waste is improved for the proposed VIORC system. These could be attributed to the partial evaporation process in the evaporator.

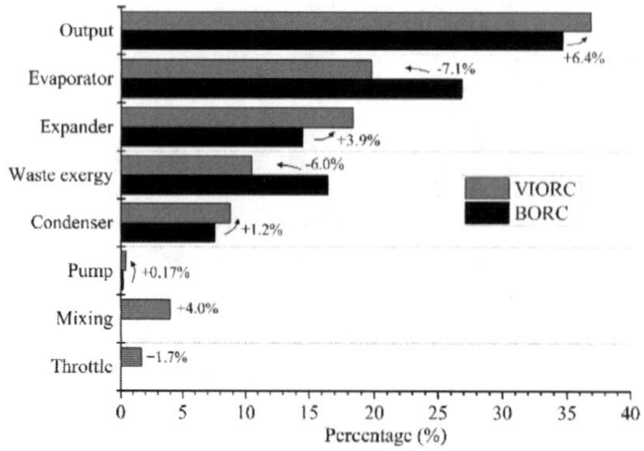

Figure 11. Exergy loss distribution of the proposed VIORC system and the BORC system.

Conclusions

In this study, a vapor injection organic Rankine cycle (VIORC) system using positive-displacement expanders was proposed to improve the cycle efficiency of the traditional ORC system. The models for thermodynamic analysis and performance evaluation of the proposed system were established. The effects of key operating parameters on the system performance were investigated. The performance of the proposed VIORC system was compared with that of the BORC system. The specific conclusions are detailed below:

1. All of the three independent operating parameters (vapor quality, evaporation temperature, and statured temperature at the pressure of point 6) in the VIORC system had an optimal value to maximize the system's net output power.
2. Under the waste heat temperature of 90 to 150 ℃, in comparison with the BORC system, the VIORC system achieved a net output power boost of 3-8%. As the waste heat temperature increased, the improvement effect became more significant.
3. The partial evaporation process in the evaporator had the main contribution to the performance improvement for the VIORC system.

It should be noted that in this paper the expander efficiency is simplified as a constant to compare the performance between the basic ORC system and the proposed system. In practice, the efficiency is largely affected by the operation conditions of expanders and ORC systems especially for the vapor injection expander. Further investigation is necessary to evaluate the proposed VIORC system's performance on the real operation conditions.

References

[1] Armaroli N and Balzani V 2007 *Angew. Chem. Int. Ed* **46** 52–66.
[2] Dai Y, Wang J and Gao L 2009 *Energy Conversion and Management* **50** 576-582.
[3] Hærvig J, Sørensen K and Condra T J 2016 *Energy* **96** 592-602.
[4] Acar H I 1997 *Energy Conversion and Management* **38** 647–657.
[5] Mahmoudi A, Fazli M and Morad M R 2018 *Applied Thermal Engineering* **143** 660-675.
[6] Lecompte S, Huisseune H, Broek M, Vanslambrouck B and Paepe M D 2015 *Renewable and Sustainable Energy Reviews* **47** 448-461.
[7] Song J and Gu C W 2015 *Energy Conversion and Management* **105** 995-1005.
[8] Song J and Gu C W 2015 *Applied Energy* **156** 280-289.
[9] Sung T and Kim K C 2016 *Applied Thermal Engineering* **100** 1031-1041.
[10] Wang Q, Wu W, Li D, Wang J and He Z 2020 *Energy Conversion and Management* **221** 113171.
[11] Li D, He Z, Wang Q, Wang X, Wu W and Xing Z 2020 *Applied Thermal Engineering* 116363.
[12] Li J, Ge Z, Duan Y Y, Yang Z and Liu Q 2018 *Applied Energy* **217** 409-421.
[13] Ho T, Mao S S and Greif R 2012 *Energy* **42** 213-23.
[14] Subha M, Shahbaz A and Sudipta D 2018 *Energy* **163** 849-862.
[15] Mago P J, Chamra L M, Srinivasan K and Somayaji C 2008 *Applied Thermal Engineering* **28** 998-1007.
[16] Ying Y and Hu E J 1999 *Applied Thermal Engineering* **19** 1173-1180.
[17] Pei G, Li J and Ji J 2010 *Applied Thermal Engineering* **30** 998-1004.
[18] Fan W, Guo S and Han Z 2019 *10th International Conference on Applied Energy* **15** 2403-2408.
[19] Yi F, Xudong Z, Jing L, Guiqiang L, Steve M, Yuanda C, Ali B, Min Y, Yousef G, Samson S and Xiaoli M 2020 *Applied Energy* **275** 115323.
[20] Sun Q, Wang Y, Cheng Z, Wang J, Zhao P and Dai Y 2020 *Renewable Energy* **147(3)** 2822-2832.
[21] Jang Y and Lee J 2018 *Energy Conversion and Management* **160** 31-47.
[22] Rashwan S S, Dincer Ibrahim and MohanyAtef 2019 *Energy Conversion and Management* **184** 416-426.

IOP Conf. Series: Materials Science and Engineering 1180 (2021) 012054 doi:10.1088/1757-899X/1180/1/012054

Analysis and Optimization of Over Compression for Rotary Cylinder Compressor Based on PV Test

Y S Hu[1,3], H J Wei[1,2,3], J Xu[1,2,3], Z C Du[1,2,3*], P L Zhang[3] and L P Ren[1,2,3]

[1]State Key Laboratory of Air-conditioning Equipment and System Energy Conservation, Zhuhai, Guangdong 519070, China

[2]Guangdong Key Laboratory of Refrigeration Equipment and Energy Conservation Technology, Zhuhai, Guangdong 519070, China

[3]Gree Electric Appliances, Inc., Zhuhai, Guangdong 519070, China

* Corresponding Author

Abstract. Rotary cylinder compressor is a novel positive displacement compressor. Theoretical analysis shows that the compressor may have large over compression loss. Based on PV test, the over compression characteristics of rotary cylinder compressor was analyzed, and the over compression loss accounts for 5.5% of the indicated power. The over compression was optimized, and the optimized over compression loss is only 0.5% of the indicated power. By further comparing the energy efficiency improvement of the whole compressor, it is found that the improvement effect is equivalent to that of the over compression, and the effectiveness of the scheme is confirmed.

1. Introduction

The over-compression of positive displacement compressor has a great influence on its performance. In view of the over-compression problem, domestic and foreign scholars have done a lot of research. Peng Xueyuan et al. [1] established a mathematical model of the discharge process of a single-tooth rotor compressor, and believed that the over-compression was inevitable in the late and middle stages of the discharge process, which would significantly increase the power consumption in the discharge process. The study of Xiong Wei et al. [2] on the compound gear compressor shows that, with a proper clearance volume, the discharge over-compression can be solved by enlarging the flow passage area of the diversion groove. Huang Weicai et al. [3] realized the synchronous monitoring of compressor rotation angle and pressure by positioning the rotation angle of the rotor compressor. Hu Yusheng et al. [4] carried out CFD simulation analysis on the internal refrigerant pressure pulsation of the rolling rotor compressor, combined with experimental research, obtained the relationship between the pressure in the compressor pump body and the rotation angle, and calculated the indicated work of the compressor.

The Rotary Cylinder Compressor (RCC) is a novel displacement compressor, its pump body mainly contains seven parts: main bearing, shaft, cylinder, piston, cylinder sleeve, spacing board, and sub bearing is shown in Figure 1. The piston reciprocates relative to the cylinder, and the cylinder moves in a circular motion relative to the cylinder sleeve. Under the joint action of the above two relative movements, RCC completes the suction, compression and discharge processes, is shown in Figure 2.

RCC has a significant advantage at the medium and low frequency, and the advantage shows more obvious with the decreasing frequency. The experimental results of the prototype fully verified the performance advantages of rotary cylinder compressor at volumetric efficiency and electrical efficiency, which was consistent with the theoretical analysis. [5] However, RCC's PV characteristics have not been studied yet.

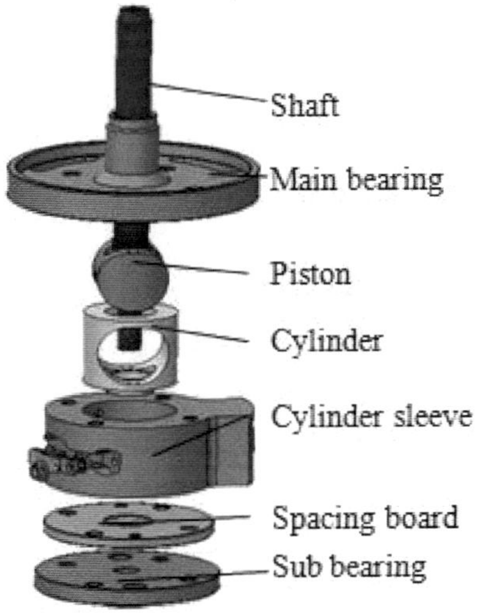

Figure 1. Structure decomposition diagram

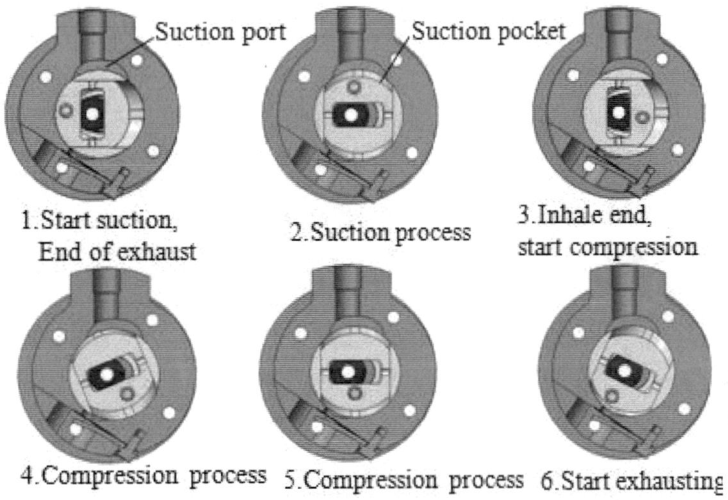

Figure 2. Schematic diagram of suction and discharge process

The starting angle of suction is defined as 0 °, 0 ~ 180 ° is the process of suction, and 180 ~ 360 ° is the process of compression and discharge. The relationship between the working chamber volume and the angle is as follows [5]

$$V = eS(1-\cos\theta) \tag{1}$$

The volume change rate is as follows

$$\dot{V} = eS\sin\theta\cdot\dot{\theta} \tag{2}$$

The effective discharge port area S

$$S = \begin{cases} \pi r^2, & \pi r^2 > 2\pi re(1-\cos\theta) \\ 2\pi re(1-\cos\theta), & \pi r^2 < 2\pi re(1-\cos\theta) \end{cases} \tag{3}$$

Discharge rate v

$$v = \frac{\dot{V}}{S} \tag{4}$$

The calculation results show that the discharge rate of the RCC compressor is too large at the end of the discharge stage, and there may be a large over-compression, which affects the compressor performance. It is necessary to study the over-compression of RCC compressor.

2. PV test principle and scheme

2.1 Test principle

The PV test system of RCC consists of compressor performance test bench, compressor, angle positioning system, and pressure test system, the compressor for PV test is shown in Figure 3. The relationship between pressure and angle (P-θ) can be obtained by testing the pressure of RCC compressor at various angles. Combined with the relationship between volume and angle, the pressure and volume curve (P-V) of the compressor can be obtained. In the PV curve, the indicated power is the area enclosed by P and V. The calculation diagram of indicated power is shown in Figure 4. The indicated power can be calculated by the following formula.

$$\mathrm{w} = \sum_{i=1}^{n} \Delta V_{\mathrm{i}} \Delta P_i \tag{5}$$

The indicated power of the compressor can be obtained and the loss of the compressor can be further evaluated.

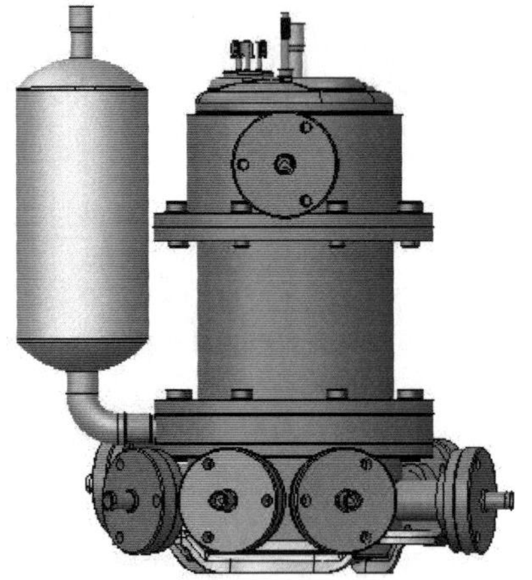

Figure 3. Compressor for PV test

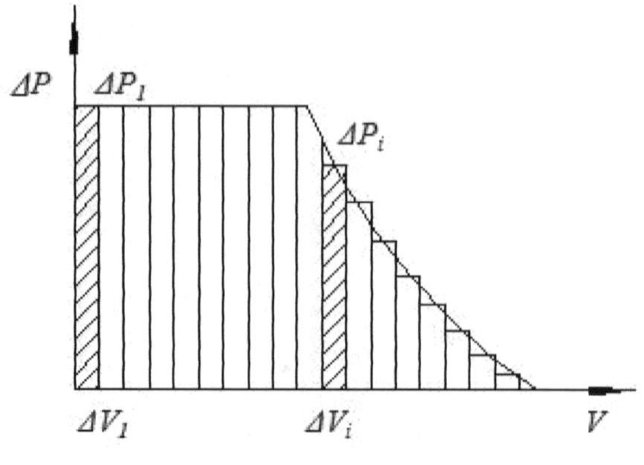

Figure 4. Schematic diagram of indicated power calculation

2.2 Angle test

A positioning plate is arranged on the top of the compressor shaft, and a gear groove with regular distribution is arranged on the positioning plate, and the displacement sensor is aligned with the gear groove plane, as shown in Figure 5. When the compressor is running, the distance between the displacement sensor and the positioning plate will change periodically, and the pulse signal will be output. The schematic diagram of pulse signal is shown in Figure 6. By analyzing the pulse signal, the running angle of the compressor can be known [3].

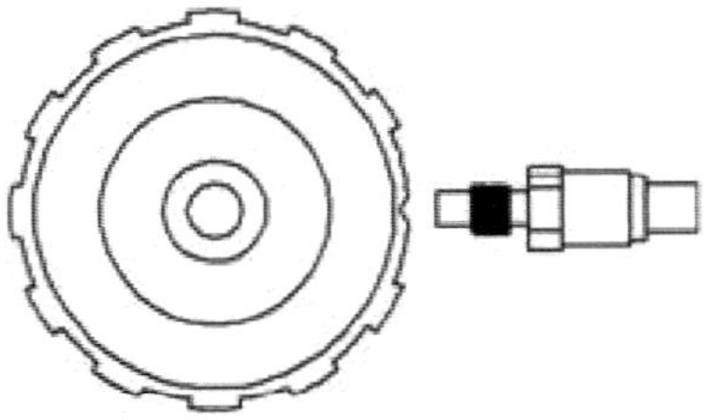

Figure 5. Schematic diagram of installation position of displacement sensor and positioning plate

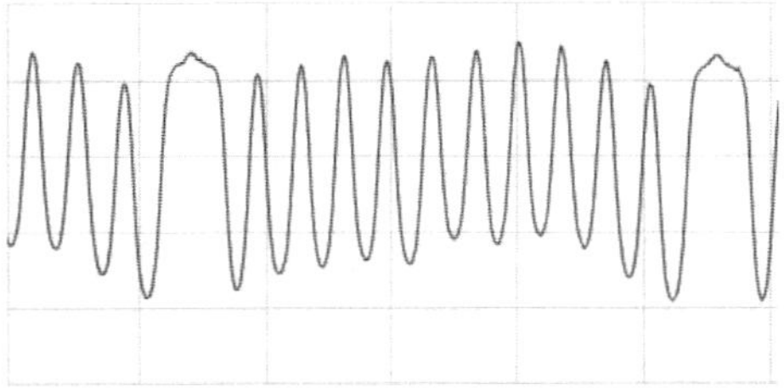

Figure 6. Schematic diagram of pulse signal

2.3 Pressure test

The pressure sensors are arranged along the radial direction of the cylinder sleeve. In order to ensure the complete pressure curve, when the pressure sensor is arranged, it is necessary to ensure that any two adjacent sensors have a certain overlap in the test angle range. The schematic diagram of the position of the pressure sensor is shown in Figure 7. In this experiment, the pressure sensor pressure test angle range is shown in Table 1. Kulite XTC-142B-190M pressure sensor is used in this experiment. Before the test, the test pressure of the sensor is compared with the theoretical pressure to ensure the accuracy of the test.

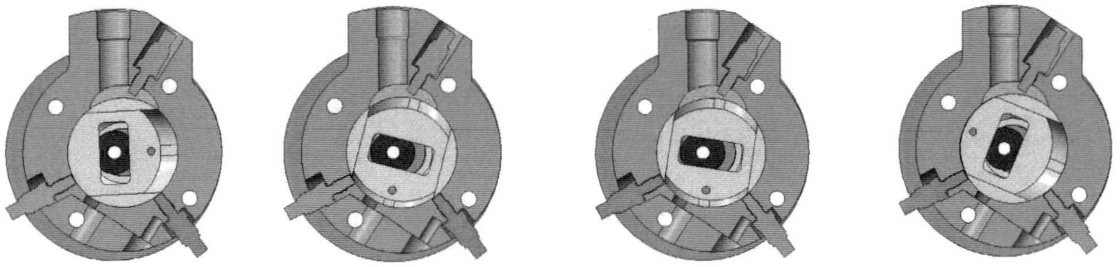

S_2 connection angle S_2 separation angle S_3 connection angle S_3 separation angle

Figure 7. Schematic diagram of the position of the pressure sensor

Table1. Angle range of pressure sensor in pressure test

Sensor	Starting angle	Ending angle
S_1	0°	180°
S_2	175°	290°
S_3	280°	390°

3. PV test analysis

Based on the PV test scheme designed above, the PV test of RCC is carried out, and the test conditions are shown in Table 2. Through the PV test, the valve opening loss, over-compression loss and suction loss of RCC compressor can be studied. This paper mainly studies the discharge over compression loss of RCC compressor, and optimizes the design of RCC compressor based on the research results; other characteristics of PV test will be discussed in detail in other papers, and this paper will only briefly describe.

Table 2. Compressor operating parameters

Refrigerant	R410A
Condensation Temperature(℃)	40
Evaporation Temperature(℃)	20
Volume(cc)	9.6
Operating frequency（Hz）	60

3.1 Sensor curve analysis

3.1.1 P-θ curve of Sensor 1

The sensor 1 is installed in the suction tank, which is used to monitor the pressure value in the suction chamber at different angles during the suction process; combined with the theoretical suction pressure, the suction loss of RCC compressor can be evaluated. P–θ curve of the sensor 1 is shown in Figure 8. According to the test results, there is an obvious pressure valley near 70°, indicating that the compressor has a certain suction loss.

3.1.2 P-θ curve of Sensor 2

The sensor 2 is mainly responsible for monitoring the pressure value in the suction chamber at different rotation angles during the compression process, and can also be used to evaluate the air-tightness between the cylinder and cylinder sleeve. P - θ curve of the sensor 2 is shown in Figure 9. In the range of 280 ° to 350 °, sensor 2 is in the clearance between cylinder and cylinder sleeve, and the pressure change in this angle range can reflect the sealing performance: the better the tightness is, the slower the pressure reduction rate is, and the higher the pressure value is at 350°.

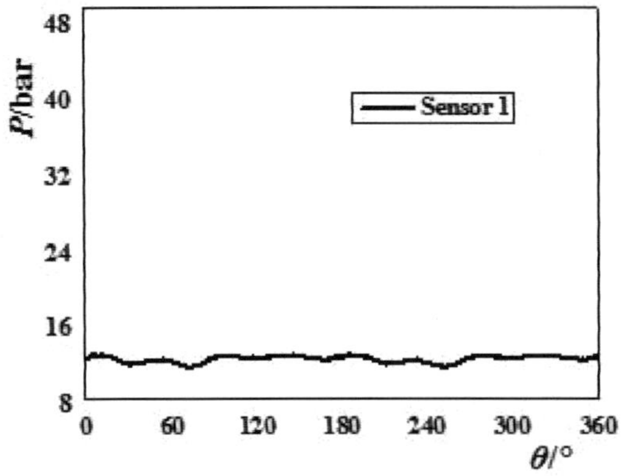

Figure 8. P–θ curve of the sensor 1

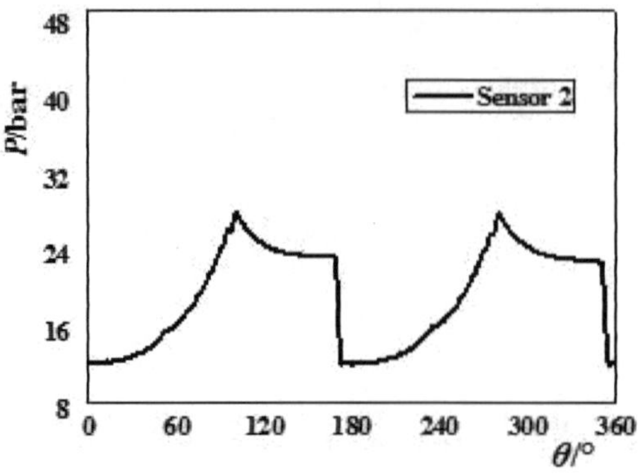

Figure 9.P–θ curve of the sensor 2

3.1.3 P-θ curve of Sensor 3

The sensor 3 is mainly responsible for monitoring the pressure state in the discharge process, which can be used to evaluate the loss in the valve opening stage and the over compression loss; at the same time, the re-expansion process of the high-pressure refrigerant in the clearance volume can be evaluated. P - θ curve of the sensor 3 is shown in Figure 10.

3.1.4 Complete P -θ curve

The pressure curves of sensor 1 (0° to 360°), sensor 2 (0° to 90° and 180° to 270°) and sensor 3 (90° to 180° and 270° to 360°) can be respectively taken to form a complete P- θ curve within a compression period. The complete P–θ curves is shown in Figure 11.

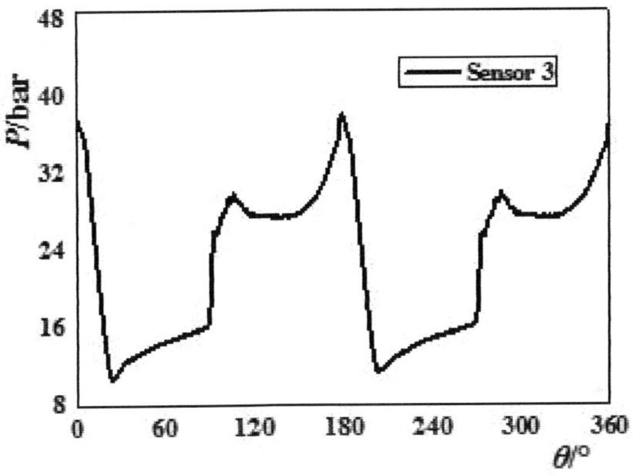

Figure 10. P–θ curve of the sensor 3

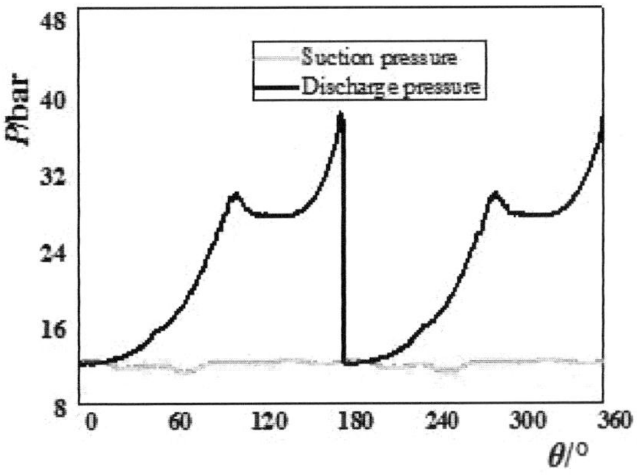

Figure 11. Complete P–θ curves

3.2 Loss calculation

According to the relationship between volume and Angle, the P–θ curve was transformed into the P-V curve to calculate the indicated work losses. The P-V curve is shown in Figure 12, and the distribution of indicated power loss is shown in Table 3. From the test results, the over-compression loss accounts for the largest proportion.

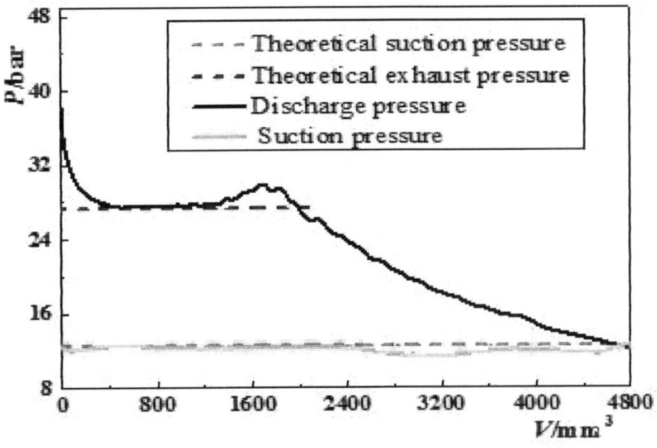

Figure 12. P-V curve

Table 3. Distribution of indicated power loss

Name	Power consumption(W)	Proportion
W	407.1	100%
$W_{1(Suction)}$	11.1	2.7%
$W_{2(Over\text{-}compression)}$	22.4	5.5%
$W_{3(Valve\ opening)}$	13.4	3.3%

IOP Conf. Series: Materials Science and Engineering 1180 (2021) 012054 doi:10.1088/1757-899X/1180/1/012054

4. Optimization design and PV test comparison

4.1 Optimization design

In the first PV test, the over-compression loss accounted for 5.5% of the indicated power, which seriously affected the performance of the prototype. Analyzing the discharge rate of the prototype, it is found that the discharge rate is larger at the end of the discharge. Enlarging the area of the discharge passage can effectively reduce the discharge rate and discharge resistance, thereby reducing the over-compression loss. Based on the above analysis, the scheme of setting auxiliary discharge port is designed. The diameter of the auxiliary discharge port is 3 mm, and the working angle range is 253 ° to 369 ° (73 °to 189 °). The schematic diagram of auxiliary discharge port is shown in Figure 13, discharge rate before and after optimization is shown in Figure 14.

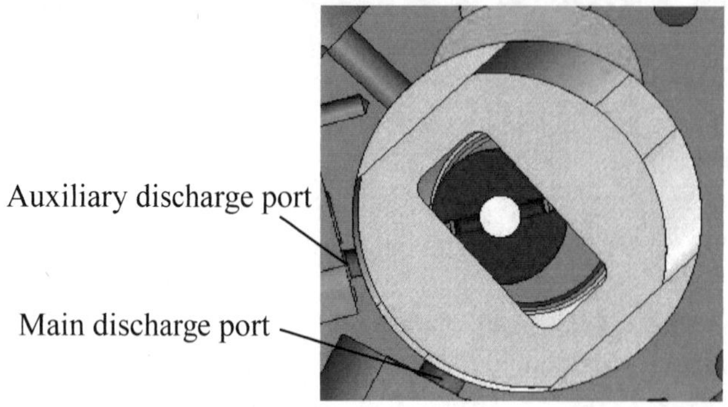

Figure 13. Schematic diagram of auxiliary discharge port

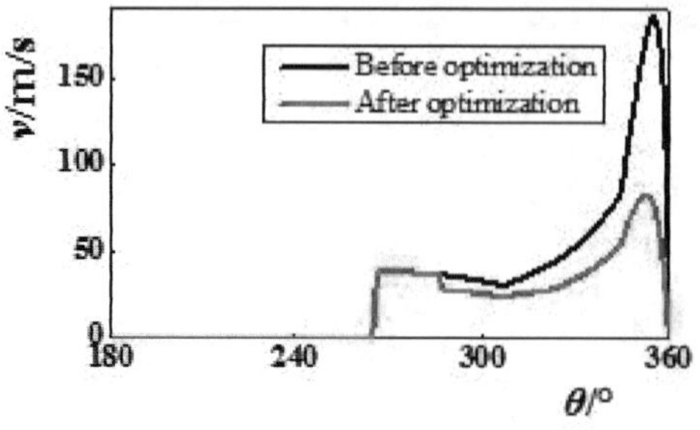

Figure 14. Comparison of discharge rate before and after optimization

4.2 Analysis of test results

With the addition of auxiliary discharge port, the pressure at the end of discharge is significantly reduced, and the over-compression is improved significantly. P–θ curve of the pressure sensor 3 before and after optimization is shown in Figure 15. P-V curve before and after optimization is shown in Figure 16. The distribution of indicated power loss before and after optimization is shown in Table 4. The calculation results show that the over-compression loss is reduced from 5.5% to 0.5%, and the COP of the prototype is increased by 9.1%.

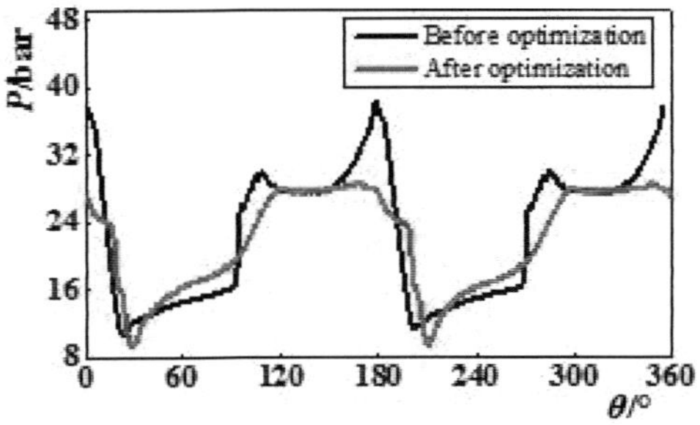

Figure 15.P–θ curve of the pressure sensor 3 before and after optimization

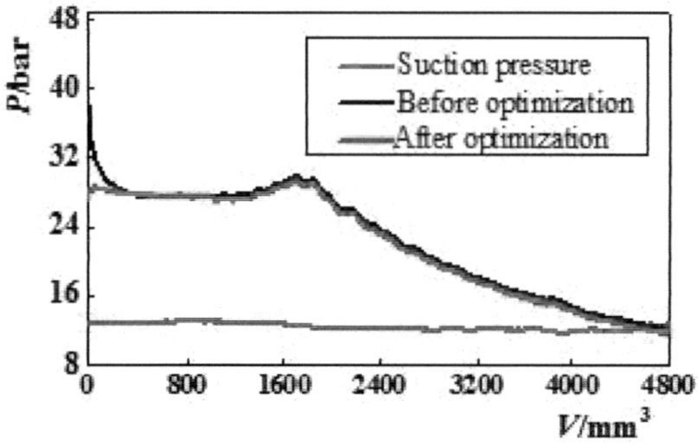

Figure 16. P-V curve before and after optimization

Table 4. Indicated power loss distribution before and after optimization

Name	Proportion before optimization	Proportion after optimization
W	100%	100%
$W_{1(Suction)}$	2.7%	2.9%
$W_{2(Over\text{-}compression)}$	5.5%	0.5%
$W_{3(Valve\ opening)}$	3.3%	3.2%

5. Conclusions

If the effective discharge area is too small, the RCC compressor will be over compressed at the end of the discharge stage, which will reduce the performance of the compressor. By setting the auxiliary discharge port, the effective discharge area at the end of the discharge stage can be increased, the discharge rate can be reduced, so as to reduce the over-compression.

6. Next step

From the PV test results, the over-compression of RCC compressor has been effectively improved, and the valve opening loss after improvement accounts for the largest proportion. The next step is to study the loss of the valve in the opening stage to further reduce the discharge loss of the compressor.

References

[1] Peng X Y, He Z L and Shu P C 1997 Research on exhaust over-compression of single tooth rotor compressor. *Compressor Technology*,6:6-8.

[2] Xiong W, Feng Q K, Peng X Y and Li X M 2004 Mathematical calculation and experimental study of compound gear-type compressor. *Journal of Mechanical Engineering*, 40:189-194.

[3] Huang W C, Yin W H and Geng J Q 2015 Experimental study on rotation angle monitoring of rotary compressor. *Refrigeration and Air-conditioning*, 15:35-38.

[4] Hu Y S and Liang S B 2011 Simulation and experimental study of refrigerant pressure fluctuation in rolling rotor compressor. *Electrical Appliances*, 8:48-51.

[5] Hu, Y S , Wei, H J , Xu, J , Du, Z C , Yang, S and Ren, L P. 2019 The theoretical and experimental research of a novel rotary cylinder compressor. *IOP Conference Series: Materials Science and Engineering*, *604*(1), 012073.

Basic design procedure for an internally geared screw compressor

Matthew Read
City, University of London, Northampton Square, London, UK
E-mail: m.read@city.ac.uk

Abstract. The internally geared gerotor type of positive displacement machine is widely used for pumping applications, but can also be used to achieve internal compression or expansion of a working fluid. A further development of this configuration is the application of helical twist to the inner and outer rotors. Previous studies have shown that this can decrease the power transferred between rotors, and increase the relative maximum flow areas at the inlet and discharge ports. There are possible benefits to this configuration when compared to conventional twin-screw machines, including reduced leakage area, stiffer rotors, and co-directional thermal expansion of rotors. However, no experimental data is currently available for this configuration, and optimising the geometry of the gerotor screw machine requires a detailed understanding of the porting, leakage flows and mechanical losses. An experimental programme is being conducted to investigate the influence of these factors on performance. This paper discusses preliminary results from geometrical analysis leading to a simple design procedure. The basic design of an internally geared air compressor is demonstrated, based on the operation characteristics of a conventional twin screw machine.

Keywords: screw, internally geared, gerotor, compressor

1. Introduction

Internally geared helical twin screw machines have been proposed for application as positive displacement compressors and expanders. The use of stationary end plates with carefully shaped holes acting as inlet and discharge ports can allow internal compression or expansion to occur using fixed profile and helical pitch for the rotors [1–4]. These machines are then analogous to conventional twin screw compressors, but have potential advantages in terms of stiffer rotors, co-directional thermal expansion of rotors, and reduced sealing line length for a given working chamber displacement. Comparing these two configurations requires a fundamental understanding of the relationship between geometric parameters including the rotor profile, maximum diameter, D, length, L, and wrap angle, Φ, and the operation of the machine characterised by the volumetric flow rate, $\dot{Q}$, built-in volume ratio, ϵ_v, fluid filling velocity, u, rotational speed, ω, and bearing loads. This paper focusses on characterising these parameters for the internally-geared machine, in order to develop a design procedure for a prototype machine.

2. Relationships between geometric and operational parameters

The current study considers only simple composite epi-hypo cycloid rotors, where the rotor profile is defined by a shape parameter, ν, and the number of lobes, N. These profiles are fully described in ref. [3]. This rotor profile can then be scaled by the maximum profile diameter, D. The swept volume of a machine is an important parameter which depends on the geometry of the rotors; once normalised by the volume of a cylinder enclosing the rotors as shown in Equation 1, it is a function of the rotor profile and the wrap angle of the helical rotor [4]. The swept volume can then be related to the ideal volumetric flow rate of the machine, $\dot{Q}_{id}$ (Equation 2). Assuming the fluid is incompressible during the filling process, the rate of change of volume divided by the port area provides an indication of the fluid inlet velocity. This value tends to infinity as the port area tends to zero, but quickly drops to an approximately constant value for most of the filling process. A further performance parameter, u^*, is therefore defined here as the minimum ideal velocity of the fluid at the inlet port during filling (Equation 3). It is possible to define a non-dimensional group, λ_1, relating rotor diameter, volumetric flow rate, and u^* as shown in Equation 4. A second non-dimensional group, λ_2, relating rotational speed, rotor length and u^* can also be defined (Equation 5). As with the normalised swept volume, $\bar{V}_{sw}$, these non-dimensional groups are only functions of the rotor profile shape (defined by ν and N, but not depending on the scaling factor D) and the helical wrap angle.

$$\bar{V}_{sw} = \frac{V_{sw}}{\frac{\pi}{4}LD^2} = f(\nu, N, \Phi) \tag{1}$$

$$\dot{Q}_{id} = V_{sw}\left(\frac{\omega}{2\pi}\right) \tag{2}$$

$$u^* = \omega \times \min\left(\frac{1}{A_{lp}}\frac{dV_{wc}}{d\phi}\right) \tag{3}$$

Non-dimensional groups:

$$\lambda_1 = \frac{\dot{Q}_{id}}{D^2 u^*} = f(\nu, N, \Phi) \tag{4}$$

$$\lambda_2 = \frac{u^*}{\omega L} = \frac{\bar{V}_{sw}}{8\lambda_1} = f(\nu, N, \Phi) \tag{5}$$

The following relationships therefore apply:

- λ_1 relates D to the ratio of volumetric flow rate to nominal filling velocity. For a given value of u^*, the volumetric flow rate is proportional to D^2.
- λ_2 relates L to the ratio of nominal filling velocity to rotational speed. For a given value of u^*, the rotational speed is inversely proportional to the length.

Basic design procedure for an internally geared screw compressor

2.1. Selection of compressor operating speed

The rotational speed of the outer rotor is related to the diameter, due to the operating limits of the bearings. The speed limit, $v_{lim} = \omega D/2$, depends on the type of bearings used; based on reference speed data for single row cylindrical deep groove and roller bearings, an appropriate rotational speed limitation is $v_{lim} = 30m/s$, while for hydrodynamic bearings much higher surface speeds of 150-250 m/s are reported.

2.2. Non-dimensional parameters as functions of rotor geometry

The non-dimensionalised values of rotor diameter and L/D ratio are shown below as function of the profile shape parameter, ν, and the normalised wrap angle, $\bar{\Phi}_o$ (defined as the actual wrap angle divided by the angle of rotation required for a working chamber volume to go from zero to maximum and back to zero), for cases when $N_o = 3, 5$ and 7.

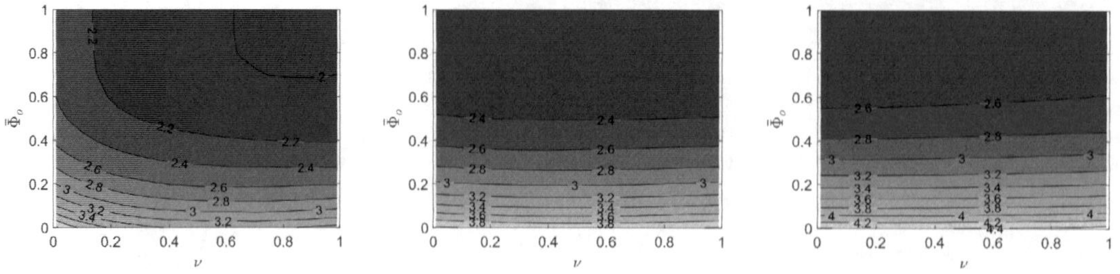

(a) (b) (c)

Figure 1: Contour maps showing $D\sqrt{\frac{u^*}{Q_{id}}} = \sqrt{\frac{1}{\lambda_1}}$ as function of ν and $\bar{\Phi}_o$, for cases when $N_o = (a)3, (b)5$ and $(c)7$

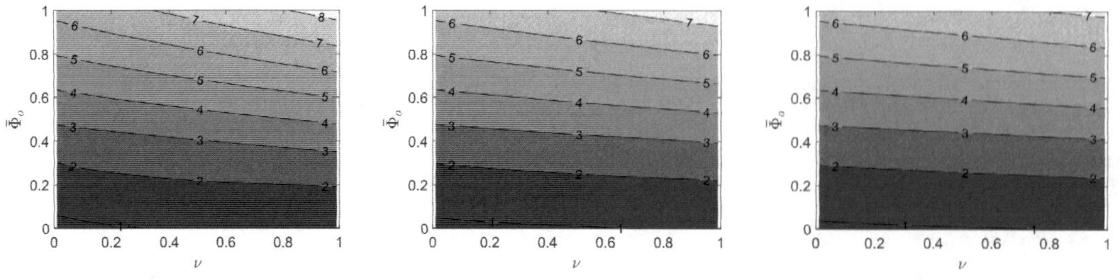

(a) (b) (c)

Figure 2: Contour maps showing $\frac{L}{D}(\frac{v_{lim}}{u^*}) = \frac{1}{2\lambda_2}$ as function of ν and $\bar{\Phi}_o$, for cases when $N_o = (a)3, (b)5$ and $(c)7$

The results shown in Figures 1 and 2 can be used to produce a further plot of the non-dimensionalised machine volume ($\propto LD^2$), which is shown below for the same cases.

Basic design procedure for an internally geared screw compressor

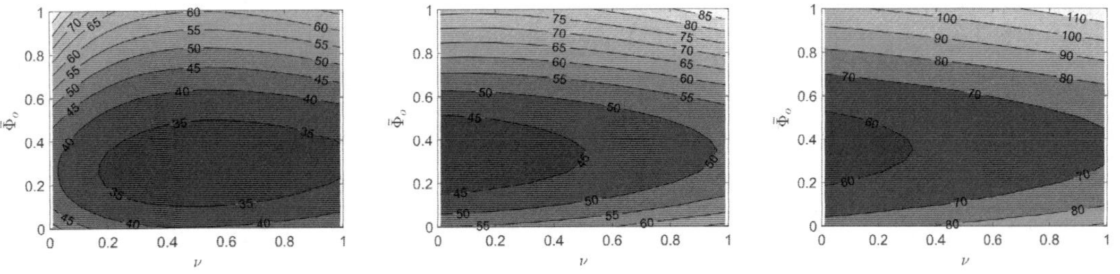

(a) (b) (c)

Figure 3: Contour maps showing $LD^2\sqrt{\frac{v_{lim}^2 u^*}{Q_{id}^3}}$ as function of ν and $\bar{\Phi}_o$, for cases when $N_o = (a)3, (b)5$ and $(c)7$

The results in Figure 3 show that the machine volume can be minimised by using lower values of N_o, and that for the values considered ($N_o = 3, 5$ and 7) the volume is minimised with a value of $\bar{\Phi}_o = 0.3 - 0.4$. This may be an important requirement for particular applications, but there are other factors which should also be considered when selecting an appropriate geometry and wrap angle.

2.3. Power transfer between rotors

The pressure of the working fluid acting on the exposed faces of the rotors results in torque applied about the axes of rotation. As these axes are offset, the torque is different during rotation. If the variation of pressure in the working chamber is known, the rotor torques due to the fluid in a single working chamber can be calculated. The effect of all simultaneous working chambers can then be found, and by considering the sign and magnitude of these net rotor torques and the gearing ratio between them, the maximum proportion of input power being transferred between the rotors, $\Pi_{\max} = \min(\max(|P_o/P_{in}|), \max(|P_i/P_{in}|)$, can be found. Note that the choice of driven rotor should be made to ensure that $\max(|P_{idle}/P_{in}|)$ is minimised in order to reduce the power transferred via rotor-to-rotor contact. Once an application is characterised by specifying the pressure ratio, this can be used to estimate a required value of ϵ_v. In this paper a simple polytropic compression cycle has been assumed with no pressure losses during filling and emptying, and discharge pressure equal to maximum chamber pressure (i.e no over- or under-compression). The inlet conditions have been taken as 1.01 bar and 288K. Using Equation 6, the polytropic exponent has been chosen to allow a temperature rise of 100K for the compression process.

$$n_{poly} = 1 + \frac{\ln(T_{dis}/T_{suc})}{\ln(\epsilon_v)} \tag{6}$$

The examples shown in Figure 4 illustrate that once the working chamber pressure is defined, this is all the information required to make an initial assessment of suitable values of ν and $\bar{\Phi}_o$ that will limit the rotor-to-rotor power transfer for particular values of N_o. In all cases, the local minima shown at lower values of ν correspond to a driven

Basic design procedure for an internally geared screw compressor

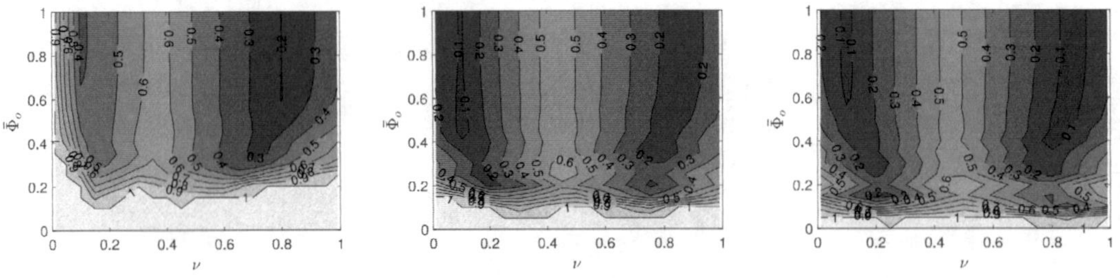

(a) (b) (c)

Figure 4: $\Pi_{\max}$ as function of ν and $\bar{\Phi}_o$, for cases with $\epsilon_v = 5$ when $N_o = (a)3, (b)5$ and $(c)7$ (note that contours are capped at a maximum value of 1)

inner rotor, while the local minima at higher values of ν correspond to a driven outer rotor.

2.4. Rotor-to-rotor leakage line lengths

The leakage paths that form between the rotors are an important geometrical characteristic of twin screw machines, as they will have a strong influence on the volumetric and isentropic efficiency of the machine. This can be characterised by considering the maximum total combined length of these rotor-to-rotor contact lines for single working chamber, $\max(\ell_{tot})$. This is only an approximate guide to leakage effects, as it does not consider other paths (such as end-face leakage), and does not consider the net effect of increasing and decreasing leakage areas between multiple working chambers during the compression process. It does however provide a useful indication of how this varies with different geometrical and operational parameters. A non-dimensional group, λ_3, can again be defined by dividing $\max(\ell_{tot})$ by the length of the helical path following the pitch circle radius of the outer rotor along its length. The value of λ_3 can be shown to always equal 2 for the epi-hypo composite rotor profiles considered in this paper; this can most clearly be seen in the case of straight rotors, where $\Phi_o = 0$, and the total length of the rotor-to-rotor contact paths will have a constant value of $2L$ throughout the compression cycle.

$$\lambda_3 = \frac{\max(\ell_{tot})}{\sqrt{\{L^2 + (\Phi_o \rho_{p,o})^2\}}} = \frac{\max(\ell_{tot})}{D\sqrt{\{(\frac{L}{D})^2 + (\frac{N_o \Phi_o}{2N_o + 4\nu})^2\}}} \tag{7}$$

$$\Phi_o = \bar{\Phi}_o \frac{2\pi}{N_o}(N_o + \nu) \tag{8}$$

$$\lambda_3 = \frac{\max(\ell_{tot})}{D\sqrt{\{(\frac{L}{D})^2 + (\frac{(N_o+\nu)\pi\bar{\Phi}_o}{N_o+2\nu})^2\}}} = 2 \tag{9}$$

It is clear from Equation 9 that the values of D and L must both be known before the value of $\max(\ell_{tot})$ can be found for specific values of N_o, ν and $\bar{\Phi}_o$. The procedure for calculating these values and assessing the suitability of a design is discussed in Section 3.

Basic design procedure for an internally geared screw compressor

2.5. Bearing forces

The net forces exerted on the rotors can be considered in terms of the net torques exerted about a rotor's global x, y and z axes. In each case the torque about one axis is due to components of the net force aligned with the other two axes. In order to assess the bearing forces acting at each end of a rotor the torques about the x and y axes must be known. If we take the example of the torque about the x axis, it will consist of a torque due to a force in the y direction ($\propto LD^2$), plus a torque due to a force in the z direction ($\propto D^3$). It is therefore only possible to evaluate these bearing loads once values for L and D have been specified.

The maximum diameter of the outer rotor profile, D, can be used as the minimum allowable bore diameter for the outer rotor bearings. The minimum diameter of the inner rotor profile, $d = D(1 - \frac{3}{N_o+2\nu})$, can be used as the maximum allowable outer diameter for the inner rotor bearings.

The axial and radial forces exerted on the inner and outer rotor are of similar magnitude, and both fluctuate with period $2\pi/(\omega_o N_o)$ during machine operation. The significantly larger size of the outer rotor bearings imposes a speed limitation on the machine, and generally leads to these being lightly loaded. Ensuring that these bearings achieve the required minimum load is therefore a key design requirement. Although the inner rotor rotates faster than the outer due to the gearing ratio, the smaller bearing diameter means that these operate well below their speed limit. In both cases the operating life of the bearing under the load conditions is a key consideration. The results for bearing loads presented for the design example in the following section therefore consider both the minimum and maximum axial force and radial forces (at the high and low pressure ends of the machine) for the rotors. The results also assume that the bearing centres are located at the corresponding end face (hp or lp) for both the inner and outer rotor. Bearing offsets can easily be applied to this calculation for more accurate calculation of bearing loads in a later stage of the design process once suitable components have been identified.

3. Design procedure

Specifying the required operational parameters $\dot{Q}_{id}$, u^*, and v_{lim} allows the required dimensions of the machine to be identified for particular values of ν and $\bar{\Phi}_o$. The initial design procedure can therefore be applied as follows:

$$D = \sqrt{\frac{\dot{Q}_{id}}{\lambda_1 u^*}} \tag{10}$$

$$\frac{L}{D} = \frac{u^*}{2v_{lim}\lambda_2} \tag{11}$$

$$\max(\ell_{tot}) = 2D\sqrt{\left(\frac{L}{D}\right)^2 + \left(\frac{(N_o+\nu)\pi\bar{\Phi}_o}{N_o+2\nu}\right)^2} \tag{12}$$

Basic design procedure for an internally geared screw compressor

(i) Assume initial values for profile shape (N, ν) and wrap angle $(\bar{\Phi})$
(ii) Calculate or lookup non-dimensional parameters λ_1, λ_2 and Π_{max} for specific profile shape and wrap angle
(iii) Specify required $\dot{Q}_{id}$ and allowable u^*
(iv) Use Equation 9 to find D
(v) Specify v_{lim} and calculate corresponding ω_o
(vi) Use Equation 10 to find L
(vii) Use Equation 11 to find $\max(\ell_{tot})$
(viii) Define fitness function based on relative importance of machine volume, Π_{max}, rotational speed, $\max(\ell_{tot})$, and bearing loads
(ix) Iterate profile shape and wrap angle to optimise fitness function

3.1. Example of geometry selection

In order to investigate the initial design of an internally geared screw compressor, a conventional twin screw machine has been used to specify the basic operating parameters, and as a basis for comparison. This machine is a typical design for a small oil-injected air compressor, and the key geometrical parameters are as follows:

Table 1: Example of conventional twin screw compressor geometry

Main/gate rotor lobe no.	4/5
Rotor centre distance	71.0 mm
Rotor length	158.0 mm
Main rotor outer diameter	101.9 mm
Gate rotor outer diameter	80.2 mm
Main rotor wrap angle	306°
Swept volume	0.720 litres/rev of the main rotor

Based on this screw compressor geometry, the software SCORG [5] has been used to assess the volume and port areas as functions of rotor angular position. For a main rotor tip speed of 40m/s (corresponding to $\omega = 785$rad/s) this results in values of $\dot{Q}_{id} = 0.0895\text{m}^3/\text{s}$ and $u^* = 23.1$m/s. The value of the bearing limiting speed, v_{lim}, assumed for rolling element bearings is 30m/s. These input values have been applied to a internally geared geometry with $N_o = 5$, resulting in the following dimensional contour plots shown in Figures 5-7. These figures show the relevant geometric and operational parameters as functions of ν and $\bar{\Phi}_o$ for this specific application.

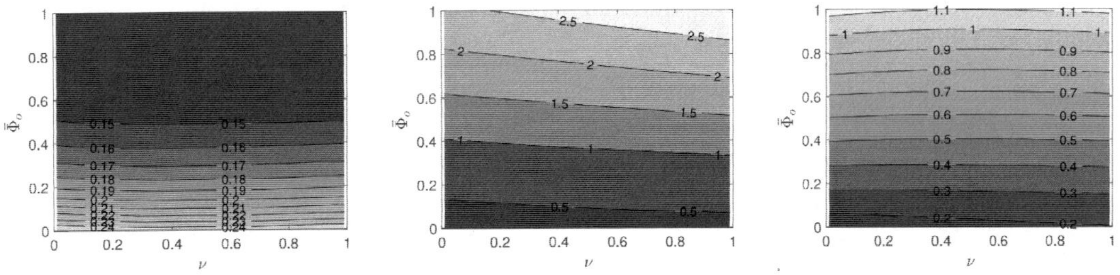

(a) (b) (c)

Figure 5: $(a) D$, $(b) L/D$, and $(c) \max(\ell_{tot})$ as functions of ν and $\bar{\Phi}_o$ for $N_o = 5$, $\dot{Q}_{id} = 0.0895\text{m}^3/\text{s}$ and $u^* = 23.1\text{m/s}$

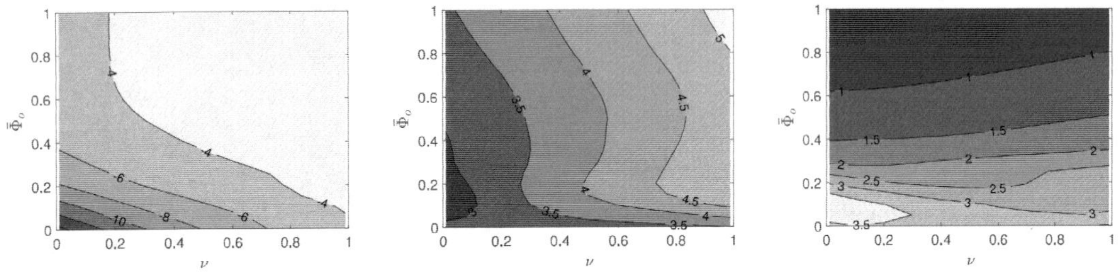

(a) (b) (c)

Figure 6: Outer rotor radial and axial minimum bearing forces: $(a) \max(|B_{z,o}|)$, $(b) \max(B_{rad,o})_{hp}$, $(c) \max(B_{rad,o})_{lp}$ for $N_o = 5$, $\dot{Q}_{id} = 0.0895\text{m}^3/\text{s}$, $u^* = 23.1\text{m/s}$ and $\epsilon_v = 5$

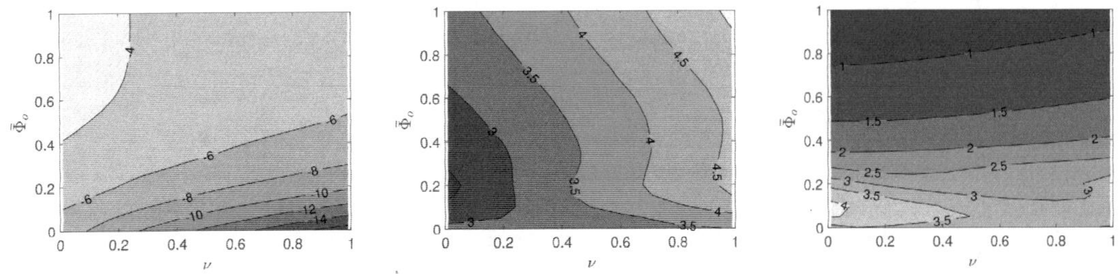

(a) (b) (c)

Figure 7: Inner rotor radial and axial maximum bearing forces: $(a) \max(|B_{z,i}|)$, $(b) \max(B_{rad,i})_{hp}$, $(c) \max(B_{rad,i})_{lp}$ for $N_o = 5$, $\dot{Q}_{id} = 0.0895\text{m}^3/\text{s}$, $u^* = 23.1\text{m/s}$ and $\epsilon_v = 5$

If the aim of the design exercise is chosen to be minimising the value of $\Pi_{\max}$ for an outer-driven configuration, the results in Figure 4 for the case when $N_o = 5$ show that this occurs when $\nu = 0.8$ and $\bar{\Phi}_o = 0.5$. This results in the geometric design parameters shown in Table 2. In this case the results suggest that, compared to the

Basic design procedure for an internally geared screw compressor

Table 2: Example of initial specification of internally geared compressor geometry

Input parameters	$\dot{Q}_{id}$	0.0895 m^3/s
	u^*	23.1 m/s
	N_o	5
	ϵ_v	5
Conditions for min power transfer	ν	0.8
	$\bar{\Phi}_o$	0.5
	$\Pi_{\max}$	0.0982
	Driven rotor	Outer
Geometrical parameters	Φ_o	209°
	D	149.4 mm
	d	81.5 mm
	L	209.6 mm
	$\max(\ell_{tot})$	588 mm
Operational parameters	ω_o	402 rad/s
	ω_i	502 rad/s
	n_{poly}	1.185
	p_{suc}	1.01 bar
	p_{dis}	6.74 bar
Bearing loads	$\min/\max(B_{z,o})$	-2.85 / -2.70 kN
	$\min/\max(B_{rad,o})_{hp}$	3.62 / 4.44 kN
	$\min/\max(B_{rad,o})_{lp}$	1.18 / 1.44 kN
	$\min/\max(B_{z,i})$	-5.74 / -5.53 kN
	$\min/\max(B_{rad,i})_{hp}$	3.45 / 4.21 kN
	$\min/\max(B_{rad,i})_{lp}$	1.33 / 1.69 kN

conventional machine, the internally geared configuration has a larger maximum rotor diameter, lower rotational speed, and longer rotors. The wrap angle of the rotors is significantly lower for the internally geared configuration, leading to significantly lower maximum total radial leakage line length.

The bearing loads and rotational speeds defined in Table 2 have been used to select appropriate bearings using the SKF 'Bearing Select' tool [6], as shown in Table 3. For both rotors, the axial loads are assumed to be supported by bearings at the high pressure (hp) end of rotors, in combination with the radial load. This is to allow the end face clearance gap to be accurately set during assembly of the machine, and in order to accommodate possible thermal expansion of the rotors in operation, the design must allow non-locating displacement on the seat for the deep groove bearing at the low pressure (lp) end of the outer rotor. For the inner rotor, non-locating displacement occurs within the cylindrical roller bearing selected. The basic rating life of the bearings (a factor only of the load and speed) has been assessed, along with the SKF rating life which is also dependant on lubrication conditions and the bearing fatigue load limit. The lubricant is assumed to be ISO 46 oil. The rated life of the hp bearings is significantly lower due to the higher combined radial and axial loads, although this could be increased if necessary through the use of multiple bearings to distribute these loads.

Basic design procedure for an internally geared screw compressor

Table 3: Rolling element bearings selected for loads identified in Table 2; bearing types are deep groove (DG), angular contact (AC) and cylindrical roller (CR).

Bearing location	-	-	o, lp	o, hp	i, lp	i, hp
Designation	-	-	61832		N207ECP	QJ306MA
Type	-	-	DG		CR	AC
Bearing bore	d_b	mm	160		35	30
Bearing outer diameter	D_b	mm	200		72	72
Bearing width	B_b	mm	20		17	19
Basic dynamic load rating	C	kN	49.4		56.0	53.0
Basic static load rating	C_0	kN	64.0		48.0	41.5
Fatigue load limit	P_u	kN	2.0		6.1	1.8
Minimum load	F_m	kN	0.731		0.271	0.117
Reference speed	n_{ref}	r/min	6300		11000	-
Limiting speed	n_{lim}	r/min	4000		12000	17000
Basic rating life	L_{10h}	h	175000	1360	$> 2e^5$	793
SKF rating life (ISO 46)	L_{10mh}	h	$> 2e^5$	63400	$> 2e^5$	12800

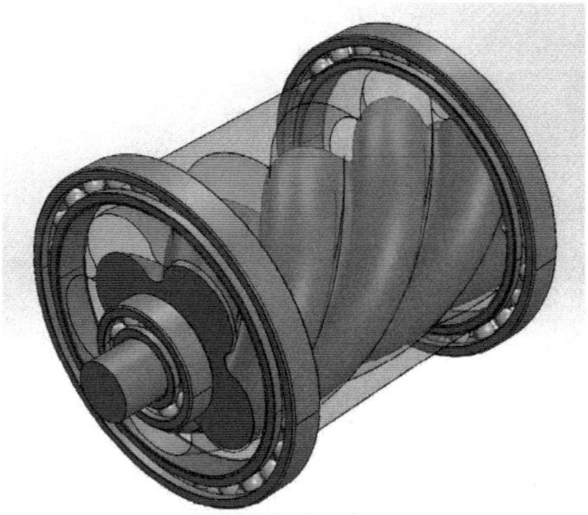

Figure 8: Model illustrating basic geometry of rotors and bearings as specified in Tables 2 and 3 (note that outer rotor is shown as transparent for clarity)

4. Conclusions

Basic design of a internally geared gerotor type screw compressor has been performed. Key operational parameters considered are ideal volumetric flow rate, nominal port flow velocity, and bearing velocity limit. Through the definition of non-dimensional shape parameters, relationships showing rotor length and diameter required as a function of profile shape and helical wrap angle are found. Once the rotor geometry is defined, the leakage line length and the bearing forces can be calculated. A design procedure is described in which the optimum geometry can be chosen based on a fitness function involving power transfer ratio, machine volume, leakage line length, operating speeds and bearing loads. A conventional twin-screw compressor is used to illustrate this design process for an equivalent gerotor-type machine. Rotor geometry and bearing selection are characterised. While this does not allow direct comparison of the performance of the conventional and internally-geared, detailed thermodynamic chamber modelling and experimental testing are planned to investigate the operation of these machines. Future work will focus on applying this method to design an experimental machine to investigate the performance of this novel compressor configuration.

References

[1] Read M, Smith I and Stosic N 2017 Internally geared screw machines with ported end plates *IOP Conference Series: Materials Science and Engineering* vol 232 (IOP Publishing) p 012058
[2] Read M, Stosic N and Smith I 2018 Port flow losses in internally geared positive displacement machines *International Compressor Conference, Purdue University, USA*
[3] Read M, Smith I and Stosic N 2020 *Proceedings of the Institution of Mechanical Engineers, Part E: Journal of Process Mechanical Engineering*
[4] Read M G, Stosic N and Smith I K 2020 *Journal of Mechanical Design* **142**
[5] Screw compressor rotor grid generator URL `https://pdmanalysis.co.uk/scorg`
[6] SKF bearing selection tool URL `https://www.skf.com/uk/support/engineering-tools/bearing-select`

A study on the kinematics of a new Schukey-type rotary compressor

B Cui[1*], K Becker[1], U Lüdersen[1], M Gottschlich[1] and S Kabelac[2]

[1] Institute for Process & Energy Engineering and Climate Protection, Hochschule Hannover, Ricklinger Stadtweg 120, 30459, Hannover, Germany

[2] Institute for Thermodynamics, Gottfried Wilhelm Leibniz Universität Hannover, An der Universität 1, 30823, Garbsen, Germany

[*] E-mail : bin.cui@hs-hannover.de

Abstract. A new type of rotary compressor, called "rotary-chamber compressor", consists of two interlocking rotors with 4 wings each, that perform non-uniform rotary movements. Both rotors have the same direction of rotation, while one rotor is accelerating, the other rotor is retarding. After surpassing a specific mark, the sequence changes and the leading rotor begins to retard and vice versa. Due to the resulting relative phase difference, the volume between the two wings is changing periodically, which allows pulsating working chambers. The technology was first introduced by its founder Jürgen Schukey in 1987. Since then, no further development on this machine is known to us except our own. In this contribution, a study on the kinematics of the rotary-chamber-compressor is presented. Initial studies have shown that changes in the kinematics of the rotors will have a direct influence on the thermodynamical variables, which, if optimized, can lead to an increased performance of the machine. Therefore, a mathematical model has been developed to obtain the performance parameters from different kinematic concepts by using numerical CFD analysis. Furthermore, additional optimization possibilities will be listed and discussed.

1. Introduction

In recent years, several novel compressor types are of interest with the focus to achieve the same or better performance levels compared to existing technologies at reduced costs, noise and vibration. They are typically rated on mass flow rate (MFR) and energy performance, given by the volumetric efficiency λ and isentropic or polytropic efficiencies [1]. Therefore, a variety of studies on novel rotary compressors [2–7], mostly on rolling piston types, have been carried out.

A new rotary type compressor called "rotary-chamber compressor" was first introduced by Jürgen Schukey [8]. He designed the machine to be operated as a compression or expansion engine for refrigeration and heat recovery applications. Since then, the interest of the machine was fading until in recent years, it has been gaining more attention to be used as a compressor technology for refrigeration cycles. The main components of the examined compressor consist of two interlocking rotors, each with four wings, which form the 8 moving chambers within the housing. The rotational movement of each rotor is specified through its own gearbox with special toothed gears, which allows the change of rotational speeds relative to each other. Both rotors rotate in the same direction, with one rotor accelerating and the other rotor retarding vice versa while the main shaft rotates constantly. As a result, the chambers fluctuate in volume, which leads to its suction, compression and discharge movement. Depending on the position of the inlet and outlet ports, the movement allows the compression of the working fluid. Furthermore, the machine is working without suction and discharge valves, since the suction and discharge process is controlled by the position of the rotor wings and ports. Figure 1 shows the main components of the rotary-chamber compressor in an exploded view.

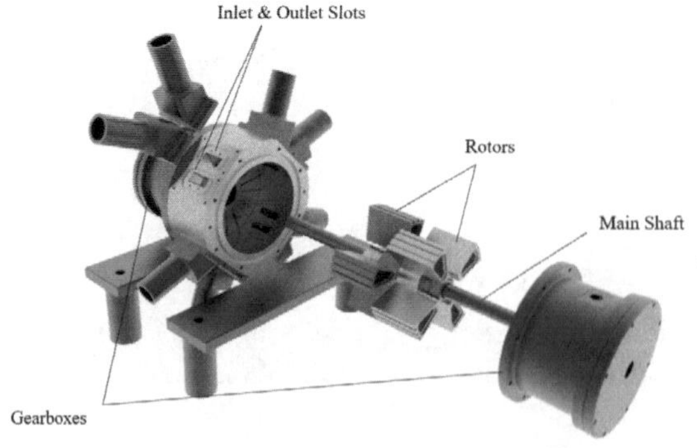

Figure 1. Exploded view of the rotary-chamber compressor.

The compression cycle is illustrated in Figure 2 with its three steps. Each cycle begins with the filling of the working chamber γ with the fluid through the inlet port (3). During the suction process, the leading rotor (1) is in its accelerating phase of its movement pattern while the complementary rotor (2) is retarding. In that case, the working chamber increases in volume and the fluid is sucked in. This step ends, when rotor blade (2) closes the inlet port during the counterclockwise rotation. In that moment, the chamber volume is at its maximum. After closing the inlet, rotor blade (2) starts accelerating while rotor blade (1) starts retarding and thus the chamber volume begins to decrease. Since the working chamber is not connected to any inlet- or outlet port, the compression takes place between rotor blade (1) and (2). After reaching its minimum volume, rotor blade (1) passes the outlet port (4) and the fluid will be discharged. From then on, rotor blade (1) starts to accelerate while rotor blade (2) starts to retard again. The discharge process ends when rotor blade (2) passes and closes the outlet port. From then on, the compression cycle starts again. The cycle takes places in all chambers simultaneously. Therefore, one revolution of the main shaft results in 32 compression cycles (8 chambers x 4 sections), which distinguishes the machine with a high volume turnover.

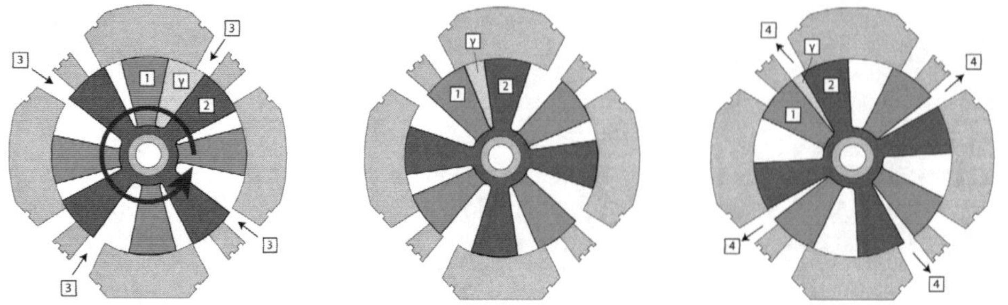

1. Suction 2. Compression 3. Discharge

Figure 2. Compression cycle in three steps.

Preliminary studies have investigated the impact of different port positions and rotational speeds of the engine [9], showing that the focus should lie on reducing the minimum clearance volume to improve the volumetric efficiency. One way to do so, is to optimize the current kinematics of the rotors. Beforehand, the impact of compressor kinematics for reciprocating and rolling piston type compressors was examined by Pandeya [10], who concludes that discharge flow losses are significantly different by different compressor kinematics due to higher flow velocity and mass density during the discharge process. Yet, further motion analysis only was done for gerotor- [11], Wankel- [12] or rolling piston [13]-compressor types.

In this paper, different concepts of the rotor kinematics will be developed and presented, starting with the current function of this machine which was obtained through incremental measurements. Afterwards, continuous and discontinuous functions are developed and examined, also with regard to whether both rotors should have the same or whether each rotor should have its own individual kinematic function. In order to take effects of the momentum of the gas into account, each kinematic function is tested in the commercial CFD-Solver ANSYS CFX to determine the changes in mass flow rate MFR, volumetric efficiency λ, internal pressure ratio Π and indicated work W_{id}. The goal of this contribution is the quantitative determination of the effect of various kinematic functions for this rotary piston-type compressor.

2. Mathematical Model /Methods

2.1. Current kinematic function and settings

The oscillating specific movement pattern for each rotor is defined by its own special toothed gears, which are installed with an offset of 45° to each other and connected to the main shaft. The main shaft itself is driven by an electric motor and rotates at constant speed. In contrary, the angular movement of the rotors are not constant, but oscillating and rotating faster and slower relative to the main shaft, which results in the accelerating and retarding movement of the rotor blades.

To determine the current kinematic function, incremental measurements of the angular velocity at the rotor blades were taken. Subsequently, the kinematic was determined by using a non-linear regression with the Levenberg-Marquardt-Algorithm to determine the coefficients a, b, c and d of eq. (1) and (2).

$$\omega_{\mathrm{R1}}(\theta) = \left\{-a \cdot \left[1 - cos\left(\frac{\pi}{45^\circ} \cdot \theta + b\right) + \frac{c}{2} \cdot sin^2\left(\frac{\pi}{45^\circ} \cdot \theta + b\right)\right] + d\right\} \cdot k_{\mathrm{b}} \cdot \frac{\omega_\theta}{\omega_{\mathrm{N}}} + \omega_\theta \tag{1}$$

$$\omega_{\mathrm{R2}}(\theta) = \left\{-a \cdot \left[1 - cos\left(\frac{\pi}{45^\circ} \cdot \theta + b + \pi\right) + \frac{c}{2} \cdot sin^2\left(\frac{\pi}{45^\circ} \cdot \theta + b + \pi\right)\right] + d\right\} \cdot k_{\mathrm{b}} \cdot \frac{\omega_\theta}{\omega_{\mathrm{N}}} + \omega_\theta \tag{2}$$

With ω_θ as the (constant) angular velocity of the main shaft, ω_{N} being the normal angular velocity of the measurement and θ as the crank angle of the main shaft, the relative changes in angular velocity can be plotted shown in Figure 3, where each rotor passes four minima and four maxima during one revolution of 360° of the main shaft crank angle. As a result, the chamber angle and volume function can be derived (see eq. (3) and (4)), showing a congruent sine-like behavior in Figure 3. The maxima or minima of the chamber volume occurs during the intersection of the angular velocity curves. This is when the compression or the suction process ends. In addition, the geometrical dimensions of the machine are represented in Table 1.

Chamber Angle function:

$$\gamma(t) = \gamma(t - \Delta t) + \int \left(\omega_{\mathrm{R1,2}} - \omega_{\mathrm{R2,1}}\right) dt \tag{3}$$

Volume function:

$$V_C(\gamma) = l \cdot \frac{\pi}{360^\circ} \cdot \gamma \left(R_{\mathrm{a}}^2 - R_{\mathrm{i}}^2\right) \tag{4}$$

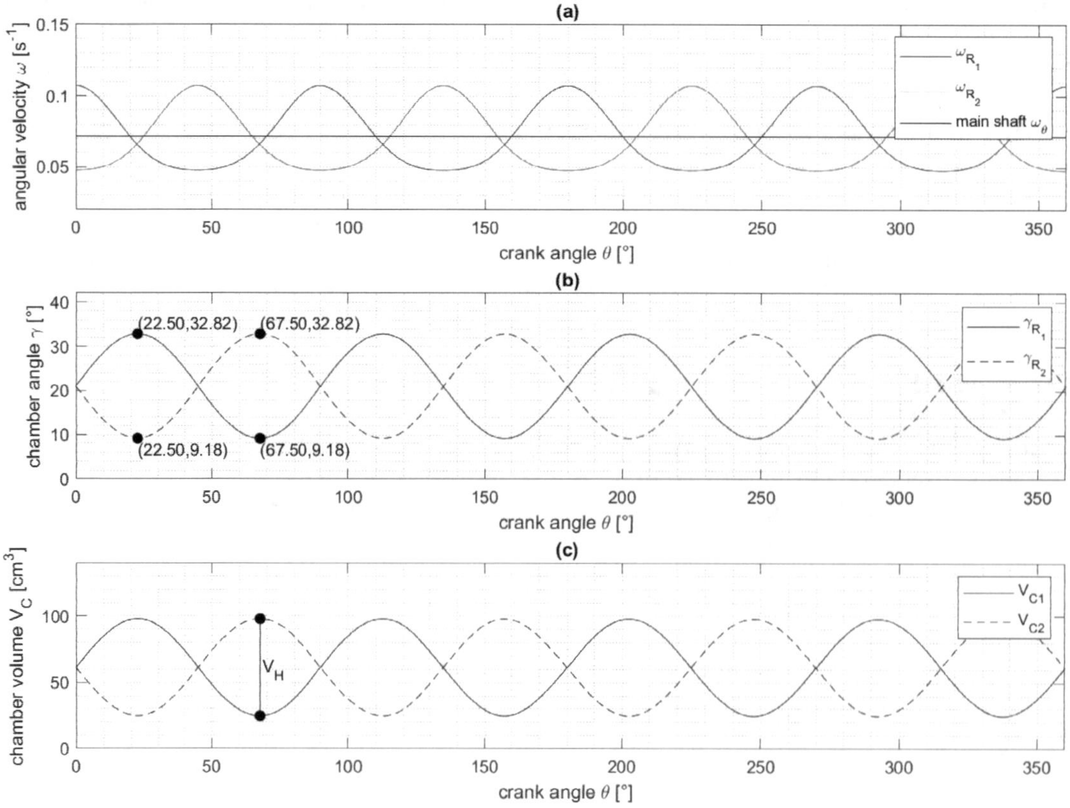

Figure 3. Function of (**a**) angular velocity ω of rotors and main shaft at $\omega_0 = \omega_N = 0.717\ s^{-1}$ =const., (**b**) chamber angle γ with $\gamma_{\max} = 32.82$ ° and $\gamma_{\min} = 9.18$ °.

Table 1. Geometry parameters and dimensions of the present compressor.

Min. chamber angle $\gamma_{\min}$	9.18°	Max. chamber angle $\gamma_{\max}$	32.82°
Min. chamber volume $V_{\min}$	28.46 cm³	Max. chamber volume $V_{\max}$	101.82 cm³
Piston displacement V_H	73.36 cm³	Total vol. turnover per revolution	2350 cm³

2.2. *Kinematic function approaches*

The current kinematic function is showing a minimum chamber angle of $\gamma_{\min}$ =9.18° which is equal to the minimum clearance angle and equivalent to the minimum clearance volume. This is due to the fact that the rotor width and rotor kinematics do not fit together perfectly. However, for a better clearance of the fluid and a possibly better performance of the compressor in general, the goal of developing new rotor kinematics is the reduction of $\gamma_{\min}$while maintaining a high volumetric efficiency with low backflow losses. Of course, it would be possible to set the minimum chamber angle to approximately 0° over a larger rotor width. As a result, however, the volume throughput itself would also decrease.

Therefore, innumerable kinematics can be developed, assuming every rotor motion can be realized. This leads to the question whether a continuous or discontinuous function is better in terms of performance parameters. It is also of interest, if both rotors should rotate with the same function but phase-shifted with periodical behavior so that every chamber fulfills the same movement, or if each rotor should have its own individual function. For each constellation, some functions have been developed, shown and categorized with their fictional name in Figure 4.

	Same function for each rotor, phase-shifted	**Individual function for each rotor**
Continous function	Amplitude, Flattened	Fourier
Discontinous function	Linear, MixLinear	Bestfit

Figure 4. Types and names of the developed kinematic functions.

For the purpose of reducing the minimum clearance angle from $\gamma_{\min} = 9.18°$, all of the developed kinematic functions should indicate a minimum clearance angle of $\gamma_{\min} = 2°$. This is a freely selected minimum clearance angle, since an angle of 0° is avoided due to safety factors, where the rotors could possibly hit each other.

Consequently, thus also increases the maximum chamber angle to $\gamma_{\max} = 40°$, which on the other hand results in a bigger piston displacement V_{H} and an increase of the mass flow rate. For a better understanding on how some of the following functions were determined, the compression cycle is subdivided into the significant angular positions, shown in Figure 5. Both rotors start in the neutral rotor position, defining chamber 1 and 2 with the left and right chamber wall positions $\phi_{\mathrm{C1/2,L}}$ and $\phi_{\mathrm{C1/2,R}}$. By taking a look at the sketch in Figure 5, following associations can be derived

$$\phi_{\mathrm{C1,L}} = \mathrm{Rotor2\_R} \tag{5}$$

$$\phi_{\mathrm{C1,R}} = \mathrm{Rotor1\_L} \tag{6}$$

$$\phi_{\mathrm{C2,L}} = \mathrm{Rotor1\_R} \tag{7}$$

$$\phi_{\mathrm{C2,R}} = \mathrm{Rotor2\_L}\,. \tag{8}$$

The suction begins when the left chamber wall $\phi_{\mathrm{C1/2,L}}$ passes the right inlet wall $\phi_{\mathrm{In,R}}$ and opens the chamber for the filling. The suction process ends when the right chamber wall $\phi_{\mathrm{C1/2,R}}$ passes the left inlet wall $\phi_{\mathrm{In,L}}$ so that the chamber is not being connected with the inlet port. The discharge process on contrary begins when the left chamber wall has $\phi_{\mathrm{C1/2,L}}$ reaches the right outlet wall $\phi_{\mathrm{Out,R}}$, allowing the fluid inside the chamber to leave through the outlet. Likewise, to the suction process, the discharge ends when the right chamber wall $\phi_{\mathrm{C1/2,R}}$ passes the left outlet wall $\phi_{\mathrm{Out,L}}$, closing the chamber for the fluid to exit. With this in mind, several continuous and discontinuous functions can be determined and optimized.

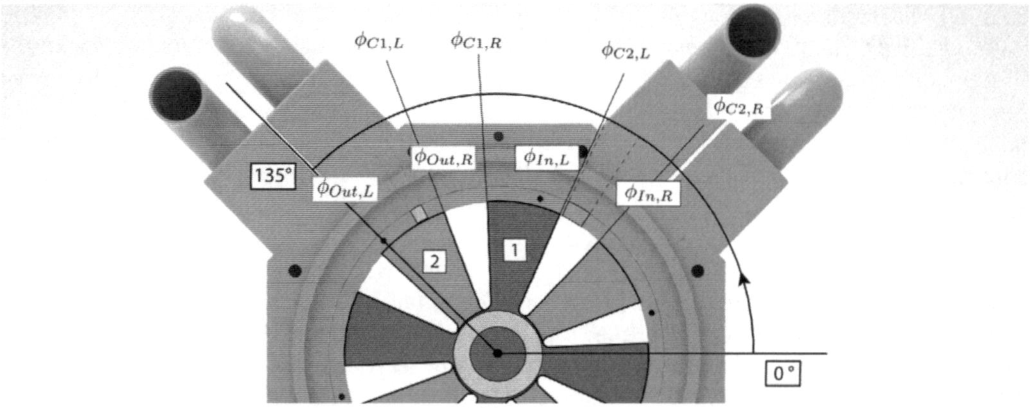

Figure 5. Position of each rotor, inlet and outlet walls in its neutral position for the development of new kinematic functions.

2.2.1. *Continuous functions*

Continuous functions have the property to be defined at every point and undergo no interruptions, jumps, or breaks and can be easily described with mathematical equations. To find and adapt functions for our optimization problem, it is vital to determine the correct coefficients. The easiest way of determining a continuous function for our problem is to take the current angular displacement function and increase the amplitude, which results in a larger amplitude of the volume fluctuation and a smaller minimum chamber angle. In the following, the first developed function is called 'Amplitude' since only the amplitude u_1 for rotor 1 and v_1 for rotor 2 was increased, compared to the current function. The amplitudes were iteratively increased in a numerical scheme until the minimum chamber equals 2°. The angular displacement functions f_1 and f_2 for each chamber wall are shown in eq. 9 and 10.

$$f_1(\theta) = u_1 \cdot \left\{ \operatorname{atan}\left(\frac{\sin\left(\frac{\pi}{45°} \cdot \theta + u_2\right)}{u_3 - \sin\left(\frac{pi}{45°} \cdot \theta + u_2\right)} \right) + \operatorname{atan}\left(\frac{sin\left(\frac{\pi}{45°} \cdot \theta + u_2 + u_4\right)}{u_3 - sin\left(\frac{pi}{45°} \cdot \theta + u_2 + u_4\right)} \right) \right\} \tag{9}$$

$$f_2(\theta) = v_1 \cdot \left\{ \operatorname{atan}\left(\frac{\sin\left(\frac{\pi}{45°} \cdot \theta + v_2\right)}{v_3 - \sin\left(\frac{pi}{45°} \cdot \theta + v_2\right)} \right) + \operatorname{atan}\left(\frac{sin\left(\frac{\pi}{45°} \cdot \theta + v_2 + v_4\right)}{v_3 - sin\left(\frac{pi}{45°} \cdot \theta + v_2 + v_4\right)} \right) \right\} \tag{10}$$

Furthermore, another continuous function was developed by the superimposition of various functions showing some flattening behaviour and thus called 'Flattened', where both rotors also have the same kinematic but phase-shifted (see eq. 11 and 12).

$$f_1(\theta) = z_1 \cdot \operatorname{atan}\left(z_2 \cdot \sin\left(\frac{\pi}{45°} \cdot \theta\right)\right) \tag{11}$$

$$f_2(\theta) = -z_1 \cdot \operatorname{atan}\left(z_2 \cdot \sin\left(\frac{\pi}{45°} \cdot \theta\right)\right) \tag{12}$$

One continuous function, where each rotor has his own individual kinematic function, has been developed and is named 'Fourier', inspired by the Fourier series.

$$f_1(\theta) = w_1 \cdot \sin\left(\frac{\pi}{45°} \cdot \theta\right) + w_2 \cdot \sin\left(2 \cdot \frac{\pi}{45°} \cdot \theta\right) \tag{13}$$

$$f_2(\theta) = -w_1 \cdot \sin\left(\frac{\pi}{45°} \cdot \theta\right) - w_2 \cdot \sin\left(2 \cdot \frac{\pi}{45°} \cdot \theta\right) \tag{14}$$

The coefficients z_1 and z_2 and w_1 and w_2 were also determined iteratively by a numerical scheme, where both values were iterated to match the minimum clearance angle of 2°. The only difference between them both is, that during the determination of the coefficients for the Fourier function only one chamber was considered, and the other chamber was being neglected according to the suction and discharge process shown in Figure 5.

2.2.2. *Discontinuous function*

Discontinuous functions indicate breaks, jumps and abrupt changes in their behaviour, mostly being described with mathematical equations for a given definition range. In this study, three discontinuous functions have been developed which indicate linear behaviour and holding phases during the rotor movement. The functions are called 'Linear', 'MixLinear' and 'BestFit' and were determined according to Figure 6. They represent the case that the transmission gears are able to perform the discontinuous functions.

At first, a best-guess function is used as the input for the optimization. From that, each timestep is obtained, where the left and right inlet and outlet walls are passed by each chamber wall. Subsequently, those are the characteristic timemarks for which the discontinuous functions will be fitted in. If the right chamber wall passed the right outlet wall, the leading rotor with the left chamber wall starts to stop in order to reach the minimum chamber angle as soon as possible. If the minimum chamber angle is reached, a holding phase occurs where the chamber angle is kept constant and is not rising until the right chamber wall has passed the left outlet wall. This is due to suppress backflows back into the chamber. If the discharge process is passed, the suction can continue. The same analogy can be adopted for the suction step, where a holding phase can be established when the maximum chamber angle is reached.

The function 'BestFit' only tries to optimize one chamber movement and neglects the other one, which results in two individual kinematic functions for each rotor. 'Linear' on the other hand tries to match both holding phases, where it assures that both rotors have the same kinematics, but their rotating phase is shifted. Due to the abrupt changes in the 'Linear' and 'BestFit' function, 'MixLinear' tries to smooth the curvature course, by using a smoother acceleration movement of both rotors at the start of the suction and compression phase.

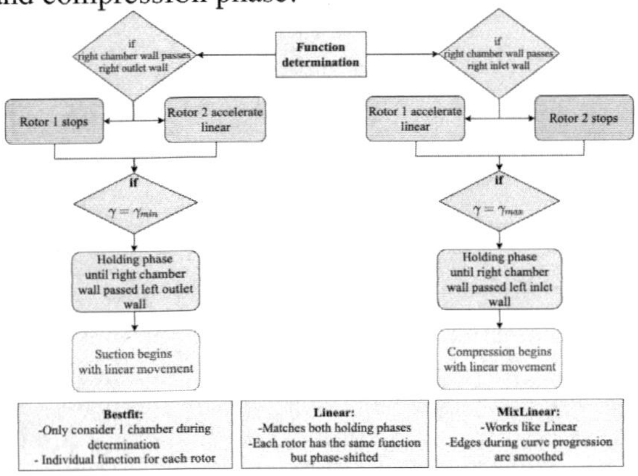

Figure 6. Flow diagram on how the discontinuous function were determined.

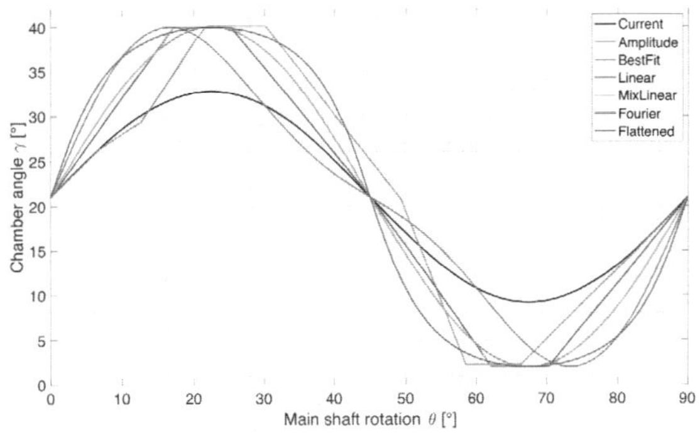

Figure 7. Chamber angle course of the developed functions during one working cycle.

Figure 7 shows the resulting chamber angles from the continuous and discontinuous functions. As shown, each function beside the current function is matching the criteria of a minimum chamber angle of $\gamma_{\text{min}} = 2°$ and a maximum chamber angle of $\gamma_{\text{max}} = 40°$. All of these functions show a quite similar behavior compared to each other except for the current function.

3. Solution Procedure

As mentioned, each function will be tested in a CFD parametric study to consider effects on the momentum of the gas. To achieve reliable numerical results, a fine mesh and time resolution is needed side by side with an adequate setup of boundary conditions. In order to calculate one full revolution of the main shaft, 2000 timesteps for each function are necessary to fulfil the convergence criteria of RMS 1E-4. The volume fluctuation and change at each timestep require a fine resolution and are modelled with the moving mesh model in CFX. Due to the limited computing time, a hexa-mesh with 4,704,593 nodes and 7,934,527 elements is used, which was proven to be accurate enough in a preliminary mesh study. The leakage flow through the annular gap between the rotor blades and the housing is also taken into account with an annular gap size of 0.1mm.

The chosen boundary conditions are derived from the application as a compressor in an air refrigeration cycle on a laboratory scale, shown in Table 2. The SST-turbulence model is chosen, since it provides accurate results for a wide range of applications and delivers a robust scheme. The characteristic movement of each rotor is described with specified angular displacement functions (presented in section 2.2) of the left and right chamber wall by using the CFX expression language (CEL).

Table 2: Boundary conditions used in simulation.

Boundary condition	Inlet	Outlet
Temperature Inlet	300.15 K	308.15 K
Pressure Inlet	1.0 bar	1.8 bar
Fluid	Air, ideal gas	
Rotational Speed	1200 rev/min	

4. Results and Discussion

The results of the kinematic study are shown in Figure 8 and Table 3. The comparison of the volumetric efficiency λ and mass flow rate MFR between each function is presented in Figure 8 (a) and (b). Both graphs intend to show, that the current function has the lowest volumetric efficiency and mass flow rate, having an efficiency of only $\lambda = 85{,}3\%$ and a mass flow rate of 0.047 kg/s. The mass flow rate of every examined function was increased to 0.08-0.085 kg/s, while the volumetric efficiency increases up to 5-10% compared to the current function. The main reason for the increasement of the mass flow rate is due to the larger displacement volume. Furthermore, 'Amplitude', 'Linear' and 'MixLinear' function indicate the biggest increase of the volumetric efficiency up to 94%, followed by the 'Flattened' and 'BestFit' function with 91% and 90%.

Figure 8 (c) shows the pressure-volume diagram, where deviations can be identified between the examined functions. Starting with the current function, the area is significantly smaller compared to the examined functions due to the lower displacement volume. However, 'Bestfit' is showing a very small are compared to the other functions which indicate that this function is probably not ideal at all. 'Amplitude' and 'Linear' is showing the largest area due to the large increase in pressure. Figure 8 (d) also shows the pressure behaviour of one chamber during the first revolution. Compared to the current function, 'BestFit' only indicates a slightly higher maximum pressure up to 4.5 bar and an internal pressure ratio of $\Pi = 3.68$. 'Flattened' and 'MixLinear' on the contrary show a quite higher maximum pressure between 7.5-8.5 bar with a pressure ratio of about $\Pi = 6.4$. 'Amplitude' and 'Linear' show a similar behaviour with a maximum pressure at 10 bar and a pressure ratio of over $\Pi = 7$ seen in Table 3. However, for a delivery pressure of 1.8bar, an internal pressure increased to nearly 10 bar results in a very low adiabatic efficiency. Therefore, port timing and design needs to be improved as well as this machine consists of no valves.

It should also be mentioned that the results from the Fourier function were disabled due to convergence problems, showing huge deviations between both chamber pressures and temperatures. The results from the 'BestFit' function also indicates, that individual functions for each rotor do not show any significant improvements compared to the equal functions for both rotors. Therefore, it can be concluded that due to the periodical design of the machine, each rotor should have the same function, but phase shifted.

All of the examined functions do show a significant improvement of the mass flow rate due to the reduction of the minimum clearance angle and also led to a better volumetric efficiency with the same displacement volumes. If we just take a look on the mass flow rate, 'Linear' and 'MixLinear' show the best result, closely followed by 'Amplitude'. However, in terms of pressure ratio, 'Amplitude' is the leading kinematic slightly followed by 'Linear'. Therefore, it can be concluded, that either one of these functions should be considered for the upcoming optimization. Which and how both of these functions are realized depends on the gear design. For cylindrical gears, the 'Amplitude' function is probably the one to choose, where stepping gears could be more suitable for discontinuous functions.

Table 3. Results of the study comparing the performance parameters of each kinematic function.

Performance Parameters	Current	Amplitude	BestFit	Linear	MixLinear	Flattened
Internal Pressure ratio Π [-]	2.72	7.64	3.68	7.13	6.44	6.42
Mass flow rate MFR [kg/s]	0.0473	0.0839	0.0796	0.0842	0.0841	0.0811
Volumetric Efficiency λ [-]	0.853	0.941	0.896	0.944	0.939	0.910
Indicated Work W_{id} [J]	9.015	40.087	23.340	43.368	38.638	34.108

5. Conclusions and future work

A new rotary type compressor named rotary-chamber compressor has been introduced. The working principle is based on two phase-shifted rotors, which are realized by special toothed gears. The construction and working principle are relatively simple and the size easily scalable. Compared to other rotary compressors, this novel compressor has a great potential for the application in refrigeration systems because of its high volume turnover. Due to the operation without oil and active valves systems, the compressor comes with low maintenance, making it sustainable.

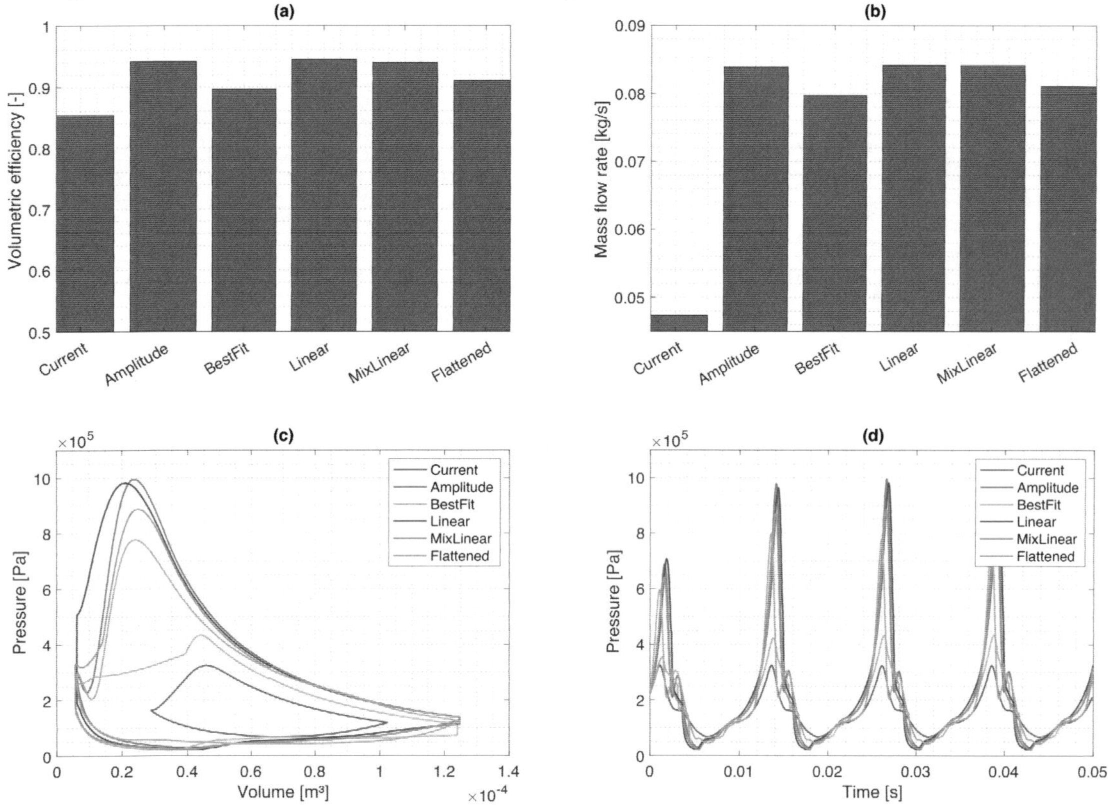

Figure 8. Results of the study at 1200 rev/min. **(a)** volumetric efficiency, **(b)** mean mass flow rate,**(c)** pressure-volume diagram, **(d)** pressure behaviour of the examined kinematics.

Different kinematic functions have been developed and tested with a mathematical model using the CFD-solver ANSYS CFX to take effects on the momentum of the gas into account. It was shown that the current kinematic function, obtained by measurements, has potential for improvements. By increasing the amplitude of the function its volumetric efficiency of $\lambda = 85.3\%$ can rise to a value of $\lambda = 94\%$. Also, by reducing the minimum clearance angle in general, all kinematics have shown an efficiency between 91% and 94% while the mass flow rate could nearly be increased to 90%. The highest volumetric efficiency was determined for the 'Amplitude' function slightly followed by the 'Linear' function. However, it is also in question which kinematic can be realized in the future since each function requires a change of the gear. Possible gear-types could be stepping or cylindrical gears. It can also be concluded that each rotor should have the same kinematic function but should be phase shifted due to the periodical design of the machine, which also can be seen in the performance parameters.

Additionally, it is not clear if changing one parameter in the 'Amplitude' is the optimum solution or if additional coefficients need to be changed as well to achieve a higher volumetric efficiency. Therefore, the next work is focusing on optimizing further parameters of the 'Amplitude' kinematic. Since the machine is still in its development phase, additional optimization possibilities do exist like determining the optimum rotor blade geometry or improving the sealing. Even though the results show great potential in the use for air refrigeration processes, the experimental validation of the presented method needs to be done before further analysis is carried out.

Acknowledgments
The work was supported by the North-German Supercomputing Alliance (HLRN). We are grateful to the HLRN supercomputer staff for their support, effort and the opportunity to use HPC ressources.

Nomenclature

CEL	CFX Expression Language	(–)
MFR	Mass flow rate	$(kg \cdot s^{-1})$
SST	Shear Stress Transport	(–)
λ	Volumetric Efficiency	(–)
ω	Angular velocity	(s^{-1})
θ	Main shaft rotation	(°)
γ	Chamber angle	(°)
Π	Internal pressure ratio	(–)
ϕ	Angular position	(°)
f	Angular displacement	(°)
k_{b}	Constant variable	(s^{-1})
W	Work	(J)

References

[1] Groll EA. Compressor technology. *Int J Refrig* 2013; 36: 1793–1795.
[2] Bradshaw CR, Groll EA. A comprehensive model of a novel rotating spool compressor. *Int J Refrig* 2013; 36: 1974–1981.
[3] Yap KS, Ooi KT, Chakraborty A. Analysis of the novel cross vane expander-compressor: Mathematical modelling and experimental study. *Energy* 2018; 145: 626–637.
[4] Na SK, Lee TM, Choi GM. Efficiency analysis of the novel twin chamber rotary compressor. *IOP Conf Ser Mater Sci Eng* 2019; 604: 012066.
[5] Yusheng Hu, Jia Xu, Zhongcheng Du. Study of Novel Rotary Cylinder Compressor. In: *International Compressor Engineering Conference*. Purdue, https://docs.lib.purdue.edu/icec/2602 (2018).
[6] Hu YS, Wei HJ, Xu J, et al. The Theoretical and Experimental Research of a Novel Rotary Cylinder Compressor. *IOP Conf Ser Mater Sci Eng* 2019; 604: 012073.
[7] Zhang Y, Madamedon M, Copeland C, et al. The roticulating concept air compressor: experimental and numerical investigation. *IOP Conf Ser Mater Sci Eng* 2019; 604: 012070.
[8] Schukey J. *Drehkolbenmaschine*. EP 0 316 346 B1, 1990.
[9] Cui B, Becker K, Gottschlich M, et al. Parameterstudie eines neuartigen Schukeyrotationsverdichter. Hannover: DKV e.V., 2020.
[10] P. N. Pandeya. Analysis of the Effect of the Compressor Kinematics on the Valve Losses. In: *International Compressor Engineering Conference*. Purdue, https://docs.lib.purdue.edu/icec/433 (1984).
[11] Beard JE. *Kinematic analysis of gerotor type pumps, engines, and compressors*. Ph.D., Purdue University, https://www.proquest.com/dissertations-theses/kinematic-analysis-gerotor-type-pumps-engines/docview/303393643/se-2?accountid=10843 (1985).
[12] R. S. Leemhuis, W. Soedel. Kinematics of Wankel Compressors (or Engines) by Way of Vector Loops. In: *International Compressor Engineering Conference*. Purdue, https://docs.lib.purdue.edu/icec/228 (1976).
[13] Yasutaka Ito, Hitoshi Hattori, Kazuhiko Miura. Numerical Analysis for Rotating Motion of a Rolling Piston in Rotary Compressors - Effective Factors for Characteristics of Rotating Motion of a Rolling Piston -. In: *International Compressor Engineering Conference*. Purdue, https://docs.lib.purdue.edu/icec/1963 (2010).

IOP Conf. Series: Materials Science and Engineering 1180 (2021) 012057 doi:10.1088/1757-899X/1180/1/012057

Design Improvements of Vane Bearing Compressor

Y S Hu[1,3], H J Wei[1,2,3], J Xu[1,2,3], P K Wan[1,2,3*], L P Ren[1,2,3] and Z Fu[3]

[1]State Key Laboratory of Air-conditioning Equipment and System Energy Conservation, Zhuhai, Guangdong 519070, China
[2]Guangdong Key Laboratory of Refrigeration Equipment and Energy Conservation Technology, Zhuhai, Guangdong 519070, China
[3]Gree Electric Appliances, Inc., Zhuhai, Guangdong 519070, China
*Correspondence: wpk_1219@163.com

Abstract. A variant of vane bearing compressor with double rolling bearing had been introduced two years ago (Hu and Xu, 2019). In order to solve the problems of complex assembly process and abnormal wear in the vane bearing compressor with double rolling bearing, this paper presents a type of improved vane bearing compressor with single rolling bearing. The improved vane bearing compressor integrates the rolling bearings and the cylinder into a rolling bearing cylinder unit. The suction t and discharge ports are placed on the end face of the main bearing. With these improvements, the number of the end face friction pairs of rolling bearing inner ring is reduced by half, the number of parts is reduced accordingly, and the assembly process is thus made simpler than that with double rolling bearing. This paper introduces the operational characteristics and makes a series of structure optimization to improve the performance of the improved vane bearing compressor with single rolling bearing. The energy efficiency of the improved vane bearing compressor with single rolling bearing is found to be 5% higher than that with the double rolling bearing.

1. Introduction

With the advantages of simple structure, easy processing, self-balance of rotor, low noise and vibration, lower torque fluctuation, low cost of manufacturing and so on, vane compressor is widely used in the fields of air compression, vacuuming, low capacity refrigeration, air conditioning equipment and automotive air conditioning.

The severe mechanical friction between the vane tip and the inner wall of the cylinder is the main disadvantage of the conventional vane compressor. In order to mitigate this problem, the optimal back pressure of the vane had been discussed and a segmented throttling back pressure chamber had been introduced [1]. Ma et al. pointed out that the frictional loss of vane tip accounts for 87.1% of the total mechanical loss of the compressor [2]. A large number of research results indicate that with the action of the appropriate back pressure which is acting on the back of the vane, the contact force between the vane tip and the inner wall of cylinder could be minimized, the friction loss can thus be reduced effectively and the performance of compressor can be improved [3-5]. Guo et al. set three back pressure states, and showed that constant back pressure is more conducive to improve the motion characteristics and the performance of compressor [6]. The above studies have focused on how to reduce the power loss at the vane tip. Therefore, the frictional loss and the reliability at the rubbing region between vane tip and cylinder are the bottlenecks limiting the wider application of the compressor.

A variant of the vane bearing compressor with double rolling bearing (VBC-D) had been introduced in 2019 (Hu and Xu, 2019) [7]. With the action of the double rolling bearing, the sliding friction between vane tip and cylinder inner wall has been translated into rolling friction. The inner ring of bearing rotates together with vane under the action of friction, which reduces the relative speed of the vane tip and the inner wall of cylinder. Therefore, the frictional power loss at the vane tip reduced. However, this brings other complications such as (i) a complex assembly process, and (ii) an unreasonably small clearance control, which is prone to abnormal wear.

In order to solve the above two issues (i) and (ii), this paper presents a new variant of vane bearing compressor with single rolling bearing (VBC-S). The VBC-S adopts a single rolling bearing structure and the suction as well as the discharge ports are placed on the end face of the main bearing. With these design changes, the rubbing regions at the end face is reduced by half, the number of parts is also reduced, and the assembly process has also been made simpler.

2. Basic structure

Figure 1 shows the pump structure for VBC-D [7]. It is composed of a main bearing, a cylinder, a sub bearing, a shaft, two rolling bearings and three vanes. The vane tip is in contact with the inner surface of the inner rings for the two rolling bearing and leaves a gap with the inner wall of the cylinder. During the operation of the compressor, the frictional force of the vane tip at the inner ring of the rolling bearing drives and rotates the rolling bearing. Therefore, the vane tip does not in contact with the cylinder and hence no friction loss between them.

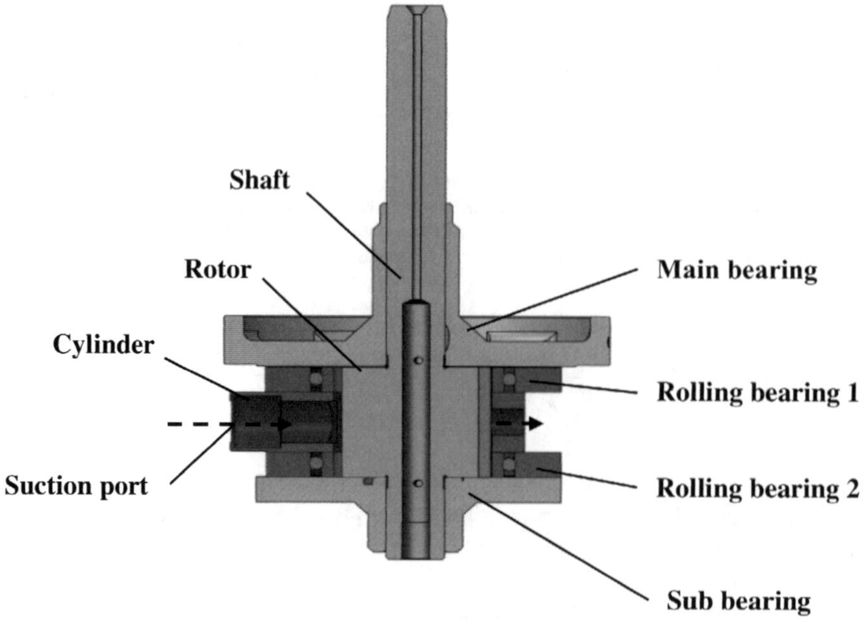

Figure 1. The pump body structure of the VBC-D

Figure 2 shows the pump structure for VBC-S. It is composed of a main bearing, a bearing cylinder, a sub bearing, a shaft, and three vanes. The rolling bearings and the cylinder are integrated into a rolling bearing cylinder unit whose inner ring rotate freely. The friction between the vane tip and the inner ring of the rolling bearing cylinder rotates the rolling bearing.

Due to the cylinder subassembly is replaced by a bearing cylinder unit, the suction and discharge ports will have to be repositioned.

As shown in Figure 2, a suction port begins with a circular channel located radially at the main bearing radially and it connects to the crescent shaped suction port which is located at the end face of the main bearing.

The discharge port is located at the end face of the main bearing. The sliding vane sweeps over the discharge port during the discharge process. If the width of the discharge port is too large, the sliding vane is not wide enough to cover the discharge port completely. When this situation occurs, the front chamber of the slide vane will be communicated with the rear chamber, resulting in an undesirable internal leakage. If however, the discharge port is too small, it will cause a serious over compression and results in significant discharge loss. Therefore, the shape of the exhaust port is designed to be the shape like rhombus to ensure the area of discharge, and the side width is less than the thickness of the sliding vane to avoid internal leakage.

In the end of the discharge process, the gap of the working chamber gets narrower. The end face cross-sectional area of this gap is too narrow for the discharge port. Therefore, this paper presents an annular discharge "channel" for the end stage of the discharge. The annular discharge "channel" is composed of a U-shaped groove located on the side of the vane slot and an annular groove is located at the main bearing. With the compressor rotating, the U-shaped groove on the rotor will connect to the annular groove of the main bearing only the end of discharge process. Thus, the high-pressure gas is discharged from the pump through the U-groove of the rotor and the annular groove of the main bearing.

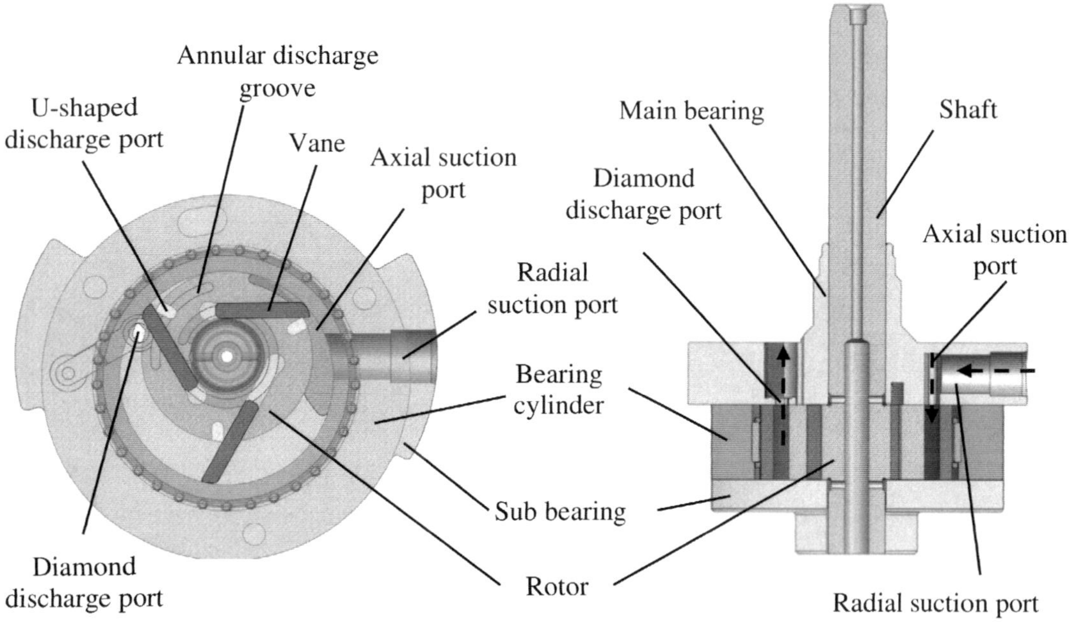

Figure 2. The pump structure of the VBC-S

3. Contrastive analysis between the VBC-D and the VBC-S

3.1 Quantity of component

The VBC-S which is proposed in this paper combines the cylinder, upper rolling bearing and lower rolling bearing into one bearing cylinder, which has two fewer parts than the VBC-D.

Compare with the VBC-D, there is no need to assemble the upper rolling bearing and the lower rolling bearing with the cylinder first in the single rolling bearing scheme and four assembly screws can be reduced. Therefore, the number of pump parts in VBC-S is reduced by six, and the actual assembly efficiency is increased by more than 30%.

3.2 Structure analysis

The suction port of the VBC-D is set in the radial direction of the cylinder. In order to meet the requirements of the suction port size, the cylinder must ensure a certain thickness. At the same time, only the inner rings of the upper rolling bearing and the lower rolling bearing are in contact with the sliding vane tip during the operation, but a certain contact area is needed to ensure the bearing capacity.

So the bearing also needs to ensure a certain height, which eventually leads to a higher overall height of the cylinder assembly.

For the VBC-S, there is no suction port and exhaust port in the radial direction of the bearing cylinder. The inner ring of the bearing cylinder is in contact with the sliding vane tip in the whole height direction to ensure the bearing capacity. Therefore, the height of the bearing cylinder can be appropriately reduced.

3.3 Mechanical loss analysis in theoretical

Based on the structural characteristics of VBC-D and VBC-S, this paper aims to compare the difference of friction loss caused by inner ring. According to the double rolling bearing scheme of VBC-D [7], there are four friction pairs between two inner rings and upper rolling bearing, lower rolling bearing, cylinder. But for the case of the single rolling bearing scheme of VBC-S, there is only one inner ring, hence the number of friction pairs is reduced to two.

Assuming viscous friction, the loss calculation of one friction pair [8] is given in Equation (1).

$$W' = \frac{\pi\eta(R_o^4 - R_i^4)\omega_{ir}^2}{\delta_{im}} \quad (1)$$

where η is the dynamic viscosity of lubricating oil, R_o is the outer radius of inner ring, R_i is the inner radius of inner ring, ω_{ir} is the angular velocity of inner ring, δ_{im} is the oil film thickness within the inner ring and the main bearing.

Under specific working conditions, this paper compares theoretical mechanical loss for VBC-D and VBC-S with the same design parameters which are shown in Table 1.

The operational conditions and related parameters are given in Table 2.

Table 3 shows that the frictional loss of the VBC-S is 81.8W lower than that of the VBC-D.

Table 1. The related design parameters

Parameters	values
η (mm^2/s)	8.31
R_o (mm)	70
R_i (mm)	60
δ_{im} (mm)	0.02-0.03

Table 2. Working conditions and related parameters

Working condition	Related parameter
Refrigerant	R410A
Condensation Temperature (°C)	54.4
Evaporation Temperature (°C)	7.2
Suction Temperature (°C)	35
Rotation Speed (r/min)	3600

Table 3. Comparison of the frictional regions and the mechanical loss

Scheme	Number of Frictional region	Predicted Mechanical loss (W)
VBC-D	4	163.6
VBC-S	2	81.8

4. Performance optimization of VBC-S

4.1 Pressure loss

This measured variation of the working chamber pressure of VBC-S is shown in Figure 3.

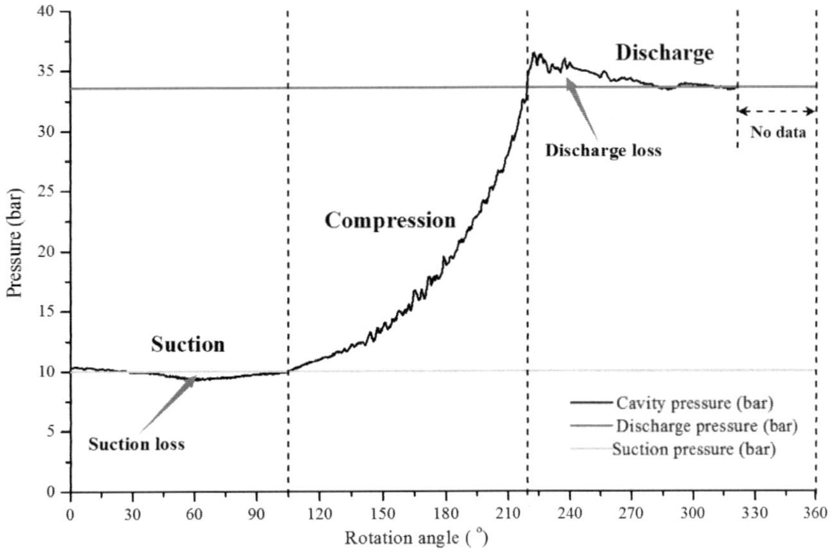

Figure 3. Variations of the working chamber pressure of VBC-S

As shown in Figure 3, there appears to have some suction loss during the suction process.

At the end of the suction process, the working chamber breaks away from the suction port and the compression process begins. The chamber pressure increases as the working chamber volume reduces. When the chamber pressure reaches that of the set discharge pressure, the discharge valve opens and the discharge process begins. Figure 3 shows that there is discharge loss indicated by over compression during the discharge process.

There is no relevant data from 320°~360° angular position due to the cavity space thus this range is too small to accommodate the pressure sensor.

4.2 Two suction structures

In view of the suction loss of the VBC-S in high operating speed, a suction structure is re-designed to locate at the main and sub bearing to increase the suction port area as shown in Figure 4.

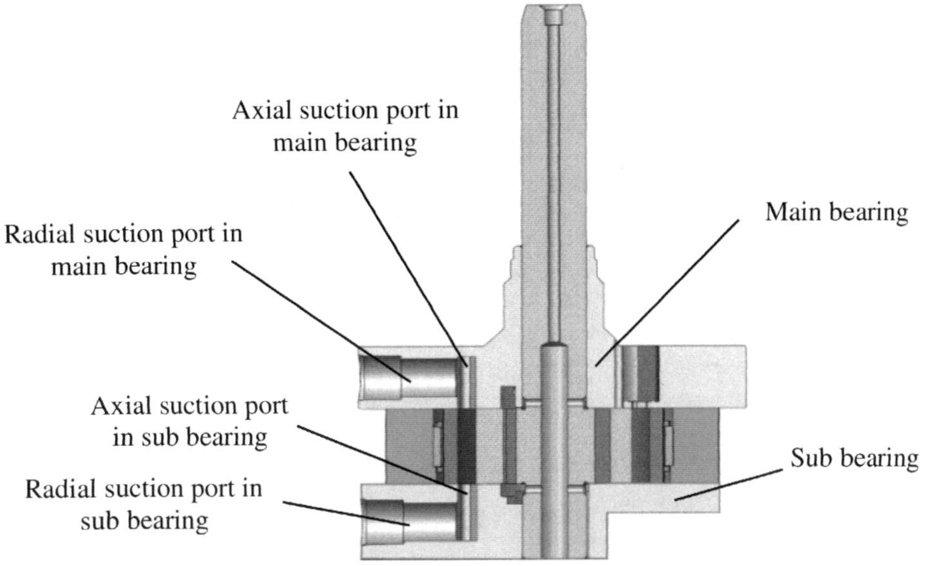

Figure 4. The pump structure of the VBC-S with two suction structures

IOP Conf. Series: Materials Science and Engineering 1180 (2021) 012057 doi:10.1088/1757-899X/1180/1/012057

The test rig schematic of compressor performance is shown in Figure 5, the main components are compressor, condenser, electronic expansion valve, evaporator, and mass-flow meter. There are some temperature and pressure sensors on pipe lines, their measuring positions are also shown in Figure 5.

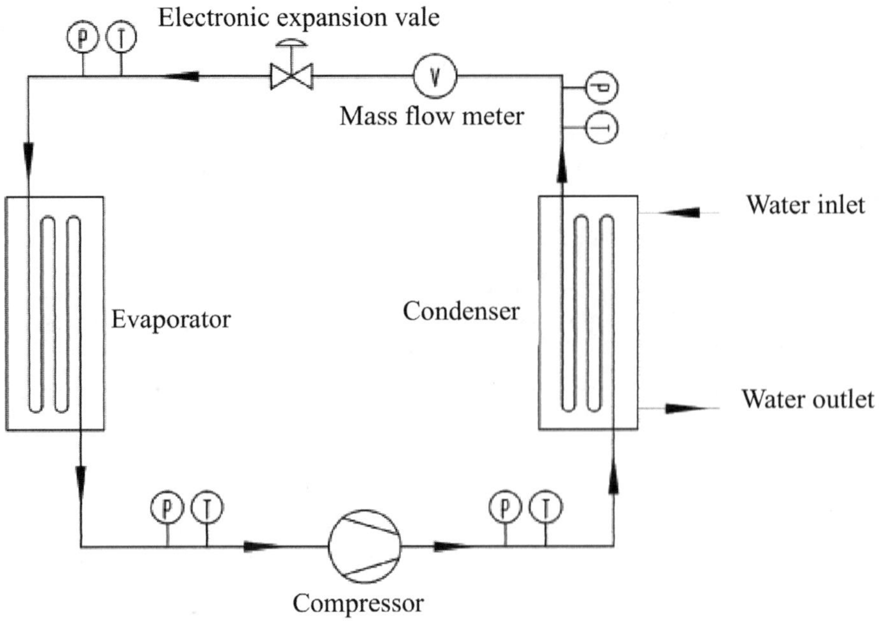

Figure 5. The test rig schematic of compressor performance

According to the measured results shown in Figure 6, as compared with that with only one suction port, the refrigerating capacity of the compressor with two suction ports is increased by 3.57%, the power consumption is increased by 2.28%, and the COP is increased by 1.26%. With the suction area increases, the amount of compressed refrigerant also increases, so the power consumption increases.

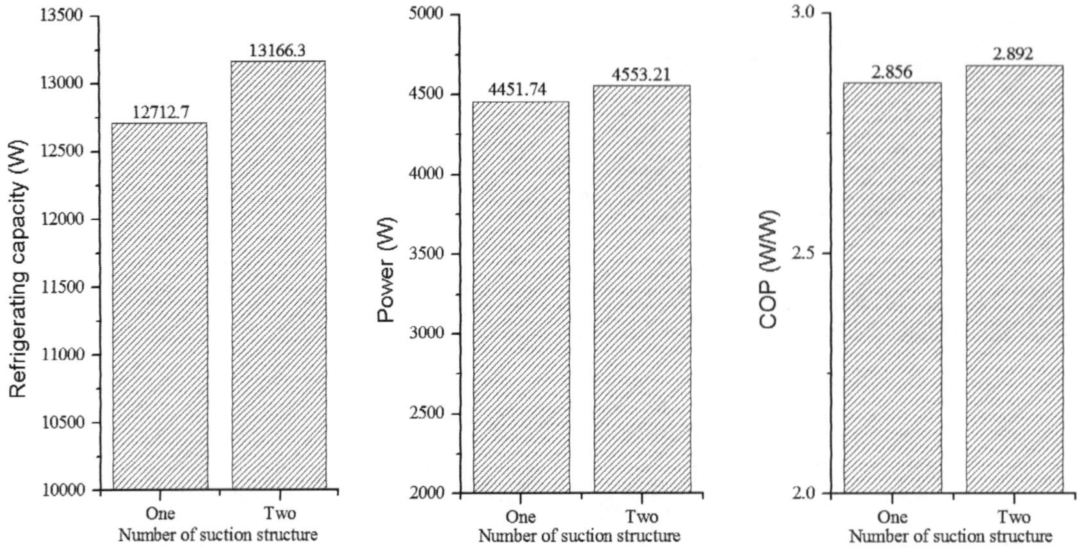

(a). Comparison of the refrigerating capacity (b). Comparison of the power (c). Comparison of the COP

Figure 6. Comparison of the performances in different suction structure

4.3 Two rhombus-shaped discharge ports

As shown in Figure 7, to reduce the over-compression at the initial stage of discharge process, two rhombus-shaped discharge ports are located in the axial direction of the main bearing to increase the discharge port area.

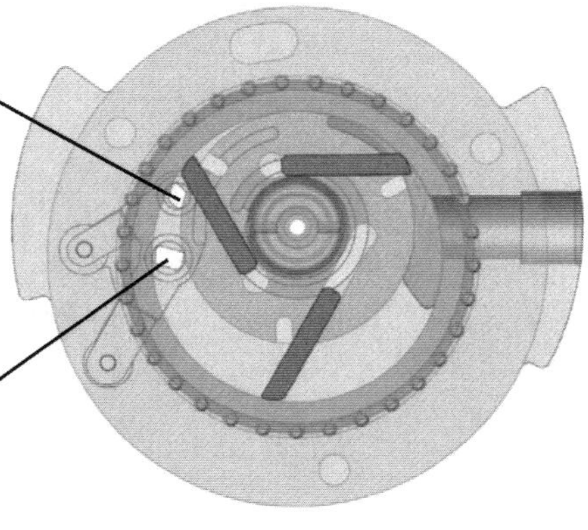

Figure 7. The pump structure of the VBC-S with two discharge ports

Figure 8 shows the comparison of instantaneous discharge velocity at the initial stage of discharge, the results show that the discharge velocity can be reduced by 47% after increasing the discharge area.

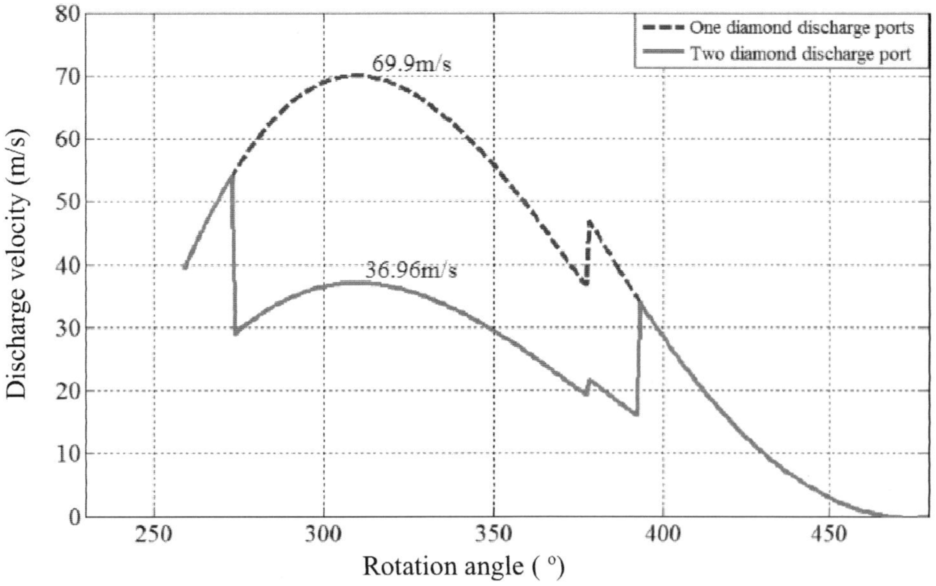

Figure 8. The instantaneous discharge velocity at the initial stage of discharge

According to the results of the performance test which is shown in Figure 9, as compared with the single discharge port, the refrigerating capacity of the compressor with double port is increased by 0.35%, the power consumption is decreased by 1.26%, and the COP is increased by 1.63%. With the reduction in over compression, the pressure difference between the front and rear cavities reduces. At the same time, the internal leakage of the pump decreases and the cooling capacity increases.

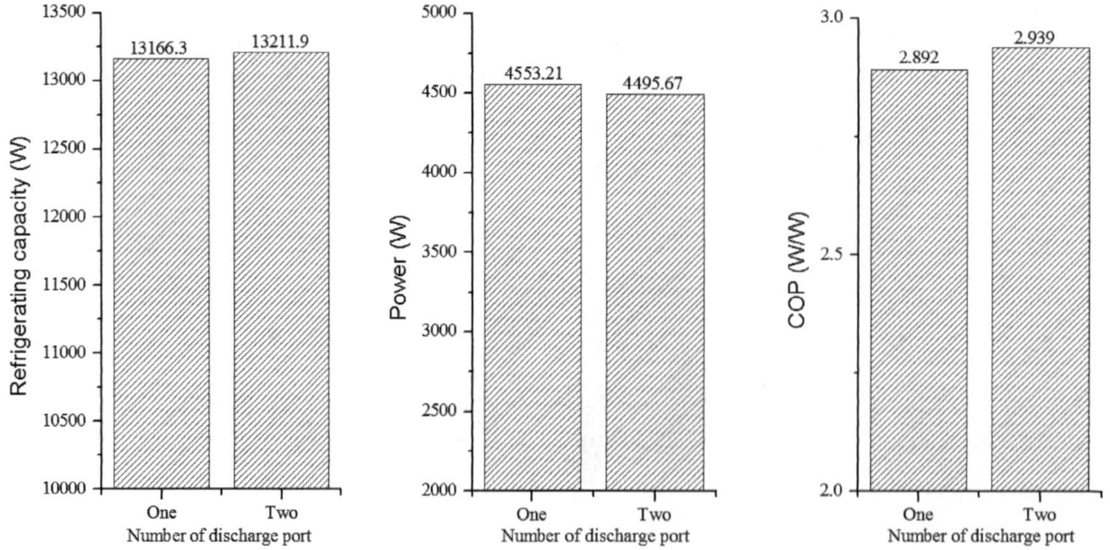

(a). Comparison of the refrigerating capacity (b). Comparison of the power (c). Comparison of the COP

Figure 9. Comparison of the performances in different discharge port number

4.4 Other structural optimization

In order to further improve the performance of the VBC-S, the bearing cylinder height, inner diameter, rotor eccentricity and other parameters are optimized on the basis of double suction and double exhaust ports.

According to the measured results shown in Figure 10, as compared with the VBC-D, the refrigerating capacity of the VBC-S increased by 2.99%, the power consumption reduces by 2.68%, and the COP increased by 5.86%.

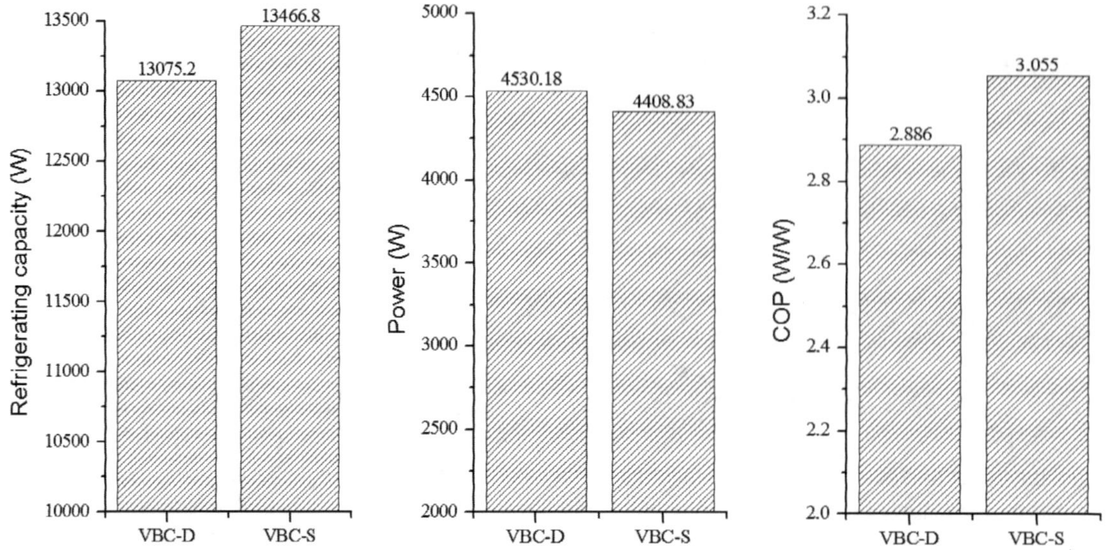

(a). Comparison of the refrigerating capacity (b). Comparison of the power (c). Comparison of the COP

Figure 10. Comparison of the performances between the VBC-S and the VBC-D

5. Reliability

After the performance test of the VBC-D [7], the pump parts were separated. As shown in Figure 11, the inner wall near the suction port of the cylinder had serious wear. The reason may be that the radial load exerted on the inner ring of the bearing by the back pressure of the sliding vane caused the clearance of the bearing at this angle to be basically zero or negative, and causes the vane tip to rub with the inner wall of the cylinder, resulting in large friction and abnormal wear.

Figure 11. The serious wear of the VBC-D cylinder

In the reliability test, VBC-S operated continuously for 300 hours under high load conditions (the evaporation pressure is 8.8 bar, the condensation pressure is 38.7 bar, the suction temperature is 30°C). Figure12 shows the wear of the pump parts after the reliability test of the VBC-S. The results show that there is no obvious wear and scratch, and the reliability is greater than that of the VBC-D.

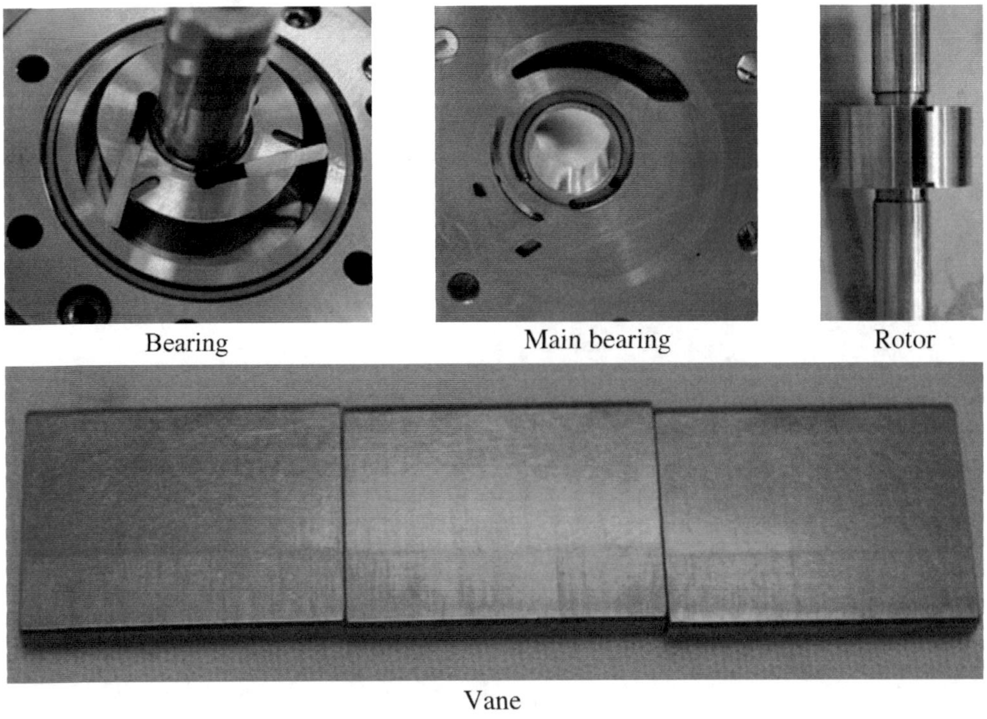

Figure 12. The wear of the VBC-S pump parts

6. Conclusion

This paper introduces an improved high efficiency variant of vane compressor (VBC-S) with a single rolling bearing. When compared to the vane compressor (VBC-D) with double rolling bearings, the following are observed:

(1) The structure of the VBC-S is simpler, the number of parts is reduced, and the height of the bearing cylinder can be adjusted more freely. And the assembly efficiency can be improved by more than 30%.

(2) Through the design of suction and discharge ports, the suction and discharge losses are reduced. After the design optimization, the energy efficiency of VBC-S is 5% higher than that of VBC-D.

(3) The results of reliability test show that under high load condition, VBC-S runs continuously for 300 hours without abnormal wear.

Reference

[1] Wu L Z and Ma G 2003 Analysing of Friction and Wear of Twin-piece Vane Compressors, *Fluid Machinery*, **31(1)**:13–16.

[2] Ma G Y and Li H Q 2001 Rotary Compressor, China Machine Press, p140-166. BeiJing.

[3] Huang Y M and Chii-Haur 1993 Effect of Radial Blades and Port Location on the RCAC System. ASME, **PVP253**:185-191.

[4] Fukuta M 1993 Vane Behaviour in Vane Compressors under Start-up Operation (1st Report, Force Acting on Vane). Nippon Kikai Gakkai Ronbunshu, B Hen/Transaction of the Japan Society Mechanical Engineers, Part B ,**58(567)**:3487-3492.

[5] Fukuta M 1994 Vane Behaviour in Vane Compressors under Start-up Operation (2nd Report, Vane Behavior with Pressure Rise in Back Chamber). Nippon Kikai Gakkai Ronbunshu, B Hen/Transaction of the Japan Society Mechanical Engineers, Part B ,**60(571)**:879-884.

[6] Guo B, Hu J H and Shu P Z 1998 Influence of Vane Slot Back Pressure on the Characteristic of Vane Motion in Rotary Vane Compressor. Chinese Journal of Applied Mechanics. 15(1):65-70.

[7] Hu Y S and Wei H J 2019 A Theoretical Study on the Novel Structure of Vane Compressor for High Efficiency. IOP Conference Series: Materials Science and Engineering 604 (2019) 012067

[8] Wen S Z and Huang P 2002 The fundamentals of tribology (2nd edition). Tsinghua Publishing House. BeiJing.

IOP Conf. Series: Materials Science and Engineering 1180 (2021) 012058 doi:10.1088/1757-899X/1180/1/012058

Current limits of CO2 compressors working in integrated mechanical subcooling cycles

Laura Nebot-Andrés*, Daniel Sánchez, Ramón Cabello, Daniel Calleja-Anta, Carlos Fossi, Rodrigo Llopis

Thermal Engineering Group, Mechanical Engineering and Construction Department, Jaume I University, Spain

* lnebot@uji.es, +34 964 718133

Abstract. CO_2 systems used in refrigeration are becoming more complex with the aim of improving their energy performance. The integrated mechanical subcooling system is being studied to enhance the COP and the cooling capacity of the plants as one of the possible solutions for hot climates in the last years.

Last studies determined their optimum working conditions where the COP is maximum. However, the compressor of the integrated mechanical subcooling cycle works out of the range of operation in some cases. This work presents the main technical limitations of this compressor in an experimental plant.

The main objective is to determine what these limits are and to highlight the need to develop new compressors for these applications. Improving the operation of these compressors will further increase the energy performance of the overall system.

1. Introduction

Carbon dioxide refrigeration systems have been the target of many of the scientific investigations in recent years, with the aim of improving classical CO_2 systems to make them more competent, especially in hot climates. All of these efforts have been driven by the F-Gas Regulation [1] which limits the use of high GWP refrigerants in many of the refrigeration applications. It is necessary to look for refrigerants with low GWP and for certain applications, CO_2 is the only candidate with low GWP that guarantees safety and is neither flammable nor toxic (A1 ASHRAE classification). Although CO_2 systems are widely used in northern Europe, in areas where the average annual temperature is higher, these systems suffer a significant decrease in performance.

Subcooling methods define a clear line of research that is gaining much weight at the moment [2]. Llopis et al. [3] compiled in a general way the effects of subcooling on CO_2 cycles and reviewed all the existing methods. Initially, with the most basic subcooling methods, such as the internal heat exchanger [4] and later with more complex systems such as dedicated mechanical subcooling DMS [5, 6], which uses a refrigerant different from CO_2 in the auxiliary cycle.

The Integrated Mechanical Subcooling (IMS) cycle, presented in this document, only uses CO_2 as refrigerant. The purpose of this system is to subcool the CO_2 at the gas-cooler outlet thanks to a CO_2 flow extracted from the main cycle and that is expanded, passing through the subcooler and being re-compressed until the inlet of the gas-cooler. The interest of this cycle is that it is simpler than the dedicated one; it has fewer components and works only with CO_2. Like DMS, this subcooling cycle increases the plant's cooling capacity and COP with respect to the base cycle [7].

The IMS system was first proposed by Shapiro's patent [8]. Cecchinato et al. [9] evaluated the system from a theoretical point of view and obtained an increase of 17.3% in energy efficiency in relation to a basic single-stage CO_2 cycle for an evaporation level of -10°C and an outlet temperature of gas-cooler at 30°C. Later, Catalán-Gil et al., [7] analyzed the thermodynamic models of the IMS and the DMS for supermarket applications, achieving annual reductions in energy consumption between 2.9% and 3.4% for warm areas and between 1.3% and 2.4% for temperate regions. Nebot-Andrés et al. [10] study the IMS system from a theoretical approach, optimizing the gas-cooler pressure and degree of subcooling, reaching increases of 15.9% for −10ºC of evaporation temperature and 35ºC of ambient temperature compared to the CO_2 cycle with intermediate exchanger [11].

Integrated mechanical subcooling represents an important interest for the improvement of CO_2 cycles, since it has great potential for improvement. This cycle was experimentally tested by Nebot-Andrés et al. [12] and its optimal conditions have been determined.

The experimental results showed that the pressure conditions and degree of subcooling for which the maximum COP values were achieved correspond to operating points that are outside the application limits of the IMS compressor. Consequently, this work has been developed to highlight the operating limits of these systems and emphasize the need to develop equipment that adapts to these operating conditions.

2. Experimental plant description

This section presents the experimental installation used to evaluate the optimal conditions of the CO_2 transcritical cycle with the integrated mechanical subcooling system. The most important details of the main components of the cycle are provided and the measurement system used in the plant is described.

2.1. Experimental plant

The scheme of the experimental plant tested in this work is shown in Figure 1 and its real plant in Figure 2. The plant is a CO_2 single-stage transcritical refrigeration system with an integrated mechanical subcooling system extracting gas at the exit of the subcooler. The main single-stage refrigeration cycle uses a semi-hermetic compressor with a displacement of 3.48 $m^3 \cdot h^{-1}$ at 1450 rpm and a nominal power of 4 kW. The expansion is carried out by a double-stage system, composed of an electronic expansion valve (back-pressure) controlling the gas-cooler pressure, a liquid receiver between stages and an electronic expansion valve, working as thermo-static, to control the evaporating process. Evaporator and gas-cooler are brazed plate counter current heat exchangers with exchange surface area of 4.794 m^2 and 1.224 m^2, respectively. The subcooler is situated directly downstream of the gas-cooler. It is a brazed plate heat exchanger with an exchange surface area of 0.850 m^2. It works as evaporator of the mechanical subcooling system and subcools the CO_2 at the exit of the gas-cooler. The mechanical subcooling cycle is driven by a variable speed semi-hermetic compressor with displacement of 1.12 $m^3 \cdot h^{-1}$ at 1450 rpm. The expansion valve of the IMS cycle is electronic, working as thermostatic.

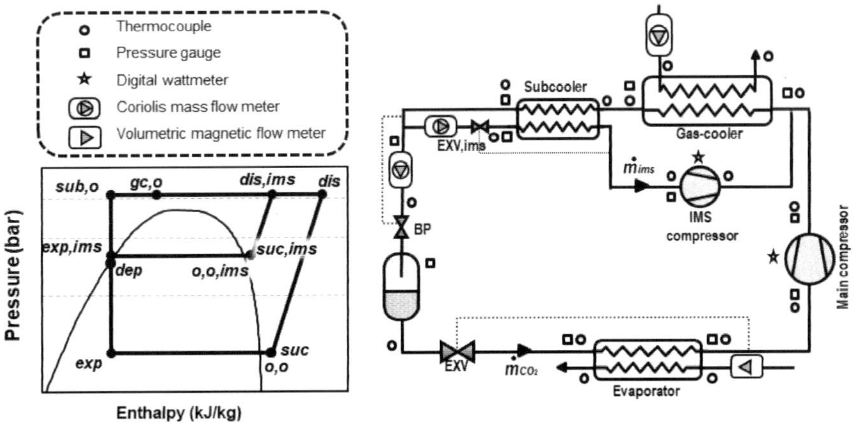

Figure 1. Plant and measurement system diagram.

Heat dissipation in gas-cooler is done with a water loop, simulating the heat rejection level. The evaporator is supplied with another loop, working with a propylene glycol–water mixture (60% by volume) that enables a constant entering temperature in the evaporator. Both the mass flow and the inlet temperature are controlled in these loops.

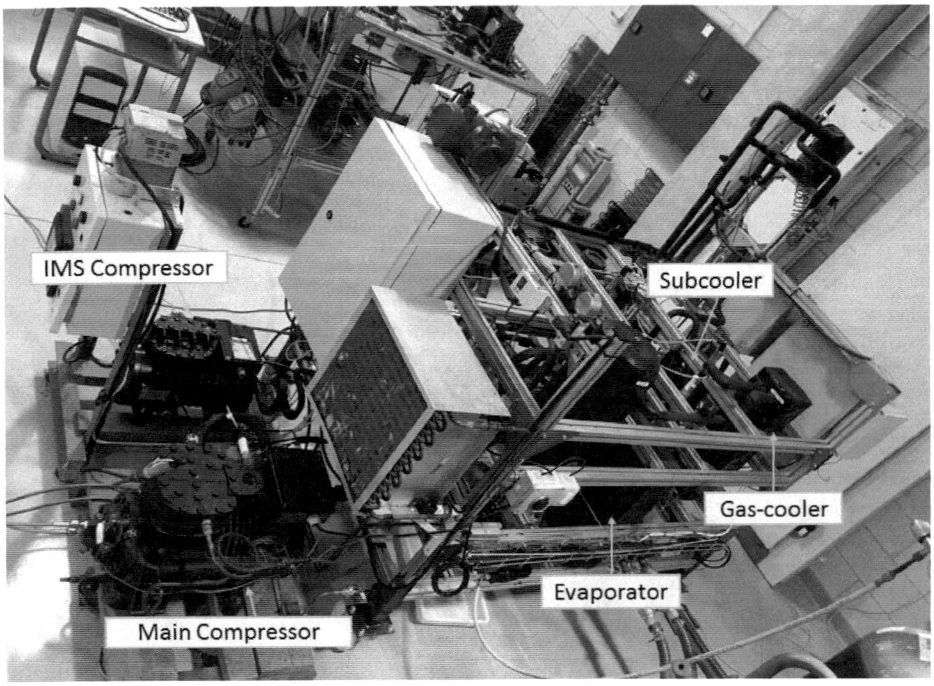

Figure 2. Experimental plant

2.2. Measurement system

The thermodynamic properties of the working fluids are obtained thanks to the measurement system presented in Figure 1. All fluid temperatures are measured by 18 T-type thermocouples. The thermocouples placed at the evaporator and at the exit of gas-cooler and subcooler are immersion thermocouples. 11 pressure gauges are installed along all the circuit. CO2 mass flow rates are measured by two Coriolis mass flow meters, as well as dissipation water flow of the gas-cooler, which is measured using another one. The flow of the other secondary fluids is measured by a magnetic volumetric flow meter. Compressors' power consumptions are measured by two digital wattmeters. The accuracies of the measurement devices are presented in Table 1.

Measured variable	Measurement device	Range	Calibrated accuracy
Temperature (°C)	T-type thermocouple	-40.0 to 145.0	±0.5K
CO_2 pressure (bar)	Pressure gauge	0.0 to 160.0	±0.6% of span
CO_2 pressure (bar)	Pressure gauge	0.0 to 100.0	±0.6% of span
CO_2 pressure (bar)	Pressure gauge	0.0 to 60.0	±0.6% of span
CO_2 main mass flow rate ($kg \cdot s^{-1}$)	Coriolis mass flow meter	0.00 to 1.38	±0.1% of reading
CO_2 IMS mass flow rate ($kg \cdot s^{-1}$)	Coriolis mass flow meter	0.00 to 0.083	±0.1% of reading
Water mass flow rate ($kg \cdot s^{-1}$)	Coriolis mass flow meter	0.00 to 13.88	±0.1% of reading
Glycol volume flow rate ($m^3 \cdot h^{-1}$)	Magnetic flow meter	0.0 to 4.0	±0.25% of reading
Power consumption (kW)	Digital wattemeter	0.0 to 6.0	±0.5% of reading

Table 1. Accuracies of the measurement devices.

2.3. Experimental test procedure

This section describes the strategy carried out to do the experimental tests in order to determine the optimal cycle conditions. Three heat rejection levels and three different evaporation temperatures have been tested.

- All tests are carried out under transcritical conditions.
- Hot sink temperatures: 25.0, 30.4 and 35.1 °C (maximum deviation of ± 0.20 °C). The inlet temperature of the water in the gas-cooler remains fixed at the stipulated values and the flow rate remains fixed at 1.167 $m^3 \cdot h^{-1}$.
- Cold sink temperatures: The propylene glycol-water inlet temperature in the evaporator and the flow rate are maintained. The temperatures tested are -1.3, 3.8 and 10.1°C (± 0.23 °C). The flow is fixed at 0.7 $m^3 \cdot h^{-1}$.
- The pressure in the gas-cooler is regulated with an electronic Back-Pressure thanks to a PID controller. In each test the pressure is varied in order to identify the optimum and reach the maximum COP conditions.
- The main compressor always runs at nominal speed (1450 rpm). The IMS compressor speed varies to modify the subcooling degree.
- The electronic expansion valves were configured to obtain a useful degree of superheating of 10K in the evaporator and 5K in subcooler.

All tests are carried out under stationary conditions of at least 10 minutes, taking data every 5 seconds, obtaining the test point as the average value of the entire test. Thermodynamic properties are determined based on data measured using Refprop v.9.1.[13].

3. Optimization of the plant and main energy results

This section details how the existence of an optimum COP has been determined experimentally, and how the plant has been evaluated to identify the optimal working conditions.

3.1. Optimization process

To obtain the maximum COP, the pressure and the subcooling values are modified following a method similar to a Simplex algorithm. When having three initial points, gas-cooler pressure or subcooling degree were increased or decreased following the trend of the previous points, to get closer and closer to the maximum COP point, until the increases achieved between the new value and the previous one were less than 1%.

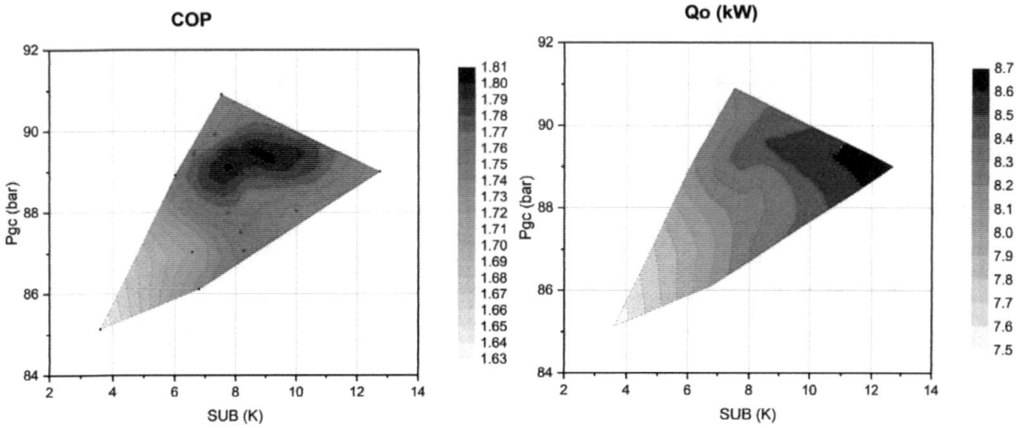

Figure 3. COP and cooling capacity as a function of gas-cooler pressure and subcooling for tw,in=35.1°C and tg,in=10.0°C.

Figure 3 shows the COP and the cooling capacity measured for water inlet conditions of 35.1°C and glycol inlet of 10.0°C, for different pressure levels and degrees of subcooling. The existence of a maximum COP is clearly observed. Regarding the capacity of the plant, it is observed that it is greater the greater the subcooling degree is. It means that the capacity of the plant can be adjusted by modifying the degree of subcooling.

3.2. COP and cooling capacity

The evolution of the COP for all the evaluated conditions is shown in Figure 4 (left) and the cooling capacity that the plant is providing under maximum COP conditions is represented in Figure 4 (right).

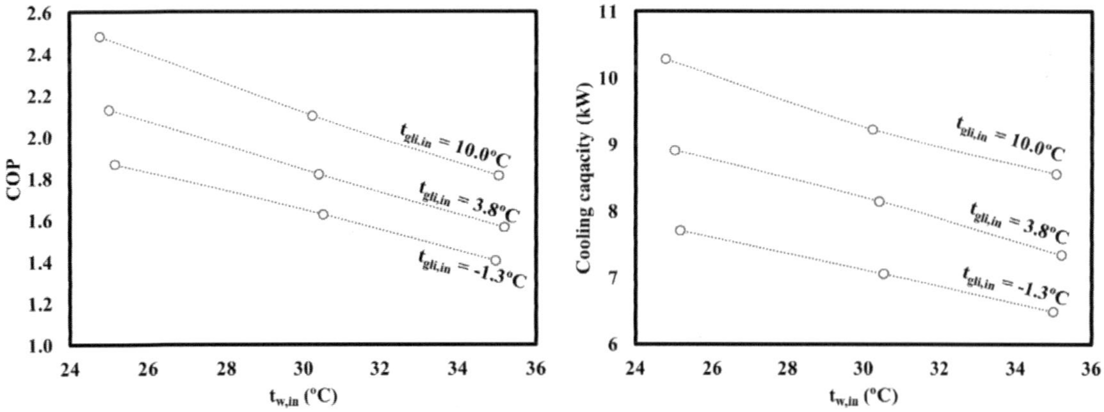

Figure 4. Experimental COP and cooling capacity of the plant.

Regarding the COP, a clear trend in its evolution can be seen, marked by the inlet temperatures of water and glycol. It can be perceived that the COP is lower when the glycol inlet temperature is lower, that is, when the evaporation level is lower. It can also be observed that, as the inlet water temperature increases, the COP decreases. The measured values range from 1.40 to 1.87 for $t_{g,in}$=-1.3°C, from 1.56 to 2.13 for $t_{g,in}$=3.8°C and from 1.81 to 2.48 for $t_{g,in}$=10.0°C.

For the cooling capacity, a linear trend can be observed depending on the hot sink temperature, reducing capacity as the inlet water temperature increases. Likewise, it is clearly seen how the capacity is greater when the evaporation level is greater. The measured values range from 6.5kW to 7.7kW for $t_{g,in}$=-1.3°C, from 7.3kW to 8.9kW for $t_{g,in}$=3.8°C and from 8.6kW to 10.3kW for $t_{g,in}$=10.0°C.

4. Operation of the auxiliary compressor at optimum conditions

As mentioned in the previous sections, this system bases its operation on the use of an auxiliary compressor that recompresses to gas-cooler pressure a part of the CO_2 mass flow that has been evaporated in the subcooler. The operation of this compressor is very particular since it must adapt its rotational speed to obtain a specific subcooling degree for which the plant's COP is maximum.

Table 2 presents the main operating parameters of the IMS compressor for the optimal points that have been obtained experimentally and presented in section 3. The presented parameters are the rotation frequency, the compression ratio, the global and volumetric performances and the suction temperature. As can be seen, the compression ratios of all the tests are very low and specifically less than 1.5 for some test cases (test points 2, 6, 7, 8 and 9). These compression ratios are outside the operating range of the compressor, which must work with a minimum compression ratio of 1.5. Regarding the frequency, we observe that all the optimal points are achieved with very low frequencies, in some cases lower than 30Hz (test points 1, 2, 4, 5, 7 and 9). For this type of plant, a smaller compressor should therefore be implemented, but the one used is the smallest existing of this type of compressors. Regarding the volumetric efficiency, the compressor operates with values between 40 and 50% while the global efficiency's values are better (except point 1) due to the low compression ratio. Finally, another operating parameter that needs to be highlighted is the temperature obtained in the suction of the compressor.

Test point	$t_{gli,in}$ (°C)	$t_{w,in}$ (°C)	τ	η_v	f (Hz)	$t_{suc,ims}$ (°C)	η_g
1	-1.3	25	1.8	0.42	29	8.4	0.52
2	-1.3	30	1.4	0.42	27	21.5	0.75
3	-1.3	35	1.7	0.47	32	20.0	0.76
4	3.8	25	1.7	0.42	28	13.9	0.72
5	3.8	30	1.5	0.44	29	20.3	0.78
6	3.8	35	1.4	0.54	33	26.1	0.79
7	10.0	25	1.4	0.40	25	20.6	0.70
8	10.0	30	1.3	0.51	30	28.0	0.86
9	10.0	35	1.2	0.42	25	33.9	0.91

Table 2. Main operation parameters of the IMS compressor.

In Figure 5 the application limits of this particular compressor in terms of gas-cooler pressure and evaporation temperature can be seen. The experimental data measured in the plant for the IMS compressor are shown in red. As can be clearly seen, although the pressure levels are correct, the evaporation temperature of the IMS cycle is too high, falling outside the compressor operation boundary.

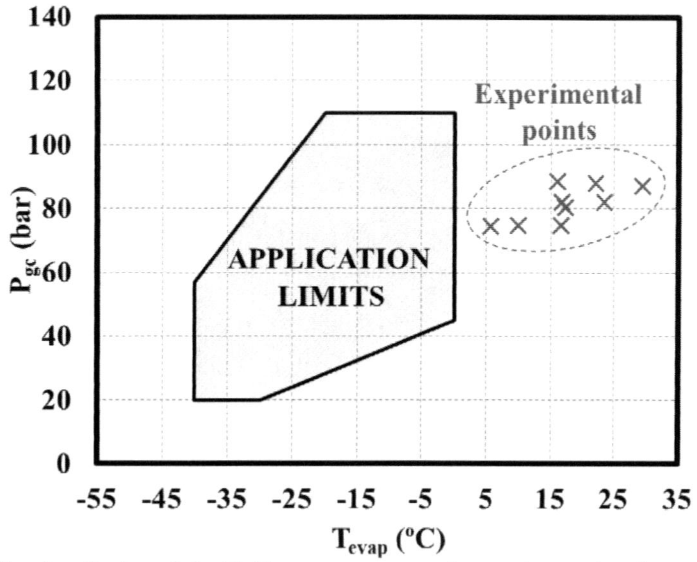

Figure 5. Application limits of the IMS compressor and experimental points of operation.

COP values presented in section 3.2 are quite lower than those obtained in previous theoretical studies [10]. The experimental COP are in average 21.5% lower than the values obtained theoretically. A large part of this reduction may be due to the work of the compressor.

5. Conclusions

This work presents the experimental test of a CO_2 transcritical refrigeration plant with integrated mechanical subcooling working at optimum conditions. The heat rejection levels of 25.0ºC, 30.4ºC and 35.1ºC and the evaporator inlet temperatures of -1.3ºC, 3.8ºC and 10ºC are evaluated under steady-state conditions. The main compressor operates at nominal speed, while the speed of the auxiliary compressor is varied to obtain the optimum degree of subcooling. It is a system with a very high potential for improvement compared to classic CO_2 refrigeration systems.

Experimental tests have demonstrated the existence of a maximum COP, obtained under optimal conditions of pressure and subcooling degree. One of the optimized parameters is controlled by modifying the auxiliary compressor's speed. However, results showed that to reach the optimum COP conditions, the auxiliary compressor must work out of the operability limits, working with high suction temperatures and compression ratios too short.

In conclusion, it would be interesting to develop CO_2 compressors for this application, making the overall system performance even better.

6. Acknowledgements

The authors thank the Ministerio de Ciencia y Tecnología (Spain) project RTI2018-093501-B-C21, the Ministerio de Educación, Cultura y Deporte (Spain), grant FPU16/00151 and the Jaume I University (Spain), project UJI-B2019-56 for financing this research work.

References

[1] European Commission, Regulation (EU) No 517/2014 of the European Parliament and of the Council of 16 April 2014 on fluorinated greenhouse gases and repealing Regulation (EC) No 842/2006., (2014).

[2] B. Yu, J. Yang, D. Wang, J. Shi, J. Chen, An updated review of recent advances on modified technologies in transcritical CO2 refrigeration cycle, Energy, 189 (2019) 116147.
[3] R. Llopis, L. Nebot-Andrés, D. Sánchez, J. Catalán-Gil, R. Cabello, Subcooling methods for CO2 refrigeration cycles: A review, International Journal of Refrigeration, 93 (2018) 85-107.
[4] J. Rigola, N. Ablanque, C.D. Pérez-Segarra, A. Oliva, Numerical simulation and experimental validation of internal heat exchanger influence on CO2 trans-critical cycle performance, International Journal of Refrigeration, 33 (2010) 664-674.
[5] R. Llopis, L. Nebot-Andrés, R. Cabello, D. Sánchez, J. Catalán-Gil, Experimental evaluation of a CO2 transcritical refrigeration plant with dedicated mechanical subcooling, International Journal of Refrigeration, 69 (2016) 361-368.
[6] R. Llopis, R. Cabello, D. Sánchez, E. Torrella, Energy improvements of CO2 transcritical refrigeration cycles using dedicated mechanical subcooling, International Journal of Refrigeration, 55 (2015) 129-141.
[7] J. Catalán-Gil, R. Llopis, D. Sánchez, L. Nebot-Andrés, R. Cabello, Energy analysis of dedicated and integrated mechanical subcooled CO2 boosters for supermarket applications, International Journal of Refrigeration, (2019).
[8] D. Shapiro, Refrigeration system with mechanical subcooling, in: https://patentimages.storage.googleapis.com/56/2a/72/a987f71d7ab41a/US7628027.pdf, Vol. US 7,628,027 B2, United States, 2009.
[9] L. Cecchinato, M. Chiarello, M. Corradi, E. Fornasieri, S. Minetto, P. Stringari, C. Zilio, Thermodynamic analysis of different two-stage transcritical carbon dioxide cycles, International Journal of Refrigeration, 32 (2009) 1058-1067.
[10] L. Nebot-Andrés, D. Calleja-Anta, D. Sánchez, R. Cabello, R. Llopis, Thermodynamic analysis of a CO2 refrigeration cycle with integrated mechanical subcooling, Energies, 13 (2019).
[11] Y. Chen, J. Gu, The optimum high pressure for CO2 transcritical refrigeration systems with internal heat exchangers, International Journal of Refrigeration, 28 (2005) 1238-1249.
[12] L. Nebot-Andrés, J. Catalán-Gil, D. Sánchez, D. Calleja-Anta, R. Cabello, R. Llopis, Experimental determination of the optimum working conditions of a transcritical CO2 refrigeration plant with integrated mechanical subcooling, International Journal of Refrigeration, 113 (2020) 266-275.
[13] E.W. Lemmon, M.L. Huber, M.O. McLinden, REFPROP, NIST Standard Reference Database 23, v.9.1. National Institute of Standards, Gaithersburg, MD, U.S.A., (2013).

IOP Conf. Series: Materials Science and Engineering 1180 (2021) 012059 doi:10.1088/1757-899X/1180/1/012059

Increasing Boiling Fluid Flowing Efficiency from Motive Nozzles of Two-Phase Ejectors

Serhii Sharapov [1], Oleh Chekh[1], Danylo Husiev[1] , Volodymyr Klymenko[2] and Oleksiy Shaparenko[3]

[1]Technical Thermophysics Department, Sumy State University, 2, Rymsky-Korsakov st., 40007 Sumy, Ukraine;
[2] Department of Mathematical Analysis and Optimization Methods, Sumy State University, 2, Rymsky-Korsakov st., 40007 Sumy, Ukraine;
[3] Department of Applied Materials Science and Technology of Structural Materials, Sumy State University, 2, Rymsky-Korsakov st., 40007 Sumy, Ukraine

d.husiev@kttf.sumdu.edu.ua

Abstract. The article contains the possibility of increasing boiling fluid flowing efficiency from expanding channels. This process takes place in the motive flow nozzle of a liquid-vapor ejector, working on the principle of thermal stream compression. Efficiency increasing by profiling the diffuser part of the nozzle. Modern industry uses nozzles, which are like de Laval nozzles, with straight walls of the diffusers. The authors suggest paying closer attention to profiling these nozzles, which might increase their efficiency and improve their gas-dynamic characteristics. For comparison, we choose a channel of a traditional form (with straight walls of the diffuser) and a channel of parabolic shape. The article contains a mathematical model to calculate the process of flowing the boiling fluid from the authors-designed channels – the peculiarities of this model that appear after changing the geometry of its streaming part. We obtain comparative analysis calculation results based on the mathematical model and the Ansys CFX workflow model. As a result of numerical calculation using the authors mathematical model and modelling in the Ansys CFX software package, it concludes that the parabolic shape of the diffuser is the most favourable. In the boiling process, the liquid central core is boiling at the optimum distance from the nozzle throat, and the flow of a stable vapor structure with the required pressure value for each regime forming at the outlet.

1. Introduction

Technological processes under vacuum are increasingly using in modern industry. But, not all processes realizing in atmospheric pressure conditions. Also, with pressure below atmospheric, we obtain a higher quality of final products, for example, in vacuum metallurgy, the amount of impurities in steel that are present during casting under atmospheric pressure is significant [1]. Main areas of application: vacuum cooling systems for biodiesel production, installations for pumping vapor-air mixture from steam turbines condensers, vacuum systems of evaporation plants, heat pumps, heating installations, etc. [2, 3].

Using the primary flow energy is one of the ways to create a vacuum of a secondary flow in vapor jet ejectors. This type of jet apparatus uses widely in many industries. It has a simple design and reliable operation. Wherein vapor jet ejectors have some significant disadvantages associated with imperfections in their workflow [4, 5].

The application of a new type of two-phase jet apparatus becomes topical, but the workflow complexities hamper their development. Recently, international research in this industry has become popular [6].

The most promising type of two-phase jet apparatus is a liquid-vapor ejector, which operates on the principle of thermal jet compression [7–9]. The principle of jet thermocompression is to use the working fluid subcooled to saturation as an active flow. Fluid starts boiling in the throat and the nozzle diffuser. The nozzle's outlet section forms a supersonic jet of a dispersed vapor-droplet structure, which injects a passive flow. The jet has a high volumetric vapor content. Compared with vapor ejectors, it has a positive effect on workflow efficiency.

The primary flow in the diffuser of the motive nozzle is the main factor determining the efficiency of a liquid-vapor ejector. Theoretical and experimental studies of the motive nozzles, with geometry close to a de Laval nozzle, were carried out. The motive nozzle has a straight wall of the diffuser geometry. Studies show the high efficiency of the process of evaporation, in which the motive nozzle velocity ratio reaches 92–97% [10, 11]. However, negative factors occur in the motive nozzle with such diffuser geometry. There is a flow separation from the channel walls, shock waves and condensation shocks, and an unstable velocity profile over the channel area [12, 13].

The above disadvantages of the motive nozzle design lead to profiling its flow part to improve the gas-dynamic characteristics.

2. Numerical Investigation

2.1. Motive nozzle design

The authors studied the issue of profiling boiling liquid nozzles. This research allowed us to determine the most appropriate diffuser forms: the logarithmic form, the parabolic form, the elliptical form, and the form calculated by the Vitoshinsky formula (Fig. 1).

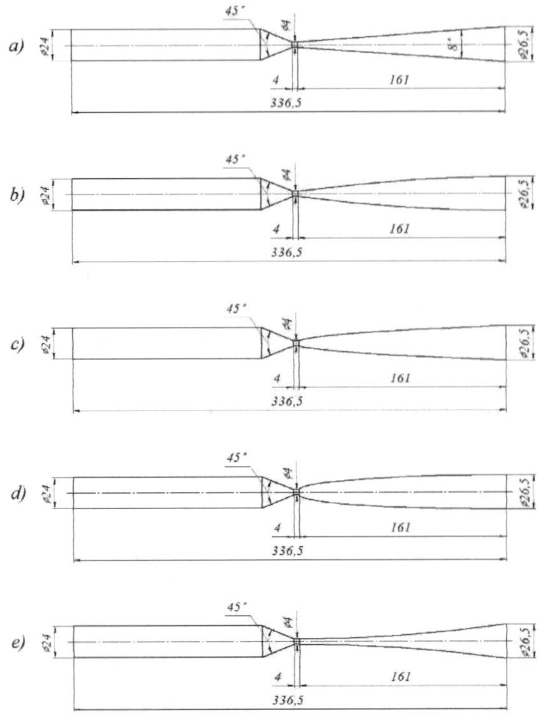

Fig. 1 Motive nozzle design:
a) – straight wall nozzle [7], b) – logarithmic form,
c) parabolic form, d) – elliptical form, e) Vitoshinsky form

2.2. *Mathematical Model*

Calculations are carrying out according to the authors method. It is basing on the mathematical model for the vacuum operation regime, which is an improved model of V. N. Marchenko for the compressor operation regime [9].

This model is basing on a dynamic model of the flow of a metastable superheated fluid. The model use the method of indirect determination of averaged flow parameters. These are the distribution of phase temperatures, velocities, vapor mass fraction, and discrete phase sizes.

The essence of this method is the joint consideration of experimental data [14]. This method includes flow rate, jet pulse, static pressure distribution, and flow visualization. Balance equations: the state of thermally metastable vapor-droplet medium, mass conservation (taking into account the phase transition), total enthalpy (first law of thermodynamics), amount of movement (impulse), entropy production (second law of thermodynamics). And the solution, together with the equations of vapor mass fraction, the channel geometry calculation, and the tangential stress on the wall.

3. CFD Simulation

3.1. *Initial parameters*

Software ANSYS CFX is used to simulate the flow of the nozzle. The standard system of equations Navier – Stokes is adopted as a mathematical model. The k–ε model is used to simulate turbulence. It is a joint solution of the Navier-Stokes equations for time-averaged constituent variables and additional equations for determining the pulsation components:

- continuity equation (mass conservation);
- equation of motion (conservation of momentum);
- transfer equation for turbulent kinetic energy;
- equation dissipation of turbulent kinetic energy.

The Rayleigh – Plesset equation describes boiling kinetics in the diffuser of the nozzle [15]:

$$R_B\frac{d^2R_B}{dt^2}+\frac{3}{2}\left(\frac{dR_B}{dt}\right)^2+\frac{2\sigma}{\rho_f R_B}=\frac{p_v-p}{\rho_f}$$

where R_B - vapor bubble radius; σ - coefficient of surface tension between liquid and vapor; ρ_f- fluid density; p_v - saturated vapor pressure; p_f - fluid pressure; p - fluid pressure around the bubble.

The nozzle is axisymmetric. The calculation mesh generation is carried out automatically. It consists of 30,000 cells approximately and is compacted in places of flow restructuring (Fig. 2).

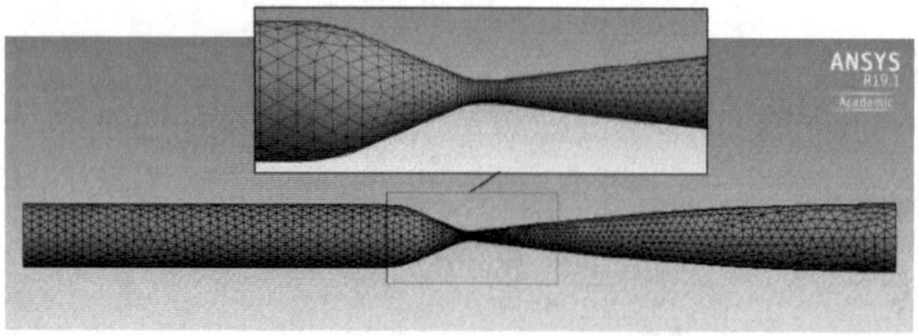

Fig. 2 Computational mesh (using the example of logarithmic primary flow nozzle Fig. 1, b)

3.2. Working fluid properties
Water using as a working fluid. Properties of water are in the ANSYS-CFX internal material database industrial standard IAPWS. Water boils while passing through the primary flow nozzle and in characteristic sections has the operating parameters shown in Table 1.
The regimes are selected in Table 1 based on the main branches of the application of two-phase ejectors. The first regime uses in binary steam turbine installations for small-scale power generation, the second regime - in installations for biodiesel production, the third - in heat pump installations for heating systems, the fourth - in vacuum evaporation installations.

3.3. Boundary conditions

Boundary conditions (Fig. 3) are defined as "Inlet" with "Bulk Mass Flow Rate", "Wall" without roughness, and "Outlet" with "Average Static Pressure". The ambient pressure is assumed to be equal to atmospheric pressure under normal conditions 101325 Pa.

Table 1. Operating fluid parameters
in characteristic sections of the motive nozzle

Regime	p_{in}, bar	t_{in}, °C	p_{out}, bar
1	35	203	1
2	20	178	0,742
3	10	150	0,502
4	3	110	0,289

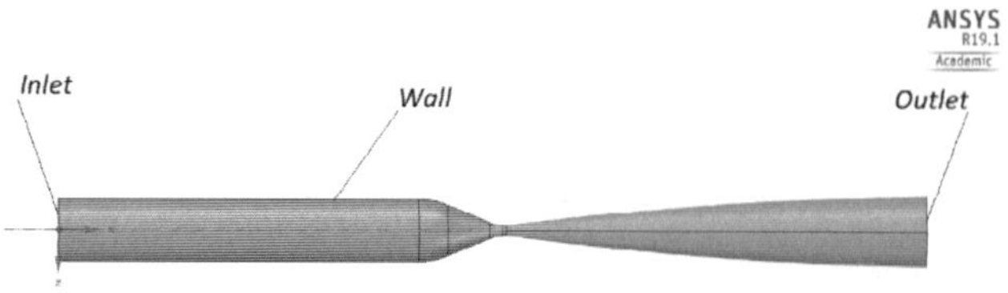

Fig. 3 Computational mesh (using the example of logarithmic primary flow nozzle Fig. 1, b)

4. Results and discussion
The mathematical model proposed by the authors makes it possible to calculate an active flow nozzle with straight diffuser walls. Therefore, to assess the feasibility of profiling, we simulated the process of a nozzle of various geometry in the Ansys CFX software.

Figures 4–20 show the results of numerical simulation of various geometry of the diffuser of the nozzles. The dependences of changes in parameters, such as pressure, velocity, Mach number, and vapor mass fraction, are determined. The efficiency of the process of the active fluid flow from these nozzles is also determined. The results of the comparative analysis are given for the first regime. On the other regimes, the results are similar.

4.1. Numbering

Figure 4 presents the results of the CFD simulation in the ANSYS CFX software the pressure distribution along with the nozzle. Figure 4, shows the pressure distribution along the length of the nozzle with straight walls depending on the initial parameters (Table 1).

Figure 5 is a graph of pressure distribution along the nozzle length, which shows the calculation results according to the method described in [5, 9] and according to the CFD simulation data obtained.

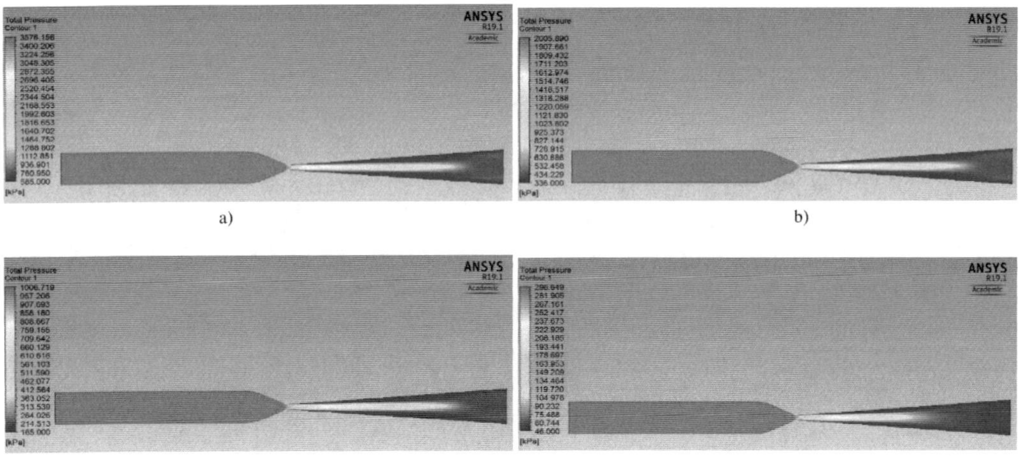

a) b) c) d)

Fig. 4 Pressure distribution in straight-wall nozzle:
a) 1st regime, b) 2nd regime, c) 3rd regime, d) 4th regime

Figures 4 and 5 show that the required pressure is reached at the nozzle outlet in all regimes. But with decreasing the initial parameters of the active flow in the nozzle diffuser, occur faster evaporation of the liquid central core. This effect has a positive impact on the completion of the boiling process.

Figures 6–8 show the CFD simulation results of pressure distributions along the length of nozzles with different diffusers (Fig. 1, b–e), depending on the initial parameters (Table 1).

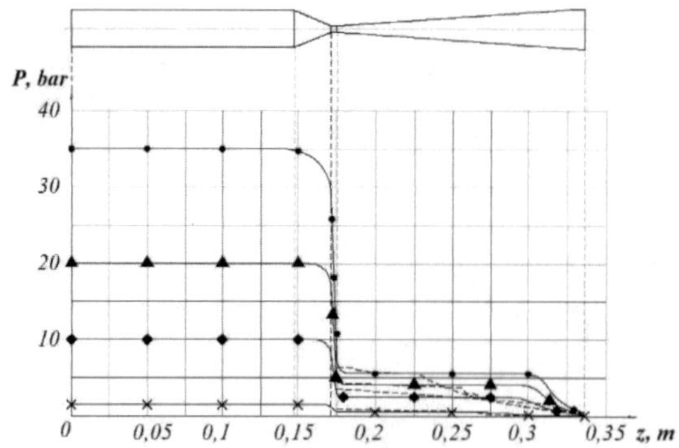

Fig. 5 Pressure distribution in straight-wall nozzle:
– – – – numerical investigation results, ————— CFD simulation results, ● 1st regime, ▲ 2nd regime, ◆ 3rd regime, × 4th regime)

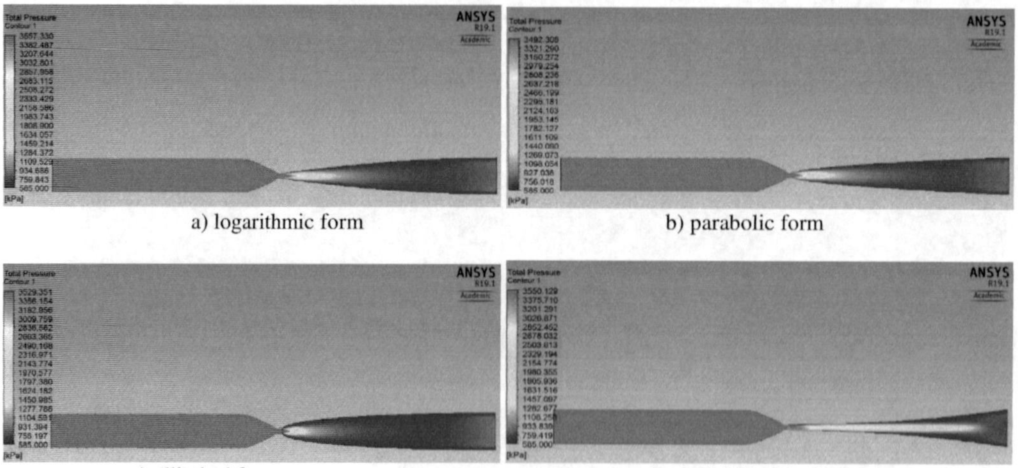

a) logarithmic form b) parabolic form

c) elliptical form d) Vitoshinsky form

Fig. 6 Pressure distribution in nozzles (Fig. 1, b–e) for the 1st regime

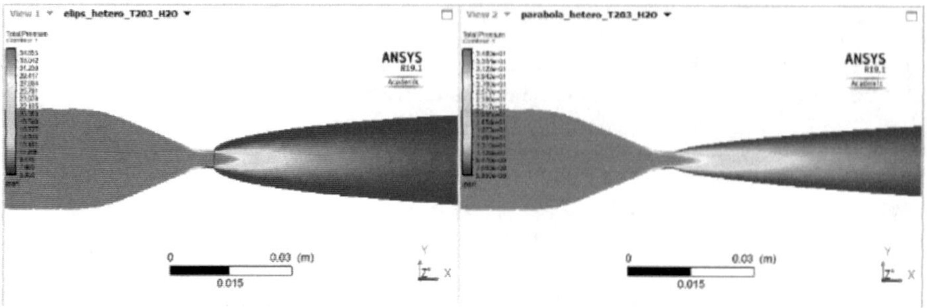

a) b)

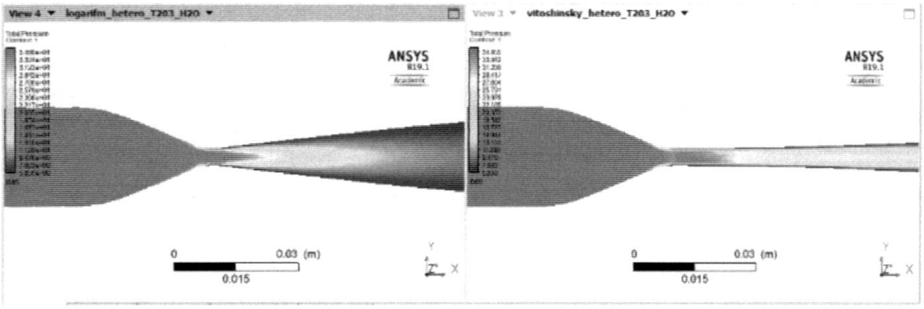

c) d)

Fig. 7 Pressure distribution in nozzles (Fig. 1, b–e) for the 1st regime

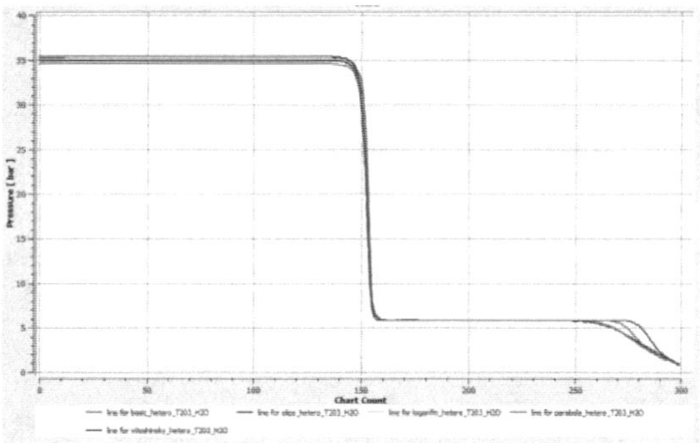

Fig. 8 CFD simulation results of pressure distribution in nozzles for the 1st regime: – straight wall nozzle, – logarithmic form, – parabolic form, – elliptical form, – Vitoshinsky form

The evaporation of the central core of the fluid occurs most rapidly in a nozzle with an elliptical form (Fig. 6, c). However, possible flow rotating in the diffuser, associated with a sudden expansion after the throat (Fig. 7, a). In the Vitoshinsky nozzle, the evaporation of the central core of the liquid occurs at a significant distance from the throat (Fig. 6, d). This effect adversely affects the process of boiling.

As shown in Figure 8 in the nozzles shown in Figure 1, b – d, the evaporation of the central core of the liquid occurs faster than that of a nozzle with straight walls of the diffuser (Fig 1, a).

4.2. Velocity and Mach number distribution

Figure 9 shows the results of a CFD simulation of a velocity change along the length of a nozzle with straight walls (Fig. 1, a) depending on the initial parameters (Table 1).

Figures 10 and 11 are graphs of the dependence of the velocity distribution and the Mach number along the nozzle length. It presents the calculation results according to the method in [5, 9] and according to the data from the CFD simulation.

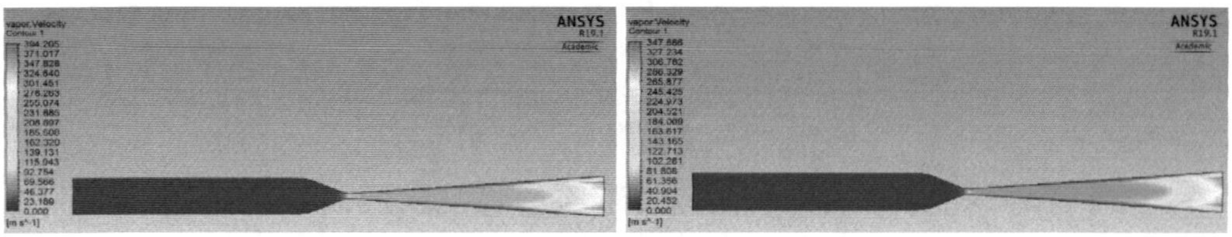

a) b)

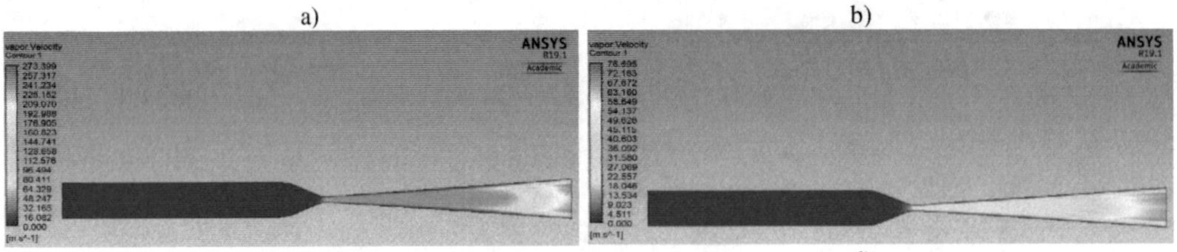

c) d)

Fig. 9 Velocity distribution in straight-wall nozzle:
a) 1st regime, 2nd regime, 3rd regime, 4th regime

Figures 9 and 10 show that with a pressure decrease in all regimes, the velocity at the nozzle outlet decreases. The faster the evaporation of the central core of the liquid, the faster the stabilization of the velocity profile.

As shown in Figure 11, a supersonic flow regime is achieving in each of the regimes, and the lower the nozzle outlet pressure, the greater the Mach number.

The difference in the graphs of the Mach number in the calculation method and CFD simulation is explained by the fact that the calculation method contains the law for changing the Mach number for a two-phase mixture, and in ANSYS CFX the Mach numbers for individual phases are considered separately.

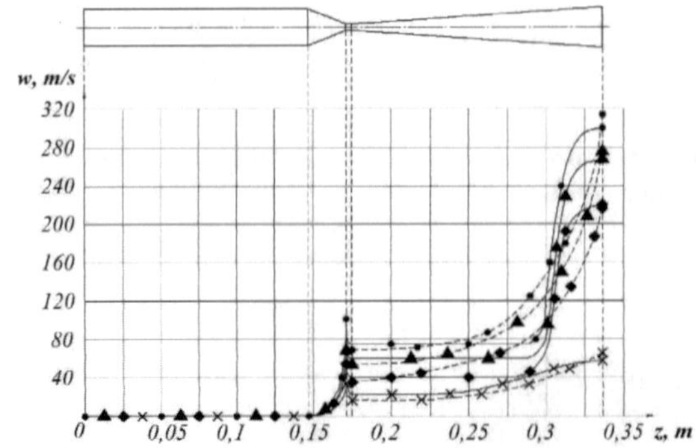

Fig. 10 Velocity distribution in straight-wall nozzle:
– – – – numerical investigation results, ———— – CFD simulation results,
● 1st regime, ▲ 2nd regime, ◆ 3rd regime, × 4th regime)

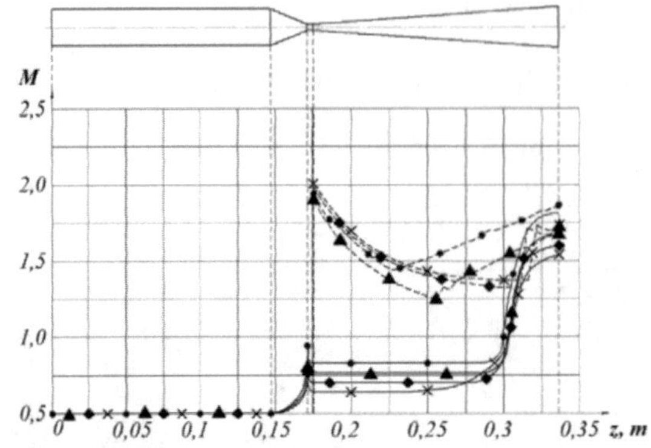

Fig. 11 Mach Number distribution in straight-wall nozzle:

– – – – numerical investigation results, ————– – CFD simulation results,
● 1st regime, ▲ 2nd regime, ◆ 3rd regime, × 4th regime)

Figures 12–14 show the results of the CFD simulation of speed variations along the length of nozzles with different geometry of the diffuser (Fig. 1, b – e) depending on the initial parameters (Table 1).
As shown in Figure 12 and Figure 13, the velocity profile stabilizes most quickly in a nozzle with an elliptical form of the diffuser (Fig. 12, c, Fig. 13, a). This effect is due to the longitudinal coordinate of the evaporation of the liquid central core. The speed of 420 m/s is reached at the nozzle outlet. In nozzles with a logarithmic and parabolic form, the velocity stabilization is approximately the same (Fig. 13 b, c). The velocity value at the outlet of these nozzles is approximately 380–400 m/s.

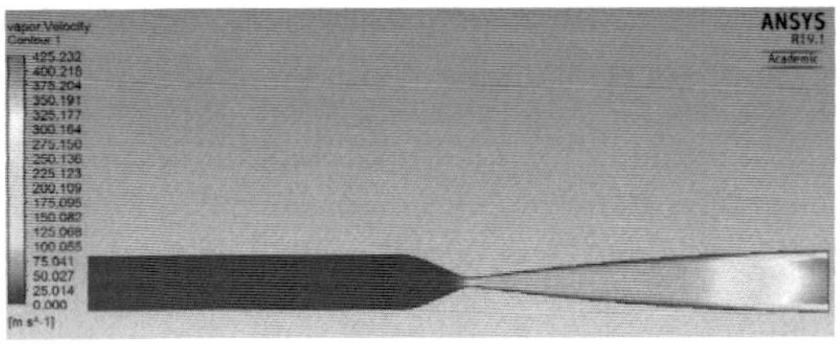

a) logarithmic form

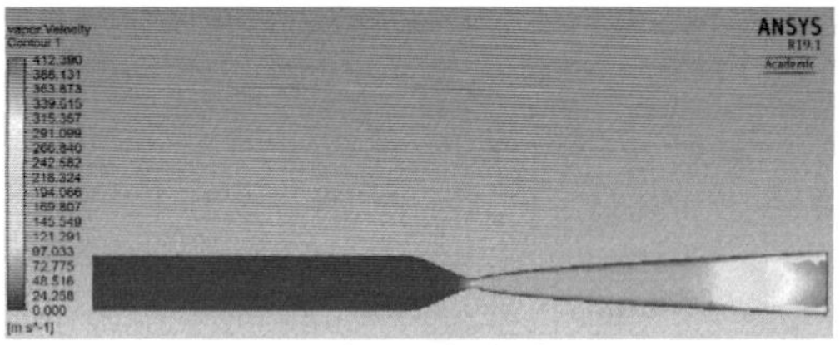

b) parabolic form

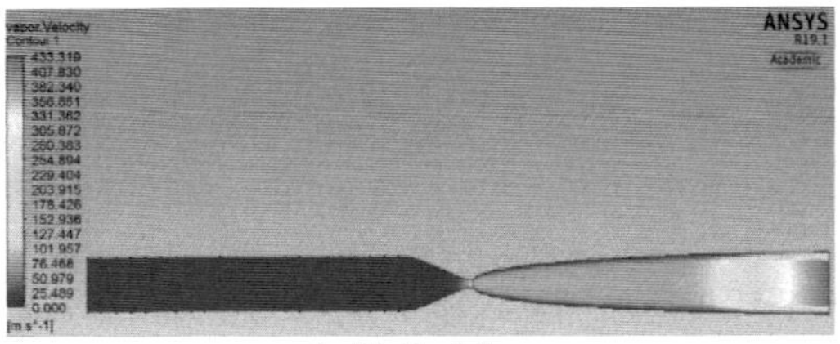

c) elliptical form

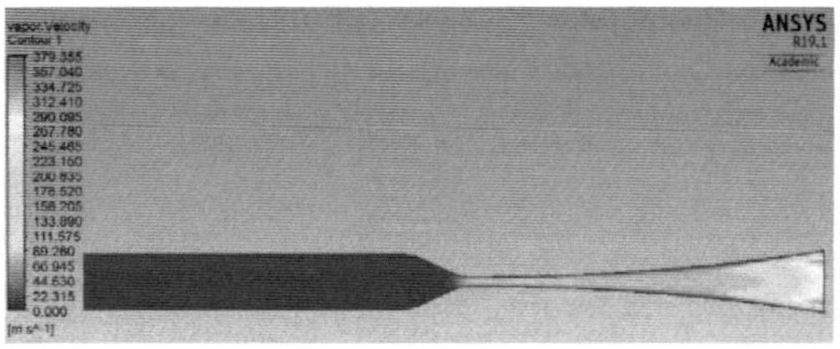

d) Vitoshinsky form

Fig. 12 Velocity distribution in nozzles (Fig. 1, b–e) for the 1st regime

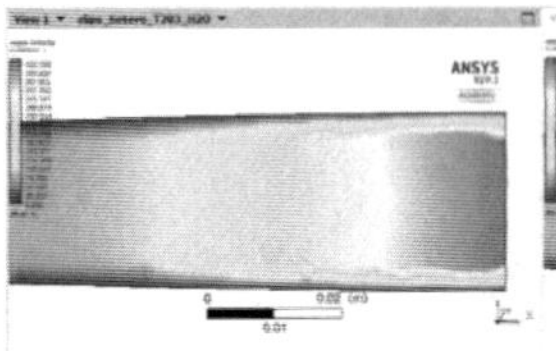

a)

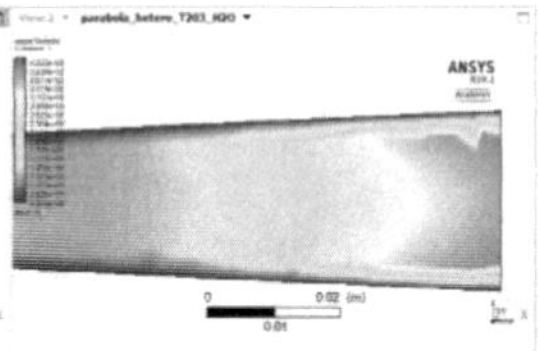

b)

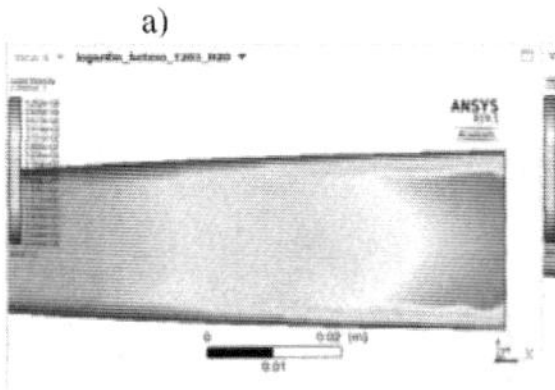

c)

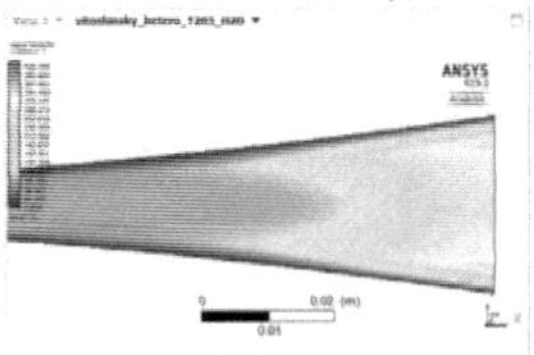

d)

Fig. 13 Velocity distribution in nozzles (Fig. 1, b–e) for the 1st regime

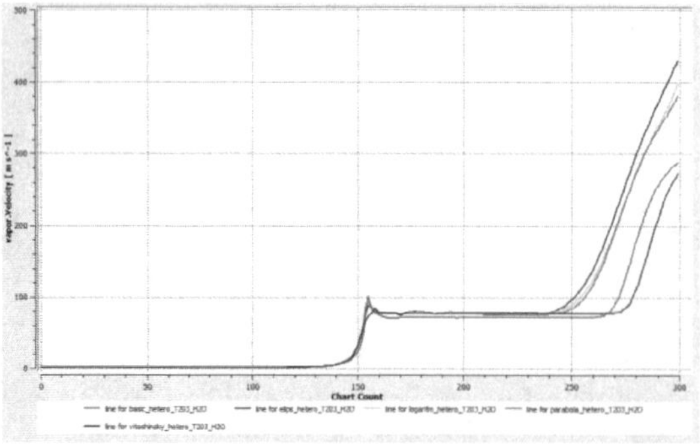

Fig. 14 CFD simulation results of velocity distribution in nozzles for the 1st regime: ———— – straight wall nozzle, ———— – logarithmic form, ———— – parabolic form, ———— – elliptical form, ———— – Vitoshinsky form

In the Vitosinsky nozzle, there is a central core of the fluid in the nozzle outlet. And there is a significant imbalance in the velocity profile (Fig 12, d, Fig. 13, d). It affects the speed of the nozzle outlet, which is 260 m/s.

In the logarithmic and parabolic nozzles, shown in Figure 1, b – d, the optimal velocity is 350–400 m/s, as shown in Figure 14. Speed reduction at the nozzle outlet is better for boiling and heat and mass transfer processes at the phase boundary.

4.3. Vapor Mass Fraction Distribution

Figure 15 shows the CFD simulation results of vapor mass fraction distribution along the nozzle length with straight walls (Fig. 1, a) depending on the initial parameters (Table 1).

Figure 16 shows the graphs of the vapor mass fraction distribution along the nozzle length. According to the method in [5, 9], the calculation results according to the data obtained from the CFD simulation.

When the initial parameters of the active flow at the nozzle inlet decrease, then the mass vapor fraction at the outlet also decreases. This effect indicates the incompleteness of heat and mass transfer processes between the vapor and liquid phases (Figure 15 and Figure 16).

Figures 17–20 presents the results of a CFD simulation of a vapor mass fraction distribution along the length of the nozzles with different geometry of the diffuser (Fig. 1, b – e) depending on the initial parameters (Table 1).

At nozzles with a logarithmic and parabolic form of the diffuser, the optimal vapor mass fraction in the nozzle outlet is 0.45–0.48 kg/kg (see Figures 17–19).

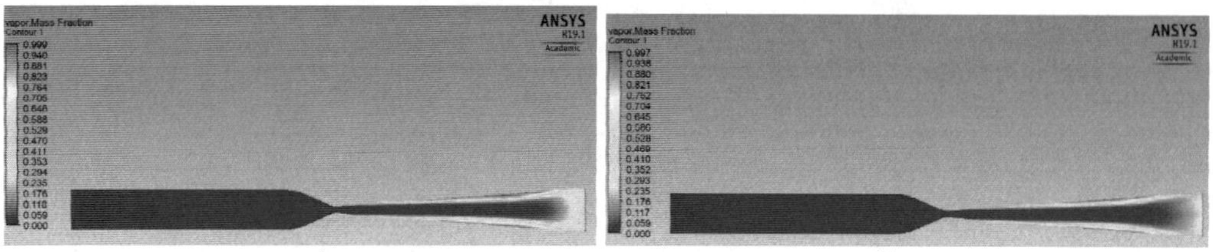

a) b)

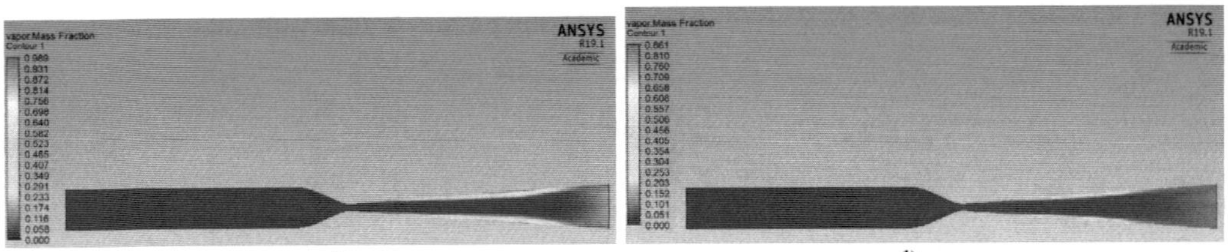

c) d)

Fig. 15 Vapor mass fraction distribution in straight-wall nozzle:
a) 1st regime, 2nd regime, 3rd regime, 4th regime

The elliptical nozzle and the Vitosinsky nozzle have many features. This sharp increase presence in the vapor mass fraction due to the feature of the geometry of these nozzles.

Thus, in a nozzle with an elliptical form, an abrupt change in the channel geometry (sudden expansion) occurs in the initial section (Fig. 18, a). The vapor mass fraction in the nozzle outlet increases and equals 0.505 kg/kg. At the nozzle with the Vitosinsky form, this area is closer to the nozzle outlet (Fig. 19, d), but the form is concave, leading to a decrease in the outlet vapor mass fraction (0.32 kg/kg).

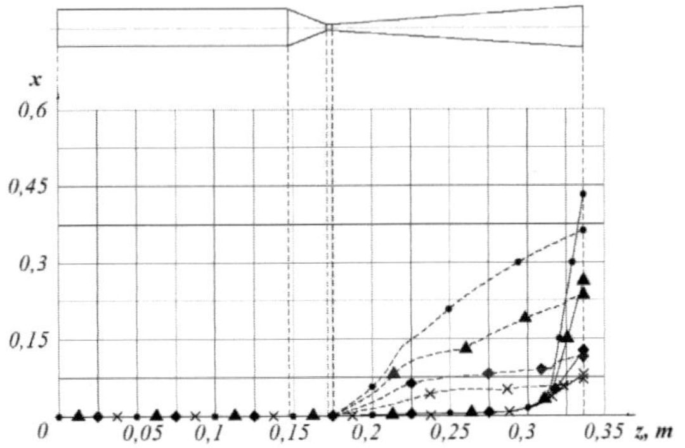

Fig. 16 Vapor mass fraction distribution in straight-wall nozzle:
– – – – numerical investigation results, ———— – CFD simulation results,
● 1st regime, ▲ 2nd regime, ◆ 3rd regime, × 4th regime)

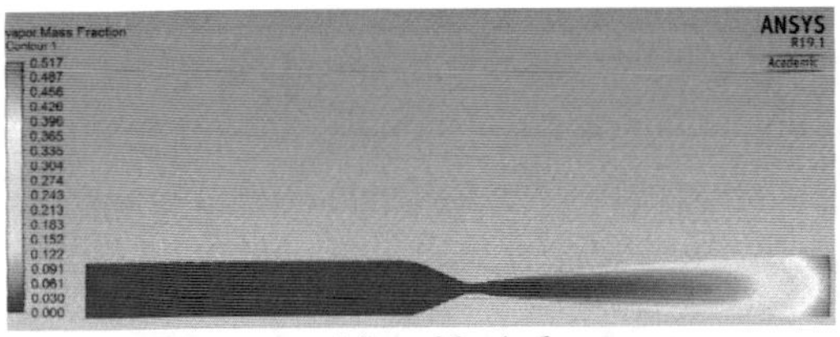

a) logarithmic form

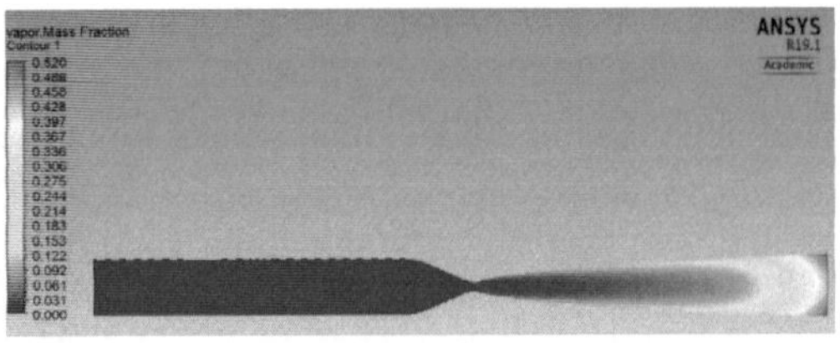

b) parabolic form

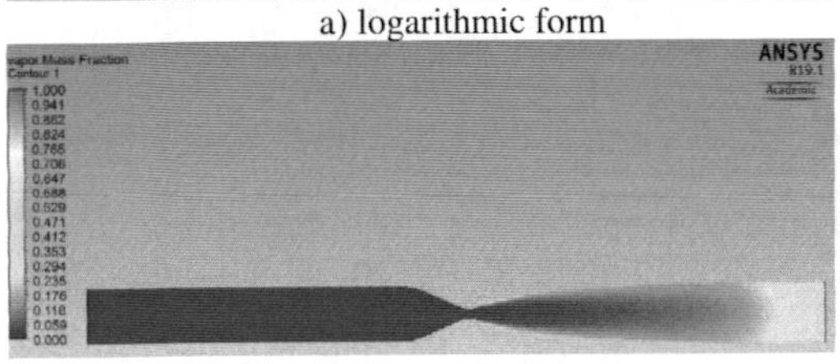

c) elliptical form

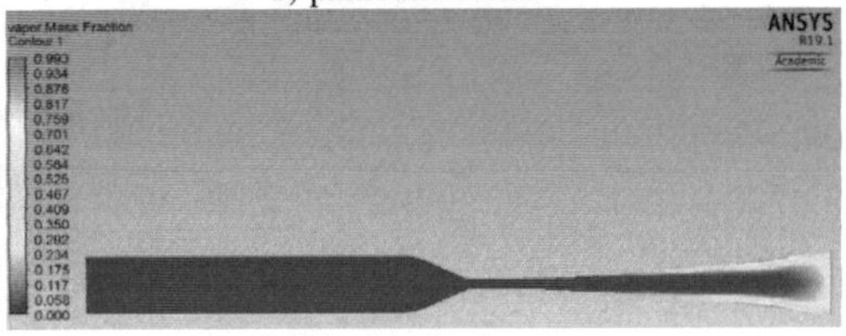

d) Vitoshinsky form

Fig. 17 Vapor mass fraction distribution in nozzles (Fig. 1, b–e)
for the 1st regime

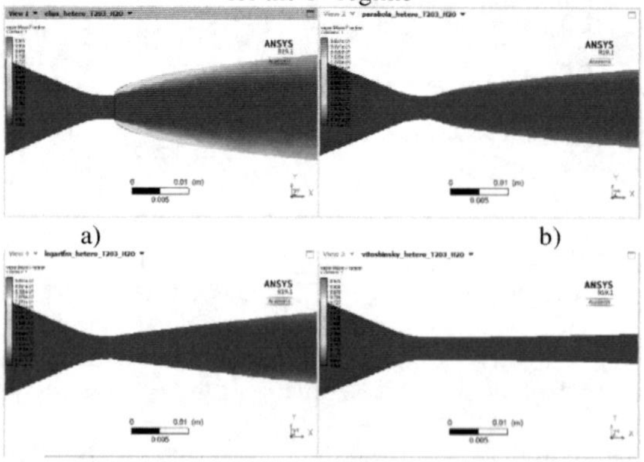

a) b)

c) d)

Fig. 18 Vapor mass fraction distribution in nozzles (Fig. 1, b–e)
for the 1st regime

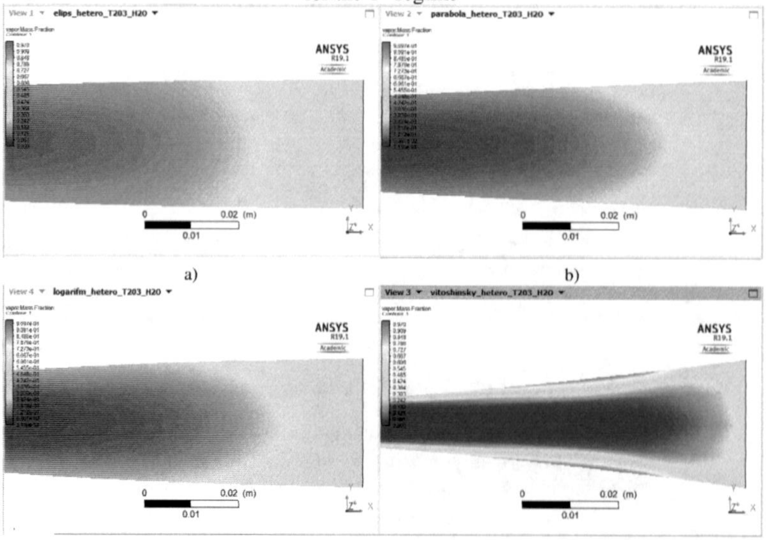

a) b)

c) d)

Fig. 19 Vapor mass fraction distribution in nozzles (Fig. 1, b–e)
for the 1st regime

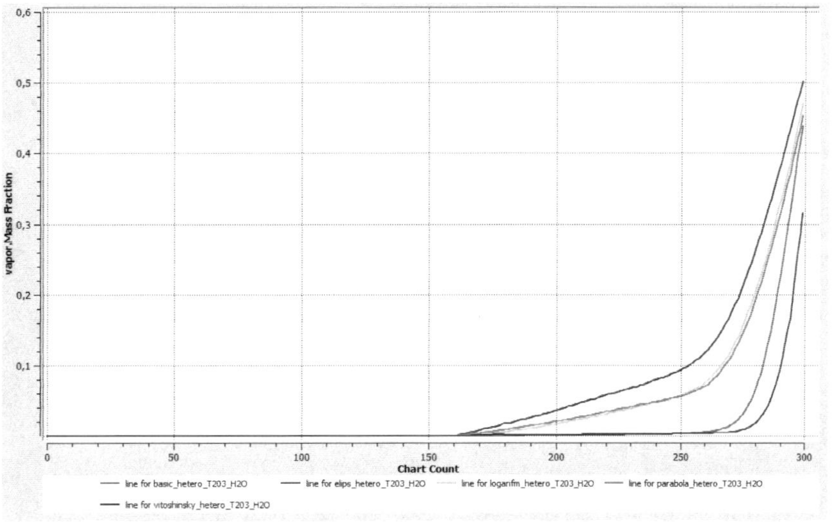

Fig. 20 CFD simulation results of velocity distribution in nozzles for the 1st regime:———— – straight wall nozzle, ———— – logarithmic form, ———— – parabolic form, ———— – elliptical form, ———— – Vitoshinsky form

In the nozzles shown in Figure 1, b–d, the most suitable vapor mass fraction consists at the level of 0.45–0.48 kg/kg (see Figure 20), which achieve in the logarithmic and parabolic form nozzles.

4.4 Efficiency of Nozzles

As a criterion determining the efficiency of fluid outflow from the nozzle, using the nozzle velocity coefficient φ_a. It is the ratio of the velocity at the nozzle outlet v_a to the isentropic velocity at the nozzle outlet v_{as}:

$$\varphi_a = \frac{v_a}{v_{as}}. \qquad (1)$$

Figure 21 shows the dependences of the nozzle velocity coefficient on the level of expansion of the working fluid in the nozzle p_{in}/p_{out}.

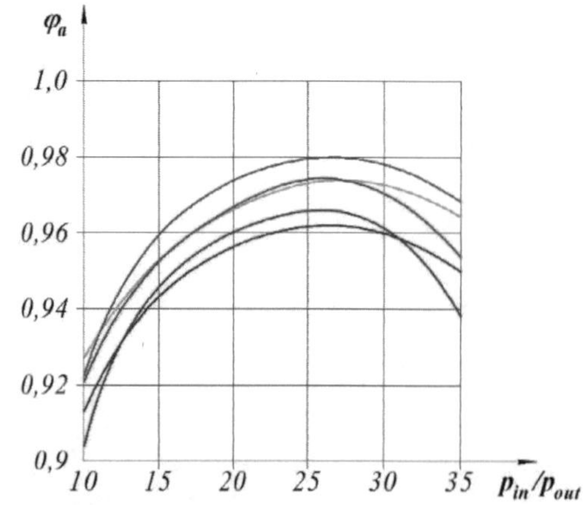

Fig. 21 The dependences of the nozzle velocity coefficient on the level of expansion of the working fluid in the nozzle p_{in}/p_{out} for the regimes presented in table 1: ——— – straight-wall nozzle, ——— – logarithmic form, ——— – parabolic form, ——— – elliptical form, ——— – Vitoshinsky form

Figure 21 shows that profiling the nozzles in which the boiling liquid flows has a positive effect. This results in an increase in the speed coefficient from the values φ_a = 0,92–0,97 to the values φ_a = 0,95–0,98. Elliptical nozzles have the highest rates of speed.

5 Conclusions

The authors made a numerical analysis and CFD simulation in the ANSYS CFX software nozzles of various geometry of the diffuser. As a result, we can draw the following conclusions:

1. Profiling the diffuser of the nozzle in which the boiling liquid flows has a positive effect on the nature of the flow process. It is possible to reduce flow separation from the channel walls, shock waves, and condensation shocks and obtain a stable velocity profile over the channel area.

2. The parabolic form is the most favorable because the central core of the liquid is boiling at the optimum distance from the nozzle throat, and the flow of a stable vapor structure with the required pressure value for each regime is formed at the outlet. Speed value at w_a = 380–400 m/s with the optimum vapor mass fraction in the range x_a = 0,45–0,48 kg/kg.

3. Profiling the diffuser of the nozzle increases the efficiency of the flow process. An increase in the velocity coefficient from a value φ_a = 0.92–0.97 for a nozzle with straight walls to the value φ_a = 0.95–0.98 for a nozzle with a parabolic diffuser it proves.

4. Vitoshinsky form nozzle is not suitable for the flow of boiling liquid. It is strongly narrow, and the evaporation of the central core of the fluid occurs at a significant distance from the throat. It also negatively affects the velocity profile and the vapor mass fraction at the nozzle outlet.

5. References

[1] Silla, H. Chemical process engineering. design and economics. New York–Basel, Marcel Decker Inc, 2003. – 483 p.

[2] Sharapov S. O. Application of jet thermal compression for increasing the efficiency of vacuum systems [Text] / S. O. Sharapov, V. M. Arsenyev, V. M. Kozin // IOP Conference Series: "Materials Science and Engineering". – 2017. – Vol. 233.
[3] Sharapov, S., Husiev, D., Panchenko, V., Kozin, V., & Baha, V. (2020). Analysis of the possibility of using R718 for a heat pump of a heating system based on a liquid-vapor ejector. Eastern-European Journal of Enterprise Technologies, 6(8 (108), 39–44. https://doi.org/10.15587/1729-4061.2020.217274
[4] Austin, F., Street, H., Applied chemical process design. Front Cover. Plenum Press, 1978 – Science – 294 p.
[5] Power, Robert B., Steam Jet Ejectors for the Process Industries. McGraw–Hill Chemical Engineering, 1993. – 498 p.
[6] Shurchkova Ju. Adiabatnoe vskipanie. Prakticheskoe ispol'zovanie – Kyiv: Naukova dumka, 1999. – 228 p.
[7] Marchenko V. Strujnaja termokompressornaja ustanovka (STK): naznachenie, princip dejstvija, termodinamicheskaja model' i rezul'taty raschetnyh issledovanij parametrov rabochego processa – Visnik Sums'kogo derzhavnogo universitetu. Serija Tehnichni nauki, 2004. – 50–60 pp.
[8] Sharapov, S., Starchenko, M., Protsenko, M., Panchenko, V., Kovtun, V. (2019). Improvement of the vacuum cooling system for biodiesel production. Journal of Engineering Sciences, Vol. 6(1), pp. F6-F11.
[9] Marchenko V. Perspektivy primenenija i rezul'taty chislennogo modelirovanija strujnoj termokompressornoj ustanovki – Visnik Sums'kogo derzhavnogo universitetu. Serija Tehnichni nauki. – 2005. – 36–40 pp.
[10] Tonkonog V Fazovye perehody v potoke zhidkosti – Trudy Kazanskogo gos. Tehn. Universiteta im. A. N. Tupoleva. – 2004.
[11] Chang, Paul K. Separation of Flow. Pergamon, 1970 – 796 p.
[12] Chekh O., Sharapov S., Prokopov M., Kozin V., Butrymowicz D. Cavitation in nozzle: the effect of pressure on the vapor content. – Advances in Design, Simulation and Manufacturing II, pp 522-530 https://doi.org/10.1007/978-3-030-22365-6_52
[13] Sharapov S. Experimental investigation of liquid-vapor ejector with conical mixing chamber / S. Sharapov, V. Arsenyev, V. Kozin // Technology audit and production reserves. – 2016. - №4/1 (30). – P. 50-55.
[14] Liao Y. and Lucas D. 2015 3D CFD simulation of flashing flows in a converging-diverging nozzle Nucl. Eng. Des. 292 149-163.
[15] ANSYS INC.: CFX–Solver Theory Guide, pp. 390 (2009)

Simulation and optimisation of opened sewage concentration air cycle combined with the heat pump under actual weather conditions

Y Zhang[1], T S Guo[1], W F Wu[1,*], C Y Li[1] and X L Wang[2]

[1] Institute of Compressor Engineering, Xi'an Jiaotong University, Xi'an, China
[2] School of Engineering, College of Science and Engineering University of Tasmania, Hobart, TAS 7001, Australia
*Corresponding author

E-mail: weifengwu@mail.xjtu.edu.cn

Abstract. Opened air cycle is a promising sewage concentration technology using heated air as the medium to absorb water from the sewage. Unlike closed cycles, the air flow is directly ejected out of the system after humidification in opened cycles. Operating conditions and the circle's efficiency are highly affected by the weather, such as the time-varying ambient temperature and relative humidity. Since the evaporating temperature relies on the ambient temperature, to restrain its dependence, the evaporator was proposed combined with the ejected air flow in this paper. Such a design could improve the efficiency of the heat pump by increasing its evaporating temperature. Simulation of the newly proposed open sewage concentration cycle was performed to investigate its thermal process and economic efficiency under actual weather conditions. Thermodynamic models of the compressor, heat exchangers and the sewage concentration chamber were established. Climatic conditions at Xi'an, China were referenced as actual weather conditions. Optimisation of the working parameters was performed to achieve higher economic index. The results demonstrated that the average monthly power consumption of the optimized cycle was 0.3508 kWh/kg. Compared with the published data of the closed cycle, the total cost of new opened cycle was nearly one-third lower.

1. Introduction

With the development of industry, sewage treatment demands are increasing year by year. It will cause environmental pollution and water waste if the sewage is charged directly. The study on economic and efficient wastewater concentration technology is important to China's industrial circles[1,2]. As a kind of efficient and energy-saving wastewater concentration technology, heat pump drying technology solves the problem of difficult structural design and easily fouling[3] in MVR[4,5] and membrane process[6]. It can be applied in atmospheric pressure and low ambient temperature, which has obvious advantages for heat sensitive sewage[7,8]. So such technology has broad prospects for application in actual industry.

In 1943, Sulzer Company first used heat pump drying technology to build dehumidification drying device[9]. Xie proposed closed liquid concentration cycle combined with the heat pump under atmospheric pressure, which provided the possibility of efficient and economic concentration of heat-sensitive material liquid[10]. Wang applied this technology to the concentration of copper pipe washing solution, which could recover 70% water in waste liquid with low energy consumption, proving the practicability of this technology[11]. Liu added bypass pipe device to the closed liquid concentration system to save energy, which successfully reduced energy consumption by 30%[12]. Saensabai conducted system simulation calculations theoretically on both opened and closed structures, and found that the theoretical energy consumption ratio of the opened structure was lower[13]. This was because in the closed cycle, circulating air needed to be dehumidified in the evaporator in order to assure the good moisture absorption later in the sewage concentration chamber. So the evaporating temperature would be lower than that of opened structures under similar operation conditions, causing the system more power-wasting. So in order to build an opened structure, the evaporator was proposed combined with the ejected air flow in this paper.

In the opened system, Yuan analyzed the influence of evaporation pressure according to the ideal model[14]. Zhou conducted an experimental investigation of the effects of spray flow rate, circulating air flow rate and trade effluent concentration on the treatment capacity[15]. Zhang studied the relationship between the gas-liquid mass flow ratio, waste liquid temperature and the concentration efficiency of the device in order to estimate the performance of the system[16]. Later, Zhang proposed the experimental correlation formula of wastewater concentration efficiency, and the maximum error between the calculated value and the experimental value was within 20% [17]. The research on opened liquid concentration system mainly focused on experiments and ideal model analysis. Realistic simulation and seasonal optimisation of systems under actual weather conditions still need further study.

The main aim of this study is to assess the thermal process and economic efficiency of an opened sewage concentration air cycle combined with the heat pump under actual weather conditions. The improved cost annual value method is applied to evaluate the investment cost, maintenance cost and operating cost of the system comprehensively. The monthly optimisation of the actual system mainly depends on operation parameters such as time-varying ambient temperature and relative humidity, the evaporating temperature and the condensing temperature of the heat pump. The influences of these operation parameters on key parameters like inlet air flow, evaporation mass flow and power consumption are studied.

2. Mathematical model of the system

2.1. System description

Opened sewage concentration air circle includes heat pump loop, air loop and sewage loop. The main process of the circle designed in this paper is shown in figure 1.

The ambient air flows through the condenser after preheating. With higher temperature and lower relative humidity, the air flow is pumped into the sewage concentration chamber by the fan. In the chamber, the wet air is hygroscopic and cooled, then becomes saturated wet air[15]. The heat is recovered by the heat exchanger 2. After separating condensate water, the air flow is directly ejected by the evaporator.

The heat exchange temperature difference $\triangle T$ is settled as 5°C. The system pressure P_{sys} is 101.325 kPa. According to Li's thesis[18], the wind speed is set as 15 m/s in which case the evaporation efficiency is larger and there will not be obvious liquid floating phenomenon. The mass ratio of inlet sewage to concentrated sewage is 6:1. *emoney* (the electricity consumption of evaporated water per unit mass) is introduced as the measurement standard of thermal efficiency of the system. *emoney* is equal to the total electricity consumption of all energy-consuming components (including compressor, fan, pump, etc.) divided by m_{zfph} (total evaporation mass rate of the sewage in the system).

2.1.1. Heat pump loop and fan. The heat pump loop is the heat energy source of the whole system. It uses R134A as the working medium. By optimizing the evaporating temperature and condensing temperature of the heat pump, the compressor power can be calculated.

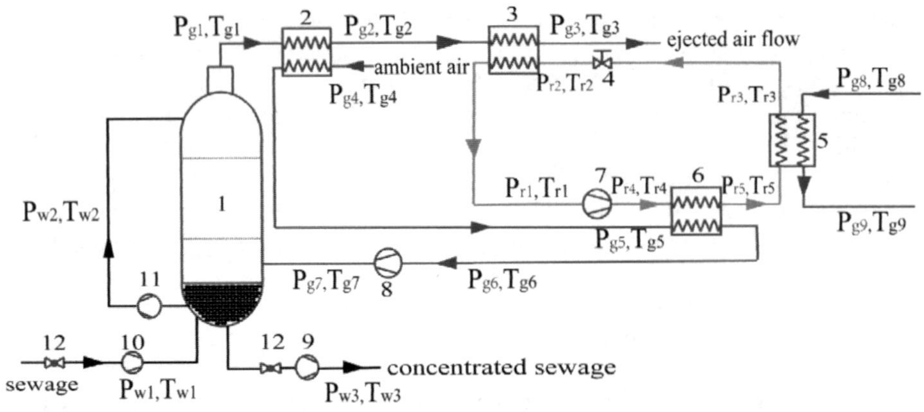

1——sewage concentration chamber 2——heat exchanger 3——evaporator 4——valve
5——air-cooling radiator 6——condenser 7——compressor 8——fan
9——drainage pump 10——inlet pump 11——circulating water pump 12——valve

Figure 1. Flow chart of opened sewage concentration air cycle combined with the heat pump.

The fan drives the air flow of the system. The total fan power needs to consider the resistance loss of the gas-liquid separator, heat exchangers and pipeline flow resistance in the circle. The resistance losses of gas-liquid separators are mostly determined by experiments, and lacks the correlation formula. Therefore, the resistance loss of gas-liquid separators is set as general number 10 kPa[19]. The loss calculation of heat exchanger, pipeline resistance[20,21] and airflow resistance in the chamber are shown in table 1.

In summary, N_{fan} (the total fan power) can be calculated as equation (1).

$$N_{fan} = Q_{air}(P_{local} + P_{onway} + P_{ht} + P_{fan} + P_{sys}) / \eta_{fan} \tag{1}$$

Where Q_{air} is the air flow rate in the pipe; the meaning of P_{local}, P_{onway}, P_{ht} and P_{fan} is shown in the table 1; P_{sys} is system pressure. η_{fan} is the fan efficiency.

Table 1. The loss calculation of pipes, heat exchangers and the chamber.

Physical meaning	Calculation formula
Pipeline resistance loss along the way P_{onway}	$P_{\text{onway}} = \frac{\delta}{d} \times \frac{v^2 \rho}{2} l$
Local resistance loss of pipeline P_{local}	$P_{\text{local}} = 0.5 \xi \rho v^2$
Total friction resistance of heat exchanger P_{htf}	$P_{\text{htf}} = P_{\text{htfl}} \times \Phi_1^2$
Liquid phase shunt resistance of heat exchanger P_{htfl}	$P_{\text{htfl}} = 2 f_1 L G^2 (1 - x^2) v_1 D^{-1}$
Coefficient of liquid friction along the heat exchanger f_l	$f_1 = 2.5 Re^{-0.3}$ [22]
Gas phase shunt resistance of heat exchanger P_{htfg}	$P_{\text{htfg}} = 2 f_{\text{g}} L G^2 x^2 v_{\text{g}} D^{-1}$
Coefficient of gas friction along the heat exchanger f_g	$f_{\text{g}} = 0.3164 Re^{-0.25}$ [23]
Friction component liquid phase apparent coefficient Φ_l^2	$\Phi_1^2 = 1 + C X^{-1} + X^{-2}$
Martinelli parameter X	$X^2 = P_{\text{htfl}} \times P_{\text{htfg}}^{-1}$
Total resistance loss of heat exchanger P_{ht}	$P_{\text{ht}} = (1.1 - 1.15) P_{\text{htf}}$
Air resistance loss in the the chamber P_{fan}	$P_{\text{fan}} = 0.5 \varepsilon v^2 \rho$ [24]

2.1.2. Sewage concentration system. The flow chart of the evaporation system is also shown in the left part of figure 1. The main components include the sewage concentration chamber and pumps.

In this paper, air is directly heated and contacted with cold sewage in the chamber, which can avoid fouling problem in the heat exchanger. For the sewage concentration chamber, the chamber temperature T_{g1} is calculated through equations of mass and energy conservation.

The inlet sewage temperature is set as normal temperature 20°C in summer and 15°C in winter. The evaporation mass rate m_{zfph} is calculated by equation (2).

$$m_{zfph} = m_{air}\left(d_{g1} - d_{g6}\right) \tag{2}$$

Where m_{zfph} is the evaporation mass rate of the system; m_{air} is the mass flow rate of dry air; d_{g1} and d_{g6} are the moisture content of point *g1* and *g6* respectively.

When calculating circulating pump power, the pressure loss of fluid flowing through the nozzle is considered. The calculation formulae are shown in table 2.

Table 2. Calculations of circulating pump power.

Physical meaning	Calculation formula
The pump power q_{pump}	$q_{pump} = 9.8HQ(n_1 n_2)^{-1}$ [25]
Injection drop $\triangle P$	$\Delta P = 0.5\rho Q^2 C^{-2} A^{-2}$ [26]
Total pump power q_{pumpt}	$q_{pumpt} = \Delta P \times Q + q_{pump}$

When the circle runs steadily, the air has reached saturation state out of the sewage concentration chamber and the saturation temperature under this pressure is not higher than the temperature of the sewage concentration chamber. In addition, heat and mass transfer driven by the pressure of water vapor in the chamber is weakened after reaching the saturation state. The dominant driving power turns into the temperature difference between sewage and air. Therefore, the temperature of air in the chamber needs to be higher than that of sewage to ensure the stable operation of the system.

2.2. Cost calculation

In this paper, the improved cost annual value method is applied for analysis. The cost of evaporated water per kilogram COST is calculated by equation (3).

$$COST = EC + \frac{IC \times CRF}{WT \times m_{zfph}} + \frac{MC}{WT \times m_{zfph}} \tag{3}$$

Where selected parameter values in the model are shown in table 3.

2.2.1. Cost calculation of Electricity energy. The electricity energy includes the energy consumed by components like the compressor, the fan, and the pump. The electricity bill *EC* is calculated as equation (4).

$$EC = ele_{price} \times emoney \tag{4}$$

Where ele_{price} is the electricity price.

Table 3. Parameter values in the economic model of the actual circle.

Parameter	Valves
System investment return years/yr	5
The discount rate /%	10
Investment recovery coefficient *CRF*	0.264
Electricity price ele_{price}/yuan·kWh^{-1}	0.8675
Annual running time of the system *WT*/h	8600

2.2.2. Cost calculation of initial investment. The initial investment *IC* includes equipment costs, installation cost, labour cost and construction cost.

According to 'Estimated Index of National Municipal Engineering Investment'[27], the cost is calculated as equation (5).

$$IC = 1.178\sum C_i \tag{5}$$

Where C_i is the price of the key equipment.

Table 4 shows the price calculation of the key equipment in the system.

Table 4. Price calculation of the key equipment.

Equipment	Purchase cost calculation
Compressor	$1.051\times 39.5m_{\text{ref}} \times \varepsilon \ln \varepsilon \times (0.9-\eta_{\text{ad}})^{-1}$ [28]
Heat exchanger	$2143\times A_{\text{ht}}^{0.514}$ [29]
Throttle valve	$114.5m_{\text{air}}$ [30]

2.2.3 Cost calculation of maintenance. According to RAZMI's paper, the maintenance cost is calculated as equation (6) [31].

$$MC=1\%IC \tag{6}$$

Where the maintenance fee *MC* includes the management fee and staff salaries.

3. System simulation

3.1. Weather data

The opened sewage concentration air cycle to be evaluated in this study is assumed to be installed in Xi 'an. Figure 2 shows the average air tem-perature and relative humidity data of Xi'an.

In the simulation, *emoney*(power consumption of evaporating water per unit mass) was selected as the optimisation target of the system. *emoney* may be affected by ambient temperature and relative humidity, as well as the evaporating and condensing temperatures of the heat pump. Since the ambient temperature and relative humidity are given, evaporating temperature and condensing temperature are chosen as optimisation variables to reach the monthly minimum of *emoney*. In this way, a four-dimensional problem is successfully simplified to a two-dimensional problem.

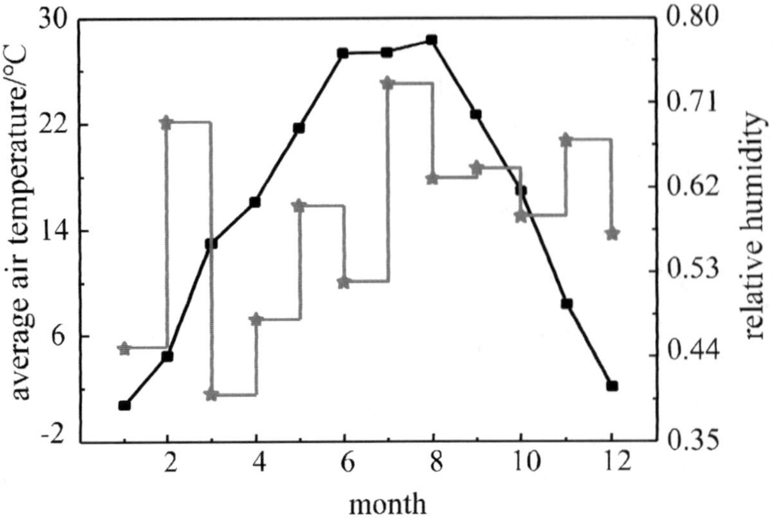

Figure 2. Air temperature and relative humidity in Xi 'an.

3.2. Optimisation model of the actual operating system

The refrigerating capacity is determined by the working conditions of the compressor. When calculating the compressor power, the adiabatic efficiency and volume efficiency are involved.

The evaporation mass of the system is no longer fixed. The air flow rate is adjusted to balance the heat exchange between the air side and the refrigerant side in heat exchangers.

The model can be expressed by equation (7). The complex method is used to solve the problem and get the corresponding monthly optimal operating conditions.

$$\min emoney\left(T_{\mathrm{evp}}, T_{\mathrm{con}}\right)$$
$$s.t.\begin{cases} 21 \le T_{\mathrm{con}} < 1.607T_{\mathrm{evp}} + 74.1, -15 \le T_{\mathrm{evp}} < 0 \\ 21 \le T_{\mathrm{con}} < 74.1, 0 \le T_{\mathrm{evp}} < 7 \\ 1.25T_{\mathrm{evp}} + 11.25 \le T_{\mathrm{con}} \le 74.1, 7 \le T_{\mathrm{evp}} \le 15 \end{cases} \tag{7}$$

Where T_{con} is the condensing temperature and T_{evp} is the evaporating temperature.

4. Results and discussion

4.1. Influencing factors of system performance

By calculation, the change of air relative humidity has small influence on the system performance compared with other variables, so this paper mainly studies on the influence of other three variables: inlet air temperature, evaporating temperature and condensing temperature.

4.1.1. Influences of inlet air temperature on system performance. Figure 3 displays change rules of key parameters in the system such as inlet air flow rate and evaporation mass flow rate when inlet air temperature is changed.

When evaporating temperature, condensing temperature is constant, refrigerant states and refrigerant mass flow rate in the heat pump loop stay constant, so the refrigerating capacity and compressor power consumption are not changed. The chamber temperature rises with the air temperature, and inlet air flow rate decreases due to constant refrigerating capacity, which leads to reduced evaporation mass flow, causing *emoney* increasing.

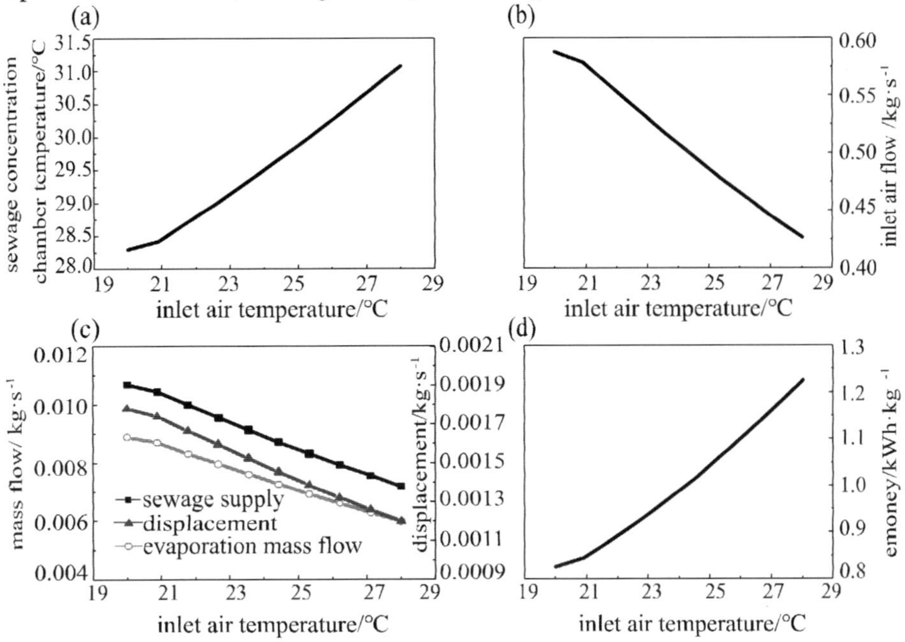

Figure 3. Influences of inlet air temperature on key parameters of the system: (a) sewage concentration chamber temperature; (b) inlet air flow; (c) m_{zfph}, sewage supply and displacement; (d) *emoney*.

4.1.2. Influences of heat pump evaporating temperature on system performance. Figure 4 displays change rules of key parameters in the system such as inlet air flow rate, evaporation mass flow rate and power consumption when the evaporating temperature is changed.

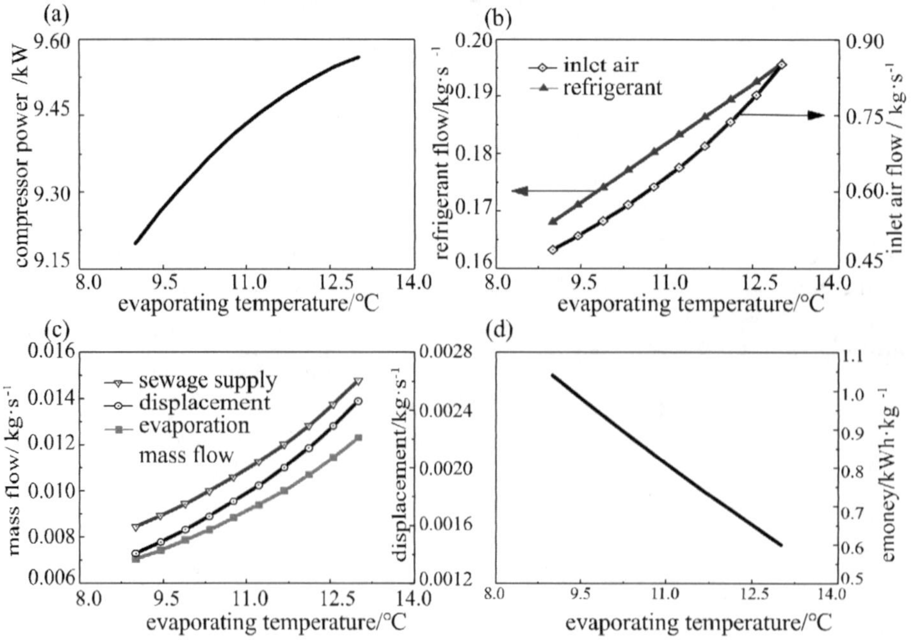

Figure 4. Influences of evaporating temperature on key parameters of the system: (a) compressor power; (b) refrigerant and inlet air flow; (c) m_{zfph}, sewage supply and displacement; (d)*emoney*.

For the refrigerant cycle, when the condensing temperature is constant and the evaporating temperature is increased, the density of inlet gas of the compressor is increased. The volumetric efficiency of the compressor is also increased, which leads to the increased mass flow rate of the refrigerant. So the compressor power consump-tion rises up. The total power consumption of other dissipative components is around 12 kW and only changes a little. So total power consumption rises with the evaporating temperature. The inlet air flow rate increases, resulting in an increase in sewage supply, evaporation mass flow and displacement. The increasing speed of system evaporation mass is faster than that of power consumption, so *emoney* decreases.

4.1.3. Influences of heat pump condensing temperature on system performance. Figure 5 displays change rules of key parameters in the system such as evaporation mass flow rate and power consumption when the condensing temperature is changed.

When the condensing temperature increases, the outlet temperature of the condenser increases, leading to the increase of the chamber temperature. For the heat pump loop, the increase of pressure ratio leads to the decrease of compressor volumetric efficiency, while the evaporation temperature remains the same, so the density of the inlet gas through the compressor is invariant. The refrigerant mass flow rate is equal to the inlet gas density times theoretical volume times volume efficiency. So the refrigerant mass flow rate drops, leading to a decline in refrigerating capacity. Therefore, the inlet air mass flow rate drops, resulting in a decrease in water supply, displacement and evaporation mass flow. The compressor power consumption increases, causing total power consumption rising, so emoney increases.

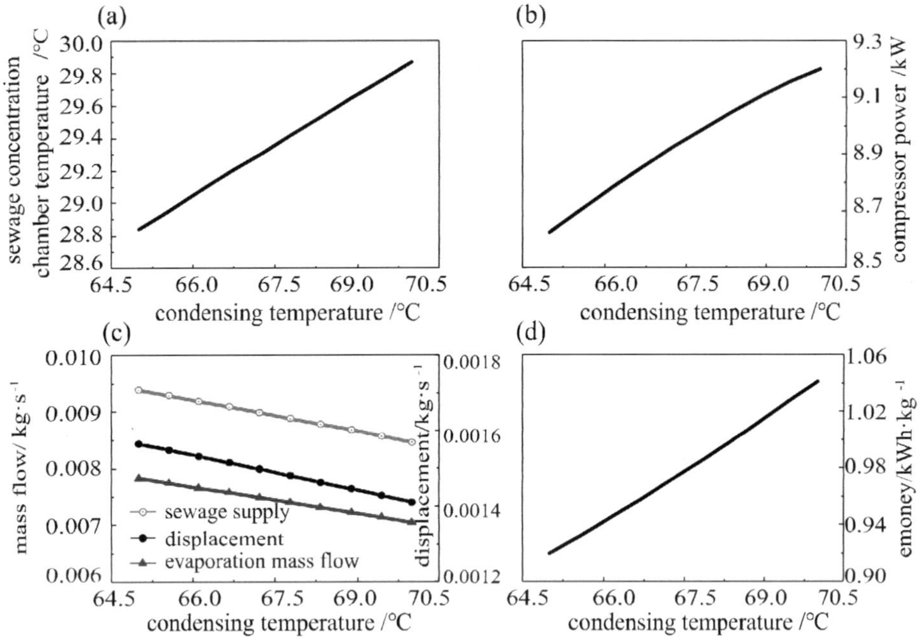

Figure 5. Influences of condensing temperature on key parameters of the system: (a) sewage concentration chamber temperature; (b) compressor power; (c) m_{zfph}, sewage supply and displacement; (d) *emoney*.

4.2. Energy consumption and water treatment

According to the analysis of the influence of the above operating parameters on the energy consumption of the system, the direction of performance optimisation of the system can be concluded. Figure 6 displays calculated *emoney* and evaporation mass flow of each month in the corresponding optimal operating conditions under actual weather conditions.

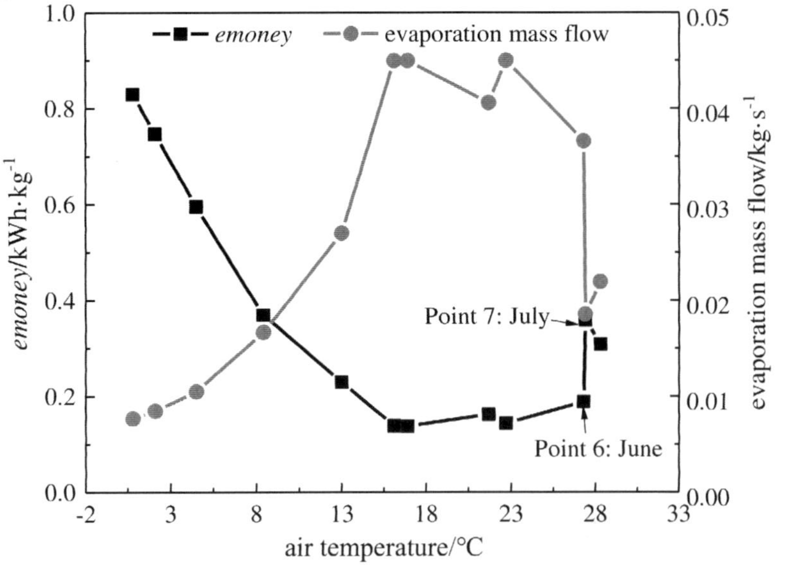

Figure 6. *emoney* and evaporation mass flow under actual weather conditions.

The average monthly power consumption of evaporated water per unit mass is 0.3508 kWh/kg. At low temperatures, the evaporating temperature becomes quite low due to the limit of the cold air temperature, which causes a surge in compressor power consumption and a decline in evaporation mass flow. And finally *emoney* is expected to soar. At high temperatures, the overall operating performance is at the high level. The decrease of evaporation mass flow leads to the increase of *emoney*. To avoid the phenomenon of liquid floating, the inlet air flow rate needs to be limited at high temperatures, leading to the decline of evaporation mass flow and the rise of *emoney*. The temperature difference of point 6 and point 7 is 0.1°C while the relative humidity difference is 21%. When temperatures are similar, *emoney* increases with increasing air relative humidity due to the rise in compressor power.

4.3. Economic analysis
Figure 7 shows *COST* under monthly optimized operating conditions can be calculated based on the economic model above.

As seen from Figure 7, *COST* reaches a minimum value in April when its temperature is 16.1°C. The tendency of *COST* changes correspondingly in agreement with *EC* (the electricity cost). *IC* and *MC* changes little when the actual system is settled. *COST* jumps in July because the electricity cost (is proportionate to *emoney*) rises up in July. The compressor power plays a leading role in the electricity cost at low temperatures while the evaporation mass flow is dominant in the change tendency of *emoney* at high temperatures.

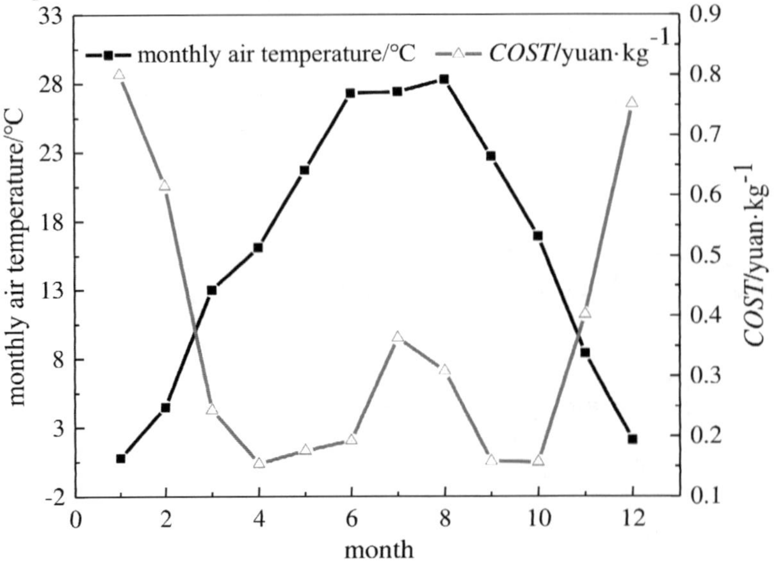

Figure 7. *COST* under monthly optimized operating conditions.

The calculated *COST* (including investment cost, electricity cost and maintenance cost) is 209,300 yuan/yr, evaporation mass flow rate is 833.6 t/yr, sewage treatment is 1000.3 t/yr, concentrated sewage is 166.714 t/yr and average evaporated water per ton costs 251.1 yuan/t.

According to the data provided in Xie's paper [32], the daily evaporated water mass is 50kg and the operation cost is about 6570 yuan/yr, so the average evaporated water cost is 360 yuan/t. So the total cost of the new opened cycle in this paper is nearly one-third lower.

5. Conclusions
The main purpose of this paper is to investigate the thermal process and economic efficiency of the opened sewage concentration air cycle under actual weather conditions. Based on the economic index, the system is optimized monthly and the following conclusions are drawn:

(1) When the evaporating temperature and condensing temperature are constant, the increase of inlet air temperature may decrease inlet air flow and evaporation mass flow, causing a drop in *emoney*.

(2) When temperatures are similar, *emoney* increases with increasing air relative humidity, causing a rise in *COST*.

(3) For the actual system, the average monthly power consumption of evaporated water per unit mass is 0.3508kwh /kg, which is relatively energy-saving.

(4) By optimizing the system monthly, the average the average evaporated water cost under actual weather conditions is 30.3% lower compared with the cost of the closed system in the literature.

The results of the analyses indicate that opened sewage concentration air cycle combined with heat pump is economically feasible for sewage concentration. This study may be improved in future work with the addition of an energy recovery device.

Acknowledgements

This paper is supported by General Project of Chinese National Natural Fund (Project No.: 52076162).

References

[1] Liu Z, Mao X and Jiang H 2019 *Chin. J. Envir. Manag.* **11** 40- 5
[2] Zuo W, Zhou Y and Ge S F 2018 *Envir. Eng.* **36** 47-51
[3] Zhang Z, Jiang H and Gong W Q 2020 *J. Xi'an Jiaotong Univ.* **54** 101-9
[4] Bian X T, Huang Y M, Guo R T and Xu D H 2019 *Inorg. Chem. Indus.* **51** 7-12
[5] Chen M Y, Lan D W and Liu Y C 2018 *Env. Prot. Chem. Indus.* **38** 19-24
[6] Das J, Rene E and Dupont C 2019 *Bioresour. Technol.* **273** 581-91
[7] Zhao Z B, Zhu B X and Li J R 2015 *Fluid Mach.* **43** 76-81
[8] Singh D V and Pedersen E 2019 *Energy Conv. Manag.* ***111*** 315-28
[9] Colak N and Hepbasli A 2009 *Energy Conv. Manag.* **50** 2180-6
[10] Xie J H, Zhou H and Chen D 2008 *Light Indus. Mach.* **26** 25-8
[11] Wang X, Liu J H and Zhao Y J 2017 *Chin. J. Refrig. Techno.* **38** 67-73
[12] Liu S C, Li X Q and Song M J 2018 *Appl. Therm. Eng.* **130** 735-44
[13] Saensabai P and Prasertsan S 2003 *Dry. Technol.* **21** 103-27
[14] Yuan J and Jing S M 2015 *Sci. Techno.* **31** 218-21
[15] Zhou X F, Liu J H and Zhang L 2017 *J. Eng. Thermal Energy Power* **32** 32-6
[16] Zhang H C, Liu J H and Xu X J 2018 *Exp. Technol. Manage.* **35** 103-6
[17] Zhang H C and Liu L C 2020 *J. Eng. Thermal Energy Power* **35** 204-9
[18] Li T Y, Dong X, Hung D L S and Li X S 2018 *Int. J. Heat Mass Transf.* **127** 244-54
[19] Lee S H, Choong S Y and Abdullah L C 2019 *Energies* **12** 41
[20] Zhang M Y 2017 *Fluid Mechanics* (Beijing: Higher Education Press) pp 333-43
[21] Yang C L 1994 *Engineering Design Manual of Plate Heat Exchanger* (Beijing: China Machine Press) pp 79-87
[22] Tao W Q 2001 *Numerical Heat Transfer* (Xi'an: Xi 'an Jiaotong University Press) pp 483-507
[23] Ceng M, Wang Q W and Qu Z G 2002 *J. Xi'an Jiaotong Univ.* **36** 237-240.
[24] Dang Z G, Zhang Z Q, Gao M and He S Y 2019 *Int. J. Heat Mass Transf.* **135** 220-34
[25] Zhao Z G, Shi J L and Zhou C H 2000 *Jour. Hyd. Eng.* **3** 14-20
[26] Liu C W and Li Z M 2000 *Drill. Prod. Techno.* **5** 3-5
[27] Wang P Z, Zhao P and Xu W P 2019 *Appl. Therm. Eng.* **162** 114321
[28] Houssainy S, Janozorgi M and Kavehpour H P 2018 *J. Energy Sto.* **18** 206-17
[29] Wang X R and Dai Y P 2016 *Energy Conv. Manag.* **126** 561-72
[30] Razmi A, Soltani M and Aghanajafi C 2019 *Energy Conv. Manag.* **187** 262-73
[31] Li Z, Tian Q and Guo W Q 2014 *Sci. Techno. Eng.* **14** 228-32
[32] Xie J H and Chen D 2008 *Chem. Equip. Techno.* **29** 8-12

IOP Conf. Series: Materials Science and Engineering 1180 (2021) 012061 doi:10.1088/1757-899X/1180/1/012061

Conjugate Heat Transfer Simulation and Cooling Optimization of the Flow inside a Screw Compressor

J F Willie

Compressors and Vacuum Pumps Systems (CVS) Engineering GmbH, Großmattstraße 14, D-79618 Rheinfelden, Germany

E-mail: james.willie@cvs-eng.de

Abstract. Computational Fluid Dynamics (CFD) as a predictive tool is playing an increasing role in the design and optimization process at Compressors and Vacuum Pumps Systems (CVS) Engineering GmbH. Design optimization using Design of Experiment (DoE) and continuous improvement of our products has led to an increase in the use of CFD and other predictive tools. This paper investigates the use of CFD combined with Conjugate Heat Transfer (CHT) to optimize the flow and the cooling inside a rotary screw compressor.

1. Introduction

Computational Fluid Dynamics (CFD) as a predictive tool is playing an increasing role in the design and optimization process at Compressors and Vacuum Pumps Systems Engineering (CVS) GmbH. Design optimization using Design of Experiment (DoE) and continuous improvement of our products has led to an increase in the use of CFD and other predictive tools. This paper investigates the use of CFD combined with Conjugate Heat Transfer (CHT) to optimize the flow and the cooling inside a rotary screw compressor. Due to the small clearances in these machines, thermal expansion of the rotors can lead to contact between the rotors and between the rotors and the housing. Determining the exact hot gap size will help reduce leakages, increase the volumetric efficiency, and help prevent contact due to thermal expansion. Using CHT to determine the thermal loads on the rotors and CFD to determine the pressure load on the rotors it is possible to find the correct hot gap sizes inside these machines using Fluid-Structure Interaction (FSI). Other critical phenomena like over and under-compression can be prevented with the help of CFD when running off design. Having localized over-compression, for example, can lead to sudden drop in pressures when the discharge port opens and this can induce shock waves that can lead to noise and pulsation. It can also lead to an increase in the power consumption of the compressor and hence to an increase in the temperature. Optimizing the suction and the discharge can lead to an increase in the volumetric efficiency and improve the cooling of the compressor. This paper presents an example of how CFD and CHT can be used in the development and optimization of screw compressors. Similar work has been reported in [1, 2, 3, 4, 5, 6]. In all but one of them, the flow considered was without heat transfer and as a result the wall temperatures were imposed as boundary conditions. In the case that considered heat transfer validation was not done with measurement. In this paper, it is shown that with the use of CFD and CHT it is possible to predict the cold and hot flows that includes heat transfer to account for the losses in the bearings, gears and sealing inside a screw compressor.

The paper begins with the description of the measurement setup. This is followed by the case setup including the description of the geometry for both the cold and hot cases. This is proceeded by the results and discussion and finally, the conclusions and future work are given.

2. Measurement setup

A picture showing the setup of the measurement test stand is depicted in figure 1. The following are measured: Volume flow rate, inlet/discharge temperature and pressure, compressor power, pressure drop (Δp) and the temperatures of the gear casing, compressor housing and the bearing carrier as well as the bearings temperatures of the main and gate rotors.

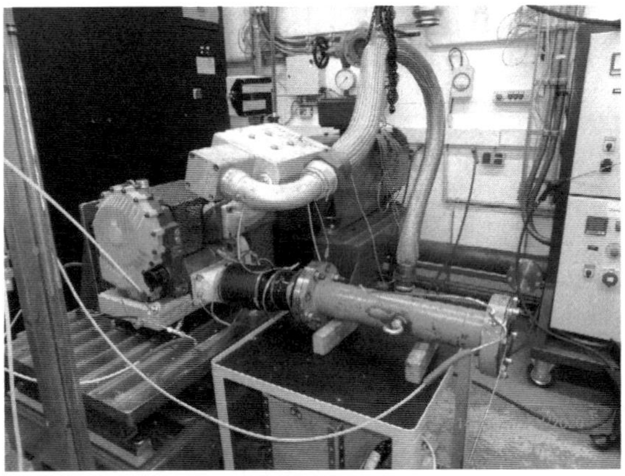

Figure 1. Picture of measurement setup in the laboratory

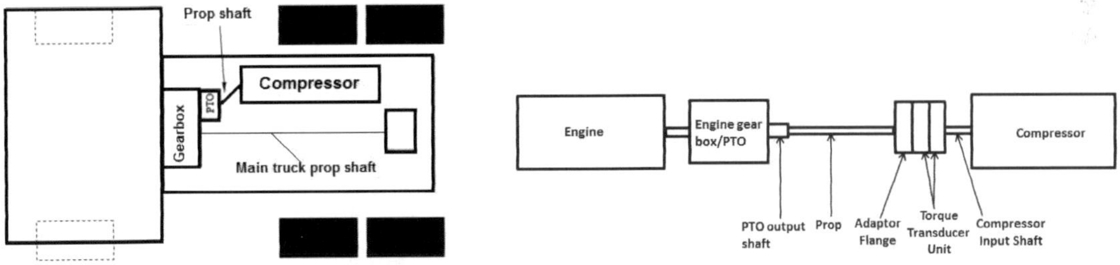

(a) Sketch of truck with compressor

(b) Schematic of the measurements setup

Figure 2. Schematics of compressor on a truck and a torque measurement setup

In real life application, the compressor is mounted on a truck and driven via a Power Take-Off (PTO) drive shown in the schematic in figure 2(a). To mimic the setup on the truck, a prop-shaft with an angle similar to that on the truck is used to connect the electric motor to the compressor as shown in figure 1. Torque is measured using the setup in figure 2(b). It is done by mounting a torque flange from the company HBM onto the compressor shaft and the signal input as a voltage to the Brüel & Kjaer (B & K) pulse noise and vibration analysis system. Microphone and accelerometers are used to correlate the noise and the vibration to the torsional response synchronized by a tachometer signal on the prop-shaft drive coupling. More information on this setup can be found in [7, 8, 9]. Dynamic pressure is measured by using high temperature pressure transducers as reported in [5].

2.1. Operating point simulated

The simulated point reported in this work is for the maximum speed of the main rotor at an input speed of 1800 rpm. The measurement point simulated is shown in table 1. In this table, the volume flow rate is normalized with the maximum designed value and the main and gate rotors speeds are normalized with the main rotor maximum design speed.

Table 1. Screw compressor test parameters used in the CFD and CHT simulation.

Parameter	Measured value
Inlet pressure (p_1)	1.00 bar
Inlet temperature (T_1)	21° C
Discharge pressure (p_2)	3.5bar g
Discharge temperature (T_2)	219°C
Normalized volume flow rate ($\dot{V}$)	0.855
Normalized female rotor speed (n)	0.67
Normalized male rotor speed (n_m)	1.0
Compressor power (P)	70.87kW

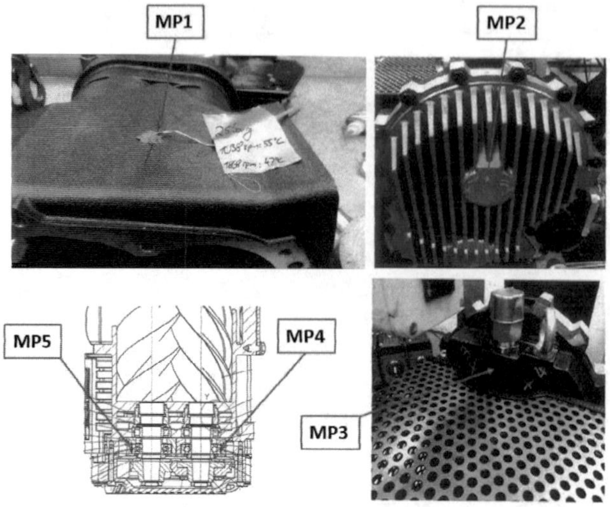

Figure 3. Picture showing temperature measurement points

To enable the simulation and the validation of the CHT case the temperatures at MP1, MP2, MP3, MP4 and MP5 in figure 3 are measured. In addition, the oil sump temperature is measured until steady state is reached, when the temperature of the oil is no longer changing. The oil used is Anderol FG XL 100 food grade oil.

3. Case setup

This section presents the case setups for the baseline (benchmark) geometry of the cold flow and coupled flow with heat transfer. This is important for the DoE part of this work because it would provide a reference for comparison when the optimized geometry of the housing is proposed. Both the cold and the coupled flows with heat transfer are simulated and validated with measurement data. The baseline geometry is denoted as geometry A in this paper. In the CFD setup, monitoring points are defined at each of the measurement points shown in figure 3. The screw compressor simulated includes: Inlet manifold, rotors (main and gate) including the rotor shafts, discharge flange, screw housing , gear casing, bearing carrier (at the non-drive end) and the oil sump as depicted in figure 4. To limit the computational cost due to the many DoE simulations performed and because the oil sump is detached from the rotor housing, the oil sump is not included in the DoE simulations. The cold flow considers only the fluid and the CHT simulations account for both the fluid and the solids. The working fluid is air and the solids are made of the following materials: The gear casing, compressor housing, intake and discharge manifolds are made from Cast Iron while the oil sump and the bearing carrier are made from Aluminum and the rotors and the rotor shafts are made from steel.

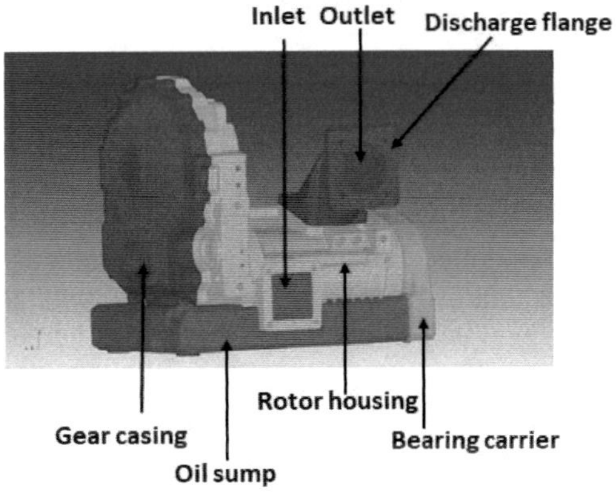

Figure 4. Baseline model (Geometry A) used for benchmarking the simulation and optimization

For the pre-processing CONVERGE Studio are used. It is important to note that the cartesian mesh used in this tool is automatic and adaptive using Automatic Mesh Refinement (AMR). Therefore, only the boundary conditions, the materials and the numerical parameters have to be specified. The simulation uses a pressure inlet and a pressure outlet. The Reynolds Average Navier Stokes (RANS) and the Pressure Implicit with Splitting Operators (PISO) solver implemented in the CONVERGE is used. To account for compressibility the equation of state for ideal gas is chosen because it depicts the behavior of air very well for moderate pressures. The mesh size was influenced by the base grid. Adaptive mesh refining based on temperature and velocity gradients and fixed embedding for the geometry are used in the setup of the AMR. For the cold flow, 8 mm base grid is used. This was decided by performing grid sensitivity analyses of the flow to ensure that the results obtained were grid independent. The mass flow rate, velocity and the volume flow at the inlet and outlet were checked for mesh independence. The gaps (Radial gap between the rotors, housing gap and axial gap) in the geometry correspond to the cold gaps. In reality and during operation, these will change when the rotors and the casing heat up and expand. To account for this in CONVERGE, sealing and porous media are used to reduce the gaps to mimic the hot gaps in the machine during operation. The sealing and porous media are varied until the predicted volume flow rate matches the measured volume flow rate. Grid sensitivity analysis is also repeated for the CHT case. In this case, the setup is done such that losses are imposed at the bearings and gears as heat fluxes using:

$$P_{sha\ t} = P_{isen} + \dot{Q} \tag{1}$$

where P_{isen} is the compressor work given by $P_{isen} = \frac{\kappa}{\kappa-1}p_1\dot{V}[(\frac{p_2}{p_1})^{\frac{\kappa-1}{\kappa}} - 1]$ and $\dot{Q}$ denote the losses from the gears, bearings, and sealing. $P_{sha\ t}$ is the motor power, κ is the ratio of the specific heats taken as 1.4, $\dot{V}$ is the volume flow rate and p_1 and p_2 denote the suction and discharge pressures, respectively. Thus, using equation (1), the losses can be estimated since $P_{sha\ t}$ is measured and P_{isen} can be determined from measured values. With these losses imposed the CHT case is simulated. The details of the setup for the CHT case can be seen in the CONVERGE User Manual [10]. Because the cooling oil is not included in the model, its effect is not accounted for. Thus, the temperature monitored at the 5 points are exaggerated when compared to measurements. The losses are therefore tuned by reducing their values and the simulation repeated until the temperatures monitored are matching those measured within a reasonable degree of accuracy ($\pm$10% of the measurement). Fixed embedding is used in the solid upto level 3 and level 3 AMR is used on the velocity-Sub-Grid Scale (SGS). The CHT tuning is done for two fluid base grids 16 mm and 24 mm and two solid base grids (2 & 4mm) with 5 fluxes applied at the bearings and the gear location and tuned. A $\dot{Q}$ value of 2kW gave acceptable results when compared to the temperature measured at the 5 monitoring points as shown in figure 5. The temperatures are normalized using the maximum simulated temperature of 451°C located at MP5 in the baseline geometry. The same normalization is used in figure 7. The highest deviation in figure 5(a) is 5.4% and it occurs at MP4 while the maximum deviation in figure 5(b) is 7.3% occurring at MP4. The values reported here were taken at an Crank Angle (CA) 1900°. Based on these result it was decided to use a base grid of 24mm in order to limit the mesh size and reduce the computational time and cost. The results of this base grid when compared to measurements is within good agreement. 4 mm was used as base grid for the solid. For the CHT setup interfaces are defined between the solids and between the fluid and the various solids to account for heat transfer. The heat transfer coefficient is determined implicitly by the CONVERGE solver. For the heat transfer between the solids and the surrounding air a value of 10 W/(m^{2}K) is used for the heat transfer coefficient (α). This value is acceptable for the lab conditions because the surrounding air is almost stagnant. Based on results of the mesh sensitivity analyses it can be said that the results reported in this paper are mesh independent and therefore physical. The simulations are done with the CONVERGE solver in parallel using msmpi with 8 cores on a Windows Workstation with an Intel (R) Xeon (R) Gold 6142, 2.6 GHz processor and a RAM of 128 GB. For turbulence modeling, the $k-\varepsilon$ model is used with the law of the wall using the standard wall function [10, 12]. Post-processing is done using Tecplot.

4. Results and discussions

This section presents the cold flow results of the baseline geometry validated with measurement data. Doing this gave us the opportunity to identify potential areas for improvement in the compressor. This is followed by the presentation of the CHT results of the baseline and the optimized geometries. Based on the results the optimized geometry was fabricated using rapid prototyping and tested to check the veracity of our model prediction. The test results are in line with what the model predicted.

4.1. Cold flow

The cold case simulated is compared and validated with measurement data as shown in table 2. The values are normalized using the maximum values for each parameter. CFD is reporting a

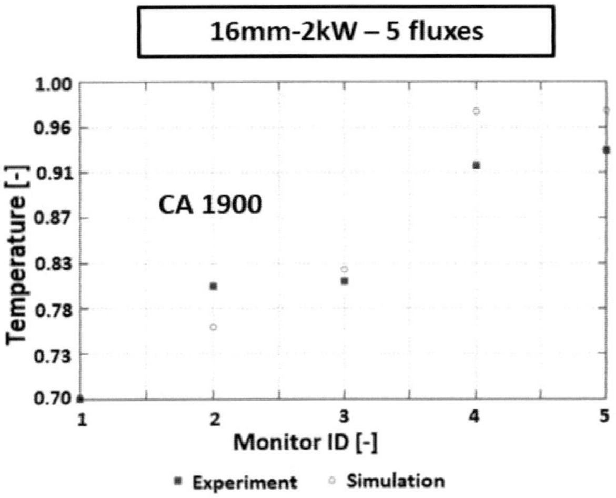

(a) Mesh sensitivity analysis for CHT base grid 16mm

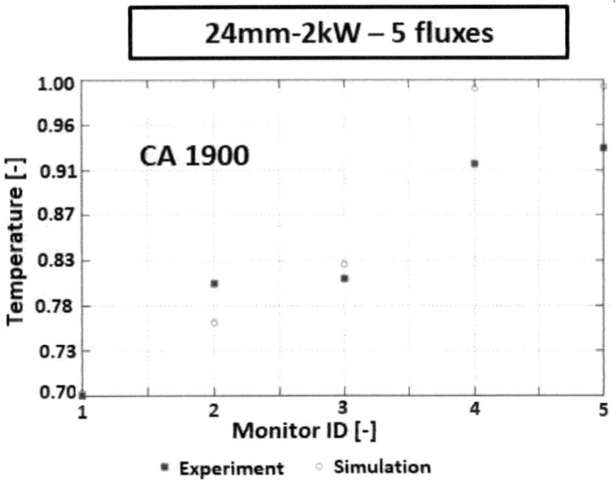

(b) Mesh sensitivity analysis for CHT base grid 24mm

Figure 5. CHT simulation of the baseline case and validation with measurement

Table 2. Comparison of CFD prediction with measurement data

Parameter	CFD	Measured/computed	% Error
Normalized mass flow rate (-)	1.0	0.93	7.3
Normalized volume flow rate (-)	1.00	0.95	5.9
Normalized compressor power(-)	1.0	0.97	2.86
Axial load on male rotor (N)	1943	1732.3	-
Axial load on female rotor (N)	220	95.64	-

larger volume flow and hence the compressor power computed in CFD is larger than measured. This is because in the measurement there is a pressure drop at the suction and this drop was not included in the CFD. The suction pressure was taken as atmospheric (See table 1). The same is true for the main and gate rotor torques. The compression work is done by the main rotor while the gate rotor transports the fluid. Hence, the main rotor torque from CFD is 48.59Nm and the gate rotor torque is 3.35Nm. The reason for the difference between the measured and predicted axial load is because the axial load is not measured but calculated using a simplified analytical model that does not consider the 3D effects like turbulence and over and under-compression. The CFD predicted axial load consider these effects. The trend is however as expected with the main rotor axial load much larger than the gate rotor.

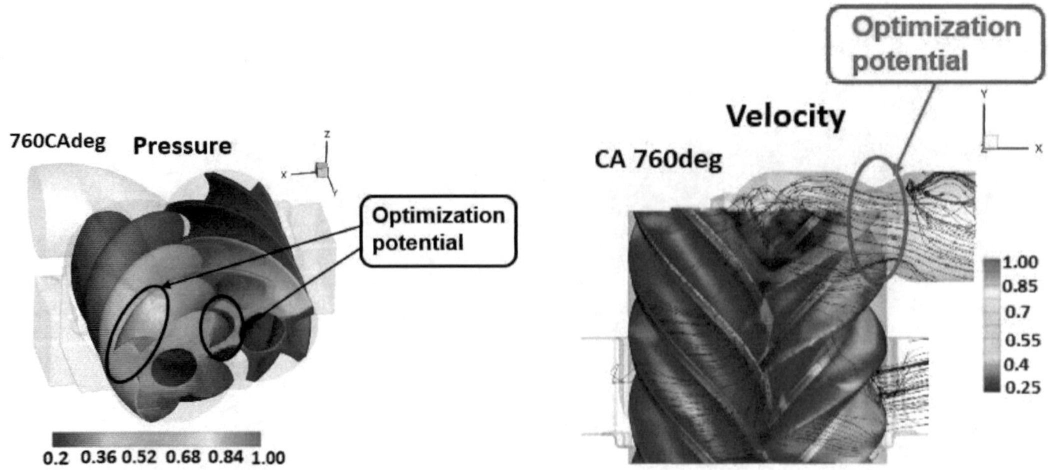

(a) Pressure contour, cold flow (b) Velocity vectors, cold flow

Figure 6. Normalized contour plots of the pressure and the velocity over the rotors

The flow field is analysed by showing the pressure and the velocity contours over the rotors as depicted in figure 6. The average velocity is 45m/s at the inlet and 30m/s at the outlet. The suction velocity is larger than the optimum value allowed for screw compressors, hence it was decided to optimized the suction, which led to an improvement in the volumetric efficiency and cooling of the oil. The pressure contour shows localized areas of over-compression (both axial and radial) close to the discharge as shown in figure 6 [11]. Monitoring points were introduced in the flow close to the discharge to confirm the 3D contour plot results in figure 6(a).

4.2. Coupled flow with heat transfer
The CHT results of the baseline case are presented here and compared to that of the optimized geometry D with the larger inlet. The monitoring points temperature are compared at simulation time 1900deg Crank Angle (CA). The discharge air temperatures are compared at 2800deg CA. Other performance parameters like the rotor torque and hence the compressor power are compared. This is a good indicator for pressure drop inside the compressor. The lower the compressor power the lower the pressure drop Δp.

Figure 7 shows the normalized temperatures of the baseline geometry A compared to the optimized geometry D relative to the measurements. The values are normalized using the maximum value in temperature occurring at MP5 in geometry A. The optimized geometry is showing lower temperatures at MP1, MP4 and MP5, which are crucial in the cooling of the compressor and the oil sump. At MP2 and MP3 the values are close. At simulation time 2800deg CA the discharge air temperature in the optimized geometry D is higher than in geometry A as expected because the air is first used to cool the compressor before it is compressed in geometry D as shown in table 3.

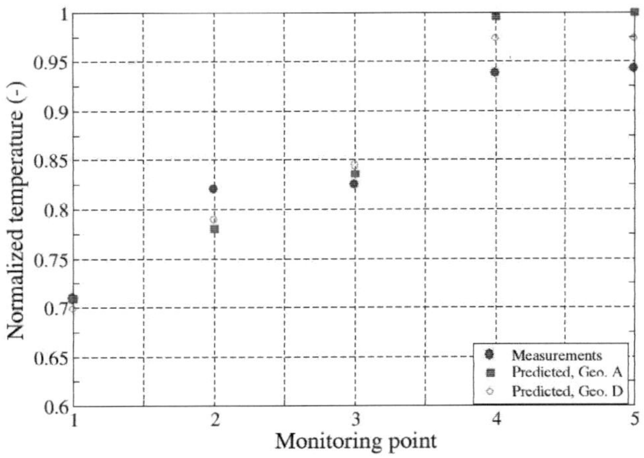

Figure 7. : Normalized predicted bearings and housing temperatures

Table 3. Normalized discharge air temperature comparison between CFD and measurement

Sim. time(CA deg.)	Measured Temp. (-)	Geometry A (Baseline) (-)	Geometry D(Optimized)(-)
2800	1.0	0.9045	0.9085

The compressor power predicted in the optimized geometry is 70.7 kW and that predicted in the baseline case is 73.05kW, which is an improvement since more of the power in the baseline case would end up as heat due to losses and this will in turn lead to an increase in temperature of the cooling oil.

The contour plots of the temperature of the housings and solid rotors using the same scale are compared as shown in figure 8. The trend clearly shows that the cooling in the optimized geometry D is better than in the baseline geometry A. Although the simulation times are different it is important to note that the position of the rotors are identical in all the plots.

4.3. Validation of the DoE results

To check the veracity of the DoE results, geometry D is fabricated using rapid prototyping and measured. Performance maps for the volume flow rate, temperature, efficiencies, etc. are determined and compared with the baseline geometry using the same cooling oil. The measurements are done at input speeds of 1000 rpm, 1200 rpm, 1400 rpm, 1600 rpm and 1800 rpm and at different discharge pressures: No load, 0.5bar(g), 1.0 bar(g), 1.5bar(g), 2.0 bar(g) and 2.5bar(g). In all the tests the same trend is observed and so we report here the results of the worst case scenarios at input speeds of 1000 rpm and 1800 rpm as shown in figure 9. The results are corrected to the standard room temperature of 20°C.

In figures 9(a) and 9(b), the results of the discharge air temperature and oil sump temperature at 1000 rpm are shown. As expected, the discharge air temperature in the optimized geometry increased because the larger volume of air is used to first cool the oil and the rotor casing before it is compressed. At full load, the temperature difference is 10°C between the baseline case and the optimized case but the discharge temperature is still within the customer specification. The higher temperature difference is also due to the higher losses and the higher leakages occurring

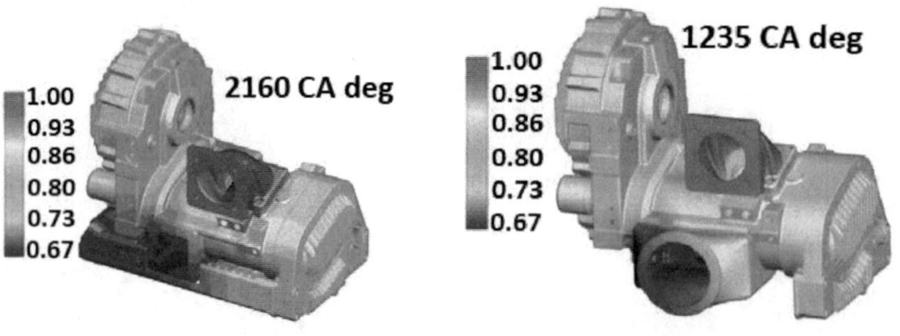

(a) Housing temp., baseline geometry A (b) Housing temp., optimized geometry D

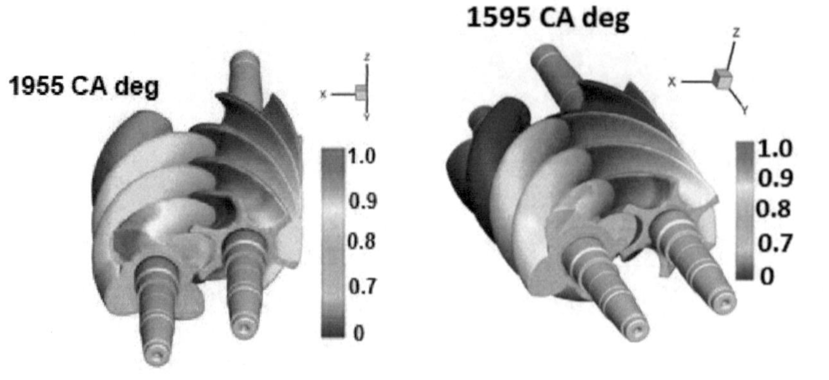

(c) Rotors temp., baseline geometry A (d) Rotors temp., optimized geometry D

Figure 8. Normalized contour plots of the temperature of the housings and the rotors

inside the compressor at the lower speed. However, the oil sump temperature at steady state in the optimized housing is 20°C less than that in the baseline machine. In figures 9(c) and 9(d), the results at 1800rpm input speed are reported. Similar to the 1000rpm input speed case the discharge air temperature in the optimized geometry increases but the increase is lower at 2°C and crucially, the oil sump temperature decreases by 30°C to 123°C, which is very close to the required oil sump temperature of 120°C from our sales department. The trend seen here is in agreement with the prediction made by the CHT results.

To understand why the looses are higher at the lower input speed and why the discharge air temperature is higher in the optimized geometry, the efficiency curves are plotted for the benchmark or baseline geometry verses the optimized housing as shown in figure 10. It is observed that in the optimized housing the design point has shifted from the optimum internal volume compression of v_i= 2.1. At 1800 rpm, the maximum point of the efficiency curve is close to the maximum point in the baseline machine. At 1000rpm the difference is more pronounced and the maximum efficiency is at π=2.5 in the baseline machine and this has shifted to π=2.07 in the optimized case. This means that a further optimization of the discharge contour to determine the optimum internal volume compression is required.

5. Conclusions

This paper reports the successful use of CFD and CHT to optimize and improve the cooling inside a screw compressor. It shows that CFD can be used as an invaluable tool in the design and optimization of screw compressors. This reduces the time and cost of development by limiting

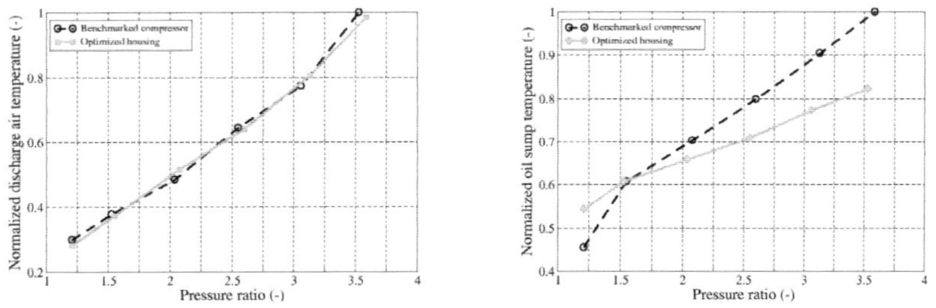

(a) outlet air temp., 1000rpm input speed (b) Oil sump temp., 1000rpm input speed

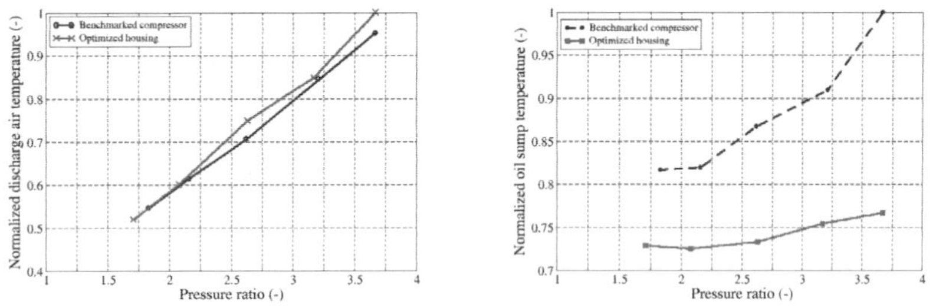

(c) outlet air temp., 1800rpm input speed (d) Oil sump temp., 1800rpm input speed

Figure 9. Normalized measured discharge air temps.

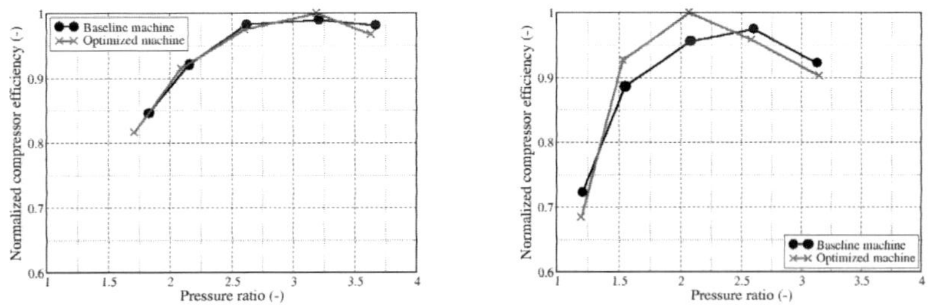

(a) Input speed 1800rpm (b) Input speed 1000rpm

Figure 10. Plots of the normalized measured compressor efficiencies

the testing and the number of prototypes required.

As future work, the optimum internal volume compression required to give the maximum compressor efficiency in the optimized geometry D will be determined using DoE and measurements and presented. FSI will be used to determine the rotors deflection and hence the hot gaps inside the machine.

Acknowledgment
The author would like to thank Timo Martschinke and Andreas Matt for carrying out the measurements. I would also like to thank Riccardo Gretter of Convergent Science for his support.

References

[1] Hauser J and Schulze-Beckinghausen P 2016 *Performance Prediction of Oil-Free Screw Compressor using Computational Fluid Dynamics* (VDMA Compressors Users Int. Forum 2016, Düsseldorf, Germany), September 14-15, pp. 303-312

[2] Rane S, et al 2015 *Analytical Grid Generation for Accurate Representation of Clearances in CFD for Screw Machines* IOP Conf. Ser.: Mater. Sci. Eng. **90** 012008

[3] Spille-Kohoff A, et al 2015 *CFD Simulation of a Screw Compressor including Leakage Flows and Rotor Heating* IOP Conf. Ser.: Mater. Sci. Eng.**90** 012009

[4] Hesse J, Spille-Kohoff A, Hauser J and Schulze-Beckinghausen P 2014 *Structured Meshes and Reliable CFD Simulations: TwinMesh for Positive Displacement Machines* (Int. Conf. on Screw Machines 2014, TU Dortmund University, Dortmund, Germany) September 23-24 2014 (VDI-Berichte 2228) pp. 297-105

[5] Willie, J 2017 *Use of CFD to predict trapped gas excitation as source of vibration and noise in screw compressors* IOP Conf. Ser.: Mater. Sci. Eng.**232** 012021

[6] Willie, J 2018 *Analytical and numerical prediction of the flow and performance in a claw vacuum pump* IOP Conf. Ser.: Mater. Sci. Eng.**425** 012026

[7] Willie J, Asal W and Sachs R 2014 *Analysis of the Noise and Vibration of a Dry Screw Compressor* (Int. Conf. on Screw Machines 2014, TU Dortmund University, Dortmund, Germany) September 23-24 2014 (VDI-Berichte 2228) pp 91-308

[8] Willie J and Sachs R 2015 *Structural and Torsional Vibration Analysis of a Dry Screw Compressor* IOP Conf. Ser.: Mater. Sci. Eng. **90** 012005

[9] Willie J and Sachs R 2016 *Structural and Torsional Vibration and Noise Analysis of a Dry Screw Compressor* (Proc IMechE Part E. J Process Mechanical Engineering 0) vol 0, DOI: 10.1177/0954408916648989

[10] CONVERGECFD Manual Series 2019 *Conjugate Heat Transfer Modeling*, (CONVERGE CFD Software) pp. 521-536

[11] Fister W 2013 *Fluidenergiemaschinen: Physikalische Voraussetzungen, Kenngrößen, Elementartstufen der Strömungs-und Verdrängermaschinen*, (Berlin: Springer-Verlag, Band 1)

[12] Pope S B 2000 *Turbulence Flows*, ed 1 (Cambridge University Press)

IOP Conf. Series: Materials Science and Engineering 1180 (2021) 012062 doi:10.1088/1757-899X/1180/1/012062

Reducing the noise from the cavity by tuning the compressor's oil level

Paulo Martins, Ph.D.

Secop Austria GmbH, Mühlwaldstraße 21, 8200, Gleisdorf, Austria

paulo.martins@secop.com

Abstract. This paper proposes a unified solution for noise of two different versions of the same compressor. First, the problem was identified as a high noise in 1kHz third-octave band from the version using R134a refrigerant and 1.25kHz band from R600a version. The first hypothesis was the noise coming from the motor through shifting frequency, which would explain the different frequency ranges for different refrigerants. However, this was debunked experimentally, and results are shown in this paper. The other hypothesis was the noise coming from the cavity. This was proved to be the problem's root cause and this paper shows the simulated and experimental results which corroborates the inference. Therefore, we propose a single solution for both versions by using an optimized amount of oil, to control the cavity volume.

1. Problem Definition and Literature Review

The object of study is a battery driven compressor that weights approximately 1.5kg, and measures 130mm wide, 110mm long and 100mm tall. Hence, its cavity volume is very small, measuring approximately 320 cm^3. This compressor's family has 2 different versions, one dedicated to run with R134a and the other with R600a, however, the same 1.4cc version was tested for both types of refrigerants, due to availability of parts.

With R134a, the noise problem was identified in between 800Hz and 1kHz 1/3rd octave bands, and with R600a, between 1kHz and 1.25kHz.

These problematic regions were also sensitive to the compressor speed. While running on 2300 RPM (hereinafter referred as "low speed"), the effect was less harsh. On the other hand, the compressors showed a higher noise while running with 4500 RPM (hereinafter "high speed").

The results shown in Figure 1 (a) and (b) are from running noise measurements with evaporator and condenser temperatures at -10°C and +55°C respectively, following CECOMAF standards. The noise was measured according to ISO-3743 for reverberant rooms, and the results are displayed in Sound Power Level (normalized for data protection).

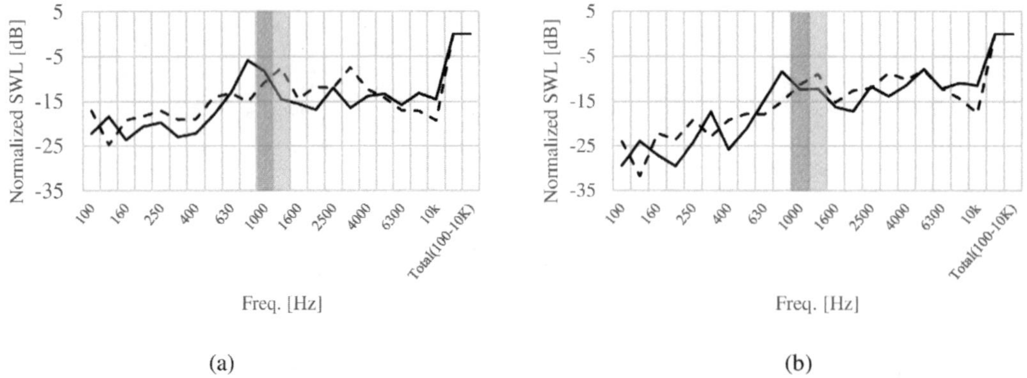

Figure 1. Critical bands at High Speed (a) and Low Speed (b), with R134a shown in full line and R600a in dashed line.

Similar problems were investigated by [1] and [2]. They approached the characterization of cavity resonances and identified the changes in oil level as significant to change the cavity's modal density in low frequencies. However, they focused on changes of shell design, which can have a high cost, depending on the maturity of the project.

On the other hand, the work of [3] awares about noise problems that could occur through transmission of vibration. Therefore, we also point out the importance of optimizing the oil level considering all different aspects of noise generation.

2. Cavity Hypothesis

In the meantime, we carried several simulations of the cavity acoustics, as an attempt to determine a correlation with noise from critical bands. ANSYS® 2019 R2 was used, the acoustic mesh consisted in both cavity and suction muffler, with elements of type FLUID221 with edges of 1mm length, hence the mesh had approx. 180 thousand elements and 292 thousand nodes. The element FLUID221, according to ANSYS® Reference [4], is a higher order 10-node solid element that with quadratic shape functions.

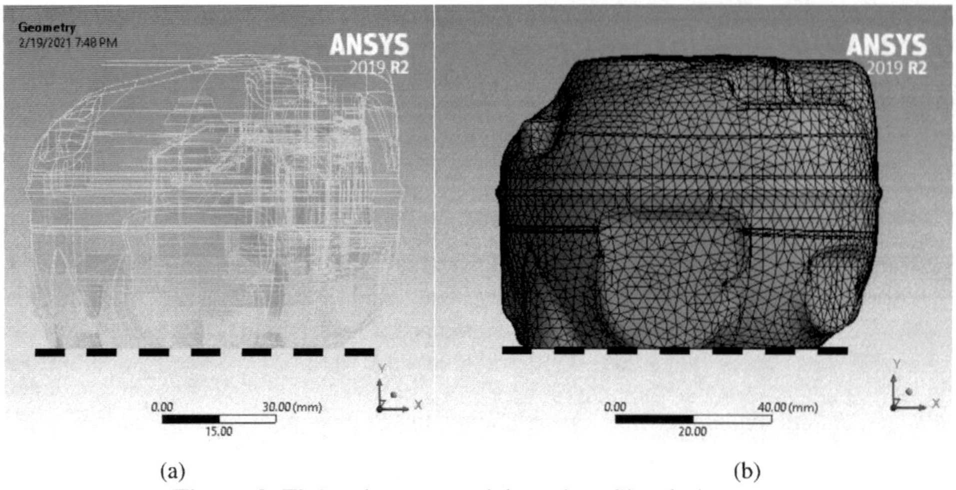

Figure 2. Finite element model used on Simulations.

An unitary surface velocity was prescribed at the suction port on the cylinder head, and an average pressure was calculated from all surfaces that had direct contact with the shell (hereinafter referred as "cavity bounds").

To obtain the proper cavity's gas-filled geometry, we used a cutting plane (identified on Fig. 2 as a black dashed line) and neglected the oil-filled volume. Afterwards, to simulate changes in oil level, we simply moved this cutting plane to the desired position.

With approximately 15ml of oil, the cavity response had an interestingly similarity to the noise of tested samples. The results below represents the simulation versus maximum and average narrow band values of all speeds, from all samples.

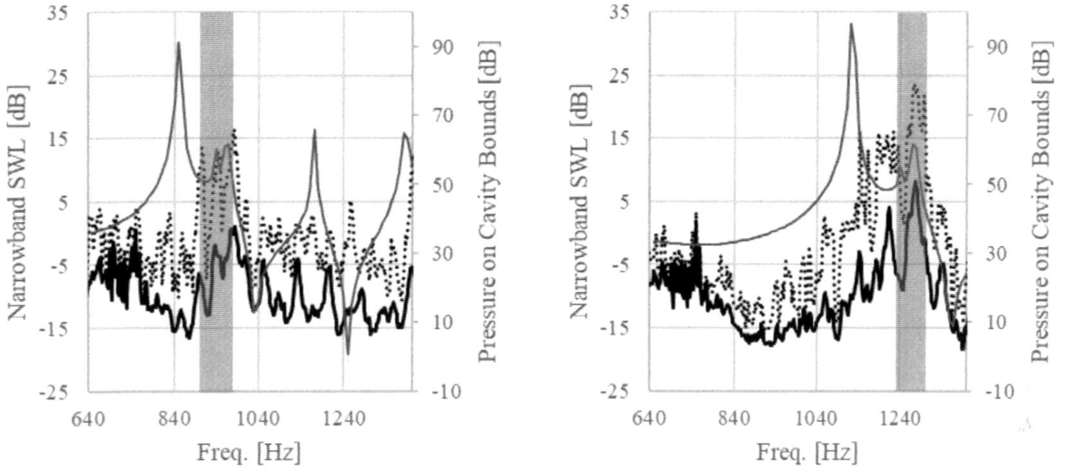

(a) (b)

Figure 3. Narrowband SWL versus pressure on Cavity Bounds for R134a (a) and R600a (b)

Therefore, because of this outstanding similarity, we started to investigate further the cavity-oil relationship.

It appears the mechanism for noise generation was highly connected to a specific "cluster of modes". This cluster was formed by 2 main modes, which high pressure regions are located on the shell, aligned with the front of the piston (see Figure 4 below). Since this is a rather flat surface, the pressure pulsation could excite the shell and, finally, generate noise.

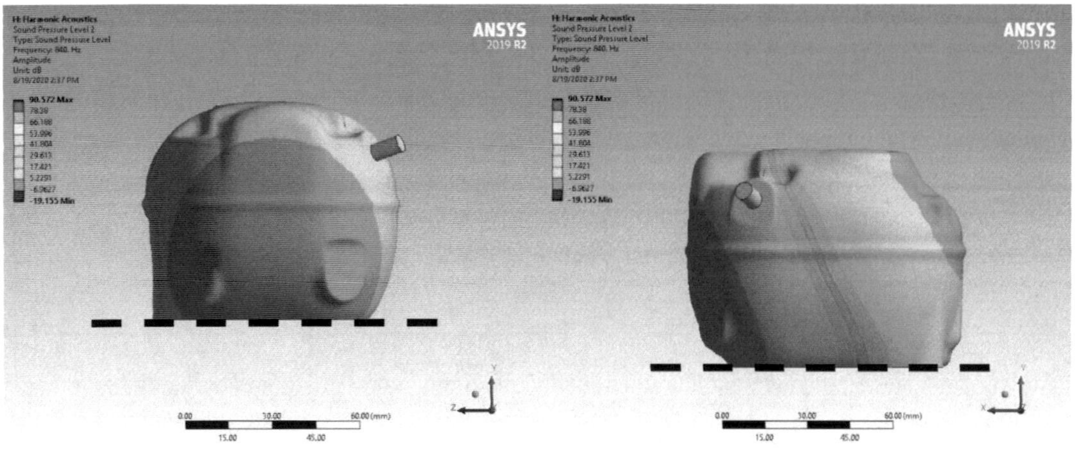

(a) (b)

Figure 4. Example of problematic cavity mode at the front of the piston (a) and on the side (b).

Finally, we performed a heuristic optimization and, when comfortable with the results, we validated the findings with another round of measurements.

3. Heuristic Optimization

After seeing the high sensitivity to oil level, we saw a strong potential for tuning it. Thus, we performed a heuristic optimization by varying the oil level, gas temperature and compressor inclination.

Simulations with oil level variation consisted in a fixed temperature for the refrigerant of 60°C, a desired frequency range of 640Hz to 1.4kHz, spaced in 76 solution intervals. We chose this coarse frequency resolution to avoid long calculation times.

Within this temperature, R134a has a speed of sound $c = 170.67\ m/s$ and density $\rho = 2.15\ kg.m^{-3}$, and R600a has $c = 225.42\ m/s$ and $\rho = 1.23\ kg.m^{-3}$.

For the geometry variation, we started with 15ml of oil, which was arbitrarily chosen as the reference plane of cut. Then, by displacing this plane upwards 2mm, 5mm and 7mm to respectively achieve 23ml, 37ml and 47ml of oil. The results of average pressure on the cavity bounds are shown below for R134a and R600a.

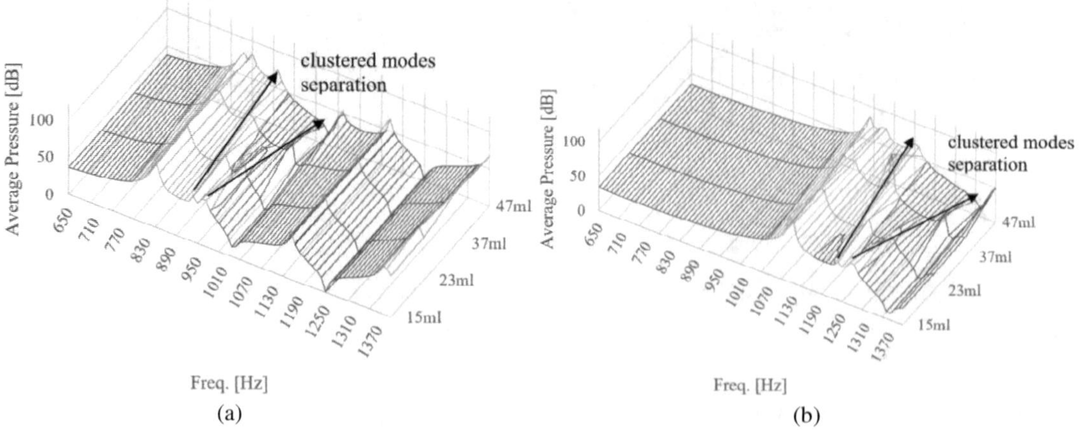

Figure 5. Average pressure on Cavity-Shell interface with oil variation for R134a (a) and R600a (b)

We can observe from the results a clear separation of the cluster of modes. The mode from higher frequency even gets filtered away by the suction muffler, as expected.

These results also show how this small compressor has a sensitive cavity; thus, further simulation was carried out to ensure safe margins before investing resources on experiments.

The next round of simulations took into consideration the effects of temperature variation over a fixed amount of oil. Hence, we considered 23ml of oil (reference plane +2mm), and we simulated for 50°C, 70°C and 90°C. The used refrigerant properties can be seen in the table below.

Gas	Temp. [°C]	c [m/s]	ρ [kg.m^{-3}]
R134a	50	168.04	2.22
	70	173.25	2.09
	90	178.25	1.97
R600a	50	222.07	1.27
	70	228.7	1.19
	90	235.11	1.12

Table 1. Refrigerant properties for Temperature Variation.

The average pressure on cavity-shell interface are shown on the figures below. The major difference is a shifting of the resonances towards higher frequencies, as expected. Also, no new resonances interactions between muffler and cavity appeared, allowing us to conclude this design is robust from the temperature variation point of view.

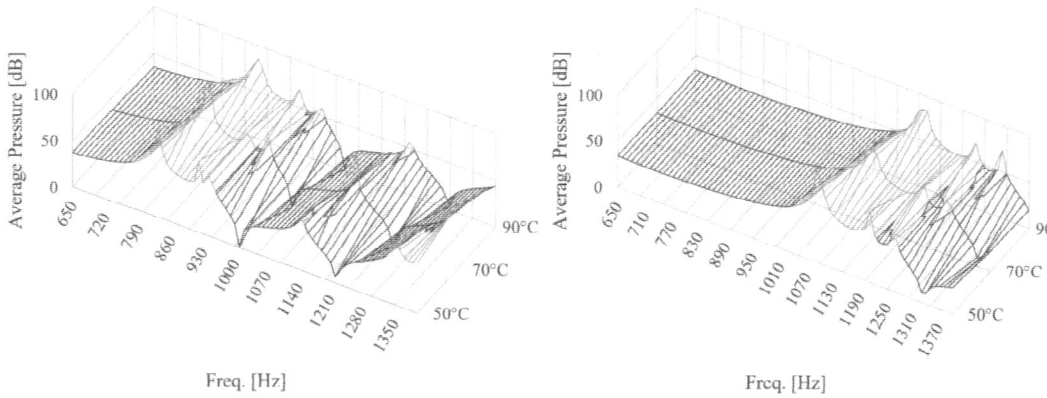

(a) (b)

Figure 6. Average pressure on Cavity-Shell interface with temperature variation for R134a (a) and R600a (b).

Finally, out of curiosity, we also simulated the effects of inclination on the cavity acoustics. Since the effects for R134a and R600a were known to be similar, we only simulated with the former. The used geometry can be seen in the Figure 7 below, and results in the next Figure 8.

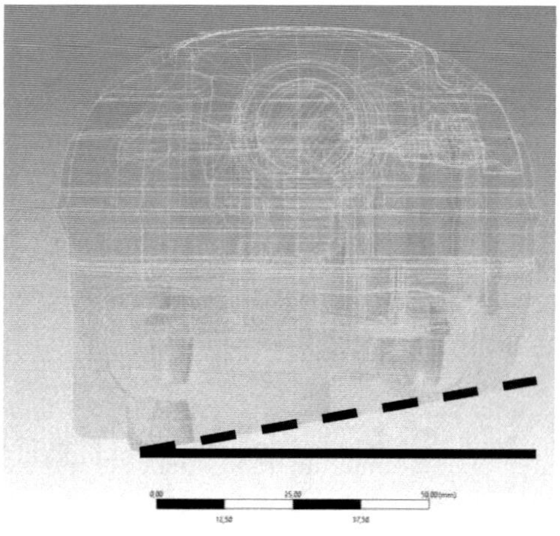

Figure 7. Geometry with oil level at 10° of inclination.

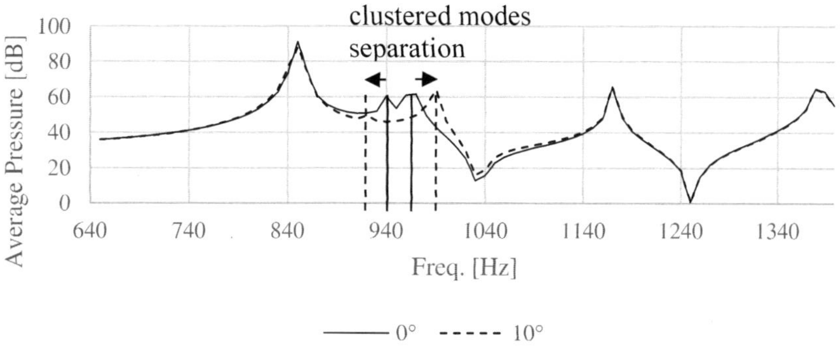

Figure 8. Average Pressure on Cavity-Shell interface with inclination variation for R134a.

IOP Conf. Series: Materials Science and Engineering 1180 (2021) 012062 doi:10.1088/1757-899X/1180/1/012062

The same separation of the clustered modes was observed, suggesting a good palliative if we are unable to re-assess the oil level, e.g., if the compressor is already assembled in the appliance and kit rework is unfeasible.

4. Experimental Validation

The experimental validation consisted in several noise measurements according to ISO-3743, with the compressor running filled with R600a at -10°C/+55°C T_{evap}/T_{cond} CECOMAF conditions.

We started with the lowest oil level, approximately 48ml which corresponds to 53g. To control the oil amount by its mass is preferable because it allows easy checking between measurements by using a simple scale. Hence, after each measurement, the compressor was weighted, and its oil level was increased by steps of approximately 2.5g. The results are displayed in Sound Power Level, but it was omitted due to data protection.

Results with compressor running at 2300 RPM, 3000 RPM, 4000 RPM and 4500 RPM are shown respectively in Figures 9 (a), (b), (c) and (d).

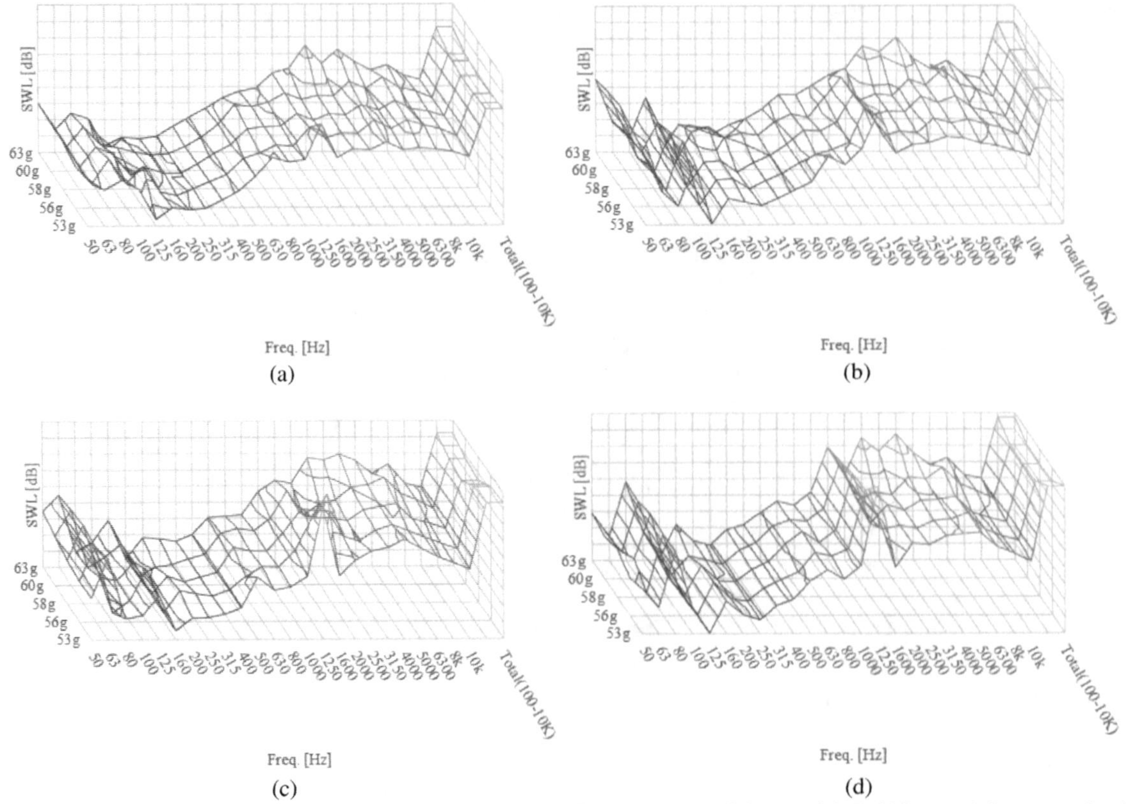

Figure 9. SWL of compressor at -10°C/+55°C CECOMAF conditions with R600a, running speed of 2300 (a), 3000 (b), 4000 (c) and 4500 RPM (d), filled with different oil levels.

We could clearly observe the appearance of a sweet spot for oil level within the cavity bands. With oil increase, the problem in cavity bands would decrease. For better visualization, we took an average of all speeds and normalized each band by its maximum, resulting on the surface plot of Figure 10.

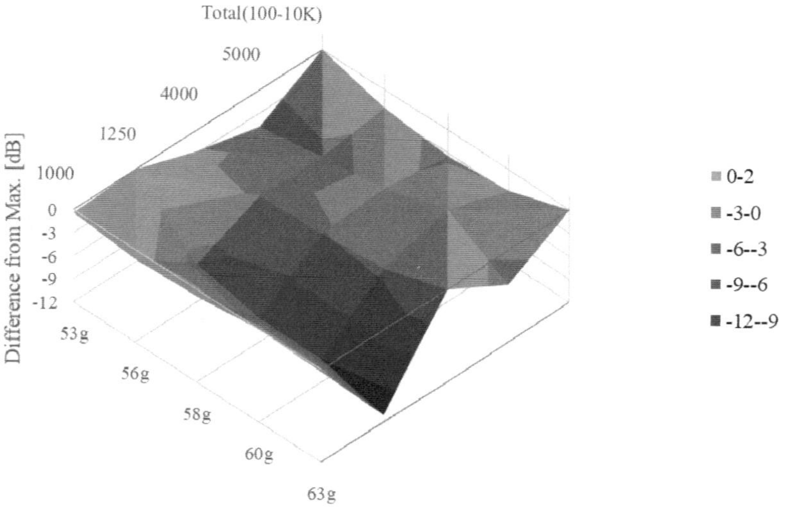

Figure 10. Average differences from maximum SWL from each critical band, from all speeds. Compressor running with R600a, -10°C/+55°C CECOMAF conditions with different oil levels.

Another interesting conclusion can be drawn by observing this plot above: Even with the decrease of noise coming from the cavity bands, the 4k and 5kHz bands showed a significant increase. This may have happened because of direct transmission of vibration through the oil if it touches the stator and rotor at the same time.

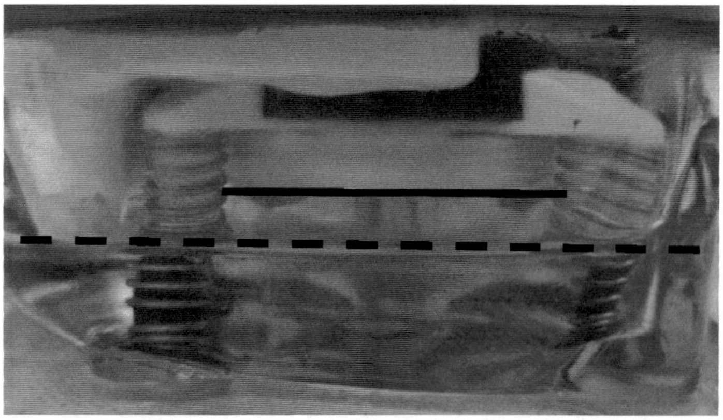

Figure 11. See-through acrylic shell shows 58g of oil (dashed line) is already dangerously close to stator (full line).

Further investigation with an acrylic shell showed the amount of 58g of oil was already dangerously close to the stator, and any inclination could make the oil surface to touch the moving parts and transmit vibration directly.

However, since this phenomenon was observed later and requires a totally different approach for understanding, it will not be discussed in the present paper.

5. Conclusion

This paper covered the issue of a single compressor having distinct high noise bands for distinct refrigerants.

The first hypothesis was the noise being influenced by the electric motor through frequency shifts with different loads. But this was quickly debunked experimentally. The test consisted in maintaining the same pressure on the evaporator and ensuring the same power consumption, so the high noise should come from the same frequency ranges. However, we observed the same problematic noise bands, even with these specific conditions.

The second hypothesis was the cavity acoustics. Simulations showed strong similarities between a cluster of modes and sound power level measurements. Simulations with oil variation, temperature variation and inclination also increased the strength of this statement.

Finally, the observations made from the simulations were confirmed experimentally by testing a compressor with R600a refrigerant, varying its oil level in steps of 2.5g, measuring its running noise with different speeds.

The correlation of oil increase *versus* cavity noise decrease was confirmed, but another harmful effect was observed at 4k and 5kHz bands. This was due to the oil level touching moving parts and transmitting the vibration directly to the shell. This effect was also pointed out by [3], however there are few studies available.

We recommend further studies regarding the direct transmission of vibration through oil, and its effect on the total noise from the compressor.

References

[1] Bucciarelli, M.; Giusto, F.; Cossalter, V.; Lio, M. Da; and Gardonio, P., "Modal Analysis of a Compressor Shell and Cavity for Emitted Noise Reduction" (1992). International Compressor Engineering Conference. Paper 922. https://docs.lib.purdue.edu/icec/922

[2] Albrizio, F.; Genoni, C.; Bianchi, V.; and Frontini, G., "Noise Reduction of Hermetic Compressor by Identification of the Gas Cavity Resonance" (1990). International Compressor Engineering Conference. Paper 753. https://docs.lib.purdue.edu/icec/753

[3] Soedel, Werner, 1936, Sound and vibrations of positive displacement compressors / by Werner Soedel. p. cm. ISBN 0-8493-7049-3

[4] Ansys® Mechanical APDL 2019 R2, help system, Chapter 7: Element Library, FLUID221, ANSYS, Inc.

Acknowledgements

The author would like to acknowledge the technical staff of Secop Austria GmbH, the “Lab Guys”, who made this research possible. Also, to thank Secop for the nurturing environment.

IOP Conf. Series: Materials Science and Engineering 1180 (2021) 012063 doi:10.1088/1757-899X/1180/1/012063

CFD analysis on the effect of discharge port geometry of the hook and claw vacuum pumps

K Theofanidis[1], **Y Lu**[2] **and A Kovacevic**[3]
[1,2,3]Center for compressor technology, City University London, EC1V0HB, UK

E-mail:[1] kyriakos.theofanidis24@gmail.com, kyriakos.theofanidis@city.ac.uk

Abstract.Due to the sharp discontinuities in the profile of hook and claw pumps which make it difficult to produce a good quality body fitted mesh, their CFD simulations are challenging. However, the cut-cell Cartesian meshing method is capable to handle the very complex moving geometries and is easy to setup. However it may lead to conservativeness issues. The aim of this paper is to investigate the influence of discharge port geometry of the hook and claw pump on its performance by using commercial software ANSYS Forte, which utilizes automatic mesh generation using the cut-cell Cartesian method. Firstly, five discharge ports with different geometries were designed. Secondly, the pumps with these different port configurations were set up and ran in the ANSYS Forte. Finally, the power and volume flow rate were compared. It was noticed that the geometry with the extended discharge port tip and the preliminary discharge port opening result in performance and reliability improvements. The simulation results can be used to further optimize the discharge port during the design of these machines.

1 Introduction

The procedure of analyzing claw vacuum pumps involves rotor profile design, using various mathematical principles and coding, CAD design and CFD simulations. Since the use of vacuum pumps is increasing rapidly over the years, properly executed analysis and innovation will be important for the evolution of them. They consist of two rotating rotors and a casing. The rotating rotors trap and move the air that is coming from the inlet towards the outlet discharge gate. Previous researches have gone very deep into the understanding of their working principle using many software packages such as SCORG, ANSYS Fluent, Star CCM+ and others. For this case another leading CFD package was used: ANSYS Forte. It is utilizing the cut-cell Cartesian mesh generation method which is not used a lot to investigate rotating machines. Thus, the accuracy and stability of this method was examined. The main component of study is the port designs, which is a not much previously researched area. The geometries of the inlet and discharge ports play a significant role in the operation of the pump, and no major investigation showcased that until now. In order to conduct this analysis, three different discharge port configuration models have been modeled and compared in terms of performance, efficiency and reliability.

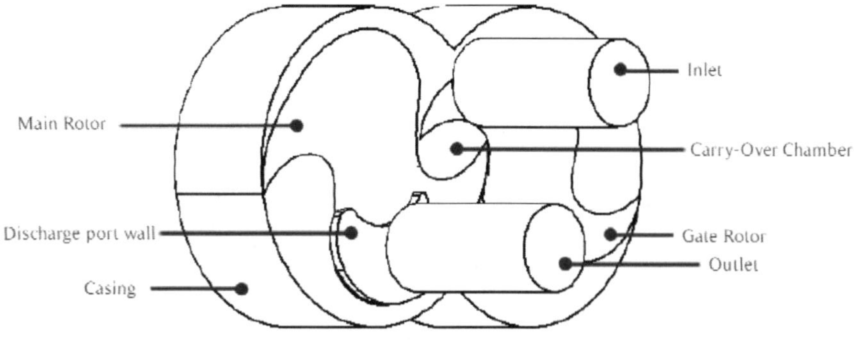

Figure 1: Claw pump model for CFD analysis

2 Background

2.1 *Vacuum pumps overview*

Vacuum pumps and compressor systems are rotary positive displacement machines that include only a few rotating and moving parts. Kovacevic et al. (2006) [1] presented the operation of positive displacement rotary machines and their principles of operation while it is one of the few books that facilitate an extensive study in CFD analysis. The pair of the moving parts are the rotors, that mesh with each other utilizing their helical groove design. Some pumps, like the one that is being investigated, do not have grooves and the rotors are pretty short and straight, while still being able to pump a significant amount of gas. These pumps are known as claw pumps. This is achieved due to the meshing geometry of them, which utilizes the mathematical approached envelope method. The two rotors are firmly surrounded by a casing [2]. In between the rotors and the casing, a radial clearance exists, in order to ensure smooth operation and minimal wear. The vacuum pumps have a working chamber that the meshing rotors form by rotating and they move it while it is filled with air from the inlet towards the outlet discharge port. The two rotors are synchronized by timing gears which are also transmitting the motor's power to move them [3][4][5]. Besides the inlet, the compression, and discharge processes, claw vacuum pumps form a special kind of process which is called the mixing process. The mixing process, is observed at the carry-over chamber which is formed in between the two rotors. The chamber accommodates gas after the discharge process and gas from the inlet process (*Figure 1*).

Stosic et al. (2005) [2] presented the main advantages of screw machines. Screw machines only consist of rotating parts and not translating ones and thus they can run on higher speeds. The contact forces are low and therefore wear is minimized. Also, the pressure inside rises and this results in less gas leakage. Furthermore, Tuo et al. (2018) [6] and Wang et al. (2015) [7] showcased some more advantages including: high reliability, stable medium conveying, small vibration, oil-free usage, and compact structure.

Regarding the disadvantages, Lu et al. (2017) [8], presenting the working process in the twin screw vacuum pumps, denoted the most important one of them: Very high rotational speeds can significantly decrease the discharge pressure whereas low speeds can lead to over consumption. Thus, speed affects the performance dramatically and may lead to failures. Except that, Kovacevic et al. (2007) [9] mentioned that cavitation and erosion at the rotor shafts and gaps also lead to damage.

2.2 *Design analysis*

The main focus in the design process of a claw vacuum pump is the generation of the rotors. The shape of the rotors defines the main performance characteristics of the pump which are related to the displacement and the leakage flows. The rotor design is achieved usually within a coding software like MATLAB or FORTRAN with the use of derived geometrical equations that define the curve plot of each rotor. Stosic et al. (2018) [10] provided a clear mathematical approach on rotor generation which used the envelope method to acquire the meshing points and meshing conditions of the rotors. Later on, a set of coordinates and their derivatives were obtained generating a variety of primary curves. The rest of the curves were calculated automatically using numerical series. For the optimization of the rotors, an investigation of each rotor feature must be conducted separately. Many studies have focused in the analysis of different aspects of the rotors with impressing results. For example, Yan et al. (2017) [11] noted that the use of involute-cycloid profiles leads to increased efficiency that results in larger mass flow rate while also larger mass flow pulsation. Also, Lu et al. (2015) [12] added that by combining a large flow cross section area with a small sealing line and a small blow hole area will lead in higher efficiency.

Lastly, Stosic et al. (2011) [13] highlighted that despite the level that rotor generation has reached, there is always more space for improvement in making lighter and more efficient rotors. Moving on from the rotors, other characteristics regarding the casing also affect the performance of the pump. These include: The port design, the axis distance, the axial clearance and the radial and interlobe gaps.

2.3 *Computational fluid dynamics (CFD)*

Computational fluid dynamics (CFD) is a way of applying numerical methods to analyze the heat transfer and fluid dynamics governing equations in order to acquire fluid flow problem solutions. In order for this to be achieved, modern computers require algebraic equations, that have been discretized from continuous differential ones, in order to calculate a solution. Under the hood of a CFD package like ANSYS Fluent and ANSYS Forte, there are a series of governing equations that tackle the procedure of calculations. These equations are basically the conservation equations which include: The continuity equation (1), the momentum conservation Navier-Stokes equation (2), the space (3) and energy (4) conservation equations.

$$\frac{\partial \rho}{\partial t} + \frac{\partial \rho u_i}{\partial x_i} = 0 \tag{1}$$

$$\rho\left(\frac{\partial \vec{u}}{\partial t} + \vec{u} \cdot \nabla \vec{u}\right) = -\nabla P + \nu \triangle \vec{u} \tag{2}$$

$$\frac{\partial}{\partial t}\int_V dV = \int_S \vec{V_b} dS \tag{3}$$

$$\rho\frac{DV}{Dt} = -\nabla q - p(\nabla \cdot V) + \rho\dot{Q} \tag{4}$$

The above equations are used to form the 2 basic approaches for a study of a flow which are: The simulation and the modeling techniques. The simulation ones, include the direct numerical simulation (DNS) and the large Eddy simulation (LES). DNS and LES, have very specific applications and their computational cost is very high. Thus, the industry has only started to use them the latest years, that the computational power has been increased, and only as a separate optional feature in CFD solvers. Nowadays, the most common approach in numerical flow studies is the modeling technique called Reynolds Averaged Navier Stokes (RANS) method. RANS is the oldest method of CFD analysis. It is very simple, it has a low computational demand and it is very efficient for turbulent flow analysis. ANSYS Forte, which is the solver used for this investigation also follows the RANS approach, aiming to simulate the averaged flow field. This approach utilizes the RANS equations themselves (5), (6) and the Reynolds stresses equation (7) presented below [14][15]:

$$\frac{D\bar{u}}{Dt} + \frac{1}{\rho}\nabla\bar{p} - (\nabla \cdot \nu\nabla)\,\bar{u} = \nabla \cdot \tau \tag{5}$$

$$\frac{\partial u_i}{\partial x_i} = 0 \tag{6}$$

$$\tau_{i,j} = \rho(u_i' u_j') = \rho(u_i u_j) - \rho(u_i)(u_j) \tag{7}$$

CFD provides a deep insight into the flow and thermodynamic behavior of positive displacement machines. As the computational technology advances, major steps forward are being made in the numerical analysis methods. Information like the mass flow rate, the rotors torque and power, the pressure distribution and velocity field can all be obtained from a CFD solver [9][11]. One of the most important aspects of a CFD analysis is the mesh quality. A good mesh will ensure accurate, efficient and robust results. A lot of CFD studies have been conducted focusing on the mesh generation and analysis. Rane et al. (2018) [16], mentioned that due to high magnitude of deformation of the domain and the geometrical complexity of the rotors, many commercial flow simulation packages are not suitable for screw machine analysis.

ANSYS Forte has a capable model for positive displacement machine analysis and the integrated mesh generation utility is useful and powerful for flows like that. The most common used CFD solvers, utilize the simple body-fitted structured mesh generation approach. A structured grid is defined as a mesh that its nodes have the same number of elements around them. It follows a specific shape pattern that is being repeated in all directions in the investigated geometry. Structured grids are most commonly seen in regular bodies with no complex geometries that allow the reproduction of the same grid (usually polygonal) shape. The rotors of claw pumps though, have very sharp and pointy orientations. This complexity of the shape, makes the meshing almost impossible when trying to use the body-fitted structured mesh approach [15]. This is where ANSYS Forte grid generation shines. ANSYS Forte uses a special kind of structured not body-fitted meshing technique called the cut-cell Cartesian approach. The cut-cell Cartesian method, as its name suggests, divides the examined area using a normal Cartesian grid which adapts to the geometry. A valuable benefit of it is the excellent near wall area treatment resulting in high accuracy.

Despite all the above characteristics, the cut-cell Cartesian method has some disadvantages. The one with the most impact, is the fact that the mesh is not generated densely enough at the leakage areas and therefore additional models are required to cover this gap. This means that for the claw pump case, the clearances between the casing and the rotors need additional mesh models and this results in a more expensive computational cost and also it might lead to inaccurate results. This is why, for some cases this method is longer and more difficult to be conducted with meshing errors often appearing. Also, it is not fully conservative making it less popular for simpler designs[17][18].

2.4 *Port design*

The ports of a pump are very important elements affecting the performance. They are responsible for bringing gas in and out of the working chamber of the pump. Yan et al. (2016) [19] in the research of numerical modeling of twin-screw pumps mentioned that the port geometry has little effect on the mass flow rate but has significant effect on power. Also, it was noted that the port shape has possible effect on pressure distribution and shaft power. Other than that, no major investigations have been made for the port geometries of a pump. Regarding the inlet, a circular shape is most commonly used. In regards to the discharge though, many alternations of the shape exist. The geometry can differ in elements that affect the angle of the rotor rotation that the discharge process begins and the discharge volume of the carry-over chamber.

3 Rotor profile design

The cycloid is a widely used curve in the pump rotor profile design. The rotor profile used in this case study consists of straight lines, cycles, cycloids, polynomial curves, and corresponding involute curves as demonstrated in *Figure 2* below. The main rotor is defined as the rotor connected with the motor. It is defined in the global coordinate system X_1, Y_1, Z_1 while the gate rotor is defined in the local coordinate system X_2, Y_2, Z_2. The first straight line curve A_1B_1, the second curve sector B_1C_1, the third cycloid curve C_1D_1 and the fifth curve E_1F_1 are defined in the global coordinate system while the fourth curve D_2E_2 is defined in the local coordinate system. The rest curves can be obtained using the theory of involute gears [20].

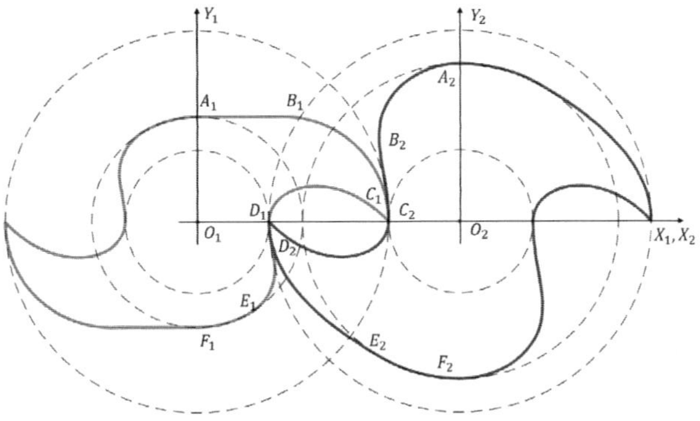

Figure 2: Rotor Profiles

To demonstrate the function of every curve, the design parameters are shown in *Table 1*. The inner and outer circle radius are r_i and r_o individually. The axis distance is $Ac = ro + ri$ and the pitch circle radius is $rp = (ro + ri)/2$.

Table 1: Rotor parameters

Parameter	Value	Parameter	Value
Rotor combination	1by1	Rotor length	25 (mm)
Axis distance (Ac)	45 (mm)	Axial clearance	0.08 (mm)
Outer circle diameter (ro)	65.42 (mm)	Radial gap	0.07 (mm)
Inner circle diameter (ri)	24.42 (mm)	Interlobe gap	0.08 (mm)

4 Port design

In order to do the discharge port comparison, a base claw pump model was designed. With this model as a starting point, 5 different port configurations (*Figure 3*) were created. Each had a different discharge port geometry. All the geometries follow the arcs and overall dimensions of the main rotor. The changes in the geometries were defined by two main aspects: The angle of discharge and the extended discharge port tip. By altering the geometry of the discharge port, the angle at which the port opens and starts to discharge changes. In this way, by rotating the main rotor at 350°, 355°, 360° and 370° and by following its lobe's arc, 4 different configurations were created. The only difference in them is the length of the outer arc, which is the factor that defines the angle that the port starts opening. With this change, the volume ratio index ($Vi = \frac{V_{max}}{V_{discharge}}$) is also changing (*Table 2*). In regards to the extended discharge port tip, the case with the angle of discharge at 360° was modified so that the port's tip is longer, enabling the carry-over chamber to be discharged in a higher rate. All the ports had as basis the port's 1 design. In (*Figure 3*) the base port 1 is denoted as the yellow hatched area whereas the other ports are denoted as different colors. The port 1 geometry is shown in the foreground in all cases so the differences are clear. Lastly, the extended discharge tip of port 2 is shown with an arrow.

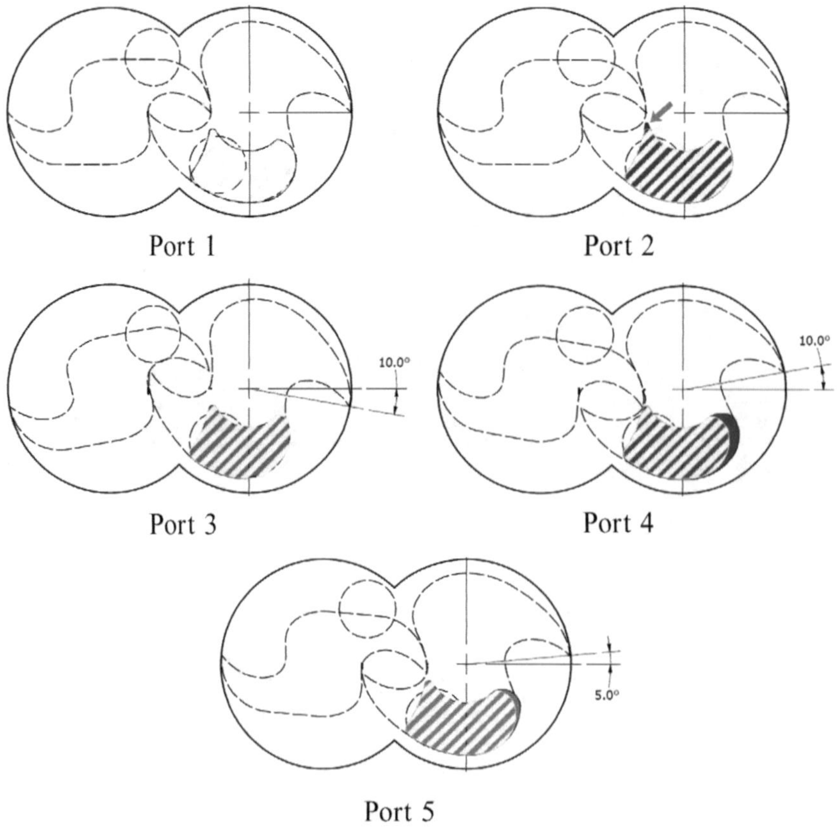

Figure 3: Port configurations

Table 2: Port characteristics

Port	1	2 (Extended tip)	3	4	5
Discharge angle	360°	360°	370°	350°	355°
Vi (Volume ratio index)	1.28	1.28	1.31	1.22	1.225

5 Simulations and results

5.1 *Set-up and boundary conditions*

ANSYS Forte is using the finite volume method (FVM), solving the governing algebraic equations. The boundary conditions and the set-up settings are listed in *Table 3* below. The working process of the claw pump was assumed to be an adiabatic one, with the operational fluid being air.

Table 3: Simulation set-up

Turbulence model	SST-k Epsilon
Inlet boundary conditions	Inlet Pressure: 1bar
Outlet boundary conditions	Outlet Pressure: 1.21 bar
Time step	10^{-5} sec

5.2 *Meshing*

ANSYS Forte's automatic mesh generation (AMG) does not require much input and processing. After defining the surfaces and their refinement methods, the only input required is the global mesh size. The global mesh size defines the resolution of the mesh, and therefore the detail of the results. A smaller mesh size, normally would give a more accurate result. But there is a disadvantage: Smaller mesh sizes require more computational time and power. This is quite an important factor, since time and power resources are limited. That is why, a mesh independence study was conducted, trying mesh sizes 0.2 cm, 0.3 cm and 0.4 cm on port 1. It was found that the 0.3 cm mesh is the most suitable, because it balances computational resources and accuracy, since the percentage difference of it comparing to the 0.2 cm mesh was only 1.96% (*Table 4*).

*The mesh size 0.4 cm was too coarse and therefore resulted in errors.

Table 4: Mesh independence study

Global Mesh size (cm)	Inlet. vol. flow rate (L/min)
0.2	396
0.3	388.2
0.5	8.6*
% difference: 0.2 vs 0.3	1.96%

5.3 *Results*

The results were validated by comparing the mass flow rates of the work of Gu et al. (2020) [3]. In this investigation a similar but not identical pump was considered. Therefore some small differences are expected. Also, the main objective is the comparison of the port geometries and not the absolute values. For the above reasons, a 10% discrepancy is reasonable and accepted. The validation data suggest an average mass flow rate of $6.63 \cdot 10^{-3}$ (kg/s). As observed from *Table 5* bellow the average % difference obtained is 9.17% which is within the acceptable range.

Table 5: Validation data comparison

$\dot{m}$ (avg)	Validation $\dot{m}$	% difference from validation
$7.3 \cdot 10^{-3}$	$6.63 \cdot 10^{-3}$	9.17%

In *Table 6* and *Figure 4* below, the results of the simulations are presented. Using ANSYS CFX-Post, the pressure, temperature and port velocity contours of all cases were plotted.

Table 6: Results Table

Port	Inlet Volume flow rate (L/min)	Power (W)	Power to vol. flow rate ratio (specific power)
1	388.2	266	0.685
2	365.4	270	0.739
3	374.4	281	0.750
4	385.6	263	0.682
5	386.4	267	0.690

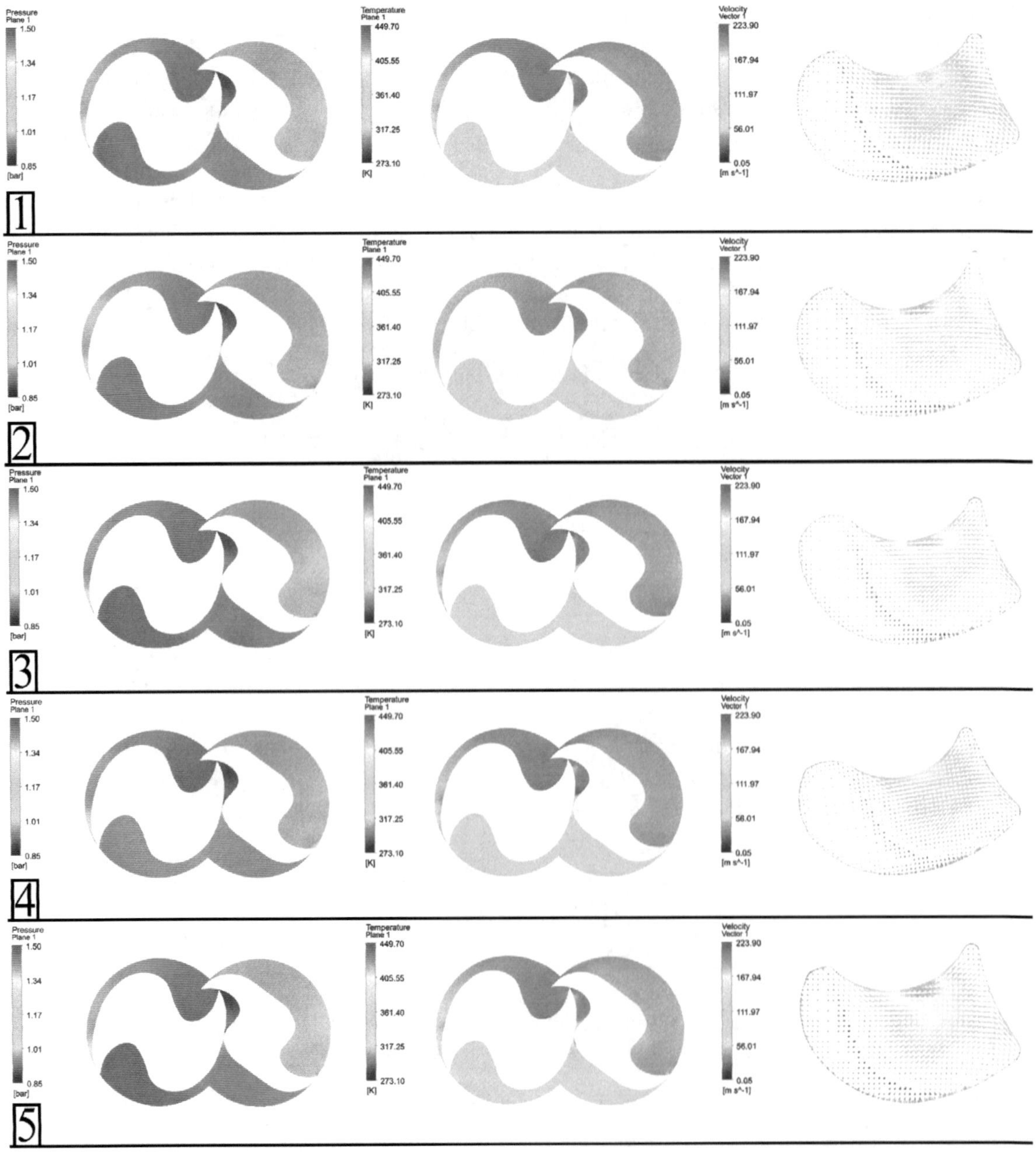

Figure 4: Pressure/Temperature/Port Velocity Contours

As it is observed from *table* 6, the case with the highest power is port 3. Despite that, the fourth case has the best performance since it has the lowest specific power. Port 3 is the least efficient with the highest specific power. Moving on to the graphs (*Figure 4*), the time step showcased above is located mid-way the discharge phase in order to showcase the effect of the port's geometry. The following observations were made:

- The pressure distribution graphs show that overall the five cases are similar. Despite that, the fourth case seems to have slightly lower pressure build up during the discharge phase, showing that the 10° early opening of the port actually helps in the discharge process. In contrast to that, port 5 which has 10° late opening angle, results in the highest pressure among all the cases. Furthermore, port configuration 2 has also relatively low pressure, showing that the extended tip of the discharge port helps in the discharge process and in the carry-over chamber off-loading. A high pressure build-up can cause reliability issues over time to the pump and therefore the lower the pressure, the better the pump will perform and thus the failure possibilities will be reduced.

- The temperature distribution graphs also show almost identical results with a small but significant difference. In ports 1,3,4 and 5 in the spot where the two rotors are very close, the temperature rapidly increases between the clearance. This shows that the extended discharge tip geometry on the second port is working effectively and therefore off-loads the carry-over chamber more efficiently. This is really important because, the high temperatures may cause the rotors to expand over time, and due to the very small clearances, it may lead to a collision and therefore a failure of the pump. Also, a high pressure and temperature loaded carry-over chamber leads to more power consumption which is not desirable.

- Finally, the velocity vectors of the five cases also show that port 4 is less loaded, with lower velocity. The maximum velocity that has been reached in port 4 is significantly lower than any other case. The reason behind that is the 10° early opening of the discharge port, showcasing again the importance of this geometry feature. Higher velocity means higher kinetic energy and thus higher power consumption.

Summarizing the results, ports 1,3 and 5 do not seem to present any positive development. In fact, port 3 has the highest specific power, and therefore the worst performance, while port's 5 highest pressure and velocity make it the most unreliable of them all. The ports that have showed positive developments are port 4 and port 2. Port's 4 early discharging, ensures the highest performance in combination with good reliability and efficiency due to the low pressure and exit velocity. From the other hand, the extended tip of discharge port 2, ensures great carry-over chamber off-loading, reducing the possibilities of failure, and therefore increasing the reliability of the corresponding claw pump.

6 Conclusion

Overall, the cut-cell Cartesian meshing method seemed to be really consistent throughout the simulations conducted. The automatic mesh generation of ANSYS Forte was very important since it was accurate and time-saving without many errors occurring. As far as the port geometry is concerned, it seems that it does not have a huge effect, but the small alterations that some features produce are very important. In detail, the 10° early opening of port 4 is resulting in an overall improvement in the performance, the reliability and the efficiency. Also, the extended discharge tip of the port 2, had a significant effect on the discharging process. The carry-over chamber's off-loading, is a major feature that maximizes the reliability of the claw pumps. Therefore port 2 might not be the best performing one by a small fraction, but is the most reliable ensuring very good carry-over chamber discharging. Generally, a detailed study and optimization of the discharge port geometry of claw pumps is a worthy procedure, since interesting and significant developments in the behavior of the pumps are observed.

References

[1] Kovacevic A, Smith I, Stosic N. Screw Compressors three dimensional computational fluid dynamics and solid fluid interaction. London: Springer-Verlag; 2006.

[2] Stosic N, Kovacevic A, Smith I. Screw compressor geometry mathematical modelling and performance calculation. Berlin: Springer-Verlag; 2005.

[3] Gu P, Xing L, Wang Y, Feng J, Peng X. Transient flow field and performance analysis of a claw pump for FCVs. International Journal of Hydrogen Energy. 2020;46(1).

[4] Kisu Im, P. (1997). Screw Vacuum Pump Having A Decreasing Pitch for the Screw Members.

[5] Lu Y, Kovacevic A, Read M, Basha N. Numerical Study of Customised Mesh for Twin Screw Vacuum Pumps. Designs. 2019;3(4):52.

[6] Tuo J, Guo B, Wu R, Zhang Y, Chen X. Study on the performance prediction of dry twin screw vacuum pump. IOP Conference Series: Materials Science and Engineering. 2018;425.

[7] Wang J, Jiang X, Cai Y. Investigation of a novel circular arc claw rotor profile for claw vacuum pumps and its performance analysis. Vacuum. 2015;111.

[8] Lu Y, Fu Y, Guo B, Fu L, Zhang Q, Chen X. Numerical Simulation of the Working Process in the Twin Screw Vacuum Pump. IOP Conference Series: Materials Science and Engineering. 2017;232.

[9] Kovacevic A, Stosic N, Mujic E, Smith I. CFD Integrated Design of Screw Compressors. Engineering Applications of Computational Fluid Mechanics. 2007;1(2).

[10] Stosic, N., Kovacevic, A., Rane, S. and Smith, I. Screw Rotor Profiles of Variable Lead Vacuum and Multiphase Machines and Their Calculation Models. 2018

[11] Yan D, Tang Q, Kovacevic A, Rane S, Pei L. Rotor profile design and numerical analysis of 2–3 type multiphase twin-screw pumps. Proceedings of the Institution of Mechanical Engineers, Part E: Journal of Process Mechanical Engineering. 2017;232(2).

[12] Lu Y, Guo B, Geng M. Study on Design of Rotor Profile for the Twin Screw Vacuum Pump with Single Thread Tooth. IOP Conference Series: Materials Science and Engineering. 2015;90.

[13] Stosic N, Smith I, Kovacevic A, Mujic E. Geometry of screw compressor rotors and their tools. Journal of Zhejiang University-SCIENCE A. 2011;12(4).

[14] ANSYS, Inc, (2020). Forte Quick Start Guide, Forte tutorials, Forte Theory manual.

[15] Zikanov O. Essential computational fluid dynamics. 1st ed. John Wiley Sons; 2010.

[16] Rane S, Kovačević A, Stošić N. Grid Generation for CFD Analysis and Design of a Variety of Twin Screw Machines. Designs. 2019;3(2).

[17] Kaczmarek, R., (2013). It'S All About The Mesh – Cartesian Cut-Cell Approach. [online] Convergecfd.com. Available at: <https://convergecfd.com/blog/its-all-about-the-mesh-cartesian-cut-cell-approach>

[18] PDM Analysis, (2020). Challenges In Grid Generation And 3D CFD Modelling Of Positive Displacement Machines. [video] Available at: <https://pdmanalysis.co.uk/webinar-series/webinar-11/>

[19] Yan D, Kovacevic A, Tang Q, Rane S, Zhang W. Numerical modelling of twin-screw pumps based on computational fluid dynamics. Proceedings of the Institution of Mechanical Engineers, Part C: Journal of Mechanical Engineering Science. 2016;231(24).

[20] Litvin F L and Fuentes A Gear Geometry and Applied Theory (New York: Cambridge university press). 2014

IOP Conf. Series: Materials Science and Engineering 1180 (2021) 012064 doi:10.1088/1757-899X/1180/1/012064

Lubricant Options for Screw Compressors Using Alternative Refrigerants

J Karnaz
Shrieve Chemical Products, LLC

E-mail: jkarnaz@shrieve.com

Abstract. Environmental concerns have created need to find replacements for HFC and in some cases HCFC refrigerants used in screw compressor applications. Like most refrigerant changes it becomes necessary to assess the best lubricant option in order to maintain system reliability and performance. Regardless of the magnitude of the changes it allows opportunity to appraise alternative lubricant chemistries that are different from what was used in the past.

This paper will investigate various alternative refrigerants being evaluated for screw compressor applications. Choices will be presented that not only maintain what was seen in the past but also offers potential to advance application ranges while using of a single lubricant with various operational designs. As always needed when evaluating lubricants and refrigerants, information on the interaction properties along with stability and material compatibility will be presented. Comparisons will be made to legacy data operation and potential improvements will be presented that are available when using combinations of alternative refrigerants and lubricants.

1. Introduction

Food preservation and processing along with building comfort heating/cooling and hot water production, are essential today to our way of life. Screw compressors are used in numerous sizes, designs, operating at various conditions with different refrigerants. This variety requires compressor lubricants of varying viscosities and chemistries to provide needed bearing protection and overall operational reliability and performance.

The lubricant must provide the needed sealing between the rotors and optimal viscosity at other bearings, minimize wear and the ability to help provide ideal performance. Like in other types of compressors, lubricant and refrigerant are going to be at some level of interaction which will drive the viscosity of lubricant and refrigerant combination; sometimes referred to as the working viscosity. The concept of not only understanding this interaction, but also ways to accurately measure, are essential tools to compressor, bearing and system design engineers to calculated bearing loads and viscosity requirements. After determining requirements for lubricant chemistry and viscosity, it is essential to examine the stability and compatibility of these multiple combinations.

The type of heat exchanger design; air-cooled or water-cooled, along with other system components and special enhancements will all dictate which lubricant is chosen for the given refrigerant.
This paper will dive into the market of screw compressor operation and examine what is required to provide lubricants to meet compressor demands and refrigerant interactions. The focus will be on evaluating lubricant options for screw compressor used in various refrigeration and air conditioning applications with newer low Global Warming Potential (GWP) refrigerants.

2. Refrigerants and Screw Compressor Operation

Screw compressors use several refrigerants in numerous applications, R-22 is a HCFC refrigerant that is still in use today and several of the common HFC refrigerants like R-134a are predominate. Screw compressors in industrial applications use several low GWP refrigerants in systems using natural refrigerants like ammonia (R-717), carbon dioxide (R-744); propylene (R-1270), propane (R-290) and others. Since some of these HCFC, natural and HFC refrigerants have history in operation and lubricants used in these applications, this paper will focus more on newer low GWP refrigerants such as those based on HydroFluoroOlefins (HFO) and blends. Figure 1 is an example of low GWP refrigerants that are used or under evaluation today. When compared to R-134a, these refrigerants either lack the equivalent capacity and/or are less efficient. Like R-134a, some refrigerants are designated as non-flammable (A1) but will not have the lowest GWP values while others will be mildly flammable (A2L) with smaller GWP values. This paper will emphasize lubricant evaluations for medium pressure refrigerants like R-1234ze(E) and R-513A as alternatives to R-134a [Kujak 2018]. Table 1 is a listing of various refrigerants of interest to the market, their chemical composition and global warming potential (GWP) values [Makhnatch 2019].

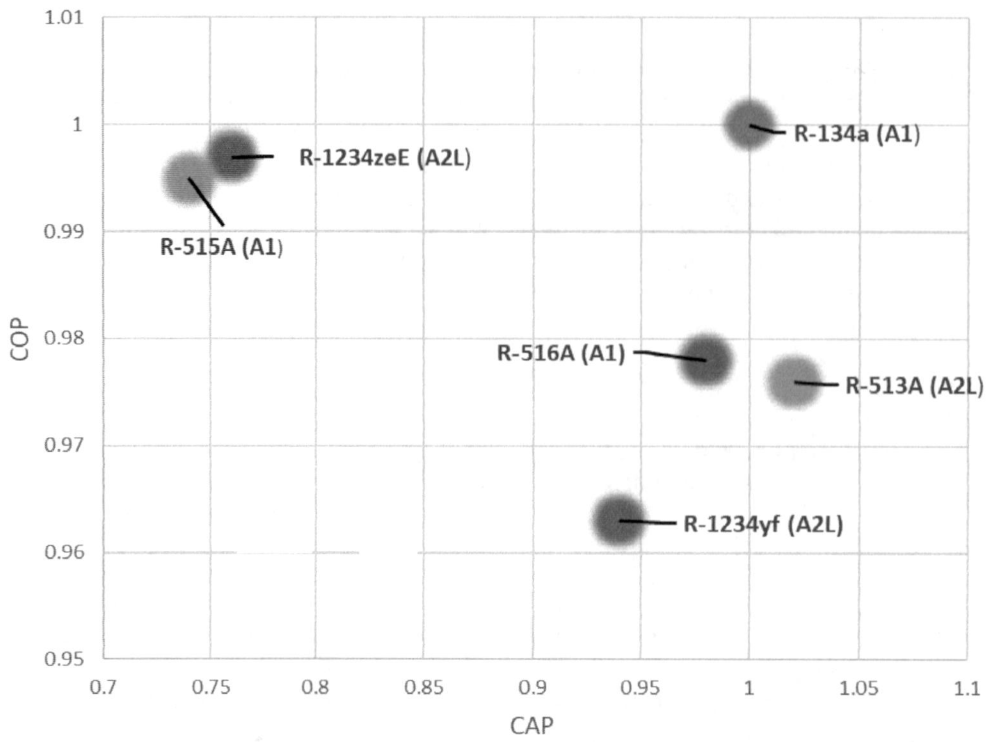

Figure 1: Performance characteristic comparison of some Low GWP Refrigerants and R-134a

Table 1: Refrigerant Composition and GWP Value

Refrigerant	R-134a	R-1234yf	R-1234zeE	R-32	R-152a	R-227ea	GWP
R-134a	100%						1430
R-1234yf		100%					4
R-1234zeE			100%				7
R-513A	44%	56%					631
R-515A			88%			12%	393
R-516A	8.5%	77.5%			14%		145

3. Lubricant and Refrigerant Interaction Consideration in Screw Compressor Operation

Measuring viscosity of a lubricant is straight forward and there are several methods, techniques, and equipment to produce reliable numbers. But when it comes to determining values of viscosity of lubricant and refrigerant mixtures, more sophisticated equipment and techniques are required. The system needs to be sealed and capable of sustaining high pressure, Jonsson (1998) describes a falling ball viscometer with sapphire viewing windows sealed in a chamber. Though lubricant and viscosity values can be obtained, the method lacks the ability to measure density in-situ, with limits on temperature and pressure ranges. The technique is cumbersome and time consuming in collecting the needed number of data points for accuracy. A more advanced method was developed by [Seeton 2010] which utilizes a closed loop system within an environmental chamber. The lubricant and refrigerant are in circulation while the chamber slowly goes through a temperature profile and collecting data for multiple temperature points, viscosity (usually from two in-line viscometers at different ranges), pressure, mass flow and density. The unit volume is fixed so vapor space correction can be done at each lubricant and refrigerant combination as the temperature and fluid density changes. Gathered data is used to generate ten-point coefficients for viscosity (cPs), density and refrigerant dilution which is used to create Daniel Plots and temperature-pressure solvers. This technique has been used to generate hundreds of plots of different lubricants and refrigerant combinations and is an essential tool when attempting to migrate to new refrigerant and lubricant combination as a screening method [Karnaz 2018]. This method can be used to predict how a screw compressor system operating at discharge conditions with an oil separator can utilize an oil cooler to generate the needed bearing viscosities as demonstrated below.

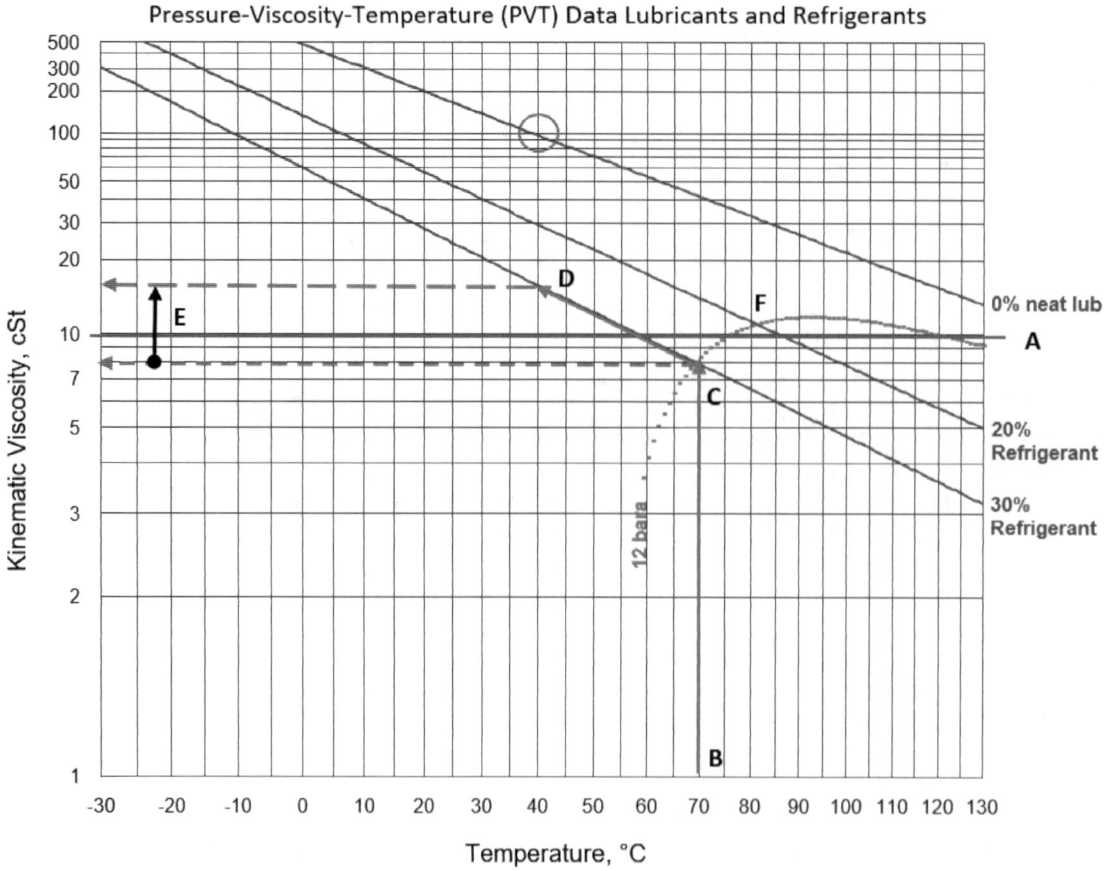

Figure 2: Example of Manipulating Working Viscosity in Screw Compressor Operation

Figure 2 is an example of how PVT generated data can be useful in showing how a lower viscosity lubricant can still meet minimum viscosity requirements in screw compressor operation. The plot in the example is an ISO 100 lubricant and Point A represents minimum working viscosity requirement for this particular system. Point B and C represent oil temperature and pressure at the discharge condition; in this case showing 30% refrigerant dilution and a viscosity of 8 cSt. Since in this example the compressor requires a minimum of 10 cSt, this lubricant does not meet the demand, but if the equipment utilizes one of several methods to cool the lubricant then this can change this acceptance. Moving from Point C to Point D we see if we cool the oil from 70°C to 40°C by keeping a constant dilution, we double the working viscosity to 16 cSt and move above the minimum. Even without oil cooling we can reach a needed viscosity if we move from Points C to F by increasing the temperature 10°C while maintaining the same pressure, the dilution will drop to 20% moving the viscosity to 11-12 cSt. These techniques may become more vital with the use of some of the HFO low GWP refrigerants that lower the working viscosity due to higher refrigerant solubility rates and other properties.

4. Lubricants for HFO Based Refrigerants to Replace R-134a

Measured miscibility interaction of lubricants and refrigerants is important to understand and could be vital for meeting energy efficiency requirements through heat transfer parameters, in addition to making sure the compressor doesn't fail due to lack of lubricant [Karnaz 2019]. As lubricant moves out of the compressor into the system, it is important to make sure lubricant is effectively returned to the compressor. Some units, like those used in screw compressors, may purposely use oil separation as a lubrication point to the compressors which could make miscibility targets not as critical. Others though will rely on certain levels of miscibility to maintain what was mentioned earlier. For R-134a used with screw compressors, POE lubricants have been a lubricant of choice; Figure 3 shows the miscibility curves for three POE lubricants of varying viscosity. For most lubricant and refrigerant combinations, as viscosity increases, the miscibility temperature will rise as seen in the figure 3; when these same lubricants are measured with R-1234zeE the miscibility points for all three viscosities moves below -60°C at all concentrations. Since moving to lower temperature miscibility results in higher solubility of the refrigerant in the lubricant, HFO-1234zeE solubility is going to be one way of reducing working viscosity which could negatively affect screw compressor lubrication.

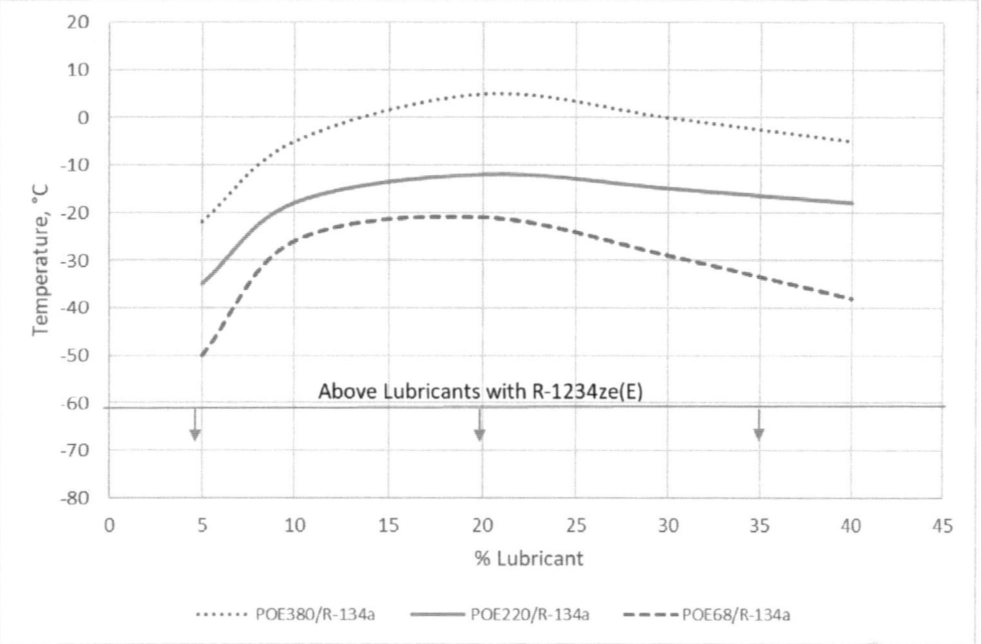

Figure 3: Miscibility of POE Lubricants with R-134a and R-1234zeE

The refrigerant over solubilizing in the lubricant is not new, screw compressor operation with R-22 is an example. When R-22 operating conditions move to the point where solubility gets too high and working viscosity is reduced in mineral and alkylbenzene based lubricants, ester-based lubricants that can reach higher base viscosity levels like 300-400 centistokes are used. To successfully develop lubricants for new refrigerants with reduced working viscosities, Shrieve has established processes to evaluate candidates in a timely and effective manner. As already mentioned, initial investigation of R-1234zeE showed the potential to have high solubility particularly in lubricants that will still maintain various miscibility levels. To begin the assessment, it is important to understand what was used in the past and for purposes of drop-in scenarios, along with evaluating minimal levels of investigation and change. Compressors will function at numerous operating parameters based on temperatures and pressures which can complicate the evaluations.

If we look back at Figure 1 and use the examples of that explanation, we can start to make comparisons of traditional POE lubricants used today with R-134a and whether these can be used in a drop-in scenario with new refrigerants. Table 2 shows viscosity at 40°C, which defines the base oil, and 70°C, then proceeds to show how much viscosity will drop with 30% R-134a. As expected, addition of any refrigerant quickly drops the viscosity of the fluid mixture. Higher diluting effects of R-1234zeE over R-134a start to appear; for R-134a the move from ISO 68 to ISO 220 is doubling working viscosity then tripling to ISO 380 but for R-1234zeE this move is closer to 1.5 times for ISO 220 and double for ISO 380. So just moving to higher viscosity of the same chemistry type might not be adequate.

Table 2: Initial Studies to Evaluate R-134a versus R-1234zeE

Lubricant	Viscosity at 40°C	Viscosity at 70C	V40 and 30% R-134a	V70 and 30% R-134a	V40 and 30% R-1234zeE	V70 and 30% R-1234zeE
POE68	65 cSt	20 cSt	5.6 cSt	3.0 cSt	4.6 cSt	2.4 cSt
POE220	213 cSt	50 cSt	9.9 cSt	4.8 cSt	7.5 cSt	3.6 cSt
POE380	370 cSt	75 cSt	15.0 cSt	9.0 cSt	9.2 cSt	4.3 cSt

Table 3 is a look at various conditions with POE lubricants and R-134a, these values will be used for later comparisons. Both POE 120 and POE 220 have been used with R-134a in many screw compressor applications. The Polyether chemistry, PE 100, is specifically designed to work with new HFO refrigerants; and as can be seen, is also an acceptable alternative to be used with R-134a. The working viscosity of PE 100 (which is a 100 cSt lubricant) falls closer to want is seen with POE 220.

Table 3: R-134a and Lubricant Working Viscosity and Dilution

	Conditions		POE 120		POE 220		PE 100	
	Temp	Pres	Vis, cSt	Dil, %	Vis, cSt	Dil, %	Vis, cSt	Dil, %
Condition A	26°C	4 bar	31	18	51	18	**42**	22
Condition B	52°C	6 bar	17	13	28	14	**31**	14
Condition C	68°C	14 bar	5	24	7	25	**11**	26

To overcome dilution and working viscosity factor of some HFO refrigerants and lubricants, Shrieve has developed lubricants that provide properties to help maintain adequate working viscosities while still maintaining traditional base lubricant neat viscosities. Figure 4 is a comparative look at a traditional POE, a modified POE and PE chemistry with refrigerant R-1234zeE. The PE chemistry is also shown with R-513A, as an example of how this one lubricant chemistry can cover multiple refrigerants. The POE 120 shown above with R-134a was not compared with R-1234zeE because it did not provide enough working viscosity at operating conditions.

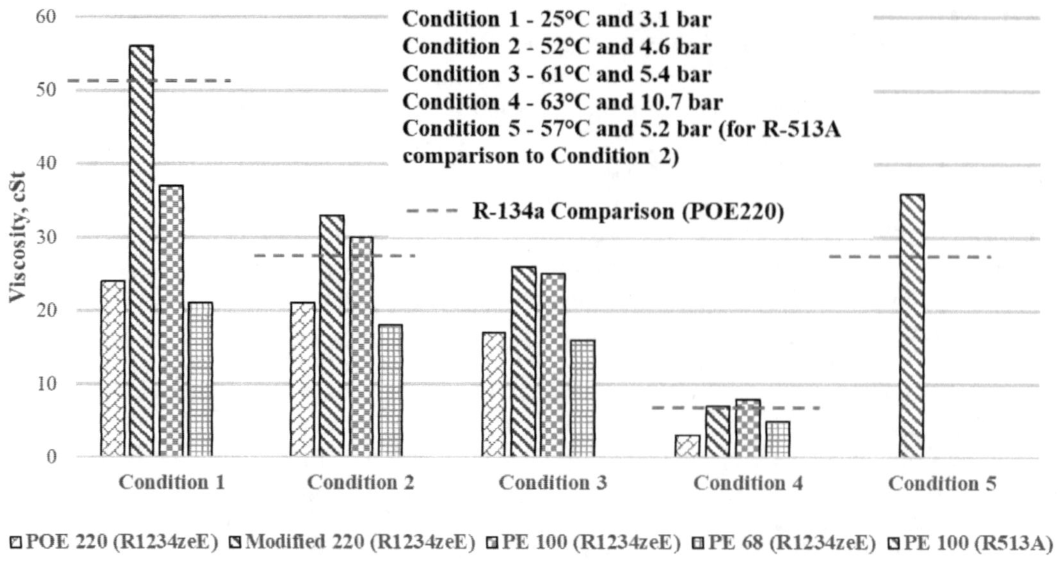

Figure 4: Evaluating Working Viscosity of Lubricants for HFO Refrigerants

Further evaluation of the data in figure 4 shows the difficulties of using current POE 220 with R-1234zeE along with benefits of using PE 100 lubricant with R-1234zeE and other HFO refrigerant alternates. For Condition 1 it can be seen how much viscosity is reduced with traditional POE 220/R-1234zeE; also, modified POE 220 with R-1234zeE and even when traditional POE 220 was used with R-134a at this condition there is too much working viscosity. PE 100 provides optimized viscosity over this condition and several other working conditions even though it has half the starting base of an ISO 220 POE. For Condition 5 with R-513A the PE 100 lubricant is also a good fit, meeting the working viscosity requirement for a screw compressor during operation.

5. Lubricant and Refrigerant Stability and Material Compatibility Testing

History has shown us when changes are made to refrigerants there is ultimately some change that needs to be made to materials used in application. This change may be caused by an incompatible situation between materials and refrigerants, lubricants, combination of the two or changes to operating conditions caused by the new working fluids. It is important to try to get an initial understanding of interactions occurring within a system by first screening various candidates. This concept is typically accomplished by testing combinations of refrigerant, lubricant, and material within an enclosed vessel. The art of hermetically sealing glass tubes with these combinations has been successfully utilized in the industry to provide a visual and analytical study of both stability and compatibility. ASHRAE Standard 97 outlines the method and expertise needed to maintain the integrity of the sealed glass when heated to high temperature which generates high pressures with the refrigerants. Other methods have been developed which use metal vessels which can also be sealed and withstand higher pressures while also utilizing larger quantities of material. Once again, a standard has been developed to provide details on this concept which is outlined in ASHRAE GPC 38.

Just as when the screw compressor industry transitioned from CFC/HCFC refrigerants to HFC refrigerants making evaluation of stability and material compatibility, a similar task is required when transitioning from HFC refrigerants to more environmentally friendly options. Table 4 and Figure 5 show examples of stability testing in sealed glass tubes of a lubricant candidate with R-1234yf containing R-513A refrigerant. This combination allows for understanding the effects of R-1234yf which will have different stability profile versus R-134a which is part of this refrigerant blend. Testing is done at elevated temperature to represent an extended time in system operation. Sometimes it becomes necessary to adjust temperatures to allow for a successful time-based evaluation, too low of temperature – no valuable kinetic information is achieved to make predictions, while too high of temperature – unrealistic mechanisms of breakdown are created. Overall, stability is achieved with this combination of lubricant and refrigerant.

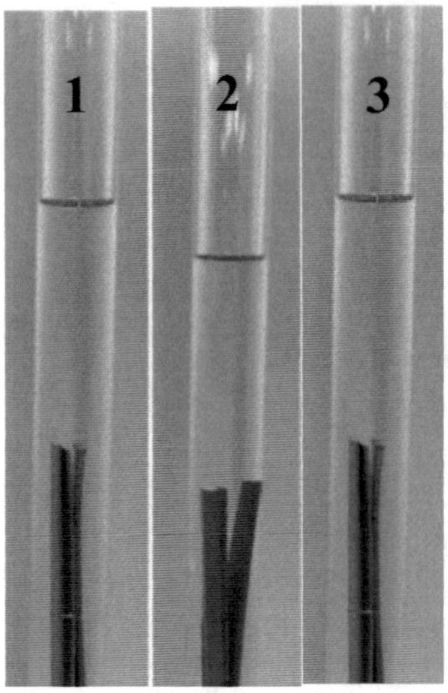

Figure 5: Fluid Appearance

Table 4: Results after Accelerated Stability Testing

Testing with Lubricant or Lubricant + Additive with R-513A at 175°C for 14 Days for Samples 2 and 3 while Sample 1 is appearance and evaluation before heat aging				
Lubricant Sample	Additive	Lubricant Appearance	TAN mgKOH/gOil	Fluoride Ion ppm
1	None (Neat Sample)	Colorless to Pale Yellow	<0.1	0
2	None	Pale Yellow to Yellow	0.2	41
3	Yes	Colorless to Pale Yellow	<0.1	0

Within the compressor and the system, there are numerous metal, plastic and elastomer chemistries interacting with refrigerants and lubricants. Figures 6 and 7 are examples of testing material at elevated conditions with various refrigerants and lubricants. Lessons learned changing from CFC to HFC refrigerant chemistries is that elastomer material will behave differently in different refrigerants and lubricants. So, it is important to understand interaction at elevated conditions. In figure 6 we see example of testing O-ring material in PE 100 lubricant with R-513A with comparison to POE 220 and R-134a. Compared to the neat oil and POE 220/R-134a, this material has no significant hardness change in PE 100 and R-513A along with no change to the acidity of the lubricant after test.

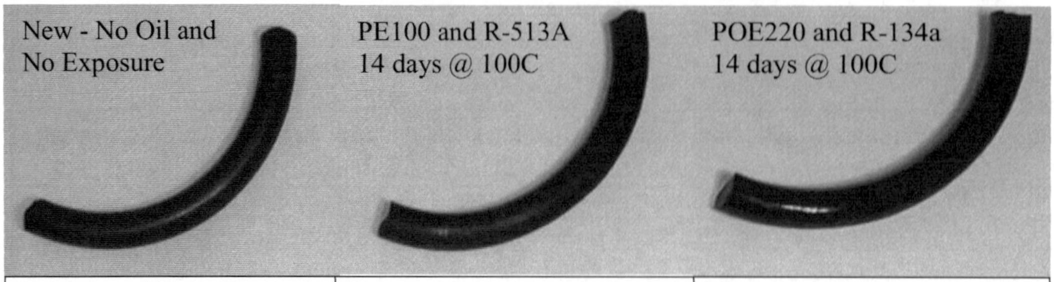

Hardness = 77	Harness = 76	Hardness = 79
TAN = 0.02 mgKOH/gOil	TAN = 0.03 mgKOH/gOil	TAN = 0.05 mgKOH/gOil

Figure 6: O-ring Material Compatibility Testing in Lubricants and Refrigerants

Figure 7 is a more detailed look at material compatibility with neoprene elastomer. Testing is done in three different refrigerants with one lubricant, PE 100, making comparison of dimensional and hardness changes. Results indicate very little change in each refrigerant and acceptability of this material.

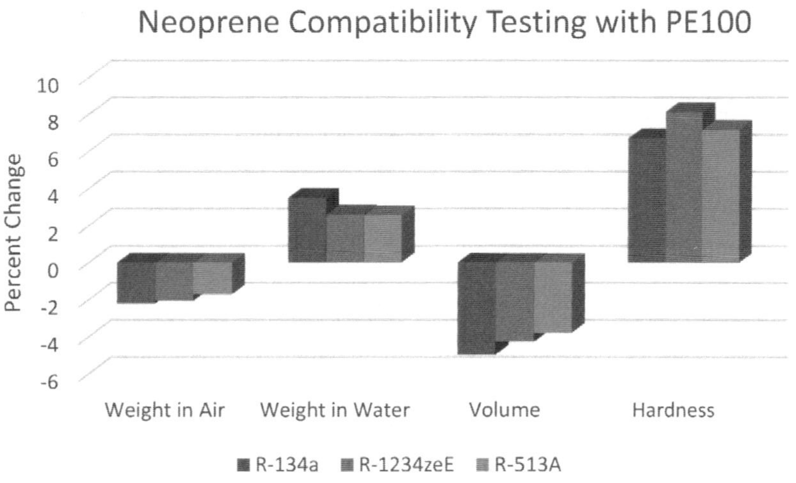

Figure 7: Elastomer Material Comparison in Various Refrigerants

6. Conclusions

Screw compressors that have historically operated with HCFC and HFC refrigerants with various lubricants, could require change to lower GWP refrigerants. Some of these refrigerant options starting to be used or investigated are based on hydrofluoroolefin (HFO) chemistry, one being R-1234ze(E). This refrigerant provides a low GWP value and with system modifications is acceptable in screw compressors replacing R-134a. Like historical transitions from one refrigerant to another, there is a need to determine lubricant options. Some of this investigation is out of necessity, like the transition from CFC to HFC (mineral oil to synthetics), and others provide opportunities to make improvements or expand other chemistry options.

Refrigerants bring different interaction parameters with lubricants, one of the newer low GWP refrigerants, R-1234zeE, creates reduced working viscosity. This is important in screw compressor operation that typically requires higher working viscosity for operation. This adjustment in the past has been accomplished by moving to higher viscosity lubricants but sometimes this still presents challenges. Trying to maintain levels of miscibility while adjusting the working viscosity to meet desired targets is more difficult with R-1234zeE.

To offset the working viscosity reducing effects of some of the HFO refrigerants, Polyether chemistry has been evaluated. Results of PE 100 lubricant with HFO refrigerant shows acceptability at various working conditions associated with screw compressors operating either in water-cooled or air-cooled applications. This lubricant is also shown to work with other HFO refrigerants like R-513A and others, in addition to working with current R-134a designed systems. PE chemistry also shows good stability and material compatibility at operating conditions with typical compressor and system materials. The Polyether base viscosity can be adjusted to lower values to provide optimized operation. There is also plenty of room to adjust to higher viscosities to accommodate severe operating conditions that could require higher working viscosity, all without changing other properties of the lubricant and refrigerant interaction.

References

[1] S. Kujak "Medium Pressure NGR Alternatives for Chillers" 13th International Symposium on New Refrigerants and Environmental Technology 2018. Kobe 2018.
[2] P. Makhnatch "New Refrigerants for Vapour Compression Refrigeration and Heat Pump Systems; Evaluation in a context of the requirements set by F-gas Regulation and the Paris Agreement Goals" KTH Royal Institute of Technology Doctoral Thesis 2019.
[3] U. Jonsson "Elastrohydrodynamic Lubrication and Lubricant Rheology in Refrigeration Compressors" Department of Mechanical Engineering Lulea University of Technology Doctoral Thesis May 1998.
[4] C. Seeton "Measurement of Refrigerant/Lubricant Thermophysical Properties" NIST 2010.
[5] J. Karnaz, C. Seeton "Evaluation of Lubricant Properties and Refrigerant Interaction" 24th International Compressor Conference at Purdue 2018.
[6] J. Karnaz "Opening up the Lubricant Toolbox" 11th International Conference on Compressors and Their Systems London 2019.
[7] ASHRAE Standard 97-2007 "Glass Sealed Tube Method to Test the Chemical Stability of Materials for Use within Refrigerant Systems". 2017
[8] ASHRAE Guideline 38 "Guideline for Using Metal Pressure Vessels to Test Materials Used in Refrigeration Systems". 2018

AUTHOR INDEX